LEAD-ZINC-TIN '80

LEAD-ZINC-TIN '80

Proceedings of a World Symposium
on Metallurgy and Environmental Control
sponsored by the TMS-AIME Lead, Zinc, and Tin
Committee at the 109th AIME Annual Meeting,
February 24-28, 1980, in Las Vegas, Nevada

Edited by

John M. Cigan
St. Joe Minerals Corporation
Monaca, Pennsylvania

Thomas S. Mackey
Key Metals & Minerals Engineering Corporation
Texas City, Texas

Thomas J. O'Keefe
University of Missouri—Rolla
Rolla, Missouri

A Publication of The Metallurgical Society of AIME

A Publication of The Metallurgical Society of AIME
P.O. Box 430
420 Commonwealth Drive
Warrendale, Pa. 15086
(412) 776-9000

Printed in the United States of America.
Library of Congress Card Catalogue Number 79-93007
ISBN Number 0-89520-358-8

Cover Design: *Judith M. Uhl*

Organization of Lead-Zinc-Tin'80 Symposium

SYMPOSIUM CHAIRMAN:

Thomas S. Mackey
Key Metals & Minerals Engineering Corporation
Texas City, Texas

SYMPOSIUM VICE CHAIRMEN:

John M. Cigan
St. Joe Minerals Corporation
Monaca, Pennsylvania

Thomas J. O'Keefe
University of Missouri—Rolla
Rolla, Missouri

Derek A. Temple
Imperial Smelting Processes Limited
Avonmouth, Bristol; England

PROGRAM COMMITTEE

Robert W. Balliett
Phelps-Dodge Brass Company
Anniston, Alabama

Raymond C. Bell
Consultant
Rossland, B.C.; Canada

William O. Gentry
Globe-Union Incorporated
Milwaukee, Wisconsin

Nassas E. Ghatas
Canadian Electrolytic Zinc Company
Valleyfield, Quebec; Canada

Thomas I. Moore
Consultant
Bartlesville, Oklahoma

Arthur E. Morris
University of Missouri—Rolla
Rolla, Missouri

Luis W. Pommier
Gulf Research & Development Company
Pittsburgh, Pennsylvania

R. David Prengaman
RSR Corporation
Dallas, Texas

Peter Tarassoff
Noranda Research Centre
Pointe Claire, Quebec; Canada

Thomas A. Theobald
National Zinc Company
Bartlesville, Oklahoma

Gary E. Welch
St. Joe Minerals Corporation
Clayton, Missouri

SYMPOSIUM COMMITTEE

R. W. Balliett
Phelps-Dodge Brass Company
Anniston, Alabama

W. R. Bechdolt
New Jersey Zinc Company
Palmerton, Pennsylvania

R. C. Bell
Consultant
Rossland, B.C.; Canada

A. Booth
AMAX Lead & Zinc Division
Clayton, Missouri

J. M. Cigan
St. Joe Minerals Corporation
Monaca, Pennsylvania

C. H. Cotterill
U.S. Bureau of Mines
Washington, D.C.

P. Duby
Columbia University
New York, New York

M. M. Fine
U.S. Bureau of Mines
Rolla, Missouri

T. M. Fusco
St. Joe Zinc Company
Monaca, Pennsylvania

W. O. Gentry
Globe-Union Incorporated
Milwaukee, Wisconsin

N. E. Ghatas
Canadian Electrolytic Zinc Company
Valleyfield, Quebec; Canada

A. H. Larson
Gould Incorporated
St. Paul, Minnesota

T. S. Mackey
Key Metals & Minerals
Engineering Corporation
Texas City, Texas

G. M. Meisel
Consultant
El Cerrito, California

T. I. Moore
Consultant
Bartlesville, Oklahoma

A. E. Morris
University of Missouri—Rolla
Rolla, Missouri

T. Nagano
Mitsubishi Metal Corporation
Tokyo, Japan

T. J. O'Keefe
University of Missouri—Rolla
Rolla, Missouri

W. Opie
AMAX Base Metals R & D Inc.
Carteret, New Jersey

R. W. Pickering
Electrolytic Zinc Co. of Australasia Ltd.
Melbourne, Victoria; Australia

L. W. Pommier
Gulf Research & Development Company
Pittsburgh, Pennsylvania

R. D. Prengaman
RSR Corporation
Dallas, Texas

M. V. Rao
NL Industries
Hightston, New Jersey

G. Reis
Vieille-Montagne
Angleur, Belgium

W. P. Roe
Asarco Incorporated
South Plainfield, New Jersey

A. W. Schlechten
Colorado School of Mines
Golden, Colorado

P. Tarassoff
Noranda Research Centre
Pointe Claire, Quebec; Canada

D. A. Temple
Imperial Smelting Processes Limited
Avonmouth, Bristol; England

T. A. Theobald
National Zinc Company
Bartlesville, Oklahoma

G. E. Welch
St. Joe Minerals Corporation
Clayton, Missouri

SESSION CHAIRMEN

R. W. Balliett
Phelps-Dodge Brass Company
Anniston, Alabama

A. K. Barbour
RTZ Services Limited
Bristol; England

S. Bergsoe
Paul Bergsoe & Son
Glostrup; Denmark

A. Booth
AMAX Lead & Zinc Division
Clayton, Missouri

J. F. Cole
Lead Industries Association, Inc.
New York, New York

N. Dreulle
Compagnie Royale Asturienne des Mines
Auby lez Douai; France

W. O. Gentry
Globe-Union Incorporated
Milwaukee, Wisconsin

N. E. Ghatas
Canadian Electrolytic Zinc Company
Valleyfield, Quebec; Canada

E. King
Gulf Chemical & Metallurgical Company
Texas City, Texas

A. H. Larson
Gould Incorporated
St. Paul, Minnesota

T. S. Mackey
Key Metals & Minerals
Engineering Corporation
Texas City, Texas

H. Maczek
Berzelius Metallhutten GmbH
Duisburg; Germany

P. R. Mead
Sulphide Corporation Pty., Ltd.
Boolaro, N.S.W.; Australia

H. L. Montague
Asarco Incorporated
New York, New York

T. I. Moore
Consultant
Bartlesville, Oklahoma

A. E. Morris
University of Missouri—Rolla
Rolla, Missouri

B. H. Morrison
Noranda
Toronto, Ontario; Canada

H. Nakamura
Mitsui Mining and Smelting Company
Oomuta, Sukuoka, Japan

E. van den Neste
Metallurgie Hoboken-Overpelt
Overpelt, Belgium

A. Norro
Boliden AB
Skelleftehamn; Sweden

L. W. Pommier
Gulf Research & Development Company
Pittsburgh, Pennsylvania

M. V. Rao
NL Industries, Incorporated
Hightston, New Jersey

W. P. Roe
Asarco Incorporated
South Plainfield, New Jersey

A. von Ropenack
Ruhr Zink, GmbH
Datteln; Germany

T. N. Rosenqvist
University of Trondheim
Trondheim; Norway

A. W. Schlechten
Colorado School of Mines
Golden, Colorado

P. Tarassoff
Noranda Research Centre
Pointe Claire, Quebec; Canada

T. A. Theobald
National Zinc Company
Bartlesville, Oklahoma

G. E. Welch
St. Joe Minerals Corporation
Clayton, Missouri

A. Young
Brunswick Mining & Smelting Ltd.
Belledune, N.B.; Canada

Thomas J. O'Keefe
University of Missouri—Rolla
Rolla, Missouri

Raold Lindstrom
U.S. Bureau of Mines
Reno, Nevada

David Taschler
Air Products & Chemicals Company
Greenville, Pennsylvania

Foreword

The metallurgical industry, reacting to a variety of outside stimuli, is in a constant state of flux. As such, in any specific area of interest, e.g., the lead-zinc industry, it is imperative to periodically have an assessment by technical experts to determine where we have been and where we are so that we may better influence where we are going. When the AIME World Symposium on the Mining and Metallurgy of Lead and Zinc was held in St. Louis, Missouri, in October 1970 (with the Extractive Metallurgy Proceedings edited by Carl H. Cotterill and John M. Cigan), it had been thirty-four years since the previous AIME volume on the metallurgy of lead and zinc was published. Today, the pressures from outside the industry are having more rapid and more profound effects on the lead and zinc industry, so it is appropriate to focus on the state of the industry after a lapse of only ten years, taking stock of the many significant changes occurring as we begin a new decade.

During the 1970's, two dominant factors have been shaping our industry, and they will continue to do so as we proceed into the 1980's. These two are the zealous concern with the environment which has resulted in a plethora of environmental laws and regulations and the rapidly escalating cost of energy. Together with the availability of suitable mineral resources, environmental concerns and energy costs are playing dominant roles in influencing our extractive metallurgy processes and the markets for our products.

In the 1970 Symposium, we focused our attention on the descriptions of total lead and zinc plants. In zinc, the new plants of the 1970's have primarily been modifications of the electrolytic plant type built in the late 1960's. In the 1970's, the ISF plants have continually modified process operations to make the existing plants more versatile and efficient. The 1970's showed no major changes in primary lead smelters. In zinc, zinc-lead, and lead plants, there was an accelerating interest in automation of processes, continuous processing, process development, and gaining an enhanced understanding of the fundamentals of processes. In the face of environmental, energy, and other pressures, a number of promising new process candidates are evolving.

For the 1980 Symposium, all papers were specifically invited after we had solicited, by means of a questionnaire, input from major lead, zinc, and tin producers throughout the world. With the Symposium Committee providing counsel, the Program Committee selected from among the contributions offered. The final sixty papers which comprise the symposium were chosen to provide high quality, properly balanced coverage of process and product topics. In contrast to the 1970 Symposium state-of-the-art total plant theme, the 1980 Symposium concentrates on a variety of technical aspects providing a detailed technical update in those areas where significant changes are taking place. The 1980 Symposium Proceedings are, therefore, a valuable complement to the 1970 Symposium Proceedings.

All papers were submitted in advance so that the Symposium Proceedings would be available at the time of the oral presentations, thereby permitting more fruitful discussions. In order to foster in-depth exchanges of ideas among symposium attendees, "Free Form" discussion sessions were scheduled covering each of the primary metal process areas. As an important adjunct to the symposium, the U.S. Bureau of Mines Metallurgy Research Center in Reno, Nevada, hosted field trips. We are indebted to the staff of the U.S. Bureau of Mines for their generosity in providing symposium participants with valuable insights into promising new extraction processes.

An integral part of this symposium is a series of technical contributions on the extractive metallurgy and applications of tin. The 1980 Symposium represents the most significant forum in decades for the presentation and discussion of tin metallurgy. This gathering of tin metallurgical experts from throughout the world will serve as a fertile seed fostering future tin technical exchanges.

Anticipating continually accelerating changes in the lead-zinc industry, by the end of the 1980's a new industry-wide metallurgical assessment will be in order. The status of one or more metal process systems related to lead and zinc should also be considered then. Having committed the contributions of the Lead-Zinc-Tin '80 World Symposium to the working minds and bookshelves of our industry, universities, and research laboratories, we charge our colleagues to now set their sights on significant future developments in planning and bringing to successful fruition the Lead-Zinc '90 and Related Metals World Symposium.

The editors wish to express their thanks for the substantial efforts expended by the authors, the Program Committee, the Symposium Committee, the Session Chairmen, the TMS Staff, and their respective secretaries and employers. The superb assistance of this fine group of dedicated people has resulted in a symposium volume which will prove time and time again a valuable contribution to the industry.

JOHN M. CIGAN
THOMAS S. MACKEY
THOMAS J. O'KEEFE

December 3, 1979
Warrendale, Pennsylvania; USA

Table of Contents

Section C—LEAD SMELTING AND REFINING

Chapter

Section D—NEW LEAD AND ZINC EXTRACTION PROCESSES

Chapter

Section E—PHYSICAL CHEMISTRY OF LEAD, ZINC, AND TIN EXTRACTION

Chapter

Section F—LEAD AND ZINC ENVIRONMENTAL

Section G—TIN EXTRACTION AND APPLICATIONS

LEAD INDUSTRIES INTO THE 1980's

John A. Wright, President

St. Joe Lead Company, Clayton, Missouri

Abstract

This keynote paper discusses the changes of the 1970's impacting the lead industry and expectations for the 1980's. Included in addition to a review of world and U.S. supply and demand is a discussion of technological considerations in the markets for lead and lead products. Particular emphasis is placed on the potential effects of government regulations on the industry during the next decade and a recommended business approach to these regulations.

Introduction

The decade of the seventies was one of great change for the lead industry. Technological innovation, coupled with a dramatic shift in market structure and new government regulations, caused virtual upheavals in many aspects of our business. I am pleased that the symposium committee invited me to prepare this keynote paper, because as we enter the era of the eighties I believe the lead industry is at a crucial juncture. The direction we take from here will depend on many factors -- many of which we cannot control -- but our business planners, researchers, engineers, production managers and others must become cognizant of the opportunities that await us, the pitfalls that must be avoided at all cost, and the means by which we can successfully improve our lot during the next ten years.

U.S. Supply/Demand Outlook

The recent flurry of activity in the lead market was the result of a complex set of interdependent events. U.S. demand was at or slightly below the high consumption rate of 1978, but supply fell below expectations. The situation in the U.S. was fueled by work stoppages in the primary sector , reduced availability of lead scrap for secondary operations, and a low level of imports due to stronger demand in the international lead market.

Compounding the marginal supply situation in 1979 was the unprecedented buying activity of the Soviet Union. Speculation as to the reasons for the large purchases include increased vehicle assembly, domestic lead production difficulties in the Soviet Union, and increased ammunition requirements due to recent Sino/Soviet tensions. Whatever the reason, the result was that the Soviets succeeded in bidding up the price of lead on the London Metal Exchange. As predicted, the Soviets were apparently unwilling to tolerate the high rate of foreign exchange in the long term, and thus LME prices began to moderate during the summer, a period when price adjustments on the international exchange traditionally occur.

The last increase in primary lead capacity occurred in the late 1960's resulting in a 37% increase in refined lead output. New mining projects have done little more than keep pace with smelter demand. The lack of commitment to new facilities stems from several factors; significant erosion of the industry's second largest market, gasoline additives, and the uncertainty of U.S. environmental regulations represent the two chief causes. With the development of new technology and the removal of other doubts concerning the viability of the lead business, I am looking for an increase in annual primary lead output in the U.S. approaching 150,000 tons over the next ten years.

Secondary lead output has in recent years been essentially dependent on scrap prices and availability. Nameplate capacity of the secondary lead producers is believed to be roughly 20% in excess of output and I project that secondary capacity will increase by as much as 200,000 tons in the next ten years as less lead is consumed for non-recoverable uses, such as gasoline additives, and more is consumed in storage batteries.

Lead consumption is monitored on a monthly basis by the U.S. Bureau of Mines. Reports filed through the cooperation of lead consumers are used by the bureau to classify consumption into 30 separate categories. While the bureau interacts with hundreds of lead consumers in compilation of its statistics, not all lead consumption is reported, and understandably so. For this reason I have listed in Table I figures for both "apparent" and reported consumption.

Overall, the outlook for both lead production and consumption appears favorable, with a 2.6% compounded annual growth rate for production and apparent consumption over the next ten years.

Consumption by major market segment is outlined in Table II, with the figures corresponding to projections of reported consumption. The category marked "SLI Batteries" refers to starting, lighting and ignition batteries, which includes the typical automotive battery. Its annual growth rate is expected to be about 2.6%. "EV Batteries" refers to the electric vehicle battery which we expect to become a significant lead consumer by 1990.

4

Table I: Outlook for the U.S. Supply/Demand Balance for Lead
(metric tons)

		1980	1985	1990
A.	Production	1,342,000	1,495,000	1,740,000
	Primary	590,000	655,000	725,000
	Secondary	752,000	840,000	1,015,000
B.	U.S. Imports	200,000	210,000	220,000
C.	U.S. Exports	5,000	5,000	5,000
D.	Net Inventory Change Decrease or (Increase)	27,000	---	---
E.	Apparent Consumption	1,510,000	1,700,000	1,955,000
F.	Reported Consumption	1,400,000	1,550,000	1,830,000

Table II: Outlook for U.S. Lead Consumption by Market Segment
(metric tons)

	1980	1985	1990
SLI & Industrial Batteries	865,000	1,010,000	1,125,000
EV Batteries	---	100,000	290,000
Gasoline Additives	175,000	86,000	68,000
Pigments	85,000	85,000	85,000
Ammunition	50,000	50,000	50,000
Solder	35,000	32,000	32,000
Cable Covering	10,500	10,000	9,000
Caulking Lead	4,500	4,000	3,000
All Other	175,000	173,000	168,000
Total	1,400,000	1,550,000	1,830,000

The use of lead in gasoline additives will be reduced by just under
5% annually, unless a major shift in policy toward lead-in-gasoline is
taken by the Environmental Protection Agency and Department of Energy.
In 1978 lead consumption for gasoline additives declined 40,000 tons and
a similar drop had been predicted for 1979. However, with the gasoline
crisis of last summer the EPA delayed implementation of the latest step
in its phase-down program until late 1980, essentially freezing gasoline
additive lead consumption at 1978 levels. Beyond 1985 the bulk of the
U.S. production of lead-based additives will be for export.

The remaining consumption categories of pigments, ammunition, solder,
cable covering, caulking lead and other are expected to display stable to
slightly decreasing consumption trends, largely due to replacement by

more economically feasible and environmentally safe materials, and by technological obsolescence.

World Outlook

The principal sources of lead mine production during the period 1957-1976 were mixed ores and lead ores. World mine production of lead from mixed ores (lead and zinc) during this period decreased from 80.7% to 62.9% of lead concentrate production, accompanied by an increase in the relative importance of lead ores from 13.1% to 28.6% of lead output. The rise in the importance of lead ores primarily resulted from new lead mine developments in the United States, Sweden and France.

Worldwide, lead mine production from mixed ores remains the dominant pattern. In Europe 67% of total lead mine production is from mixed ores; in Canada 92.2%; in Japan 48%; in Mexico 87.6%; in Peru 82%; in Australia 99.5%. In the U.S. only 10% of the lead mine production is from mixed ores, while 86% is produced from lead ores. It is possible that future world lead concentrate production will be dampened by a continuing depression in the world zinc market. U.S. lead concentrate production should remain strong throughout the next decade.

As shown in Table III, world refined lead production and consumption is forecast to grow at a compounded annual rate of 1.8% over the next ten years.

Table III: World Supply/Demand Balance for Refined Lead
(Primary and Secondary, 000 metric tons)

	1980	1985	1990
Production	3,800	4,155	4,500
Apparent Consumption	3,750	4,155	4,500

Projected mine development in Canada, Australia, Ireland and South Africa should increase their contribution to concentrate output. Canada appears to be the most active with two new mine developments scheduled to start production during the second half of 1979. One of these, Gays River owned by Esso Minerals Canada, has ore reserves estimated at 10.9 million metric tons of 2.78% Pb and 4.23% Zn. Yava Mines, owned by Barymin Exploration, has projected ore reserves of 1.1 million metric tons of 5.6% Pb. During the next five years, other anticipated Canadian developments include Grum, owned by Cyprus Anvil (estimated reserves of 25.2 million metric tons of 3.1% Pb) and Arvik, owned by Cominco (estimated reserves of 23 million metric tons of 4.3% Pb).

In Australia, production at the Que River Mine, owned by Cominco, is expected to commence by 1984. Ore reserves are estimated at 6.8 million metric tons. Completion of Broken Hill South's Cobar Mine is expected by 1983 and should add 260,000 metric tons of combined Cu/Pb/Zn ore to current production.

Bula Mines Ltd. has applied for planning permits for a lead/zinc site located near Tara mine at Navan, Ireland. Production is currently scheduled to begin in 1981, yielding 14,500 metric tons of lead per year at full capacity.

6

African concentrate output should show a marked increase by mid-1980, with the opening of the Aggenays mine. The mine will have an eventual annual capacity of 90,000 metric tons of contained lead, but should only produce about 25,000 tons in 1980.

Technological Considerations

One of the major technical developments of the 1970's is the maintenance-free battery. About 40% of the automotive batteries produced last year used non-antimonial grid alloys, and recent products such as precision rolled lead strip will expand the production of maintenance-free batteries using highly automated processes.

The trend toward smaller cars has resulted in a trend toward lighter original equipment batteries. When another wave of downsizing fever hits Detroit in the mid-80's, auto battery lead requirements will be again reduced. Other advances in battery technology may serve to increase battery efficiency and reduce lead requirements, but the lead-acid battery should not be appreciably different from its present form.

With the lead-acid battery being responsible for 61% of all lead consumption, the lead industry is perhaps in a vulnerable position. The subject of technological obsolescence cannot be dismissed out of hand, but the lead-acid battery system has inherent advantages over any contender, including: technical simplicity, raw material economics, ease of manufacturing, performance stability, and minimal maintenance. Coupled with these distinct advantages are the facts that the major capital investment in lead-acid battery facilities has already been made, and that the lead-acid battery is a known entity to distributors, mass merchandisers and consumers.

One key to our lead supply/demand projections is the development of a viable electric vehicle market. The EV industry at present can be politely termed a "cottage industry" with some 20 or more manufacturers in the U.S. and many others in Europe and Japan. However, recent powertrain and storage battery advancements make the futuristic concept of a million EV's humming down our streets a reality which is just around the corner. Research on multi-rotor motors and 1000 cycle deep discharge batteries suggest that a range of 150 to 200 miles and battery life of four to five years may be commercial developments within one year. Luxury features such as on-board air conditioning are even being considered.

During the next few years many questions will be answered about whether the American people can integrate the concept of a limited-range vehicle with their love of the internal combustion automobile. Certainly there is a need for the fuel conservation feature of EV's, and driving statistics support the appropriateness of EV's entering the mainstream of U.S. transportation. Nearly 35% of all driving involves commuting to and from work, with an average driving distance of less than 10 miles. Furthermore, 91% of all trips taken in an automobile in the U.S. involve distances under 21 miles. This is a strong endorsement for the market potential of electric-powered autos in the second and third vehicle markets. Dramatic increases in the price of gasoline have removed all doubts that electrics are cheaper to operate. Initial capital costs may involve premiums of 25% to 100% today because of the low volumes of EV's being produced in custom and small mass-production operations. With proper scale-up, premiums could conceivably be reduced to the range of 10%, giving electrics a clear economic edge. Most of the major U.S. auto producers have announced plans to market EV's by 1985, with the first generation design most likely to incorporate advanced lead-acid batteries.

To assist in demonstrating the viability of lead-acid powered electric vehicles, St. Joe and other corporations have added units to their company fleets employing state-of-the-art EV technology. We acquired three converted Dodge OMNI autos for our fleet last October and have an on-going market evaluation program in the St. Louis metropolitan area. Our vehicles, are powered by 16 heavy duty lead-acid EV batteries and feature up to a 70 mile range at 45 mph. Top speed is rated at 65 mph and I am particularly impressed with the vehicle's acceleration performance and ease of handling. Recharge costs in our location are typically less than a penny a mile.

At our lead smelter we have successfully implemented into regular service an electric-powered, full-size cargo van. The vehicle travels a daily regime of 15 to 20 miles in servicing and collecting air samples from remote environmental monitoring stations. The key to the use of electrics in industrial applications is the proper matching of service needs to vehicle capabilities.

The additional energy conservation feature of EV's is that recharging normally occurs during non-peak hours. In addition, in St. Louis the EV is a petroleum conserver when charged at any hour, as 98% of our electricity is generated at coal-fired stations.

Another future development involving utilities will be the use of storage batteries to shave peak-power generating requirements. Storing energy produced during off-peak demand periods for use during periods of high demand is being tested under a government program in New Jersey. The lead-acid battery is not the most likely candidate for this application, as several advanced battery candidates in a fixed or non-mobile installation, where space and auxiliary support units are not restricted, have superior operating characteristics. The R&D work on the load leveling concept will likely continue into the 1990's before we see any significant commercialization.

Government Regulations

Time magazine perhaps overstated the government regulation problem when it reported, "The growing number of federal rules and regulations . . . seem to float out from Washington as casually as children blow soap bubbles."

We recently took a survey at St. Joe to determine just how many agencies impact our business. It turned out that the job would have been easier had we asked what agencies do not affect us.

Murray Weidenbaum, Director of the Center for the Study of American Business at Washington University, in his book The Future of Business Regulation, succinctly describes how government regulation has been a subtle, yet onerous, barrier to productive economic activity in this nation. Breaking the cost of regulations into five areas, he discusses:

1. The cost to taxpayers to support government regulators,

2. The cost to the consumer in the form of higher prices to cover the extra expense of producing goods and services due to government regulations,

3. The cost to workers due to jobs eliminated by government regulations,

4. The cost to the _economy_ resulting from the loss of smaller enterprises which simply cannot afford the burden of government regulations, and

5. The cost to _society_ due to the reduced flow of new and better products and a reduced growth in the standard of living.

We all see the direct costs of government regulations. In 1978 the cost of government mandated safety and environmental control equipment added $666 to the price of a new auto. Each new house nationwide has an average $2000 added to its cost due to government regulations. That is about $4 billion nationwide.

Other important direct costs we can easily single out are the budgets of the federal agencies and bureaus which must administer the regulations. This figure was $4.8 billion in fiscal 1979, up 115% in just five years.

There are many indirect costs of regulations as well. The paperwork involved with regulations is an expensive time consuming preoccupation with reports, applications, questionnaires, and replying to agency requests. The work is mundane and unproductive, and the delays caused by the governmental process is becoming more and more severe. The mining industry in the U.S. has encountered unexplained delays in obtaining permits to explore for minerals on government lands available for commercialization. A two or three year delay on a drilling permit is not an unusual circumstance.

The sum of all direct and indirect costs of federal government regulations, ignoring the nuisances of state and local costs, has been estimated by the Center for the Study of American Business to be $66 billion annually -- split out at $3 billion for taxpayers and $63 billion for business. In other words, each dollar appropriated by Congress for government regulation results in $20 of expense to the private sector.

This 20:1 leverage brings about what Dr. Weidenbaum calls the "induced" costs of government regulations. It should be obvious that business does not have a bottomless pocket, and that its capital resources are limited. But the necessity of devoting a larger and larger portion of each capital dollar into non-productive compliance with government regulations can only serve to have a negative impact on our innovation and development. Is it any wonder that growth in worker productivity in our nation has suddenly come to a virtual standstill.

Before making any suggestions as how we should tackle this **insidious** problem, let me specifically focus on the problems facing the lead industry. Our most pressing regulatory headaches are the OSHA workplace standard for lead, the EPA Ambient Air Quality Standard for Lead, and various hazardous wastes regulations.

Both the OSHA and EPA lead standards, promulgated in late 1978, are under litigation, with court decisions expected sometime in 1980. The OSHA standard involves drastic reductions in workplace exposures to lead and decreased reliance on the use of respirators to accomplish these goals. The EPA standard involves an ambient air fenceline measurement, without regard to background concentrations from non-industrial sources. The EPA lead-in-air limit is believed to be far more stringent than required to provide an adequate margin of safety. Both the OSHA and EPA emphasis is on the use of costly engineering controls, but with dichotomous goals; OSHA's prescription is for venting of the workplace while EPA's aim is to keep emissions contained in the plant. In each case we believe that the technology does not now exist which would permit compliance by either primary or secondary smelting and refining operations.

What in effect EPA and OSHA are advocating is a total rebuilding of the lead industry to satisfy a pair of incompatible regulations with highly questionable limits. While an environmentally safe process has not been yet developed which would meet EPA and OSHA standards, the capital requirements to rebuild the industry could easily exceed $1 billion.

Proposals for government regulations covering hazardous waste disposal and hazardous spills have come from several corners since the LOVE canal incident in the state of New York. The Carter Administration proposed the creation of a "superfund" to pay for the cleanup of spills and wastes involving oil, hazardous materials and hazardous wastes. While it is difficult to argue against the need for a mechanism to react promptly in the case of an environmental accident, this fund would circumvent procedures already contained in laws such as the Clean Water Act, and would require funding not from government or producers of waste products or disposers of hazardous wastes, but from raw material suppliers in the mining and chemical industries. Lead, for example, would be initially assessed $2 per ton of production, with potentially higher "tax" levies as the needs of the program were developed. We support the position of the American Mining Congress in favor of a program limited to provide for the immediate cleanup and stabilization of imminent hazards to public health caused by past inadequate waste disposal practices. However, we believe that the cost of cleanup to affected individuals and companies should be drawn from public funds and not assessed to industrial concerns totally unrelated to the nature of the incident.

A further concern of our industry is the regulatory manipulation of hazardous wastes. At the end of 1978, the EPA proposed a "toxicant extraction procedure" (TEP) to define whether or not a material is suitable for disposal in a sanitary landfill, and this proposal would have labeled as "hazardous" any material that failed the test. The EPA proposal was immediately attacked by many parts of the scientific community as being a poor test, selected by EPA without proper input from professionals in the field. The test is not sufficiently discriminating, as too many materials failed the test -- from oak leaves to concrete. Our concern is that lead mining tailings and blast furnace slag do not pass the test. This would call for them to be labeled as "hazardous materials" even though the experience of the past decade has demonstrated that these materials are environmentally safe.

Another manner in which government regulations will continue to impact the lead industry is the promulgation of rules regulating our end-use markets. The most notable example is the EPA phasedown schedule for the reduction of lead in gasoline. The rules set in the early seventies would have seen essentially all lead eliminated from U.S. gasoline production in ten years. The disturbing part of the EPA phasedown program is that the manufacturer of unleaded gasoline wastes crude oil at a time when the U.S. cannot tolerate an increased burden on this limited resource. The savings of crude oil brought about by the EPA's delay last year in the phase-down program is significant; 350,000 barrels of oil per day worth $2.5 billion per year. It is likely that complete conversion to unleaded fuel could cost American consumers an additional $7.6 billion per year. The EPA's phase-down program will effectively eliminate the lead industry's second largest market, except for a residual export business by domestic gasoline additive producers. However, the problems of U.S. refineries in converting to unleaded production coupled with the domestic energy crisis have resulted in numerous delays and modifications of the phase-down program. Thusfar, increased demand by battery and other consuming industries has roughly kept pace with the decline in lead consumption for gasoline additives, and so the lead

10

industry has not yet experienced the trauma of a vast oversupply situation. However, in anticipation of such a supply imbalance in future years the primary industry still has no announced plans for any new mine/mill/smelter operations and the secondary lead industry has experienced a marginal increase in nameplate capacity in the past few years.

The options available to our industries in dealing with problems involving government regulations are rather limited. It would be impractical for us to crawl back into the tortoise shells we occupied as conservative traditionalists during the first sixty years of this century. It would be equally imprudent for us to steadfastly and stubbornly oppose all regulatory efforts. What is needed by our industry is the adoption of a "common sense" approach to regulation.

One side of the common sense approach involves the need for social responsibility in the business world. Many volumes have been produced on this subject, and most business schools now feature in their required curriculums one or more courses on the subject of "Business and Society". But an important aspect of social responsibility which must not be ignored is the role of business in opposing over-zealous regulation, regulation which goes far beyond what is reasonable for the business community to do in promotion of common good, regulation unnecessarily stringent to the degree that business is irreparably harmed.

Confrontation with environmentalists and special interest groups is not the strategy I am advocating. Confrontation is the most effective tool of the self-serving special interests. What I do propose is that industry take advantage of its inherent skills and resources -- not just its financial resources and top management -- but the largely untapped resources of its employees, customers and vendors. Soliciting the assistance of the special interests internal to our business community is a study in effective communications. It has been suggested that the personal impact of 1000 deaths 10,000 miles away can be equated with 100 deaths 500 miles away and to 1 death a block away. Such is the role of perception in mustering support for an attack upon unjust regulation. It does little good for me to provide exhortations to the media like "over 3000 direct and 10,000 indirect jobs would be lost if the lead industry in Missouri were to shut down". But tell an employee who has lived and worked and raised a family in a smelter community, without health incident, that the EPA is threatening to impose rules which are not technologically feasible and may take away his chance to earn a livelihood, and that man will do more than make some innocuous gesture. He is likely to write and telephone his congressman and senators, he may even pressure his union into a position to at least not support the implementation of the rule, and he will coerce members of his family, neighborhood, social organizations, religious community and others to take similar action.

Vendors and customers will often take positive steps in helping to fight the battle which threatens a viable customer/supplier. Often this can result in a broadening of the base of support into other states or at least other congressional districts.

Upper management personnel and technical staffs can similarly solicit support from service organizations in the affected communities, and from trade associations, utilities transportation suppliers and other groups with a vested interest in the health and vitality of the industry.

An additional resource we should not overlook is state government. While the federal authorities function to set rules and regulations it is agencies within state government which are often required to establish implementation plans and interact directly with industry. It is at this level that the dilemma of satisfying federal rules vs. jeopardizing jobs comes into play. Nothing can substitute for good rapport with state and local officials.

Summary

It will be interesting to look back ten years from now on the lead industry projections I have submitted to you today. With the metamorphis taking place in our consuming segments and market structure, with new opportunities like electric vehicles and load leveling on the horizon, and with the continuing threat of even more restrictive and debilitating government regulations, we in the lead industry do find ourselves at an important crossroads. As you may have gathered I am very bullish on the prospects for our industry. I am convinced that our engineers, planners and management personnel are well equipped to tackle the problems I have outlined, and that the era of the eighties will be challenging, at times very frustrating, and in the end quite rewarding.

ZINC IN THE 1980's

James L. Broadhead

President
St. Joe Zinc Company
Pittsburgh, Pennsylvania

The zinc industry has just passed through perhaps the most tumultuous decade in its history--a period encompassing shortages as well as oversupply, record profits and losses, unprecedented demand followed by market stagnation, and the beginnings of new eras of governmental regulation and energy consciousness. The conditions that now exist offer difficult challenges to all of us associated with the businesses of mining, smelting, or consuming zinc. There is severe over capacity, slow growth in demand, loss of critical markets, and prices far too low for a reasonable return on invested capital. As bleak as conditions now look, however, there are favorable opportunities for our industry that, if vigorously and intelligently pursued, will enable us to overcome the difficult problems confronting us and restore our industry to good health. This is the challenge I present to you today.

Before discussing the factors likely to affect our future, it would seem appropriate to reflect on the industry as it existed at the time of the last World Symposium and to review what has taken place in the intervening years. I'll begin by describing the conditions that prevailed at the time of the last World Symposium held in St. Louis in October, 1970.

The 1960's

The 1960's were good years for the zinc industry. Consumption grew rapidly and mine and smelter production increased to meet demand.

Consumption

Free world zinc consumption grew at a compound annual rate of over 6%, reaching 4.06 million tons* in 1969 (Table 1). Major areas of growth included Japan and the United States with growth rates of 13.6% and 4.9%, respectively.

The large improvement in Japanese zinc consumption arose from the rapid construction of a sophisticated industrial infrastructure requiring such products as galvanized sheet roofing, siding, and ducting, and galvanized structural members for

---

*Unless otherwise noted, all references in this paper to "tons" shall mean metric tons of 1,000 kilograms or 2,204.62 pounds avoirdupois.

Table 1 FREE WORLD ZINC METAL CONSUMPTION

(Thousand Metric Tons)

	1960	1969	1970	1971	1972	1973	1974	1975	1976	1977	1978 (Est)
North America											
U.S.	806	1,236	1,082	1,107	1,265	1,348	1,220	840	1,022	1,041	1,051
Canada	49	108	96	101	117	135	128	150	134	121	145
Mexico	23	44	48	42	49	61	57	63	72	60	72
Sub Total	878	1,388	1,226	1,250	1,431	1,544	1,405	1,053	1,228	1,222	1,268
Europe	1,102	1,458	1,430	1,428	1,550	1,711	1,671	1,254	1,472	1,462	1,666
Japan	189	595	619	628	717	815	695	547	699	665	733
Rest of World	217	619	626	644	709	807	815	728	761	776	868
TOTAL	2,386	4,060	3,901	3,950	4,407	4,877	4,586	3,582	4,160	4,125	4,535

Source: ILZSG

14

highways, railroads, and electrical and telephone transmission systems. Much of the steel produced in the rapidly expanding Japanese mills was exported, so that zinc consumption growth that might have been experienced in Western Europe and the United States took place in Japan.

In the United States, zinc consumption benefited from strong growth in both galvanizing and die casting. Galvanizing was spurred by the construction of ten large continuous galvanizing lines. Every major American steel company built a modern efficient line capable of consuming 9,000 tons of zinc per year. Zinc consumption for continuous galvanizing sheet increased from 178,000 tons in 1960 to 244,000 tons in 1969, a compound annual rate of 3.6%. A major factor in this growth was the selection of prepainted galvanized sheet for preengineered buildings. The galvanized coating provided excellent protection against corrosion while the paint finish added an additional element of protection as well as improved aesthetics.

An even bigger success story in the United States during the 1960's was the use of zinc die castings, which grew at an annual rate of 6%. The automotive market was particularly important in this growth. To refresh your memory, it was common for an intermediate size automobile to weight 4,000 pounds and to contain 80 pounds of zinc die castings. The 1968 leader in zinc usage, for example, was the Chrysler Imperial 4-door hardtop, which contained 207 pounds of zinc die castings. This included a 27 pound grill, a 22 pound instrument panel assembly, 11 pounds of taillight housings, and 13 pounds of rear door glass channels.

Mining

The large growth in free world consumption was accompanied by a corresponding increase in world mine production, fostered primarily by the discovery of large ore bodies in Canada (Table 2). The startup of Brunswick Mining and Smelting at Bathurst, New Brunswick, in 1964, Pine Point Mines in the Northwest Territories in 1965, Ecstall Mining Ltd. in Ontario in 1967, and Anvil Mining Corp. Ltd. in the Yukon in 1969, increased Canadian zinc mine production by approximately 800,000 tons of zinc in concentrates per year, establishing Canada as the preeminent zinc mining nation. Other major mine developments in the free world were the Yeoun Wha mine in South Korea and the Mogul and Tynagh mines in Ireland.

Toward the end of the 1960's the United States zinc mining industry began to experience difficulties. Previously significant reserves in Montana, California, and Nevada were downgraded due to low metal prices, increased mining costs and, in some instances, a lack of operational smelting facilities within a reasonable distance. Similarly, zinc mining essentially ceased in Kansas, Oklahoma, and southwest Missouri, an area that earlier had been the leading zinc producer in the United States. At the same time zinc ore reserves in Pennsylvania, New Jersey, and New York increased modestly as a result of expanded mine development activities. The net result was a loss of approximately 6.35 million tons of zinc reserves from previously existing zinc mines in the United States. Fortunately, the newly discovered Viburnum, Missouri, lead belt contained approximately 9 million tons of by-product zinc reserves. Thus, in the aggregate, United States zinc reserves grew by 2.65 million tons.

Smelting

Smelter production in the free world (Table 3) was able to keep up with rapidly increasing demand and, by 1969, free world capacity (Table 4) was approximately 4.67 million tons. The horizontal retort process, first commercially adopted in 1800, accounted for 15% of that total capacity; the electrolytic process introduced in 1915, accounted for 56%; the vertical retort process, developed in 1930, contributed 14%; the electrothermic process, commercially introduced in 1936, added 4%; and the Imperial Smelting Process, a newcomer in 1950, accounted for 11%.

15

Table 2   FREE WORLD MINE PRODUCTION OF ZINC

(Thousand Metric Tons)

	1960	1969	1970	1971	1972	1973	1974	1975	1976	1977	1978 (Est)
North America											
U.S.	395	502	484	456	437	435	448	424	438	407	305
Canada	308	1,096	1,136	1,134	1,129	1,227	1,159	1,052	979	1,067	1,029
Mexico	271	252	267	265	271	271	262	220	268	265	257
Sub Total	974	1,850	1,887	1,855	1,837	1,933	1,869	1,696	1,685	1,739	1,591
Europe	481	745	731	705	726	752	784	801	857	946	1,047
Japan	157	269	279	294	280	263	240	253	259	275	274
Rest of World	846	1,057	1,186	1,207	1,337	1,361	1,349	1,546	1,552	1,632	1,707
TOTAL	2,458	3,921	4,083	4,061	4,180	4,309	4,242	4,296	4,353	4,592	4,619

Source: ABMS

16

Table 3   FREE WORLD PRIMARY ZINC METAL PRODUCTION

(Thousand Metric Tons)

	1960	1969	1970	1971	1972	1973	1974	1975	1976	1977	1978 (Est)
North America											
U. S.	787	1,008	866	769	641	605	575	450	486	430	457
Canada	238	423	418	373	476	533	438	426	472	494	496
Mexico	53	83	81	83	84	72	137	150	171	175	171
Sub Total	1,078	1,514	1,365	1,225	1,201	1,210	1,150	1,026	1,129	1,099	1,124
Europe	889	1,359	1,381	1,273	1,471	1,573	1,713	1,454	1,584	1,673	1,642
Japan	181	712	676	716	804	843	850	698	742	775	768
Rest of World	252	501	562	563	637	632	649	575	640	700	753
TOTAL	2,400	4,086	3,984	3,777	4,113	4,258	4,362	3,753	4,095	4,247	4,287

Source: ILZSG

17

Table 4 FREE WORLD PRIMARY ZINC METAL PRODUCTION CAPACITY

(Thousand Metric Tons)

	1969	1979	1990
North America			
U. S.	1,085	655	655
Canada	505	635	760
Mexico	90	200	275
Sub Total	1,680	1,490	1,690
Europe	1,775	2,330	2,510
Japan	635	965	965
Rest of World	575	935	1,185
TOTAL	4,665	5,720	6,350

By the end of the decade, it was apparent that the horizontal retort process was technically obsolete and closedowns in Europe were already underway. These closings and the rapid growth in free world zinc consumption spurred new smelter construction. By the end of the 1960's, 540,000 tons of additional zinc refining capacity--mostly electrolytic--was scheduled for construction, nearly 60% of which was in Europe.

Outlook at the Time

Because the decade of the 1960's was a stable, long lasting period of sustained growth, the common expectation was that future economic growth would also be steady and vigorous. Commentators in the late 1960's and early 1970's generally looked at future zinc consumption with confidence, predicting compound annual growth rates ranging from 3.5% to over 5%.

The 1970's

Economic Influences

The rapid growth predicted for the 70's did not materialize. While 1973 was a boom year for all major metals, on the whole the world economy in the 1970's has not been as robust and as conducive to increased metal consumption as in the prior decade. The O.E.C.D. Index of Industrial Production, which includes the major European nations, Canada, the United States, Japan and Australia, grew at a compound annual rate of 3.3% from 1970 through 1978 as compared to a 5.6% rate during the immediately preceding period. Furthermore, most of the growth was achieved by 1974, the major recession of 1975 reducing the Production Index so significantly that recovery was not experienced until 1977. Annual growth since 1974 has averaged only 2%.

Two events stand out as having a profound and long lasting impact on industrial production. The first, the October 1973 Arab oil boycott, proved to be the forerunner of a series of production and price actions by OPEC members that would bring the price of oil in July 1979 to ten times its price at the start of the 1970's. The rapid escalation in oil prices focused attention on wasteful energy practices and led to a shift to more energy efficient designs, processes, and applications. The second event was the organization in December 1970 of the U.S. Environmental Protection Agency. The EPA and its counterparts in other countries have had a pronounced effect on capital costs and operating costs of almost all industrial plants in the developed nations of the free world. Both events had a dramatic impact on zinc consumption and production.

Consumption

During the period 1969 through 1978, free world consumption of slab zinc grew from 4.06 million tons to 4.54 million tons (Table 1), an increase of only 1.2% annually. The growth occurred disproportionately among the major consuming areas, with Japan and Europe growing at rates of 2.3% and 1.5%, respectively, United States consumption declining at 1.8%, and the other free world nations growing at 3.9%. These disappointing statistics reflect a lower per capita consumption of zinc, even as our economies grew, and a gradual loss of importance of zinc as an element of industrial production.

The reduced growth has occurred in all major market segments. In galvanizing, highway construction is less than in recent years, especially in the United States; industrial construction is down as the economies of the developed nations have been stimulated by greater expenditures on consumer goods and services rather than by the capital projects that previously provided substantial markets for zinc; and galvanizers

19

have been more efficient in their use of zinc. The galvanizing lines installed in the late 1960's, for example, decreased zinc consumption per pound of steel by about 15% through more efficient coating control systems.

In the die cast area, there has been a significant reduction in the use of zinc in automobiles in the United States due in part to design and safety changes, but primarily due to weight reduction efforts in response to governmentally mandated fuel economy standards. Because zinc weighs almost as much as iron, it is a natural target for weight reduction. I referred earlier to the typical intermediate size 1968 model American automobile, which contained nearly 80 pounds of zinc die castings. In 1979, the average American car used slightly less than 28½ pounds of zinc die castings.

In brass, consumption has been adversely affected by decreased industrial construction, reduced military usage, and by substitution of plastic piping and hardware.

Mining

World mines in the 1970's were more than adequate to satisfy the modest increase in zinc consumption (Table 2). Through the 1970's Canada maintained its position as the dominant miner of zinc, accounting for more than 25% of production. Large new developments included the Mattabi Mines in 1972 and Sturgeon Lakes Mine in 1975. The major developments in Europe were the Navan deposit in Ireland and the Rubiales mine in Spain. Several mines in the free world closed in 1977 and 1978 because of excessively low zinc prices resulting from excess capacity.

Smelting

While there were initially some worries concerning the adequacy of supply in the 1970's, those concerns evaporated by the middle of the decade. Disappointing consumption, combined with the large increase in smelting capacity, resulted in a severe supply/demand imbalance. In 1978, the average free world smelter utilization was less than 75%, considerably lower than that experienced 10 years earlier.

Of particular interest is the marked shift in the geographical distribution of zinc production facilities. The greatest change, of course, was in the United States (Table 5) where ten domestic zinc plants closed in 1969 or thereafter. Of these ten smelters, one electrolytic plant was eventually renovated and reopened and one pyrometallurgical plant was replaced by a new electrolytic plant at the same location. During the same period one new plant, an electrolytic facility, was constructed at a greenfield site. In the aggregate this represented a loss of approximately 430,000 tons of annual productive capacity. The United Kingdom was the only other major industrial nation to experience a severe loss in productive capacity, while Germany, Italy, and Japan increased productive capacity significantly.

The failure of the U.S. industry to maintain capacity sufficient to satisfy U.S. demand appears to be the result of a complex interaction of government policies and industry attitudes. There is little doubt that there is a larger burden of governmentally created uncertainties and disadvantages on U.S. producers than on their competitors and that foreign governments do more to help and less to hinder their mining and metals industries than does the United States. Price controls imposed during a period of rising world demand and prices, and a one million ton government stockpile overhanging the market are but two examples of government action in the United States that acted as a suppressant to new capacity. On the other hand, in some cases, American management appears to have been unwilling to take reasonable risks of the type assumed by their foreign counterparts.

20

Table 5   U. S. PRIMARY ZINC SMELTING INDUSTRY

1969-1979

Company	Location	Type*	Capacity**	Remarks
Amax	Blackwell, OK	HR	82,000	Closed - 1973
American Zinc	Dumas, TX	HR	53,000	Closed - 1971
American Zinc	Sauget, IL	E	76,000	Closed - 1971; renovated and reopened by Amax - 1973
Anaconda	Great Falls, MT	E	147,000	Closed - 1972
Anaconda	Anaconda, MT	E	79,000	Closed - 1969
Asarco	Amarillo, TX	HR	48,000	Closed - 1975
Asarco	Corpus Christi, TX	E	100,000	
Bunker Hill	Kellog, ID	E	95,000	
Eagle Picher	Henrietta, OK	HR	36,000	Closed - 1969
Jersey Miniere	Clarksville, TN	E	80,000	New plant - 1978
M & H Zinc	Meadowbrook, WV	VR	41,000	Closed - 1971
National Zinc	Bartlesville, OK	HR/E	50,000	Electrolytic replacement - 1975
New Jersey Zinc	Depue, IL	VR	64,000	Closed - 1971
New Jersey Zinc	Palmerton, NJ	VR	82,000	
St. Joe Zinc	Monaca, PA	ET	180,000	

* HR   Horizontal Retort
  E   Electrolytic
  VR   Vertical Retort
  ET   Electrothermic

** Annual Capacity in Metric Tons

The 1970's saw the increasing predominance of the electrolytic process. This process now comprises more than 75% of zinc production facilities, as opposed to approximately 56% in 1969. The growth was caused primarily by the environmental problems, high energy costs, and manpower requirements of the vertical and horizontal retort processes; the trend toward special high grade zinc in the early 1970's, especially for die cast applications; and significant improvements made in the electrolytic process. These improvements include lower operating costs achieved through automation, particularly in the cell house, larger continuous units, and higher recovery due to introduction of hot acid leach processes for the extraction of zinc from ferritic calcines that had previously been discarded.

The 1980's

Current Trends

Historically, the consumption of zinc has been closely related to the health of the world economy; consumption has been high in periods of substantial growth and low during economic recessions. Thus, zinc has tended to be an especially cyclical commodity and, although the cyclical lows have been painful, a producer could gain some solace from the knowledge that good times had always followed such lows in the past. Unfortunately, as we have seen, there appears to have been a negative change in this situation in recent years that has deprived producers of this small measure of comfort. Economic growth in the free world has slowed, and the continuing oil crisis appears to offer little prospect of an early return to the strong economic growth previously enjoyed. Furthermore, the economies of the developed nations are increasingly reliant on consumer producers and services rather than on the capital projects that traditionally have been the source of so much zinc demand. Also, there has been a structural dislocation in the zinc market due to the partial loss of the large U.S. die cast automotive market.

If zinc consumption were to continue to grow at the rate experienced since 1969, more than ten years would be required for demand to reach 90% of present free world capacity. Adding the new capacity that we expect will be coming on stream (Table 4) increases the time required beyond the year 2,000. These depressing statistics are not presented as a projection of what will actually happen. We all know that markets do not grow or decline at uniform rates and that the zinc business has been and will probably continue to be unpredictable. Certainly the favorable results of the 1960's were not an accurate guide to the 1970's. These figures are presented only to emphasize the seriousness of our current plight and to focus attention on the compelling need to alter the present supply/demand imbalance. Failure to change the situation in a reasonable period of time will have serious, adverse economic consequences to the entire industry.

Consumption

Zinc Characteristics - It would seem appropriate to begin an examination of the prospects for increased consumption by reviewing the characteristics of zinc that determine its usefulness in the market place. Zinc has a low melting point and can be cast in fine detail into complicated shapes. In addition, it is anodic to steel and has a low, uniform rate of corrosion. Zinc is also easily transformed into various grades of zinc oxide having useful chemical, electronic, optical, and thermal characteristics.

Zinc is not a miracle metal, however, and is not without its disadvantages. It is relatively heavy, having almost the same density as iron and almost three times the density of aluminum. Also, it has a low strength to weight ratio.

Traditional Uses - Traditionally, zinc's low, uniform corrosion rate and its high position in the electromotive series of metals have led to its use in galvanizing. Its low melting point and excellent castability have led to its use in die casting, the major applications being in builders' hardware and automobile components. Its oxides have been used for the manufacture of rubber, paints, ceramics, chemicals, and photocopy paper. In addition, its favorable characteristics when alloyed with copper account for its use in the brass and bronze business.

Unfortunately, there are several available substitutes for zinc in most of its applications: aluminum, plastic, and magnesium products can be substituted for zinc die castings; aluminum, paints, weathering steels, and plastics can replace zinc for corrosion protection of steel in many applications (and corrosion-protected steel can be supplanted by other materials); stainless steels, plastics, and aluminum can be substituted for brass; and oxides of titanium and magnesium as well as other materials can replace zinc oxide in some chemical and pigment applications.

We have found in recent years that zinc has lost position in the automotive die cast market to plastics and aluminum because of its inherent weight disadvantages. It has also lost opportunities in other markets to previously existing or newly developed products that have some of zinc's favorable characteristics and that have been more aggressively marketed.

Growth Opportunities - A job for all of us in the zinc industry, quite obviously, is to mount effective research, development, and marketing programs that will capitalize on the favorable characteristics of zinc, while avoiding or minimizing its unfavorable traits. We will be supported in our efforts by the growing trends in our society toward increased conservation of non-renewable natural resources and demand for longer product life. Zinc itself is highly corrosion resistant and is very effective in protecting steel, the major structural material. Thus, it seems ideally positioned to respond to the desire for conservation.

We will also be helped in certain applications by the urgent need to conserve energy. The conversion of zinc ore in the ground to a pound of refined slab requires only 1/4 to 1/3 the energy required for the comparable production of a pound of aluminum or plastic. Even on a volume basis zinc is less energy intensive than aluminum, its major competitor in some applications. The low melting point of zinc also results in energy savings in fabrication. Furthermore, zinc's position in the electromotive series makes it useful for electrochemical applications that can divert energy consumption away from certain scarce or costly fossil fuels.

The ability to use these trends and the other favorable characteristics of zinc to our advantage lies with us. What is needed is the commitment to vigorous research, development, and marketing efforts rather than mere wishful thinking.

Thin-Wall Technology - A good example of what can be done in research and market development is thin-wall die casting. This technology, developed in response to the automotive industry's designing out zinc die castings because of their weight, capitalizes on zinc's excellent casting characteristics while minimizing its weight. In the past, a zinc die cast product gained its strength from the dense, chilled skin, the center section of the casting providing little or no mechanical properties. The thickness of the die casting depended not on the functional requirements of the part, but on processing technology. Through extensive research conducted by ILZRO, procedures were developed that virtually eliminated the non-effective center of a die cast component, thereby saving 30-50% of the casting weight. These so called "thin-wall" die castings have helped slow the rate of loss in the automotive area and have actually recaptured several parts previously lost. More research and development of this type and of the type done by CSIRO in Australia is necessary if we are to have strong market growth in the future.

Corrosion-Protection - I have already mentioned the trend toward longer product life and increased conservation of natural resources. The United States National Bureau of Standards issued a report in 1978 stating that corrosion in the

23

United States alone cost an estimated $70 billion in 1975 of which $10 billion was avoidable. It seems safe to assume that most of the loss was due to rusting of steel, the major structural metal in our society. Coating steel with zinc is a proven method of protection against corrosion and could offer us excellent opportunities for growth. Looked at another way, there are about 425 million tons of steel produced in the free world each year, less than 5% of which are galvanized. An increased market penetration of only 1% would increase annual free world zinc consumption by 290,000 tons, about 6½% of 1978 consumption.

The automobile market alone seems to offer great potential. The quest for lighter vehicles to conserve energy is leading to the use of thinner gauge steel. These body parts will require corrosion protection to maintain product life, thereby offering attractive opportunities for increased usage of zinc-based coatings on one or both sides of the steel. This trend toward greater automotive corrosion protection may be accelerated by governmental action. There is already Canadian legislation concerning the life expectancy of cars, and in the United States, the Department of Transportation has called on automakers to improve rust warranties and the National Highway Safety Administration recently forced Fiat to buy back cars from consumers because of excessive body corrosion.

We should keep in mind, of course, that these and other opportunities in the corrosion resistance area are not lost on our competitors and we can count on aggressive marketing programs by producers of other metals and organic coatings. It is heartening, therefore, to see the vigor with which the American Hot Dip Galvanizers Association is pressing its "FutureZinc" program and the enthusiasm with which it is being supported by the Zinc Institute. It is essential that we also intensify fundamental research in the galvanizing area in order to better understand the governing features of the galvanizing process and the controllable factors that influence the type of galvanized coating that is developed.

The Building and Construction Committee of the Zinc Institute has targeted another area in the corrosion resistance market offering significant opportunities for zinc usage. The committee has found that with little or no technological changes, galvanized steel studs and joists can be directly substituted for traditional wooden products in residential construction. Factors such as price, availability, and quality of lumber products are beginning to favor the consistent quality of galvanized steel in this large market. The builders of the world, however, are unlikely to beat a path to our doors, and an effective educational program is required if this market is to come even close to achieving its large potential. A major effort along these lines will be made by the Zinc Institute in connection with the January 1980 National Association of Homebuilder's Show in Las Vegas.

Other Growth Opportunities - There appear to be other growth opportunities in areas capitalizing on zinc's favorable characteristics that could alter the unattractive consumption figures determined by extrapolating present conditions and trends. Builder's hardware, for example, is now a major market for zinc die casting. In this market, the weight of such products as door and window hardware, locks and keys, furniture and cabinet hardware, and bathroom and plumbing fittings is not a disadvantage, and the excellent castability of zinc is a distinct benefit. Furthermore, the ability to chromeplate or simulate copper, brass, and pewter finishes adds versatility to the use of zinc.

One must also consider the large potential new markets in nickel zinc, zinc chlorine, and zinc bromine batteries for electric vehicles and utility load leveling. Electric vehicles were relatively plentiful in the 1920's, but were made obsolete by the improved performance of gasoline engines. While conventional gasoline and diesel powered automobiles still perform better than electric vehicles, improvements in battery technology and the shortage of oil may provide a fresh opportunity for this quiet and non-polluting method of transportation. Even limited production of zinc battery powered vehicles could increase zinc consumption noticeably. For example, about 220 pounds of zinc are incorporated in the nickel zinc batteries required to propel a single automobile. Were such batteries ultimately to capture a market for one million vehicles, the zinc required would be more than 100,000 tons.

Keep in mind, however, that for the past ten years breakthroughs in battery technology and vehicle performance have been announced almost annually at symposiums like this. Also remember that there are a large number of battery systems competing for the prize. I suspect that the system supported with the best research and development program will ultimately prevail. I'm proud to say that St. Joe's subsidiary, Energy Research Corporation, is one of several companies actively working on the development of prototype nickel zinc batteries for vehicle propulsion.

Zinc foundry alloys also offer potential for increased zinc consumption. This series of zinc-aluminum alloys can be substituted for bronze, iron, and aluminum castings in certain applications using traditional foundry equipment for melting and casting. The low melting temperature of zinc offers considerable energy savings and relatively fume-free operations, and the resulting product is easily machineable with virtually no porosity. Thus far, zinc foundry alloys have been used for machine housings, lock hardware, pullies, cams, and bearings, to name but a few applications.

Lest you be tempted to regard the potential market as relatively minor, keep in mind that the U.S. foundry industry shipped almost 18.6 million tons of castings in 1978. Even a 1% share of that market would exceed the 1978 consumption of primary zinc by the U.S. brass industry. A 2.5% share would exceed the entire 1978 production of zinc die castings. Efforts are being made by a major alloyer and several zinc producers to promote this family of zinc alloys. As in the case of almost all other applications for zinc, however, more companies must be actively involved in research and development.

Mining

Although there may be temporary shortages, it appears likely that sufficient feed materials will be available to satisfy foreseeable demand. Many zinc mines are operating at less than full capacity and some recently closed mines would undoubtedly be reopened if justified by more reasonable prices. Furthermore, a number of new mines are on the horizon, including Howard's Pass and Arvik in northern Canada, Detour River in Quebec, several central Tennessee properties, the Crandon ore body in Wisconsin, and the Gamsberg Mine in South Africa. Zinc is a relatively abundant mineral in the earth's crust and substantial additional mining capacity could be developed given reasonable advance notice. A disturbing trend in the United States, however, is the legislative withdrawal from possible mineral exploration of vast areas of relatively unexplored territory, eliminating what might possibly be the best ore bodies within our borders.

We should keep in mind that there has been a gradual reduction in the overall economic quality of zinc ore in the developed nations as ore grades decline, production goes deeper, and mines are established in more remote locations. This would appear to favor the increased processing of secondary materials. The production of zinc from scrap requires less than 30% of the energy of primary smelting, and recycling is increasingly the subject of the governmental tax incentives. According to the U.S. Bureau of Mines, only 27% of the zinc secondaries available in the United States are actually recycled. Thus, work is needed on developing better systems of scrap collection as well as more efficient and environmentally acceptable recovery technologies. In 1978, the U.S. Congress enacted a 10% Recycling Investment Tax Credit that could encourage developments in these areas.

Smelting

Factors Influencing Capacity - Increasing demand will be of little value to zinc producers if it is accompanied by large capacity increases as it was in the recent past. It would seem appropriate, therefore, to examine factors that may influence capacity. One such factor is costs--both the costs of building a new facility and the costs of operating it.

Capital Costs - We are all aware of the significant increase in the cost of new zinc plant construction. Only ten years ago a 100,000 metric ton zinc refinery could be constructed in North America for approximately $80 million or $800 per ton of installed capacity. It now appears that the cost in July 1979 dollars to construct the same plant is closer to $185 million or $1,850 per ton of installed capacity. This significant increase reflects not only the rapid rise in the cost of equipment, but an increase in construction labor rates. In addition, the higher long-term interest rates now in effect increase the rate of return required on any new project.

Operating Costs - Operating costs have also risen dramatically over the period we are discussing. Ten years ago the direct operating costs of producing a pound of zinc, including purchased feed, was approximately 10¢ per pound for a modern electrolytic plant. The cost for a new plant today is closer to 30¢ per pound, despite increased automation and improved recovery techniques. Among the most important reasons for this large increase are energy, labor, and government regulation.

Prospects for Increased Capacity - While it is true that competing materials are being affected in varying degrees by most of these same cost increases, the inhibiting effect on new zinc plant construction seems likely to be greater. The present over capacity, stagnation in consumption, and low prices are generally not perceived as offering the opportunity for an adequate return on the necessary new capital. These factors, as well as the continuing energy crisis and the general perception of zinc as a mature and unglamorous commodity, have caused a sense of uneasiness and uncertainty greater than that existing in the businesses of our competitors. Therefore, it seems unlikely that there will be large increases in zinc plant capacity in the free world in the next several years.

Overall, we project the net addition of 630,000 tons to free world zinc smelting capacity by 1990, which is equivalent to an average capacity increase of approximately 1% per annum. If this figure is correct, zinc consumption would have to increase at an annual rate of nearly 1.3% merely to retain the present level of overcapacity. Much greater growth would be required to justify operations at reasonable levels of capacity utilization.

Process Development - If new plants are constructed, it seems likely that they will be of substantial capacity in order to obtain the advantages of lower unit capital and operating costs. One could also anticipate larger operating units within the plants, such as larger roasters and cathodes, to facilitate computer control and reduce labor costs. Also, it is possible that the trend to electrolytic plants will be slowed and that more attention will be focused on new systems that rely on fossil fuel directly. The development of such systems, however, is likely to take substantial periods of time.

The incentive for renewed interest in pyrometallurgical processes comes primarily from opportunities for lower energy costs. First, there is an inherent inefficiency in the two-step process of converting fossil fuel to electricity and then using that electricity to extract zinc. Second, it seems likely that power costs, at least in the United States, will increase at an even greater rate than the cost of fossil fuel. There has been an increasing tendency to shift the cost burden of electrical power generation from the individual user to the commercial consumer, and our failure to develop more nuclear power and coal-fired generating plants will result in a shortage of generating power that will inevitably lead to a rash of construction projects. The capital costs for such projects will be included in the rate base of all consumers, thereby escalating costs of purchased electrical power.

Other reasons favoring a return to pyrometallurgical processes include lower capital costs, the ability to treat some complex concentrates more efficiently, the growing market trend toward galvanizing grades (the natural product of pyrometallurgical processes), a superior ability to utilize secondary materials, and the growing environmental problems connected with disposal of electrolytic leach residues. In addition, it may also be easier for some developing nations, because of the absence of an industrial infrastructure, to operate pyrometallurgical facilities. The major problem inhibiting construction of pyrometallurgical plants appears to be compliance with environmental standards, development of alternatives to coke as the primary fuel source, and the integration of instrumentation and automation into the operating procedures. These are areas requiring intensive research and development.

Need for Research, Development and Marketing

It is traditional to end addresses of this type by expressing confidence in the future of the industry. Present circumstances make that a difficult tradition to follow. We are beset with serious problems, including excessive productive capacity, slow growth in demand, loss of significant markets, and an apparent retreat in world economic growth, long the driving force of zinc demand. Furthermore, the zinc industry has yet to demonstrate a clear commitment to finding the answers to these problems. In preparing for this talk I found that most of the market opportunities previously discussed were the same growth prospects mentioned by commentators ten years earlier. These markets have not materialized to any great extent, in part because they have not been vigorously pursued. Whether or not they materialize in the future depends in large part on the intensification of joint and individual research, development and marketing activities by all of us associated with the zinc industry. The challenges to our traditional and prospective markets are certain to continue. Our competitors are well organized and well financed and can be counted on, not only to resist expansion of zinc consumption, but to seek further penetration into markets not already taken from us.

I am encouraged, however, by what I perceive as a new spirit of activism and enthusiasm apparent in our trade organizations. Most producers appear to realize that they are not selling a commodity for which there are assured markets and that more than lip service need be paid to new product and market support activities. Thus, I'll end my presentation by saying I am hopeful that world zinc producers, recognizing the need for vigorous efforts, will make more substantial commitments of manpower and funds to research, development, and marketing. Should my hopes be realized, then I will indeed be confident in the future of the zinc industry.

ENERGY USE IN ZINC EXTRACTION

Herbert H. Kellogg
Henry Krumb School of Mines
Columbia University
New York, NY 10027

The energy requirements for zinc production by modern electrolytic and Imperial Smelting Process flowsheets are evaluated, based on data reported by fourteen plants. For production of one ton of SHG zinc the process fuel equivalents are 50.1×10^9 J for the electrolytic process and 44.8×10^9 for the ISP. For one ton of prime western zinc the values are 49.9×10^9 J for the electrolytic process and 37.5×10^9 J for the ISP.

Potential improvements that might result in energy savings for conventional zinc production processes are discussed. Energy requirements for the concept of a novel low-energy pyrometallurgical zinc process are estimated to be less than 30×10^9 J per ton of SHG.

Introduction

This study consists of an analysis of the energy requirements for the two principal methods of zinc extraction—the electrolytic process, and the Imperial Smelting Process (ISP). The nature and magnitude of the various energy inputs to these established processes suggest modifications that might reduce the energy-intensity of zinc production, and such modifications, along with a concept for a radically new zinc extraction process, are briefly considered.

A number of studies of energy utilization in zinc extraction already exist in the literature (1, 2, 3, 4), so that one may question the need for yet another. What I hope to add to existing knowledge on the subject is greater detail, consistency of basic assumptions which makes possible valid comparisons between processes, and data drawn from current operations of many plants. In preparation for this study the author sent a questionnaire to ten electrolytic and four ISP plants, located throughout the world, requesting detailed information on their use of energy. All of these questionnaires were completed and returned, and they comprise an invaluable data bank for this study. The companies that supplied the data deserve our recognition and thanks for their efforts in compiling the data and their willingness to release it.

The method adopted here for comparison of zinc processes is similar to that used by the author for comparison of energy use in copper smelting (5). The energy requirements for a hypothetical new plant are calculated for an assumed set of uniform local conditions (concentrate analysis, nature of fuel, degree of sulfur capture as acid, etc.). The hypothetical plant always incorporates the best modern practice, as far as energy saving is concerned, but does not include any improvements that are not already practiced on a commercial scale. Forthcoming, but not fully proven, developments are relegated to the latter sections of this paper.

The process fuel equivalent, PFE, is calculated for each process according to the formula (6)

$$PFE = F + E + S - B$$

F is the energy content of all fuels used by the process; E is the fuel equivalent of electrical energy used—assumed here to be 11.1×10^6 J/kwh (10500 BTU/kwh); S is the fuel equivalent of major consumable supplies (reagents, fluxes, oxygen, etc.); B is the fuel equivalent of byproducts and useful surplus heat. All units are metric, and ton stands for 1000 kg (except where specified as short ton (2000 lbs.)).

The energy content of fuels is taken as their gross heating value. The exception to this is metallurgical coke, a manufactured fuel, for which the energy value chosen is 36.6×10^9 J/ton (31.5×10^6 BTU/short ton). This value was calculated by Battelle-Columbus (1), based on a full energy analysis of the heating value of the coal used, energy for firing coke ovens and energy credits for the several byproducts produced.

Sulfur Recovery as Acid

Both of the zinc processes considered herein result in oxidation of most of the sulfur in the zinc sulfide concentrate to gas containing SO_2. In fluid-bed roasting, used by the electrolytic zinc process, the roaster gas may initially contain 10-12% SO_2; after cooling, cleaning and dilution with sufficient air to permit acid manufacture the gas strength is typically 7% SO_2. The ISP flowsheet employs updraft sintering of the concentrate, usually with recirculation of weak gases to the sinter machine, and also produces a gas strength of about 7% SO_2 for acid manufacture.

29

Although other alternatives, such as production of liquid SO_2 or elemental sulfur, exist for the disposal of these gases, acid production is the standard of the industry, worldwide. In this study, the double-contact acid process is employed for both the electrolytic and ISP flowsheets because the tail gas from this process (~500 ppm SO_2) will meet the U.S. EPA new source performance standards. In some locations in Japan even stricter local regulations require reduction of exhaust gases to about 50 ppm SO_2; to meet such standards, scrubbing of acid-plant tail gas with lime or zinc oxide slurry is employed.

In a previous study (5) estimating equations were given for the energy requirement for acid production (single and double-contact) as a function of the SO_2 strength of the clean dry gas. For a gas of 7% SO_2 these equations yield 111 kwh/ton acid (100%) for single-contact and 156 kwh/ton acid for double-contact. These values include the energy used in pumps and fans for the gas-cleaning circuit (scrubbers, mist Cottrell, drier, etc.) as well as the contact plant itself and the pumping of cooling water.

As a check on the accuracy of the estimating equations, I have reports from five plants using the single-contact process that are in fair agreement (range, 81-113 kwh/ton acid, average 96 kwh/ton acid). The average for these plants is lower by 15% (15 kwh/ton) than that calculated by the estimating equation. In contrast, reports from four plants using the double-contact process were in poor agreement with each other (69-149 kwh/ton acid) and yield about the same average energy consumption as the five single-contact plants. It seems likely that some of the respondents to the questionnaire failed to include the energy for gas cleaning, or pumping of cooling water, in their reports for the energy for acid making, thus resulting in abnormally low values of total energy consumption.

Double-contact acid making certainly uses more energy than the single-contact process, for the additional pumping and fan energy required in the second stage. For the purposes of this study I have selected the unit energy value of 135 kwh/ton acid (100%) for the double-contact process. This value is 15% lower than the value calculated from the previously used equation, a compromise suggested by comparison of the equation with the reported values for single-contact plants.

Acid making is the only unit process common to both electolytic and ISP flowsheets. The energy requirements for all other unit processes are analyzed under the sections devoted to these flowsheets. Acid is properly considered a byproduct of zinc production. As such, a byproduct credit should be allowed, according the PFE formula given in the Introduction. In conformity to the earlier study on copper smelting (5), the small byproduct credit for acid has not been applied here. For those who wish to include this credit, the Battelle-Columbus study (1) calculates the unit energy for sulfuric acid to be 0.96 x 10^9 J/ton (0.83 x 10^6 BTU/short ton).

Electrolytic Zinc

The roast, leach, electrowin flowsheet clearly occupies the leading role in production of zinc. Not only does this process now account for about 75% of world zinc production, but the fraction of zinc production produced by electrowinning has grown steadily over the past two decades. The two new zinc plants built in the U.S. within the past few years have both employed this process. The electrolytic zinc process represents, therefore, the standard to which other processes must be compared with respect to metallurgical performance, environmental problems, quality of product and energy requirements.

The analysis of the zinc concentrate chosen for both the electrolytic and

ISP flowsheets is : 52.% Zn, 1.5% Pb, 8.0% Fe, 32.% S. The relatively high
iron content of this concentrate requires that one of the new hot-leaching
methods (resulting in an iron residue of jarosite, goethite or hematite) be
employed in order to achieve a high recovery of zinc. For purposes of this
study I have assumed the use of the Vieille-Montagne (V-M) goethite leach system (7). This method is well established in three plants, and possesses a
minor advantage over the jarosite method because of the smaller amount of iron
residue requiring environmentally acceptable disposal. The difference in
energy consumption between the various leaching methods makes little difference to the total energy consumption for the flowsheet, as will be apparent
at a later point. Table I summarizes basic assumptions and major material
flows for the production of 1 ton of special high grade zinc. The energy requirements for the flowsheet are estimated by separate analysis of six sections as follows: roasting, acid making, leaching and purification, electrolysis, melting, casting and dross treatment, and miscellaneous.

Table I. Material Balance for Electrolytic Zinc
(V-M goethite leach)

Conc. Anal.: 52% Zn, 1.5% Pb, 8.0% Fe, 32.% S
Overall zinc recovery: 96%

Material	Tons/ton SHG zinc ingot
Concentrate	2.00
Acid (100% H_2SO_4)	1.78
Lead-silver residue	0.29
Goethite residue	0.36
Cathode zinc*	1.075

*Assumes 2% formation of dross on melting (dross
returned to process) and 5% of molten zinc used to
produce zinc dust for electrolyte purification.

Roasting

This unit of the flowsheet is intended to include the auxiliary operations
of receiving, storage and handling of concentrate, air compression for the
roaster, waste-heat boiler, cyclones, hot electrostatic precipitator and hot-
gas fan. Of the two roaster designs commonly used, V-M/Lurgi and Dorr, the
former, which has direct feed of solid concentrate, and employs cooling coils
in the roaster bed, is capable of recovering 1.0 to 1.2 tons of steam per ton
of concentrate from the waste-heat boiler plus the cooling coils. The Dorr
Fluo Solids roaster, with concentrate fed as a slurry, recovers only 0.8-0.9
tons of steam per ton of concentrate, but has the advantage of producing a
calcine with lower amounts of sulfate sulfur.

The two energy items of importance for this flowsheet section are total
electrical energy used (for air compression, fan, materials handling, electrostatic precipitator, etc.) and the recovery of surplus steam available for use
in leaching and electrolyte purification. The table below summarizes the data
reported from nine plants (six V-M/Lurgi and three Dorr) for this study.

The lower energy utilization and greater steam recovery of the V-M/Lurgi
system make it the choice for this study, and the unit energy values adopted
are those listed above under Selected Values. Converted to the basis of 1 ton

of SHG ingot zinc, these selected values become: <u>120 kwh/ton SHG and 2.1 tons surplus steam/ton SHG.</u>

	Range	Average	Selected Values
Electricity, kwh/t conc.			
V-M/Lurgi	50- 86	63	60
Dorr	56-120	88	
Surplus steam, t/t conc.			
V-M/Lurgi	0.9-1.2	1.07	1.05
Dorr	0.8-0.9	0.85	

Acid Plant

The unit energy for acid making has already been discussed. It only remains to multiply the unit energy (135 kwh/ton acid) by the rate of acid production from Table I (1.78 tons acid/ton SHG) to calculate the total energy value of <u>240 kwh/ton SHG ingot zinc.</u>

Leaching and Electrolyte Purification

This section of the flowsheet includes all of the leaching and precipitation steps employed in the V-M goethite process (neutral, hot and superhot leach, reduction of iron, neutralization and goethite precipitation), as well as cold and hot purification employing the V-M reverse-antimony method. Both leaching and purification processes are continuous. Products are purified pregnant electrolyte, Cu-Cd sludge, Co sludge, Pb/Ag residue, goethite residue and sulfur residue that is returned to roasting.

Electric energy is used for pumping, agitation, air compression, filtration, ball-milling of residue, etc. Energy is also used as steam for heating the hot leach and purification steps. The average value from two plants employing the V-M goethite process was 177 kwh/ton cathode Zn for the total electric consumption. This value is close to that reported by plants using other continuous leaching methods (conventional or jarosite), where the range was 167-224 kwh/ton cathode zinc. The value of 177 kwh/ton cathode zinc will be accepted for this study; when multiplied by the factor (from Table 1) of 1.075 tons cathode/ton SHG ingot this yields <u>190 kwh/ton SHG ingot zinc.</u>

Steam consumption for plants employing the goethite process was 1.86 tons/ton cathode for one plant, and 1.9 tons (summer) to 2.5 tons (winter) per ton cathode for another plant. The difference in steam consumption between summer and winter operation was also reported by a plant using jarosite precipitation (2.2 tons (summer) to 3.1 tons (winter) per ton cathode). In general, however, it was difficult to understand the wide variations in steam consumption reported (conventional leach: 0.33 to 1.4 tons steam/ton cathode; jarosite leach: 0.67 to 3.28 tons steam per ton cathode) even taking climate into consideration. A more detailed analysis of solution volumes, temperatures, heat losses, indoor vs. outdoor leaching tanks, etc. would be necessary for better understanding of this subject.

For this study, the average steam consumption over the cycle of a year will be taken as (1.9 + 2.5)/2 = 2.2 tons steam/ton cathode, or <u>2.4 tons steam/ton SHG ingot zinc</u>. The part of this steam requirement not satisfied by

32

surplus steam from the waste-heat boiler is, therefore,

$$2.4 - 2.1 = 0.3 \text{ tons steam/ton SHG ingot}$$

This additional steam will be generated in a fuel-fired boiler operating at 85% boiler efficiency, with a calculated fuel rate (gross heating value) of 3.2 x 10^9 J/ton steam. The fuel consumption for this purpose is, therefore:

$$0.3 \text{ tons steam} \times 3.2 \times 10^9 \text{ J/ton steam} = \underline{0.96 \times 10^9 \text{ J/ton SHG ingot}}$$

An additional small energy requirement for this plant section is fuel to dry the lead/silver residue prior to shipment. Two plants reporting this item yield an average value of 0.5 x 10^9 J/ton ingot zinc. The total fuel energy for raising steam and drying of Pb/Ag residue is, therefore, $\underline{1.46 \times 10^9 \text{ J/ton}}$ $\underline{\text{SHG ingot}}$. The fuel equivalent of reagents used will be considered at a later point. The energy used for treatment of cadmium sludge for recovery of this metal is properly charged against cadmium, and will not be included here.

Electrolytic Tank Room

In addition to the electrowinning cells, this plant section includes the auxiliary operations of circulation of electrolyte to the cooling towers, and handling and stripping of cathodes. Electric energy for electrolysis is the single largest contributor to the energy for the electrolytic zinc flowsheet.

Of the ten plants reporting data for this study the range in D-C energy used for electrolysis was 3100 to 3475 kwh/ton cathode, with an average of 3261 kwh/ton cathode. It is now common practice in many plants to operate at different current densities during the day--typically, 300 amps/m^2 at peak power periods when the unit power cost is highest, and up to 700 amps/m^2 during periods when the unit power cost is low. One plant reports the extreme of 50 amps/m^2 (just enough to prevent redissolution of the cathodes) used for several hours each day when the power cost is highest. Such variations in current density may be necessary to minimize power bills, but they do not result in the most efficient operation from the viewpoint of energy consumption.

For the purpose of this study I have chosen a unit energy for electrolysis of $\underline{3185 \text{ kwh D.C./ton cathode}}$. This is lower than average, but well above the minimum reported. It corresponds to operation with 90% current efficiency and a total cell voltage of 3.5 volts, a combination that can usually be achieved by operating at about 450 amps/m^2. Modern silicon rectifiers with 98% efficiency are now commonly used, and their use here would result in an A.C. power consumption of 3185/.98 = $\underline{3250 \text{ kwh A.C./ton cathode}}$.

Energy for circulation of electrolyte and handling and stripping of cathodes varies from plant to plant, in the range 37 to 157 kwh/ton cathode. The high value of 157 kwh/ton was reported by one plant for the summer months (winter value was 108 kwh/ton) in a hot, humid region where cooling towers operate at low efficiency. For this study the average of seven plants reporting--$\underline{80 \text{ kwh/ton cathode}}$--will be accepted.

Total electric energy for the tank room operations is therefore 3250 + 80 = 3330 kwh/ton cathode, or 3330 x 1.075 = $\underline{3580 \text{ kwh/ton SHG}}$ ingot zinc.

Melting and Casting

This unit operation, typically conducted by melting cathodes in an electric induction furnace, also includes the auxiliary operations of casting and treatment of melting dross to recover a dry dross that is returned to the process. Plants supplying data for this study report 107 to 133 kwh/ton cathode for the induction furnace energy, with an average value of 118 kwh/ton cathode. For dross treatment and casting the results are more variable, three plants reporting values of about 10 kwh/ton cathode, and two plants with values of 40 and 55 kwh/ton of cathode. The reasons for these large differences are unknown. For this study, the values of 115 kwh/ton of cathode for melting and 10 kwh/ton of cathode for casting and dross treatment are chosen, so that the total energy requirement is 125 kwh/ton of cathode or 134 kwh/ton SHG ingot.

For production of dross in melting and the amount of zinc used for production of zinc dust, the reported values were 1.2 to 3% of cathode weight (average 2.0%) as dross and 2.5 to 10.% of cathode weight (average 5.8%) for zinc dust production. The accepted values for this study are 2.% dross and 5.% for zinc dust, which values correspond to 1.075 tons cathode/ton SHG ingot given in Table I.

Miscellaneous

Two additional plant functions are included in this section—1) energy requirements for office, laboratory, shop and maintenance; 2) waste-water treatment and disposal of goethite residue in plastic-lined dumps. Based on very limited evidence from one plant, the energy consumption for these purposes may amount to 20 kwh/ton SHG ingot for the former, and 50 kwh/ton SHG ingot for the latter, for a total of 70 kwh/ton SHG ingot. Any uncertainty in these values is compensated in the comparison of electrolytic zinc with ISP zinc by use of the same estimate in both cases.

Finally, I include in this section the energy equivalent of consumable reagents for the process. Rough estimates for the amounts and energy equivalent of these are given below.

Reagent	Kg/ton SHG ingot	Energy/kg	10^9 J/ton SHG ingot
Zinc dust	50.	made in process	0.0
Lime	16.	9×10^6 J/kg	0.14
NH_4Cl	0.8	54×10^6 J/kg	0.04
MnO_2	2.0	6×10^6 J/kg	0.01
Sb_2O_3	3.0	60×10^6 J/kg	0.18
		Total	0.37

The only items on this list that require comment are the zinc dust and lime. The flowsheet of the hypothetical plant incorporates the energy for zinc dust production, so that no additional charge should be made against the process for this reagent. Lime is used at various points in the gas-cleaning circuit prior to acid making, and in waste-water treatment for neutralization of acidity. The total of 16 kg of lime per ton of zinc is an actual value for one plant where the information was available. The exact amounts and types of reagents used can be expected to vary widely from plant to plant, depending on specific impurities in the concentrate and the degree of water treatment practiced. The total energy equivalent of reagents is increased from 0.37 $\times 10^9$ J, shown above, to 0.5 $\times 10^9$ J/ton SHG ingot in the energy summary to account for a number of minor reagents and chemicals not included in the list above.

Summary for Electrolytic Zinc

Table II gathers together the energy inputs for the six sections of the electrolytic zinc flowsheet. The total of 50×10^9 joules/ton SHG zinc probably has an accuracy of $\pm 2 \times 10^9$ joules, although existing plants are likely to show larger positive deviations because of the failure to incorporate one or more energy-saving features assumed in this study. Substitution of the

Table II. Process Fuel Equivalent for Electrolytic Zinc (SHG)*

Plant Section	Amount	x	Unit Energy	10^9 J/ton SHG ingot
Roasting:				
Electric	120 kwh	x	11.1×10^6 J/kwh	1.33
Acid Plant				
Electric	240 kwh	x	11.1×10^6 J/kwh	2.66
Leaching & Purification				
Electric	190 kwh	x	11.1×10^6 J/kwh	2.11
Fuel	(see text)			1.46
Electrolytic Tank Room				
Electric	3580 kwh	x	11.1×10^6 J/kwh	39.74
Melting & Casting				
Electric	134 kwh	x	11.1×10^6 J/kwh	1.49
Miscellaneous				
Electric	70 kwh	x	11.1×10^6 J/kwh	0.78
Reagents	(see text)			0.50
Total				50.07

*See Table I for basic assumptions and material flows. 2.4 tons of steam are required for leaching and purification, but most of this (2.1 tons) is generated in roasting.

jarosite process for the goethite process should affect the total by less than $\pm 1 \times 10^9$ joule. Conventional leaching, applied to the high-iron concentrate assumed here, would reduce the energy for leaching, and show a byproduct credit for surplus steam, but the recovery of zinc would drop to about 90%, thus increasing the tons of concentrate required to produce 1 ton of ingot and the energy for roasting and acid making. Moreover, the leach residue would be rich in zinc, requiring some additional processing (Waelz kiln, or lead blast furnace plus slag fuming) in order to complete the recovery of zinc, lead and silver.

The estimated total from Table II is compared in Table III with other estimates for the total energy requirements for electrolytic zinc. For this purpose the estimates of electrical energy requirements by other authors are all converted to equivalent fuel at the unit value used here (11.1×10^6

joules/kwh), and English units are converted to metric units where necessary. Among these estimates, the one that is in most serious disagreement with the present study is that of Meisel (4), who offers an estimate of total electric energy requirements of 4960 kwh/ton SHG (4500 kwh/short ton SHG) that is nearly 15% larger than data available to the author. Moreover, Meisel charges the energy for 1.7 tons of steam per ton SHG, which appears to overlook the generation of surplus steam in roasting. The estimate of Battelle (1) also appears to be high, and part of the reason for this is their charge for natural gas in drying of concentrate and in roasting--requirements that do not pertain to modern zinc roasting. Gordon's (3) estimate for a modern plant I believe to be low because of underestimation of the electrical energy requirements beyond those for electrolysis; for this purpose his estimate is 525 kwh/ton SHG, compared to 754 kwh/ton SHG found in this study.

Table III. Comparison of Energy Estimates for Electrolytic Zinc

Source	10^9 J/ton SHG ingot
This study - modern plant	50 ± 2.
Battelle-Columbus (1) - average U.S. plant	57.0
Gordon (3) - modern plant	46.2
Gordon (3) - old plant	53.2
Hopkin and Richards (2) - modern plant	50.3
Meisel (4) - modern plant	60.5

Imperial Smelting Process

The Imperial Smelting Process (ISP) is the only other zinc producing process, besides the electrolytic process, that has shown growth in productive capacity over the past two decades. Today there are eleven plants using this process in ten different countries, and plans for two additional installations. In 1977 the ISP accounted for a little over 10% of the world slab zinc production.

The ISP produces both zinc and lead, typically in the proportion about 0.5 ton of lead/ton zinc. The co-production of these metals is both an advantage and a limitation of the process. An advantage, because the lead and silver content of zinc concentrates can be recovered by a single process, whereas the electrolytic process must rely on retreatment of lead/silver residue in a lead blast furnace; this consideration becomes especially important for treatment of zinc-lead ores that do not respond well to differential flotation, and yield zinc concentrate with high lead levels. The ISP can also advantageously treat a wide variety of drosses, residues and oxidic materials of varying Zn/Pb ratio, which are not amenable to the electrolytic process. Co-production of lead is a limitation for treatment of clean zinc concentrates, where lead production is not desired. Despite encouraging tests with feeding the ISP with hot briquettes made from low-lead zinc calcine (2) this proposed version of the "low lead" ISP remains untested on 100% briquette charge.

A second important difference between the ISP and electrolytic zinc lies in the quality of the metal produced. The electrolytic process produces very pure zinc, most of which meets the specifications for Special High Grade (Grade I). The direct product of the ISP is Prime Western (Grade IV), and, to produce the higher grades of zinc, an ISP plant must employ additional capital and operating expense for refluxing columns*. It is only fair to point out

---

*The ingenious vacuum distillation process (8), which operated successfully at the Swansea plant before this plant was shut down, is not considered in this study because of the lack of current commercial use. It produced intermediate grades of zinc.

that direct production of Prime Western is not an unalloyed disadvantage for the ISP, particularly in today's market where the producer of electrolytic zinc must debase his SHG zinc (~41¢/lb) with purchased lead (~48¢/lb) in order to produce PW zinc (~39¢/lb).

The differences between ISP and electrolytic zinc, with respect to co-production of lead and zinc, and quality of product, require special considerations for a comparison of energy requirements for the two processes. Two different energy comparisons will be made: one will compare electrolytic zinc (SHG) with the ISP plus refluxing columns to produce SHG zinc; the second comparison will be between electrolytic zinc debased with lead to produce prime western zinc, and the ISP producing prime western zinc directly. Since either version of the ISP also produces lead, a byproduct energy credit for lead production will be employed. The energy charge for lead (for debasing) or byproduct credit (for ISP) is based on the Battelle-Columbus energy study for lead production in the blast furnace (1). Refined lead would be used for debasement, and the Battelle study gives a value of 31.1×10^9 J/ton (26.8 x 10^6 BTU/short ton) for this product. The analysis of the ISP flowsheet will only follow the lead to the stage of drossed bullion, so the energy credit for lead should correspond to the energy required in the lead blast-furnace route to convert lead concentrate to drossed bullion (with acid production from the sulfur); for this purpose the Battelle study gives 14.0×10^9 J/ton (12.1 x 10^6 BTU/short ton).

Basic assumptions and major material flows for the ISP flowsheet are summarized in Table IV. The same zinc concentrate used for the electrolytic flowsheet is employed, plus a lead concentrate with a relatively high zinc level (7% Zn). These concentrates are proprotioned to yield 0.5 ton of drossed lead bullion for each 1 ton of PW zinc produced. If these concentrates average about 0.4% Cu, then about 42 kg of copper dross (assumed to be 22.% Cu, 58% Pb) will also be produced for each ton of PW zinc.

The energy requirements for the ISP flowsheet are analyzed under the following sub-headings: sinter plant, acid plant, furnace plant, refluxing plant and miscellaneous.

Table IV. Material Balance for ISP Flowsheet

Zinc conc: 52.% Zn, 1.5% Pb, 8.% Fe, 32.% S

Lead Conc: 7.% Zn, 66.% Pb, 4.% Fe, 16.% S

Recovery of Zinc*: 96.%

Recovery of Lead*: 96.% (sum of drossed bullion plus copper dross)

Material	tons/ton of PW zinc
zinc concentrate	1.872
lead concentrate	0.774
coke (dry basis, 88% C)	0.955
limestone flux	0.060
drossed lead bullion	0.500
copper dross (22% Cu, 58% Pb)	0.042
Acid (100% H_2SO_4)	2.09
P.W. zinc	1.00

*Unaccounted losses may reduce these recoveries by 1 to 3%.

Sinter plant

This plant section includes the receiving, storage and handling of concentrates, sinter mixing system, updraft sintering machine, gas recirculation system, crushing, sizing and handling of sinter and sinter returns, and the ventilation system for control of particulates and fume in the sinter plant. For the four plants reporting data for this study the electric energy used in this section ranged from 153 kwh to 199 kwh per ton of PW zinc. The average value of 170 kwh/ton PW zinc will be accepted for this study.

Fuel is also used in this section for the sinter-machine ignition furnace and for motorized vehicles in amounts ranging from 0.43 to 0.88 x 10^9 J per ton of PW zinc. Once again the average value of 0.62 x 10^9 J per ton of PW zinc is accepted for this study.

Acid Plant

The same unit energy for acid production (135 kwh/ton acid) used for the electrolytic flowsheet is adopted here. For production of 2.09 tons of acid (Table IV) per ton of PW zinc, the total energy requirement for the acid plant is 2.09 x 135 = 282 kwh/ton PW zinc.

Furnace Plant

This plant section includes the coke preheater, blast preheater, blast furnace, lead splash-condenser, cooling launders for separation of zinc from lead, gas-cleaning system for the condenser off-gas, forehearth, slag granulation, drossing of lead bullion and ventilation for hygiene control.

Four energy-accounting items are of importance for this plant section: electrical energy (for blast compression, gas cleaning, ventilation, pumping of condenser lead and a variety of materials handling operations), coke requirement for the blast furnace, miscellaneous fuels (for the forehearth, pilot burners, standby operation, etc.), and the generation and use of condenser gas--a low-grade fuel, containing about 21% CO, and usually referred to as LCV (low calorific value) gas.

For electrical energy used there is good agreement among the four plants reporting data on a value of 365 kwh/ton PW zinc, and this value is adopted here. For miscellaneous fuels, three plants reporting data give a range of 0.64 to 1.45 x 10^9 J/ton PW zinc; an average value of 0.9 x 10^9 J/ton PW zinc is employed in this study.

Coke for the blast furnace, like the electrolysis energy for electrolytic zinc, is the single predominant use of energy in the ISP process. The amount of coke used will depend on many interrelated factors such as coke quality and carbon content, preheat temperature of blast and coke, quality of the sinter and condenser efficiency. The coke consumption adopted here is based on what ISP operators consider to be typical good practice, expressed as 0.84 tons carbon burned per ton of zinc (PW) liquated from the cooling launder. With typical metallurgical coke quality in the U.S., containing 88% carbon (dry basis), this yields a coke consumption of 0.84/.88 = 0.955 tons/ton PW zinc. This is in good agreement with the value used by Hopkin and Richards (2) (1.03 tons wet coke (7% moisture), or 0.958 tons dry coke per ton PW zinc). With a unit energy value of 36.6 x 10^9 J/ton coke (see Introduction), the energy associated with coke in the ISP becomes:

$$.955 \times 36.6 \times 10^9 = \underline{34.95 \times 10^9 \text{ J/ton PW Zn}}$$

The fuel value of LCV gas is about 2.7×10^6 J/Nm^3 (73 BTU/SCF). This low-grade fuel is used by all ISP plants for preheating of the blast air and coke. These uses consume only 60-80% of the total LCV gas generated, and many plants flare the surplus gas. Some plants are using surplus LCV gas, on an experimental basis, as a supplementary fuel for the ignition furnace and in reflux columns. One plant has a fully established boiler-turbine-generator unit, fired by surplus LCV gas to co-generate steam (for sale) and electric power for use in process. For this study the surplus LCV gas will be assumed to be used entirely for generation of electric power.

Data reported for this study yield the following average values for the disposition of LCV gas in this flowsheet:

	LCV gas, Nm^3/ton PW zinc
Total gas produced	4550
Used for blast preheat (900°C)	2250
Used for coke preheat (800°C)	500
Surplus LCV gas	1800

Use of this very low-grade fuel in a small boiler-turbine unit should result in production of electric power at the rate of about 1 kwh per 8.5 Nm^3 of LCV gas. The total credit for power generated by 1800 Nm^3 of surplus LCV gas is, therefore, 1800/8.5 = 212 kwh/ton PW zinc. This calculation is semi-quantitatively verified by the reported values from the one plant where 9.2 Nm^3 of gas is used to generate 1 kwh plus 2.0 kg of saturated steam in a co-generation scheme. For any ISP plant that can sell saturated steam to a neighboring plant, the co-generation of steam and power would afford a larger energy credit than that employed here for generation of power alone.

Refluxing of PW to SHG Zinc

All ISP smelters reflux some portion of their production of crude zinc to produce the higher grades, including SHG. The crude zinc is fed to the lead columns, and the volatilized zinc and the cadmium is condensed and refluxed a second time in cadmium columns. The bottoms from the cadmium columns are SHG zinc. The bottoms from the lead column are enriched in lead and iron; after liquation to remove excess lead and iron, and treatment with sodium metal for removal of arsenic and antimony, the partly purified metal is cast as prime western. Metal volatilized from the cadmium column (~20% Cd) is treated for cadmium recovery in a small cadmium refining column.

The use of electric energy in this section is relatively minor, about 25 kwh/ton SHG zinc. The process uses considerable fuel, however, and the four plants reporting give a range of 6.7×10^9 - 9.0×10^9 J/ton SHG zinc, with the high values associated with plants of smaller capacity, and visa-versa. For the purposes of this study the value of 7.0×10^9 J/ton SHG will be used. This is the average value for the two largest reflux plants reporting data.

Miscellaneous

The items to be included here are the energy equivalents of reagents and fluxes employed, the credit for production of byproduct lead, and the electric requirements for office, laboratory, shop, maintenance and waste-water treatment. For the electric energy requirement the same estimate used for the electrolytic zinc flowsheet, 70 kwh/ton PW or SHG zinc, is adopted.

Consumption of reagents and fluxes is estimated as follows:

Reagent	Kg/ton zinc	Energy/kg	10^9 J/ton zinc
limestone flux	60.	0.35×10^6 J/kg	0.02
lime	20.	$9. \times 10^6$ J/kg	0.18
NH_4Cl	1.0	$54. \times 10^6$ J/kg	0.05
sodium metal	1.0	$107. \times 10^6$ J/kg	0.11
			0.36

As was done for the electrolytic flowsheet, the total energy equivalent for reagents will be increased to 0.5 x 10^9 J/ton PW or SHG zinc. The unit credit for byproduct lead has already been discussed--14 x 10^9 J/ton drossed lead. For one-half ton of byproduct lead the credit is, therefore, 7.0 x 10^9 J/ton PW or SHG zinc.

Summary for ISP

Table V summarizes the energy inputs to the ISP flowsheet, including re-fluxing of crude zinc to produce one ton of SHG zinc. The only other study of energy for this process, known to the author, is that of Hopkin and Richards (2). Their energy values for production of one ton of SHG zinc are 48.1 x 10^9 J to produce one ton PW zinc (+0.5 ton lead), plus 6.6 x 10^9 J for re-fluxing to SHG zinc, less 8.0 x 10^9 J credit for byproduct lead, which calcu-lates to 46.7 x 10^9 J/ton SHG. This agrees reasonably well with the total re-ported in Table V (44.8 x 10^9 J/ton SHG). The agreement is fortuitous, in part, because some of their basic assumptions differ from those used here. For example: they charge the process for the fuel value of the sulfide concentrates used, whereas this is not done here or in most other studies of energy consump-tion in metal production; their unit energy values for coke, oil and electric energy are slightly lower than those used here. Regarding the actual amounts of fuel, coke and electricity used, the analysis by Hopkin and Richards agrees fairly well with the present study for fuel and coke, but their total electric requirement (600 kwh) for production of one ton of PW zinc (plus 0.5 ton lead) is far less than that shown in Table V (887 kwh). Although they claim to take acid-plant requirements into account, their low value of total electric re-quirements suggests that they may have omitted or underestimated this item. The unusual result that their total energy requirement for one ton of SHG zinc (46.7 x 10^9 J) is _higher_ than that found here (44.8 x 10^9 J), yet their elec-tric energy consumption is 32% _lower_ than found here, results from their prac-tice of charging the process for the fuel value of concentrates used.

Comparison of Electrolytic and ISP Zinc

Comparison of Tables II and V shows that the ISP plus refluxing flowsheet produces SHG zinc for about 10% lower consumption of energy resources than the electrolytic process. The energy advantage of the ISP increases when the comparison is made for prime western zinc, as shown in Table VI. Direct pro-duction of prime western zinc by the ISP requires 25% less energy than debase-ment of electrolytic zinc with lead. The lower energy requirement of the py-rometallurgical flowsheet comes as no surprise, since for many other metals the pyrometallurgical methods offer even greater energy advantages than shown here for zinc.

Despite its energy advantage, the ISP plays a secondary role in zinc pro-duction compared to the electrolytic process. At least one important reason for this is the difficulty and expense associated with control of particulates and fume in the work-place atmosphere and plant emissions for the ISP. The

Table V. Process Fuel Equivalent for ISP Production of SHG Zinc*

Plant Section	Amount x Unit Energy	10^9 J/ton SHG ingot
Sinter Plant		
Electric	170 kwh 11.1 x 10^6 J/kwh	1.89
Fuels		0.62
Acid Plant		
Electric	282 kwh 11.1 x 10^6 J/kwh	3.13
Furnace Plant		
Electric	365 kwh 11.1 x 10^6 J/kwh	4.05
Misc. fuels		0.90
Coke	0.955 tons 36.6 x 10^9 J/ton	34.95
Credit for:		
surplus LCV	212 kwh 11.1 x 10^6 J/kwh	(-2.35)
Refluxing Crude Zinc to SHG Zinc		
Electric	25 kwh 11.1 x 10^6 J/kwh	0.28
Fuel		7.00
Miscellaneous		
Electric	70 kwh 11.1 x 10^6 J/kwh	0.78
Reagents (see text)		0.50
Credit for:		
0.5 t lead		(-7.00)
Total		44.75

*See Table IV for basic assumptions and material flows.

ISP is basically a zinc-making process, but, as discussed above, the current technology is tied to co-production of lead, and the lead splash-condenser remains at the heart of the process (even if a "low-lead" ISP, based on hot-briquetted zinc calcine, were perfected). Current ISP plants already put considerable effort into environmental control of dust and fume in the sinter plant, furnace plant and refluxing plant; if the new EPA and OSHA standards for lead in the ambient atmosphere and the work-place environment become effective, an ISP plant in the U.S. would require design changes that achieve far more rigorous environmental control.

In comparison, the electrolytic process has only minor environmental problems. Fluid-bed roasting, unlike sintering, can be a tightly-closed process so as to avoid fugitive dust and gas. The waste solutions and slurries from leaching and electrolysis can be collected, confined and treated so as to

Table VI. Production of Prime Western Zinc by the ISP and Electrolytic Processes

ISP	10^9 J/ton PW zinc
Total from Table V	44.75
Less energy for refluxing (Table V)	-7.28
Total for prime western zinc	37.47
Electrolytic	
0.99 tons electro. zinc x 50.07 x 10^9 J/ton	49.57
0.01 tons refined lead x 31.1 x 10^9 J/ton	.31
Total for prime western zinc	49.88

permit recycle of water to the process or discharge to the environment. Such treatment is costly, but not prohibitively so.

Disposal of leach residue (jarosite or goethite) is the most difficult environmental problem facing the electrolytic plant. The amount of such residues is large--0.6-0.9 tons of jarosite, or 0.3-0.5 tons of goethite per ton of zinc --they are finely divided, voluminous, contain soluble salts, and are unstable in the natural environment. Today they are impounded in plastic-lined ponds, but this method consumes considerable land area, and a better disposal method needs to be found. Some method for using these residues (either to recover the zinc and iron values, or as an additive to some high-volume product like cement) would prove the best alternative. The worst case would be the need to furnace-treat the residue at high temperature to produce an environmentally stable clinker or slag suitable for land disposal. For the 0.36 tons of goethite residue, from the electrolytic flowsheet in this study, I estimate that furnace treatment might require an additional 2 x 10^9 J/ton of SHG zinc--an energy (and cost) requirement to be avoided, if possible.

Retorting Processes

There still remain several vertical retort plants (N.J. Zinc Co.) and electrothermic retorts (St. Joe Minerals Corp.) in the U.S. and elsewhere in the world. Energy consumption for these processes has been analyzed in the Battelle-Columbus study (1). The Battelle results have been recalculated to eliminate the energy for mining and beneficiation of ore, and to remove the credit for acid production, so as to make the values comparable to those in this study. On this basis, the PFE values are 60.4 and 10^9 J/ton zinc for the vertical retort, and 72.2 x 10^9 J/ton zinc for the electrothermic process. Neither of these values includes energy for refluxing of zinc, so that the product includes a mix of prime western and intermediate grades of zinc. The best comparison of these values, therefore, is with the values in Table VI, and we see at once that the retort processes are far more energy-intensive than either the ISP or electrolytic flowsheets. Indirect heating through retort walls (vertical retort) or with electric energy (electrothermic retort) is inefficient. These processes suffer further from the need for considerable excess carbon as a reducing agent, and from the need for sintering or bri-quetting the retort charge. With the present outlook for energy prices the future of these process is not bright.

Promising Process Improvements

In this section I consider potential improvements to the electrolytic and ISP flowsheets that can be expected to decrease the energy-intensity of these processes. In both cases there are innumerable opportunities to make minor improvements that will save energy. I refer to such items as better insulation of steam pipes and heated vessels, more efficient motors properly sized for the intended service, pipe sizes adjusted to reduce pressure drop, maximum heat recovery from exhaust gas streams and many similar changes. These improvements are important, but my focus will be on more fundamental changes in the process itself.

Electrolysis with Methanol

Table II shows that electrolysis energy accounts for 79% of the total energy used to produce electrolytic zinc. Reduction of this energy component offers rich rewards, but sixty years of development and experimentation with different electrolytes, different current densities, smaller anode-cathode distances, and different anode materials have not changed the electrolysis energy to any significant extent.

Just this past year my colleague at Columbia, Professor Paul Duby, and his graduate student, Paul Vining, have discovered a most promising approach to significant lowering of the electrolysis energy for zinc as well as other metals. They have devised a means to anodically decompose methanol, dissolved in the zinc electrolyte at a concentration of 2-3%, to carbon dioxide according to the reaction:

$$H_2O + CH_3OH \rightarrow CO_2(g) + 6H^+ + 6e$$

This anode reaction possesses a far lower positive potential than the decomposition of water to oxygen used in conventional electrowinning, with the result that the cell potential is greatly reduced. In laboratory experiments they were able to operate a zinc electrowinning cell with a current density of 400 amps/m^2 and total cell voltage of 1.9 volts to produce good quality zinc deposits for a calculated expenditure of about 1800 kwh/ton cathode. Prof. Duby and Mr. Vining are seeking patent protection, and details of the method cannot yet be divulged.

Of course, the drastic reduction in electrical energy compared to conventional cell operation is not obtained free--either in dollars or in energy accounting terms. Methanol is consumed by the anode reaction and must be replenished, and it requires both dollars and energy to make methanol. However, rough calculations of these dollar and energy costs indicate a net saving of about 30% in the energy requirements for methanol electrolysis, compared to conventional operation. The dollar cost shows a break-even when power cost is about 2-2.5¢/kwh, with dollar savings increasing rapidly as unit power costs rise above 2.5¢/kwh.

To date, experimentation has been limited to a small laboratory scale. Many unexpected problems may arise when the method is tested on a larger scale. It has my vote, however, for the potential of a significant reduction in the energy total of Table II.

Changes in the ISP

Hopkin and Richards (2) have thoroughly analyzed potential improvements in this process, so that I will only make a few comments on their study. One major improvement that they foresee is the replacement of updraft sintering by fluid-bed roasting followed by hot-briquetting to produce feed for the blast furnace. If briquette quality is such that the blast furnace operation is not adversely affected, this change could result in the ability to operate with very low lead in the charge (essentially a zinc blast furnace, rather than a zinc-lead blast furnace) and the elimination of the hygiene problems surrounding the sinter plant. If this method is applied to low-lead feed materials, so that little or no lead bullion is produced, I believe the energy saving would be minor (~1. x 10^9 J/ton zinc), because, in this case, the large additional credit for surplus steam generated in roasting (~6.7 x 10^9 J/ton zinc) is more than cancelled by the loss of credit for byproduct lead (~7.0 x 10^9 J/ton zinc). In other words, by making the ISP an all-zinc process one forfeits the advantage that has long been claimed--that the lead produced by the process is obtained free of any extra energy requirement. A small energy credit would still result from the substitution of more efficient roasting for sintering.

If, on the other hand, fluid-bed roasting and hot-briquetting can be applied to the usual raw materials, having a zinc-lead ratio of about 2 then the energy saving could approach 8. x 10^9 J/ton zinc, because the lead credit would still pertain, and the steam credit from roasting as well.

Hopkin and Richards (2) point out that a large fraction of the energy employed as coke in the ISP leaves the system as warm water from the cooling launders (about 6.3 x 10^9 J/ton zinc), and they discuss difficulties as well as possible methods by which this energy could be used or recovered. I agree as to the importance of saving this energy, and would add the thought that the entire launder system needs redesign to greatly minimize the exposure of molten lead to the atmosphere if the ISP process is to be capable of meeting stricter environmental standards. If both environmental control and energy savings are linked together in redesign of an enclosed cooling launder, it would seem possible that the energy saved could compensate for the cost of environmental control.

Radical New Technology

For many years I have believed in the possibility of a radically improved pyrometallurgical process for zinc production that would use less energy than existing processes and simplify problems of environmental control, without requiring excessive capital investment. My leading candidate for this role would start with fluid-bed roasting of concentrate to produce calcine, acid, and surplus steam for sale or power production. This is the well established first step of the electrolytic flowsheet for which Table II indicates an energy expenditure of 3.99 x 10^9 J/ton zinc (including acid production) and a credit for 2.1 tons of steam. The steam credit would be 2.1 x 3.2 x 10^9 = 6.7 x 10^9 J, so that the roasting-acid making step would show an overall credit of (6.7 - 3.99) x 10^9 = 2.7 x 10^9 J/ton zinc.

As in the ISP, carbon monoxide derived from carbon would be the reducing agent for zinc oxide. To avoid two liabilities of the ISP--the need to agglomerate the feed, and the need for very expensive lump coke--I would mix finely-divided carbon (coke breeze or petroleum coke) with the finely-divided calcine. To reduce the volume of gas produced, enrich this gas with zinc, and save energy, I would employ oxygen (98% O_2) rather than air for burning the carbon.

The finely-divided feed mixture (with fluxes necessary to form slag from the gangue) would be reacted with oxygen in one of several possible high-intensity reactor designs--a flash-smelting unit (9), a cyclone furnace (10), or a submerged-tuyere unit similar to a slag-fuming furnace (11). All of these reactor types are capable of very high specific capacities and very rapid reaction rates for finely-divided solids. The patent of Derham (12) covers a cyclone furnace applied to the zinc oxide-carbon reaction.

The gas resulting from the reactions would be similar to that from ISP, except for the absence of nitrogen, and calculations indicate the possibility of a gas analyzing 15% Zn, 14% CO_2, 64% CO, remainder H_2, H_2O and N_2. This gas would be condensed in an ISP lead splash-condenser with about one-half the rate of lead circulation currently used for lean blast-furnace gas (~7% Zn). The condenser off-gas would be a much richer fuel (82% CO + H_2) than LCV gas (~21% CO) from the ISP, and would be burned in a boiler to produce surplus steam or electric power. Condensed zinc could be recovered either by vacuum distillation, or by the conventional cooling launder plus refluxing.

This concept has been criticized (2) as having "no energy advantage" relative to the ISP. Although this "process" is only a concept, and no one can be sure, at this time, that it can be made to work as conceived, I will try to show here that the concept does possess a very large energy advantage. The calculated performance figures given below were derived by a detailed heat and mass balance, performed with the help of a computer, for smelting of zinc oxide calcine (70% Zn, 3.5% Fe, 1.8% Pb) with coke breeze (83.5% C, 15% ash), commercial oxygen (98% O_2), and flux (Fe_2O_3). The reactor heat losses were estimated for a water-cooled-wall reactor protected by a frozen slag layer. Results for a typical calculation follow, based on production of 1 ton of PW zinc from a conventional lead splash-condenser and cooling launder (90% efficiency):

Reaction temperature: 1415°C

Gas Analysis: 63.9% CO, 14.3% CO_2, 15.0% Zn, 5.4% H_2, 0.3% H_2O

Zinc lost to slag: 1.3% of total

Material flows	tons/ton PW zinc
Calcine	1.587
Coke breeze	1.275
Flux	.027
Oxygen	1.419
Slag	.372
Condenser off-gas	2100 Nm^3
Off-gas heating value (net)	10.2 x 10^6 J/Nm^3 (275 BTU/SCF)

An energy analysis of this part of the total flowsheet shows the following:

Material	Amount	x	Unit Energy	10^9 J/ton PW zinc
Coke breeze	1.275 ton	x	24.4 x 10^9 J/ton	31.11
Oxygen	1.419 ton	x	4.0 x 10^9 J/ton	5.68
Electric	150 kwh	x	11.1 x 10^6 J/kwh	1.67
Credit for condenser off-gas:				
1) as steam	6.6 tons	x	3.2 x 10^9 J/ton	(−21.12)
or				
2) as electric	1280 kwh	x	11.1 x 10^6 J/kwh	(−14.21)
			Total with steam credit	17.34
			Total with electric credit	24.25

Two features of this concept should be noted. It does consume a large amount of carbon (more than the ISP), but it consumes this as low quality coke-breeze (much lower unit-energy value and cost than metallurgical coke), and it

45

produces a very large amount of relatively rich fuel gas, and this is responsible for a large energy credit. It might be possible to sell this gas, of medium heating value, or it could be used to produce steam, electricity or both for sale. If the roasting plus acid-making energy (a credit or 2.7×10^9 J/ton zinc) is added to the smelter energy (24.25×10^9 J/ton zinc, with electric credit) and the energy for refluxing from Table V (7.28×10^9 J/ton zinc) the total process energy for 1 ton of SHG zinc is about 28.8×10^9 J. Sale of the fuel gas, or co-generation of steam and electric energy, would lower this energy value even further.

The author would have liked to consider the energy requirements for other potential new zinc production concepts, but unavailability of data and lack of time prevented this. Important among such omissions are the Sherritt-Gordon direct pressure leaching of zinc sulfide (13) and the U.S. Bureau of Mines work on electrolysis of molten zinc chloride (14). The Kivcet process (15), although primarily a producer of lead, also recovers zinc; comparison of this new commercial process with conventional production of lead and zinc must await the availability of firm production data.

Acknowledgment

The writer is deeply indebted to the following companies for their contributions to the energy questionnaire on which this study is based: Cominco, Ltd., Electrolytic Zinc Co. of Australia, Ltd., Hachinoe Smelting Co., Ltd., Imperial Smelting Processes, Ltd., Jersey Miniere Zinc Co., Korea Zinc Co., Ltd., Mitsubishi Metal Corp., Mitsui Mining & Smelting Co., National Zinc Co., Outokumpu Oy, Sulfide Corp. Pty., Ltd., Sumiko I.S.P. Co., Ltd., Texasgulf Canada, Ltd., and Société des Mines et Fonderies de Zinc de la Vieille-Montagne. The information they supplied for this study was an invaluable aid, but the use to which I have put it is entirely my responsibility.

References

1. "Energy Use Patterns in Metallurgical and Nonmetallic Mineral Processing, High Priority Commodities," Battelle-Columbus Laboratories, Columbus Ohio, Nat. Tech. Inf. Ser. PB 245 759/AS, 1975.

2. W. Hopkin and A.W. Richards, "Energy Conservation in the Zinc-Lead Blast Furnace," Journal of Metals, 30 (11) (1978) pp 12-17.

3. A.R. Gordon, "Improved Use of Raw Material, Human and Energy Resources in the Extraction of Zinc," pp 153-60 in Advances in Extractive Metallurgy 1977, M.J. Jones, ed.; Inst. Min. and Met., London, 1977.

4. G.M. Meisel, "New Generation Zinc Plants, Design Features and Effect on Costs," Journal of Metals, 26 (8) (1974) pp 25-32.

5. H.H. Kellogg and J.M. Henderson, "Energy Use in Sulfide Smelting of Copper," pp 373-415 in Extractive Metallurgy of Copper, Vol I, J.C. Yannopoulos and J.C. Agarwal, eds.; TMS-AIME, New York, 1976.

6. H.H. Kellogg, "Energy Efficiency in the Age of Scarcity," Journal of Metals, 24 (6) (1974) pp 25-29.

7. R. Knobler, T.I. Moore, R.L. Capps, "New Zinc Electrolysis and Residue Treatment Plant of the National Zinc Co," Erzmetall, 32 (3) (1979) pp 109-116.

8. C.F. Harris, "Production of Low-Lead Zinc from the Imperial Smelting Furnace," in Proceedings of Fourth International Conference on Vacuum Metallurgy,

Iron and Steel Institute of Japan, The Japan Institute of Metals, and the Vacuum Society of Japan, 1974.

9. M.C. Bell, J.A. Blanco, H. Davies and R. Sridhar, "Oxygen Flash Smelting in a Converter," Journal of Metals, 30 (10) (1978) pp 9-13.

10. C.A. Maelzer, R. Biquet, R.F. Derksen, "Suspension Smelting of New Jersey Residues," Applied Energy, 1 (1975) pp 107-118.

11. H.H. Kellogg, "A New Look at Slag Fuming," Engineering and Mining Journal, 158 (3) (1957) pp 90-92.

12. L.J. Derham, "Extraction of Zinc," U.S. Patent 3, 271, 134, Sept 6, 1966.

13. B.N. Doyle, I.M. Masters, I.C. Webster and H. Veltman, "Acid Pressure Leaching of Zinc Concentrates with Elemental Sulfur as a Byproduct," Eleventh Commonwealth Mining and Metallurgical Congress, Hong Kong, 1978, Paper No. 56.

14. D.E. Shanks, F.P. Haver, C.H. Elges, M.M. Wong, "Electrowinning Zinc from Zinc Chloride-Alkali Chloride Electrolytes," U.S. Bur. Min Report of Invest., No 8343, 1979.

15. A.P. Sychev, "Oxygen Electrothermal Processing of Lead Concentrates in the Kivcet CS Equipment," Tsvet. Met., 50 (8) (1977) pp 8-15.

ADVANCES IN LEAD, ZINC AND TIN TECHNOLOGY -

PROJECTIONS FOR THE 1980's

T.R.A. Davey

Metallurgy Department, University of Melbourne, Australia

Progress in the technology of lead, zinc and tin extraction over the last ten years is reviewed, and forecasts are made of developments to be expected over the next decade. Some of the major problems currently requiring attention are outlined.

The lead blast furnace, so long supreme as the lead metal producer, is now being challenged by direct smelting processes, which offer cost savings, freedom from dependance upon high quality metallurgical coke, and greatly simplified pollution control measures. Two new continuous lead refining processes have emerged: for decoppering and debismuthising.

Electrolytic zinc processing has forged ahead of all competition for high-grade concentrates, with improvements in the treatment of primary leach residues, by jarosite, goethite or hematite processes. The Imperial Smelting process is expanding into the processing of mixed lead-zinc-copper concentrates, residues, fumes and miscellaneous low-grade materials. The Kivcet process offers a substitute for slag-fuming of copper or lead slags for zinc recovery.

Tin smelting reverberatories are beginning to be replaced - by the TBRC* to date - and both smelting and mineral processing are being partially replaced by fuming processes. Centrifuging of drosses, and vacuum distillation, are being introduced into tin refineries' flowsheets, reducing the need to up-grade concentrates before smelting, and offering a pyrometallurgical alternative to electrolytic refining of heavily contaminated crude tin.

* Top Blown Rotary Converter

Introduction

In their review of process research on lead and zinc extraction, at the St. Louis Symposium in 1970, Davey & Bull(1) concluded that "the major developments over the rest of this century will be improvements in the performance of existing processes and equipment, through the application of process control". While this has proved accurate to date for zinc, it has certainly not been so for lead - nor would it have applied for tin.

Economic difficulties at times during the past decade for all three metals have prevented developments which might otherwise have been expected, or even caused abandonment of successful developments - as, for example, vacuum dezincing of I.S.F. lead at Swansea. The enforced provision of large amounts of capital for environmental protection measures has necessitated curtailment of capital expenditures on many research and development projects which would otherwise have been pursued vigorously.

The uncompromising attitude of the U.S. E.P.A. (which unfortunately often tends to be followed in other countries) in setting and insisting on unrealistic standards for gaseous emissions, has forced many producers to spend considerable resources on radically new, low-temperature possibilities for technological development, many of which will pose far greater environmental hazards in the disposal of liquid or solid wastes, as well as being much higher in capital and energy requirements. The end result may well be a movement of metal production away from developed to under-developed countries, where a more realistic balance may be possible between conservation of environment, capital resources, and energy sources - without the public debates dominated by emotional and populist propaganda which clouds the issues in most developed countries.

Thus the prediction of developments over the next decade involves political factors as much as economic, scientific and technological. Since lead and zinc occur predominantly as sulfides (and even tin oxide, cassiterite, is often associated with pyrite) the disposal of sulfur frequently poses a greater problem than the production of the metal *per se*. Although in the longer term sulfur will be a resource requiring conservation, for the next decade it will be in potential over-supply, and whereas in a planned economy it would be a simple matter to reduce the mining of native sulfur, and use by-product sulfur from metals or coal, this is not at all easy in free-enterprise economies. Nevertheless Governments should be made aware of the desirability of this course of action, lest the world's resources continue to be grossly misused. For example, it is much more desirable to produce sulfuric acid for immediate use, than to make calcium sulfate for dumping, and possibly rendering hard very large tracts of ground waters, as well as being difficult to revegetate. Even worse is the unnecessary consumption of large amounts of fuel or energy to make elemental sulfur, for safe storage. Disposal of sulfur from sulfide concentrates as sulfuric acid by-product enables use to be made of the fuel value of sulfur (and iron) in what is a fairly high-grade fuel. These values are lost in many of the alternative processes proposed to avoid the production of SO_2, and such processes are high energy consumers.

Perhaps it is not too much to hope that the U.S. E.P.A. and other government agencies will in the near future face up to the real problems associated with sulfur recovery and disposal. There are more rational and practical bases for control of emissions(2), and it should not be beyond the wit of economists to devise a rational scheme to encourage sulfur recovery with optimum use of resources in free enterprise economies. Also, it would be a tragedy if critical energy shortages forced the abandonment of reasonable environmental protection standards rather than the promotion of economically feasible recovery of by-product sulfur.

Lead

Undoubtedly the major development of the last decade has been the emergence of Kivcet as a mature process for production of lead by direct smelting of sulfide concentrates, with simultaneous production of zinc electrothermically from the slag, if warranted(3,4). Although Russian scientists consider that their engineers have proceeded very slowly and conservatively in scaling up from small pilot units, very soon several plants of capacities in the range 50-100 thousand tons per annum of crude lead will be brought into operation.

The Kivcet furnace (see Figure 1) combines the Outokumpu method of combusting fine sulfide concentrate (in a vertical tower, with almost pure oxygen, plus fluxes and returned dusts) with the St. Joe method of generating zinc vapour from the slag (using electrodes suspended in the slag settling bath, plus some fine coke for extra reduction) and the Imperial Smelting lead-splash condenser to produce liquid P.W. grade zinc - or alternatively, burning the zinc vapour to form oxide suitable as feed for an electrolytic plant. (The zinc oxide is an excellent neutralising agent.)

Much of the success of the Kivcet process is attributed to the successful development of (i) a water-cooled under-flow wall separating the SO_2-bearing gases in the smelting chamber, and the zinc-bearing gases generated from the slag; (ii) an electrostatic filter which continuously returns oxide dust and fume to the smelting shaft; and (iii) hermetically tight seals between the electrodes and the furnace top, so that no leaks occur to the atmosphere.

In contrast to the suspension reactor of the Kivcet type, other direct smelting processes being developed rely on a converting or submerged combustion technique, the reacting gases being blown into a liquid bath, or jetted on to it at high speed, so that in either case very intensive mixing of the bath occurs. These include the QSL process (Queneau-Schumann-Lurgi) in which oxygen is blown into the bath through the centre of a double tuyere, or Savard-Lee-type injector, surrounded by a shielding gas of SO_2 or hydrocarbon, so as to prevent too-rapid refractory wear - which was the cause of St. Joe's failure to smelt successfully in a converter in the 1960's. Like Kivcet, QSL uses nearly pure oxygen, but instead of electrical energy input, coal or coke breeze and propane are burnt to make up the heat balance. It is described in detail elsewhere in this symposium.

The Sirosmelt process(5) may also be applied to lead smelting, as it has been to tin. This involves injection of air and fuel (solid, liquid or gaseous) into a liquid bath by means of lances. As with the QSL process, the concentrate and fluxes are added in pelletised form. Since simple, air-cooled lances are used, instead of tuyeres, the reactor used need not be tiltable.

The TBRC is being used to smelt sulfidic or oxidic materials(6,7), and it is claimed that the rotary action of the furnace gives good mixing. In fact, this is true only while the charge is solid or pasty. Once liquid, the mixing must be provided by jetting of gas on to the bath surface. This gas may, of course, be oxidising, neutral or reducing, and can be varied at will during the course of a charge cycle.

The supersonic top-blowing technique pioneered by Wuth at Berlin, and being developed by Klöckner-Humboldt-Deutz, offers an alternative to lancing or tuyere-injection, and may also shortly enter the field of possible new lead smelting techniques.

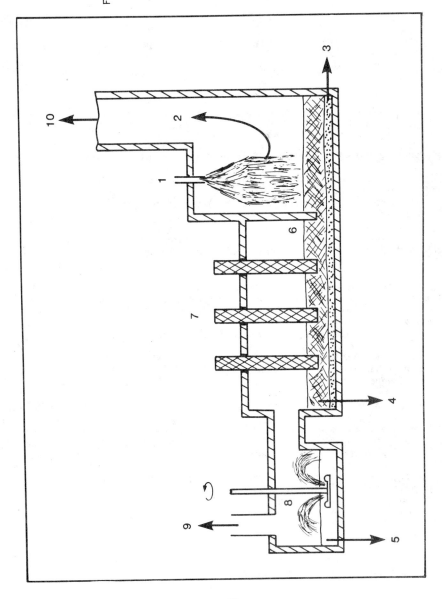

Fig. 1 - KIVCET PROCESS

1. Concentrate burner
2. Fume-laden gas
3. Lead tap
4. Slag tap
5. Zinc tap
6. Water-cooled
   under-flow wall
7. Electrodes
8. Lead-splash
   condenser
9. CO off-gas
10. To electrostatic
    precipitator

51

All of the above are capable of being sealed fairly completely from the working atmosphere, and so can meet the most stringent environmental control regulations.

The major problems in direct lead smelting are that PbS is very volatile (and hence temperatures and gas volumes must be kept as low as possible) and the metal/slag equilibrium is such that two (preferably counter-current) stages are required to produce both a slag low enough in lead to discard, and a lead low enough in sulfur to refine. The relationship between these quantities (valid only in the low percentage ranges) is(7):

$$\%S \text{ in lead} = \frac{610 \cdot p_{SO_2}}{(\%Pb \text{ in slag})^2} \cdot 2 \tag{1}$$

This form of relation follows from the approximate equilibrium

$$(S)_{\text{in lead}} + 2(O)_{\text{in slag}} = (SO_2)_{\text{in gas}} \tag{2}$$

$$\text{with } K = \frac{a_{SO_2}}{a_S \cdot a_O^2} \tag{3}$$

and putting the approximations S% in lead $\propto a_S$, and Pb% in slag $\propto a_O$.

Neither Kivcet nor QSL are operated in truly counter-current fashion, and so both produce an intermediate slag which is very high in lead, requiring coke breeze or coal to reduce its lead content to acceptable levels for discard.

Kivcet requires a feed of dried concentrate, and hot waste gases are utilised for the drying operation. QSL feeds concentrate as pellets, and it is claimed that the pelletising costs are no greater than drying costs.

Large Kivcet units are expected to operate with a consumption of 650 kWh per ton of lead produced, and 76 kg coke breeze. QSL is estimated to require 95 kg coal per ton of lead. A limitation of the QSL process is that the slag's zinc content must be kept below 9% Zn, in order to prevent zinc volatilisation which would add unnecessarily to recirculating flue dust.

Capital costs will be less for QSL than for Kivcet, but maintenance costs will be higher, as converter re-lining will certainly be required at fairly frequent intervals.

If a truly counter-current two-stage direct smelting process for lead sulfide concentrates is developed, then in the first stage a high-sulfur lead will be produced, in equilibrium with a low-lead slag, suitable for discard or sending to slag fuming; and in the second stage sulfur will be lowered in the lead by converting or by reaction with returned oxide flue-dusts, while the high-lead slag passes back to the first stage for reduction of its lead content. The sulfur content of the lead produced does not need to be eliminated completely in the second stage, as the Davey-Port Pirie continuous decoppering unit(8) can also be run as a desulfuring unit, and any low-copper mattes produced could be recycled to the smelting unit.

Future developments we may expect include the establishment of the Kivcet process throughout the world as one of the standard procedures. (It is thought to be favoured for early introduction at Port Pirie and Trail.) Presumably it will not be long before the electrothermic portion of the process is replaced in some instances by direct combustion of fuel. In this event the zinc would be produced as oxide, rather than as P.W. metal.

We may also expect to see a trial, on extensive pilot plant scale, of the QSL process, and similarly of processes involving lancing (either submerged or top-jetted) instead of tuyeres, which can be used with a stationary furnace, a suitable design of which should give long refractory life. A further possibility, not yet under serious consideration, as far as is known, would be a stationary converting-type furnace with tuyeres and oil-filled plugs, as used at Messina in South Africa for copper converting(9).

These processes will be two-stage, because of the slag/metal equilibria mentioned above, and if counter-current in principle will require no reducing agent, although some carbon or hydro-carbon *fuel* will be required. Oxygen-enrichment rather than preheating will be used as much as possible to keep gas volumes low, to minimise lead volatilisation and gas filtering costs, and to provide a high SO_2-tenor gas for acid manufacture.

It is also to be expected that the TBRC will make some headway in the near future as a batch process for smaller producers, but that in the long run it will be realised that exactly the same processing can be achieved in a simple tilting (or even stationary) top-blown or lanced vessel, without the very expensive provision of both tilting and rotary mechanisms.

It can scarcely be expected that the blast furnace will disappear in the foreseeable future, and we may expect the wider acceptance of oxygen-enrichment or preheating of blast. This will open the way to steam injection at the tuyeres, as recommended so long ago by Lumsden(10) but so far not attempted in the primary lead blast furnace, although adopted with remarkably successful results in secondary lead blast furnaces(11,12): greatly increased furnace speed, improved consistency of performance, and decreased coke and iron consumption.

Developments in secondary lead smelting may well see a trend towards smelting of whole batteries in blast furnaces, with after burners and flash melting of dusts. This will generally be a more hygienic and environmentally satisfactory operation than breaking of cases and separation of constituents, although chlorine from PVC separators poses problems. Prevention of fume leakage from blast furnaces will probably always be more effective than from rotary furnaces. A swing to blast furnaces will throw a higher load on to refining, since it will not then be possible to smelt in two stages, with a limited amount of reduction in the first stage, to produce a soft lead and an antimonial slag, followed by production of a high-antimonial lead in a second reduction stage.

Use of the Kivcet process for treating battery scrap requires that the batteries be crushed, followed by separation of PVC for **recovery or** use of the crushed cases as reductant. The sulfur is recovered as SO_2 and does not require iron for matte formation , as in the blast furnace. This application of Kivcet may well also have a future. The use of mineral separation techniques to separate crushed battery constituents is also a feature of Tonolli blast furnace processing.

Refining

Recent developments include improvements in de-coppering, desilvering, bismuth dross processing, and new techniques for the removal and recovery of tellurium and selenium. As well as these batch processes, continuous developments include decoppering, debismuthising and fractional crystallisation for simultaneous removal of silver and bismuth.

About a decade ago it was discovered by accident(13) that a mixture of pyrite and elemental sulfur is effective in decoppering soft lead which contains less than the 0.02% Ag or 0.10% Sn required to "catalyse" the normal decoppering process using sulfur alone(14). This permits the effective decoppering of lead such as that from the Missouri belt which contains only about 0.006% Ag and no tin. With this procedure, it is not necessary to stir the lead at a temperature close to the freezing point - in fact the operation is more effective at higher temperatures(15).

Asarco have also found two-stage processing with NaOH and pyrite to be effective in decoppering, with the further claimed advantage that SO_2 evolution is avoided(16).

A recent development significantly improves the batch desilverising operation, reducing the labour, fuel and zinc requirements, and improving the consistency of final silver values in the desilverised lead(17).

The Port Pirie continuous decoppering unit has run into difficulties, and operations are temporarily suspended for modifications. Possibly the mixer-settler units were made too small for stable operations - even small deposits of dross accretion caused major blockages. It can confidently be expected that continuous decoppering will be established during the next decade as a sound operation. Since oxidation of the surface of sulfide particles is now believed to play a significant role in the process(15), it may be necessary to abandon the concept of a cascade of kettles, and re-design the process to utilise only one stirrer kettle.

Continuous debismuthising in two stages (a first stage to about 0.01% Bi or less, with Ca and Mg additions, followed by a second stage to about 0.001% Bi with Sb additions) was successfully developed on pilot scale at Port Pirie(6,18) and a full-scale plant is now being built. The advent of this process now removes the last impediment forcing some refiners to prefer electrolytic refining to pyrometallurgical. Although fine debismuthising by means of antimony dates back to 1936(19), until recent investigations in Australia(20) the procedure was not well understood, and was difficult to conduct in practice. The modern continuous version is claimed to be much more economical than the batch process.

Continuous refining operations following a concept of this writer, utilising mixer-settler units for the stages of decoppering, softening, desilverising and final refining, were developed in a brief pilot plant campaign some 17 years ago, but the work remained unpublished until recently(21). The concept offers numerous advantages over the Williams-Port Pirie softening and desilverising processes, being lower temperature and easier to mechanise and automate. It may be expected that one of the large lead refiners will take up this development in the next decade or so.

Smirnov(22) has long advocated the adoption of continuous refining stages in Russian refineries, and large pilot plants have been successfully operated, but no completely continuous lead refinery has yet been built in Russia.

Recent developments in by-product treatment include Asarco's simplification of the processing of bismuth crusts, by a controlled oxidation which produces a bismuth alloy free from the Ca and Mg reagents, and partially freed from lead(23). It can be expected that the Leferrer-Penarroya vacuum dezincing furnace may make further head-way in replacing Faber-du-Faur retorting of silver-zinc alloy crusts, and possibly the change to induction heating suggested by Gerlach et $al.$(24) will accelerate the trend to vacuum distillation, as it lessens the sensitivity of the process to disturbance by oxide drosses.

The current movement to lower antimony content of battery grid metal (a lowering from 5-6% Sb to 2-3% Sb, or even to nil in the case of the Ca-lead alloys) will bring about changes in the by-product treatment of antimonial slags or drosses. Harris processing produces a sodium antimonate, used for enamelling, and so will not be affected by changes in battery demand specifications, but other refineries may be faced with the prospect that the demand for antimonial lead is far short of their potential production, and will have to consider instead production of antimony trioxide, the demand for which as a flame-retardant additive could well increase above the present growth-rate of 10-15% p.a., as a result of safety legislation. The Sb_2O_3 will be fumed from antimony metal baths or from antimonial slag, under appropriate conditions of temperature and oxygen potential, and must be of a high degree of purity (about 99.9% Sb_2O_3) and fine-ness (average particle size 1 μ), the latter requirement being met by shock-chilling of the vapour at a suitable rate.

A most appropriate use of hydrometallurgical techniques is illustrated by two processes for the treatment of copper dross. Berzelius dissolve their (predominantly sulfidic) dross in sulfuric acid and electrowin the copper, returning lead sulfate for smelting. I.S.P* and C.R.A** have developed an ammonia-ammonium carbonate leach(25) to treat their copper drosses, which are very low in sulfur, followed by solution purification and solvent extraction of the copper, stripping with sulfuric acid and producing high-grade copper sulfate. The leach residue is returned to the smelter at the sinter plant. These procedures take advantage of the ability of hydrometallurgy to make a clean separation at low temperatures; the solutions are recycled and the solid residues return directly to the smelter, thus avoiding the pollution problems which usually plague hydrometallurgical processes.

Zinc

In **contrast to areas of** lead and tin, no new processes for zinc have emerged in the past decade, except for a theoretical possibility not so far tested in practice.

Very considerable improvements have been made in both the electrolytic and Imperial Smelting (I.S.) processes, and progress will continue unabated for the next decade, as both are far from having exhausted their capacity for improvement.

A notable advance in electrolytic zinc technology is the almost universal adoption of fluid-bed roasting of zinc calcines. It was demonstrated by Jorgenson & Mumme that the low-melting phase sometimes encountered is due to pyrite, not to lead(26).

The jarosite, goethite, and haematite processes have reached maturity, and permit the recovery of 10-15% additional zinc formerly lost in the primary leach residues. The jarosite process has the advantage of removing sulfate as well as iron from the circuit, thus reducing the requirement of lime for neutralisation purposes. The difficulties of disposal of jarosite (due to its significant content of toxic heavy metals Zn, Pb, Sb, As, Cu, Ni, Hg) have led to a significant effort to reduce these, and the E.Z. Coy.*** of Australia has succeeded in developing a two-stage jarosite precipitation process, the neutralisation and cooling being performed separately, so that the heavy metals largely enter a preliminary precipitate which may be returned so that they leave the circuit in the residue destined for a lead smelter.

  * **Imperial Smelting Processes**
 ** **Conzinc Riotinto Australia**
*** **Electrolytic Zinc of Australasia**

Although they do not remove sulfate, the goethite and hematite
processes produce a product suitable for feeding to the iron blast furnace,
and so solve the problem of iron residue disposal, if transport from the
zinc to the iron works is feasible, and if the toxic heavy metal content is
low enough to be acceptable to the iron maker. In practice the toxic impurity
contents have usually been too high.

Improvements continue in purification procedures, with continuous
processing and automation, amongst which the Vieille Montagne and Outokumpu
fluid bed zinc dust reactors are prominent. In the future we may look to the
introduction of solvent extraction for solution purification.

Automatic stripping of zinc cathodes, together with automated mechanical
handling and cleaning of electrodes, has been widely introduced. Although
continuous electrolysis involving fluid bed or stirred electrodes is still
a dream of the future, no doubt progress will be made in this direction over
the next decade.

Use of induction furnace melting of zinc cathodes, and automatic casting
and bundling machines, have become almost universally adopted.

Along with mechanisation and automation there emerges the possibility of
increased computer control of operations, and this can be expected to become
commonplace in the next decade.

A further tantalising possibility is the development of new electrolytes
- perhaps organic salts - to reduce the energy requirements for electrolysis.

Present I.S. process history is also a record of continual and gradual
improvement. Giant steps were taken recently in the understanding of sinter
structure's bearing on its performance in the zinc-lead blast furnace(27,28).
An ideal structure is one comprising a lattice of needle-like crystals of
ZnO (zincite) which do not soften or melt or become reduced until low down in
the furnace shaft, with other constituents in the form of a glass between the
network, including lead which becomes reduced relatively high in the furnace.
When a network of high-melting crystals - zincite or franklinite ($ZnO;Fe_2O_3$)
or mellilite ($2CaO;ZnO;2SiO_2$) does not form, then much lower-melting silicates
constitute the matrix, and the sinter softens high in the furnace shaft and
causes numerous problems: hang-ups in the shaft, unevenness of gas
distribution, variability of reduction and generally slow production rates
and high zinc content of slags. With an understanding of the correct sinter
types to aim for, goes improved capability to treat lower grade and more
complex materials. The I.S. furnace is now established as a most satisfactory
unit for treating complex Zn-Pb-Cu materials, whether as concentrates or
intermediate products, even of comparatively low-grade. This trend can be
expected to continue, the I.S.F. becoming a scavenging unit for low-grade
complex materials, just as the lead blast furnace has always been for low-
grade leady materials.

The substitution of a portion of the sinter charge by hot-briquetted or
binder-briquetted material has added to the proportion of returned or recycled
materials that may be treated in the I.S.F., including fine oxide drosses
from an I.S.F. or refinery plants, oxide flue dusts (including material
reclaimed from iron blast furnace fumes), coke fines, and other residues from
outside plants.

Selenium, as well as cadmium, is now recovered from sinter plant flue
dust(27).

56

Cowper stoves have replaced alloy steel recuperators for preheating of blast, and enabled much higher blast pre-heat temperatures to be achieved (approaching 1000°C instead of 700°C).

The improvement of the lead-splash condenser efficiency to a stage where it can cope with gases containing H_2O has permitted the injection of oil into the tuyeres, to replace portion of the coke, which has been economic to date - but may not be for much longer, unless waste oil products can be used. There may, however, be a case for using natural gas in some locations. If abrasive-resistant tuyeres and conduits can be developed, a further possibility would be the injection of coke fines or powdered coal, with the object of reducing the amount of costly metallurgical coke consumed by the I.S. process.

Further developments at the bottom of the furnace include the use of downwardly-sloping tuyeres in an attempt to perform some dezincing from the slag pool there - to recover additional zinc without using appreciably more reducing agent. It is too early to say whether this development will be successful, but it is obviously well worth extensive effort.

There are three major drawbacks of the I.S. Process: the need for expensive preparation of charge by sintering; consumption of coke, which is an expensive form of fuel and reductant; and the relatively impure product - PW grade. This last disadvantage was actually overcome to a great extent by the development of a vacuum dezincing plant producing zinc of about 0.02% Pb content at Swansea in the 1960's. By an incredible decision this plant was scrapped when the Swansea plant was closed down in 1970. Since then all zinc produced in I.S.F.'s has been PW grade, and any production in excess of demand for this grade has had to be refined by the New Jersey refluxing process, which is very expensive and a high energy consumer.

The heat generated in the I.S.F. condenser by condensation of zinc and cooling of the non-condensible gases is practically sufficient to re-distill the zinc twice over, and we may look to a vacuum refluxing process development to produce zinc of the highest grades without further expenditure of energy for the separation of less volatile metals (Pb, Cu, Fe, etc.) and only a small amount of energy to separate cadmium.

There are also moves underway to develop partial freezing or crystallisation techniques to refine zinc from some of its impurities, but these do not have the merit of utilising the otherwise waste energy of the condenser, and also fail to separate some important impurities - e.g. Cu and Sn. The principles and apparatus employed are similar to those previously described for refining lead(12) and tin(29). Although the energy consumption and capital costs are moderately low, the process has only limited promise because it fails to remove copper and tin, and it is difficult to see how it could compete with vacuum distillation techniques when these are developed.

Recent developments in the direct smelting of copper and lead have no counterpart so far in the metallurgy of zinc. It had been generally considered that phase relations in the Zn-S-O system precluded the direct production of zinc from ZnS, but several years ago the writer pointed out to Yazawa that his calculations on this sytem indicated that, under rather specific conditions, direct smelting of ZnS to Zn is theoretically possible, and might be practically attainable. Independently Yazawa et al, (30) and Davey and Turnbull(31,32) investigated the possibilities of such a direct smelting process, the former by a free-energy-minimisation technique without using a computer, and the latter using a computer program so that much more extensive calculations were possible, enabling a hypothetical smelting process, using a typical Australian zinc concentrate (with the customary impurities) and a typical coal for fuel, to be optimised.

It was shown(31,32) that a type of flash smelting operation could be carried out at a temperature of about 1600K, burning powdered or crushed coal and a zinc flotation concentrate, to yield recoveries of about 95-97% Zn as vapor, in a gas of composition quite similar to I.S.F. gas, but containing about 8% SO_2 in addition. The coal consumption (about 0.4 t/t of zinc produced) is much less than the I.S.F. coke consumption, and is in any case a much cheaper fuel. The zinc loss is of course sensitive to the concentrate/coal/air ratios, but these are not so critical as to make a process unpractical. It is not yet known whether a gas containing so much SO_2 can be cooled in such a way as to yield most of its zinc content as liquid, without reversion to ZnO or ZnS. However, the ability of the I.S.F. lead-splash condenser to prevent reversion due to reaction of zinc vapor with CO_2 or H_2O, gives some grounds for optimism that it could also cope with SO_2. The relative rates of reversion of zinc vapor with CO_2, H_2O and SO_2 are being determined experimentally in Australia to assess the probability that efficient condensation of zinc would be possible, using such a process. In this event, pilot scale development of the smelting process will be undertaken.

If it proves impossible to condense zinc vapor with high recoveries from an SO_2-containing gas, then direct zinc smelting will prove impossible, and consideration should be given to the alternative new method for smelting zinc-sulfide concentrates as proposed by Kellogg(33,46). This involves dead roasting of sulfide to oxide and subsequent reduction of oxide to metal, using fluidised beds or suspension reactors. Such a process would be expected to offer certain advantages over the I.S. process, since the roasting operation should be cheaper than sintering, handling of the calcine to the smelting operation should be much simpler, and no expensive coke would be required as fuel and reductant.

Tin

A generalised flowsheet for tin production from ore to finished metal is shown in Figure 2. Ten years ago the almost universally used flowsheet consisted of mineral processing to a high-grade concentrate (over 60% Sn) followed when necessary by roasting/leaching to reduce the contents of "dirty" elements (Pb, Bi, As and Sb) before two stage smelting. Only exceptionally was the second-stage slag fumed for additional tin recovery. If it contained substantially above about 2% Sn, it might be subjected to yet a third smelting stage, to produce a tin-iron alloy for return to the first stage. The discard slags from second-stage smelting rarely contained in fact less than 1% Sn as was usually claimed in the literature.

During the past decade, an increasing proportion of lode tin has been mined, due to the gradual exhaustion of sources of alluvial tin. It has become increasingly difficult to obtain high-grade tin concentrates at high recovery from lode material and not only is the average tin content decreasing, but the concentrations of associated sulfide impurities (particularly of Pb, Bi, As and Sb) are increasing.

Rather than lose increasing amounts of tin by attempting to up-grade the concentrates, there has been a trend towards fuming processes (item (2) in Figure 2) which can give a medium-grade concentrate (40-50% Sn) at high recovery (over 90%) rather than 50% or less recovery of about 60% Sn concentrates by mineral processing. Furthermore, elimination of S, Pb, Bi, As and Sb by mineral processing cleaning stages often also entails

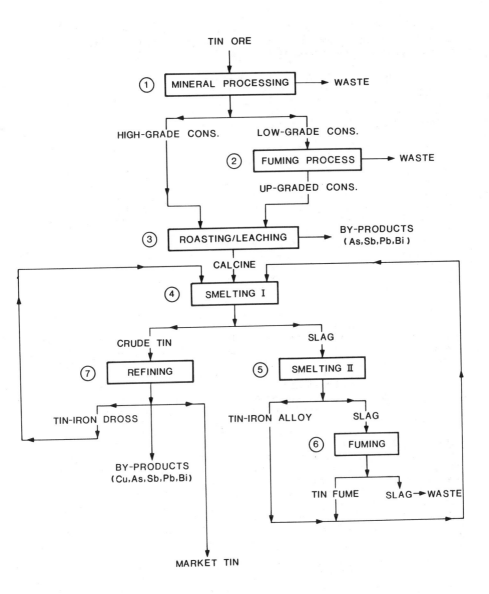

Fig. 2 - GENERALISED TIN FLOWSHEET

unacceptable tin losses, and tends to be replaced by treatments at the smelter: roasting/leaching (item (3) in Figure 2) before smelting, or refining processes on the metal after smelting (item (7) in Figure 2).

Roasting (in air, with or without addition of NaCl for chloridising) followed by leaching (in hot water or dilute acid) had long been practised at Williams Harvey's works, and is no innovation. However the increasing need to cope with greater impurity levels has led to new developments in refining, both electrolytic and pyrometallurgical.

In an endeavour to recover more of the fine tin produced by ever finer grinding to liberate cassiterite (especially from intimately associated sulfide minerals) flotation has been widely introduced, not merely to float sulfides away from cassiterite concentrates, but to float cassiterite from the gangue. Although the production of this flotation concentrate can boost tin recovery significantly (by 20% or more) the product is very low-grade (around 20% Sn content) and calls for new methods of treatment - although today the problem is sometimes being stalled by shipping the concentrates to classical smelters, where their impurities are diluted by blending with high-grade concentrates. In the future they will be more and more up-graded by fuming, and the impurities recovered as by-products by roasting/leaching or refining processes. (The "dirty" elements tend to accompany the tin into the fume.)

Fuming may, in favorable cases, replace mineral processing altogether, to produce a concentrate directly from the ore(34). Although it is generally reckoned that a tin content of around 7% or more is required to justify fuming, this does not apply if the ore itself contains sufficient combustible materials for autogenous smelting - and this can be so for highly pyritic or pyrrhotitic materials, whose iron and sulfur constitute a high fuel value. The concept of matte smelting was originally proposed for recovery of sulfur (and copper and noble metals as by-products) from pyritic residues(35). The pyrite, together with fluxes of silica and lime if necessary, is injected with air into a bath of FeS matte, whereupon the sulfur is evolved partly as sulfur vapor and partly as SO_2. The iron is oxidised and combines with silica to form a fayalite slag. The SO_2 may be reduced by coal burnt above the bath, and elemental sulfur recovered from the gases, or alternatively the elemental sulfur vapor is burnt to SO_2 and used for sulfuric acid manufacture. This process concept overcomes the problems of stickiness encountered in attempting to smelt pyrite in a shaft furnace, and avoids loss of sulfur in a discard matte, as in the Orkla process. Copper and gold may be recovered by accumulation in the matte bath and withdrawing it periodically for converting in a separate vessel. Tin and other volatile impurities may be recovered as fume.

Reverberatory furnaces had been used for nearly a century for both stages of smelting (I for concentrates, II for slags) and while concentrates were very high-grade, the amount of slag produced was very small, so that there was no great disadvantage in tapping it from the first smelting stage, granulating, stockpiling and later feeding it mixed with additional reducing agent and flux, to the second-stage smelt. When the concentrates are low-grade, however, the slag make preponderates over metal production, and this procedure becomes unacceptable because of high handling, stock-piling and reheating costs. The reverberatory furnace is not suitable for performing the second-stage smelt immediately after the first stage, because the flux and coke can not be mixed into the molten slag within the furnace. Consequently attempts were made to substitute rotary for stationary reverberatory furnaces, but the degree of mixing achievable was not sufficient, and the temperatures attainable were not high enough for satisfactory second-stage smelting(36). The successful carrying out of two

stages of treatment in one vessel without removing the charge was later
achieved at Texas City Tin Smelter, using a TBRC(37), and (experimentally) in
Australia by Sirosmelt, a submerged combustion development(5), which avoids
the high capital cost of a combined rotary and tilting furnace, and achieves
very intensive mixing by the use of lances.

The second stage of smelting produces not crude tin, but a tin-iron
alloy, which must be returned to the first smelting stage, where iron
substitutes for coke as reducing agent, and enters the slag. The lack of
stirring in a reverberatory furnace limited the amount of tin-iron alloy that
couldbe circulated,' and made it difficult for the smelter superintendent to
calculate charge compositions that kept the circulating iron in balance,
before Davey & Floyd published the physico-chemical basis of the two-stage
smelting process, and showed that accumulation of hard-head (the tin-iron
alloy) need never be a problem if it is caused to react with first-stage
slag(38).

The optimisation of combined smelting and fuming operations by a
computer program was examined by Davey & Flossbach(36) who concluded that,
certainly in Bolivia, and probably under a wide range of conditions, the
second-stage smelt should be omitted when slag fuming is practised.
Klöckner-Humboldt-Deutz came to the same conclusion, and built the ENAF tin
smelter at Oruro with a primary smelting stage (in a reverberatory) followed
by a sulfide fuming stage, injecting oil, air and pyrite into a slag bath
in a water-cooled shaft furnace(39). Thus the tin content of the primary
slag is recovered as an oxide fume instead of a tin-iron alloy, and very
little iron is recycled to the first-stage smelt. (However, the tin-iron
dross from the refining kettles must still be recycled.) The Oruro
reverberatory furnaces were the first built with a non-porous hearth for tin
smelting, although this has always been common practice for all other metals.

The new Texas City plant, already mentioned(37) also eliminates the
second-stage smelt. After tapping crude tin from the TBRC, the primary slag
remains in the furnace, and $CaCl_2$ is added to fume off $SnCl_2$, which is scrubbed
and reacted with lime to produce SnO_2 for return to the primary smelt, and
regenerate $CaCl_2$ for re-use.

It will be surprising if any further smelters are built with reverberatory smelting for low- or medium-grade concentrates. The future clearly
belongs to the intensive reactors, of which the TBRC has reached the stage
of commercial exploitation and promotion. Since the stirring achieved by
lancing or super-sonic top-blowing is much more effective than the mechanical
stirring produced by furnace rotation, and since the capital and maintenance
costs of non-rotary vessels are much lower than those of the TBRC, we may
expect eventual replacement of TBRC's by plant based on lancing (Sirosmelt)
or super-sonic top-blowing (being developed by K.H.D.).

Lurgi have re-introduced electric furnace smelting at the Berzelius plant,
and for a new smelter being built in Thailand(40). The attraction at the
former site is doubtless that the small amount of waste gas saves on gas
cleaning costs. Electric furnacing has long been used at Phuket for second-
stage smelting, where the slag is used to collect columbium, and has a very
high melting point. It also has a long history of use in both stages of
smelting (instead of reverberatories): at Trail during World War II, and
at small smelters in Brazil, South Africa and Japan(41). It must be an
attractive alternative for smelting small quantities of high-grade
concentrates in locations where electrical power is cheaper than hydrocarbon fuels, and has the added attraction, for new installations, that staff
can be trained by Lurgi at their prototype Berzelius plant.

A very interesting recent proposal for alkaline leaching of fumed tin uses Na_2S + NaOH, followed by electrowinning of tin with an alkaline electrolyte, whereby the reagents are regenerated(42). The long-established wet-way techniques of Harris refining of lead are used to remove most of the common metallic impurities (Pb, Sb, As, Bi) from solution, and very pure tin cathodes have been produced in laboratory tests. This procedure would appear to have a bright future if satisfactory means are developed for dealing with residues and by-products. Although naturally occurring cassiterite is inert to chemical attack, the SnO_2 produced by oxidising fumed SnO or SnS apparently dissolves readily in alkaline solution.

The increased emphasis on impurity removal has, naturally, led to developments in refining, both electrolytic and pyrometallurgical. The grade of product at Vinto has been substantially improved by the substitution of phenol-cresol sulphonic acid for cresol sulphonic acid electrolyte, and numerous minor operating modifications(43). The improvements are so significant that it is intended to increase electrolytic refining capacity at Vinto and reduce the extent of fire-refining(44).

However, considerable advances in pyrometallurgical refining have been achieved by the adoption of Russian-developed centrifuges and vacuum distillation units. The former permit fully automated and hygienic removal of drosses from a kettle of tin or other low-melting metals and so facilitate the removal of iron, copper, arsenic, antimony and bismuth, which may be removed by precipitation reactions. The vacuum distillation unit likewise hygienically and without labor removes Pb, Bi, As and Sb from tin(6,45). These two units are rapidly being adopted throughout the world, and appear to meet the needs of the immediate future as far as can be foreseen.

Conclusions

The reader will have noticed this writer's comparative neglect of many hydrometallurgical proposals put forward as a panacea for present troubles. This is because of a strongly held conviction that, as long as energy consumption is an important consideration, new processes must utilise the fuel value of sulfide concentrates in order to be feasible, since a great deal of energy has already gone into producing what is in fact a high-grade fuel. There are also serious problems involved in disposing of fine-grained residues containing toxic impurities. Furthermore, solid-liquid separation processes are expensive in capital and energy costs.

Hydrometallurgy does have a place in the future scheme of things if solution mining can replace conventional mining, milling and smelting processes. It will also retain a place where it can provide a high degree of selectivity in making separations at low temperatures - especially in electrowinning of high-grade zinc from high-grade concentrates, and electrorefining of both tin and lead, as well as treatment of complex by-product materials such as copper-lead drosses, or impure tin-bearing fumes. The survival of such processes as those associated with the names jarosite, goethite and hematite will depend in the long run on their ability to produce waste end products which can go into smelting operations, whose only solid residue to be disposed of consists of an inert silicate slag.

Present problems in lead and zinc metallurgy are heavily centered around the need to seal the reactors effectively to prevent atmospheric discharges of toxic or obnoxious emissions. In the long term, these spell the doom of processes including sintering operations, and call for the development of small, intense reactors, with any intermediate products conveyed pneumatically or as liquids between sealed vessels. The Kivcet process typifies a designer's response to these needs.

The use of suspension reactors rather than fluid beds can be expected for developments where fritting would otherwise upset the process, and control of temperatures within low limits is not required. Suspension versus submerged smelting versus top-jetting techniques will no doubt generate continuing controversy, probably resulting in a place for all three in future developments.

The need for a new pyrometallurgical zinc process is becoming widely recognized, and either a direct smelting process, as previously outlined, or a two-stage pneumatic process(46), or both, may shortly be developed. The new process will, however, be of limited application if it cannot accept a wide range of complex input materials, recovering numerous by-products, and producing a very pure zinc product (e.g. by vacuum distillation) without a large additional energy consumption.

Although tin refining now seems to be well catered for, there is room for development of new continuous lead refining techniques, especially designed for low temperature, highly automated operation. It is imperative that a vacuum distillation process, including refluxing, be developed to produce zinc of the highest grade, utilising the otherwise waste energy of condensation of zinc produced pyrometallurgically.

Given reasonable economic conditions, there is every reason to expect a decade of more rapid advance in lead, zinc and tin technologies than ever experienced before.

References

1. T.R.A. Davey and W.R. Bull, "Process research on lead and zinc extraction"; pp.975-996, Vol.1, & pp.1008-1029, Vol. 2, in *AIME World Symposium on Mining and Metallurgy of Lead and Zinc*, D.O. Rausch *et al.* ed.; AIME, N.Y., 1970.

2. S.A. Zabel, "Comparative controls for sulfur oxides in four countries", pp.314-324 in *International Symposium on Sulphur Emissions and the Environment*; Soc. of Chem. Ind., London, 8-10 May, 1979.

3. E. Müller, "Smelting of lead concentrates by the Kivcet process", *Erzmetall*, 29 (1976), pp.322-327 (in German).

4. K.B. Chaudhuri and G. Melcher, "Comparative view on the metallurgy of the Kivcet-CS and other direct lead smelting processes", *C.I.M. Bull.*, 71 (799) (1978), pp.126-130.

5. J.M. Floyd and D.S. Conochie, "Reduction of liquid tin smelting slags", *Trans. (Brit) I.M.M.*, 88 (1979), pp.C114-127.

6. T.R.A. Davey and G.M. Willis, "Lead-zinc-tin extractive metallurgy review", *Journal of Metals*, 31(5), (1979).

7. T.R.A. Davey, "Developments in direct smelting of base metals - a critical review", Paper 6 in *International Symposium on Chemical Metallurgy*; Indian Inst. Metals, Bombay, Jan. 1979.

8. W. Peck and J.H. McNicol, "An improved process for continuous copper drossing of lead bullion", *Journal of Metals*, 18 (1966), pp.1027-1032.

9. J. Newton and C.L. Wilson, "Metallurgy of Copper"; J. Wiley & Sons Inc., N.Y., 1942, p.183.

10. J. Lumsden, assgd. to National Smelting Co., "Lead blast furnace processes"; *U.S. Patent* 3,243,283, March 29, 1966.

11. K.D. Libsch and M.V. Rao, assgd. to N.L. Industries Inc., "Secondary lead smelting process"; *U.S. Patent* 4,115,109, Sept. 19, 1978.

12. T.R.A. Davey and G.M. Willis, "Lead-zinc-tin extractive metallurgy review", *Journal of Metals*, 30(4) (1978), pp.12-19.

13. W. Gibson, formerly Smelter Supt., Moloc, Missouri - personal communication.

14. T.R.A. Davey and J.V. Happ, "The decoppering of lead, tin and bismuth by stirring with elemental sulfur", *Proc. Aust. I.M.M.*, No. 237, 1971, pp.63-70.

15. T.R.A. Davey, G. Jensen and R. Segnit, "Decoppering hard lead with sulfur and pyrite" - to be published; *Australia-Japan Extractive Metallurgy Symposium*, Aus. I.M.M., Melbourne, 1980.

16. Y.E. Lebedeff and W.C. Klein, assgd. to Asarco, "Process for decoppering lead", *U.S. Patent* 3,694,191, Sept. 26, 1972.

17. T.R.A. Davey, "Improved process for refining metals by drossing procedures"; *Aust. Patent* PD9176, June 13, 1979.

18. J.D. Iley and D.A. Ward, "Development of a continuous process for the fine debismuthising of lead", pp.133-139 in *Advances in extractive metallurgy 1977*, M.J. Jones ed.; Instn. Min. & Met., London, 1977.

19. J.O. Betterton and Y. Lebedeff, "Debismuthising lead with alkaline earth metals", *Trans. AIME*, 121 (1930), pp.205-225.

20. J. Moodie, "Debismuthising of lead", *M. App. Sci. Thesis*; University of Melbourne, 1976.

21. J.F. Castle and J.A. Richards, "Lead refining: current technology and a new continuous process", pp.217-234 in *Advances in extractive metallurgy 1977*, M.J. Jones ed.; Instn. Min. & Met., London, 1977.

22. M.P. Smirnov, "Lead Refining"; Metallurgiya, Moscow, 1977. (In Russian)

23. J.E. Casteras and C.D. Martini, "Method of treating debismuthising dross to recover bismuth"; this Symposium, 1980.

24. J. Gerlach, B. Kociuk and R. Kammel, "Vacuum distillation of rich crusts", *Metall*, 32 (1978), pp.879-883.

25. W. Hopkin, "Processes for treatment of zinc-lead and lead blast furnace copper drosses", TMS-AIME Paper S lection A 76-99 (1976).

26. F.R. Jorgensen and W.G. Mumme, "Pyrite as an agglomerating agent during the fluid bed roasting of zinc concentrate"; *TMS-AIME Paper Selection* A78-5 (1978).

27. P.R. Mead, "Treatment of a diverse range of low grade feed materials in the ISF"; this Symposium, 1980.

28. R.J. Holliday and P.R. Shoobridge, "A study of the composition, micro-structure and softening characteristics of some zinc-lead sinters", pp.311-321 in *North Queensland Conference, 1978*; Aust. I.M.M., Melbourne, 1978.

29. T.R.A. Davey and G.M. Willis, "Lead-zinc-tin extractive metallurgy review", *Journal of Metals*, 29(3) (1977), pp.24-30.

30. A. Yazawa, T. Kyomizu and S. Katoh, "Direct zinc distillation by the high temperature oxidation method - theoretical study of its possibility", *Bull. Res. Inst. Min. Dressing & Metallurgy*; Tohoku Univ., 33(1) (1977). (In Japanese).

31. T.R.A. Davey and A.G. Turnbull, assgd. to C.S.I.R.O., "Production of zinc from sulphide concentrate"; *Aust. Patent* PD 4973 (1978).

32. T.R.A. Davey and A.G. Turnbull, "Direct smelting of zinc sulfide concentrates", to be published in *Australia-Japan Extractive Metallurgy Symposium*; Aust. I.M.M., Melbourne, 1980.

33. H.A. Kellogg, "Conservation and metallurgical process design",(Wernher Memorial Lecture), *Trans. (Brit) I.M.M.*, 86 (1977), pp.C47-58.

34. K.A. Foo and J.M. Floyd, "Development of the matte fuming process for tin recovery from sulphide materials"; this Symposium, 1980.

35. T.R.A. Davey, assgd. to C.S.I.R.O., "Production of sulphur from pyrites"; *Aust. Patent* PC1417, 1975.

36. T.R.A. Davey and F.J. Flossbach, "Tin smelting in rotary furnaces", *Journal of Metals*, 24 (1972), pp.26-30.

37. E.B. King and L.W. Pommier, "The future of the Texas City Tin Smelter", in *International Tin Symposium, La Paz*, Nov. 1977 (Microfiche available from Centro de Documentacion, P.O. Box 8686, La Paz, Bolivia.)

38. T.R.A. Davey and J.M. Floyd, "Slag-metal equilibria in tin smelting", *Proc. Aust. I.M.M.*, No. 219 (1966), pp.1-10; No. 223 (1967), pp.75-82.

39. H. Weigel and D. Zetsche, "Vinto tin smelter - unusual methods to smelt difficult ores". *World Mining*, 27(8) (1974), pp.32-39.

40. M. Schmidt and A. Selke, "Extractive tin metallurgy", *Lurgi Information* 5/79 (1979), pp.22-24.

41. M. Gamroth, S. Wirosoedirdjo and P. Paschen, "The state of plant technology in tin metallurgy", *Metall*, 31, 1977, pp.999-1004 (in German).

42. L.W. Pommier and S.J. Escalera, "Processing of tin from impure raw materials", *Journal of Metals*, 31(4), (1979), pp.10-12.

43. J. Lema and L. Montoya, "Electrolytic refining of tin at Vinto", in *International Tin Symposium*, La Paz, 1980 (in Spanish).

44. F. Huanca, ENAF, Vinto - personal communication.

45. Anon, "Metal refining vacuum apparatus", Licensintorg, Moscow, 1974.

46. H.H. Kellogg, "Energy use in zinc extraction", this Symposium, 1980.

"METALLURGIE HOBOKEN-OVERPELT PROCESS FOR ROASTING

ZINC CONCENTRATES IN A FLUID BED"

by R. DENOISEUX
R. WINAND
H. WILLEKENS
L. VOS

METALLURGIE HOBOKEN-OVERPELT S.A., BELGIUM

Abstract

A particular zinc roasting process was developed some time ago, by
Metallurgie Hoboken-Overpelt (MHO) in Belgium, aiming to produce a
suitable feed for horizontal retorts without previous Dwight-Lloyd
sintering.

When the obsolete thermal plant was to be replaced by a modern electrowinning plant it became obvious that the roasting process did also
fit perfectly the hydrometallurgical prerequisites.

This process was characterised by the treatment of hard pellets in
a specially designed fluid bed roaster.

Its main features are highlighted with regard to their impact on
roasting and hydrometallurgical treatment of zinc bearing materials
as well as on heat recovery, sulfuric acid production from the
process gases and general efficiency.

I.- Introduction

The roasting process that is now used by Metallurgie Hoboken-
Overpelt (M.H.O.) at its Overpelt Division has to be seen in
its historical context. The name of the company resulted from
the merger of two companies, the non-ferrous activities of
which were complementary : on the one hand, Metallurgie Hoboken
specialized in Cu, Pb, Co and precious metals and, on the
other, Compagnie des Métaux d'Overpelt, Lommel et Corphalie
specialized in zinc and cadmium.

After the second world war, when plant activities were resumed,
the latter company used only horizontal retort furnaces for
zinc production. These furnaces required both dense and porous
granulated feed that was easy to reduce with carbon monoxide.
Because of the limited volume of the retort furnaces all
"ballast" of the charge had to be avoided, which led to a
reduction in the production capacity.

The roasting process that was used in 1946, the Rigg-Overpelt
process, combined pre-roasting in rotating multiple-hearth furnaces and a Dwight Lloyd sintering machine; this combined unit
necessitated intensive operating and maintenance manpower.

A first prototype of a new roasting furnace, called "fluid-bed
semi-flash", to treat unsorted ores was tested from June, 1949
to July, 1950; it had a daily capacity of 35 t of zinc concentrates. Fig. 1 shows a vertical section of this semi-industrial
unit. The furnace is fed laterally with crude ores, but roasts
only a small proportion of its charge in the fluid-bed - hence
its name. Only 40 % of the calcine was recovered by the overflow of the fluid bed, the finest particles, dragged away by
the gases and roasted in suspension (flash), being recovered
in the cyclone near the furnace (about 55 % of the calcine)
and the remainder being collected in an electrostatic precipitator.

FLUID BED SEMI-FLASH ROASTER

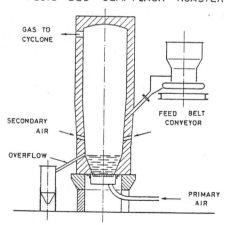

FIG.1

Tests with this prototype furnace were financed by a research partnership that included, among others, Union Minière and Dorr Oliver; the latter company wanted to determine how the fluid-bed method could be applied to zinc concentrates.

The furnace was not at that stage equipped with a waste heat boiler and some difficulties were encountered in treating the concentrates with high Cu and Pb contents, which at that time constituted an important part of the supply to the works. Moreover, since desulphurization of the dusts, removed and collected in the cyclone and the Cottrell remained unsatisfactory, despite the large volume of the flash roasting zone, recycling of these dusts would have been metallurgically unavoidable. It is not impossible that these difficulties could have been overcome, but the new orientation at that time stemmed from the fact that it was hard to find a granulation method that could make the powdery calcine suitable for treatment in the horizontal retorts.

From 1951, tests were begun on a second prototype furnace, of rectangular section, to treat blendes granulated with zinc sulphate to act as a binder. Successful results were achieved rapidly and the first boiler in Europe - perhaps even in the world - to recover heat from the sulphur-containing gases was completed and tested at the same time.

The industrial installation of two granulation-roasting lines to supply the zinc works of Overpelt and Lommel was commissioned in February, 1954.

After a fairly lengthy investigation the hydrometallurgical process was chosen to replace the retorts.

Pilot-plant tests showed that pelletized roast was very suitable for the hydrometallurgical route. Accordingly, there was no need to change the roasting method, not least because it was known that all electrolytic refiners tended to fine-grind the calcines to be leached.

Two and a half year after the commissioning of the new electrolytic plant M.H.O. decided to increase the capacity of the existing roasting furnaces; the excellent properties of the process in feeding the tankhouse since May, 1974, were confirmed on the industrial scale.

2.- Principles of M.H.O. roasting process

Before a detailed description of the process and its technical characteristics is given its main features can be seen from the flow sheet shown in Fig. 2. Basically, the metallurgical process for the production of zinc metal from sulphur-containing concentrates or crude blendes necessitates two separate steps :

- Elimination of sulphur, with the aim of converting sulphides into oxides and preventing the formation of zinc sulphate, as well as converting the maximum possible part of the sulphur into H_2SO_4.
- Transformation of the roasted or calcined blende into zinc metal.

The chemical reactions that areuused in zinc hydrometallurgy are well known. This process is strongly influenced
by the physical quality and the chemical composition of
the calcine. Since the M.H.O. process yields several favourable properties from that point of view, we shall only
deal with this phase of the process insofar as this roasting process is of particular interest to the calcine user.

Fig. 2 shows the circuit of materials and fluids to the electrolytic zinc plant. Three sections of the flow-sheet
- preparation of the charge,
- homogenizing, granulation and drying, and
- roasting, waste heat boiler and electrofilter
are presented in detail in Figs. 3-6.

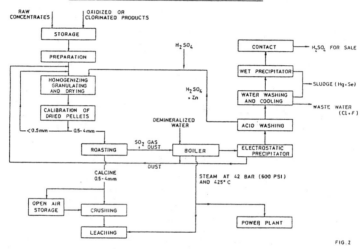

PELLETIZING , ROASTING AND ACID PLANT FLOWSHEET

FIG.2

In the process the charge to the roasting furnaces is granulated and dried, zinc sulphate acting as the binder for the granulation. Zinc sulphate was selected to avoid the addition of
ballast to the charge to the zinc plant, since it decomposes
during roasting. The size of the pellets was fixed at 0,5-4mm.

Maximum elimination of sulphur, leaving only traces of zinc
sulphide and sulphate in the calcine and ensuring the best zinc
recovery, requires four conditions :

- the calcine should not be extracted by "overflow" of the
  fluid bed
- the calcine should not be extracted near the point at
  which the crude feed is introduced to the furnace

72

- roasting should be completed in a part of the furnace where the 02 content is sufficient
- roasting should be completed in preheated air

A detailed description of how these four conditions are met will be given later, but at present it can be said that pre-heating is achieved by direct contact in counter-current flow of combustion air and hot calcine, which is ipso facto cooled before leaving the furnace.

To maintain a granulated charge in the fluidized state it is necessary to use in the furnace a space rate higher than 2.5 m/sec., which allows a very high furnace productivity : more than one tonne/h/m^2 of grate area.

As a direct consequence of the granulation of the charge the proportion of dusts removed by the gases of the roasting furnace is remarkably small.

3.- Detailed description of plant and features of M.H.O. process
----

In this section the intention is to clarify the principles that led to the conception of the process and to its development by means of the following equipment description.

3.1 Preparation and pelletizing of roasting charge (Figs.3 and 4)

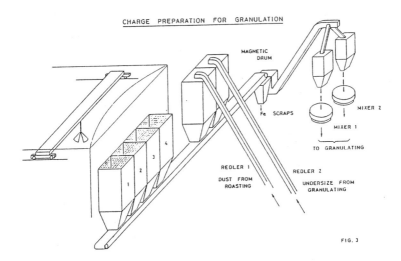

CHARGE PREPARATION FOR GRANULATION

FIG. 3

The concentrates are unloaded by an overhead crane into a group of four bins, which allows the simultaneous intro-duction of four different charge components. Above these bins a coarse grate eliminates foreign substances that are harmful to the subsequent handling system.

Besides the raw materials, the charge consists of recycled dusts from the roasting plant and the "undersize" of the granulation. These dry and fine returns are sent by

73

Redler conveyors to two storage bins. Feed regulators unload the charge components from the six bins on to a conveyor belt, the head drum of which is equipped with a magnetic separator. The proportioned charge is taken by a bucket-elevator to the two bins that are placed above the mixers.

The downstream installation of the Overpelt pelletizing and roasting plant consists of two equivalent production lines. The split in two units is merely a matter of flexibility. In the mixers (Fig. 4), ahead of both granulation lines, the zinc concentrates, the more or less oxidized products and the sulphuric acid are homogenized, the reaction of the two latter components of the mixture forms the granulation binder.

Oxidized products needed for this operation might be :

- dusts collected under the boiler or in the electro-filter
- oxides from the conditioning of zinc ashes
- calamines or blendes pre-roasted in, for example, multiple-hearth furnaces.

All the sulphuric acid that results from the washing of sulphurous gases is recycled to the mixers, and commercial acid is used only for make-up.

MIXING _ DRYING _ PELLETIZING

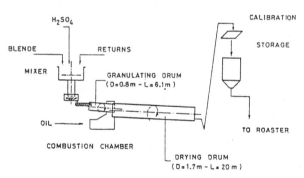

FIG. 4

When leaving the mixers, the paste, containing some 3 % of sulphate sulphur, is fed into a group of two rotary furnaces, where it is pelletized and dried at the same time.

The pelletized charge is then sized and the 0.5 - 4 mm size is directly supplied to a bin near the roasting furnaces by a belt conveyor, and undersize (0,5 mm) is returned by a Redler to the mixers ahead of the granulation lines.

3.2 Roasting furnace with its heat recovery boiler and dry electrofilter

3.2.1 Roasting
Careful charge preparation yields an extremely homo-

74

geneous and closely sized feed, freeing the roasting
process of any constraints with regard to particle
sizing, moisture or analytical requirements.

Belt conveyors move the crude and still hot granules
from the screening station to the loading bin of the
fluid-bed roasting furnaces.

ROASTING PROCESS

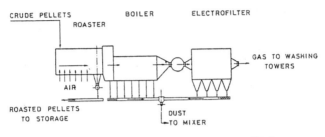

FIG. 5

Figs. 5 and 6 show that the furnaces consist mainly of
a vertical refractory lined shaft. Over some 85 % of
their length the bottom of the furnace consists of an
air distribution grate.

FLUID - BED ROASTER

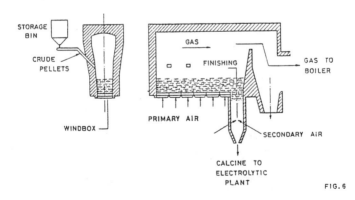

FIG. 6

The balance of the bottom is the aperture, feeding the
"finishing" bin for cooling the calcine and completing
the roasting. Calcine extraction by overflow, which is
a harmful system for the quality of the roasted ores,
is thus avoided.

Earlier tests had shown significant differences with

75

regard to the extent of sulphur removal between extraction by overflow of the fluid bed and by underflow via the finishing bin. An average figure of these differences is 0.25 % in sulphur sulphide and 0.35 % in sulphur sulphate.

One reason for this situation is a systematic 8 to 10 % difference in density : the crude pellets are lighter than the calcine and therefore tend to float at the upper level of the fluid bed. In addition, and very important for the low sulphate content of the calcine, the atmosphere in the upper part of the finishing bin has to be noted. It is well known that sulphates decompose easier in gases with a very low SO_2 content.

The upper part of the finishing bin receives

- from above, through the aperture in the grate, calcine at about 980°C, the roasting of which is practically complete (this aperture is at the opposite end to the feed point)

- from below, preheated air, which prevents any entry of sulphurous gases from the fluid bed to the bin : the atmosphere is very rich in oxygen and, hence, very poor in SO_2.

Therefore, in this part of the finishing bin, the four conditions for good roasting and a high decomposition of sulphates ideally prevail.

Fig. 5 is very similar to that which could have resulted from the further development of the 1949 prototype - the semi-flash furnace. The present furnace, however, has the following particularities :

- rectangular shape, which imposes a progressive longitudinal movement of the material

- very high ascension speed of the gases; the "space rate" is generally higher than 2.5 m/sec. by virtue of the granulation of the charge; furnace production is at least 1 $t/h/m^2$ grate area.

- the furnace arch is about 9 m above ground level

- the proportion of calcine extracted at the bottom of the furnace represents about 80 % of the total; the balance, i.e. the fine, which has a high sulphate content, is recycled to the granulation

- the calcine is cooled by counter-current flow in the "finishing" bin (see Fig. 6) by part of the combustion air used in the furnace : this air is therefore well preheated and almost complete finishing of the roasting is ensured. The yield $\dfrac{S \text{ in acid}}{S \text{ of raw materials}}$ will thus be optimal and the calcine will have the highest possible ZnO content

- the roasting temperature is very high and often exceeds 1000°C.

- behind the boilers and the dry electrofilters, roast gases are first washed with sulphuric acid : this washing cleans the gases from particulates and also traps the SO_3, the acid so produced being sent to the granulating section.

Crude pellets (Fig. 6) are fed laterally and above the fluid bed by a device that automatically regulates the flow of material according to the prevailing temperature in the mass. Moreover, this device is completely sealed and requires no manual control. Inspection holes for observation of the bubbling of the mass are also placed in the side wall. When leaving the furnace, gases immediately enter the heat recovery boiler, followed by a dry electrofilter. After acid washing and in line with the usual processes gases are cooled by water spraying in a vertical tower and then reach the wet electrofilters and the contact plant.

Fluidization of screened and homogeneous pellets is characterized by significant bed stability. Moreover, since the granules are loaded dry and hot, ignition occurs very quickly. The level of the fluid bed may be maintained at about 1 m between the grate and the upper average level of the bed. This limits the blowing pressure of combustion air at 0.11 bar or 1100 mm H_2O for 85 % of the air flow. The remaining 15 % of the air needed for the finishing, which in the first instance is used to cool the calcine, requires 0.25 bar or 2500 mm H_2O. In total, the average pressure will be about 0.13 bar or some 1300 mm H_2O, corresponding to a very small power consumption.

The amount of dusts is about 20 %.

In addition to the ease and rapidity of the starting operations, the M.H.O. roasting furnace has the following features :

- the virtual elimination of the elements Cl, Hg, Se and F because of the homogeneity of the fluid bed and the prolonged residence time of the full production of roasted ore in the fluid bed at an average temperature of some 1000°C

- the opportunity to treat, without problems, charges that are particularly rich in Pb and Cu and such other troublesome elements as MgO, BaO, SiO_2.

An example of an assay of a Canadian concentrate that is processed without blending, despite its extreme fineness, and the corresponding calcine composition are given in Table I.

Table I : Canadian concentrates

Grain size	Typical screen analysis, %		Typical analysis, %	
	U.S. Series		Raw product	Calcine
> 124 μm	on 120	0,25	Zn 52	62
124/44 μm	120/325	3.95	Pb 2.1	2.5
< 44 μm	thru 325	95.80	S_{total} 34.3	0.5
			$S_{sulphate}$ 0.5	0.45
			$S_{sulphide}$ 33.8	0.05

It is assumed that the possibility to treat mixtures with very high Pb and Cu contents results from the fact that fluidization is very stable by virtue of the particle sizing of the charge, its constancy and its homogeneity. Moreover, the strong agitation of the mass of the bed, because of the high space rate, avoids the danger of sticking.

African concentrates – for example, those from Kipushi with 6 % Cu and Tsumeb with 10 % Pb and 3.5 % Cu – have been treated during long campaigns without causing accretions on the furnace walls. Assays for these materials are given in Table 2.

Table II

	Cu concentrate Kipushi		Pb-Cu concentrate Tsumeb	
Content	Raw, %	Roasted, %	Raw, %	Roasted, %
Zn	52.2	63.5	50.8	58.5
S_{total}	32.5	0.35	29.2	1.3
$S_{sulphate}$	–	0.30	–	1.2
Pb	1.6	1.9	8.9	10.1
Cu	4.6	5.4	3.4	3.8
SiO_2	1.6	2.1	0.8	0.9
Cd	0.36	0.43	1.15	1.25

Since the calcine is granulated, there is no objection to storing it in the open air over longer periods, which removes the need for calcine storage bins. Additionally, because of the granulation and the straight forward and complete elimination of chlorine, it is possible to treat significant quantities of zinc ashes as well as oxidized and chlorinated products with no previous dechlorination step.

The M.H.O. roasting furnace is very compact as a result of its specific roasting capacity (higher than 1 tonne/ h/m^2 of grate area) and the small proportion of flue dusts : the flash volume is further reduced in that the dusts are recycled to the granulation. The height of the furnace from grate to arch is about 5.5 m.

As a consequence of these features its construction price and the heat losses are low.

The heat balance of the furnace is also improved in three ways :

- by drying the charge : each kg of water used in the roasting should, in fact, be vaporized and the corresponding steam will leave the waste heat boiler at a cost of 2890 kJ (690 kcal)

- by recovering the sensible heat of the calcine by counter-current air circulation at the finishing stage : this recovery is the more important in that 80 % of the calcine is extracted in this way

- by feeding the charge straight from the dryer and passing to the furnace at an average temperature of 80°C.

On the other hand, decomposition of the $ZnSO4$ binder is an endothermic reaction that consumes heat. On the whole, the furnace fed with dried pelletized charge enables the concentration of the combustion heat of the crude blends in a recoverable state in the boiler. In fact, the exhaust gases pass immediately to the boiler, where cooling to 290°C produces maximum steam for energy or heating purposes.

3.2.2 Waste heat boiler

Roasting furnaces operating at M.H.O. are equipped with forced circulation boilers of the La Mont type, which comprise vertical bundles of tubes that can be replaced easily. These boilers are directly connected to the furnaces.

As was pointed out above, dust production is drastically reduced because of pelletization of the charge, and it is obvious that this is very beneficial to the service life of the bundles. After 25 years of use, those for the Overpelt boilers are still in operation. Consequently, maintenance expenses for these boilers are low and state authorities have recognized this fact.

Legally, the periodic inspection of boilers in Belgium has to be performed every year, but, for M.H.O., inspections have been extended to 18 months, which is the maximum time that can be permitted. During this 18-month period the furnaces run continuously without interruption or shut down for cleaning.

Clearly, the general reduction in the frequency and
duration of shutdowns has a favourable influence on the
efficiency of manpower and plant productivity, and this
minimizes the erection costs of a plant.

3.2.3 Purification of gases before washing

When a pre-dried charge is roasted, the dry electrofilter
operation is never disturbed by clogging of the electrodes, since the gases are dry and condensation is
avoided. There is no interruption for manual cleaning of
the electrodes.

3.2.4 Transport, storage and grinding of calcine

Since most of the sensible heat of the pelletized calcine
(see earlier) is recovered before it leaves the furnace,
there is no objection to handling it by open chain
conveyors or by ordinary trucks, and its storage in open
yards poses no problems. Grinding may be carried out
immediately without intermediate cooling.

3.3 Washing and cooling sulphurous gases

After the dry electrofilter, and before reaching the wet
electrofilters, gases must be washed and cooled. Generally,
such washing is performed with water. The water acidifies and
picks up dusts and impurities, and its ultimate destination is
the cause of problems and expense.

In the M.H.O. roasting process washing is done by spraying with
rather concentrated acid (40 % H_2SO_4). The acid used for
washing then passes to the pelletizing stage, where it forms
the ideal binder because of its reaction with ZnO. Entrained
dusts are simultaneously reintroduced and, as a result, all the
zinc contained in the ores passes to the calcine and to the
electrolytic plant.

Besides the economic and ecological advantages of the zinc
recovery, the process makes SO_3 collection and cooling of the
sulphur gases possible.

3.3.1 SO_3 collection

The first washing tower merely uses in situ produced acid,
which is recycled to the granulation mixers. This acid
collects the SO_3 that is always present in the roast
gases, especially those that emanate from a waste heat
boiler. SO_3 collection is not sufficient to meet the
granulation needs and fresh acid has to be added, but the
use of acids of a commercial quality is avoided. In turn,
this makes it possible to employ lower-grade acids that
are not suitable for sale – for example, black acids,
residual acids, etc.

3.3.2 Cooling

Usually, after washing, sulphurous gases are cooled

adiabatically with water in a separate tower. In turn,
the cooling water is then cooled in an exterior
circuit and recycled.

It is easier to cool the gases and the quantities of
water to be condensed are smaller, when the humidity
of the gases is drastically reduced by feeding the
furnaces with dried pellets.

Bleed offs are made periodically on the water-cooling
circuit. Figures obtained at Overpelt during operation
indicate about 1 ton of water for 10 tonnes of treated
concentrates, but this volume relates to elimination
of chlorine rather than of condensation water.

4.- Quality of calcine and its influence on hydrometallurgical process
----

4.1 General

It is presumed that the electrolytic plant is technically
advanced, producing the metals contained in the ferrites
by a hot leach in a sufficiently acid medium.

Specific questions related to the different separation
processes of the iron dissolved by this leach, such as the
Goethite, Jarosite and Hematite processes, are not dealt
with here as such a survey would go far beyond the present
subject.

Only the influence of the quality of the calcine on the
leaching and the retreatment of the by-products will be
considered - quality of the calcine means both its physical
properties, its desulphurization and its chemical composition.

4.2 Leaching the calcine

It has been said that, in addition to its high zinc content,
which is useful at the neutralizion stage and allows the
losses of zinc and secondary metals to be minimized, pelletized calcine contains very little sulphur - more particularly, sulphur sulphide. It is difficult to define precisely
to what extent this ZnS is lost in hydrometallurgy, but it
may be said that 30 % is not recovered.

It has also been noted that the high roasting temperature
and the longer residence time of the calcine at this temperature ensure so complete an elimination of Cl, that virtually none remains in the pelletized calcine. This
explains why the Cl content of the solutions to be electro-
lysed does not exceed 60 mg/l at Overpelt, despite the
introduction of substantial quantities of zinc ash and
the complete absence of electrolyte bleed-off during five
years of operation.

Moreover, the pelletized roast ores also contain very little
sulphate. This property has a particular value in that it
greatly facilitates the acidification that is required for
the different operations of the ferrite leaching.

Without jeopardizing the sulphate balance of the electrolysis
it is possible to obtain a lead sulphate residue with high Pb

and Ag contents.

4.3 Reprocessing by-products

The aim of reprocessing hydrometallurgical by-products is to recover valuable secondary metals, and this may be commercially important, especially in regard to custom treatment.

In addition to the advantage of the process in respect to silver-bearing lead sulphate production, similar conclusions may be drawn for copper and the other by-products of the purification. Remembering that this pellet roasting process enables the treatment of blendes with a high copper content, the process permits a zinc plant to produce more cements with high Cu contents.

5.- Flexibility of M.H.O. process

It was mentioned that the downstream installation of the Overpelt pelletizing and roasting plant consists of equivalent lines. The split into two units was merely for flexibility in that there is never a complete interruption of zinc plant supply when a roasting furnace is stopped, nor an interruption of the flow of sulphur gases to the contact plant. Obviously, if such flexibility were not required, it would be advisable to install only one line, which would, of course, involve a lower investment.

But, if flexibility were the main reason for the selection of two units, it must be said that there were no metallurgical or other construction considerations that necessitated that selection. Doubling the capacity of the furnaces or of the granulators, for example, poses no problems.

Each line has its own possibilities for adaptation to production needs :

- the space rate, calculated at 1000°C, can vary from 2.4 to 3.6 m/sec. : accordingly, production at the maximum is about 50 % higher than at the minimum space rate
- temperatures in the bed vary from 880°C to 1060°C, depending on the required quality of the calcine for the leaching steps.

If even greater flexibility is needing to cut production in the event of a recession, for example, a very simple change of the upper sieve at the calibrating section of the pelletizing stage enables the furnace to be fed with pellets 0.5-3 mm size rather than the standard size (0.5-4 mm).

With the smaller feed size it should be possible to reduce the space rate to 2 m/sec without disturbing the stability of the fluidization.

It was stated earlier that the furnace at Overpelt is very compact and that the level of fluid bed may be kept low (about 1 m.) : this explains why starting operations are easy and rapid.

The calcine starting mass, which is, of course, cold, is very low - this is more pronounced in that the air distribution grate over 85 % of the length of the furnace is divided into separate windboxes to allow starting operations over half the furnace length. Some 5 tons of calcine are used to start a furnace of 275 t/day capacity.

This particular feature of the air distribution grate is also
very attractive to the user in the event of a failure of the power
supply to the plant. In 95 % of cases the bed during this failure
remains gently on the grates and, when power is restored, air
blowing is sufficient to restart the fluidization. If the breakdown lasts too long, or when the calcine is to sticky because the
breakdown occurs at high temperature, it is not possible to re-
start the fluidization and some agglomerates will be produced at
the charging end of the furnace. Because of the design of the
furnace and its rectangular size, a six-hour period will suffice
to remove the agglomerates and to return the furnace to normal
operating conditions.

During 25 years of operations at the Overpelt plant such a
defluidization has occurred only twice.

6.- Future developments

The hydrometallurgical zinc extraction process is far from new,
but considerable progress has been made in the last few years.

The first zinc calcine leaching systems yielded residues that were
either reprocessed by the pyrometallurgical route or, and this
happened frequently, heaped up. Under these conditions the type
of roasting was of lesser importance than the conditions during
the roasting itself.

With the appearance of different acid leaching processes requirements for the quality of the calcine became increasingly severe
in connection with improving the rapidity of solubilization, the
filterability of the solutions and the extraction yield of the
metals.

These elements are very sensitive to different parameters, which
can easily be controlled in a process in which the reproducibility
of the results is guaranteed by careful charge preparation and by
the elaborate automation that the process allows at the roasting
furnace.

It is a tendency of modern metallurgical practice, and one that
can only be increased with the evolution of metallurgy, for
processes to be selected that aim at reducing manpower and
increasing equipment efficiency.

The M.H.O. roasting process is in line with that expected
evolution since it meets these criteria and standards.

7.- Conclusions

The principles of the roasting process of a granulated and dried
charge have been described, together with the advantages to the
user because of the three specific features of the process :
- drying : favourable heat balance; rapidity of ignition; no
  cleaning of the dry electrofilters; no disposal of spent
  wash liquors
- pelletizing : few dusts removed by the gases, thus giving

long life of the boilers; the possibility of treating
charges with high Pb and Cu contents as well as chlorinated
products; no restraint on roasting as far as the choice of
the mixtures is concerned, which offers the opportunity to
introduce the raw materials at the most judicious point in
the circuit; high specific production of the furnaces

- stability of fluidization - by virtue of careful charge
  preparation, which enables, despite its very small power
  consumption, a calcine of excellent quality to be obtained
  without chlorine and with high Zn and a very low sulphur
  contents.

The process allows a high utilization factor of the equipment,
with a minimum of stoppages and no accretions on the walls in
contact with dusty gases.

From a metallurgical point of view roasting of the pelletized
charge is also attractive hydrometallurgically - the sulphate
balance is handled with ease, there is a particularly high and
economic recovery of by-products and the losses of zinc and
secondary metals are reduced.

From an economic point of view it would be useful to have a
comparative survey of the available types of roasting processes
with regard to investment costs and operating expenses, but
such a comparison is beyond the scope of this presentation,
which is deliberately limited to the technical features of the
process that is used by M.H.O. It is clear, however, that
M.H.O. proceeded with the roasting process applied previously
by Compagnie des Métaux d'Overpelt because both the granulation
and drying operations, apparently unnecessary for roasting,
do represent a good investment.

If zinc metallurgy continues along the hydrometallurgical route,
the advantages of the M.H.O. process are readily apparent.
Current research indicates that there remain considerable
possibilities for the adaptation of this roasting process to
this wet metallurgy, with which it has been associated for
only five years. Said possibilities exist also to adapt
the process to the steady evolution of the ecological,
economic and social conditions in which our companies must
exist.

8.- Acknowledgement

The authors wish to thank the company's general management
for permission to publish this paper, and their colleagues
for valuable assistance in its preparation.

THE BOLIDEN - NORZINK MERCURY REMOVAL PROCESS
FOR PURIFICATION OF ROASTER GASES

Georg Steintveit
Norzink AS
Odda, Norway

Abstract

Mercury as an impurity in zinc concentrate was previously mainly traced as metallic particles in the gas purification system of roaster gases for sulphuric acid manufacture.

Although part of the mercury also arrived into the product acid, the metal was not a matter of concern as a detrimental contamination until the beginning of the nineteen **seventies.**

At Norzink we started the development work on the mercury problem in 1967. After consideration of several possible processes, we decided to develop a mercuric chloride washing process whereby the remaining metallic mercury in the otherwise clean roaster gases was removed before the gases entered the drying tower.

The first installation at our 250 tons per day fluid bed roaster was started in 1973. A second unit for our newest 350 tons per day roaster was commissioned in 1975. By this process we produce 98 - 98.5 percent sulphuric acid with a mercury content of 0.3 ppm from zinc concentrate containing 350 ppm Hg.

Introduction

The raw materials for zinc manufacture normally contain numerous impurities which have to be removed during the processing of the metal. Some of these, such as copper, cadmium, lead and silver, have resulted in valuable byproducts, others like iron, cobolt, nickel, arsenic, antimony, are undesirable, but have had a great influence on the metallurgical processes of the zinc extraction.

Mercury is also one of the elements normally present as an impurity in zinc sulphide concentrates. Contents from 5 to 350 ppm are often found, but concentrates with up to 3000 ppm are occasionally used in zinc production. Such high mercury containing ores may moreover be referred to as a source for mercury production.

By treating zinc concentrates with the moderate mercury contents mentioned above, the problem for the zinc metallurgist was previously mainly the nuisance of removing the metallic mercury sludges and its possible corrosion effects on the standard gas purification equipment between roasting and the sulphuric acid plant.

Some of the mercury arrived in the product sulphuric acid but as specifications on the impurity previously did not exist, this was no matter of concern for the marketing of the acid.

This situation changed gradually when it was acknowledged that mercury had to be considered as one of the serious environmental poisons. In Scandinavia a report was published (1) showing that the mercury compounds used as fungicides in agriculture via water would concentrate in the fauna. The mercury content in fish might thus increase from the normal 0.044 to 0.167 ppm (2), to fatal levels.

At Norzink our ore supply came from overseas and from Norwegian and Swedish mines with mercury analyses of 20 to 40, 30 to 40 and 20 to 500 ppm respectively. In the sixties it became apparent that we would have to depend merely on Scandinavian sources, where the higher mercury bearing Swedish ores would be the predominant supply. Accordingly a development of a mercury removal system became a major requirement.

We realized that the available analytical methods to determine traces of mercury and identification of its different compounds were not sufficiently sensitive, and we had initially to develop adequate and practical sampling and analytical procedures.

This was the situation back in 1967, the year when the Swedish National Institute of Public Health established the safety level of mercury in fresh water fish to 1.0 ppm. Analyses of fish taken in the fjord not far from our plant showed about 3 ppm Hg. Somewhat later the Japanese Reports on diseases in Niigata and Minamata (3), (4), were known and they greatly disquieted the authorities and the public.

Liberation of Mercury During Roasting

The desulphurization of the zinc concentrate is carried out in fluid bed roasters at temperatures between 900 and 950 degrees centigrade. The greater part of calcine is recovered in the steamboiler hot cyclone and dry electrofilter. The roaster gas is discharged at temperatures 300 to $350^{\circ}C$ from the boilers, operating at 40 kp/cm^2 and 270 to 300°C from the dry electrostatic precipitators.

The mercury is assumed to be present in the concentrate as mercury sulphide, which during the roasting is transformed to metallic mercury and mercury oxide. Depending upon the presence of selenium some mercury selenide is found, together with minor amounts of mercury sulphide being carried over.

It is our experience that 95 to 98% of the mercury in the concentrate is following the roaster gas into the washing system. The remaining 2 to 5 per cent is found in the calcine.

The first stage in the conventional washing of the roaster gases is the adiabatic cooler, which may consist of a normal washing tower or at newer plants, a venturi type scrubber. The temperature at the exit is determined by the moisture content and the temperature of the entering gases and is in our plant 55 to 60°C.

Dependent on the mercury content in the roaster gases entering the washing system, part of the mercury is condensed here.

However, the main condensation of mercury and mercury compounds takes place in the indirect gas cooler with a discharge temperature of 30 to 38°C, followed by the precipitation of the remaining particulate metallic mercury and of particulate mercury compounds in the wet electrostatic precipitator.

We have found that 60 to 80 per cent of the mercury entering the washing system is élminiated here. The percentage of mercury removal in the standard washing and precipitation installations is reduced at relative low mercury contents in the ore, but we find often higher contamination of mercury in the product acid than should be expected considering the lower content of mercury in the feed. At gas temperatures of 30 to 38°C at the inlet of the drying tower we found 30 to 40 mg Hg per m^3, corresponding to 100 ppm of Hg in the sulphuric acid. This about relates to the equilibrium mercury vapour pressure at the temperature conditions mentioned.

Our plant experience was further that mercury contents up to 200 ppm in the product acid were found when treating zinc concentrates with from 100 to 350 ppm Hg. This was due to supersaturation and carry over of particulate matter.

The situation regarding mercury contamination was thus very serious, particularly so because the scrubbing and cooling water of our contact plant, started in 1965, operated as an open system with discharge of 1200 m^3 per day of contaminated water directly into the fjord.

By closing this system and installing indirect gas cooling, the excess water to be purified and filtered was reduced to 30 to 50 m^3 per day. Simultaneously we limited the ore supply for roasting as a short term provision to concentrates with maximum 50 ppm of mercury.

The Removal of Mercury from the Roaster Gases

We had stopped mercury pollution to the sea, but the complete removal of mercury from the roaster gases was a much more complex problem.

Our sulphuric acid was mainly used for fertilizer production, and we considered it essential to comply with our customers' requirements of 1 ppm.

Different process possibilities were discussed and studied jointly with the Swedish company Boliden.

The following seven possible process alternatives were more or less thoroughly evaluated.

- Addition of sulphur compounds to the roaster gas
- Addition of selenium compounds to the roaster gas and to the washing liquid
- Washing of gases with concentrated sulphuric acid at high temperatures
- Filtration of the gases through impregnated activated carbon
- Filtration of the gases through selenium containing filtermedia
- Precipitation of the mercury in the acid and separation of the precipitates from the same
- Washing the gases with special chemical containing solutions

Our conclusion from the development work were:

1-2. Both sulphur, sulphur compounds and selenium added directly in the gas had some effect, but did not sufficiently remove mercury. The selenium was also too expensive to be applied as an addition.

3. Washing the gases at temperatures around 180°C with concentrated sulphuric acid effectively oxidized metallic mercury and precipitated the metal as mercury sulphate. At such high temperatures the moisture in the gas was not retained and a procedure working on this principle did not dilute the concentrated acid applied as scrubbing liquid.

In the first hand this process seemed to have the advantage to remove the mercury directly after the dry electrostatic precipitation. It became clear, however, that inevitable presence of minor quantities of chloride in the feed to the furnace would result in a carry over of mercury chloride compounds that might call for additional removal stages.

The system was studied in pilot plants in 1967-68. We estimated, however, both the investment and operation cost too high to proceed with the industrial development of this alternative.

The system was later developed industrially at the Kokkola plant of Outokumpu in Finland (5, 6).

Two other processes based on filtration of the gases through activated filtermedia were, however, brought to final industrial development. Both processes are quite efficient and specially suitable for cleaning gases with a relatively low content of mercury (7).

4. The impregnated activated carbon process was worked out on laboratory scale at Norzink (8), while the industrial installation was developed at Boliden.

5. The unit with selenium activated filtermedia was developed from laboratory up to full plant scale at Boliden (9).

6. Several methods of precipitation of insoluble mercury compounds directly in the product acid were tried. We found, however, that a practical process had to be based upon some dilution of the acid, and this did not suit our requirement.

7. A special gas washing process to remove the mercury vapour from the roaster gases down to 0.05 to 0.1 mg per Nm^3 wad patented by Boliden (10). The process applied was solutions of halogenides and semihalogenides, and the laboratory scale tests looked very promising for the conditions in our zinc roasting and adjacent sulphuric acid plant.

We decided at Norzink to develop an industrial installation, and chose to work with the cheap and simple mercuric chloride solution as washing liquid in connection with our 250 tons per day fluid bed roaster.

The Boliden - Norzink Mercury Removal Process

The process principle is based upon the following reactions:

$$Hg^{2+} + Hg^{o} = Hg_2^{2+} \qquad \text{(I)}$$

$$Hg_2^{2+} + 2Cl^- = Hg_2Cl_2 \qquad \text{(II)}$$

$$Hg^{2+} + nCl^- = HgCl_n^{2-n} \qquad \text{(III)}$$

$$HgCl_n^{2-n} + Hg^{o} = Hg_2Cl_2 + (n-2)Cl^- \qquad \text{(IV)}$$

The system $Hg^{2+} - Cl^-$, mercuric chloride, is an efficient oxidation agent for the mercury vapour Hg^{o} (I, IV). The chloride precipitates the monovalent mercury to Hg_2Cl_2, mercurous chloride (II). Partly the chloride forms complexes, $HgCl_n^{2-n}$, with the mercuric ions (III), thus acting as inhibitor for the reduction of Hg^{2+} with SO_2.

In the system using this procedure, Hg^{2+} as a reactant, is continuously consumed and must be replenished. This is easily accomplished by reoxidizing part of the precipitated mercurous chloride by chlorine.

$$Hg_2Cl_2 + Cl_2 = 2HgCl_2 \qquad \text{(V)}$$

The gas washing stage for removing mercury from roaster gases is illustrated in Fig. 1. Wash solution is sprayed into a scrubber tower and meets the gas which is to be cleaned in counter current flow. The tower is equipped with a mist eliminator to prevent droplets of wash solution from being carried over. Part of the wash solution from the bottom of the tower goes back into circulation and part goes into a sludge separator. The solution from the separator is returned to the scrubbing liquid circuit or is diverted from the process, while part of the sludge which consists primarily of Hg_2Cl_2, is transferred to an oxidation plant, where Hg_2^{2+} is oxidized to Hg^{2+}. This is returned to the process. The rest is removed from the system as a raw material for mercury production.

The mercury removal scrubbers are included in the gas cleaning system of our sulphuric acid plants. The residue separated from the combined gas cleaning system contains 50 to 60 per cent Hg. Fig. 2 shows the flow diagram of the mercury removal plant. The first unit treating 30 000 Nm^3 roaster gases per hour was started in 1973 and the second unit for 40 000 Nm^3 gas per hour in connection with our 350 tons per day fluid bed roaster in 1975. Both units have two washing towers operated in parallel to make it possible, if necessary, to clean one of the twin towers and still have the mercury removal plant in operation. The experience up to now, however, is that no deposit is formed in the scrubbers and future plants may be designed with one single tower only. Fig. 3 and 4 show the scrubbers for our 250 and 350 tons per day roasters respectively.

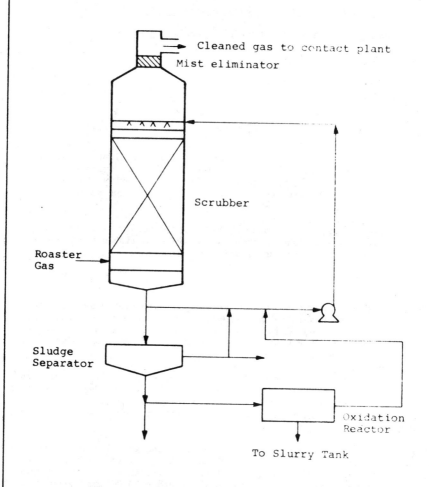

FLOW DIAGRAM OF BOLIDEN - NORZINK
MERCURY REMOVAL PROCESS

Cleaned gas to contact plant

Mist eliminator

Scrubber

Roaster Gas

Sludge Separator

Oxidation Reactor

To Slurry Tank

Fig. 1

FLOWDIAGRAM OF ROASTING AND GAS PURIFICATION SYSTEM

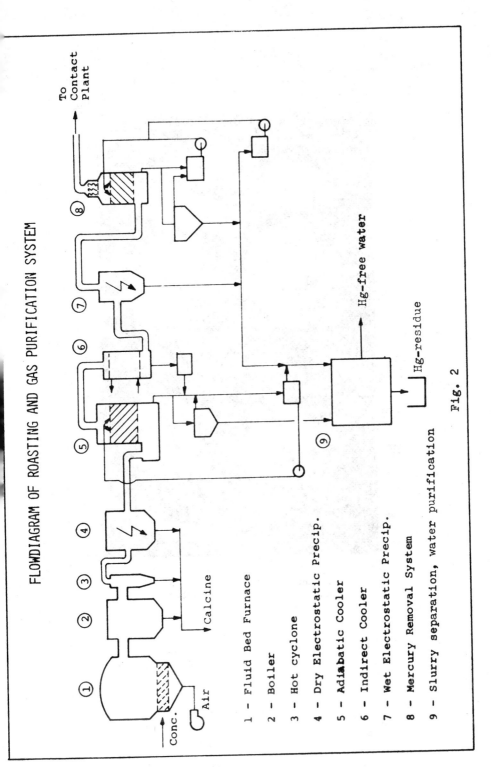

1 - Fluid Bed Furnace

2 - Boiler

3 - Hot cyclone

4 - Dry Electrostatic Precip.

5 - Adiabatic Cooler

6 - Indirect Cooler

7 - Wet Electrostatic Precip.

8 - Mercury Removal System

9 - Slurry separation, water purification

Fig. 2

Mercury removal scrubbers for
the 250 tons per day roaster

Fig. 3

Mercury removal scrubbers located after
the wet electrostatic precipitator for
the 350 tons per day roaster

Fig. 4

BOLIDEN NORZINK MERCURY REMOVAL SYSTEM
Performance figures during April - June 1979.
ppm Hg in zinc concentrate and in product acid.

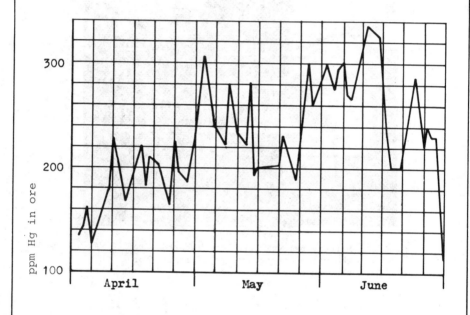

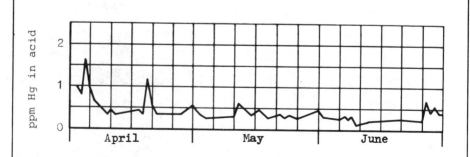

Fig. 5

The mercury is removed from the gas at temperatures of 30 to 38°C, which are the normal conditions after the washing system in contact plants.

The scrubbing solution applied at these temperatures creates no problem regarding construction materials, and the general maintenance is very low.

In Fig. 5 is given the daily variations of ppm Hg in the zinc concentrate roasted and in the product acid for the months April, May and June 1979. The mercury level in the acid obtained from concentrates up to about 350 ppm is only 0.3 ppm or even lower. This implies a mercury content in the purified gas of 0.05 to 0.1 mg/Nm^3, which represents a total purification efficiency of 99.9 per cent.

Without the mercury removal system our acid would contain up to 200 ppm.

Costs

The extra equipment installed in 1973 and 1975 for the mercury removal plants, including the scrubbing system, tanks, separation units, piping and pumps amounted to about US $300 000 for each system.

The operation and maintenance costs are about 10 cents per ton monohydrate. Of this one third represents the cost of chlorine and other chemicals.

The power required for pumps and fans is 7 kWh per ton of acid.

One man working part time on dayshift 5 days a week carries out the regeneration of the active washing solution. The supervision is made by the sulphuric acid plant operator.

Conclusion

After 6 years of service we may conclude that the Mercury Removal System developed has met or even surpassed the requirements originally specified.

The purity of the fjord water as regards mercury is the same as in the North Sea.

The process has attracted attention outside our organization and is up to now licensed to three other companies.

References

(1) Borg K. 1958, Proc. VIII Nord. Veterinarian Congress 394. Helsinki.

(2) Raeder & Snekvik 1941, Proc. Royal Norwegian Society 1941.

(3) Niigata Report; Report on the cases of mercury poisoning in Niigata (Ministry of Health and Welfare. Tokyo 1967).

(4) Minamata Report; M. Kutsuma (Ed.) Minamata disease. Study Group of Minamata Disease. (Kimumoto University, Japan 1968).

(5) J. Rastas, E. Nyholm, J. Kangas, Mercury recovery from SO_2-rich smelter gases. Reprint E/MJ 1971 by McGraw-Hill Inc.

(6) A. Kuivala, J. Poijarvi, Sulphuric Acid washing removes mercury from roaster gases. E/MJ October 1978, 81-84.

(7) O. Sundström, Mercury in Sulphuric Acid,Sulphur, **116, 37-43 (1975)**.

(8) Norwegian Patent Nbr. 125 716.

(9) Swedish Patent Nbr. 33 75 98.

(10) Swedish Patent Nbr. 360.986, US Pat. Nbr. 3.849.537.

THE JAROSITE PROCESS - PAST, PRESENT AND FUTURE

V. Arregui, A.R. Gordon and G. Steintveit

Respectively: Asturiana de Zinc S.A., Madrid,
Spain, Electrolytic Zinc Co. of A'asia Limited,
Melbourne, Australia, and Norzink AS, Odda,
Norway.

Abstract

For many years the recovery of zinc in the electrolytic process was
restricted in those cases where there was a significant proportion of iron
in the zinc concentrates treated. The need for an economic means of
removal of high concentrations of iron from solution was met by the
development of several processes of which the jarosite process has
proved to be the most widely adopted. The jarosite process was
developed independently in Europe and Australia in the 1960's and
adopted rapidly in the works of the inventors and many other zinc plants.
At the present time its use is enabling a very significant improvement in
efficient use of zinc concentrates, together with improved recoveries of
associated metals such as lead, silver, gold, copper and cadmium.

The process has been further improved to give even higher
recoveries of zinc and other metals and it is now possible to produce a
purer jarosite which facilitates recovery of its iron content as well as
making it more acceptable for disposal by other routes.

Future zinc plants are envisaged in which zinc recoveries of over
98 per cent should be possible together with high recoveries of other
non-ferrous metals and production of an iron-bearing product suitable as
raw material for iron manufacture.

Introduction

The hydrometallurgical treatment of zinc sulphide concentrates has always been concerned with the separation of zinc from iron. This arises from the fact that most zinc concentrates contain iron, some in significant proportions. The iron content would lie normally between 5 and 12 per cent.

Apart from processes involving the direct leaching of concentrates, such as that proposed to be installed by Cominco in Canada (1), it is necessary that concentrate be oxidised prior to hydrometallurgical treatment. In most modern roasting processes much of the iron present in the zinc concentrate forms zinc ferrite ($ZnO \cdot Fe_2O_3$). This occurs whether the iron is in solid solution in the zinc sulphide mineral or is present in a separate iron mineral. It is possible to roast zinc concentrate using the sintering process (2) and the proportion of iron which forms ferrite may be small. However, sintering is rarely used as a step preparatory to hydrometallurgical processing. Even if it were, there would still be the need for a process for removing soluble iron from solution.

In most electrolytic zinc plants a residue containing zinc ferrite is formed where roasted concentrate ("calcine") is leached in spent electrolyte at a temperature of about $50-60^{\circ}C$ (the so-called "neutral" leach). Formerly, this neutral leach residue was subjected to a mild acid leach, to dissolve excess zinc oxide, and subsequently either:

(a)   stockpiled

(b)   fed to a lead blast furnace, as at Cominco's Trail operations, zinc being recovered from the slag as zinc oxide

(c)   treated in one of a variety of pyrometallurgical devices in which reduction was used to form lead and zinc which volatilised and were then reoxidised and recovered as crude oxides.

All these methods are mentioned in the report of the 1970 AIME Symposium (3). The same volume, however, described the new types of hydrometallurgical processes which had become available just prior to 1970. These included the jarosite and goethite processes. Subsequently the hematite process was developed in Japan and applied at the plant of the Akita Zinc Company (4). This paper reviews the development of the jarosite process, its present use and likely future developments.

Historical

The problem of recovering zinc from electrolytic zinc plant residues is as old as the electrolytic process itself. Ralston (5) and Goldbright and Niconoff (6), report early investigations on the subject.

Tainton (8) reported that most of the zinc and iron in the residues dissolves readily in sulphuric acid solution, provided there is an excess of acid, and the temperature is close to the boiling point of the solution. The problem was to precipitate iron from solution in a form which could be readily separated from the solution. Tainton precipitated the iron at high temperature at high pH apparently with reasonable success. The form of the iron precipitate in that case was probably iron hydroxide or goethite or a mixture. The precipitate was not particularly easy to separate from the solution and Tainton used the costly and cumbersome Burt pressure filters for this purpose.

At about the same time, the Anaconda Company developed a process (9) in which residue was leached in hot sulphuric acid, so that most of the iron and zinc dissolved, followed by a second step in which the ferric sulphate in the solution was reacted with zinc ferrite to give further solution of zinc from the ferrite and precipitation of iron as a basic sulphate. The resulting residue, containing iron, plus lead, silver and gold, was treated in a lead smelter. Mitchell (10) of the Anaconda Company proposed that the basic iron sulphate be mixed with acid and roasted at a temperature where zinc sulphate was stable and the basic iron sulphate unstable. The roasted product was to be leached so that zinc and other soluble sulphates - which could include those of sodium and potassium for example - would be returned to the zinc circuit while the residue would contain iron oxide, plus lead sulphate, silver and other insoluble materials.

Anaconda apparently did not adopt this procedure fully and did not persist with the treatment of residue by leaching, possibly because, in the absence of sufficient complexing ions such as potassium, sodium or ammonium, the ease of formation and the insolubility of the basic iron sulphate would have been low.

Experimental work along similar lines was carried out at the EZ research laboratories over a considerable period in the 1920's. During this work, it was apparent that under certain conditions, zinc ferrite was acting as a neutralising agent - that is the ferrite was being attacked so that the zinc was going into solution and, at the same time, iron was precipitating as a basic sulphate. However, the role of complexing ions was not suspected or understood.

Many companies experimented with various types of processes for the treatment of residues in the 1930's to 1950's. Both Norzink and EZ operated selective sulphation processes at various times and Norzink also operated a type of fuming furnace (7 ,11).

However, efforts continued to develop a leaching process which would enable a good recovery of zinc and other metals such as cadmium and copper, a good recovery of lead and silver in a saleable product, and a ready separation of iron from solution with little loss of zinc.

These objectives were attained as a result of studies at Norzink, EZ and Asturiana, and the jarosite process became a practical reality. The development at Asturiana was associated particularly with the problem of using zinc sinter in the manufacture of electrolytic zinc. An important factor which was understood as a result of these developments was the role played by complexing ions in the formation of insoluble basic iron sulphates.

Jarosite and its Precipitation

The natural mineral called jarosite was apparently first recognized at Barranco Jaroso in Murcia, Spain, according to Dana (12) and given the formula $K_2Fe_6(OH)_{12}(SO_4)_4$.

The mineral in which sodium is present instead of potassium is called natro-jarosite. Another related mineral is called carphosiderite in which the potassium ion is replaced by H_3O^+. Some authors have stated that minerals designated carphosiderite are in most cases actually natro or sodium jarosite (13). Haigh (14) reported that the x-ray pattern for artificially prepared carphosiderite did not correspond to that published up to that time as that of carphosiderite, and concluded that the then published pattern was that of natro-jarosite.

Other possible jarosite type compounds contain ions of lead, silver or rubidium replacing potassium ions. It has also been reported that calcium forms a jarosite type compound (15) but Haigh reported that attempts to confirm its formation were unsuccessful.

It also seems that solid solutions form readily particularly those in which there is partial replacement by hydronium ion (H_3O^+) of one or other of the other ions. It is also likely that compounds with more than one of the "jarosite forming" ions in the crystal lattice may form.

Phipps and Hutchison (16) reported a study of the factors affecting the formation of jarosite and stated that the rate of precipitation of ferric iron as ammonium jarosite from acid zinc sulphate solutions near $100^\circ C$ increased with:

(a) increasing concentration of ammonium ion

(b) decreasing concentration of acid

(c) increasing surface area of jarosite particles present

In addition, their results indicated that the concentration of ferric iron had little effect, except at low concentrations of iron when the rate decreased, and that increasing sulphate ion concentration resulted in the formation of jarosite of smaller size. The interesting fact was also noted that the "jarosite" formed was non-stoichiometric. Even with a ratio of ammonium to ferric ions in solution much in excess of the ratio in stoichiometric jarosite, the ratio in the precipitate was less than stoichiometric.

The beneficial use of jarosite particles as "seeds" in the precipitation step as reported by Hutchison and Phipps has been practised in several plants, including that of Texasgulf, Canada (17), and is a feature of the Spanish Patent No. 407,811 of Asturiana de Zinc (47).

Rastas, Fugleberg and Huggare (18) noted that with a solution containing less than half of the stoichiometric amount of sodium, the precipitate formed contained some glockerite ($Fe_4SO_4(OH)_{10}$). At higher concentrations of sodium, the precipitate was all jarosite, the proportion of sodium present being about half the stoichiometric amount. The jarosite contained hydronium ions as well as sodium ions.

The precipitation of jarosite has also been studied by Getskin, Margulis, Belsekeeva and Yaroslavtsev (19) who noted that the precipitation of jarosite at $90^\circ C$ was more effective with potassium ions than with ammonium ions, with sodium ions being the least effective of the three. These results agreed with the results of the work of Haigh (14) carried out at higher temperatures. The Russian authors also noted the effect of increased concentrations of zinc sulphate on the "dispersion" of the jarosite and also the benefit of "seeding" to increase the rate of precipitation, especially if sodium or potassium were being used. These results are generally in agreement with those of Hutchison and Phipps.

Steintveit (20), in a paper published earlier than some already referred to, dealt with much of the basic chemistry of the precipitation of iron, including the effect of the addition of potassium, rubidium, ammonium and sodium ions. The results were generally similar to those reported above with rubidium being between potassium and ammonium ions in effectiveness.

Studies of the precipitation of jarosite were also carried out in Spain by Limpo et al (21). Their results agreed generally with those of others, except that they found that iron concentration affected the precipitation rate significantly. They noted the effect of seeding also. Rastas et al (22) also discussed the kinetics of the reaction finding relative effectiveness of the ions of sodium, potassium and ammonium to be generally similar to that noted by others such as Steintveit, and also noting again the favourable effect of "seeding" by jarosite.

Other papers of interest dealing with either thermodynamic or kinetic aspects of jarosite formation are those by Masson and Torfs (23), McAndrew et al (24), and Kwok and Robins (25).

Development and Use of the Process

The development and use of the jarosite process has been well documented. The early patents were those of Norzink AS (26), Asturiana de Zinc S.A. (27) and Electrolytic Zinc Company of Australasia Limited (28). Steintveit and co-workers at Norzink have published a series of papers dealing with the theory of the process and the application at the Norzink plant (20,29,30,31,32).

Likewise, authors from EZ have published descriptions of the EZ plant and its operation, as well as general articles on the jarosite process (16,33,34,35,36). Several other general reviews have been published (37,38). The EZ plant is of special interest since, as well as treating residue currently produced, the plant also treats residue from a large stockpile at a rate of about one-quarter to one-third of the total residue treated. Other authors have written general papers dealing with the process (39,40,41,42).

Gordon and Pickering reviewed some of the applications of the process (37) but for ease of reference, some of the flow sheets will be reproduced here. Figure 1 shows one of the simplest ways in which the process may be applied.

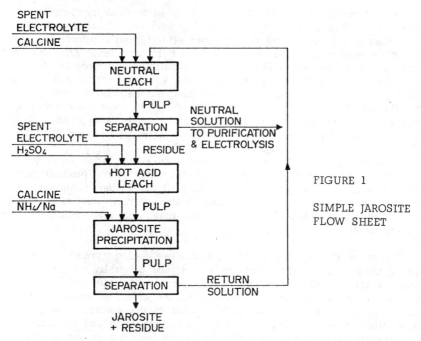

FIGURE 1

SIMPLE JAROSITE
FLOW SHEET

Referring to this flow sheet calcine is leached in a mixture of spent electrolyte and acid solution returning from the jarosite precipitation step. In this "neutral" leach step, zinc oxide dissolves, and any soluble iron precipitates carrying with it impurities harmful to electrolysis. This precipitate and other materials insoluble under the neutral leaching conditions - chiefly zinc ferrite - form the "neutral leach residue". The residue and solution are separated, the latter proceeding to purification and zinc electrolysis, the former to hot acid leaching where it reacts with a mixture of spent electrolyte and sulphuric acid at a temperature of 95°C. The resulting pulp is then partly neutralised by the addition of calcine, and jarosite is precipitated in the presence of complexing ions, commonly sodium or ammonium. The pulp is separated usually by thickening and filtration into the "jarosite" residue and solution which is returned to the neutral leach.

In considering the developments of alternative flow sheets it is interesting to notice that there have been two lines of development. One line has been concerned with flow sheets which would give the high recoveries of lead and silver together with high recoveries of zinc and other metals such as copper and cadmium. The second has been concerned with maximising the extraction of zinc and other metals excluding lead and silver and minimising investment when raw materials do not contain significant amounts of lead and silver.

Flow Sheets Incorporating Lead and Silver Recovery

Multistage Leaching

Lead and silver compounds are insoluble at all stages and exist at maximum concentration in the hot acid leach stage. Maximum recovery of lead and silver is dependant upon minimum use of calcine in the jarosite precipitation stage. Also, in order to produce a residue with the maximum concentration of lead and silver, it is desirable that the extraction of zinc and iron be maximised in the hot acid leach step and that the reprecipitation of iron as jarosite be avoided. To achieve this, Sitges and Arregui (43) proposed the multistage countercurrent hot acid leach (Figure 2) procedure in which neutral leach residue is leached in a first stage solution returned from a second leaching stage in which the residue from the first stage is leached in hot strong acid solution. Residue separated from the second stage is low in zinc and iron and consequently high in lead and silver. The process may be extended to further stages, and although it would not normally be worthwhile using more than two stages, three stages have been used in at least one plant.

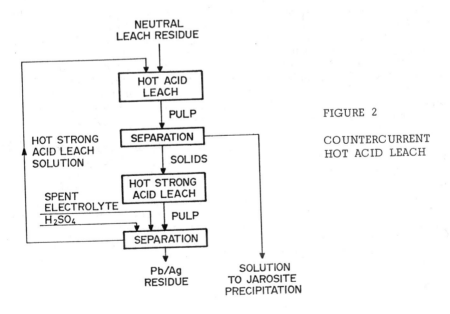

FIGURE 2

COUNTERCURRENT
HOT ACID LEACH

Pre-neutralisation

Another procedure proposed for the increasing of recovery of zinc, copper and cadmium as well as lead and silver is the pre-neutralisation step. Solution leaving the hot acid leaching step is usually at a reasonably high acid concentration. For jarosite precipitation to proceed quickly and efficiently the acid concentration must be reduced to a lower level. This is usually achieved by adding calcine, the residue from which, containing zinc, copper and cadmium in the ferrite portion as well as the associated silver and gold, joins the jarosite and is lost. The pre-neutralisation step is interposed between the hot acid leach and jarosite precipitation, and the acid content of the solution is partly neutralised with calcine, the residue from which may then be separated and returned to the hot acid leach where zinc, etc. plus iron is dissolved and lead and silver joins the lead-silver residue. The partly neutralised solution then proceeds to jarosite precipitation in which the remainder of the required neutralising material is added.

Jarosite acid wash

The recovering of zinc and other associated metals can also be enhanced by using the "jarosite acid wash" due to Steintveit (44). Since the amount of ferritic residue present in the jarosite will depend on the iron content of the calcine, it is clearly advantageous to use a low iron calcine in the jarosite precipitation step. It should be noted that zinc sinter may not contain ferrite, the zinc and iron contents being highly soluble. Sinter is therefore an excellent neutraliser for jarosite precipitation and has been so used by Asturiana de Zinc S.A. and at Societe de Prayon S.A. in Belgium. When sinter is used, the zinc dissolves, remaining in solution, while the dissolved iron forms jarosite. Steintveit noted that the jarosites were stable at acidities at which the zinc ferrites dissolve at a reasonable rate. The jarosite acid wash then consists of a stage in which the precipitate is reacted with hot acid in which zinc and other metals **from** calcine residues dissolve. Iron also dissolves, and part of this may **form** additional jarosite. After separation of the remaining solids, which still contain the lead and silver from the calcine used in jarosite precipitation, the solution may be returned to the pre-neutralisation or the jarosite precipitation step.

Steintveit (30) and Steintveit and Dyvik (31) discussed various flow sheets incorporating combinations of the steps already discussed, including arrangements wherein the major part of the calcine was treated in a high acid leaching step instead of a neutral leach step. This procedure derived from an adaptation of batch leaching operations (45, 46). Use of a high acid leach for calcine can, with sufficient process steps, give good extractions of zinc and other metals together with reasonable recovery of lead and silver, but the volume of plant required is much greater than in plants using the procedures starting with a neutral leach step.

Steintveit's published calculations show the possibility of extractions of up to 98% of the zinc content of calcines with about 10% iron, with about 90% extraction of copper and cadmium, and up to 80% recovery of lead and silver.

Whether seeking such high extraction is warranted is entirely a matter of economics, taking into account the additional capital and operating costs required for the more extensive plant arrangements, and the additional metal recoveries achieved thereby.

Two-stage precipitation of jarosite

Another important development is the procedure developed by Asturiana de Zinc S.A. in which jarosite is precipitated in two stages (47) Figure 3.

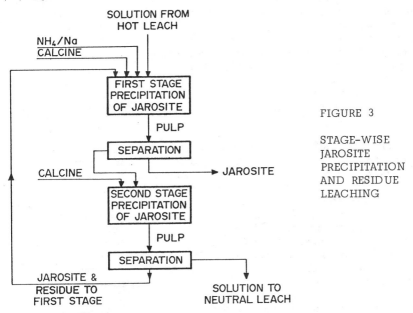

FIGURE 3

STAGE-WISE JAROSITE PRECIPITATION AND RESIDUE LEACHING

In the first stage solution containing soluble iron and acid is reacted with some calcine and solids returned from a second stage. These solids contain jarosite and zinc ferrite residue from calcine added in the second stage. The acidity in the first stage is sufficiently high to dissolve the ferrite, so that the solids separated after this stage, which represent the final jarosite waste, consist of insoluble materials plus jarosite with very little ferrite. The solution from the first stage is still moderately high in soluble iron and passes to the second stage where calcine is added to reduce the acidity to a low figure and so precipitate most of the iron. The resulting precipitate is separated and returned to the first stage, the solution proceeding to the neutral leach. By this means also high extraction of zinc can be obtained. In addition, recycle of jarosite and other solids from the second to the first stage has a beneficial effect on the jarosite precipitation rate and the physical properties of the jarosite. An alternative form of this process for use when recovery of lead and silver is not important will be referred to later.

<u>Flow Sheets for Minimum Investment without Lead and Silver Recovery</u>

<u>Zinc ferrite as a neutralising agent</u>

The reaction between zinc ferrite and sulphuric acid is a vital feature of all the leaching methods for the treatment of electrolytic zinc residues.

$$ZnO \cdot Fe_2O_3 + 4H_2SO_4 \longrightarrow ZnSO_4 + Fe_2(SO_4)_3$$

This is the essential reaction of the hot acid leaching step. The kinetics of this reaction has been the subject of some study, much of it not published. Rastas, Fugleberg and Huggare discuss the reaction briefly (18), stating that the rate is reaction, not diffusion controlled, and that the other significant variables are temperature and specific surface area of the ferrite. Over a moderate range acidity had a small effect.

Later work was reported by Rastas et al (22) in which they studied more fully the influence of surface area on the rate of solution of ferrites.

Norzink have studied the relation between roasting conditions and reactivity of calcine. The reactivity is studied by leaching the calcine at a constantly maintained acidity and temperature using automatic apparatus. Variation in roasting conditions may cause variation in the particle size distribution of the calcine and resulting ferrite, as well as of the proportion of ferrite formed. In addition, it is known that the structure of ferrites may differ depending on their rate of cooling, for example, and it is possible that their leaching characteristics may be affected also. Experimental leaching of various residues has shown significantly different rates, and it is suspected that the presence of various impurities, for example silica, may also influence the reaction.

It is evident however that the ferrite is a neutralising agent, albeit a much more leisurely one than zinc oxide. The precipitation of jarosite results in the generation of sulphuric acid, so:

$$3Fe_2(SO_4)_3 + 10H_2O + NH_4OH \longrightarrow 2NH_4 \cdot Fe_3(SO_4)_2OH_6 + 5H_2SO_4$$

combining the ferrite leaching and jarosite precipitation reactions we have:

$$3ZnO \cdot Fe_2O_3 + 10H_2O + 2NH_4OH + 7H_2SO_4$$
$$\longrightarrow 2NH_4Fe_3(SO_4)_2(OH)_6 + 3ZnSO_4.$$

Provided, therefore that the acidity of the solution is sufficiently high to give a reasonable rate of leaching of ferrite, and yet not high enough to prohibit jarosite precipitation the above reaction will proceed. It normally proceeds slowly because of the above constraints. That such a reaction was possible was indicated in EZ's original patent (28).

Considerable experimental work was carried out on this reaction in EZ's laboratories over several years and it was proven to be practicable. The reaction proceeded more rapidly using potassium rather than ammonium, with which long reaction times proved necessary. The procedure was not adopted at EZ because separation of lead silver residue was desired. Steintveit indicated (20) in 1970 that a variety of neutralising agents might be used instead of zinc calcine and gives an example of the use of ferric oxide. The use of zinc ferrite in this manner has been adopted in Outokumpu's use of the jarosite process at Kokkola, Finland (48).

Flow sheets for minimum investment

Mention has already been made of Asturiana's two-step jarosite precipitation procedure in which ferrite is used as a neutralising agent in jarosite precipitation. The same Asturiana patent (47) describes a simplified procedure for use when lead and silver recovery is of no interest, as shown in Figure 4.

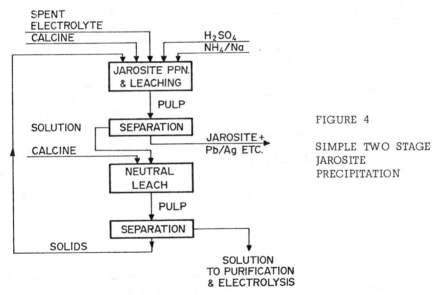

FIGURE 4

SIMPLE TWO STAGE
JAROSITE
PRECIPITATION

There are two reaction stages - one in which the conditions of temperature and acidity are such that zinc ferrite is attacked at the same time as jarosite is precipitating, and another in which the iron in solution from the first stage is precipitated as jarosite plus hydroxide. The solids from this latter stage are recycled to the hot acid leaching stage where they assist in "seeding" jarosite precipitation. In this stage also, any ferrites in the solids are dissolved and any iron dissolved forms jarosite.

Numerous other flow sheet arrangements are possible, but two of particular interest will be mentioned. One of these is shown in Figure 5 and it is a flow sheet of great simplicity and effectiveness in giving good extractions of zinc, cadmium, etc. Lead and silver are incorporated in the jarosite residue. In this flow sheet, neutral leach residue is fed directly to the jarosite precipitation step, where the neutralising material it contains, including zinc ferrite, contributes to the precipitation reaction. Solids from the jarosite step are recycled to the hot acid leach where ferrite leaching is completed, thus combining the jarosite acid wash and hot acid leach in one step (49).

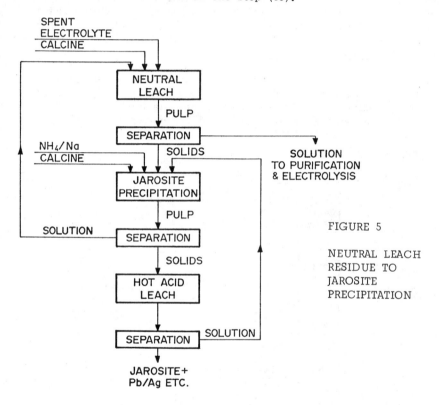

FIGURE 5

NEUTRAL LEACH RESIDUE TO JAROSITE PRECIPITATION

Another arrangement is shown in Figure 6 in which neutral leach residue proceeds to the hot acid leach, the solution from which goes to jarosite precipitation and all the pulp from jarosite precipitation is returned to the neutral leach, so that the solids from jarosite precipitation enter the hot acid leach with the neutral leach residue (50).

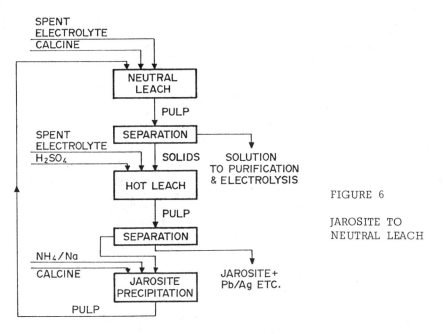

FIGURE 6

JAROSITE TO
NEUTRAL LEACH

In choosing a particular flow sheet it is necessary to consider the characteristics of the calcine available for treatment and whether production of a lead silver residue is economically justified. From a knowledge of the leaching characteristics of the calcine, which should be determined by test, it is then possible to estimate the conditions in all parts of the circuit which will give the minimum amount of equipment for a desired extraction of zinc.

Use of the Jarosite Process

The jarosite process is thus well accepted and has been very widely adopted in the electrolytic zinc industry. Plants using the process, and their nominal capacities, are:

	Nominal Capacity tonnes per year
Asturiana de Zinc S.A., Spain	200,000
Budelco, Holland	150,000
Canadian Electrolytic Zinc, Canada	205,000
Compagnie Royale Asturienne des Mines, France	120,000
Electrolytic Zinc Co. of A'asia Limited, Australia	210,000
Hemijska Industrija Zorka, Yugoslavia	30,000
Korea Zinc, Korea	50,000
Met Mex Penoles, Mexico	100,000
Norzink AS, Norway	85,000
Outokumpu Oy, Finland	160,000
Preussag-Weser-Zink, Germany	110,000
Sulfacid, Argentina	30,000
Societe de Prayon, Belgium	60,000

Texasgulf, Canada   120,000
Toho Zinc, Japan   140,000

In addition the process will be used in the Minero Peru plant at present under construction and planned to have a capacity of 100,000 tonnes per year.

Thus, about one-quarter of the world's zinc is produced in plants using the process.

The energy and utility requirements have been discussed in reference 38.

The process has also been incorporated in proposed processes not yet adopted. For example, reference 77 mentioned United States patent 3,879,272 which describes a cupric chloride leach of copper concentrates, with discard of iron from solution as jarosite. Several other patented procedures in the treatment of complex concentrates or zinc raw materials have incorporated the process (51,52,53,54,55,56,57,58,59).

Handling and Treatment of Lead Residue

As indicated earlier, the recovery of a residue containing lead and silver may be justified if these elements are present in significant quantities.

The composition of the residue from the hot leaching stage is a function of the composition of the concentrates treated.

Residue as produced at EZ's plant had the composition as follows for the year ended June 1978:

Lead	22.0%
Zinc	10.0%
Iron	12.7%
Calcium Oxide	3.5%
Sulphur as Sulphate	7.3%
Elemental Sulphur	1.4%
Silica	8.2%
Alumina	2.3%
Silver	665 g/tonne
Gold	4.5 g/tonne

The chief compounds present are lead sulphate, zinc ferrite, jarosite, calcium sulphate, silica, elemental sulphur.

Such a material can be accepted as a small part only of the feed to a lead blast furnace or Imperial Smelting furnace since it does not have any significant fuel value. Therefore, it must be mixed with sulphide concentrates or with other fuel prior to the normal sintering operation. Additionally, its low lead content means that only a limited proportion can be incorporated in the sinter feed without lowering the lead content of the sinter to an unacceptable level. Therefore attempts have been made to upgrade the residue.

Asturiana de Zinc S.A. have developed a flotation process for producing silver concentrate and a material enriched in lead (60); Vieille Montagne (61) have carried out similar work and installed a plant for the purpose. They have also (62) developed a process for recovering silver from the silver concentrate.

An entirely different approach was taken by EZ and Mitsubishi (63) and also by Norzink (64) in which the residue was smelted in electric furnaces. The resulting processes, though using similar equipment are different procedurally. Both produce a lead bullion containing lead and silver.

Handling and Treatment of Jarosite Residues

These residues contain jarosite, zinc ferrite, gypsum, silica, etc. A typical composition is as follows:

	%		%
Zinc	5	Jarosite	79
Lead	2	$PbSO_4$	3
Cadmium	0.05	$ZnSO_4$	6
Copper	0.3	$ZnO.Fe_2O_3$	6
Iron	24	ZnS	1
Sulphur	12	SiO_2	2

Small amounts of other impurities are usually present, depending on the raw materials treated.

It has been possible to dispose of some of this kind of material in the sea and disposal on land is practised also. It is necessary for the land disposal area to be lined with a membrane impermeable to solutions which may contain, for example, zinc and cadmium. These could cause problems in surrounding areas if they were not contained. In some cases, drains have been provided to collect any inadvertent leakage which is returned to the pond by pumping.

Because of the environmental limitations applying to disposal at sea or storage on land, much effort has been spent on finding ways of treating or using jarosite. Several chemical treatments have been devised with a view to recovering the non-ferrous metals and iron in a useful form. Some of these are summarised in the table below:

Process	Reference
Sulphating roast followed by leaching of soluble salts and production of pure iron oxide	65
Thermal decomposition and production of iron oxides	66,67,68,69,70
Hydrothermal decomposition to form iron oxide and recover soluble salts	71,72

111

Solution of metals as chlorides and recovery from
solution, with recovery of iron oxide as pigment   73,74,75

Electric furnace smelting   64

Another proposed method of disposal is recovery of part of the iron
content as iron oxide or iron and the remainder as zinc sulphate plus
ammonium sulphate for use in mixed fertilisers (76). Mixing with acid
prior to use as a fertiliser has also been suggested (77).

It has been proposed that jarosite sludge be mixed with bitumen
emulsion, whereby the solids and bitumen separate from the solution
which contains soluble zinc and other metals. The bitumen jarosite
mixture can then be used as an additive to asphalt paving (78).

Other efforts have been directed at making the material more readily
stockpiled preferably in free standing heaps in which all the metals are
insoluble. Methods proposed include mixing with calcium oxide (79),
portland cement, or other chemicals. Use of jarosite in the manufacture
of bricks, plaster, and other building materials has been investigated,
with little promise so far because of its physical properties, sulphur
content, non-ferrous metals content, or the small amount which would
be used.

It is evident therefore that a major advance would be the production
of jarosite with a low content of non-ferrous metals, preferably such that
the iron content could be readily recovered.

Possible Production of Purer Jarosites

Earlier discussion has mentioned that one of the lines of development of the process has been towards increasing the recovery of zinc
while at the same time increasing the recovery of lead and silver. This
can be achieved by replacing calcine as a neutralising material in the
jarosite step, thereby avoiding having residue from the calcine present in
the jarosite residue. If a lead-free zinc oxide were available, it would
be a suitable replacement for calcine, as there would be no residue from
the zinc oxide to contaminate the jarosite.

Use of limestone for neutralisation

An alternative to calcine as neutralising material is limestone.
This adds gypsum to the jarosite residue, increasing sulphate removal
from the solution circuit beyond that removed by jarosite. Such sulphate
removal may be useful as it allows an increase in the amount of strong
acid added to the acid leaching stage, which either allows a more
intensive leach, or an increase in the capacity of the plant which can
be utilised if stockpile residue is available.

EZ have developed a means of using limestone indirectly. The
Selective Zinc Precipitation (SZP) process (80,81) was developed as an
alternative to electrolytic stripping of zinc from spent electrolyte prior
to discard of solution for the purpose of elimination of soluble impurities

such as manganese, magnesium, chlorine from a zinc plant circuit. In this process, zinc sulphate solution is reacted at temperatures between $50^{\circ}C$ and boiling point with limestone at a pH of about 6.5 so that basic zinc sulphate is precipitated.

$$3CaCO_3 + 4ZnSO_4 + 13H_2O \rightarrow ZnSO_4 \cdot 3Zn(OH)_2 \cdot 4H_2O + 3CaSO_4 \cdot 2H_2O + 3CO_2$$

By this means, the zinc content can be reduced to about 1 g/l with impurities remaining in solution.

A second application, discussed extensively by Matthew et al (82) is in decreasing wash water requirements by precipitating basic zinc sulphate from the filtrate from the washing of jarosite or other residues and re-using the de-zinced solution for washing the residue. In either type of application, the basic zinc sulphate is available as a neutralising material, and it is used advantageously in the jarosite precipitation step, in which case, the amount of calcine required is reduced. With a sufficiently large SZP plant, calcine may be completely replaced by basic zinc sulphate. In that case a jarosite residue is obtained which is free of lead and silver and containing only zinc in entrained solution. It does, however, contain gypsum equivalent to the limestone used in forming the basic zinc sulphate, and this increases the bulk of the residue considerably. However, recoveries of zinc, copper, cadmium, lead and silver are very much increased and the jarosite residue is more acceptable environmentally.

Figures 7 and 8 show two flow sheets incorporating the two means of using the SZP procedure.

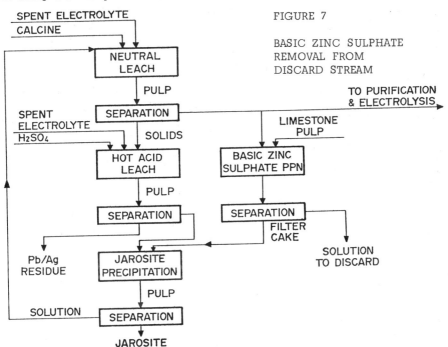

FIGURE 7

BASIC ZINC SULPHATE
REMOVAL FROM
DISCARD STREAM

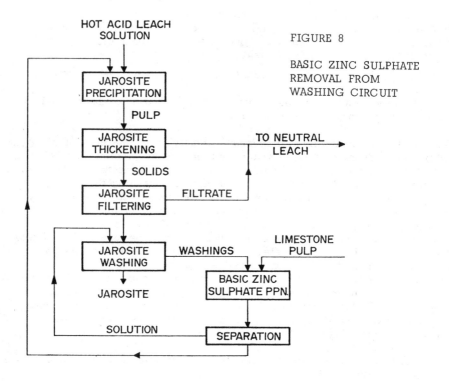

HOT ACID LEACH
SOLUTION

FIGURE 8

BASIC ZINC SULPHATE
REMOVAL FROM
WASHING CIRCUIT

Jarosite precipitation without neutralisation

Independent work in several parts of the world has been carried out in an endeavour to diminish the amount or avoid the use of neutralising agent in the jarosite step (83,84,85,86).

At EZ, extensive pilot plant testing of possible means of avoiding the use of neutralising material has been carried out and is continuing. Reference 82 describes a number of possible procedures and flow sheet arrangements, all of which will not be discussed here. The procedures used involve the adjustment of the composition of the solution prior to jarosite precipitation. This may be achieved by several means. For instance, by lowering the temperature of the solution from the hot acid leaching stage, it is possible to neutralise the acid present down to an acidity of about 5 g/l without premature precipitation of jarosite. This arises from the increased solubility of jarosite at lower temperatures (refer paper by Pickering and Kershaw, ref. 87) and from the slower rate of reaction. Undissolved solids may then be removed and returned to the hot acid leach, the solution passing to jarosite precipitation, being heated to 95°C prior to or in the precipitation step.

Another means of adjusting the solution composition consists of diluting the solution from the hot acid leaching step with zinc sulphate solution low in iron and acid.

It is possible to use various combinations of cooling and pre-neutralisation and dilution.

The solution, after adjustment of the composition by one or other of the possible means is then brought up to a temperature of about 95°C. This may be achieved, for example, by heat exchange with solution from the hot acid leach, which is cooled thereby, or by the use of steam directly or in a heat exchanger. The hot solution is allowed to react in the presence of the usual complexing ions. It has also been found that recycle of jarosite assists in the precipitation of the iron. As jarosite forms, acid is generated, and eventually the reaction virtually ceases. The amount of iron precipitated depends on the initial concentrations of ferric iron and free acid. If these are, say, 15 g $Fe^{3+}/$ litre and 5 g H_2SO_4/l, the iron may be reduced to 5 g/l, while the acidity will rise to about 20 g/l.

Advantages of producing a "clean" jarosite are:

(a) overall recoveries of zinc, cadmium, copper, lead and silver in calcine are significantly increased.

(b) the weight of jarosite is reduced.

(c) jarosite residue should be more acceptable environmentally.

The chief disadvantage is that extra equipment is required – including heat exchangers - and the size of some plant items must be increased.

Estimates have been made comparing a zinc plant using the "normal" process and the new procedures and producing 100,000 tonnes of zinc per year. The results are:

	New Procedure	Normal Process
Overall zinc recovery	99%	96.2%
Concentrate (53.6% Zn) t.p.a.	188,452	193,937
Fraction of calcine to jarosite ppm	0	29%
Lead residue t.p.d.	40	29
Jarosite produced t.p.d.	119	129
Analysis of jarosite residue **Total Zn %**	1.25	7
Water Soluble Zn %	1.0	1.0
Fe %	34.1	32.5
Pb %	0.05	3.16
Cd %	<0.001	0.06

Because the hot acid leach step has to treat more neutral leach residue, this part of the plant is larger for the new process. Also, the jarosite precipitation plant is larger because the rate of precipitation is lower. It should be noted that a small amount of neutralizing material in this step can make a significant reduction in the size of plant required and enables a lower level of iron to be reached in the solution. The use of some basic zinc sulphate from an SZP plant in conjunction with the solution composition adjustment procedures could result in a very economical plant. Figures 9 and 10 show two fundamental ways in which the new procedure may be used.

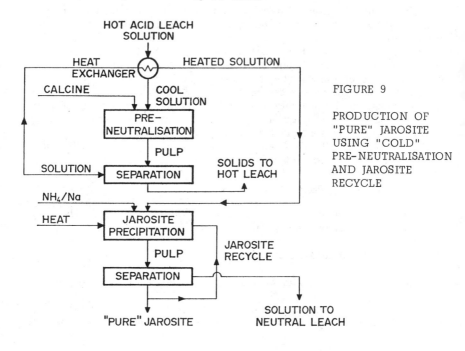

FIGURE 9

PRODUCTION OF "PURE" JAROSITE USING "COLD" PRE-NEUTRALISATION AND JAROSITE RECYCLE

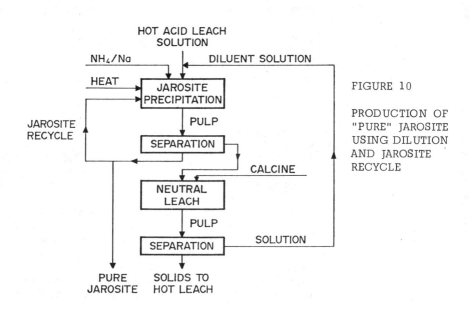

FIGURE 10

PRODUCTION OF "PURE" JAROSITE USING DILUTION AND JAROSITE RECYCLE

The "clean" jarosite produced by the new procedures may allow the production of material suitable for iron manufacture. Steps which have been tested with promising results include calcining, sintering and thermal hydrolysis. In all cases iron oxides are formed.

Summary

Since its development more than a decade ago, the jarosite process has been widely accepted especially in zinc hydrometallurgy so that a large part of the world's zinc is now produced in plants using the process. It has made an important contribution to extending the use of zinc raw materials, as well as allowing increased recoveries of other metals from zinc concentrates.

Recent developments point to the possibility of improved processes in which the recovery of all the valuable metals in zinc concentrate would be practically complete. At the same time, a jarosite residue would be produced which should be more readily disposed of or possibly converted into a raw material suitable for iron manufacture.

References

(1)   B.N. Doyle, I.M. Masters, I.C. Webster, and H. Veltman, "Acid Pressure Leaching Zinc Concentrates Produces By-Product Elemental Sulphur", Presented at Eleventh Commonwealth Mining and Metallurgical Congress, Hong Kong 1978.

(2)   H.J. Stehli, "Sintering Zinc Ores", Transactions AIME (121) (1936) pp. 374-386; K.D. McBean, "Sintering" in "Zinc", Reinhold 1959, Ed. C.H. Mathewson, pp. 155-163.

(3)   C.H. Cotterill, and C.J.M. Cigan (Editors) AIME World Symposium on Mining and Metallurgy of Lead and Zinc, Volume II (1970).

(4)   S. Tsunoda, I. Maeshiro, E. Emi, K. Sekine, "The Construction and Operation of the Iijima Electrolytic Zinc Plant", TMS Paper No. A 73-65, AIME; T.H. Ohtsuka, T. Yamada, H. Abe and K. Aoki, "Progress of Zinc Residue Treatment of the Iijima Refinery", TMS Paper No. A 78-7, AIME.

(5)   O.C. Ralston, "Electrolytic Deposition and Hydrometallurgy of Zinc", McGraw-Hill 1921.

(6)   G.L. Oldbright, and D.P. Niconoff, "The Recovery of Zinc from Ferrite Compounds in the Electrolytic Zinc Process", **Mining & Met. Investigations, Univ. of Utah, Paper No. 6, 1929.**

(7)   A.R. Gordon, R.W. Pickering and O.G. Woodward, "Fluid Roasting of Zinc Plant Residue at Risdon, Tasmania". Paper presented at Aust. I.M.M. Symposium on Fluidisation, Adelaide, Australia 1961.

(8)   U.C. Tainton and L.T. Leyson, "Electrolytic Zinc from Complex Ores", Trans. AIME (LXX) (1924) pp. 486-522.

(9)   Albert E. Wiggin and Russel B. Caples, "Electrolytic Zinc Practice at Great Falls and Anaconda", Engineering and Mining Journal, 128(8) (1929) pp. 319-324.

(10)   W.E. Mitchell, "Treating Zinc Concentrate and Plant Residue", United States Patent No. 1,834,960, lodged April 25, 1930.

(11)   J. Van Oirbeck, "Treatment of Residues from Electrolysis of Zinc", Trans. AIME (121) 1936 pp. 693-701.

(12)   E.S. Dana, "A Textbook of Mineralogy", Wiley New York 4th edition, p. 769.

(13)   A.A. Moss, "The Nature of Carphosiderite and Allied Basic Sulphates of Iron", Mineralog. Mag. (31) (1957), p. 407.

(14)   C.J. Haigh, "The Hydrolysis of Iron in Acid Solutions", Proceedings of the Australasian Institute of Mining and Metallurgy No. 223, September 1967, pp. 49-56.

(15)   D.P. Serdyuchenko, "Calcium Jarosite", Doklady Akad. Nauk S.S.S.R. (78) (1951) pp. 347-350.

(16)   R.F.S. Hutchison and P.J. Phipps, "Formation and Particle Size of Jarosite", Aust. Inst. of Mining and Metallurgy Tasmania Conference May 1977, pp. 319-327.

(17)   F.S. Gaunce, et al, "The Ecstall Story - The Electrolytic Zinc Plant", CIM Bulletin, May 1974, pp. 116-124.

(18)   J. Rastas, S. Fugleberg, and T.L. Huggare, "Treatment of Iron Residues in the Electrolytic Zinc Process", Paper presented at AIME Annual Meeting, Chicago, February 1973.

(19)   L.S. Getskin, E.V. Margulis, L.I. Belsekeeva, and A.S. Yaroslavtsev, "A Study of the Hydrolytic Precipitation of Iron from Sulphate Solutions in the Form of Jarosites", Izv. Vyssh. Uchebn. Zaved. Tsvetn. Metall. (6) (1975) pp. 40-44.

(20)   G. Steintveit, "Precipitation of Iron as Jarosite and its Application in the Wet Metallurgy of Zinc", Erzmetall (23) (1970) pp. 532-539 - See also Norwegian Patent 126,679.

(21)   J.L. Limpo, A. Luis, D. Siguin and A. Hernandez, "Kinetics and Mechanism of the Precipitation of Iron as Jarosite", Rev. Metal. Cenim (12) (1976) No. 3 pp. 123-134.

(22)   J. Rastas, S. Fugleberg, L.G. Bjorkqvist, and R.L. Gisler, "Kinetics of Ferrite Leaching and Jarosite Precipitation", Paper presented at Zinc Section of GDMB Innsbruck, May 18, 1978.

(23)   N.J.J. Masson and K.J. Torfs, "Treatment of Leach Residues from Zinc Electrolysis" Erzmetall (22) (1969) pp. B.35-B.42.

(24)   R.T. McAndrew, S.S. Wang and W.R. Brown, "Precipitation of Iron Compounds from Sulphuric Acid Leaching Solutions", CIM Bulletin January 1975, pp. 101-110.

(25)   O.J. Kwok and R.G. Robins, "Thermal Precipitation in Aqueous Solutions", International Symposium on Hydrometallurgy, AIME 1973 pp. 1033-1080.

(26)   G. Steintveit, "Process for the Separation of Iron from Metal Sulphate Solutions and a Hydro-metallurgic Process for Production of Zinc", Norwegian Patent No. 108,047 lodged April 30, 1965.

(27)   F. Sitges and V. Arregui, "A Process for the Recovery of Zinc from Ferrites", Spanish Patent No. 304,601 lodged October 2, 1964.

(28)   C.J. Haigh and R.W. Pickering, "Treatment of Zinc Plant Residue", Australian Patent No. 401,724 lodged March 31, 1965.

(29)   G. Steintveit, "Electrolytic Zinc Plant and Residue Recovery, Det Norske Zinkkompani A/S", World Symposium on the Mining and Metallurgy of Zinc and Lead, St. Louis 1970, Vol. II, pp. 223-246, AIME.

(30)   G. Steintveit, "Treatment of Zinc Leach Plant Residue by the Jarosite Process", IMM Symposium on Advances in Extractive Metallurgy, London 1971.

(31)   G. Steintveit and F. Dyvik, "The Integrated Jarosite Process in Electrolytic Zinc Production", TMS Paper No. A.72-4, AIME.

(32)   G. Steintveit, "Trends in the Optimization and Control in Zinc Hydrometallurgy, I. Chem. E. Symposium Series No. 42 Manchester, April 2, 1975.

(33)   C.J. Haigh and R.W. Pickering, "The Treatment of Zinc Plant Residue at the Risdon Works of the Electrolytic Zinc Company of Australasia Limited", World Symposium on the Mining and Metallurgy of Lead and Zinc, St. Louis 1970, Vol. II, pp. 423-448, AIME.

(34)   J. Wood and C. Haigh, "Jarosite Process Boosts Zinc Recovery in Electrolytic Plants", World Mining, September 1972, pp. 34-38.

(35)   J.T. Wood, "Treatment of Electrolytic Zinc Plant Residues by the Jarosite Process", Australian Mining, (65) (1) (1973) pp. 23-27.

(36)   C.J. Haigh, I. Oakes, and J.S. Garrigan, "Residue Treatment Plant Practice at Risdon, Tasmania", in Aust. I.M.M. Conference Papers, May 1977, pp. 339-349.

(37)   A.R. Gordon and R.W. Pickering, "Improved Leaching Technologies in the Electrolytic Zinc Industry", Trans. AIME (6B) (1975) pp. 43-53.

(38)   A.R. Gordon, "Improved Use of Raw Material, Human and Energy Resources in the Extraction of Zinc", IMM Symposium on Advances in Extractive Metallurgy, 1977, pp. 153-160.

(39)   Anon, "Zinc Leach Residues - A Mine on the Doorstep", Metal Bulletin Monthly, April 1971, pp. 6-10.

(40)   P. Druelle, "The Precipitation of Jarosite and its Application in the Hydrometallurgy of Zinc", Metallurgie (XVI) (3) (1976) pp. 148-153.

(41)   K. Mager, "Technical and Economic Aspects of the Treatment of Leach Residues of Zinc Electrolysis", Ertzmetall (29) (5) 1976 pp. 224-229.

(42)   J. Perlinski, "A Survey of the Methods of Treatment of Zinc Bearing Slimes", Rudy i Metale Niezclazne (17) (9) (1972) pp. 441-443.

(43)   F.J. Sitges and V. Arregui, "Process of Recuperation of Zinc from Ferrites", Spanish Patent No. 385,575 lodged November 14, 1970.

(44)   G. Steintveit, "A Process for the Treatment of Jarosite Residues", Norwegian Patent No. 123,248 lodged October 2, 1969.

(45)   K.H. Heino, R.T. McAndrew, and N.E. Ghatas, "Process for Leaching Zinc from Zinc Calcine", Canadian Patent No. 900,180 lodged April 23, 1970 (assigned to Electrolytic Zinc Company of Australasia Limited).

(46)   A. von Roepenack and H. Wuetrich, "Process for the Recovery of Zinc from Zinc and Iron-containing Materials", German Patent No. 1,948,411 lodged September 25, 1969 (assigned to Det Norske Zinkkompani A/S).

(47)   F.J. Sitges and Vicente Arregui, "Process for Recovering Zinc from Ferrites", Spanish Patent No. 407,811 lodged October 20, 1972.

(48)   J.K. Rastas, T.L.J. Huggare, and S.P. Fugleberg, "Hydrometall-urgical Process for the Recovery of Zinc, Copper and Cadmium from their Ferrites", Finnish Patent Application A10/73, lodged February 12, 1973.

(49)   "Process for the Preparation of an Aqueous Zinc Sulphate Solution for Manufacture of Electrolytic Zinc", Holland - Patent No. 77,05374 lodged May 16, 1977.

(50)   M.F. DeGuire, A.L. Hannaford, and L. Harris, "Electrolytic Zinc Recovery from Manganiferous Concentrate", World Mining (31) (10) September 1978, pp. 92-99.

(51) E. Peters, "Process for the Treatment of Complex Lead-Zinc Concentrates", United States Patent No. 4,063,933 (Canada 256,145 lodged July 2, 1976).

(52) Sherritt Gordon Mines Limited and Cominco Limited, "Process for the Treatment of Metal Sulphides for the Recovery of Non-Ferrous Metals", Belgian Patent No. 859,857 (Canada 285,127, lodged August 19, 1977). "Process for the Purification of Iron as Jarosite", Belgian Patent No. 869,856 (Canada 285,090 lodged August 19, 1977).

(53) L. Landucci et al, "Treatment of Zinc Plant Residue", United States Patent No. 3,976,743, (United Kingdom 40112/74 lodged September 13, 1974).

(54) International Nickel Company of Canada Limited, "Improvements in the Separation of Iron from Solutions or Pulps of Zinc", Belgian Patent No. 828,218 (Canada 207,567 lodged August 22, 1974).

(55) L.S. Getskin, A.S. Yaroslavtsev, V.M. Piskunov, V.A. Grebenyuk, A.U. Usenov, B.K. Plastinkin, Y.M. Mironov, B.A. Sysoev, "A Method of Processing Oxidised Materials containing Zinc and Iron", U.S.S.R. Certificate of Authorship (21) 1884964/02, lodged February 16, 1973.

(56) L. Harris and R.D. MacDonald, "Process for Removing Ammonium Sodium or Potassium Ions from Metal Sulphate Bearing Solutions", South African Patent No. 773,869 - U.S.A. 703,189 lodged July 7, 1976.

(57) M.J. Meixner, "The San Telmo Process - Hydrometallurgical Procedure for the Recovery of Copper and Zinc from Sulphide Ores", Erzmetall, 30 (5) 1977, pp. 204-208.

(58) "The Nitric Sulfuric Process for Recovery of Copper from Concentrate", Paper presented at AIME Annual Meeting, Denver 1978.

(59) H. Martens and K. Wolf, "Iron Removal from Zinc Sulphate Solution by Two Stage Precipitation at a Controlled Precipitation and Addition of Neutralising Agents", D.D.R. Patent No. 125,827 published May 18, 1977 (lodged March 26, 1976).

(60) Asturiana de Zinc S.A., Spanish Patent No. 411,058, lodged January 27, 1973.

(61) Annual Report of Societe Generale Belgique 1977, p. 89.

(62) F.J.J. Bodson, "Process for Recovering Silver and Possibly Gold from a Solid Starting Material Containing Said Metals", Belgian Patent No. 847,441, lodged October 19, 1976.

(63) T. Suzuki, H. Uchida, R. Pickering and I. Matthew, "Electric Smelting of Lead Sulphate Residues", TMS-AIME Paper A-77-21, 1977.

(64)   G. Steintveit, T. Lindstad, and J.K. Tuset, "Smelting of Lead-
         Silver Residue and Jarosite Precipitate", IMM Symposium on
         Advances in Extractive Metallurgy, London 1977.

(65)   Outokumpu Oy, "Process for the Purification of Residues Containing
         Basic Sulphates, Hydroxides of Iron and Ferrite of Zinc". Belgian
         Patent No. 779,613 lodged February 21, 1972 (application in
         Finland No. 508/71 February 22, 1971).

(66)   Societe de Prayon S.A., "Iron Oxide Based Pigment and Process for
         Preparing Such a Pigment", Australian Patent Application No.
         87973/75.

(67)   Frank Lawson, "Calcination of Basic Ferric Sulphates", South
         African Patent Application No. 762,867 (Australian Patent Application
         No. PC.1584, May 14, 1975) assigned to Monash University.

(68)   V. Karoleva, G. Georgieo, and N. Spasov, "Dissociation of
         Potassium Sodium and Ammonium Jarosites", Thermal Analysis Vol. 2
         - Proceedings of the Fourth ICTA Budapest 1974 pp. 601-610.

(69)   W. Kunda and H. Veltman, "Decomposition of Jarosite", Paper
         presented at AIME Denver Meeting 1978.

(70)   P. Druelle and A. Van Ceulen, "A Process for Treating Iron Rich
         Residues of Industrial Origin Resulting from the Hydro-Metallurgical
         Processing of Ores", Australian Patent Application No. 21017/17
         (France January 9, 1976).

(71)   J. Rastas, "Process for Preparation of a Raw Material suitable for
         Iron Production", United States Patent No. 3,910,784 (Finland
         Application No. 294/73, February 1, 1973).

(72)   Preussag A.G. Metall, "Process for the Preparation of Ferricginous
         Residues from Electrolytic Zinc by Aqueous Treatment Under Pressure",
         Belgian Patent No. 855,541.

(73)   Noel Druelle, "Process for Treatment of Basic Sulphate Residues
         from the Sulphuric Acid Treatment of Minerals", French Patent No.
         2,278,626 lodged July 15, 1974.

(74)   A. Van Ceulen, "Process for the Recovery of Metal Chlorides from
         Jarosite Sludges Resulting from the Sulphuric Acid Processing of
         Ores", United States Patent No. 4,070,437 (France Application No.
         73/00293 January 7, 1975).

(75)   Compagnie Royale Asturienne des Mines and Commissariat L'Energie
         Atomique, "Process for the Treatment of Electrolytic Zinc Residues
         by Recovery of Contained Metals", Belgian Patent No. 846,588
         (France application September 26, 1975 No. 75,29484).

(76)   D.A. Jackson, "Process of Treating Electrolytic Zinc Refining Jarosite Residues", United States Patent No. 3,871,589 lodged November 19, 1973. "Electrolytic Zinc Refining Process including Production of By Products from Jarosite Residues", United States Patent No. 3,937,658 lodged December 24, 1974.

(77)   G.R. Hogstrom, J.L. Stroehleln, and J. Ryan, "Sulphuric Acid plus Mining Residue equals Promising Iron Fertiliser Material", Sulphur Institute Journal, Summer 1976, pp. 12-15.

(78)   Shell International Research, "Regeneration of Mineral Sludges containing Soluble Salts", German Patent No. 2634735 (France 7524233 lodged August 4, 1975).

(79)   W.E. Poornet (Societe de Prayon), "Process for Stabilizing and Consolidating Residues Comprising Metal Compounds", Australian Patent Application No. 19809 November 19, 1976.

(80)   G.A. Major, and I.G. Matthew, Australian Patent No. 429,078, lodged May 25, 1970.

(81)   P.G. Hall, A. Johnson, and I.G. Matthew, "The Development and Application of the Selective Zinc Precipitation Process for Controlling Impurities in Electrolytic Zinc Plant Circuits", Aust. Inst. of Mining & Metallurgy Conference, Tasmania, May 1977, pp. 299-307.

(82)   I.G. Matthew, O.M.G. Newman, and D.J. Palmer, "Water Balance and Magnesium Control in Electrolytic Zinc Plants Using the EZ Selective Zinc Precipitation Process", AIME Annual Meeting, 1979.

(83)   Electrolytic Zinc Company of Australasia Limited, "Process for the Precipitation of Iron as Jarosite from Sulphate Solutions", Belgian Patent No. 866,866, (Australian Patent Application No. PD.0030 lodged May 9, 1977).

(84)   L.S. Getskin, V.A. Grebenyuk, A.S. Yaroslavtsev, A.U. Usenov, and Yu.M. Mironov, "Study and Industrial Assimilation of Hydrometallurgical Processing for Zinc Cake", Tsvetny Metally, February 1976 pp. 17-19 (English Translation February 1978 pp. 17-18).

(85)   L.S. Getskin, A.S. Yaroslavtsev, V.M. Piskunov, A.U. Usenov, Yu.M. Mironov, "Study of Hydrometallurgical Processing of Zinc Cakes", Sb. Nauchn. Tr. Vses. Nauchno-Issled Gornometall. Inst. Tsvet. Met. 1975, 25, pp. 218-222.

(86)   H. Martens and K. Wolf, "Process for the Removal of Iron from Acidic Zinc Sulphate Solutions", D.D.R. Patent No. 126074, published June 15, 1977 (lodged June 6, 1976).

(87)   R.W. Pickering, and M.G. Kershaw, "The Jarosite Process - Phase Equilibria", Paper presented at AIME Annual Meeting, 1980.

THE ELECTROLYTIC ZINC PLANT OF JERSEY MINIERE ZINC COMPANY

AT CLARKSVILLE, TENNESSEE

Lewis A. Painter
Technical Vice President
Jersey Miniere Zinc Company

Jersey Miniere Zinc Company's 81,670 metric ton per year electrolytic zinc plant at Clarksville, Tennessee went on stream on November 6, 1978 with feed of concentrates to the roaster. Full production of the plant was achieved on January 24, 1979.

The roasting and acid plant facilities were designed and constructed by American Lurgi. The hydrometallurgical and electrometallurgical facilities are based on technology of Vieille-Montagne of Belgium, on basic engineering by Mechim of Belgium and on detailed engineering by Kaiser Engineers. Construction was performed by Daniel Construction Company.

The facilities are completely new as there was no prior operation at the site, consequently, all services had to be provided.

The process was designed for treatment of low iron content concentrates from company owned mines, consequently, no residue treatment was used. The plant is highly instrumented and automated, including centralized processs control, automated filtration of purification solutions and automated cathode handling.

Introduction

The Jersey Miniere Zinc Company was formed as a joint mining and refining venture between The New Jersey Zinc Company, a wholly owned subsidiary of Gulf + Western Industries, Inc., and Union Zinc Company, a wholly
owned subsidiary of Union Miniere, S. A. of Brussels, Belgium.

Mining properties located in the general area of Carthage, Tennessee
(about 50 air miles - 31 kilometers, east of Nashville, Tennessee) were
discovered and partially developed by The New Jersey Zinc Company prior
to the formation of Jersey Miniere Zinc Company. Concentrates from these
low iron zinc ores were to provide the feed for the electrolytic zinc refinery projected for construction at Clarksville, Tennessee located on the
Cumberland River, about 40 miles (25 kilometers) northwest of Nashville.
The design of the electrolytic refinery was based on the receipt of these
concentrates.

The technology for the 90,000 short tons (81,670 metric tons) per year
plant was obtained primarily from Belgium, Germany, Norway and the United
States. Vieille-Montagne of Belgium provided basic technology for the
hydrometallurgical and electrometallurgical sections of the plant with
Mechim, also of Belgium, providing basic engineering for the Vieille-
Montagne technology. Lurgi of Germany, through American Lurgi, provided
technology, engineering and construction services for the roasting plant
and the acid plant. Norzink of Norway provided technology for the mercury
removal section of Lurgi's roasting and acid plant. Kaiser Engineers of
Oakland, California provided detailed engineering, site services and overall project managment. Auxiliaries were a part of Kaiser Engineers' engineering obligation. Construction services were provided by Daniel International Corporation of Greenville, South Carolina.

Ground was broken on July 8, 1976 and the first concentrate was fed
to the roaster on November 6, 1978. Feed of roasted zinc concentrates to
the leaching plant commenced on November 12, 1978 and electrolyzing power
was applied to the tank house on November 23, 1978. Full power, at 375
amperes per square meter, was applied to all circuits in the tank house on
January 24, 1979.

Prior to plant start-up, roasting tests were performed by the Lurgi
Company in Frankfurt, Germany and the main hydrometallurgical and electrometallurgical processes were tested in a pilot plant at G+W's research
facility at Palmerton, Pennsylvania. The pilot plant at Palmerton was
dismantled and reassembled at the Clarksville site where it was placed in
service and verified the first purified solution before it was fed to the
tank house. As a result of these measures, and of very exacting initial
process controls, the first metal produced was at good current efficiency
and of SHG quality.

Plant Description

Concentrate Receiving

Although the basic design of the plant was based on the receipt of
concentrates by barge from company owned mines, facilities were included
to unload concentrates by either truck or railroad cars. The railroad
unloading facility has the capability of handling bottom dump, flat
bottom gondolas or box cars. As the company mines are not yet supplying
the full refinery feed, the railroad and truck unloading facilities

125

Figure 1

Figure 1

Plant View Legend

1. Concentrate Sampling and Transfer Tower
2. Concentrate Storage Building
3. Concentrate Screening and Tranfer Tower
4. Water Treatment Plant. Producing Process, Potable & Deionized Water
5. Sulfuric Acid Storage Tanks for 98% Acid, 98% Acid and Oleum
6. Railroad and Truck Unloading Building
7. Auxiliary Boiler and Deionized Water Storage Tank
8. Roaster
9. Gas Cleaning, Including Mercury Treatment
10. Acid Plant and 200' Stack
11. Cooling Towers (4 cells for R & A, 2 cells for Zinc Refining)
12. Calcine Storage Silos
13. Neutral Leach Thickener
14. Lime Unloading, Storage, Slaking and Holding Tanks
15. Weak Acid Leach Thickeners
16. Leach Filter Building for Rotary Vacuum Filters and Auxiliaries
17. Air Compressor Building
18. Maintenance and Warehouse
19. Waste Treatment Plant
20. Leach Residue Pond
21. Recyclable Pond
22. Non-Recyclable Pond
23. Purification
24. Central Control Building
25. Cadmium Plant (Rated at 357 metric tons per year)
26. Central Laboratory
27. Cell Room Including Anode Cleaning and Casting
28. Sanitary Waste Treatment Plant
29. Metal Services Building, Including Dross Handling and Zinc Dust Production
30. Electrical Switching Yard - TVA Area
31. Electrical Switching Yard - JMZ Area
32. Change House and Industrial Relations Buildings

handle an important portion of the concentrate receipts.

The barge unloading facility is located on the Cumberland River about one-half mile from the main plant site. The stationary bucket unloading unit is designed for an unloading rate of 300 short tons (272 metric tons) per hour. Concentrate, hoisted from the barge, is dumped into a 55 cubic yard (42 cubic meters) feed bin equipped with dust control equipment. A 36 inch (.91 meter) overland conveyor belt, equipped with a belt magnet and a scale, transports the concentrate through a transfer and sampling building to the concentrate storage building. The overland belt conveyor is also equipped with a turnover device to insure that there is no concentrate loss over the half mile of flood plain which extends from the river to the railroad embarkment at the main plant site.

Sampling equipment for the barge unloading system consists of a primary and secondary sample cutter. There is no automatic sampling system for concentrates delivered by rail or truck as this type of shipment was originally felt to be of very small magnitude.

Incoming concentrates are stored in a 200 feet by 128 feet (61 meters by 39 meters) concentrate storage building equipped with two 25,000 CFM (42,500 m^3/hr.) bag houses. The storage area is divided into three bins, one of 7,500 short tons (6,806 metric tons) and two of 3,750 short tons (3,403 metric tons) capacity. Under special conditions, using front end loader stacking, the building has housed 25,000 short tons (22,686 metric tons) of concentrate. Concentrates are distributed to selected bins by a 30 inch (0.76 meter) tripper conveyor.

Blending of concentrates for transportation to the roaster feed bin is accomplished with front end loaders of 3 cubic yards (2.3 cubic meters) capacity. Concentrate is dumped by the loader in a 6 cubic yard (4.6 cubic meter) feed hopper from which it is transferred to a 30 inch (0.76 meters) belt conveyor system, which transports the concentrates to a transfer and screening tower ahead of the roaster feed bin. A twin deck screen in the transfer tower separates the two inch trash for discard and 3/4 inch (1.9 centimeter) concentrate lumps for size reduction.

Roasting and Acid Plants

The Lurgi roasting plant is single train and is designed to treat 466 short tons (423 metric tons) per day of concentrate and recycled dross, producing 400 short tons (363 metric tons) per day of calcine, containing less than 1.5% sulfates and 0.25% sulfides. The Lurgi acid plant, also single train, is designed to produce 500 short tons (454 metric tons) per day of 100% sulfuric acid as 95%, 98% or oleum with stack gases below 300 ppm SO_2. Included in the acid plant is a mercury removal plant designed to produce sulfuric acid containing less than 1 ppm of mercury. The combined roasting and acid plant is designed to produce the tonnages noted above on a normal operating day assuming 335 operating days per year.

Actual average production rates required to support the refinery on a 365 day per year basis are 421 short tons (382 metric tons) per day concentrates, 367 short tons (333 metric tons) of calcine and 377 short tons (338 metric tons) of 100% sulfuric acid.

Concentrates are received from concentrate storage in a 500 short ton (454 metric ton) plastic lined feed bin. Concentrate from the bin is metered and weighed by a Merrick belt scale before discharge to a table feeder, which in turn, discharges into two 16" belt slingers. Normal operation involves the use of both slingers to provide optimum distribution although

128

Table I

Concentrate Analysis

Element	Design	Typical Mix
Zn	62.0%	62.0%
Pb	0.47	0.31
Fe	1.5	1.42
Cd	0.4	0.32
Cu	0.25	0.06
Cl	0.01	0.01
SO_2	0.48	0.55
Al_2O_3	0.09	0.09
CaO	0.52	1.28
MgO	0.42	0.68
As	0.0018	0.0018
Sb	0.0012	0.0012
Hg	20 ppm	6 ppm
Mn	$<$ 0.01	0.015
In	0.001	0.015
Sn	$<$ 0.002	0.003
Ge	0.04	0.015
Co	0.01	0.01
Ni	0.01	0.01
Ag	8.9 gm/MT	9.0 gm/MT
Au	0.34 gm/MT	$<$ 0.34 gm/MT

Table II

Typical Residue Analysis (Percent)

Element	Leach Residue	Copper Residue	Cobalt Residue
Zn Total	15.0	6.0	15.0
Zn Water Sol.	2.5	1.0	2.0
Cd	.26	5.0	5.0
Cu	.34	24.0	1.0
Pb	2.0	15.0	20.0
Fe	15.0	.05	.10
Sulfide S	1.5	--	--
Co	--	.001	.4
Ni	--	.2	.3
Ge	--	.04	.06
Sb	--	.05	.05

one may be used for extended periods of time in the event of a slinger failure.

The turbulent bed roaster has a grate area of 696 square feet (64.66 square meters). It is equipped with five bed coils, each with a surface area of 42.8 square feet (4 square meters). Due to the low heat value of the concentrates from Jersey Miniere Zinc Company mines, four of the five cooling coils were removed prior to start-up. More recently, the use of higher iron content concentrates has required the installation of an additional coil for a total of two now in service.

The roaster air blower, with a maximum capacity of 24,800 CFM (42,160 m^3/hr.), can be driven by either an electric motor or a steam turbine.

Calcine leaves the roaster either as bed overflow or carry-over into the waste heat boiler. The waste heat boiler cools the SO_2 bearing gas and its load of calcine, generating about 40,000 pounds per hour (18,160 kg per hour) of steam at 570 psi (39 atmospheres). The boiler tubes suspended in the waste heat boiler are automatically cleaned with an Oschatz type rapper. The calcine from both the roaster overflow and waste heat boiler is cooled to about 100°C in a rotary drum cooler.

Gas from the waste heat boiler is further cleaned in two parallel cyclones before entering an electrostatic precipitator where the dust load is reduced to less than 460 mg/m^3. Gases leaving the electrostatic precipitator contain about 9.5% SO_2.

All calcine from the roaster bed overflow, the waste heat boiler, the cyclones and electrostatic precipitator are passed through an airswept ball mill where the size is reduced to less than 120 microns. This ground calcine is transferred to the calcine storage silos by two Fuller Kinyon pneumatic conveying systems in parallel.

The gases from the electrostatic precipitator are scrubbed in a venturi type scrubber and further cooled and humidified in a Star, or internal vane, cooler. The clean, cool gas is then passed through two wet gas electrostatic precipitators in series before entering the Norzink mercury removal tower where mercury in the gas stream is reacted with a mercuric chloride solution to form insoluble mercurous chloride which is then settled, filtered and removed from the circuit.

The mercury free gas stream is dried with 93% sulfuric acid in a venturi drying tower before passing through the SO_2 blower ahead of the double catalysis, vanadium catalyst converter and heat exchanger system. There are four catalyst beds in the converter, each with a gas to gas heat exchanger to remove the heat generated by the SO_2 to SO_3 conversion and to heat the incoming cold gas feed before entering the converter. The SO_2 bearing gas is largely (about 90%) converted to SO_3 in passing through the first two catalyst beds and this gas stream goes through an oleum and intermediate absorption tower before re-entering the last two catalyst beds for final conversion. The gas, exhausted to the atmosphere through a 200 foot (61 meter) stack, contains less than 300 ppm of SO_2.

The three grades of acid, 93%, 98% and oleum, are produced from the drying tower, the final tower and the oleum tower respectively. The acid plant has the capability of producing up to 500 STPD (454 MTPD) of either 93% or 98% acid. Up to 150 STPD (136 MTPD) of this capacity can be as oleum. There are five acid storage tanks capable of storing about one month's production. Acid loading facilities are provided for truck, rail and barge.

ROASTER PLANT

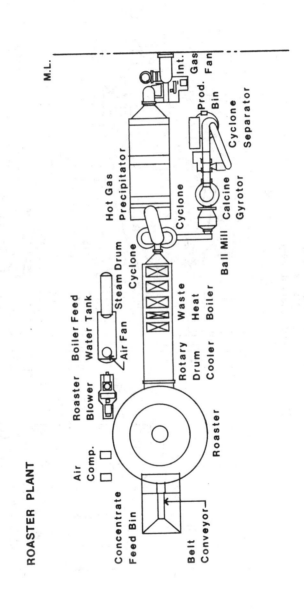

Concentrate Feed Bin

Air Comp.

Roaster Blower

Boiler Feed Water Tank

Air Fan

Steam Drum

Cyclone

Hot Gas Precipitator

Belt Conveyor

Roaster

Rotary Drum Cooler

Waste Heat Boiler

Cyclone

Ball Mill

Calcine Gyrotor

Int. Gas Fan

Prod. Bin

Cyclone Separator

M.L.

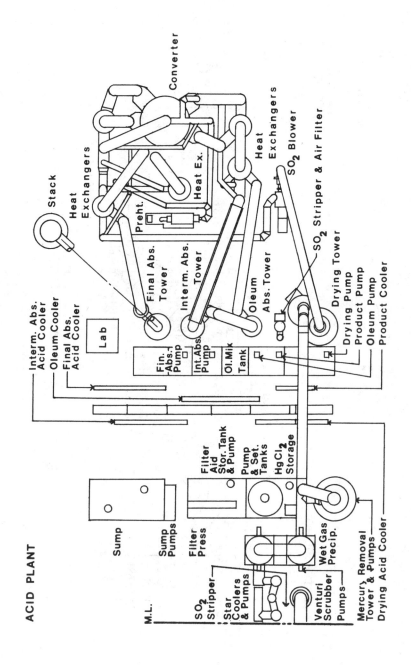

ACID PLANT

Converter

Heat Exchangers

Stack

Heat Exchangers

Heat Exchangers

SO₂ Blower

SO₂ Stripper & Air Filter

Preht.

Heat Ex.

Final Abs. Tower

Interm. Abs. Tower

Oleum Abs. Tower

Drying Tower

Lab

Interm. Abs. Acid Cooler

Oleum Cooler

Final Abs. Acid Cooler

Fin. Abs. Pump

Int. Abs. Pump

Ol. Mix Tank

Drying Pump

Product Pump

Oleum Pump

Product Cooler

Sump

Sump Pumps

Filter Press

Filter Aid Stor. Tank & Pump

Pump & Set. Tanks

HgCl₂ Storage

M.L.

SO₂ Stripper

Star Coolers & Pumps

Venturi Scrubber Pumps

Wet Gas Precip.

Mercury Removal Tower & Pumps

Drying Acid Cooler

Leaching

The leaching plant at the Clarksville refinery is designed for low iron concentrates; consequently, it does not include a residue treatment section. The flow sheet consists of a neutral leach which serves to bring most of the zinc into solution but finishes with some unreacted zinc oxide, thereby providing a purifying function. The residue from the neutral leach thickeners is then subjected to a weak acid leach ending with a pH of about 3.0.

The presence of higher than normal amounts of germanium in the design concentrate feed to the plant made it necessary to add ferrous sulfate and manganese dioxide to the neutral leach circuit in order to adequately precipitate the germanium, which is troublesome in the tank house. An addition of lime in the final stages of the weak acid leach is necessary in order to prevent excessive recycle of germanium to the neutral leach, while at the same time, providing an easily filtered thickener underflow.

Calcine from the roasting plant is stored in four concrete silos with storage capability of about three weeks at design rates of calcine usage. The calcine from the silos is moved by a Redler conveyor system to a feed bin at the head of the leaching circuit. Calcine is metered from the feed bin into one of two 30 cubic meter slurry tanks where it is mixed and partially reacted with spent electrolyte.

SIMPLIFIED LEACH CIRCUIT

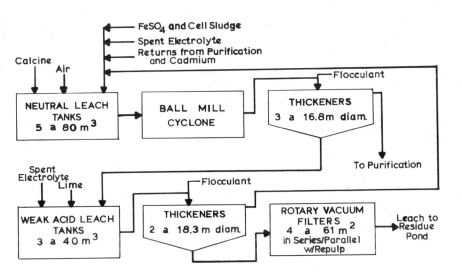

133

Overflow from the slurry tanks enters a series of five, 80 cubic meter leaching tanks maintained at a final pH of about 5.0 by addition of spent electrolyte. Sufficient ferric sulfate solution is added to maintain 1 gpl of iron in the leaching reaction. Oxidizing air is injected into each of the tanks and air is also added for solids transport at the internal air lifts.

Discharge from the final neutral leach tank is pumped through a cyclone into three, 55 foot (16.8 meter) diameter neutral leach thickeners arranged for either three thickeners operating in parallel or for the two thickeners operating in parallel feeding a final thickener. The cyclone underflow is ground to less than 75 microns in a Hardinge ball mill.

Neutral leach thickener overflow is sent by launder to the purification section. Neutral leach thickener underflow is pumped to three, 40 cubic meter weak acid leach tanks arranged in series. In these tanks, air is used only in the internal air lift systems. Spent electrolyte is added to the first tank of the series to complete the dissolution of the remaining zinc oxide. Lime slurry is added to the succeeding tanks to reprecipitate most of the germanium carried over in the neutral leach solids and redissolved in the first part of the weak acid leach.

The discharge from the weak acid leach tanks is pumped to two, 60 foot (18.3 meter) diameter thickeners in parallel. Thickener overflow is returned to the neutral leach circuit and thickener underflow is pumped to two, 660 square foot (61 square meter) area, stainless steel, rotary, vacuum filters in parallel. Cake from these filters is repulped, then filtered again in another set of two, stainless steel, rotary, vacuum filters also of 660 square feet of filter area. The final washed acid leach residue cake is then pumped to a Hypalon lined residue pond for storage or eventual sale. The residue pond is designed for about eight years of storage at full plant operating capacity.

Purification

The basic features of the purification section include: 1. The use of a water slurried zinc dust with 1% lead content; 2. The use of a cold or unheated bulk purification stage in which the zinc dust precipitates both cadmium and copper, followed by a hot, steam heated, purification in which zinc dust and antimony oxide precipitate the elements deleterious to the operation of the electrolytic cells; 3. The use of highly automated, pressure leaf filters; 4. The use of intensive releaching and filtering of both cold and hot purification residues for recovery of zinc and cadmium; and 5. The use of a final settling tank for protection against particulate carry-over prior to sending the purified solution to the tank house storage tanks.

Neutral leach thickener overflow is fed by way of a 120 cubic meter storage tank to two, 50 cubic meter, cold purification tanks. Slurried zinc dust containing about 1% lead is added to the first cold purification tank. Both copper and cadmium are precipitated in this stage and the slurry is pumped through a bank of three, stainless steel, 600 square feet (56 square meter) pressure leaf filters. These filters have a programmed cycle of recycling filtrate until a satisfactory precoat has been formed and solution clarity achieved, filtering until the filter cake build up is approaching a maximum, stopping filtration, sluicing off the filter cake and discharging it. The cycle is then repeated. The cake from these pressure filters is termed the copper-cadmium cake. It is subjected to further dewatering filtration on a small, 175 square feet (16 square meter) drum

SIMPLIFIED PURIFICATION CIRCUIT

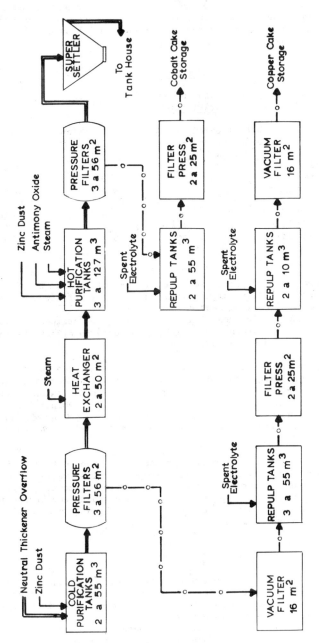

➤		MAIN PROCESS SOLUTION FLOW
•—•—○		RESIDUE FLOW
➤		OTHER FLOWS

filter, then repulped and leached with spent electrolyte in a series of three, 55 cubic meter, steam heated tanks which provide about a 24 hour reaction time. At the end of this reaction, most of the cadmium and zinc have been dissolved leaving a copper rich residue which is removed in a plate and frame filter press. The filtrate from the filter press is sent to the cadmium recovery plant. The cake from the filter press is repulped and re-leached in two small tanks with a reaction time of 8 hours to remove the last traces of cadmium. This slurry is then filtered on a 175 square feet drum filter and the recovered copper-rich filter cake is stored for sale.

Table III

Typical Solution Analysis
Purification

Chemical	Neutral Leach Thickener Over-flow	Cold Purification Filtrate	Purified Solution
Zn - grams per liter	150	150	150
Cd - milligrams per liter	500	10	0.2
Cu - milligrams per liter	0.20	0.5	0.2
Sb - milligrams per liter	0.15	0.01	0.01
Ni - milligrams per liter	0.5	N/A	0.01
Ge - milligrams per liter	0.3	0.1	0.2
Co - milligrams per liter	0.8	0.4	0.03

Filtrate from the cold purification automatic filters is stored in a holding tank from which it is pumped through two, 540 square feet (50 square meters) stainless steel, spiral heat exchangers to three, 127 cubic meter, hot, purification tanks. The reaction temperature is maintained at about 90°C. Slurried zinc dust and antimony oxide are added to effect the removal of nickel, cobalt and residual germanium.

Discharge from the hot purification is filtered in three automated pressure leaf filters of design identical to the cold purification filters. Cake from the hot purification filters is repulped and reacted with spent electrolyte in two, 55 cubic meter tanks at a pH of about 4.0 to solubilize the excess zinc dust without redissolving the impurities. The leached, hot purification cake is then filtered in a plate and frame press and the cake is stored for sale.

Filtrate from the hot purification presses is laundered to a 40 foot (12 meter) diameter, conical settling tank (super-settler) where particulates are settled out and returned to the purification cirucit. Overflow of the settling tank is acidified in a separate tank to reduce crystalization and plugging in the pipe lines and is then pumped to two cooling towers in the cell house area.

Cadmium Plant

Cadmium bearing solution from the main plant purification section is subjected to a zinc dust addition in two, 80 cubic meter tanks which serve as both cementation and wash tanks. The precipitated cadmium is settled and the solution decanted and returned to the zinc leaching circuit. The cadmium cake is subjected to two water washes before being manually removed from the tank and allowed to drain and partially oxidize for up to 24 hours.

SIMPLIFIED CADMIUM RECOVERY CIRCUIT

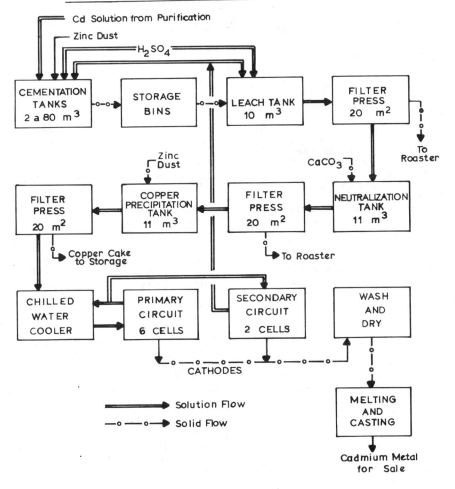

The washed and partially oxidized cadmium cake is then leached in a 10 cubic meter tank with concentrated sulfuric acid and spent electrolyte from the cadmium electrolysis cells. The resultant slurry is filtered on a plate and frame filter press. The press cake is then releached in a second 10 cubic meter tank, filtered through a plate and frame press and this final cake is returned to the concentrate storage bin for use as roaster feed.

The cadmium bearing filtrate, at about 180 gpl cadmium from the first cadmium sponge leaching tank is neutralized with limestone in an 11 cubic meter tank to remove traces of iron and arsenic. The slurry is filtered on a plate and frame filter with the cake being returned to the roaster feed and the filtrate being laundered to an 11 cubic meter copper precipitation tank. Zinc dust is added to the copper precipitation tank and the resultant slurry is filtered on a plate and frame press with the filtrate being laundered to a 26 metric ton cadmium purified solution storage tank.

The cadmium electrolysis section consists of eight concrete PVC lined cells of the same dimensions as the zinc electrolytic cells. Instead of the 48 cathodes and 49 anodes used in the zinc circuit, only 20 cathodes and 21 anodes are used in the cadmium circuit. The cells are operated at 60 amperes per square meter at full load. Six of the electrolytic cells are in the primary electrolysis circuit which maintains a cadmium concentration in the overflow of 50 gpl and an acidity of 130 gpl. External water coil cooling is used in the primary circuit. The two electrolytic cells in the spent circuit maintain a cadmium concentration in the overflow of 12 gpl and an acid concentration of 160 gpl. Cooling in the spent circuit is supplemented by cooling coils in a cell. All of the cooling water is chilled by refrigeration.

A 48 hour stripping cycle is employed. The stripped cadmium sheets are washed and dried before melting under resin in a two ton, gas fired, melting furnace. Molten cadmium is cast for shipment into stick and ball shapes.

Tank House

Purified solution at approximately 154 gpl zinc from the purification section is cooled to 30°C in one of two Hamon-Sobelco forced draft, fibre reinforced, plastic cooling towers of the spray feed type. The towers are of open construction with no packing. The cooled solution is stored in two 1500 cubic meter, purified solution tanks which provide for about 24 hours of storage. Purified solution from these storage tanks is fed to the main tank house distribution launder.

The tank house is provided with 208 concrete, PVC lined cells arranged in eight rows of 26 cells. Each cell is equipped with 48 aluminum cathodes of 2.6 square meter plating area and 49 lead-silver anodes. The anodes were continuously cast in such a way that the submerged portion contains a lead alloy with 0.75% silver, while the portion above the solution is chemical lead with no silver. The cell room is divided into two electrical circuits of four rows each. It is also divided into three plating circuits: a primary circuit of six rows, a secondary circuit of two rows and an intermittently used stripping circuit of six cells. Each plating circuit has its own cooling system consisting, in the cases of the primary and secondary circuits, of pumping the cell overflow at about 43 gpm per cell over a total of eleven Hamon-Sobelco forced draft cooling towers of the same design as the purified solution cooling towers. Normally, three of these cooling towers are in stand-by service. A circulating hot water cooling system is employed to clean the cooling towers and solution lines.

138

SIMPLIFIED ZINC ELECTROLYSIS FLOWSHEET

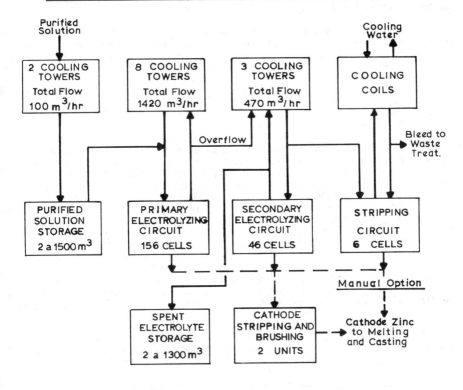

Typical Tank House Solution Analysis

Chemical	Primary Circuit	Secondary Circuit	Stripping Circuit
Zn - grams per liter	90	65	20
H_2SO_4 - grams per liter	128	175	244
MgO - grams per liter	20	20	20

In the case of the six cell stripping circuit, an external, water cooled, lead coil installation is employed for electrolyte cooling. Cell temperatures are maintained at 32°C.

Electrolyte feeding the primary cells is a mixture of purified solution and circulating spent electrolyte of about 96 gpl zinc and 121 gpl sulfuric acid. Electrolyte feeding the secondary cells contains about 71 gpl zinc and 169 gpl sulfuric acid and is made up of spent electrolyte from the primary circuit and secondary circuit recycled spent electrolyte. Spent electrolyte from the secondary circuit is returned to the leach plant at about 65 gpl zinc and 177 gpl sulfuric acid. The stripping circuit is used as required to remove magnesium from the plant solutions. In this circuit, the

139

zinc level of the bleed electrolyte is reduced to about 20 gpl and the sulfuric acid concentration is raised to 244 gpl before being discharged to the waste treatment plant.

Feed of electrolyte to each cell is by collapsible hose from an overhead launder with each cell row being fed by a single launder. The design current for the tank house is 375 amperes per square meter of cathode plating area but the name plate rating of the silicon rectifiers is 56,000 amperes which is equivalent to about 450 amps per square meter. At design current density, the plating cycle is 48 hours. This requires that one half of the tank house is stripped every day. Load optimizing equipment is employed to maintain a set plant electrical demand load by raising or lowering the tank house electrolyzing power consumption.

Cathode handling and stripping is done automatically by two overhead cranes, which serve alternate tank house rows, and two stripping and brushing units. The complete cycle encompasses a programmed series of operations: the lifting of 24 cathodes from an operating cell by an overhead crane, transporting the cathodes to a cathode car which carries the cathodes from the tank house area to the stripping area adjacent to the electrolyzing area, transferring the cathodes from the cathode car to the stripping machine by an overhead transfer crane. The stripping machine strips the zinc from the aluminum cathode by the action of a combination of chisel and separating knives. The zinc sheets are lowered and stacked for weighing on a conveyor beneath the stripping machines. The stripped aluminum cathode is automatically wire brushed, returned by transfer crane to a cathode car and from there by crane into an open position in an operating cell. The cathode handling and stripping operation is carried out by a programmable controller which can be manually over-ridden if necessary.

Cells and anodes are cleaned on a 50 day cycle with the anodes being removed for cleaning by automated high pressure water jets and straightened by a piston activated slapper.

Equipment is installed for melting and casting of the lead-silver anodes. All of the original anodes were manufactured with this equipment.

Metal Service (Melting, Casting, Dross Handling and Zinc Dust)

Stacks of cathode zinc weighing approximately 1.3 metric tons are moved by fork lift from the two cathode sheet conveyors beneath the cell room stripping machines or from a cathode zinc storage area in the metal casting building to automatic furnace charging equipment. This equipment consists of a hoist, which receives the stack of cathode sheets, raises them to the elevation of charging tables, moves them to a tilting table, which then charges the stack into two charging chutes above the single melting furnace.

The melting furnace is heated by four, 550 kW, air cooled inductors. Molten zinc flows from the furnace into a pump well from which it is pumped into the various casting operations by air operated hot metal pumps.

Metal may be pumped by way of launders to any one of the following operations:

1. Direct to the slab casting machine for casting special high grade slabs.

2. To an agitated, induction heated, two ton, 50 kW furnace pot where alloying agents such as lead and aluminum may be added continuously. The furnace overflows into a

launder feeding the slab casting machine.

3. To stationary, water cooled, 2400 pound (1,089 kilogram) jumbo molds.

4. To special shape casting.

5. To a lined ladle for transfering molten zinc to the zinc dust plant.

The 160 mold casting machine is capable of casting and stacking 18 metric tons per hour of 24 kilogram zinc slabs. Finished stacks of 40 slabs each are allowed to cool before banding.

The jumbo molds are water cooled and the tops of the 2400 pound (1,089 kilogram) jumbos are also cooled by a manual water addition.

METAL SERVICES AREA

The melting furnace is equipped with air-operated drossing doors. Ammonium chloride is added continuously along with the cathode sheets being charged to the furnace and by hand, as needed, into the melting furnace. Dross from the melting furnace is taken by dump carts to a dross recovery unit. The dump carts are hoisted and dumped across a grizzley into a cooling drum. Cooled dross from the cooling drum passes into an air swept ball mill.

Zinc oxide is swept out of the ball mill, trapped in a bag house, and then returned to the roaster. Zinc prills discharged from the ball mill are returned to the melting circuit.

Molten metal from the melting furnace is transported by fork lift in a refractory lined ladle to a hoist in the zinc dust area. The ladle is hoisted and the zinc poured into a 10 ton induction holding and melting furnace of 200 kW capacity. An air driven pump in the holding furnace pumps the metal into a circulating launder, which maintains a constant level of molten zinc in three gas-fired, graphite, blowing crucibles. A stream of molten zinc issuing from a small hole at the bottom of the crucibles is atomized with 90 psi (6.3 kg/cm^2) air. The atomized zinc is caught in an expansion chamber and is passed over vibrating screens to produce two grades of zinc dust (-75 microns and +75 - 200 microns) for use in the purification and cadmium sections of the plant.

Auxiliaries

The JMZ refinery at Clarksville is a green field plant and all services had to be provided. Power is received from the Tennessee Valley Authority at 161 KV. Water is pumped to a water treatment plant from the Cumberland River. A sanitary waste treatment plant is provided. There is no natural gas service and there are propane storage facilities provided.

The water treatment plant is capable of treating 900 gallons per minute (204 m^3/hr.) of river water and supplying the plant with process water, potable water and boiler feed water.

The waste treatment plant receives two types of waste streams: one capable of being treated and clarified and then returned to the plant as process water and one which must be treated to meet state effluent standards but which may not be returned to the plant as process water. The cell house bleed stream containing approximately 20 grams per liter of magnesium is an example of non-recyclable waste. Recyclable waste treatment consists of reaction with lime, pH adjustment with sulfuric acid, clarification and filtration. The non-recyclable system employs lime neutralization to pH 9.5 to 10.0, thickening and clarification in an 85 foot (26 meter) diameter thickener and pH adjustment with sulfuric acid. The pH adjusted thickener overflow is discharged to the river and the thickener underflow is pumped to the non-recyclable residue pond.

The machine shop (46 X 52 meters) is equipped to provide automotive repairs and servicing, instrumentation services, welding, machining and pipe fabrication. It is not designed to provide construction services nor is it designed to handle large scale repairs. A small engineering section is housed in the machine shop office.

The main warehouse (52 X 37 meters) is immediately adjacent to (really a part of) the machine shop building. Additional warehousing is provided in another unheated building on the site.

The laboratory building (52 X 15 meters) contains all of the facilities required for analysis of concentrates, calcine, liquid and solid process streams and final products. Spectrographic analysis for slab zinc production, under the supervision of the central laboratory, is carried out in a small room in the metal services building.

The administration building (52 X 15 meters) provides space for the general manager, the technical manager, accounting, purchasing, payroll and computer. The Jersey Miniere Zinc Company management offices are in Nashville.

Other buildings on the site are:

1. The central control building which contains a large
    motor control area on the first floor and the Foxboro
    Video Spec control system on the second floor. The
    control system monitors and controls all processes
    from the calcine bins through the cell room spent
    electrolyte and purified solution storage tanks. The
    roasting and acid plant, the cell room and the cadmium
    plant are under their own control systems.

2. The compressor room, which contains four 2500 SCFM
    compressors and an instrument air compressor.

3. An industrial relations and first aid building,
    (23 X 18 meters).

4. A guard house and emergency vehicle building.

5. A change house (31 X 21 meters).

6. A lubrication storage building.

7. A paint and plastic shop.

8. A stand-by boiler building housing a 60,000 pound
    (27,240 kg) of steam per hour, oil-fired boiler
    normally kept on the line and capable of handling
    all plant steam requirements in the event the roaster
    waste heat boiler is off the line. The roaster waste
    heat boiler has sufficient capacity to handle the
    plant load.

THE CANADIAN ELECTROLYTIC ZINC SULPHATE SOLUTION
PURIFICATION PROCESS AND OPERATING PRACTICE
A CASE STUDY

David D. Rodier
Assistant   Manager
Canadian Electrolytic Zinc Limited
Valleyfield, Quebec, Canada

Canadian Electrolytic Zinc Limited successfully developed and installed a continuous purification process in its plant as part of its recent 1975-76 expansion of capacity to 218,000 metric tons per year.

This paper will describe the new process, what objectives were established prior to its institution and how these objectives were achieved. The process is continuous using atomized zinc dust and antimony trioxide for cobalt removal. It produces a combined copper-cadmium-cobalt cake which is releached, resulting in a copper filter cake and a cadmium sulphate solution which is subsequently treated to recover cadmium.

Historical Background

Canadian Electrolytic Zinc (C.E. Zinc) was established in 1961 by five Canadian mining companies to treat the zinc concentrate that was produced by their newly developed zinc-copper mines in Northwestern Quebec and Ontario. Construction started in 1962 in Valleyfield and the original plant, having an annual capacity of 63,000 tons, was commissioned in 1963. In 1966, the plant's annual capacity was increased to 127,000 tons and included the expansion of both batch leaching and purification sections, an addition of one electrolytic circuit, two 32 m^2 Lurgi turbulent bed Roasters, the first installed in North America, and a Monsanto single absorption acid plant having 400 tons per day capacity.

The Leaching process of the plant was converted to a batch Jarosite process in 1970. The Burt pressure filters used to separate the leach residue from the Leach pulp, were replaced with thickeners in 1972. The last expansion, completed in 1976, raised C.E. Zinc's capacity to 218,000 tons per year and included such modern features as a third 72 m^2 Lurgi turbulent bed Roaster, a second Monsanto acid plant of 470 tons daily capacity, a continuous jarosite leaching section which utilizes the Outokumpu Oy Conversion Process, a continuous purification section (the subject of this paper), a new mechanized Cellhouse having an annual capacity of 100,000 tons of cathode and a third induction furnace in the casting section.

C.E. Zinc is currently ninety percent owned by Noranda Mines Limited and continues to be managed by Noranda. The remaining minority share is owned by Kerr Addison Mines Limited.

Process Objectives

The original purification process, consisting of a two stage standard arsenic trioxide, copper sulphate, zinc dust cementation batch process, was developed by Asarco's Corpus Christi's plant. Cobalt and copper were removed in the First Stage while cadmium removal occurred in the Second. This process involved two filtration steps, producing separate filter cakes, and the use of thirteen (13) 1.07 meter squared (42 x 42 inch) Shriver plate and frame filter presses. Note each filter press had a filter area of 64 square meters. Changing the purification process was considered an essential part of the 1975-76 expansion to achieve the following objectives:

1. Minimize the capital cost of process changeover by optimizing the use of existing equipment.

2. Effect the change without interrupting the operation of the existing plant.

3. Improve manpower productivity, especially in the area of filtration.

4. Increase the production rate by sixty percent (60%) without increasing zinc dust consumption.

5. Eliminate or at least drastically reduce the consumption of toxic reagents such as arsenic trioxide or antimony trioxide.

6. Improve the resultant copper cake quality.

7. Most importantly, for continuity of plant operation, establish a simple, non-selective, easy to control, reliable process which would guarantee the purified zinc sulphate solution ("Neutral") quality.

Original Batch Process Description

Before discussing C.E. Zinc's current process, it would be useful to include a brief description of the Corpus Christi process.

One of the five First Stage batch tanks would be filled with impure zinc sulphate solution, hereafter called Impure Feed, the steam coils of the tank would be turned on and the solution heated from 70°C to 90°C while the tank was being filled, which took approximately one hour. When the tank was half filled, a preliminary shot of zinc dust was added to cement the majority of the copper in solution. When the tank was filled, the first additions of copper sulphate and arsenic trioxide were made with a second shot of zinc dust. A half hour later, another shot of zinc dust was added and this repeated each 30 minutes, normally for a total of two and one half hours, until the cobalt was determined to be below 0.2 milligrams per litre using the Nitroso-R colour test. The batch was then pumped through a minimum of two filter presses, the filtrate going to a second stage tank and the filter cake to a storage bin and subsequently to the Noranda copper smelter.

At the start of a second stage batch, a small amount of copper sulphate (25 kilograms) was added. Zinc dust shots were added every half hour until the cadmium was determined to be out, using a qualitative test. The batch was then pumped through filter presses to the Check Tanks. Here, a sample was taken of the filtrate to be checked for cobalt, cadmium and solids. When these tests proved acceptable, the solution batch was pumped, via the neutral cooling towers, to the Neutral Storage tanks from where the solution was pumped, upon demand, to the Electrolysis section.

pH was controlled by spent electrolyte additions and checked manually.

The batch purification equipment included ten (10) process tanks with. agitators, eleven (11) filter presses in Purification and two (2) in cadmium cake leaching. This equipment was distributed as shown in the following Table I.

Table I. Batch Process Equipment

	First Stage	Second Stage	Check	Neutral Storage
TANKS				
Number	5	5	4	3
Diameter	9 m	9 m	9 m	14.8 m
Height	3.2 m	3.2 m	3.2 m	3.2 m
Shape	All cylindrical flat bottomed			
Capacity	200 m^3	200 m^3	200 m^3	550 m^3
Material	All wood stave – Douglas Fir			
AGITATORS	Lightning			
Hp	75	75		
rpm	45	45		
Type	Single turbine			
Material	All wetted parts 316L S.S.			
FILTERS				
Number	7	4		
Size (filter area m^2)	1.07 m (64)	1.07 m (64)		
Type	Shriver plate and frame			
Material	Everdur	Everdur plates, fiberglass frames.		

146

To complete the overall picture, the Return Cake leach section, that is the cadmium cake leaching section, included the following equipment:

Return Cake Leach Equipment

TANKS

Number	2
Diameter	5.8 m
Height	2.8 m
Shape	Cylindrical flat bottomed
Capacity	72 m^3
Material	316 S.S., lead lined, acid brick lined

AGITATORS

Hp	7
rpm	22
Type	sweep arm
Material	Everdur

FILTERS

Number	2
Size	1.07 m (42 inch) (64 m^2 filter area)
Type	Shriver plate and frame
Material	Everdur plates and Fiberglass frames

Canadian Electrolytic Zinc Continuous Process
Zinc Sulphate Solution Purification

Keeping in mind the objectives for the new process, it was decided to develop a continuous process which would maximize the utilization of the existing equipment, minimize zinc dust addition, use more productive liquid-solid separation techniques, produce a single combined cake of all impurities and reduce, to a minimum, the exposure of our employees to either arsenic or antimony.

Several months of bench scale and pilot plant scale testing of the process ensued, culminating in our current procedure. In general terms, the process consists of three stages; the first stage, using recycled zinc dust from the second and third stages, cements all of the copper and cadmium, and a significant portion of the cobalt in the Impure Feed in a combined cake which is settled from the solution in two 30 foot diameter thickeners. The thickener overflow solution, which has been, up to this point, maintained at 70 to 75°C is then heated with steam, using Alfa-Laval spiral heat exchangers, to 95°C while being pumped to the first tank of Second Stage. This stage removes the remaining cobalt in solution and is achieved in four tanks in series using antimony trioxide (Sb$_2$O$_3$) added as a slurry to the first tank. Atomized zinc dust is added to first, third and last tank of this stage. The solution containing the zinc dust-cobalt solids is pumped through a cyclone stand, the overflow going to the Third Stage, and the solid-rich underflow recycled to the First Stage. The Third Stage is essentially for final solution "clean-up" and serves mainly as a pumping stage for the final filtration. A small amount of zinc dust is added to prevent resolution of the cobalt, cadmium and copper. The pH is controlled at 3.6-3.8 to maintain filter rates. The solution filtration is performed using Shriver filters having Polypropylene plates and frames. The filter medium is Polypropylene cloth and Kraft paper. The filtrate is checked for impurities and pumped from the Check Tanks to the Neutral Storage tanks.

147

The filter cake is repulped using paddle mixers and repulp tanks, the solids being recycled to the First Stage.

The First Stage Thickener Underflow is pumped, using air lifts to a Pulp Storage Tank. This pulp is then filtered, using one of two Dorrco Vacuum Drum filters, the filtrate recycled to First Stage purification and the cake conveyed by screw conveyor to one of two Return Cake Leach tanks. The cake is leached in batches with a sulphuric acid content of 300 grams per litre and at 90 to 95°C. At the end of the batch, when the pulp has reached a pH 4.0, the cobalt in solution is recemented using arsenic trioxide, and if required, zinc dust. The overall batch time is up to six hours with the cobalt control step taking the last hour. The pulp is then pumped through Shriver filters, the filtrate transferred to the cadmium section and the cake dropped into the copper cake bin to await shipment to the copper smelter. Normally, three batches per day are sufficient to leach all of the purification solids.

(Please refer to Figure I)

Table II. Process Parameters –
C.E. Zinc Purification Process

First Stage

Temperature	72 ± 2 ° C
pH	~ 4.2 not controlled, as received from Leach section
Resident time	2 hours
Solid Loading	5 to 7 gpl
Flow Rate	250 to 300 m^3/hr

Second Stage

Temperature	95°C
pH	$3.85 \pm .15$
Resident time	4 hours
Solid Loading	3 to 5 gpl
Zinc dust addition	20–30 kg per minute
Sb_2O_3 addition	1.5 to 2.0 mg per 1 or g/m^3

Third Stage

Temperature	85 to 90 not controlled (° C)
pH	$3.75 \pm .15$
Resident time	45 minutes
Zinc dust addition	3 kg per minute

Return Cake Leach

Initial Acid	300 gpl
Final pH	4.0
As_2O_3 Addition	50 kg per batch
Zinc dust addition	100 kg per batch – variable

148

FIGURE I
CANADIAN ELECTROLYTIC ZINC
PURIFICATION-PROCESS

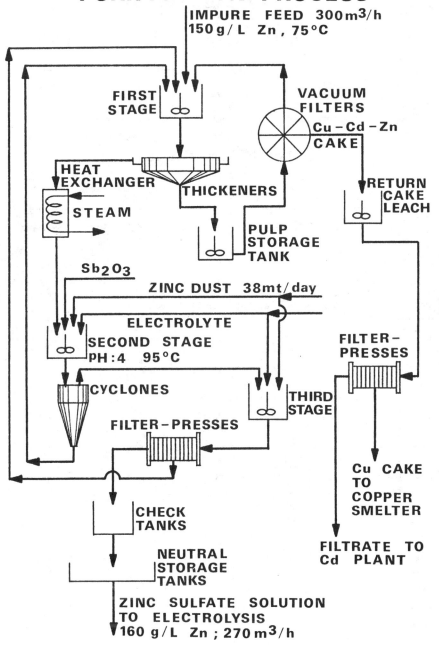

IMPURE FEED 300m³/h
150 g/L Zn, 75°C

FIRST
STAGE

VACUUM
FILTERS
Cu – Cd – Zn
CAKE

HEAT
EXCHANGER
STEAM

THICKENERS

RETURN
CAKE
LEACH

PULP
STORAGE
TANK

Sb₂O₃

ZINC DUST 38mt/day

ELECTROLYTE
SECOND STAGE
PH:4   95°C

FILTER-
PRESSES

CYCLONES

FILTER-PRESSES

THIRD
STAGE

Cu CAKE
TO
COPPER
SMELTER

CHECK
TANKS

NEUTRAL
STORAGE
TANKS

FILTRATE TO
Cd PLANT

ZINC SULFATE SOLUTION
TO ELECTROLYSIS
160 g/L Zn ; 270 m³/h

Table III. Process Typical Analyses

SOLUTIONS	Zn gpl	Cu mgpl	Cd mgpl	Co mgpl	Sb mgpl
Impure Feed	145	833	570	15	–
First Stage Overflow	148	< 1	<10	7	–
Second Stage Overflow	150	<.1	<.1	.1	–
Third Stage Filtrate	155	<.1	.2	.1	<.02

SOLIDS	%Zn	%Cu	%Cd	%As
1st Stage Thickener Underflow	45	7	6	–
2nd Stage Cyclone Underflow	75	2	2	–
3rd Stage Cake	85–90	2	2	–
Cu Cake after Return Cake Leaching	10–15	45–60	3–5	1

Table IV. Process Equipment

	First Stage	Thickener	Second Stage	Spare	Third Stage	Check	Neutral
Tanks and Thickeners	Tanks	Thickener	Tanks	Tanks	Tanks	Tanks	Tanks
Number (normally operating)	3 (2)	2 (2)	5 (4)	1 (0)	1	3	3
Diameter (meter)	9	9	9	9	9	9	14.8
Height (meter)	3-3.7	3	4.0-3	3.7	3.2	3.2	3.2
Cascading	Yes	Yes	Yes	Possible	No	No	No
Shape	All tanks flat bottomed cylinders						
Material –							
316 S.S.	1	2	3	1	–	–	–
Wood stave	2	–	2	–	1	3	3

Agitators

	First Stage	Thickener	Second Stage	Spare	Third Stage	Check	Neutral
Hp	75	5	2 @ 100 3 @ 75	1 @ 75	75	–	–
rpm	45	1/7	2 @ 78 3 @ 45	1 @ 45	45	–	–
Type	single turbine	rakes	single turbine	single turbine	single turbine	–	–
Material	316L	316L	316L	–	316L	–	–
Cyclones	–	–	14 (8-12)	–	–	–	–

Filters

	First Stage	Thickener	Second Stage	Spare	Third Stage	Check	Neutral
Number	–	–	5	–	–	–	–
Size (filter area m^2)	–	–	1.07 m (64)	–	–	–	–
Type	–	–	Shriver plate and frame				
Material			Polypropylene				
Paddle mixers			4				

Repulpers : 15 hp – 90 rpm – 22 m^3

Table V. Return Cake Leach Equipment

	Pulp Storage Tank	R.C.L. Tanks
Number	1	2
Diameter	5.8 m	5.8 m
Capacity	72 m^3	72 m^3
Shape	All flat bottomed cylindrical	–
Material	316 S.S.	304 S.S. – Lead plus acid brick lined.

Agitators

Hp	20	25
rpm	68	68
Type	sweep arm	single turbine
Material	316 S.S.	Alloy 20

	Pulp Filtrate	R.C.L. Filtration
Service		
No. (normally operating)	2 (1)	3 (2)
Type	Dorrco Vacuum Drum	Shriver Plate and Frame
Filter area m^2	9.1	64
Material	316 S.S.	1 Polypropylene recessed plate. 2 Everdur.

Process Start-up Problems and Subsequent Developments

1. Solution Transfer Hydraulics

The hydraulic design for the cascade system of tanks and launders was inadequate and at start-up would not permit operating rates in excess of sixty percent (60%) of design without solution overflowing the tanks and launders. Following re-evaluation of the system, the launder system was modified to provide greater hydraulic head differences between tanks in both the First and Second Stages. An interim "fix" was applied to the First Stage by installing 30.5 cm diameter siphons between the tanks to supplement the launders. This temporary solution could not be applied to the Second Stage because of the elevated temperatures of the solution. The changes in the launder system also resulted in the First Stage being pumped to the thickeners rather than flow by gravity as originally planned. It would be recommended that for any similar installation, the hydraulic design be verified experimentally or failing that, be quite conservative.

2. Thickener Operation

The feed to the parallel thickeners required control to equalize the distribution. When it became evident that this stage would be pumped, a distributor-surge box was installed with an overflow equipped with a coarse screen to catch any tank accretions which could otherwise block the thickener cone discharges. It was found that the original distributor box was too large and quickly filled with solids blocking the feed pipes. It was subsequently replaced with a smaller box having only the lump catching basket.

The solid settling also required improvement. This was achieved by adding small amounts of flocculent, Percol 156, a non-ionic polymer, to the thickener feed, installing feed wells in the thickeners to deflect the solution feed flow downwards and by improving the underflow pumping. The underflow was originally designed to use Allis Chalmers 2 x 2 SRL pumps. The excessive capacity of these pumps overloaded the Pulp Storage Tank with low gravity pulp. If the pumps were operated only intermittently, thickener cone blockages and dirty overflows resulted. These problems were solved by using a simple air lift arrangement controled by electronic timers which permit the operator to vary the frequency of pumping cycle thus maintaining the desired minimum underflow gravity of 1500. It was found that the key to good operation of the process is the maintenance of clear thickener overflows, otherwise the presence of copper and cadmium in the solution will hinder the cementation of cobalt. Normally, over half of the cobalt is removed in First Stage.

3. Cyclone Operation

The cyclo-pak installed to perform the liquid-solid separation between Second and Third stage consists of fourteen (14) cyclones arranged in a ring with a common feed pot and underflow discharge box. Initially, the cyclones were fabricated of rubber-lined mild steel and had adjustable underflow apexes. The linings and apexes failed with increasing frequency due to the cutting action of the sharp zinc dust particles and the high solution temperatures. They were therefore gradually changed to 316 Stainless Steel with fixed apexes. The rubber-lined header required changing after two years of operation when the rubber lining peeled from the mild steel. The underflow box required redesign to maximize flow rates and eliminate dead spots, where the chemically active solids would quickly cement, causing blockages. Ultimately, a by-pass line was installed to allow for servicing of the cyclo-pak.

Without the by-pass, production would be interrupted each time service would be required. Currently, eight to twelve (8 to 12) of the fourteen (14) cyclones are in service at any one time with the others on stand-by. The number used is related to plant flow rates and is controlled according to the pressure in the header.

4. Tank Agitation

During the start-up period, it was difficult to realize the objectives in zinc dust consumption. One of the most severe problems experienced was in the agitation of the tanks in Second Stage. While in the former batch process, the zinc dust additions were distributed throughout ten operating tanks, the new process required the bulk of the additions to occur in the first tank of Second Stage. The original agitator was inadequate resulting in rapid accumulation of zinc dust in the tank. This was solved by: installing more powerful agitators, having a higher horsepower and rpm in the first two tanks of Second Stage, changing the addition point of zinc dust from the tank perimeters to the centre by means of a screw conveyor, and by raising the steam coils up the sides of the tank from their original position at the bottom of the batch tanks where they tended to interfere with the agitation by providing an area for solid accumulation. All of these steps, coupled with the improved hydraulic design of risers and launders eliminated tank build-ups and concurrently lowered zinc dust consumption by making its use more effective.

5. Zinc Dust Addition

The original zinc dust feeders, open helical volumetric B.I.F. feeders, tended to block if any large particles of zinc dust either accumulated on their feed screen or the pipe in which the helical coil turned. The operator had no indication of the time elapsed of a zinc dust feed interruption and would haphazardly make large shot additions to compensate for what he judged to be the missing zinc dust feed. This problem was solved by installing a Wallace-Tiernan weigh belt which had a continuous chart recorder indicating the rate of zinc dust addition. This had two immediate results, the belt feeder could accommodate the occasional large piece of zinc dust without blocking and it provided the operator with a permanent record of zinc dust additions enabling him to compensate, if necessary, and thus optimize the zinc dust addition. This installation of the first weigh belt has been conservatively estimated to have reduced zinc dust consumption by one percent of cathode.

6. Third Stage Cake Handling

As originally installed, the paddle mixers could not endure the hard "tomb-stones" of zinc dust in the collected Third Stage filter presses. They required extensive modifications and are now sufficiently robust to provide over six months operation without repair. The repulp tanks had blockage problems requiring modification of their discharge feed lines to their pumps. It should be noted, however, that as the zinc dust consumption was lowered by all of the previously mentioned improvements, the whole question of solid handling simultaneously improved.

Process Reliability

There has only been one major incident of production interruption due to the process and that occurred in November 1976 when inadvertently, copper, in the form of filter cake particles, was passed to the Neutral Storage Tanks and subsequently to the Electrolysis. Since that time, no plant shutdown has been the result of the Purification, although occasionally a filtration problem may occur, due usually to a momentary upset in the Leach section. These occurrences result in a flow restriction of twenty-five percent (25%) for sixteen (16) hours duration, two to three times annually.

In addition to reliability, simplicity was one of the original objectives achieved by the process. Operating experience has shown that two operators, who control solution flows and conduct process tests, assisted by two filter press cleaners, can adequately operate the Purification section. The critical control points are First Stage Thickener overflow clarity, temperature and pH control in Second Stage and pH control in Third Stage. Second Stage zinc dust additions are adjusted to meet current cobalt levels. Filtration is critical only to assure that no solids pass through the filters to the Check Tanks.

An assistant Cadmium Operator operates the Return Cake Leach Section including the underflow pumping of the First Stage Thickeners. He is assisted in press cleaning of the R.C.L. presses by a cadmium labourer for no more than two hours per shift.

Safety and Hygiene

The original batch process consumed considerable quantities of arsenic trioxide which was shoveled by the operator into five batch tanks. This meant that there were five areas in which arsenic drums were stored.

In contrast to this, the current process adds antimony as a pumped slurry to the first tank of Second Stage. The slurry is prepared by weighing, in an enclosed glove box, the necessary antimony (3300 grams) which is dropped into a ventilated mixing tank filled with demineralized water and then, when thoroughly mixed, dropped by gravity into the pump tank below. The slurry is then pumped by a positive displacement pump to the first tank of Second Stage.

The arsenic continues to be added to two Return Cake Leach batch tanks which have forced ventilation. The work floor area of these tanks and the pump area below these tanks is continuously monitored by an automatic arsine detector.

In addition to this, all employees who manipulate either reagent are monitored by a quarterly Arsenic-Antimony medical protocol coordinated by the company Medical Director. These operators are also provided with a special set of lockers and clothes to ensure that they do not wear contaminated clothing while eating lunch or taking their breaks. Periodic educational safety meetings to review the dangers of using both reagents are conducted by the technical and supervisory staff, of the department, with the operators in this section.

Measurements of stack emissions from both the Second Stage purification and RCL tanks indicate that arsine levels are maintained below 200 micrograms per cubic meter in the tank ventilation stacks are usually below 36 micrograms per cubic meter.

Table VI. Results

Comparison of Batch and C.E.Z. Continuous Processes

	1973	1974	1978	(1st Half) 1979
Zinc Production	134,950	122,280	159,410	103,385
Impure Feed Solution				
Cu mgpl	350	450	602	570
Cd mgpl	460	493	758	833
Co mgpl	6.5	6.0	10.3	15.0
Leach Zinc Extraction	91.7	91.5	97.6	96.7
Consumption:				
Zinc Dust – tons	8,231	9,575	14,071	7,408
Percent of Cathode	6.6	7.1	8.2	6.7
Sb_2O_3 kg	–	–	3,410	2,510
Kg per ton of Cathode	–	–	.020	.022
As_2O_3 kg	129,942	130,059	37,606	42,000
Kg per ton of Cathode	.87	.98	.22	.37

	1973	1974	1978	(1st Half) 1979
CuSO$_4$ kg	497,284	346,232	–	–
Kg per ton of Cathode	3.3	2.6	–	–
Neutral Solution:				
Co mgpl	.11	.13	.17	.18
As/Sb mgpl	.02	.02	<.02	<.02
Copper Cake Analysis:				
% Cu	33.0	40.0	42.4	45.9
Cd	4.0	3.4	3.7	3.5
Zn	19.7	16.1	12.3	11.7
As	3.0	3.0	1.5	1

Discussion of Results and Conclusion

Referring to the original objectives and the results of Table VI one can see that all of the objectives were met. This can be summarized as follows:

1. Capital Cost was minimized as the new process, which although increasing capacity by sixty percent (60%) only required the installation of two 9 meter thickeners, a pump tank, three spiral heat exchangers, a cyclo-pak, four paddle mixers and two repulp tanks, launders, a 72 m^3 pulp storage tank and two Dorrco vacuum filters. Original equipment used were the tanks, filter presses and some of the pumps. No building extension was required.

2. There was no total interruption of operations by construction except for a period of 11 days where an extended wood stave tank failed.

3. Productivity was improved with the purification section employing only four men per shift versus the original five while increasing production by sixty percent.

4. Zinc dust consumption is currently at the same percentage of cathode produced as with the previous process in spite of the fact that cobalt levels are twice what they had been. The improved leach extractions are responsible for the current higher impurity levels of the Impure Feed solution.

5. Copper cake quality has improved markedly with a 5.9% increase in copper content and a 4.3% combined decrease in cadmium and zinc contents and a decrease of 60% in Arsenic content. Further improvement in this is expected with the conversion of all RCL presses to recessed plates.

6. The process has allowed C.E. Zinc to reduce the combined arsenic-antimony consumption by 55.2% based on a kilogram per ton of cathode basis. (Note 1978 showed a 72.4% decrease). In addition to this, the antimony is handled in a very secure manner and the arsenic is used in a very limited area monitored by a continuous arsine meter, developed by the Noranda Research Centre.

7. The process, in spite of some growing pains at start-up, has proven very reliable, easy to operate and has for all practical purposes eliminated the historical concern of Zinc Plant Operations of neutral solution quality maintenance. The process also has the advantage of being designed so that it is amenable to on-stream impurity analysis and fully automatic control, a benefit of its relative simplicity of control.

Acknowledgements

I wish to personally thank all those who made this paper possible and specifically thank Mr. N.E. Ghatas and M.R. Toivanen for their suggestions, the staff of Noranda Research Centre for their technical assistance and all those in Hydrometallurgical Operations who persevered during the demanding start-up period to make the C.E. Zinc process the success it is today. Special thanks are also in order to Mrs. G. Quenneville without whose help this paper would not have been possible.

References

Canadian Patent No. 1046288
Purification of Zinc Sulphate Solutions
Ghatas, N.E. - Noranda Mines Limited
Date of Issue - January 16, 1979

SOLUTION PURIFICATION AT THE KOKKOLA ZINC PLANT

Sigmund Fugleberg[1], Aimo Järvinen[2], Ville Sipilä[3]

[1] Senior Research Metallurgist
Metallurgical Research Centre
Outokumpu Oy, Pori, Finland

[2] Zinc Plant Metallurgist
Outokumpu Oy, Kokkola, Finland

[3] Superintendent Electrolytic Zinc Refinery
Outokumpu Oy, Kokkola, Finland

Abstract

The electrolytic zinc plant in Kokkola commenced operations in 1969 with a capacity of 90,000 t/a. In 1974 the zinc plant was expanded, and now the annual capacity is 160,000 tons.

At the beginning, normal hot arsenic-zinc dust purification was used for the removal of cobalt, and zinc powder purification for the removal of cadmium. The process was a batch process.

Immediately after the start-up, investigations of the purification of the zinc sulphate solution were instituted. As a result of these investigations, also the purification process was changed in connection with the expansion.

The solution purification comprises three stages. In the first stage, the copper is removed with zinc dust in a continuous process. The copper residue is separated in thickeners and the solution continues to the cobalt and nickel removal. These metals are removed with zinc dust and arsenic trioxide in an automatic batch process. The solution is filtered on filter presses, and the clear solution goes to the third stage in which the cadmium is removed in fluidized bed reactors using a zinc dust bed. The solid material is removed from the purified solution in hydrocyclones. After this, the solution is ready to be fed to the cell house.

As a result of the purification process, residues with high metal contents are produced (copper residue with 80 % of Cu, 0.2 % of Cd; cadmium residue with 90 % of Cd). With a raw solution containing 1.8 g/l of Cu, 0.42 g/l of Cd, 25 mg/l of Co and 23 mg/l of Ni, the consumption of zinc dust has been ca. 3.5 % of the cathode production.

Introduction

In the electrowinning of zinc it is of the utmost importance to have a pure solution in the electrolysis as certain elements, even if only amounting to micrograms per liter, may cause hydrogen evolution and redissolution of the zinc deposit. Thorough purification of the zinc sulfate solution before the electrolysis is therefore a must in the process, and it was not until reliable purification methods were developed that electrowinning of zinc was possible on a technical scale.

Much research work has been done in this field, and nowadays there are different possibilities of choosing a purification system that will give a sufficiently pure solution. All systems use cementation with zinc dust, possibly in connection with other reagents, such as arsenic trioxide (1), antimony trioxide or antimony metal (2), and thus the impurities are mostly separated from the solution as a metallic precipitate. For cobalt that cannot be cemented with zinc dust without additives, chemical precipitation as insoluble salt with α-nitroso- β-naphthol or xanthate is also used to some extent.

Though the primary purpose of the purification is to produce a pure solution and no other factors can be allowed to upset this, there are still possibilities for economical considerations when choosing the purification system. And as the impurities removed are valuable metals like copper, cadmium, cobalt and nickel, the market value of these elements in the precipitates produced, as well as the cost of zinc dust and other reagents, have to be taken into account.

Outokumpu has developed and patented (3, 4) a purification system in which also the economical aspects have been considered. In the process, the zinc dust consumption is brought to a minimum and the valuable metals can be obtained in a form that makes further refining easy at a low cost.

The Old Purification

When the zinc plant in Kokkola commenced operations in 1969, the solution purification used was the standard "hot arsenic-zinc dust" purification for Co removal followed by Cd removal with zinc dust only (5). The process was operated batchwise.

Cobalt Removal

In the beginning the concentrates treated were quite clean, and the raw solution contained only about 5 to 7 mg per liter of Co and 2 to 3 mg per liter of Ni. With these low impurity levels the system worked satisfactorily as far as a reliable production of pure solution was concerned. Later on when concentrates and different wastes with high Co and Ni contents had to be treated, the drawbacks of the system appeared more clearly.

The "hot arsenic-zinc dust" precipitation was carried out according to standard performance, i.e. the tank was filled up, As_2O_3 added, followed by zinc dust additions until the Co test showed less than 0.2 mg per liter. To get the Co removed, this performance demands quite high amounts of Cu to be coprecipitated; normally about 300 to 400 mg per liter of Cu is required. Such an amount of Cu is quite often the normal Cu content of zinc raw solutions, and so also in Kokkola, and therefore all Cu was precipitated in the Co removal step, even if the Cu content was some times higher. The result was that all the Cu in the solution came out of the process as a precipitate

158

containing 7 - 10 % of As, which in Kokkola made the value of Cu close to zero.

Another drawback of this Co removal is the quite high surplus of zinc dust that has to be used. Most of this surplus is leached according to the reaction:

$$Zn + 2 H_2O \rightarrow Zn^{++} + 2 OH^- + H_2 \tag{1}$$

and thus the solution becomes more basic. If the amount of zinc dust reacting according to the above reaction is high enough, the alkalinity of the solution increases until it reaches the point where $Zn(OH)_2$, or more correctly, basic zinc sulfates start to precipitate. At this point the zinc dust is passivated and the cementation reactions will cease. When the Co and Ni contents of the raw solution were high, a higher zinc dust surplus had to be added, and it happened quite frequently that the "passivation point" was reached before the Co was sufficiently removed. At this point further addition of zinc dust was useless. Acid has first to be added to lower the pH, and not until then could the zinc dust addition be continued. Such an operation was, besides being zinc dust consuming, also highly time-consuming and could cause loss of production.

In the "hot arsenic-zinc dust" purification the aim is not to precipitate Cd in this step. This can be attained as the high temperature and the presence of arsenic and other impurities lower the hydrogen overpotential on Cd to such a degree that precipitation is inhibited. The high zinc dust demand, which could be up to 4 g per liter led, however, to substantial coprecipitation of Cd, and in addition to this, the Zn content of the cake was high. Data on the old purification are given in Table I. These figures represent the average, but occasionally Cd could be up to 2 %, which represented a loss of about 10 %.

Problems with the redissolution of the precipitate were also encountered. This could occur if a batch had to stand some time before going to filtration, and it indicates that the Co compound formed was not very stable in water solution.

Cadmium Removal

After the Co removal step the slurry was filtered on filter presses. The solution went to the following step where Cd was cemented with zinc dust. This step was also batchwise operated, and when Cd in the solution was low enough, the batch was filtered on filter presses and the "Cd cake" was transferred to the Cd plant.

This operation needed also a high zinc dust surplus, which resulted in a precipitate with only about 20 to 25 % of Cd. The precipitate was very fine with a large surface area which easily caused redissolution of the cake.

The big reactors of 200 m^3 gave also problems with build-up of cake on the bottom and thus they had to be frequently cleaned.

Results from this step are also given in Table I. Fig. 1 shows the flowsheet of the old purification.

Table I. Data on the Old Purification System (January 1972)

Analyses of Solution	Unpurified ZnSO$_4$ Solution	Purified ZnSO$_4$ Solution
Zn	147 g/l	152 g/l
Cu	490 mg/l	0.1 mg/l
Cd	350 mg/l	<1.2 mg/l
Co	32 mg/l	0.4 mg/l
As		0.02 mg/l
Analyses of Solid Material	Waste from I Stage (Cu Precipitate)	Waste from II Stage (Cd Precipitate)
Zn	15 %	55 %
Cu	29 %	1 %
Cd	1 %	22 %

Chemical Consumption	I Stage	II Stage	III Stage (Continuous)
Zn Dust (Total Consumption 44.3 kg/t Zn)	26.1 kg/t Zn	17.7 kg/t Zn	0.5 kg/t Zn
As$_2$O$_3$	2 kg/t Zn		
CuSO$_4$		0.57 kg/t Zn	

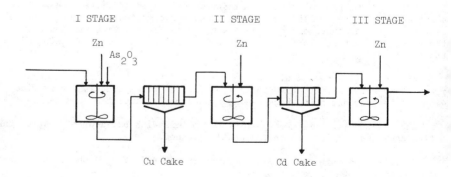

Figure 1. Old Purification Process

The New Purification

Review of Research Work

Right from the start-up of the plant, investigations on the purification of the raw solution were started. In 1974 when the plant was expanded to a capacity of 160,000 tons of zinc per year, the investigations had advanced so far that the purification was changed into a new system consisting of Cu removal, Co, Ni removal and Cd removal.

In spite of the many drawbacks of the old system, it had also some advantages compared with other possibilities. The greatest advantage was the fact that Cd was precipitated from a "clean" solution. This made it possible to improve the Cd cementation technique and gave a clean cake, which made the Cd plant simple. As this was achieved by using arsenic and as also arsenic proved to be quite a reliable additive in the removal of Co and Ni, it was decided to see what could be developed with arsenic as reagent in the Co cementation.

Studies were started to get better understanding of the mechanism of the Co and Ni precipitation using As_2O_3 and zinc dust, with special emphasis on the role of Cu in the cementation. If the amount of Cu in this step could be lowered, it would be possible to precipitate at least some of the Cu contained in the raw solution in a preceding step, and thus a pure copper product without arsenic could be obtained.

The first results showed clearly that a high Cu content was beneficial to the Co cementation and that it was not possible to get Co precipitated completely from a Cu-free solution or a solution with low Cu using the "old" procedure. This can be seen from Figure 2.

Co and Ni were precipitated in tests with higher Cu but in the cementation residue only metallic Cu and Cu_3As, but no Co or Ni compounds, could be identified by X-ray diffraction. As Co and Ni could be precipitated only

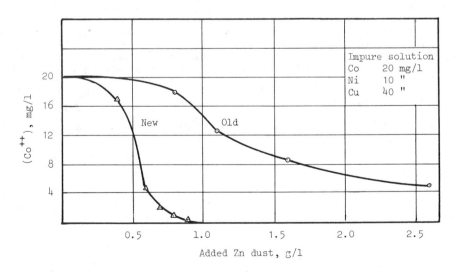

Figure 2. Cobalt Cementation Tests Comparing the Old and the New Procedure Used in Kokkola

in the presence of both Cu and As, it was assumed that Co and Ni were built into the Cu$_3$As precipitate. In this way the benefit of high Cu could be explained as coprecipitation of Co and Ni with Cu$_3$As and thus a high amount of Cu could seem unavoidable. When examining the precipitates by microprobe in order to locate Co and Ni in the Cu$_3$As, it was surprisingly found that Co and Ni were not mixed with Cu compounds to a great extent but that Co and Ni were in separate compounds with As and that this compound was substantially Cu-free. The Co, Ni, As particles were large and well defined also in a precipitate containing as much as 60 % of Cu and only about 2 % of Co and 1 % of Ni. This led to the conclusion that Co (and Ni) are separated out as a pure As compound, which later on has proved to be CoAs (or quite close to this stoichiometric compound), and thus Cu could not be inevitable in the Co cementation.

These findings were used in a new Co cementation procedure which showed that Co could be successfully precipitated with no Cu at all in the solution. Later results showed, however, that small amounts of Cu in the solution are beneficial. This amount is, however, negligible compared with the old system; the same concentration of Cu as of Co will give very satisfactory results. The result of a cementation run of Co according to the new procedure can be seen from Figure 2.

With the new procedure, all Cu in excess of about 50 mg per liter can be removed prior to the Co removal. This means that about 90 % of the Cu can be obtained as a "clean" Cu precipitate in normal solutions.

The development work on the Cd removal started from the fact that cementation of Cd on Zn dust from the pure solution after the Co removal was accompanied by very little side reactions, such as H$_2$ evolution. Therefore this simple cementation system ought to be ideal to run in a fluidized bed where the solution is run through a bed of Zn dust.

The main problem when developing this system was to solve how to maintain the fluidization of the bed. When Cd is cemented on Zn, it shows a high tendency to form dendrites (6), and when these have grown to a certain extent, the zinc grains with the attached dendrites will easily stick to one another. At this stage agglomeration will start and result in a consistency of the bed comparable with wet snow, and no fluidization is possible. To overcome this, the Cd has to be precipitated as a smooth and dense layer on the zinc grains. This is achieved if the precipitation rate per unit of surface area is kept sufficiently low (7), the effect being the same as in electrodeposition with low current density.

From the rate formula

$$\frac{dc}{dt} = k \cdot A_{Zn} \cdot C \qquad \text{or rearranged}$$

$$\frac{dc/dt}{A_{Zn}} = k \cdot C \qquad \text{where}$$

dc/dt = the cementation rate of Cd

k = the rate constant

A_{Zn} = the surface area of zinc dust present

C = the Cd concentration in bulk solution

it can be seen that the Cd concentration in the bulk solution must be kept low to achieve a low "precipitation density" (= $\frac{dc/dt}{AZn}$). In the system developed, this is obtained by having a large amount of zinc dust in the first bed and maintaining a good agitation of the bed. Under these conditions Cd is deposited as a dense, smooth and well adhering layer on the surface of the zinc particles and the bed is kept well dispersed. Pictures of the deposit are shown in Figure 3. It appears clearly that practically all Cd in the layer is formed around the zinc grains and very little finely divided material can be seen.

Although the zinc dust is covered with a dense layer of Cd, this does not affect the rate of cementation. This is seen from Figure 4 giving the result of a test, where a solution containing 300 mg of Cd per liter was run through a bed of zinc dust. Samples were taken of the outcoming solution and of the bed material at different times. Not until the Cd content of the bed has reached about 80 % does the Cd content of the outcoming solution start to increase which indicates a lower rate of cementation. This decrease in the rate is probably not caused by the hindered diffusion through the Cd layer but by the fact that the finer Zn grains are completely consumed at this stage, and thus the total effective surface area is decreased.

Application in the Plant

Copper Removal

The raw solution is continuously pumped to a tank of 200 m^3 giving a retention time of about one hour. Zinc dust is added to lower the Cu^{++} content to about 50 - 80 mg/l, after which the solution flows to two thickeners of 9 m in diameter each (old purification reactors). Cement copper is removed as underflow and washed on a drum filter before it is sent to the copper smelter. Some is also used in the copper sulfate production for fertilizers. The quality of the precipitate is shown in Table II.

In this step copper is precipitated according to the following reactions:

$$Cu^{++} + Zn^{\circ} \rightarrow Cu^{\circ} + Zn^{++} \qquad (2)$$

$$Cu^{++} + Cu^{\circ} + H_2O \rightarrow Cu_2O + 2 H^+ \qquad (3)$$

The process can easily be run so that a high portion of Cu_2O is precipitated, which means that the zinc dust consumption in the step is low. The conditions are maintained so that only very small amounts of hydrogen gas are evolved, which is favourable for a low zinc dust consumption.

Also some chloride is removed from the solution according to the reactions:

$$Cu^{++} + Cu^{\circ} + 2 Cl^- \rightarrow 2 CuCl\downarrow \qquad (4)$$

$$Cu_2O + 2 HCl \rightarrow 2 CuCl\downarrow + H_2O \qquad (5)$$

If the copper content of the solution is high, the chloride removal may be substantial. The content of Cl in the cake has been about 0.2 %, but this will vary depending on the Cl and Cu levels in the solution.

Cobalt Removal

The overflow from the Cu thickener goes to the Co removal, which is still batch-operated, but all operations are fully automated so the need for man-

163

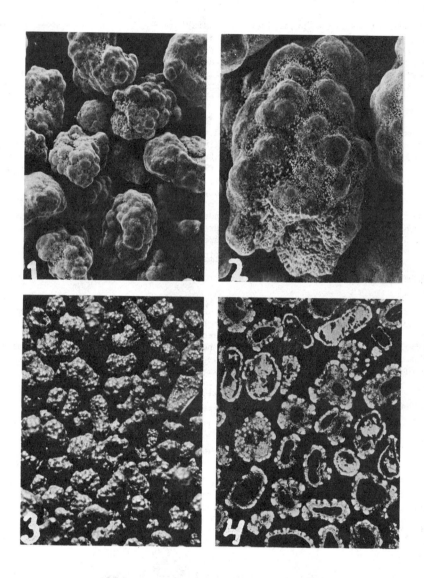

Figure 3. Photographs of cadmium cementate from
the fluidized bed reactors

1 SEM photograph   magn. 100 x
2   "   "   "   300 x
3 Photograph   "   60 x
4 Polished specimen photograph   "   60 x

SEM = Scanning electron microscope

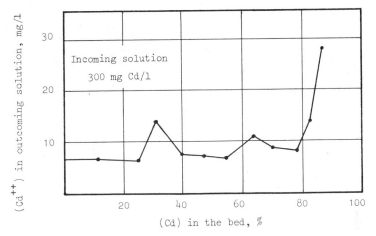

Figure 4. Performance of the fluidized bed cementation of cadmium
with zinc dust showing how the "cementation efficiency"
depends on the cadmium content of the bed material.

Table II. Data on the New Purification System (March 1979)

Analyses of Solution	Unpurified ZnSO$_4$ Solution	Purified ZnSO$_4$ Solution
Zn	140 g/l	142 g/l
Cu	1.3 g/l	<0.05 mg/l
Cd	0.4 g/l	0.4 mg/l
Co	33 mg/l	0.15 mg/l
Ni	30 mg/l	<0.05 mg/l
As		0.02 mg/l
Analyses of Solid Material	Waste from I Stage (Cu Precipitate)	Waste from II Stage (Co Precipitate)	Waste from III Stage (Cd Precipitate)
Zn	5 %	8.5 %	8.8 %
Cu	68 %	45 %	
Cd	0.2 %	0.4 %	88 %
Co		3.4 %	
Ni		3.0 %	
As		15 %	
Chemical Consumption	I Stage	II Stage	III Stage
Zn Dust (Total Consumption 32 kg/t Zn)	12 kg/t Zn	13 kg/t Zn	7 kg/t Zn
As$_2$O$_3$		1.9 kg/t Zn	

power is comparable to a continuous process. The equipment comprises 5 tanks of 200 m^3 and 7 filter presses of 80 m^2 each, 3 of which are in operation at the same time. The presses are opened only in day shift by two men who are the only operators working with filtration in the whole purification step. In the old system and with a production of 90,000 tons of Zn per year there were 2 workers in continuous shift for this purpose.

The most important reactions taking place in this step are the following

$$Co^{++} + As^{+++} + 2.5\ Zn^{0} \rightarrow CoAs + 2.5\ Zn^{++} \tag{6}$$

$$Ni^{++} + As^{+++} + 2.5\ Zn^{0} \rightarrow NiAs + 2.5\ Zn^{++} \tag{7}$$

$$3\ Cu^{++} + As^{+++} + 4.5\ Zn^{0} \rightarrow Cu_3As + 4.5\ Zn^{++} \tag{8}$$

$$Cu^{++} + Zn^{0} \rightarrow Cu^{0} + Zn^{++} \tag{9}$$

$$2\ H_2O + Zn^{0} \rightarrow H_2 + Zn^{++} + 2\ OH^{-} \tag{10}$$

Hydrogen evolution cannot be avoided under the prevailing conditions but it has greatly diminished from the level in the old system. This appears as a substantially lower zinc dust consumption, which can be seen by comparing the data in Tables I and II.

The lower zinc dust consumption gives a smaller increase in pH, which means that there is no danger of precipitating basic salts during normal operation. Therefore the process is not so sensitive to fluctuations in the Co content of the solution and considerably higher impurity levels can be handled without any intermediate pH regulation.

The low zinc dust consumption together with the low pH which can be maintained throughout the purification cycle give a precipitate with a low content of unsoluble Zn and Cd; Zn being 2 - 3 and Cd below 0.3 %. It also appeared that although the Zn content of the precipitate is much lower than in the old process, the Co compounds in it are much more stable against oxidation and the precipitate is not redissolved even if it stands for hours in the solution.

Cadmium Removal

The filtrate from the Co stage is run through a series of 5 fluidized bed reactors, each containing Zn dust beds through which the solution flows. Since a coarse deposit as shown in Figure 3 is obtained, it has been possible to replace the filter presses after this stage by hydrocyclones. These are much easier to operate than filter presses and both investment and operating costs are considerably lower. The efficiency is also good, as the normal Cd level of the cathode zinc is 5 ppm.

Table II shows the consumption figures and the analysis of the cake. The flowsheet of the purification in Kokkola at present is seen from Figure 5.

166

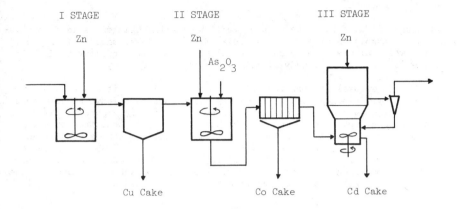

I STAGE   II STAGE   III STAGE

Zn   Zn   Zn

As_2O_3

Cu Cake   Co Cake   Cd Cake

Figure 5. New Purification Process

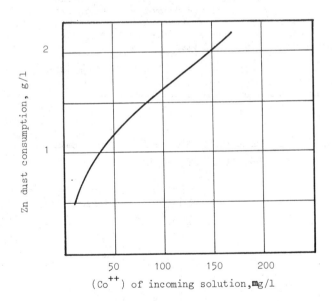

Zn dust consumption, g/l

(Co^{++}) of incoming solution, mg/l

Figure 6. Zinc dust consumption as a function of the cobalt
concentration of the raw solution in a continuous
pilot run using Outokumpu's new cobalt removal.

167

Comments on the System

There is no filtration of the overflow from the thickeners in the Cu removal in Kokkola. As the overflow contains quite a large amount of Cu precipitate, which dilutes the Co cake, this is of rather poor quality in Kokkola. Filtration of the overflow will greatly improve the quality.

The Co removal is operated batchwise in Kokkola but it can also be run continuously. This has been confirmed by pilot runs in which also solutions with higher Co and Ni contents were tested. The curve in Figure 6 showing the zinc dust consumption as a function of the Co content of the feed solution was obatined in a continuous run with three reactors in series. As_2O_3 was added only to the first reactor, whereas Zn dust was added to all three reactors. In this run, solutions with up to 200 mg of Co and 290 mg of Ni per liter were purified. The amount of Cu in the solution was about equal to the sum of Co plus Ni. The following precipitate was obtained:

Co	Ni	Cu	As	Zn	Cd
12 %	12 %	25 %	40 %	2 %	0.3 %

In laboratory tests still higher levels of Co and Ni have been cemented. The result of such a batch is given in Table III. In this test the stoichiometric zinc dust demand was about **29** g/l.

At the plant in Kokkola the Tl content of the concentrate and thus also of the solution is so low that no precautions have to be taken. If there are higher levels, Tl can be cemented in the Cd removal. Tl shows a similar behaviour as Cd in Figure 4 except that Tl starts to go through the bed at an earlier stage. This means that the Cd removal must be run at lower Cd contents of the bed than the present 90 %. If there is 10 mg/l of Tl, the bed in the Cd removal has to be changed when it reaches about 65 % of Cd. Tl can then be removed from the process in the Cd plant.

The new system was built in an existing plant and this has, of course, influenced the setup of the system. For this reason the solutions to the problems were not optimal ones. Improvements have therefore to be considered

Table III. Results of a Cementation Test in Laboratory
Using the New Co Removal Principle

Time min.	Zn g/l	Ni g/l	Co g/l	Cu g/l	Remarks
0	130	7.5	0.9	5	15 g of As_2O_3 + 30 g of Zn dust/l added
20		1.02	0.9	<0.005	
35		0.12	0.61	"	
60		<0.005	0.19	"	
120		0.01	0.28	"	4 g of Zn dust/l added
180		0.005	0.008	"	

in the future, of which the following can be mentioned:

- Filtration after the Cu removal.

- Use of automatic pressure filters.

- Installation of treatment of the Co cake including recycling of As.

- Changing of the Co removal into a continuous process.

- Automation of the system.

The development work on most of these projects is in progress or has already been completed.

Summary of Achievements

- Low zinc dust consumption. The figures for 1978 are only twice the stoichiometric amount needed.

- Pure Cu cementate that only needs to be washed to be a valuable raw material.

- Possibility of producing a concentrated Co and Ni precipitate. This precipitate can easily be treated to recycle As and produce a valuable Co concentrate.

- Concentrated Cd cake that gives a concentrated solution and accordingly a small Cd plant. 600 tons/year of Cd is now produced in Kokkola in equipment planned for the production of 200 tons from the old cake.

- No filtering after the removal of cadmium.

- Excellent selectivity in the different stages, i.e. less than 1 % of cadmium is lost with the Cu and Co cakes, about 90 % of the copper comes out as a clean end-product when having an average solution, small zinc losses with the cake.

- Investment costs competitive with other systems, especially if also the Cd plant is considered.

Figure 7. Solution purification. The reactors of the I and II
stage in the foreground, the III stage in the background.

Figure 8. Reactors from the III stage.

References

1. U.S.P., 1,332,104.

2. A. Grevel: Wiss. Veröff. Siemens-Konzern, 13 (1933), pp. 61 - 71.

3. U.S.P., 3,979,266.

4. U.S.P., 3,954,452.

5. T-L. Huggare, A. Ojanen, A. Kuivala: International Symposium on Hydro-metallurgy, AIME 1973, p. 786.

6. D.M. Chizhikov: Cadmium, Pergamon Press, 1966, pp. 148 - 149.

7. Ibid., p. 169.

RETREATMENT OF ZINC PLANT

PURIFICATION PRECIPITATE

Ernie R. Hamilton

Texasgulf Canada Ltd.

Electrolytic zinc plants conform to the same general flowsheet. After the leaching step which takes zinc and impurities into solution, there follows a purification step, which removes the impurities, producing a theoretically pure zinc sulphate solution for electrolysis.

Purification methods are generally similar in that the impurities, commonly copper, cadmium, cobalt, nickel, etc. are cemented from solution in one or more stages by the addition of zinc dust and arsenic or antimony. These cemented residues are treated to recover by-product cadmium metal and unreacted zinc dust. A copper residue is also produced. It contains the bulk of the other impurities as well as the arsenic or antimony which was added as reagent in the purification step. This residue is normally treated in a copper smelter.

Due to the high cobalt content of the feed to the Texasgulf zinc plant, a relatively large amount of arsenic must be added as reagent to effect the necessary degree of cobalt removal. As a result of this high arsenic demand the copper residue produced contains a level of arsenic (6%-8%) which reduces its value as a smelter feed to almost nil.

This paper describes Texasgulf's development and subsequent operation of a process to remove the arsenic from the copper residue in order to increase its value. The development of the process as well as operating data are discussed. Flowsheet modifications which have considerably improved performance and profitability are also described.

Introduction

In April of 1972 Texasgulf Canada Ltd. (formerly Ecstall Mining Limited) commissioned an electrolytic zinc plant at Timmins, Ontario, Canada. The plant was designed to produce 109,000 tonnes per annum of zinc metal. The yearly by-product production includes 220,000 tonnes of 93% sulphuric acid, 450 tonnes of 99.99% cadmium metal, 15,000 tonnes of lead-silver leaching residue (5% Pb, 1,100 gm/tonne Ag) and 1,200 tonnes of copper cement (50% Cu).

The plant uses conventional practices. Fluid bed roasting of concentrates is followed by leaching and residue treatment via the integrated jarosite process. A hot copper arsenic first-stage purification for copper, cobalt, nickel and metalloid removal precedes the cold zinc dust second-stage purification for cadmium and residual arsenic removal. Electrolysis of the purified solution as well as melting and casting of the zinc are conventional. The purification residues are treated in the cadmium plant for recovery of unreacted zinc, cadmium metal and copper cement.

Description of the Original Cadmium Plant

First- and second-stage purification residues from the zinc circuit are fed to separate reaction tanks in the cadmium plant. The process takes advantage of the fact that the major metals in the purification residues are in their elemental state. The two residues are treated separately to dissolve the less noble elements in sequence, relative to their positions in the electrochemical series. The process recovers zinc as zinc sulphate solution which is returned to the zinc leach circuit, cadmium as 99.99+% metal and copper as a cement product. Provision was also included to produce a cobalt residue. Due to the low grade and high cadmium content, this cobalt residue is recycled and the cobalt exits the plant in the copper cement.

The basic process is shown in figure I. First-stage residue is acid leached to dissolve the zinc and cadmium. After filtration and washing, the residue is stockpiled for treatment in a copper smelter. The typical composition of this residue is 50% Cu, 3.5% Zn, 1.5% Cd, 1.5% Co, 6.5% As. The acid leach filtrate is then mixed with second-stage purification residue in one of two zinc leaching tanks. Zinc dust present in this residue precipitates cadmium, cobalt, and other metals while being solubilized as zinc sulphate. The resulting solution, which is essentially stripped of metals other than zinc, is decanted to the zinc leaching circuit. More metallic zinc is present in the second-stage purification residues than is required to precipitate the noble metals from the copper leach solution. This zinc must be removed as the objective of this processing step is to make a separation of cadmium and zinc. Spent electrolyte, from the zinc circuit, is used for this purpose. The rate of acid addition is regulated to avoid uncontrollable hydrogen generation. Because cadmium is more noble than zinc, a reasonably clean separation of the metallic phases is possible. This is demonstrated by the analysis of the typical leach products: zinc leach residue - 5% Zn, 35% Cd, 0.25% Co, 2.5% Cu; zinc sulphate solution - 100.0 gpl Zn, 0.50 gpl Cd, 0.025 gpl Co.

After several decants of zinc sulphate solution have been carried out the cadmium rich residue is dissolved. Cadmium dissolution is completed utilizing water and 93% H_2SO_4 to produce a cadmium rich liquor by filtration on a plate and frame filter press. The typical analysis of the filtrate is 145 gpl Cd, 50 gpl Zn, 0.8 gpl Co. The residue which contains appreciable quantities of Cu, Cd and Ni is returned to the copper acid leach tank where the Cu and Ni exit with the copper cement.

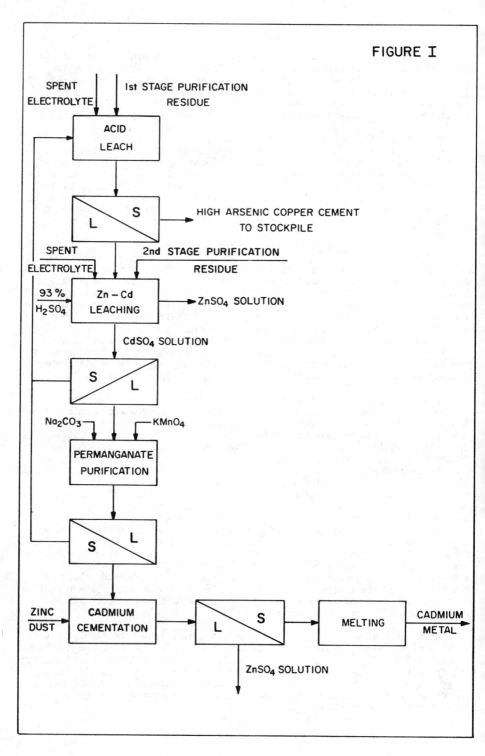

FIGURE I

Table I

Daily Material Balance of the Original Cadmium Circuit

Material Description	Units of Metal Contained (Kg)				
	Zn	Cd	Cu	Co	As
Feed Materials					
First-stage purification residue	1000	75	2000	75	210
Second-stage purification residue	4650	1300	100	0	0
Total Feed	5650	1375	2100	75	210
Products					
Zinc sulphate solution	5520	30	0	0	0
Cadmium metal	0	1300	0	0	0
Copper cement	130	45	2100	75	210
Total Product	5650	1375	2100	75	210

The cadmium leach solution is then purified by reaction with potassium permanganate to precipitate a cobalt rich residue which is separated by filtration. The 20% Cd, 15% Zn, 2% Co residue is recycled to the copper acid leach tank, where the zinc and cadmium are recovered in the acid leach filtrate while the cobalt exits the plant with the copper cement.

The filtrate from the permanganate purification is a high purity cadmium solution with a typical analysis of 145 gpl Cd, 55 gpl Zn and less than 1 ppm of Co and Ni. The solution is treated with SHG zinc dust to produce a high quality 99.5% Cd sponge which is briquetted and melted under caustic soda to produce 99.99+% cadmium metal. The solution after cementation contains 0.5 gpl Cd and 135 gpl Zn, which allows it to be sent directly to the zinc leaching circuit.

Table I shows a material balance around the original cadmium plant on a daily basis.

Development of the Copper Retreatment Process

One of the problems facing Texasgulf, during the initial years of operation, was that the high arsenic content of the copper cement made it unattractive to custom copper smelters. The high smelter charges on this material, high arsenic penalties, high freight charges and metallurgical losses of zinc and cadmium, for which no payment was received, made it very attractive for Texasgulf to develop a process which would remove the arsenic from the copper cement to upgrade the sales value.

In 1974 work commenced on a laboratory scale to develop a process by which the high arsenic copper cement produced by the cadmium plant could be treated to remove the contained arsenic and thus increase the sales value.

The process which emerged from this development work is shown in figure II. It was successfully demonstrated in a series of plant scale tests in 1975 during which time considerable design data was obtained. Approval was received to construct the full scale plant and in the fall of 1977 the plant was commissioned. The capital cost of the plant was U.S. $1,100,000. In September 1977, U.S. Patent #4,049,514, which covers the process was issued. The patent is assigned to Texasgulf Canada. The process, as shown in figure II, is described as follows.

The high arsenic copper cement produced by acid leaching and filtering the first-stage purification residue is repulped and pumped to the caustic leaching tank. Leaching is carried out with 50 gpl caustic soda at a 10-20% solid loading of high arsenic copper cement. Vigorous air addition provides the necessary air oxidation to dissolve the arsenic contained in the copper cement. The caustic leach is complete in 6-8 hours, the end point being determined by the arsenic content of the leached solids. The copper cement is filtered, washed and air blown, before being dumped to a storage bin for shipment to a smelter. Arsenic extractions are in the range of 95%. The copper cement produced has a typical analysis of 65% Cu, 3% Zn, 1% Cd and less than 0.5% As.

The filtrate, which contains 20-30 gpl Na and 9-12 gpl As, is treated in the copper arsenate precipitation tank with $CuSO_4.5H_2O$ to precipitate the arsenic as copper arsenate. After filtration the copper arsenate residue is slurried in a repulping tank which recycles the copper arsenate to the purification plant as a reagent for first-stage purification.

176

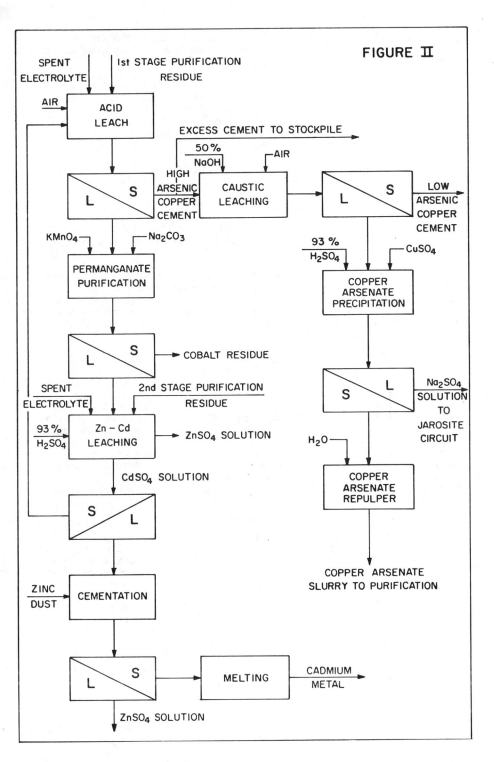

FIGURE II

177

The analysis of this reagent residue is typically 25% Cu and 27% As. The filtrate containing 20-30 gpl Na is recycled to the leaching plant as reagent for jarosite precipitation.

Provision is also installed to add air to the acid leaching step to improve the extraction of cobalt, zinc and cadmium from the high arsenic copper cement which forms the feed to the caustic leach. This results in a high level of cobalt in the acid leach filtrate. This makes it necessary to perform the permanganate purification on the acid leach filtrate prior to adding it to the zinc leaching tanks.

This permanganate purification is not required if air oxidation is not utilized to extract the cobalt from the high arsenic copper cement.

Initial Operation of the Process

The start-up of the plant was exceedingly smooth. The knowledge gained from the in-plant demonstrations aided greatly both in operator training and design of the plant.

In order to minimize start-up problems, the process was commissioned without air addition to the acid leach. This eliminated the need for production of a cobalt residue. The air oxidation - cobalt removal step has proven very difficult and, as a result, has never been operated economically. The elimination of this stage in the process affects the economics in a minor manner only.

In order to minimize the entrainment of arsenic rich solution in the copper cement product, a repulp washing system was installed. This step proved to be a severe bottleneck in the process as both the filtration rates of the repulped cement and arsenic elimination efficiency were poor. As a result, the arsenic content of the product copper cement remained above 1%. It was discovered that careful management of the caustic leach filtration step resulted in the copper cement forming a 10-20 mm layer of cake on the filter paper in the plate and frame filter press. This adherence of the cement to the filter media has allowed the introduction of water directly into the filter press feed lines to effect a very efficient follow-through washing step. Subsequent air blowing produces a low moisture product with minimal solubles entrained. The need for the difficult repulping stage was thereby eliminated, and the product quality exceeded our best expectations.

All other areas of the plant operated as designed. Within two months the plant was required to operate at 130% of the design rate, due to an increase in the copper content of the concentrate fed to the zinc plant. Since commissioning, the plant has continually been operated above design rate. Table II shows the design material balance of the cadmium plant with copper retreatment.

Improvements to the Process

A. Copper Sulphate Production

The Texasgulf zinc plant is located immediately adjacent to Texasgulf's Kidd Creek concentrator. The concentrator consumes 1.25 tonnes per day of copper ion as an activator for the flotation of sphalerite.

Table II

Design Material Balance of the Cadmium Plant
After Installation of the Copper Retreatment Plant

Material Description	Units of Metal Contained (Kg)				
	Zn	Cd	Cu	Co	As
Feed Materials					
First-stage purification residue	1000	75	2000	75	210
Second-stage purification residue	4650	1300	100	0	0
Copper sulphate	0	0	200	0	0
Total Feed	5650	1375	2300	75	210
Products					
Zinc sulphate solution	5525	30	0	0	0
Cadmium metal	0	1300	0	0	0
Copper cement	100	30	2100	0	15
Copper arsenate	0	0	200	0	195
Cobalt residue	25	15	0	75	0
Total Products	5650	1375	2300	75	210

This copper ion had been historically purchased in the form of five tonnes of $CuSO_4.5H_2O$ daily at a delivered price which was approximately double the market price for copper metal. The net smelter return on the low arsenic copper cake produced by the copper retreatment process was approximately 2/3 of the market price of copper metal. In early 1978, the relative values of copper to Texasgulf were, therefore, U.S. $0.40/lb. as copper cement and U.S. $1.20/lb. as $CuSO_4.5H_2O$. Obviously, it was very attractive to upgrade the value of the copper cement by a factor of three by converting it to copper sulphate. It was determined in the laboratory that the low arsenic copper cement produced by the caustic leach was quite easily dissolved with air oxidation in sulphuric acid. Residual arsenic, which dissolved from the copper cement, re-precipitated as copper arsenate when the pH was raised at the end of the leaching step. A separation by filtration produced a copper sulphate solution with a typical analysis of 80 gpl Cu, 6 gpl Zn, 2 gpl Cd, 3 gpl Co and less than 200 ppm As. The zinc, cadmium and cobalt levels of the copper sulphate solution are acceptable as they precipitate in the high pH water of the milling circuits.

As previously mentioned, the need for the repulp washing stage on the product copper cement had been eliminated. The repulping tanks had been located so as to receive, repulp and refilter the product copper cement. Thus, it remained only to install ventilation, acid lines and air sparging to convert these tanks to the production of copper sulphate solution. Other modifications included storage and loading systems and the purchase of a rail tanker which is used to transport the solution to the concentrator reagent storage tanks. The capital cost of these modifications was U.S. $120,000.

In the period August, 1978, when copper sulphate production commenced, to July, 1979, 337 tonnes of copper ion was produced for use by the concentrator.

B. Production and Use of Zinc Arsenate

When copper arsenate slurry became available in sufficient quantity to use as purification reagent on a plant scale, we encountered some difficulty in effecting cobalt elimination consistently. It was determined that the pH of the solution had to be maintained in a lower range using copper arsenate slurry as reagent than when using arsenic trioxide. This minor modification to the purification procedure allowed purification to proceed normally. The lower pH, however, was reflected by an increase in zinc dust consumption of 1-2% of cathode plated.

During this period, it was noted that the analysis of the copper arsenate residue varied widely with respect to copper while maintaining a relatively constant 27% As. It was also noted that the total of the zinc content plus the copper content was constantly around 25%. It was hypothesized, and later proven, that the copper arsenate was co-precipitating with zinc arsenate.

The zinc was reporting to the precipitation tank in the caustic leach filtrate. Several laboratory tests were conducted, and it was determined that zinc arsenate could be produced and utilized as a purification re-agent in the same manner as copper arsenate.

The economics of this discovery were good, as approximately U.S. $800.00 per day worth of $CuSO_4.5H_2O$ crystals were required for the production of copper arsenate. The copper sulphate produced for the concentrator was not suitable for this purpose, due to the high cobalt content of the solution. Both the zinc and acid required to carry out the precipitation of zinc arsenate were readily available in the form of spent electrolyte.

The use of zinc arsenate on a plant scale was commenced in September, 1978. The trend of high zinc dust consumption, due to the low pH required in the first-stage purification, continued to be a problem. At this point, it was discovered that, by addition of acid, the zinc arsenate could be redissolved after separation from the caustic leach solution. Utilizing dissolved zinc arsenate as purification reagent eliminated the need for operating at low pH levels and, as a result, the zinc dust consumption was reduced to levels consistent with those obtained utilizing As_2O_3 as reagent. An application to patent the use of zinc arsenate as purification reagent has been filed by Texasgulf.

C. Copper Cementation

The copper retreatment plant was designed to produce two tonnes per day of copper as copper cement. This design rate is equivalent to impure solution containing 0.5 gpl copper while the plant is operating at design zinc capacity.

When the plant was commissioned the zinc plant was operating at only 70% of capacity, however, the copper in the feed had increased so that the copper content of the impure solution was in excess of 1.0 gpl. This resulted in it being necessary to operate the copper retreatment plant at 130% of design rate. When the zinc plant returned to 100% production rate, it became necessary to bypass a portion of high arsenic copper cement. This material was added to the 3,200 tonne stockpile of high arsenic copper cement to await future upgrading.

To eliminate this problem, and create capacity to treat the stockpile, a copper cementation stage is being installed in advance of the first-stage purification. This stage will precipitate approximately 75% of the copper from the impure solution by stoichiometric zinc dust addition. The resulting cement is a high quality 90% Cu material containing no arsenic, less than 1% Zn and less than 0.5% Cd.

This cementation stage will reduce the load on the copper retreatment plant by 75%, resulting in free capacity which will then be used to treat the stockpiled high arsenic copper cement. Arsenic, cadmium, zinc and a saleable copper cement or copper sulphate will be recovered. At the time of writing, this process was in the design and installation stage. Both plant scale tests and pilot plant runs have shown it to be a reasonably easy modification. No difficulties in operating the process are anticipated.

Figure III shows the plant configuration at present. Table III gives a daily materials balance of the plant as it is currently operated.

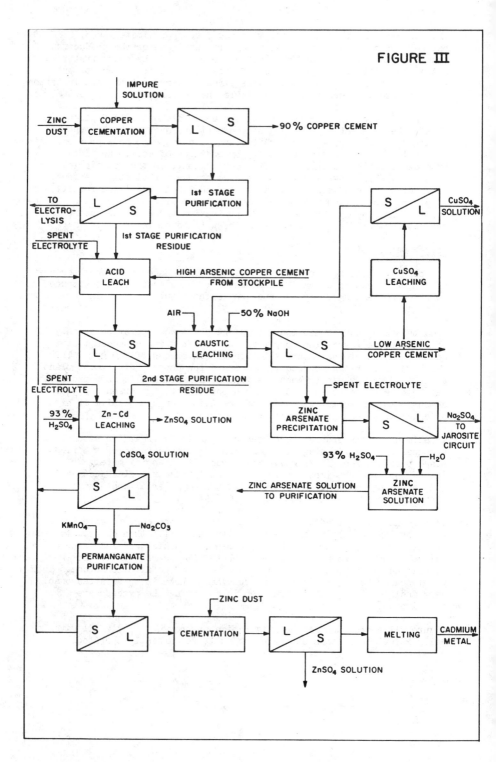

FIGURE III

Table III

Design Material Balance of Plant
After Installation of Copper Cementation

Material Description	Units of Metal Contained (Kg)				
	Zn	Cd	Cu	Co	As
Feed Materials					
Copper in impure solution to be removed in copper cementation stage	0	0	3000	0	0
Copper acid leach residue from stockpile	35	15	500	15	65
First-stage purification residue	1000	75	1000	75	210
Second-stage purification residue	4650	1300	100	0	0
Zinc sulphate (for arsenic precipitation)	275	0	0	0	0
Total Feed	5960	1390	4600	90	275
Products					
Zinc sulphate solution	5560	30	0	0	0
High grade copper cement	35	5	3000	0	0
Copper cement	15	5	350	15	10
Cadmium metal	0	1320	0	0	0
Zinc arsenate	265	0	0	0	265
Copper sulphate solution	85	30	1250	75	0
Total Products	5960	1390	4600	90	275

The Future

What does the future hold for the copper retreatment plant?

The installation of the copper cementation stage has provided the capacity to recover, in a saleable form, the copper from the impure solution, regardless of the concentration. In the future, the treatment of electrostatic precipitator dusts from Texasgulf's copper smelter will increase the copper production of the zinc plant to eight tonnes per day of copper.

The high price of cobalt makes recovery of the 75 Kg per day entering the purification plant attractive. We are pursuing several avenues for the recovery of cobalt. Several process changes which recover cobalt in a 15-20% residue are being investigated. Some plant testing has been carried out.

The proximity of the zinc plant to our copper smelter and refinery, as well as the concentrator, make processes such as the production of copper sulphate economic. Work is currently underway to integrate a bleed stream from the copper refinery into the copper retreatment plant. If successful, a crystalization plant in the refinery will not be needed, and the zinc plant will be in a position to recover nickel and cobalt from this bleed stream.

Conclusions

In addition to upgrading a virtually unsaleable residue to a valuable product, the copper retreatment process has shown several other benefits. The losses of zinc and cadmium per unit of copper produced are reduced. Reagent costs are reduced due to the recycle of arsenic ion. Installation of the process in a zinc plant utilizing the jarosite process allows easy disposal of the sodium sulphate solution produced by the process.

Table IV shows the operating data of the plant for the year ending July 31, 1979.

Table IV

Plant Operating Data August, 1978 - July 31, 1979

Copper in feed to copper retreatment plant	905 tonnes
Copper assigned to high arsenic stockpile for future retreatment	348 tonnes
Copper produced as copper sulphate solution	337 tonnes
Copper produced as saleable low arsenic copper cement	220 tonnes
Arsenic recycled as reagent to first stage purification	75 tonnes

Acknowledgements

The author wishes to thank Texasgulf for permission to publish, and the staff at Timmins for their assistance in preparation of this paper.

References

1. United States Patent Number 4,049,514

2. F.S. Gaunce et al, "The Ecstall Story - The Electrolytic Zinc Plant", The Canadian Mining & Metallurgical Bulletin, Vol. 67, No.745, May, 1974, pp. 68-77.

CONTINUOUS MONITORING
OF ZINC ELECTROLYTE QUALITY AT COMINCO
BY CATHODIC OVERPOTENTIAL MEASUREMENTS

R.C. Kerby
C.J. Krauss
Cominco Ltd.
Trail B.C.
Canada

In Cominco's electrolytic zinc operations at Trail, B.C., the electrolyte is extensively purified of harmful impurities prior to the electrowinning stage to ensure efficient extraction of high purity zinc metal. In addition, reagent additions are made to the electrolyte to optimize the electrodeposition process and to minimize acid mist evolution.

A voltammetric technique, based on cathodic overpotential measurements, was developed to provide rapid, quantitative measurements of the quality of electrolyte being supplied to the zinc electrowinning operations both before and after reagent additions. An automated instrument using this technique was developed for in-plant use. Further refinement of the technique has resulted in the development of a continuous, on-stream analyzer.

Both instruments have been undergoing extensive testing for in-plant use in controlling reagent additions and monitoring zinc electrolyte quality. Results to date indicate that these instruments can play an important role in optimizing the zinc electrowinning process.

Introduction

Cominco's electrolytic zinc plants are located on the Columbia River at Trail, British Columbia (Figure 1). The zinc plants use conventional roast-leach-electrowinning technology (1, 2). A modernization program is currently in progress which will expand production capacity from 250,000 tonnes to 270,000 tonnes of zinc per year. As part of this program, the present electrolytic zinc plants are being replaced by a new, highly mechanized plant.

Zinc electrowinning is very sensitive to the presence of impurities in solution (3). Metallic ions such as antimony, germanium, cobalt, and nickel can cause extensive re-solution of the electrodeposited zinc, leading to a decrease in the cathode current efficiency of zinc production. Cadmium, lead and copper ions co-deposit with zinc to give an impure zinc product. Zinc electrodeposition is also sensitive to the presence of many organic compounds, which can cause spongy zinc deposits or extensive dendritic growth. However, some organic compounds, eg., animal glues, can be used at specific concentrations both to control the re-solution of the electrodeposited zinc caused by metallic ion impurities and to give level zinc deposits (4).

Those metallic ions detrimental to the zinc electrowinning process are removed to low concentrations by extensive purification processes. At Cominco, the purification process consists of three stages. The first stage involves iron oxide precipitation in which iron, antimony, arsenic, germanium and indium and other impurities are removed to low concentrations in the zinc electrolyte. The second stage involves zinc dust purification in which atomized zinc is used to remove cadmium, copper, nickel, cobalt and antimony from solution. The third stage involves removing gypsum from solution by passing the electrolyte through cooling towers to cool the electrolyte from approximately 60°C to 30°C.

After purification, the zinc electrolyte is sent to the zinc tankrooms for the electrowinning of zinc. At Cominco, animal glue is added to the electrolyte to moderate the effects of impurities in solution on the electrodeposition of zinc and to provide level zinc deposits (4). Meta-para-cresol is added in conjunction with animal glue to provide a foam layer on top of the electrolyte in the cells so as to control acid misting (5). Barium carbonate is added to help reduce lead concentrations in the cathode zinc, and potassium antimony tartrate is added as required to improve cathode current efficiency and to allow easier stripping of zinc from the aluminum cathodes.

Several techniques are used to determine whether the electrolyte purification processes have removed metallic ion impurities to concentrations low enough to have minimal effect on zinc electrodeposition. These include assaying on a semi-continuous basis for a number of the impurities, as well as monitoring on the pH, temperature and clarity of the electrolyte throughout the processes. The assaying techniques used to monitor these impurities are often slow, complex and costly, thus limiting their usefulness in process control. In addition, they do not take into full account the synergistic effects of combinations of impurities on zinc electrodeposition.

Assaying for animal glue concentration in cell electrolyte to determine whether the correct glue concentration is present is extremely difficult, consequently glue content measurements are seldom taken.

Several types of rapid testers have been proposed for determining the quality of zinc electrolyte relative to its effect on zinc electrodeposition

Fig. 1 – Electrolytic Zinc Plants at Cominco Ltd., Trail,
British Columbia.

(6 - 9). These testers, although useful in providing information on high concentrations of impurities in electrolyte, are not being used by Cominco. In our opinion they do not appear to provide a continuous, reliable measurement of electrolyte quality prior to its addition to cell electrolyte, or a rapid indication that the correct amount of glue has been added to cell electrolyte to moderate impurity effects on zinc electrodeposition.

A program was initiated to develop a continuous, on-stream analyzer capable of monitoring both the quality of zinc electrolyte relative to its effect on zinc electrowinning, and the addition rate of animal glue to cell electrolyte to control electrolyte impurity effects.

Basis Of The Method

Impurities such as antimony and cobalt in zinc electrolyte will often cause extensive re-solution of cathode zinc deposits when present above certain critical concentrations. Additions of animal glues are used to moderate this effect. The interactions of impurities and animal glues on zinc electrodeposition were extensively studied at Cominco using laboratory pilot cells. Typical results are shown in Figure 2. It was found that for all concentrations of electrolyte impurities normally encountered during zinc electrowinning, there are optimum glue concentrations which maximize cathode zinc current efficiencies. As the total impurity concentration in electrolyte increases, the optimum glue concentration required to maximize current efficiency also increases. An example of this relationship is shown in Figure 3, where the shaded region defines the optimum current efficiency of zinc deposition at various concentrations of antimony and glue in solution and low concentrations of other impurities.

Solution impurities such as antimony and cobalt and solution additives such as animal glue also affect the cathodic polarization associated with zinc electrodeposition. A convenient technique for measuring these polarization effects is cyclic voltammetry (10). In cyclic voltammetry, a controlled potential is applied to an electrolysis cell and the resulting electrolysis current is displayed against the controlled potential. A typical cyclic voltammogram is shown in Figure 4. The line BC of the voltammogram corresponds to the activation overpotential associated with initial zinc deposition onto the aluminum substrate. Increasing glue concentration increases the activation overpotential, whereas increasing concentrations of impurities such as antimony, germanium, cobalt and nickel, which are deleterious to zinc electrowinning, result in decreasing activation overpotentials.

The results of a number of activation overpotential measurements at various glue and antimony concentrations in zinc electrolyte are illustrated in Figure 5 where the shaded region shows the optimum current efficiency for zinc deposition as referred to previously in Figure 3. These results indicate that Cominco hot zinc dust-purified electrolyte with glue to antimony concentration ratios giving activation overpotential measurements of 100 to 110 mV will also give maximum cathode zinc current efficiencies.

This relationship between activation overpotentials and optimum zinc production was found to hold for a variety of impurities in and glue additions to zinc electrolyte. The relationship between the measured activation overpotential for zinc electrolytes and the corresponding current efficiency of zinc deposition is shown in Figure 6.

Activation overpotential measurements are most sensitive to those impurities which are the most deleterious to zinc electrowinning current

190

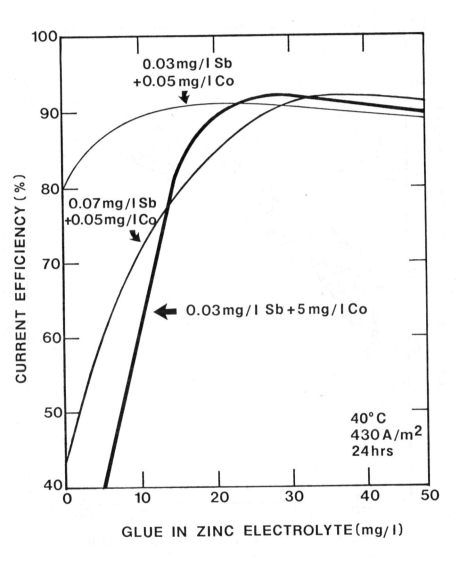

Fig. 2 - Effect of glue additions on the current efficiency
of zinc deposition from acid zinc electrolytes
(55 g/l Zn, 150 g/l H_2SO_4) containing antimony
and cobalt.

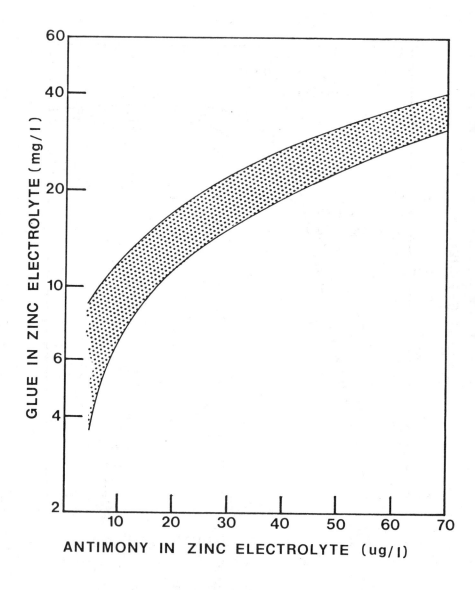

Fig. 3 - Region of maximum current efficiency of zinc production (shaded region) as related to antimony and glue concentrations in Cominco acid zinc electrolyte (55 g/l Zn, 150 g/l H_2SO_4). Electrolysis tests were carried out at 430 A/m^2 and 40°C for 24 hours.

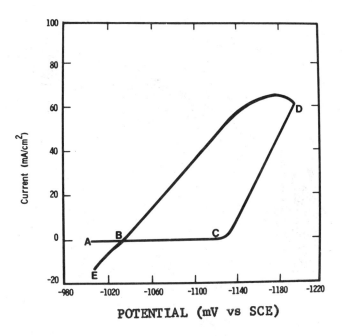

Fig. 4 - Cyclic voltammogram for acidified zinc sulphate electrolyte. (A) start of recorded signal, (B) crossover potential, (C) potential at which zinc deposition is first observed, (D) reversing potential, (E) end of recorded signal.

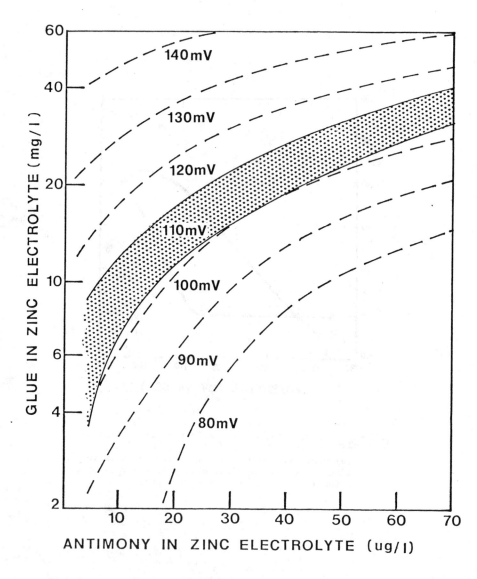

Fig. 5 - Relationship between maximum current efficiency of zinc
production (shaded region) and the associated activation
overpotential of zinc electrodeposition onto an aluminum
substrate from Cominco acid zinc electrolyte (55 g/1 Zn,
150 g/1 H_2SO_4) containing glue and antimony in solution.

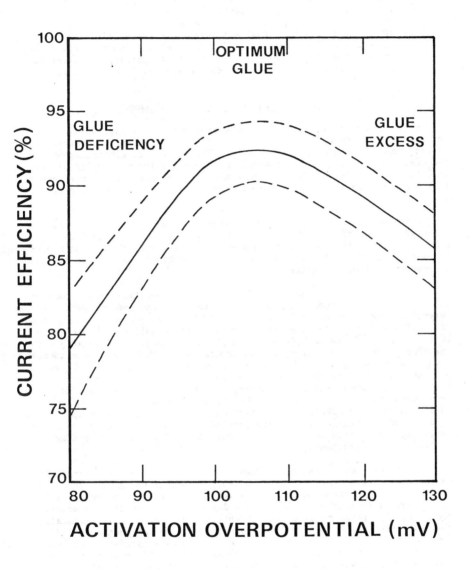

Fig. 6 – Relationship between the current efficiency of zinc
production from Cominco acid zinc electrolyte and the
associated activation overpotential of zinc electrodeposition. Also indicated are the glue concentrations
relative to impurity concentrations in the electrolyte.

efficiency, eg., antimony and germanium. Electrolyte impurities such as cadmium and lead, which have little effect on zinc electrowinning current efficiency, also have little effect on the activation overpotential (11), whereas cobalt and nickel have the most effect when impurities such as antimony and germanium are also present.

These results demonstrate that activation overpotential measurements can be used to detect harmful levels of impurities in zinc electrolyte and also to indicate the correct levels of animal glue additions required to moderate impurity effects and optimize the cathode zinc current efficiency.

Development Of Automated Plant Instruments

Having shown that the activation overpotentials of zinc deposition onto an aluminum substrate could be used to provide a rapid and accurate indication of zinc electrolyte quality relative to zinc electrodeposition, the next step was to develop a simple and reliable instrument for plant use. A cyclic voltammetric technique was not used because it would provide more than the required information and requires a skilled technician to operate it.

A voltammetric technique using a single voltage scan through the potential region of interest (ie., line BC of Figure 4) was used as the basis for one of the two automated plant instruments which were developed. For identification purposes this instrument was given the name Activation Overpotential (AOP) Analyzer.

One requirement for accurate activation overpotential measurements is that the aluminum cathode have a fresh surface for each measurement. It was determined that aluminum foil could be used for this purpose, with a fresh strip of aluminum foil being placed in the cathode holder prior to every test. Once the fresh cathode had been placed in the test cell along with the electrolyte to be tested, the instrument would follow a test sequence when a button was pushed. The voltage scan was recorded on an X-Y recorder and the activation overpotential read from the recorded voltammogram. This instrument underwent extensive testing for in-plant use, but was judged to be unacceptable because it did not meet the requirements of a continuous, on-stream analyzer requiring a minimum amount of operator attention. However, it did confirm that activation overpotential measurements could be used to assist in controlling the electrolyte purification and electrowinning stages of the zinc electrowinning circuit.

The second method which has been developed for in-plant activation overpotential measurements uses a controlled current technique, in which the cell current is controlled at a low level (0.4 mA cm^{-2}) and the resulting activation overpotential recorded on a digital voltmeter or on a suitable recorder. By operating at a set current, the voltage scan is eliminated and the instrument can be made to measure cathode potential continuously. To overcome the problem of providing a fresh cathode surface an aluminum wire is continuously passed through the cathode holder. Electrolyte is pumped from the process stream which is being sampled, through the cell past the moving cathode wire and back to the process stream. The prototype instrument was given the name Continuous Electrolyte (CE) Analyzer to differentiate it from the AOP analyzer.

In-Plant Measurement of Zinc Electrolyte Quality

The instruments have shown themselves to be ideally suited to monitoring glue additions to cell electrolyte to give optimum current efficiencies and zinc deposit levelling. Normal practice has been to add glue at a constant rate to the recirculating cell electrolyte regardless of the concentrations of impurities present. Typical results of this type of operation are shown in Figure 7, in which the weekly averages for the current efficiencies of zinc deposition are compared to their corresponding activation overpotential readings. The results are similar to those shown previously in Figure 6, in which current efficiencies are maximized at activation overpotential readings of 104 to 110 mV. Those electrolytes with lower overpotentials have a deficiency of glue relative to the electrolyte impurity content, while higher overpotentials indicate too much glue was present relative to the electrolyte impurity concentrations. The activation overpotential measurements can be used to indicate what glue addition rates are required for optimum zinc production during the electrowinning process.

The activation overpotential measurements were used to control glue additions to the cell electrolyte to maximize the current efficiencies of zinc electrowinning. Typical results are shown in Figure 8, where daily activation overpotentials of cell electrolyte are shown both for constant glue additions to electrolyte and for controlled glue additions based on maintaining the activation overpotential measurements of the electrolyte at between 100 mV and 110 mV. Controlled glue additions eliminated the periodic low current efficiency periods due to fluctuating impurity levels and thereby gave a better overall zinc productivity from the electrowinning cells.

Activation overpotential measurements can also be used to monitor the impurity content of the zinc solution after each purification step and therefore can be used by the operators as an aid in controlling the purification processes.

The Cominco zinc dust purification process for zinc electrolyte is a continuous process which occurs in two stages. Following calcine leaching with spent acid, iron purification and solution clarification, the neutral clarifier overflow solution is adjusted to approximately 1 mgL^{-1} antimony, heated to 75°C and added, along with atomized zinc, to the first of a series of five stirred tank reactors. The slurry from the first stage is filtered in Sperry filter presses. The filtrate from the first stage is added, along with atomized zinc, to another stirred tank reactor. The slurry from this second stage is filtered on Shriver filter presses. The Shriver filtrate is pumped to a cooling tower and clarifier for gypsum removal. The clarifier overflow goes to storage or directly to the zinc electrowinning cells as electrolyte feed.

The primary control parameters used in zinc dust purification are the pH and temperature of the reacting solutions, and the purity of the filtered solutions. Cadmium assays are taken once every four hours on Sperry and Shriver filtrates, while Sb, Ge, Co and Ni are assayed for on 24 hour composite samples. Activation overpotential measurements can be used to advantage to monitor for impurities such as Sb, Ge, Co and Ni which do adversely affect zinc electrowinning current efficiency. As mentioned earlier, activation overpotential measurements provide only limited information on cadmium concentrations in zinc electrolyte and this is of little consequence because cadmium does not adversely affect zinc electrowinning current efficiency.

The zinc dust purification process was monitored at three points along the system, ie., the Sperry filtrate, the Shriver filtrate and the gypsum

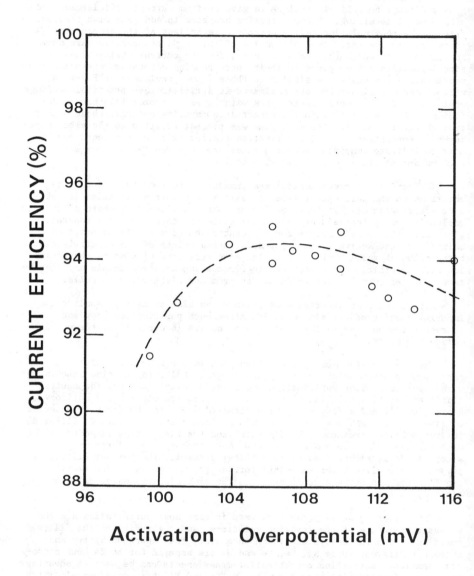

Fig. 7 – Weekly averages of current efficiencies of zinc
production of some cells at Cominco, Trail, versus
the activation overpotentials of the cell electrolyte.

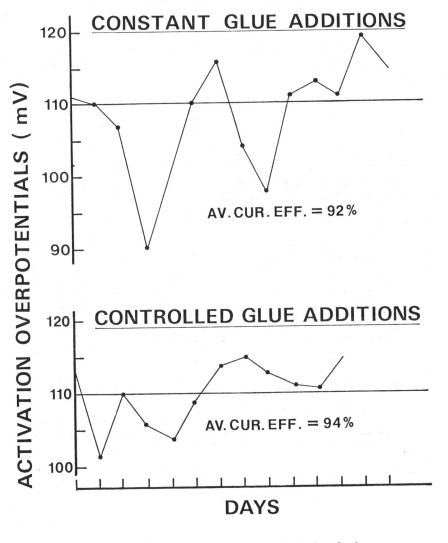

Fig. 8 – Variations in activation overpotentials of zinc
deposition onto aluminum for Cominco cell electrolyte,
as determined on a daily basis. In the case of constant
glue additions, glue was added at 15 mg/l irrespective of
impurity concentration in electrolyte. For controlled
glue additions, the rate of glue addition was varied to
maintain the activation overpotential measurements close
to 110 mV.

clarifier overflow after the cooling towers. Typical measurements obtained
from Shriver filtrate are given in Table 1 where activation overpotentials
are compared to the solution assays. The activation overpotentials decreased
with increasing impurity levels in solution, in particular those of germanium
and antimony.

Table 1   Measurement of the Quality of Zinc Electrolyte
           Obtained From Zinc Dust Purification
           (ie., After Shriver Filter Presses)

Date	Activation Overpotential (mV)	Germanium (mgL^{-1})	Antimony (mgL^{-1})	Cobalt (mgL^{-1})	Cadmium (mgL^{-1})
July 12	84	0.03	0.02	0.5	0.13
July 8	81	0.03	0.03	0.4	0.23
July 9	75	0.04	0.02	0.3	0.32
July 10	72	0.05	0.03	0.4	0.30
July 11	72	0.06	0.03	0.6	0.22

A combination of activation overpotential measurements and cadmium
assays of the zinc electrolyte undergoing purification allows the plant
operators to make a judgement as to the source of problems that may be
occurring. Examples of some of the problems that might be indicated and
their suggested corrections are given in Table 2. The application of activation overpotential measurements to process control in the zinc dust purification process is presently undergoing extensive testing.

The primary control parameters in the iron purification stages are the
pH, the iron precipitated, the temperature of the reacting solutions, and
the clarity and iron content of the neutral thickener overflow solution.
Also of importance is the adequate removal of electrolyte impurities such as
antimony, germanium and arsenic. Monitoring of these impurities is normally
limited to germanium assays on 24 hour composite samples of the neutral
thickener overflow electrolyte. Continuous monitoring of these impurities
in electrolyte would be advantageous in controlling the process.

Activation overpotentials were determined for several of the 24 hour
composite samples of electrolyte (Table 3). The activation overpotentials
decreased whenever there was an increase in germanium and antimony concentrations in the neutral thickener overflow electrolyte. The use of activation overpotential measurements to assist in the control of the iron oxide
precipitation process is presently under study.

Table 2. Suggested Changes To A Zinc Dust Purification
Process, Based On Combination Of Overpotential
Measurements Plus Cadmium Assays Of Processed Electrolyte

Overpotential (mV)	Cadmium (mgL^{-1})	Indicated Process Problem	Suggested Process Change
High (greater than 80 mV)	Low (less than 0.2 mgL^{-1})	Process operating satisfactorily	None
High	High (greater than 0.2 mgL^{-1})	Cadmium removal problem	1. Check for air in-leakage. 2. Check for excess iron in solution feed. 3. Check quality of zinc dust. 4. Check if temperature too high.
Low (less than 80 mV)	Low	Co, Ni, Sb, Ge removal problem	1. Check iron purification to limit impurities in solution feed. 2. Check quality of zinc dust. 3. Increase zinc dust and copper additions. 4. Increase temperature.
Low	High	Problem with removal of all impurities, perhaps high input of impurities from previous stage	1. Correct iron purification so as to limit impurities being fed to this stage. 2. Increase atomized zinc additions, check quality. 3. Check for iron in solution feed. 4. Increase purification time. 5. Increase Cu additions while decreasing temperature.

Table 3. Quality Of Zinc Electrolyte Obtained
From Neutral Thickener Overflow

Date	Activation Overpotential (mV)	Germanium (mgL^{-1})	Antimony (mgL^{-1})
July 1	26	0.13	0.08
17	27	0.11	0.07
11	28	0.19	0.07
19	28	0.12	0.08
13	30	0.12	0.04
10	30	0.15	0.07
12	30	0.16	0.06
7	31	0.11	0.04
14	31	0.11	0.04
20	32	0.08	0.03
21	32	0.09	0.04

Acknowledgements

The many helpful discussions with Dr. T.J. O'Keefe of the University of Missouri-Rolla on both the basic and applied aspects of activation overpotential measurements in zinc electrolytes are gratefully acknowledged. The assistance of personnel from Cominco's Technical Research, Technical Development, Instrumentation and Zinc Operations are also gratefully acknowledged. It should be noted that Cominco has a proprietary interest in procedures discussed in this paper and patent applications have been filed.

References

(1) "Cominco Zinc Department," Canadian Mining Journal, 75(5) (1954) pp. 262-282.

(2) "Trail Operations: Metal For World Markets," Eng. Mining Journal, Sept. (1973) pp. 115-120.

(3) G.C. Bratt, "A View Of Zinc Electrowinning Theory," Pap. Tasmania Conf, pp. 277-290; Australas. Inst. Min. Metall., Parkville, Aust. (1977).

(4) R.C. Kerby, H.E. Jackson, T.J. O'Keefe and Yar-Ming Wang, "Evaluation of Organic Additives For Use In Zinc Electrowinning," Metallurgical Transactions B, 8B (12) (1977) pp. 661-668.

(5) D.J. DeBiasio and C.J. Krauss, "Electrolytic Production of Zinc," Can. Patent 978,137, November 18, 1975.

(6) T.R. Ingraham and R.C. Kerby, "A Meter for Measuring the Quality of Zinc Electrolytes," Can. Met. Quart. 11 (2) (1972) pp. 451-454.

(7)   A.W. Bryson, "Solution Quality Analyzer for Zinc Sulphate Electrolyte," pp. 507-511 in Proc. 2nd IFAC Symp. (Automation in Mining, Mineral and Metal Processing) held in Johannesburg, Sept. 13-17, 1976; S. African Council for Auto. and Comp., Pretoria (1976)

(8)   S. Mukae, "Method For Judging Purity Of Purified Zinc Sulphate Solution Used For Electrolytic Production Of Zinc," U.S. Patent 4,013,412, March 22, 1977.

(9)   A.P. Saunders and H.I. Philip, "Automated Instrument For Measuring The Effects Of Impurities On Cathodic Current Efficiency During The Electrowinning Of Zinc," NIM Report No. 1906, September 2 (1977).

(10) B.A. Lamping and T.J. O'Keefe, "Evaluation of Zinc Sulphate Electrolytes by Cyclic Voltammetry and Electron Microscopy," Metallurgical Transactions B, 7B (12) (1976) pp. 551-558.

(11) D.J. MacKinnon, J.M. Brannen and R.C. Kerby, "Effect of Cadmium on Zinc Deposit Structures Obtained From High Purity Industrial Acid Sulphate Electrolyte," J. Applied Electrochem., 9 (1979) pp. 71-79.

NEW VIEILLE-MONTAGNE CELLHOUSE AT V.M. BALEN PLANT BELGIUM

Yves de Bellefroid

Manager of VM Balen Plant

Roger Delvaux

Chief Metallurgist of VM
Balen Plant

Abstract

In order to replace the old manual cell houses still operating at the Balen plant, Vieille-Montagne has constructed a new 150,000 metric ton cathodes/year electrolytical cell house, that went on stream in August 1979.

The concept of this cell house is based on the experience acquired by the company in running the first jumbo size cathode automated cell house since 1969 and in extrapolating the acquired results as much as possible, so as to obtain a house with still more reduced investment costs and a better productivity.

The characteristics are as follows :

- 4 rows of 35 sandwich cells containing each 86 cathodes of 3,2 sq.m,

- amperic nominal density 410 A/sq.m. ensuring a daily production of 2.98 t per cell.

- at the head each pair of rows is equipped with a simplified stripping machine for cathodes, composed of 3 chain conveyors taking charge of 2 stripping units and a brushing one.

- conveyors and cells are interconnected by 2 mobile cranes.

As for the investment costs, the used surface will only represent 55 % of the automated first generation houses and the electrodes weight will be brought down by ± 8 %.

The manpower required will be cut down to 0.6 hour/ton cathodes, i.e. some 20 % of what was needed in the manual cell houses.

1. Introduction

During the AIME World Symposium at St Louis in 1970 our Company presented a paper dealing with the whole Balen plant of Vieille-Montagne. (1)

It has been mentioned then that the Balen plant had put on stream in 1969 a completely new type of cell house, based on totally new plans in view of a full automation of all handling and stripping activities.

Let us point out that this cell house was originally planned for a yearly output of 60,000 t of zinc.

Due to the favourable results obtained, the current density has been increased from 340 to 415 Ampères per sq.meter, thus enabling a year production of 75,000 t cathodes in this installation. Automated handling and stripping raised the productivity up to 30 t zinc cathode per man-shift, i.e. four times the productivity level of manual stripping.

As this unit gave perfect satisfaction in use, an important step was made in electrolytical zinc manufacturing. The fame of this installation was spreading fast over the whole world. Many a producer decided in favour of our techniques when planning new electrolytical units to be built during the last decade.

As a matter of fact this new type of automated cell house with jumbo 2.6 sq.m. cathodes can now be found in

3 plants in Belgium,
1 plant in the Netherlands,
1 plant in France,
1 plant in Austria,
1 plant in Mexico and
2 plants in the United States of America.

Together they represent a total production of more than 750,000 t zinc per year, which means the major part of the new production capacities for electrolytic zinc installed throughout the world since 1970. Other units of the same type are under construction or on the drawing board.

Furthermore, other plants that were already in production when our new automated installation came on stream, have been heavily inspired by our techniques, without asking directly for our assistance to automate those cell houses.

It goes without saying that the new Balen tankhouse has been the starting signal for an advanced mechanization of most electrolytic installations in the world. This automation became a must, since labour charges went up steeply, specially in Europe, and the unions insisted more and more strongly on the easement of working conditions.

Encouraged by this success and under pressure of the aforementioned economic circumstances, aggravated by the economic crisis with a steep drop of zinc quotations, Vieille-Montagne has decided in 1977 to mechanize completely the zinc electrolysis at Balen Plant by building a second automated cell house in replacement of all manual installations still in operation but out of date, having a capacity of approximately 100,000 tons/y.

As the barrier of the human scale, blocking the cathodes dimensions due to manual stripping, has given way, a road was opened to new optimum dimensions. So it was decided to start a new era in the conception and the optimalization of the electrolytic technology.

In 1979 a new tankhouse went on stream with a capacity of 115,000 t zinc cathodes per year, that can be lifted up to 150,000 t/year increasing the number of cathodes per cell.

First I intend to summarize the data concerning the first generation automated cell house in order to compare them afterwards with the subject of this paper, namely the completely new house of V.M. Balen.

2. Automated cell house with jumbo cathodes of 2.6 sq.m. submerged area.

This automated cell house, with a year capacity of 75,000 t zinc cathodes, comprises 168 cells arranged in 6 parallel rows of 28 cells each. (fig. 1).

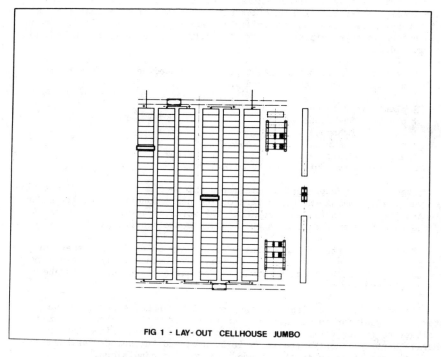

FIG 1 - LAY-OUT CELLHOUSE JUMBO

All cells are electrically in series and contain 44 cathodes each with a 2.6 sq.m. submerged area, i.e. 114 sq.m. useful surface per cell.

We adopted a maximum current density sufficiently low to be able to maintain a deposit time of 48 hours.

The average amperic yield is 90 % with a voltage of 3.38 Volts, giving for the deposit an electrical energy consumption of approx. 3,150 kWh A.C. per ton deposit zinc.

Distance between cathodes centre to centre is 90 mm.

Cooling of solutions is done outside in F.R.P. cooling towers with forced ventilation.

The circulation flow is fifteen times greater than the input of the purified solution to be electrolyzed.

The solutions' zinc content at the cells outlet is maintained at approx. 50 gr/liter; given 300 gr/l SO4 in the solution this corresponds with an acidity up to 200 gr/liter H2SO4. As far as the solution goes all cells are connected in parallel and the electrolysis is done directly in one step, at an end acidity and a residual zinc content permitting the solution to return directly to leaching without a second electrolysis step.

This simplification became possible because of the following items :

1) systematic brushing and washing of aluminium cathodes after each stripping, thus improving notably the deposits quality.

2) progress made in purifying solutions.

3) important recirculation of solutions obtaining thus a great homogeniety of cell's acidity and temperature.

In order to facilitate the placing of the cathodes during handling and also to avoid the bus-bars alongside the cells, those cells were jointly placed along the long side. Thus cathode and anode header contact are shaped to fit in each other.

The tankhouse disposes of two double stripping machines, placed in the adjoining hall. Each stripping tandem is composed of a 2 branch caroussel (fig. 2).

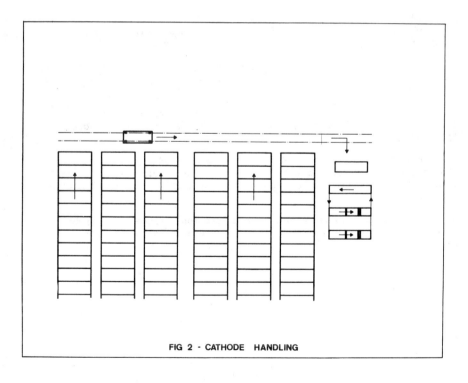

FIG 2 - CATHODE HANDLING

The chain conveyors n°1 and 2 form the two parallel branches, feeding 2 stripping units and 2 brushing ones in series with the stripping machines.

Three more chain conveyors and 6 transferring devices linking the conveyors together close the caroussel's circuit. The five chains are electrically driven, where all other movements are done by pneumatic cylinders.

Each tandem machine assures the stripping of 3 cell rows with cathode contacts situated on the same side.

A transfer car comes and goes between those rows and a stripping machine, while two cranes transport the cathode with or without zinc, one from cell to transfer car and the other from stripping unit's stocking chain to transfer car.

On the lower level each stripping tandem disposes furthermore of an automatic stacking installation for zinc sheets and removal of the stacks of sheets. The whole of the operations is monitored by a V.M. programmed electronic device.

The daily 48 hours deposit stripping of 82 cells is done in one shift by:
4 men at the 2 stripping machines
1 craneman controlling the 2 cranes in the rows.
1 lift truck driver to remove zinc sheets stacks to melting and
1 labourer
which means a total of 7 men per day, enabling an output of approx. 30 ton/man - shift.

The grooved lead anodes containing 0.75 % of silver are cleaned every six weeks with water under a 100 bars pressure and straightened by an automatic machine with a capacity of 4 to 6 cells a day.

At the same time the cells are vacuum cleaned in order to eliminate manganese sludge. Those operations are mostly done while keeping the cells on stream.

All anode transport operations are conducted manually with the above mentioned equipment during the afternoon shift. It takes 6 men to execute those operations.

Including supervision, solution's control, cleaning of pipes, launders and evaporators, detection and elimination of short-circuits, the total number of men necessary to conduct a cell house will be an average of 26 men per day.

The producing labour will be 1 hour per ton cathode i.e. approx. 3 times less than in a manual house. The maintenance labour is approx. 0.35 hour/ton cathodes.

3. New automated cell house with super-jumbo cathodes of 3.2 sq.m of submerged area.

When working out this new unit, we were guided by the following basic ideas:

a) to construct a unit   at least able to replace completely our manual cell houses, which means to have a minimum capacity of some 100,000 t of zinc cathodes per year and allowing a capacity extension up to 150,000 ton per year.

b) taking advantage of our experience acquired during the last years in conducting our automated cell house mentioned above in order to bring down the necessary investment costs by adapting the optimum size and number of cathodes per cell and furthermore the number of stripping machines and auxiliary machines needed.

c) in order to cut down investment costs an effort was made to simplify the installations.

d) production and maintenance labour costs had to be brought down by increased electrodes' dimensions, decreased number of cells and rows and by simplifying the installations.

A. Basis

As basic   characteristic we had imposed on ourselves a current density of approx. 400 Amp./sq.m. enabling a 48 hours stripping. Maintaining 90 mm distance between cathodes, the possibility to increase the electrode's headbars section and specially the acquired experience with cathodes' contacts have brought us now to an amperage per cathode of 1,300 Amp. maximum.

From there on the submerged cathode surface has been fixed on 3.2 sq.m., this corresponding in comparison with a jumbo cathode, with a 15 cm increase in length and 10 cm in width.

So the useful dimensions are: width 1 m, submerged height 1.6 m.

The use of the existing silicon rectifiers imposed 140 as **the number** of cells.

The number of cathodes per cell resulting from this imposition has been fixed on 86. We think that it will be possible to increase this up to 100 in another installation.

In the final stage the foreseen cathodic surface per cell will be 275 sq.m. (For the moment this is limited to 66 cathodes, i.e. 211 sq.m. per cell). Consequently the nominal amperage will reach 113,000 Amp.

B. Lay-out

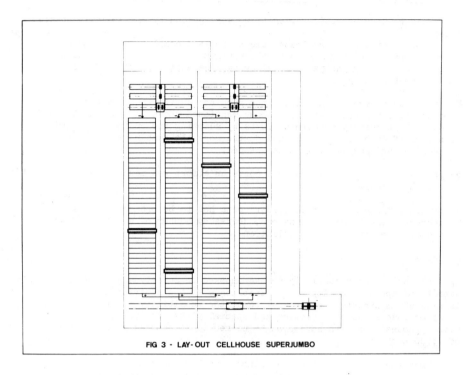

FIG 3 - LAY-OUT CELLHOUSE SUPERJUMBO

The 140 cells containing each 86 cathodes of 3.2 sq.m. and 87 anodes are disposed in 4 rows of 35 sandwich cells (fig. 3).

A double stripping machine at the head of each pair of rows can treat daily one row's cathodes, thus corresponding with the 48 hours deposit time.

Facing the stripping machine, at the other end of the rows, is a tunnel for the anode's transfer car directly in line with the washing - straightening machine for anodes. All cells are electrically in series, the row's order in series being 1 - 3 - 2 - 4 in order to obtain an identical cathodes'orientation in each pair of rows, i.e. all cathodes treated by machine 1 with contacts facing south and inversely for machine 2,

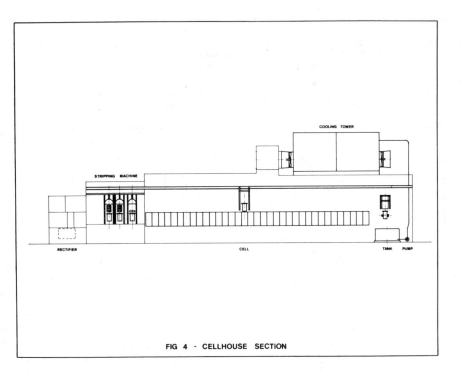

FIG 4 - CELLHOUSE SECTION

Fig. 4 shows a longitudinal section through a row of cells.

One end of the house is reserved for electric current supply, transformer rectifiers, with on top offices and control room; and stripping, stacking and evacuation of zinc.

The other end contains everything concerning the anode's and solution's circulation, i.e.:

- on the upper flooring the cooling towers disposed in such a way that part of the necessary air can be taken from the tankhouse,

- on the bottom flooring the circulation tanks and pumps,

- and on the intermediate flooring, past the anodes handling tunnel, a passage for the transfer of the crane handling the anodes in the different rows.

The total occupied surface of approx. 4,000 sq.m. represents not more than 55 % of the occupied surface in a first generation jumbo cathode tankhouse having an identical capacity.

C. Cells

In order to cut down the necessary surface, the cost of the cells and the electrodes heads and the corrosion problems, cells are no longer independent tanks with intermediate spacing, but 2 concrete upside down T's put together.

So each wall is common to two cells (fig. 5).

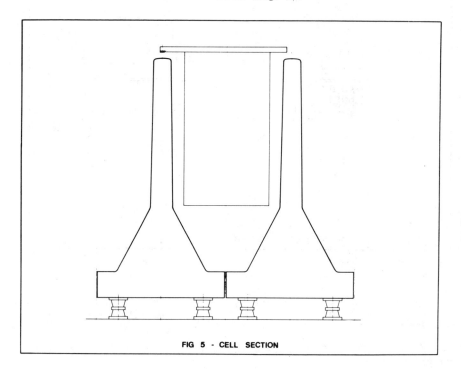

FIG 5 - CELL SECTION

Insulation between cells, only done by pvc-linings, is more than sufficient to resist the small potential differences between 2 cells. So the electrode heads are shortened by approx. 15 cm.

In order to facilitate the vacuum evacuation of manganese sludge from the cells, without removing the electrodes and consequently without time or productivitiy losses, the cell's bottom is V-shaped with a slight slope in the length direction.

The cell's overflows take the full length of a small side of the cells to avoid changing solution levels, that can cause zinc deposits ending in a sloping edge which makes stripping uneasy.

D. Electrodes

The Al cathodes are composed of a 99.5 % aluminium sheet measuring 1,000 x 1,740 x 7 mm thickness surmounted by a hard aluminium profile at 0.5 % MgSi, measuring 1,490 x 60 x 25 mm argon weld automatically on both sides of the sheet.

Two steel lifting hooks are set in the aluminium cathode head bars and an exploded Cu-Al contact is welded under the head end.

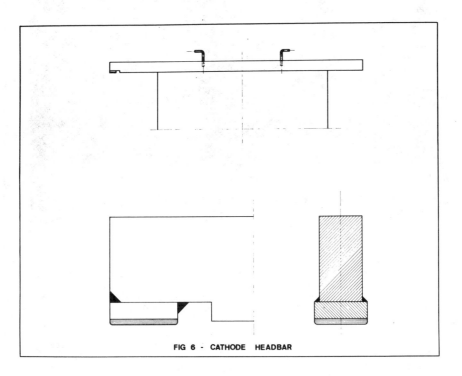

FIG 6 - CATHODE HEADBAR

213

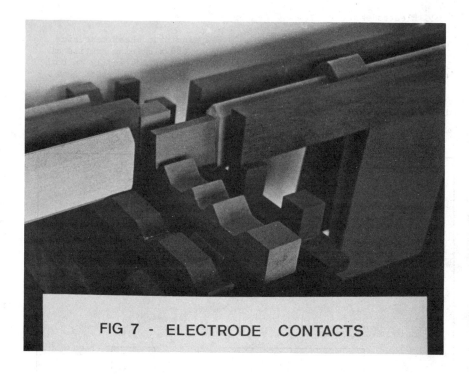

FIG 7 - ELECTRODE CONTACTS

The Al sheets are grooved on both large sides in order to improve the fastening of 2 plastic strips in clip form. Those are furthermore sealed to the sheet by silicon rubber. Thus the aluminium corrosion under the strips is eliminated, maintenance is cut down and tearing loose of strips during cathodes passage in the brushing machine is avoided.

The grooved 14 mm thick lead anodes surmounted by a 15 x 55 mm lead covered copper head are moulded in one piece. The silver content of the anode's lead has been brought down to 0,5 %. They measure 940 x 1,705 mm and weight 220 kg.

Six polypropylene clip type insulators equip each anode in order to maintain the distance between anode and cathode.

For easy handling by the very same lifting grab that handles cathodes without first removing those cathodes as it is done with the jumbo-type, anodes have 2 stainless steel hooks fixed to the copper bar, and cathodes and anodes are positioned, on their contact side, on a toothed copper bar placed on the cell wall as represented in fig. 7.

E. Stripping

There are 2 stripping machines, one at the head of each pair of cell rows (fig. 8).

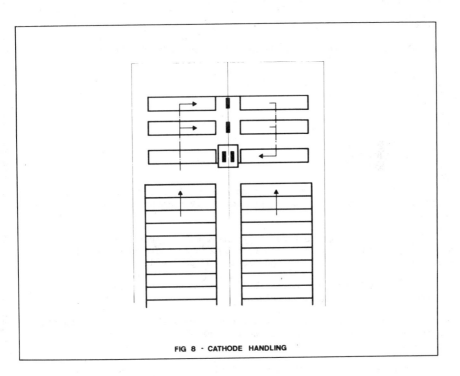

FIG 8 - CATHODE HANDLING

This disposition simplifies considerably handling of cathodes and still more so, while the cathodic surface per cell is greater. Indeed for the 2.6 sq.m. jumbo cathodes and with a year capacity of 150 000 t zinc, six stripping machines would be needed with this disposition, whereas the disposition in an annexed house would do with 4 machines and 4 transfer cars.

Each machine is simply composed of 3 chain conveyors, two feeding the 2 parallel stripping units and the third one feeding the double brushing unit, while serving also as a return loop for the cathodes in the alignment of the cell's crane.

Cell crane nr 1 transfers the zinc **laden** cathodes between cells and stripping chains 1 and 2 and returns the stripped and brushed cathodes from chain 3 to the cells.

Crane 1 bis, travelling on the road of the second row, is then used to transfer the stripped cathodes from chains 1 and 2 to brushing chain 3.

The following day all chain travelling directions are inversed to secure the second row's stripping.

So each set is composed of:

- 3 chain conveyors with direct acting on two extremities
- 2 stripping units
- 1 double brushing unit
- 2 lifting and translation cranes
- 1 stacking unit.

Are no longer needed :

- transfer cars
- 2 transversal chains
- 6 transferring devices.

This simplification leads to a reduced cost of stripping machines and programmer and furthermore to a less important maintenance and an easier conduct.

The actual guillotine type stripping machines are identical to those of the first generation and allow both automatic and controlled stripping every 24 seconds, i.e. 300 cathodes/hour and per tandem.

A good visibility and accessibility make the accidental manual interventions easy and without risks.

Thanks to perfect cathode guidance and precise positioning, cranes can operate completely automatic, being supervised at a distance by the machine men.

Automatic water sprinkling of contacts and headbars assures a good current repartition between electrodes and minimum potential drops on contacts.

Situated on the lower flooring, the two stacking units of both tandems deliver by chain conveyors to the same lift where stacks of zinc sheets are put at the disposal of a lift truck on the loading level of the melting furnaces. This lift truck circulates without danger in its own lane.

Taking into account the foremen, the lift truck driver, stops and cleaning time of the machine, global stripping requires 0.23 hour/ton zinc cathodes, i.e. 40 % less than the first generation machines (see complete table nr 3).

F. Solutions

As for the solution each row forms an independent loop consisting of: (fig. 9)
- 1 double cooling tower
- 1 feed manifold towards cells
- 1 cell's outlet manifold
- 1 circulation tank
- 2 circulation pumps towards evaporators.

216

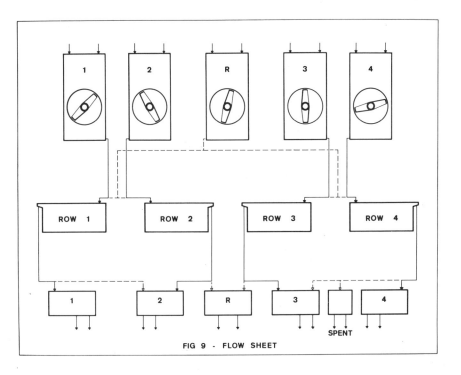

FIG 9 - FLOW SHEET

Each circuit has a feeding of fresh purified solution and the surplus volume is drained by simple overflow of the circulation tank into a dispatching spent tank.

Each loop is conceived in such a way that pipes are as short and as straight lined as possible in order to facilitate maintenance.

The cell's row can be bypassed by the loop for a weekly 24 hours water washing of evaporators and main pipes in order to prevent gypsum clogging.

A spare loop evaporator-tank-pump can then be connected to inlet and outlet of the solutions of the concerned cell's row.

The cell's feeding collector of 500mm diameter is in flexible rubber, thus enabling at each distribution stop the elimination of gypsum incrustations by simply bending the pipes. The individual feeding tubes are calibrated and are feeding the cells without adjustment. Only one adjustment at the collector's head maintains a constant pressure in same.

This simplification cuts down maintenance and untimely bath level changes that disturb the stripping rhythm.

The evaporators consist of double empty towers in F.R.P. with an horizontal section of 10 x 5 m, placed back to back and with each on top

217

a network of 72 stainless 317 pulverizers and 3 beds of droplet demisters in pvc waves. Each tower has a 428,000 cu. m./h fan with a 50 HP two speed motor.

The solutions' output is 400 cu. m./h per tower enabling for a 5° C difference an evacuation of 2,100,000 kcal/h.

G. Anodes maintenance

This is an extremely important operation for the smooth running of the electrolysis and we made a real effort to install a high productivity installation with an intense frequency of anodes cleaning.

Every day one row out of two is at disposal for anodes maintenance. Furtheron and due to the hooks on anodes and intermediate bus-bars, the anodes become independent of the cathodes and can be replaced according to the same principle as cathodes, even during stripping.

A 6 t capacity crane with special grab in three sections enables lifting in one go of 22 anodes (approx. 5 t), spaced 18 cm, i.e. one anode out of two. In that way current is not cut off on any cathode.

Four trips between cell and anode car are necessary to replace all dirty anodes of one cell by clean straigthened anodes coming from washing-straightening.

An automatic program assures the car's move to the anode treatment, the high pressure water washing (more than 100 bars), the anode straigthening and the return to the anode crane.

A team of 3 men stands for the anodes treatment of 3 to 4 cells per shift and also for the evacuation of the MnO2 sludges vacuum cleaned while the cell remains on stream.

The cleaning cycle could thus be brought down to 3 - 4 weeks with 2 shifts a day, sunday excepted and nevertheless labour costs were cut down for this operation from 0.21 h/t zinc cathodes to 0.13 h/t zinc cathodes, meaning a gain of 39 % (see complete table nr 3).

Due to this more preventive systematic treatment the costly intervention on individual anodes in order to eliminate short circuits is strongly diminished.

In case of inspection or repair of a supply pvc cell lining, it is possible to short circuit a cell by replacing the cathodes by a set of comb forming bars substituted to the cathodes.

This short circuiting is done by cutting tension at the electrolysis terminals. This comb being on stream it becomes possible to take out the anodes of the concerned cell and to empty the cell itself while stripping operations in the adjacent cells can continue quite normally.

H. Summary of the compared characteristics

Table 1 shows the main characteristics of automated tankhouses of · both first and second generation.

Table 1

Cathodes	units	jumbos	superjumbos	differences
capacity	t/year	150,000	150,000	
total current	A	47,500	113,000	
current density	A/sq.m.	415	410	
deposit time	h	48	48	
cells		330	140	- 58 %
cathodes/cell		44	86	
surface/cathodes	sq.m.	2.6	3.2	
cathodic surface/cell	sq.m.	114	275	+141 %
prod./cell/day	t	1.25	2.98	+138 %
stripping machines		4	2	- 50 %
prod./mach. h	t	17	20,8	+ 22 %
flow/cell	m3/h	6	23	
area	sq.m.	7,280	4,000	- 45 %

I. Compared investment costs

Table 2 hereafter shows compared investment costs as per May 1979 in million U.S. dollar.

Following metal prices were used as base :

Ag   242.4 $/kg

Pb   1.03 $/kg

Al   1.97 $/kg

Table 2

	units	jumbos	superjumbos	difference
electrodes	million $	6.6	6	- 9 %
immovable	"	30.2	21.6	-28,5 %
total	"	36.8	27.6	-25 %
investment/t cath.y	"	245	184	

J. Compared working costs

Table 3 hereafter shows compared costs for productive and maintenance labour needed for both types of installation.

Table 3

		jumbos	superjumbos	difference
stripping	h/t	0.38	0,23	- 40 %
cell+anode cleaning	"	0,21	0,13	- 38 %
control	"	0,12	0,06	- 50 %
others	"	0,29	0,17	- 41 %
total productive	"	1,00	0,59	- 41 %
maintenance	"	0,35	0,21	- 40 %
total	"	1,35	0,80	- 41 %

K. Capacity limit

The capacity limit of such an installation is estimated at 200,000 t zinc cathodes/year by bringing up the number of cathodes per cell to 100 and by raising the number of cells per row up to 42, this meaning a maximum in view of the time taken by the crane for one trip. It goes without saying that investment costs will be most advantageous at this level.

It is possible though to construct units smaller than 100,000 t/year in an economic way. In that case the installation will only count 2 rows and 1 stripping machine.

In order to cope with the problem of cathodes placed differently in the two rows, two methods are available :

a) an identical disposition of cathodes in both rows, but this makes the investment rather costly as bus-bars have to be foreseen along both rows,

or

b) to evacuate daily the cathodes in stock on the stripping machine to a rack and replace them by reversely oriented cathodes. This solution requires some more space and lowers slightly the stripping productivity, but asks for very little supplementary investment.

4. Conclusion

As a conclusion we can state that the double aim to cut down investment and to boost productivity has been attained by working on the same parameters :

a) reducing of number of electrodes and
                  number of cells and
consequently number of handling and stripping machines,

b) simplifying dispositions and machines, but with higher capacities for the latter.

Moreover those points mentioned above made working conditions better by reducing manual interventions and number of points to be controlled and by making different machines more easily accessible.

- - - - - -

Reference :
(1) Jean André, Roger Delvaux, Production of Electrolytic Zinc at the
                              Balen plant of S.A. Vieille-Montagne
                              AIME, Lead and Zinc 1970 Vol.II chapter 6.

COMPARISON OF CELLHOUSE CONCEPTS IN ELECTROLYTIC ZINC PLANTS

G. Freeman
A. Pyatt
The SNC Group

Summary

This paper describes the evolution of the electrolytic zinc cellhouse to
its present day automated form. Designs of the various automated cellhouses
are described with emphasis upon cathode handling and automated stripping.
Cost and operating parameters are tabulated.

Introduction

In October 1969, the Balen zinc plant of Vieille Montagne brought on stream a 60,000 tpy electrolytic zinc plant that introduced the automated cellhouse. The zinc industry, for a number of reasons, had recognized the need to mechanize the stripping of zinc. Many companies had applied themselves to the expensive and time consuming task with various degrees of success.

It is correct to say that the development of mechanical stripping of zinc cathode ushered in a new era of technology in the zinc industry. This new era resulted, in many instances, in large tonnage plants replacing many of the scattered small tonnage operations previously popular in the industry. The mechanical stripping of zinc cathodes has permitted the plant owners to resolve many labour and environmental problems while at the same time enjoy the benefits of lower unit costs that allow zinc to be the competitive metal it is today.

The object of this paper is to describe firstly the mechanical stripping systems available and secondly the mechanical cathode handling systems that are or may be associated with the mechanical stripping systems. In addition, this paper will describe six systems from the aspects of capital cost, maintenance, and manpower requirements. As zinc plants, either existing or planned, have different requirements, it is a further objective of this paper to present comparative data on the systems so as to inform the industry of the scope of the choices available.

Conventional Cellhouse

The conventional cellhouse as we know it today has evolved from the original designs of the Anaconda Mining Company, the Consolidated Mining and Smelting Company (COMINCO) and the Electrolytic Zinc Company of Australasia. These three companies are considered the pioneers and brought their plants into production between 1914-1918. The Anaconda plant in Great Falls, Montana was divided into six electrolyzing units of 156 cells each. The electrolyte was introduced into the first cell of the six cell cascade, the overflows being cumulative. The Consolidated Mining and Smelting Company operates their electrolyzing cells in units of 144 cells. Each unit is made of 4 groups of 36 cells arranged in 4 cascades of 9 cells each. Electrolysis at the Risdon plant is carried out in units of 144 cells. The electrolyte is fed to the cells individually in each cascade, the overflows being cumulative. This brief outline describes the general concept of cell arrangement as shown in the first generation of electrolytic zinc cellhouses. The next generation of electrolytic cellhouses was introduced in the Corpus Christi plant of Asarco in 1941. This plant of Asarco's introduced some new concepts in cellhouse design, arrangement and operating philosophy that have been translated into the present generation of automated cellhouses.

The zinc electrolytic cellhouses as described above were designed for the manual stripping and stacking of zinc cathode sheet. For this reason, the electrode size, cell size and zinc deposition times were designed for a man to accomplish the tasks of stripping and stacking to a reasonable limit of his capability. The conventional cellhouse uses aluminum cathodes of approximately $1.1m^2$ for zinc deposition. Further it utilizes cast concrete-lead or plastic lined cells, and has cells arranged in rows of 14 or more. There is generally an overhead trolley arrangement to facilitate the moving of cathodes from the cell to the stripping area. The process of cathode washing, stripping and stacking is done manually.

The conventional cellhouse as defined by the 1941 Corpus Christi plant, has individual cell feeds for circulating spent with cell overflows per row being cumulative. The spent is cooled exterior to the cells. The spent electrolyte is recirculated with some portion of the return spent sent to the leaching plant. The remainder of the spent is cooled, fortified with purified zinc sulphate solution and the mixture returned to the cell circulation system. The conventional cellhouse, with manual stripping is the highest employer of labour of any section of a zinc plant. The zinc cathode stripped per man day varies from extremely low values to a high of 10-11 tons.

The physical arrangement for the conventional cellhouse as portrayed in Photo 1 and Photo 2 shows the concept of the design.

Table 1 details some information on conventional cellhouses.

Photo No. 1 - Conventional Cellhouse at
Canadian Electrolytic Zinc Ltd.

TABLE I CONVENTIONAL CELLHOUSE DATA

	CORPUS CHRISTI	RUHR ZINC	CANADIAN ELEC. ZINC	OUTOKUMPU OY	TEXAS GULF	COMINCO	BUNKER HILL
CELL ARRANGEMENT	Row	Row	Row	Row	Row	Cascade	Cascade
CATHODES PER CELL	–	40	36	–	40	32	–
CATHODE SIZE M^2	1.1	1.1	1.1	1.1	1.1	1.1	1.1
CURRENT DENSITY AMPS/M^2	500–700	500–600	500–700	500–700	500–700	400–600	N.A.
METHOD OF STRIPPING	Manual	Auto	Manual	Auto	Manual	Manual	Manual
DEPOSITION TIME	24 hr.	24 hr.	24 hr.	24 hr.	24 hr.	24 hr.	24 hr.

Photo No.2- Cascade Cellhouse at Cominco Ltd.
Trail, B.C.

The conventional cellhouse portrayed in Photo 1 does allow for mechanization-automation. However, it is not generally considered practical to automate the cascade cell arrangement of Photo 2.

The Requirements for Mechanization

A prerequisite for the design and installation of an automated or mechanized cellhouse is the ability to consistently produce good quality cathode zinc with few if any sticky cathodes.

The effective application of zinc technology shows in the quality of the zinc produced in the electrolytic section of a zinc plant. The quality of the cathode zinc deposit may reflect the application of process technology as far back as the roasting section, however, the cathode morphology and purity generally reflects the level of performance of the purification process. In addition to effective purification processes, the cellhouse must be maintained in good mechanical condition and electrolyte composition and distribution well controlled. The conventional cellhouse because of the number of cathodes, cells, etc., presents control problems that result in high labour requirements to carry out a variety of functions. The size and weight of the zinc deposit is such that it can be handled by the average worker (stripper). The stripper in a cellhouse exerts considerable impact on the cellhouse production as he exercises the "now automated" functions of cathode alignment, flash plating when necessary, anodizing, washing, and electrode plate rejection.

226

Cathode purity and current efficiency is dependent more upon the purification processes and the circulating spent composition than the performance of the stripping personnel. The spent electrolyte temperature, additives, and distribution pattern within the facility further influence the quality of cathode zinc production. Another area that influences production efficiency is the atmosphere over the cells. Poor environmental conditions, essentially the level of acid mist in the air has a negative influence on worker's productivity.

The recurring problems of sticky cathodes and variations of zinc deposit characteristics were not solved easily. The solution of these two problems was required in order to proceed with the automation of cathode stripping. The increasing cost of labour, and the increasing problems of labour in the 1950's and 60's caused zinc plant operators to consider the problems of automation more seriously. Within this time period, technical advances within the zinc industry were made. Developments such as electrolyte quality control meters, magnetic flow meters, cell temperature sensors and others, improved the control over the zinc plant processes. Electrolytic zinc practice began to move from the labour intensive mode to the present concepts of control and automation. In-depth studies were carried out by a number of firms, all generally orientated to the mechanical stripping of zinc cathodes. The attainment of a steady state morphology of zinc deposition onto aluminum sheets (electrodes) was obtained through the diligent application of known zinc technology, This accomplishment laid to rest to a large extent the problem that mechanized the automated systems faced in the handling of poor quality cathode zinc and sticky cathodes.

The confidence of the plant operators ability to produce cathode zinc sheets with repetitive characteristics of adhesion to the aluminum plate and mechanical stability has made the automatic stripping of zinc a reality.

Definition of Cellhouse Types

The term "automated cellhouse" has been used to describe not only the fully automated cellhouses but also the mechanized units. The mechanized cellhouse is defined as a unit where the cathode zinc stripping and stacking is automatic but a reasonable manual input to the mechanical mode of cathode removal and repositioning from the cells is still employed. In general, new cellhouse installations are of the fully automated type. The modernized or upgraded cellhouses, because of the many factors of cost, area configuration, and other criteria, prescribes that they be of the mechanized type. Automatic cathode stripping systems have been developed at Vieille Montagne, Mitsui, Akita, Toho Zinc, Cominco and Porto Marghera. With the exception of Cominco at this time, the other designs have been commercially marketed around the world. Six cellhouses are described.

Fully Automated Cellhouse

Vieille Montagne, S.A.

The V.M. cell room was designed and developed initially for a new 60,000 tpy facility at Balen, Belgium. The cell room designed by V.M. for its licensees are variations of the Balen design. The Balen cell room consists of 168 "Jumbo" cells arranged in six lines, each of 28 cells. The cells are lined with high density polyethylene "Paraliners"; each cell contains 45 anodes and 44 cathodes, each of $2.6m^2$ effective area.

The lower edge of the "Jumbo" anodes is equipped with "drop-shaped" plastic insulators to maintain cathode spacing.

"Jumbo" electrode handling is fully automated. A travelling crane removes every other cathode, 22 at a time, and delivers them to an automated rail transfer car. Then, the crane is reset and picks up 22 stripped cathodes from the transfer car. The crane returns the stripped cathodes to the cell from which the plated cathodes were removed. Prior to being lowered into position, the stripped cathodes are "anodized" by being briefly contacted with the adjacent anodes. Upon completing replacement of each alternative cathode of a 28-cell line, the travelling crane is automatically reset for removal and replacement of the cathodes left in place at first, or moved to the next cell line being serviced.

Photo No 3 - Stripping Machine
at National Zinc Co.

The automated rail transfer car delivers the 22-plated cathodes to the travelling crane of the stripping aisle, and is loaded with 22 stripped cathodes. The stripping aisle crane serves the cathodes stripping and anode pickling and straightening machines. Each alternate 28-cell line is served by one of the two travelling cranes, a transfer car at each end of the cell lines and a cathode stripping machine and anode pickling and straightening machine in a stripping aisle at each end of the cell room.

Eight of the 168 "Jumbo" cells are always down for two days of cleaning. All electrodes are removed and the anodes are pickled and straightened. Thus, 1, 760 cathodes are stripped every 48 hours by each of the two-unit automated stripping machines. This is accomplished in one 8-hour shift with two men on duty at each of the two stripping circuits.

228

Photo No 4 - Cell Crane at
National Zinc Co.

Fully automated operation permits the cell room foreman to observe the operation from the central control room and maintain audio communications with 2 two-men stripping crews. Production rate per man shift is 28 tons cathode zinc from the "Jumbo" cell room compared with seven tons from the "standard" cell rooms.

V.M.'s records indicate an average of ten "sticker" cathodes being encountered each day; they require manual stripping. "Jumbo" cell room current density is 410 A/m^2.

A variation of the design used in Balen, was incorporated in the design of other zinc plants such as that of the National Zinc Company. The stripping machines are in-line with the cell rows so that the crane acts as the only conveyor of cathodes between the cells and stripping machine. At National Zinc, each pair of cell rows is served by two double guillotine stripping machines and one double brushing machine. Thus, for the four cell rows, there are four guillotine stripping machines. With a 48-hour cathode deposit, two rows of 32 cells are stripped each day.

There are three cranes, two for normal operation and one spare. These cranes are designed to handle 22 alternate cathodes and 23 anodes from each cell. The cranes are manually operated and travel on trolley beams over each cell row. Alternate cathodes (22 total) are first removed from a cell. This doubles the current density in the remaining 22 cathodes in the cell for the duration of the crane cycle necessary to return 22 washed, stripped and brushed cathodes to the cell.

The cathodes are dipped in a washing tank containing heated water at the end of the cell row to remove the acid from the zinc deposit and introduce cleavage between the zinc and aluminum due to the difference in coefficient of expansion of zinc and aluminum. The crane then transfers the 22 cathodes to one or the other stripping machine feed conveyor which feeds cathode individually to the machine.

Before moving the cathodes into the stripping machine, the crane operator lifts the 22 cathodes which have been previously stripped and moves them to the brushing machine conveyor.

Then the crane lifts the 22 cathodes which have previously been brushed and returns them to the cell from where they were picked up at the start of the crane cycle and replaces them. The crane then is moved to the next adjacent cell and in the same order, repeats its cycle except that the cathodes are deposited in the cathode conveyor of the alternate stripping machine. The crane will repeat this cycle until the first 22 cathodes in the same row of 32 cells have been handled. Then the cathode grab is cranked to the position where the remaining 22 unstripped cathodes can be handled.

The cathodes are fed, one at a time, by the stripping machine feed conveyor. When positioned in the stripping machine the cathodes are raised by a pneumatic cylinder to the position to initiate the stripping of the zinc deposit. Two starting and two plowing knives contact the cathode blank above the deposit.

The stripping is initiated by the starting knives on each side of the cathode. Stripping is effected by raising the cathode fully by the pneumatic cylinder. If the deposit is reluctant to strip, the cathode is lowered in place and the cycle repeated.

Stripped zinc falls into a chute and is alternately stacked by an oscillating carriage chute underneath the stripping machine. Each two cathodes are transferred to a stacking table shared by the two stripping machines. Each package of 44 zinc cathodes is lowered onto a chain conveyor designed to feed each stack to an elevator for transfer to the cathode melting plant storage. A fork truck operator weighs and stacks the cathodes in the casting plant.

The brushing machine brushes two cathodes at a time between pairs of stainless steel wire brushes irrigated by water sprays to remove sulfate and aluminum powder.

With this cathode handling system there is no cathode transfer car since all of the transfer of cathodes through the system is handled by one crane operator. There are six operators used for stripping. Two for the cranes, two strippers and two helpers. Stripping requires six to seven hours per day during the day shift, seven days per week.

The V.M. design of automated cell house is the fully automated type. There are several variations of the basic design successfully installed around the world.

Automatic Stripping & Mechanical Handling of Electrodes

There are the four basic automatic cathode stripping machines. It has been the practice of many zinc plants to incorporate the automatic stripping of zinc with a partially or wholly manually attended cathode handling system. The four automatic stripping machines and their associated mechanical handling systems will be described by plant or company.

Akita Zinc

Photo 5 gives a general view of the Akita stripping machine. Fig. 1 shows the cellhouse layout and Fig. 2 a schematic of the mechanical handling system.

Photo No.5-View of Stripping Machine
at Akita

As shown in Fig. 1 the electrolysis section consists of three circuits each consisting of cells arranged according to the Walker System. A mechanical stripping machine and a gantry crane is installed in each circuit. Two operators are stationed in each circuit, one of them operates the crane and the other engages in watching of the stripping machine. Every 48 hours, 24 electrodeposited cathodes which correspond to half of the cathodes in a cell are lifted at one time from one cell and transferred to the stock conveyor of the stripping machine by the 2.5 tons gantry crane. After that the electrodeposited cathodes are settled on the stock conveyor, and then aluminum cathodes after stripping are transferred from the arrange conveyor and inserted in the cell by the crane. The electrodeposited cathodes on the stock conveyor are fed continuously to the stripping machine. The conveyor has the capacity of holding 64 cathodes and its length is about 5m. The electrodeposit is washed away by spray washing on the way to the stock conveyor. The stripping machine is operated ten seconds in treating one cathode. The stripped zinc sheets fall to the stacking machine set under the floor. The stripped aluminum cathodes are sent to the subsequent conveyors by trolley conveyor.

The aluminum cathodes which are judged by the stripping machine operator to be impossible to strip of the deposited zinc are transferred via trolley conveyor to the disordered conveyor. The cathodes on the disordered conveyor are transferred to the pickling tank and immersed in electrolyte. After 10 hours, these cathodes are transferred to the polishing machine in order to make the surface smooth. The polished cathodes are transferred to the make up conveyor by the crane in order to make up the rejected cathodes

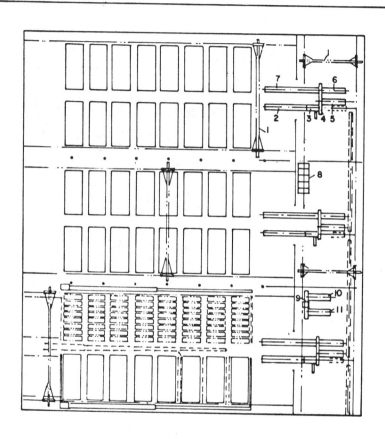

I	CRANE
2	STOCK CONVEYER.
3	STRIPPING MACHINE
4	TROLLEY CONVEYER
5	DISORDERED CATHODE CONVEYER
6	MAKEUP CONVEYER
7	ARRANGING CONVEYER
8	PICKLING TANK
9	CATHODE POLISHING MACHINE
10	POLISHED CATHODE CONVEYER
11	STOCK CONVEYER

FIGURE I   CELLHOUSE LAYOUT AT AKITA

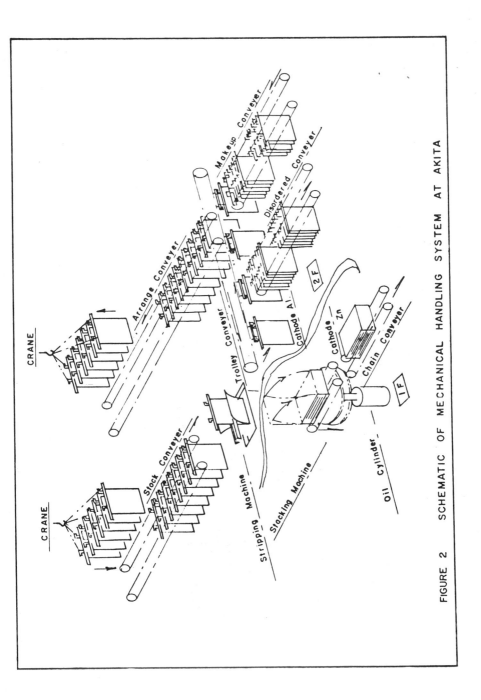

FIGURE 2   SCHEMATIC OF MECHANICAL HANDLING SYSTEM AT AKITA

to the disordered conveyor. The aluminum cathodes are sent to the arrange conveyor either from the trolley conveyor or from the make up conveyor for delivery to the cells in good order with a fixed center distance. The length of the conveyor is approximately 7m. The stripped zinc sheets are stacked automatically in bundles (Max 1.8 tons/unit) and sent to a next conveyor. All equipment incorporated in the mechanical stripping machine system is driven by electric motors, hydraulic power or compressed air. The stripping machine is equipped with its own control panel, operation panel and power switch board. If necessary, it is possible to operate the whole stripping system manually from the operation panel.

Mitsui Mining & Smelting Co. Ltd. - Hikoshima Zinc Plant

The Hikoshima cell room consists of 288 primary and 36 scavenger cells arranged in 18 lines (each of 18 cells). Each cell contains 33 cathodes and 34 anodes; the cells are rubber-lined. Effective cathode area is approximately $1.8m^2$. The sides and bottom of the cathodes are edged with plastic insulating strips.

A monorail hoist tended by an operator removes 11 cathodes from one of the cells. The cathodes are washed with water-jets, picked up by a mono-rail crane on an elliptic track, and delivered to one of two stripping machines, each tended by an operator.

At the present slab zinc capacity of 60,000 tpy, current densities are 300 A/m^2 during week-day peak hours and 500 A/m^2 during off-hours; stripping cycle is 32 to 48 hours. Electrolyte temperature ranges from 35° to 38^oC; purified solution is cooled in one and circulating electrolyte in 11 TCA (turbulent contact absorber) cooling towers, built on license from Universal Oil Products. A spare tower is provided so that the neutral solution tower may be cleaned every 15 days and the circulating electrolyte towers every 45 days.

Photo No.6-Stripping Machine -Mitsui Mining & Smelting Co.Ltd.

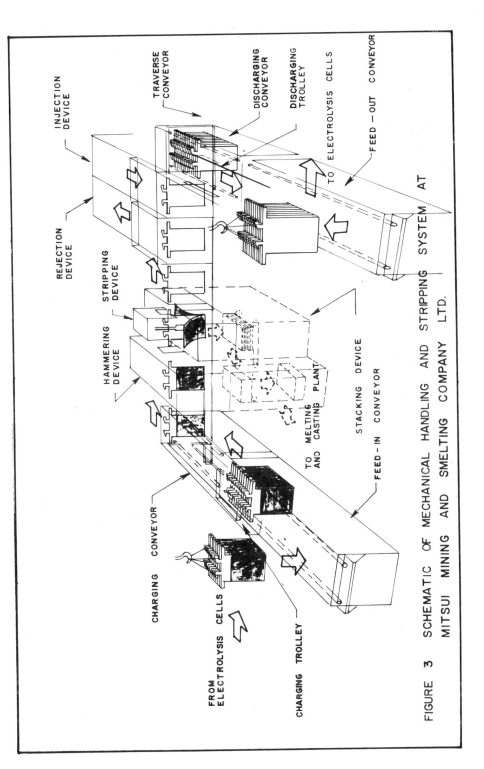

INJECTION DEVICE

REJECTION DEVICE

STRIPPING DEVICE

HAMMERING DEVICE

TRAVERSE CONVEYOR

DISCHARGING CONVEYOR

DISCHARGING TROLLEY

TO ELECTROLYSIS CELLS

FEED—OUT CONVEYOR

CHARGING CONVEYOR

FROM ELECTROLYSIS CELLS

CHARGING TROLLEY

STACKING DEVICE

FEED—IN CONVEYOR

TO MELTING AND CASTING PLANT

FIGURE 3   SCHEMATIC OF MECHANICAL HANDLING AND STRIPPING SYSTEM AT MITSUI MINING AND SMELTING COMPANY LTD.

In the cathode stripping machine, the cathodes are then stored in the charging conveyor, shifted into traverse conveyor, sent into the hammering station and the subsequent stripping station, where stripping of the deposited sheets is carried out. The stripped cathode is then shifted to the discharging conveyor and finally sent to the feed-out conveyor.

The travelling crane hoists pick up eleven (11) deposited cathodes each from the cells and the crane carries the cathodes to one of the transfer car trains staying at the end of the cell row; each train consists of two (2) transfer cars, receiving cathodes from the respective crane hoists. There is also another train staying nearby to feed the stripped cathode mother plates in turn to the cells by using the same travelling crane. These operations are done manually by crane operators.

The cathodes loaded on the transfer cars are sent to the other end by their built-in chain conveyors, where they are then picked up by circular trolley hoists; two (2) hoists are connected into a pair so that they can handle two (2) cathode bundles (11 cathodes x 2 - 22 cathodes) at a time. This is also done manually by a hoist operator.

The circulator trolley hoists will carry the cathodes to above the Wash Bath where the cathodes are lowered into the bath for washing and lifted again to resume travelling toward the cathode stripping machines. This is done automatically.

When the trolley hoists arrive at the cathode stripping machine, the bundles are lowered into the feed-in conveyor of the stripping machine one by one by a hoist operator. The hoists which have now become empty pick up the stripped cathode mother plates and return them toward the transfer cars.

Canadian Electrolytic Zinc Ltd.

The new cellhouse of the Canadian Electrolytic Zinc plant has incorporated the Mitsui automatic zinc cathode stripping machine with their own design for mechanical handling of electrodes. The handling and stripping system is shown schematically in Figure 4.

The development work on a mechanized cathode handling and stripping system began in 1969 with the intention of mechanizing this labour intensive work. As the system evolved through the development stages the priorities changed from the development of a stripping technique to a fast handling of cathodes to and from a stripping area without sacrificing the simplicity and reliability of the system.

At this point, application of materials handling equipment proven in the heavy steel, aluminum and automotive industries became essential. Cathode stripping machines, developed and operated by Mitsui Mining and Smelting Co. in their Kamioka and Hikoshima zinc plant were selected. Further, these machines could be adapted to convert manual stripping in their other conventional cellhouse. There are two cellhouses, a conventional unit with manual stripping producing 130,700 tpy zinc, and a new cellhouse producing 73,500 tpy.

The new mechanized cellhouse contains 336 cells arranged in two (2) electrical circuits of 14 rows of 12 cells each. Each cell contains 45 lead anodes and 44 aluminum cathodes. The submerged area of the electrode is $1.6m^2$. The designed current density of 375 A/m^2 should yield a 90% current efficiency on a 48 hours deposition period. There are 3 stripping machines, however the stripping is carried out by 2 machines at a time.

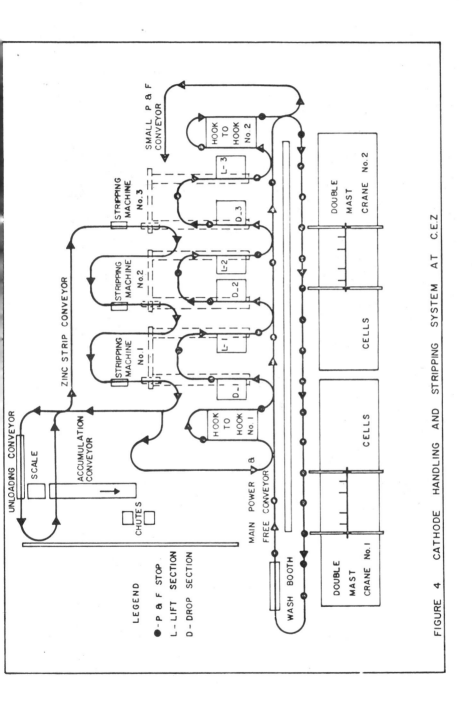

FIGURE 4 CATHODE HANDLING AND STRIPPING SYSTEM AT C.E.Z

ine third machine is standing by.

Cathode stripping is done in an area isolated from the cellroom. Cathodes are transported to this area with an overhead conveyor running the entire length of the building. This conveyor is fed by a crane which handles two rows of cells simultaneously. All operations of the cathode handling and stripping system are fully automatic, except for the very final steps of cathodes insertion and removal from the cells.

Photo 7 - Double mast crane at CEZ

Photo 7 shows one of the two double mast cranes. Both cranes can handle the cathodes in and out of the cells. However, while one is used for feeding stripping machines, the other is used for cell cleaning. Sets of 22 cathodes are put on the overhead conveyor which operates on the power and free principle. They are conveyed to a washbooth where they are washed with water at 80°C for 1 to 2 minutes. After washing they proceed to the stripping area where they arrive dry in order to avoid corrosion of the machines. At the entrance of a stripping machine, the sets of 22 cathodes are lowered and the grab mechanism releases the cathodes on a loading conveyor. The cathodes are now handled individually. To start with, the cathode passes a hammering station. It is hammered for 1 to 2 seconds in order to loosen the zinc deposit from the aluminum sheet. Stripping wedges which work on the whole length of the cathode ensure the complete separation. Stripping is helped by means of small low pressure air jets on each side of the sheet. The zinc plates are then stacked and positioned on an elevator.

Once some 25-35 zinc plates have been stacked, an elevator which is part of the stripping machine lowers it to a stack transfer area below the stripping machines. Here the stack is transferred to a scissor lift table, from which it is picked up by a standard overhead conveyor. This conveyor is equipped with heavy double C-hooks. The cathode stack is then transported to the holding area prior to melting. An electronic in line scale accumulates the weight of each stack. At the end of the accumulation conveyor, a fork truck builds bundles out of three stacks and charges them directly into the melting furnace or stores them on the charging floor. The charging of the bundles into the furnace can be mechanized but it is not so

238

at this time.

At present, cathode zinc production from the new cellhouse is being maximized to benefit from higher productivity. Deposition time varies from 40 to 42 hours and some 8,500 cathodes are stripped daily. The deposition time will be increased through further optimization of the operating conditions. The stripping of these 8,500 cathodes is done on two shifts per day, seven days a week. Present labour requirements include a cathode handling and stripping crew of four men, who are being replaced during their breaks by two additional men. These two additional men are also involved in handling reject cathodes to the cathode maintenance area, in cathode buffing and also in steaming and aligning cathode contacts on the cells.

The cathode buffing is done on a three month cycle. Stickiness of the zinc deposit has not been a problem. Only 0.9% of the cathodes are rejected for reasons of stickiness or edge strip and head bar maintenance.

The CEZ mechanization of cathode handling has proven to be an excellent system to orderly schedule and program the stripping of zinc cathode, cathode weighing, delivery of electrodes to and from electrolyzing cells to accomplish a high degree of cellhouse automation.

Toho Zinc Co. Ltd.

Toho developed at its Annaka zinc plant, west of Takasaki (122 km. from Tokyo), an automated cathode stripping system as a modification to four circuits of an existing conventional cell room. There are two rows (each of 36 cells) per circuit; each cell contains 32 conventional cathodes of $1.25m^2$ effective area. No insulated edging is provided for the cathodes. Current density is 400 A/m²; stripping cycle is 48 hours. Thus 4,608 cathodes are stripped in two shifts per day allowing ten seconds per cathode at 6.5 hours net time worked per shift. A monorail loop serves the cells, cathode wash rack and stripping machine. Sixteen cathodes are removed, washed and replaced every 160 seconds and two cathodes are stripped every 20 seconds. Removal of the cathodes from the cells and their replacement in the cells is manually controlled. All other operations of the system are automated. The circuit is manned with a crew of two; a hoist man and a lead operator who monitors the entire operation.

Cominco

The new pilot Cominco cellhouse including automated zinc cathode stripping and cathode handling is described as follows:

The cells are arranged in two rows of fifteen at the same elevation. A walkway runs the full length between the two rows of cells. The circulation system design employs piping rather than launders wherever possible. Cell overflow is recirculated through the cells in closed circuit at a rate of about twenty times the flow rate of neutral feed solution.

Cathodes are transported to and from the stripping machine by crane carriage. It is of 6-tons capacity, bridging both rows of cells. The cathode carriage is supported by two cables and restrained by four fixed and four floating guide wheels. The carriage lifts twenty-four cathodes double spaced in half-cell sets. The crane bridge has two Harnischfeger P and H auxiliary hoists of 3-tonne and 1-½ tonne capacity. These auxiliary hoists are used to transport stripping machine reject cathodes, in groups of up to thirty, to holding racks. The racks are moved by forklift to maintenance and brushing area. Brushing is carried out on one cathode at a time using an air hoist and a mechanical steel brush.

The new plant cathode zinc stripping technique is based on the Monteponi and Montevecchio patented hinged edge piece. The Cominco designed stripping machine includes a walking beam capable of receiving twenty-four unstripped cathodes from the crane, a cathode transfer mechanism, a cathode reject system, a walking beam for handling twenty-four stripped cathodes and a cathode zinc stacker.

Photo 8 — Stripping Machine
at Cominco Ltd

The power supply consists of two 33,000 ampere silicon diode recti-fiers paralleled to give a combined capacity of 66,000 amperes at 125 volts. This combination of two rectifiers provides considerable flexibility for operating over a wide range of current densities.

The cells are made of reinforced concrete with an FRP liner. The cell dimensions are 4.5 m by 1.1 m by 1.8 m deep, and the bottom of the cell extends into two hoppers. The hopper-bottom design was used to extend the time between cell cleanings and reduce cleaning labour. The cells are supported by steel I-beams on concrete columns. The footings are protected by a sloped acid-brick floor.

The cathodes are of all-aluminum construction except the contact tip, which is fabricated from copper. Two lugs are welded on top of the header bar to facilitate handling by crane.

Each cathode has a hinged edge piece (based on Monteponi and Montevecchio Canadian Patent 948157) and two edge sticks. The edge sticks are made from polypropylene and swivels on an aluminum pin. Cathode sheets are 89 by

152 cm. There are forty-eight cathodes per cell, spaced at 90mm center to center and each cathode has a submerged area of $2.3m^2$ and weighs approximately 27 kg.

Photo No. 9 - Cell room layout at Cominco Ltd.

Each anode for the new plant weighs about 214 kg. The anode header consists of a 130 cm by 5.7 cm by 1.9 cm copper bar coated with 6 percent antimonial lead. An 84 cm by 150 cm sheet of 0.75 percent Ag-Pb alloy is lead-burned to the header bar. The sheet has three bottom grooves for attaching bottom spacer clips. (Cominco, Canadian Patent 970718).

The electrode contact system is designed for 1200 A/contact. The system employs a spool-type contact bar (patent pending). Both the anodes and cathodes have inverted V-notch copper contacts. Contact is made on at least two points. This type of contact system is unique in that it fixes geometry in all directions.

The new cell is designed so that one end of the anode header is fixed by the spool contact bar. The other end of the anode header is fixed by an insulator block with restricting ridges. All electrodes hang independently and can be removed individually.

The cathode position is controlled at five points. One end of the cathode header bar is fixed by the spool contact bar. The other end of the bar is positioned between anode top spacers. The cathode sheet is guided into the cell by the top anode clip and held in place by three bottom spacer clips. The new Cominco cellhouse is being designed to produce 300,000 stpy of cathode zinc.

Table 2, Comparison of Operating and Cost Parameters, collects data pertinent to the cellhouses described.

Table 2, Comparison of Operating & Cost Parameters

	V.M.Design as at Balen Plant	V.M. Design as at Nation. Zinc	COMINCO	MITSUI	AKITA	C.E.Z.
Immersed Cathode Area (m²)	2.6	2.6	3.0	1.7	1.62	1.6
Current Density (Ampt/m²)	410-420	410-420	400	600	250-540	510
Current per Contract (Amps)	1070-1090	1070-1090	1200	1000	875	820
Cathode Deposition Time (Hours)	48	48	72	32-48	48	36
Capital Cost of Cathode Handling and Stripping System for 100,000 MTPY of Slab Zinc Production (US$) *5	4×10^6	2.4×10^6	N.A.	-	1.7×10^6	2.9×10^6
Maintenance Cost for Cathode Handling and Stripping Equipment per Annual Metric Ton of Cathode (US$/mt)	-	-	-	0.91	1.1	1.0
Building Area Required for Cellhouse Cathode Handling and Stripping for a 100,000 MTPY of Slab Zinc Production (m²)	1500	930	N.A.	1400*1	5112*2	2500*3
Building Area per Annual Metric Ton of Cathode Zinc (m²/met)	.014	.0087	-	.056*2	.05*2	.07*2
Capacity (Metric Tons of Cathode/Hour)	13-14	13-14	-	-	48	19*4

* See notes on following page.

Notes concerning Table 2 on the preceding page.

*1   This covers the area for the transfer cars, circular trolley hoists, cathode stripping machines and buffing machines.

*2   Includes building area of whole cell room.

*3   Area quoted is the area occupied by the conveyor system, stripping machines and the strip charging areas. This could be reduced with a different design of cell house configuration.

*4   Capacity calculated on 16 hours per day operation. Normally, two machines are operating with third on standby. Average capacity per machine is 9.5 metric tons per hour.

*5   Capital cost adjusted to 1979 Dollars.

Discussion of Factors for Design and Selection - Cellhouses

Cellhouse Building

There are several factors that influence the dimensions and design of the electrolyzing building. Cell size, arrangement and location of the stripping machines are pertinent factors. A quantum change in cell size has taken place and cells that accomodate 44 or more aluminum cathodes from 1.6 to $3.0m^2$ in size are becoming common. By way of illustration, the Balen plant of V-M when compared to an equivalent unit [1] having the stripping machines in line with the cell rows lowers the building space for 100,000 mtpy of zinc from 1,500 m^2 to 930 m^2, or from 0.014 m^3/mt to 0.0087 m^2/mt. These values assume that electrolyte cooling is outside of the building, as is the melting and casting. Vieille Montagne further states that the latter machine and cell arrangement lowers the capital cost per annual metric ton of zinc from U.S. \$37.4 to \$22.4. Similar reductions of capital cost (same plant size basis) result for the cathode handling system installation. The Cominco plant is being designed from grass roots and will similarly show economies of building requirement. The conventional or cascade type cellhouse by its cell configuration required approximately 4,500 m^2 for 100,000 tons of zinc per year.

A cellhouse should ideally be of low profile construction so as to incorporate a ventilation system. The ventilation system is important to provide a satisfactory atmosphere for both man and machine. The problems of corrosion are dramatically reduced in a properly ventilated cellhouse.

Type of Stripping Machine

The selection of the type of stripping machine is usually based on cost, stripping speed, and the physical damage to cathodes that may be expected. All of the stripping machines described earlier are capable, with modification, of stripping jumbo size cathodes. The economics of capital cost, machine maintenance and projected electrode damage must be considered. The cycle time per cathode for each machine defines the number of machines to be used and the hours per day the stripping cycle will cover.

The Balen machine operates at a cycle time of approximately 15 seconds, the other machines are generally operated at 10 seconds or less. When considering the machine cycle time, the size of the cathode must also be considered. For 100,000 tpy of zinc, Balen uses two stripping machines, CEZ has available 3 machines, Cominco will use 3 machines for 300,000 tpy of zinc.

Ampere Loading

The development of better contacts (copper-aluminum), has allowed the plant designers to move from the low values of 300-400 amperes/contact to the 1,200 amps/contact that Cominco will be using. The use of extruded aluminum headed bars is an improvement in the material of construction for electrodes. The extruded aluminum bar is mechanically stronger than the cast bar and also presents a more impervious surface to corrosion. In general, the type of copper-aluminum contact has moved away from the threaded type to the explosion bonded type.

(1) Communication - J. André, V.M.

244

There are various types of contacts; however, the explosion [1] bonded
type has gained much popularity recently in Canada, the U.S. and Germany.

Cell Construction

The electrolyzing cell is generally constructed of reinforced concrete.
Wood is also employed. The cell is lined with either chemical lead or a high
density polyethylene (paraliner). To facilitate cell cleaning, several cell
designs are available with bottom drains. The practice of vacuum cell
cleaning is widely used, however, the recent trend is to cell bottom drains.

Conclusions

As can be seen from the description of the automatic stripping and
cathode handling systems, there is a reasonable selection of cellhouse-
layout-stripping and handling arrangements available. The basic decision
on the approach to new cellhouse construction is the selection of the
mechanical handling system and the cell arrangement. Although, there may
be some economies of capital investment with regards to the stripping
machines, the major impact on the capital cost of a cellhouse is the
cathode handling equipment. At this time, the choice of stripping machines
is limited and this situation is likely to continue. The development of
cathode transport systems appears to be on-going. In the future, we can
expect even more sophisticated arrangements. The utilization of computers
to program the mechanical systems has been introduced in Balen. More
computer controlled installations are expected as new zinc cellhouse
construction takes place. In some plants the existing systems are undergoing continual development to tailor the systems to the individual plant
needs.

The technical developments in the area of zinc hydrometallurgy have
resulted in processes that are capable of consistently assuring good quality
zinc cathodes at elevated current efficiencies. This presented the opportunity
to optimize the materials handling operations in the electrolysis section
of the zinc plant. There has been much investigation into such criteria
as contact loading, power consumption per expended hours of deposition,
materials of construction for electrodes, electrolyte composition and
distribution, cell house ventilation plus many other aspects of the
electrowinning of zinc.

The results of the work that brought cellhouse operation to today's
level, is shown by the reduction in zinc cathode stripping labour. For
example, V-M reports 6.7 tons zinc stripped/manday for manual stripping
and 22.5 tons/manday for automated stripping in the Balen Plant. This order
of magnitude in labour reduction can be expected in a modern plant.

In addition to saving labour, plant operators have been able to work
on improving the working atmosphere both over the cells and within the cellhouse facility.

In summary, the zinc electrolytic cellhouse designer now has several
choices of automatic stripping machines available plus several proven
mechanical handling systems. The opportunity to design for full automation
of operations or partial automation is available. In closing, credit must

(1) Limpak Industries - Mississauga, Ontario, Canada.

go to the original designers of the equipment that presently gives much
benefit to the plant operators and to the people who work in the plants.

Acknowledgements

The authors would sincerely like to thank the companies and people who so
generously submitted published and other information that aided so much
in making this paper possible. Particular thanks goes to Mr.J. André of
Vieille Montagne for his assistance; to Canadian Electrolytic Zinc,
Mitsui Mining and Smelting, Cominco, Dowa Mining, and National Zinc for their
contributions.

We thank Mr. M. Watts (SNC) for his assistance in assembling and proof-
reading of the contents of this paper.

Some of the data for this publication was extracted from published sources
within the zinc industry and these sources are not individually
acknowledged.

THE ZINC-LEAD BLAST FURNACE INTO THE 1980s

C.F. Harris, A.W. Richards and A.W. Robson

Imperial Smelting Processes Limited,

Avonmouth, Bristol, England

At present there are thirteen zinc-lead blast furnaces operating in eleven countries. In 1978, the last full year for which records are complete, eleven of these furnaces produced approximately 650,000 metric tons of zinc and 300,000 metric tons of lead. The standard size furnace, with an upper shaft area of 17.2 m^2, more than doubled its annual capacity from 35,000 to 80,000 metric tons of zinc, and 17,000 to 40,000 metric tons of lead in the fifteen years up to 1978.

New techniques are being incorporated into existing furnaces which will further improve their standard of performance, recovery and profitability. These include a new method of controlled charging using ring and plug bells, bent tip tuyeres to increase gas/slag interaction, and the use of an auxiliary process to treat copper dross from lead bullion for copper recovery. Hot briquetting of oxides, residuals and drosses is an effective means of utilizing cheaper residual zinc-containing materials and has opened the way to the profitable concept of the zinc, rather than zinc-lead, blast furnace in which fluid-bed roasting of zinc concentrates will be followed by hot briquetting and smelting.

The blast furnace has proved to be the zinc and lead process most conservative of primary energy in existence and offers the basis of a flexible and viable non-ferrous metals smelting complex.

Predictions are made of the improvements expected in operating efficiency in the 1980s as a result of present and future development plans.

Introduction

Fluctuations in the demand for zinc and lead are familiar to all concerned in their production. In recent years there has been a world over-capacity for zinc production whilst the demand for lead has remained firm. Under the circumstances zinc producers in general have not been attempting production records but have been paying more attention to the efficiency of their processes. Nonetheless in 1978, the last full year for which records are complete, eleven of the twelve zinc-lead blast furnaces which were operating produced approximately 650,000 metric tons of zinc and 300,000 metric tons of lead equalling the previous highest annual total of 1974, the last full year before the present poor economic climate. The twelth furnace situated at Shaoguan in China was commissioned recently and a thirteenth furnace, the second at Miasteczko in Poland, was still being built at the time the data were assembled.

Production Data

Zinc and lead bullion production data for the eleven furnaces from their start-up dates is shown in Table I with design capacities for all thirteen furnaces. These latter figures are varied due to their dependence on the design of the particular complex in terms of acid and sinter plant capacity and the size of the blower and other ancillary equipment. The data for Copsa Mica are mainly estimated, whilst those for Avonmouth, where the present furnace was commissioned in 1967, are inclusive of earlier production since 1951. It will be noted that the present capability of the first furnace at Miasteczko has already overtaken the design capacity of the second furnace now being built.

Metallurgical Data

This is summarized in Table II which gives the mean and the best annual result for each of the parameters considered over a period of approximately 5 years. Most of the parameters are self explanatory but a few require further clarification. The interpretation of campaign duration can be difficult because the definition of a furnace campaign is not the same at all locations. Whatever the definition non technical matters such as industrial disputes and producer restrictions also play a part. It is suggested that the data presented here be used rather to illustrate that campaign duration is of significant length on all furnaces than to draw comparison between one operation and another. Condensation and separation efficiency defines the efficiency of recovery of zinc in slab metal from the gaseous zinc leaving the furnace shaft. The ISP carbon estimation is the percentage of carbon actually used compared with a consumption predicted from a simplified heat balance. The lower the numerical value the better the result which is affected by the reactivity and size of the coke used as well as by other operational factors. If comparison is made with previously published information (1, 2) it will be observed that the carbon/zinc ratio in charge has been reduced at a number of locations whilst the condensation and separation efficiency has increased and the ISP carbon estimation has been reduced at most locations. This illustrates the effectiveness of various measures taken to improve efficiency.

A description of the chemistry of the process has been given in a recent paper which analysed in detail the energy consumption of the zinc-lead blast furnace(3). The conclusion reached that of all operating zinc processes the blast furnace is the one most economical in the use of primary fossil-fuel energy, has been supported by the independent survey of H.H.Kellogg presented to this Symposium (4).

Table I.   Imperial Smelting Furnaces - Production Data

Site	Country	Company	Start-up Date	Total Production (end 1979) 10⁶Tonnes Zn	Total Production (end 1979) 10⁶Tonnes Pb	Present design capacity Tonnes Zn	Present design capacity Tonnes Pb
Avonmouth	U.K.	Commonwealth Smelting Ltd.	1951/ 1967	1.15	0.48	100,000	45,000
Cockle Creek	Australia	Sulphide Corporation Pty. Ltd.	1961	1.01	0.48	68,000	30,000
Copsa Mica	Romania	Intreprinderea Metalurgica de Metale Neferoase	1966	0.55	0.32	50,000	30,000
Duisburg	Germany	Berzelius Metallhütten- Gesellschaft m.b.H.	1965	0.94	0.44	85,000	40,000
Hachinohe	Japan	Hachinohe Smelter Co.Ltd.	1969	0.71	0.33	76,000	35,000
Harima	Japan	Sumiko ISP Co. Ltd.	1966	0.67	0.31	66,000	30,000
Kabwe	Zambia	Nchanga Consolidated Copper Mines Ltd.	1962	0.48	0.42	35,000	20,000
Miasteczko I	Poland	Huta Cynku "Miasteczko Slaskie"	1968	0.5	0.22	62,500	30,000
Miasteczko II	Poland	"	1980	-	-	60,000	32,000
Noyelles Godault	France	Sociéte Minière et Metallurgique de Penarroya	1962	1.11	0.46	105,000	45,000
Portovesme	Italy	SAMIM S.p.a.	1972	0.27	0.11	72,500	33,000
Shaoguan	P.R.China	Ministry for Metallurgical Industry	1978	-	-	30,000	11,000
Titov Veles	Yugoslavia	Zletovo	1973	0.26	0.13	72,000	40,000

Table II. Summarized Operating Data

	Avonmouth 5 y mean	Avonmouth Best	Harima 5 y mean	Harima Best	Kabwe 5 y mean	Kabwe Best	Cockle Creek 5 y mean	Cockle Creek Best	Duisburg 5 y mean	Duisburg Best
Campaign length-days	586	715	705	866	447	1506	609	829	1030	1224
Pb:Zn ratio in input	0.46	0.49	0.45	0.48	0.72	0.82	0.53	0.61	0.45	0.46
C:Zn ratio in input	0.77	0.75	0.77	0.75	1.0	0.86	0.76	0.74	0.74	0.71
Slag:Slab Zn ratio	0.67	-	0.65	-	2.6	-	0.90	-	0.67	-
Zn in slag - Assay %	8.4	7.7	7.3	7.0	8.2	7.7	7.2	6.8	6.9	6.2
Distrbn.% new metal	5.5	5.4	4.6	4.4	15.6	10.9	6.4	6.0	4.4	3.9
Carbon burning rate tonnes/24 hr EFB (1)	292	305	166	171	151	158	177	189	206	209
Metals production tonnes/24 hr EFB (1)										
Slab zinc	334	353	194	205	112	149	211	219	245	253
Lead bullion	144	164	88	93	95	104	103	113	115	117
Condensation & Separation Efficiency %	87.5	90.1	93.4	94.0	85.8	93	90.6	91.4	89.9	92.4
ISP Carbon Estimation	73.5	72.0	75.8	72.1	86.9	78.5	67.3	66.7	69.4	65.2
Zinc Recovery % of new input	93.0	93.5	94.7	95.0	74.2	86.9	92.1	92.7	93.9	94.6

Table II cont'd. Summarized Operating Data

	Hachinohe		Noyelles-Godault		Miasteczko		Portovesme		Titov Veles	
	5 y mean	Best	5 y mean	Best	5 y mean	Best	5 y mean	Best	5 y mean	Best
Campaign length–days	895	1017	487	781	341	354	429	652	153	308
Pb:Zn ratio in input	0.43	0.44	0.41	0.43	0.45	0.46	0.45	0.46	0.51	0.57
C:Zn ratio in input	0.76	0.74	0.67	0.66	0.85	0.82	0.82	0.78	0.81	0.78
Slag:Slab Zn ratio	0.57	–	0.73	–	0.97	–	0.66	–	0.76	–
Zn in slag – Assay %	7.1	6.9	8.5	7.2	7.1	6.9	6.9	6.7	7.8	6.5
Distrbn.% new metal	4.1	4.0	6.0	4.8	6.5	5.8	4.6	4.5	5.6	3.7
Carbon burning rate tonnes/24 hr EFB (1)	188	188	224(2)	245(2)	193	196	179	199	159	170
Metals production tonnes/24 hr EFB (1)										
Slab zinc	227	231	283(2)	303(2)	197	200	194	218	167	179
Lead bullion	105	107	108(2)	121(2)	85	86	79	87	83	100
Condensation & Separation Efficiency %	92.3	92.6	87.7	90.5	90.4	91.8	88.7	90.0	89.1	92.5
ISP Carbon Estimation	73.8	71.8	66.1	65.0	79.8	77.9	78.4	75.4	80.1	79.4
Zinc Recovery % of new input	94.7	94.8	93.5	94.0	90.8	92.0	94.0	94.1	92.9	94.8

(1) EFB = Equivalent full blast   (2) Tonnes/24 hrs actual production

251

Furnace Performance

Not all the furnaces listed have the same nominal size. The Avonmouth furnace is significantly larger than the standard furnace; the Noyelles-Godault furnace has been enlarged to increase its capacity; the Kabwe furnace was built larger than standard to compensate for the effect of altitude and the Harima furnace was built smaller than standard because of a lower production requirement. The shaft areas of the different furnaces are tabulated in Table III.

Table III.   Furnace Shaft Area

Site	Shaft Area
Avonmouth	27.1 m^2
Noyelles-Godault	24.6 m^2
Kabwe	19.6 m^2
All other furnaces (i.e. Standard furnace)	17.2 m^2
Harima	15.3 m^2

It naturally follows that with the exception of the Kabwe furnace, the larger furnaces would be expected to have larger capacities. Also, of course, plants were built with differing specified capacities which reflected the quality of the feed materials, the needs of the various companies and the state of development of the Imperial Smelting process at that time. Current performance is dictated by these early decisions and by more recent decisions on increasing capacity as well as by market considerations.

Data illustrating past and present performances with predictions for the future are given in Table IV. The first column is based on typical records for 1963 (5) and set alongside this information is the expected performance of a standard 17.2 m^2 furnace operating unfettered by restriction at this present time. It should be noted that the annual zinc production figure of 80,000 metric tons was achieved at Duisburg in 1973 so that this could be regarded as conservative in view of subsequent development. Also included in Table IV are predicted performance data for the early 1980s based on unrestricted production employing known developments which are in the course of adoption. The remainder of the paper deals with plant developments which are the justification of these predictions and with the more speculative performance predictions contained in a fourth column which could apply to the later eighties. Since smelter profitability is affected by factors other than the listed performance indices other process improvements are discussed.

Operating Improvements

Over the last 16 years it is clear that many operating modifications and improvements have been made resulting in the much higher output of the standard furnace. These have been achieved by developments initiated by the operating companies and by the licensing company.

The increase in furnace capacity has resulted largely from an increased blowing rate in turn dependent on improved charge preparation, a redesign of furnace shape and reduced number of tuyeres (6). Carbon efficiency has increased in line with increased air preheat, and this effect will be discussed

252

Table IV.   Standard Furnace (17.2 m^2 Shaft Area) Performance
Past, Present and Future

	Based on Data		Prediction	
	Typical 1963 furnace	Unrestricted 1979 furnace	Early Eighties	Late Eighties
Annual zinc tonnes	35,000	80,000	90,000	100,000
Annual lead tonnes	17,000	40,000	49,500	60,000
Pb:Zn ratio in input	0.49	0.50	0.55	0.60
C:Zn ratio in input	0.75	0.75	0.70	0.70
Slag:slab zinc ratio	1.00	0.65	0.65	0.65
Zinc in slag - Assay %	8.0	7.0	6.0	6.0
Distribution as % new metal	7.3	4.5	4.0	4.0
Carbon burning rate tonnes/24 hrs EFB	115	210	215	235
Metal production tonnes/24 hrs EFB				
Slab zinc	120	256	280	310
Lead bullion	58	128	155	185
Condensation & Separation efficiency %	85.0	92.0	93.5	94.0
ISP carbon estimation %	72.0	69.0	65.0	65.0
Zinc recovery % of new input	91.7	94.5	95.5	95.5
Lead recovery % of new input	94.0	95.0	95.0	95.0
Carbon burnt/tonne slab zinc	0.96	0.82	0.77	0.76

in more detail in a later section of this paper.   Condenser and separation efficiency have increased by virtue of condenser improvements resulting from laboratory studies with models (7) and by the use of immersible coolers in the launder system (6).

Present Plant Developments

In addition to the operating improvements outlined, which have occurred over the past several years, further developments have been studied and incorporated in some of the operating plants.   These can be summarized under the following headings:

(a)   Extending the range of raw materials to be treated and the charge preparation techniques.

(b)   Improving fuel economy via improved charge distribution, the use of alternative fuels, and heat recovery.

(c)   Uprating the performance of shaft and condenser.

(d)   Upgrading the products of the process.

The major advance in the first category is the development of the hot briquetting process for oxidic zinc and lead materials.   Many operators have available sources of oxidic zinc and lead in the form of mineral silicates and carbonates, Waelz oxides and other secondary oxides and drosses. A limited proportion of these oxides may be incorporated into the normal sintering process, but to allow greater flexibility and a higher proportion of oxide to be treated the preparation of a suitable agglomerated furnace burden was desired.   This has been achieved by hot briquetting, a process in which fine oxidic feeds can be agglomerated at 500-700°C in briquetting rolls without the use of added binders.   Full details of the pilot plant development work on this process are described in a separate paper (8) and operating plant has been installed at three locations.

The importance for future development is that a fluid-bed sulphide roast could link with the hot briquetting process allowing recovery of the roasting heat as in existing practice with low lead sulphidic zinc feed and providing an alternative to sintering.   The overall process could then be substantially for zinc only rather than zinc-lead feeds for locations where lead concentrates are not available or lead smelting is not required.   As a resource economy hot briquetting could utilize a total secondary oxide feed giving either a zinc or a zinc-lead smelter as required.

An additional gain from the adoption of such a process would arise under the second and third categories of the development programme.   A uniform sized charge should result in metallurgical benefits such as better carbon utilization, higher blast intensity and improved condensation.

Similar advantages to these are also being obtained under the second and third headings resulting from the adoption of a new ring and plug charging bell design (9).   The benefit of this technique is that whereas with the normal charging bell a 'hill and valley' charge profile may form, with the ring and plug bell system there is a measure of control over the distribution of sinter and coke in the furnace shaft and the facility to smooth out the charge profile, thus equalizing gas velocities over the furnace cross-section. This in turn improves fuel economy and reduces the carryover of materials from the furnace thus aiding condenser efficiency.   Further, it has been found that by controlling the disposition of coke and sinter within the furnace a coke-rich centre can aid economical working, whilst coke-rich walls, though less economic in working, have the advantage of maintaining a cleaner furnace with less accretion growth.   Furnace operation can thus be programmed to optimize the overall return.

Again, under the second and third categories, it can be reported that two major advances have been achieved on the Japanese furnaces in relation to

(1)   lowering the C/Zn ratio, that is the amount of carbon needed per unit of zinc metal cast

(2)   the recovery of heat from furnace waste gases (10).

The first relates to the drying of the air blast.   Moisture in air can reduce the carbon efficiency of the blast furnace by lowering the temperature of the raceway flame.   Previous measurements at the Sulphide Corporation smelter at Cockle Creek, N.S.W., Australia had indicated the degree of the effect when account was taken of the variability of C/Zn ratio over wide seasonal variations in humidity. Now both Japanese operators of the process have installed plant to dry the air blast. Their results, having done so, are in line with the previous predictions from Sulphide Corporation, that each 1% of water vapour removed from the blast air is equivalent to a reduction of 0.024 in the C/Zn requirement. This represents about a 3% reduction in coke usage for each 1% of water vapour. (At 1979 prices worth about $4 per tonne zinc cast). Obviously this is a procedure which would be economic only in areas where high humidities are encountered.

With respect to the second item, the recovery of heat from furnace waste gases (LCV gas), it is common practice to use some 60-70% of the output to preheat the blast air and to preheat coke. On one site some of the remainder is used to heat a cadmium reflux column, and at another site a steam boiler is fired both to generate steam and to drive a turbine coupled to the main air blower. The excess LCV gas (from 15-30% depending on the site) is flared to waste.

However at the Hachinohe smelter in Japan some of this excess gas is used to fire a steam boiler - turbo alternator set which generates about 3 MW of electrical power - approximately one-half of the total electrical requirement of the whole smelting complex. Other smelters may well follow this example which at the present time, when energy costs are rising faster than other costs, is becoming an economic proposition.

In the third category of present plant developments, leading to an up-rating of the shaft and condenser performance, can be mentioned first investigations which indicate that heat and mass transfer in the hearth area of the furnace can be increased substantially by the use of more steeply inclined tuyeres. Difficulties in incorporating such tuyeres into existing installations gave rise to the development of a bent tip tuyere which gave similar results in the laboratory investigations (7, 11). Trials on this new design of tuyere are now taking place and if the results on operating furnaces are in line with the predictions made from the laboratory studies then a reduction in carbon/zinc ratio approaching 5% should be achieved.

Modifications to the condenser inlet giving more streamlined flow have been made recently at some of the operating furnaces which have had a significant effect on improving condenser efficiency. Also a design modification to the condenser offtake has reduced carryover of lead with the exit gases (12).

The fourth category of plant development relates to the upgrading of furnace products. By far the most important advance in this respect has been the emergence of processes for the recovery of copper. The blast furnace is a minor copper producer in that on average during the last year the operating furnaces have each treated about 1000 tonnes of copper, the exact figure varying from furnace to furnace depending on the copper content of concentrates and feed used. The copper reports as a hearth product in solution in lead bullion together with small amounts of tin, silver, gold, antimony and arsenic. It is drossed from the bullion in mainly metallic form at up to 40% in admixture with lead. The dross has a low commercial value and the development of a hygienic process to recover saleable copper has been a major objective.

Earlier development work at Avonmouth centred on a process of sulphuric acid leaching followed by electrolysis for copper. This process has been adopted at the Duisburg furnace where high quality cathode is made (13). Two major problems are the engineering of the leach system and the control of impurities which dissolve with the copper, consequently development work has continued on an ammonia leach process which is considered to be more universally acceptable (14). This new process has three interacting circuits:

(a) Leaching - copper is selectively dissolved in aqueous ammonia-ammonium carbonate solution with added air or oxygen.

(b) Solvent extraction - copper from the leach solution is selectively transferred to a kerosene solution of an organic extractant.

(c) Stripping - copper is transferred from the extractant solution to aqueous sulphuric acid from which it may be recovered as cathode copper by electrowinning or as copper sulphate by crystallization.

Pilot plant operation in 1974 at 30 kg/day cathode copper has resulted in a full plant design being established and two operating companies (Sulphide Corporation at Cockle Creek, Australia and CSL at Avonmouth, U.K.) are building commercial plants which will be commissioned in 1979 and 1980 respectively. In Japan a similar plant is being built at Takehara to treat the copper dross arising at the Hachinohe zinc-lead blast furnace.

Future Possibilities and Improvements

It is certain that the trend of development and improvement will continue into and throughout the 1980s and a number of topics are considered here where it is predicted that major advances will occur in the next few years. Several of these are concerned with energy savings.

(a) Fuel Injection

This technique is used widely in the iron blast furnace to reduce fuel costs rather than energy per se and has been very successful in an era when hydrocarbon costs were low compared with coke. The practicability of oil injection on the zinc-lead blast furnace has been proved by Sulphide Corporation at their Cockle Creek smelter (15) and up to 10% replacement of coke has been achieved. Fears expressed on the probable adverse effect of water vapour in exit gases on condenser operation were not realized with up to 3-4% of $H_2O(g)$. Higher injection rates of oil have now been precluded by the rise in oil prices and it is unlikely that hydrocarbon fuels will play a major role in the foreseeable future. Solid fuel injection, as employed by a number of iron blast furnace operators is a possibility for future practice although doubts have been expressed on the economics of installation of complex milling and injection systems. In this field zinc companies will almost certainly benefit from the increased attention which the iron and steel industry will be giving to this problem.

Also under the heading of fuel injection it is worth mentioning the prospects of the recirculation to the furnace of some low calorific value gas (LCV) which leaves the blast furnace condenser after removal of zinc, and contains, typically, 11% CO_2, 22% CO and 67% nitrogen. As mentioned previously after utilizing this gas as a fuel for preheating coke and air-blast, and for other in-plant duties, a residual excess of at least 15% of the gas is flared to waste. The indications are that recirculation of this quantity of gas, with higher than normal preheat could result in a 10% increase in the amount of zinc volatilized per unit of coke consumed (16).

256

(b) Blast Preheat

One factor responsible for productivity and output increases since the
early days of operation has been the increase in blast preheat.   This is
also pertinent in connection with fuel injection discussed above.   At an
early date it was found that an increase in blast temperature of 100°C from
say 600°C to 700°C, lowered the C/Zn ratio by 0.02.   In achieving this the
coke burning rate is increased but the zinc production rate increases by more
than an equivalent amount.   Typical figures were 3% and 5% respectively.
Recent work on standard furnaces operating at the rate of 80,000 tonnes per
year zinc output have shown a similar reduction in C/Zn ratio on increasing
preheat from 850°C to 950°C.

A theoretical analysis of the benefits of preheat (17) has indicated
that the effect on further reduction of C/Zn is likely to be limited above
1000°C, but this effect has not been shown positively, as yet, on an operating
furnace.   The effects of such a limitation may be overcome by establishing
endothermic reactions in the tuyere zone of the furnace, for example by the
initial impact of fuel injection.

It is pertinent at this point to make some mention of oxygen enrichment
of blast.   A variation in oxygen content will affect the peak temperatures
attained in the gas in the tuyere zone in much the same way as preheat of the
blast.   From this point of view oxygen enrichment must be compared economically with preheating air and furnace trials have shown that enrichment is
unlikely to be viable while there is surplus furnace gas available and
suitable preheating equipment for increasing blast temperature.   Enrichment
could play an important role, however, in conjunction with fuel injection,
and it is probable that in the next few years furnace trials will be undertaken on this development.

(c) Alternative Fuels

In all metal winning processes the fraction of total costs debited to
fuel has increased in recent years.   This is true in respect of coke as with
all other fuels and there has been world-wide activity in the development of
'formed coke' processes which greatly extend the range of coals which can be
utilized to make a final product comparable with metallurgical coke in
physical and chemical properties.   Parallel advances in coke-oven technology
have given a new lease of life to the conventional process and put back the
date of commercialization of the newer processes.   Trial work has indicated
that there should be minimal problems involved in converting the zinc-lead
blast furnace to operation with a suitable formed coke, and such practice
may be common before the end of the eighties.   Whether or not this occurs
depends to a large extent on what happens in the steel industry.   Consumption
of coke on a world-wide basis in zinc blast furnaces is only about 1% of that
in the iron blast furnace and here (again) it is reasonable to assume that
the zinc industry will benefit from necessary developments in the iron industry.
An alternative fuel is not, of course, necessarily a cheaper fuel, and there
is no implication that formed coke will be cheaper than conventional coke.
A future possibility could be the linking of a formed coke plant with a zinc
blast furnace removing the necessity of a separate coke preheater with
consequent minor fuel savings.

New zinc processes are often postulated which make use of 'cheap' coke
breeze or 'low priced' chars.   It is not clear why any other form of coked
coal should be substantially cheaper than metallurgical coke on a long term
basis.   With the advances made in the recycling of coke breeze in the coke-
oven process and the environmental constraints placed on any coking process
it would seem that reliance on the continuation of a supply of 'cheap' breeze

or char depends on that material being an otherwise low-demand by-product of a normal coking process.   Such material is not generally available in the U.K. and in countries where it is available now, careful surveys should be made of future trends in the coke industry before placing reliance on its continuity.   Should it be considered, contrary to our present expectations, that coke breeze is a fuel with a future then it would be worthwhile extending research work into the briquetting of that fuel with suitable binders in an attempt to establish a viable alternative fuel for the zinc blast furnace.

(d)   Complex Materials

The next few years will see increasing quantities of low grade, complex and secondary materials in the feed to the blast furnace.   This area has been well documented by P.R. Mead in the present Symposium (18) and by Adami, Firkin and Robson recently (19), so will not be further discussed here apart from underlining the fact that benefits ensue for plant profitability.

(e)   Copper Recovery

The treatment of copper dross arisings for copper recovery, mentioned previously as a current development, has potential for the handling of much larger quantities of copper in charge.   From the theoretical viewpoint there is no reason why the copper content of lead bullion should not increase significantly above the level of 5-8% which is a typical range of present practice.   Above 8 to 10% copper in bullion the marginal recovery of further copper additions should be high.   (The activity of copper in solution in lead increases only slightly above these concentrations, so that no further increase in copper loss in slag should occur).

The major difficulties in handling higher quantities of copper in bullion are likely to be practical ones associated with the premature formation and precipitation of the large bulk of dross.   To avoid such problems and to enable the handling of higher copper a method of recirculation of de-copperized bullion to the furnace or forehearth has been proposed (20).

The Roasting Process

Improvements to the charge preparation of the blast furnace have been considered within the context of the hot briquetting process, but this of course is associated only with oxidized materials.   Advantages could accrue to the blast furnace complex if the sintering operation of roasting sulphidic materials could be replaced by a fluidized-bed process.   There is no problem here when considering a low lead operation – the technology is already in existence.   As operated at present, however, the fluid-bed process is limited in the amount of lead which can be tolerated.   It is recognised that an important step in widening the appeal of the blast furnace as a zinc-lead operation will be the development of such a fluid-bed process incorporating lead.   Development work now in hand may lead to the solution of this problem during the next decade.

Conclusion

There are continuing suggestions being made of possible new pyro-metallurgical processes for zinc.   These processes, on paper, are always claimed to be more conservant of energy, more hygienic and less costly than the existing processes even though they are normally tied to high grade concentrates.   However, it is the view of the authors' that step-wise changes in the method of operating the blast furnace are more easy to accomplish, far less costly and more likely to be successful than the development of a new process.   There will be gradual evolvement of the following:

-   mixed and variable furnace feed

-   lower coke/zinc requirement

-   use of formed coke or briquetted chars

-   waste heat recovery

-   higher copper recovery

    as well as zinc only facility where required

These measures, among others, will lead to the performance data specified in the fourth column of Table IV as we proceed into the 80s.

References

1. A.W. Robson, "Growth and Development of the Imperial Smelting Process", paper presented at the G.D.M.B., Huttenausschuss fur Zink, Eindhoven, May 1974.

2. G. Binetti, J.Koteski and D.A. Temple, "Combined Zinc and Lead Smelting: Recent Practice and Developments", Journal of Metals, 27 (9) (1975) pp. 4 - 11.

3. W. Hopkin and A.W. Richards, "Energy Conservation in the Zinc-Lead Blast Furnace", Journal of Metals, 30 (11) (1978) pp. 12-17.

4. Herbert H. Kellogg, "Energy Use in Zinc Extraction", paper presented at the 109th AIME Annual Meeting, Las Vegas, 1980 (This Symposium).

5. S.E. Woods and D.A. Temple, "The Present status of the Imperial Smelting Process", Trans.I.M.M. 74 (6) (1965) pp.297-318.

6. C.F. Harris, "Process Development and the Zinc-Lead Blast Furnace", paper presented at the 17th Annual Conference of Metallurgists of C.I.M. Montreal, Canada, August 1978.   To be published in C.I.M. Bull Autumn 1979.

7. C.F. Harris, A.W. Richards and D.A. Temple, "Process Modelling in the further development of the zinc-lead blast furnace", paper presented at the Eleventh Commonwealth Mining and Metallurgical Congress, Hong Kong, 1978.   I.M.M. London.

8. C.F. Harris and A.W. Richards, "Hot Briquetting of Oxidic Zinc and Lead Materials as a Feed for the Zinc-Lead Blast Furnace", paper presented at the 104th AIME Annual Meeting, New York, 1975.

9. A.J. Self, M.W. Gammon and C.F. Harris, "The Development and Use of Variable Charge Gear on a Zinc-Lead Blast Furnace", paper presented at the 107th AIME Annual Meeting, Denver, 1978.

10. H.Nakagawa, Y. Sugawara and K. Nakayama, "Energy savings in the Zinc-Lead Blast Furnace Complex", paper presented at the 109th AIME Annual Meeting, Las Vegas, 1980.   (This Symposium).

11. M.W. Gammon, "Heat and Mass Transfer in the Tuyere Region of a Zinc-Lead Blast Furnace - Model Studies", pp. 47-52 in Advances in Extractive Metallurgy, I.M.M., London, 1977.

12. British Patent Application, 45975/78

13. K. Tack, "Hydrometallurgical reprocessing of copper dross at the Berzelius Metallhütten-GmbH, Duisburg". (In German) Erzmetall, 29 (4) (1976), pp.276-279.

14. W. Hopkin, "Processes for Treatment of Zinc-Lead and Lead Blast Furnace Copper Drosses", paper presented at 105th AIME Annual Meeting, Las Vegas, 1976. TMS Paper A76-99.

15. G.R. Firkin and R.F. Still, "Injection of oil through the tuyeres of an ISF", paper presented at Symposium on Extractive Metallurgy, University of New South Wales, Sydney, Australia, Nov.1977 - joint meeting with Sydney Branch of A.I.M.M.

16. British Patent 1 470 722.

17. Private Communication, John Lumsden, Imperial Smelting Processes, Aug. 1973.

18. Peter R. Mead, "Treatment of a Diverse Range of Low Grade Feed Materials in the ISF", paper presented at the 109th AIME Annual Meeting, Las Vegas, 1980. (This Symposium).

19. A.O. Adami, G.R. Firkin and A.W. Robson, "Treatment of Complex Materials and Residues in the Imperial Smelting Process", paper presented at Joint Meeting of I.M.M. and G.D.M.B. "Complex Metallurgy '78", Bad Harzburg, Sept., 1978.

20. British Patent Application 7916151.

ENERGY SAVINGS IN THE ZINC-LEAD

BLAST FURNACE COMPLEX

H. Nakagawa*, Y. Sugawara** and K. Nakayama**

* Mitsui Mining & Smelting Co., Ltd.
Former General Manager of Hachinohe Smelter

** Hachinohe Smelter, Hachinohe Smelting Co., Ltd.

Abstract

The progress in energy savings over the last ten years is reviewed and the major developments in this field achieved during this period are described, namely, (a) reductions in coke consumption by use of low reactivity coke, (b) increases in blast temperatures, and (c) **reductions** in moisture content in the blast, (d) reuse of condenser (L.C.V.) gas previously wasted, e.g., the construction of the L.C.V. gas burning power plant, and (e) improvements in energy efficiency at the zinc refinery by the development of new large trays for zinc refluxers.

Finally, the over-all energy requirement of the smelter has been estimated for comparison purposes with other alternative processes.

The wide range of raw materials handled at the smelter is also briefly mentioned from the view-point of resource economy.

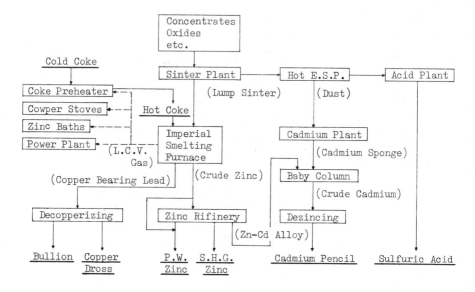

Figure 1. Flowsheet of Hachinohe I.S.F. Complex

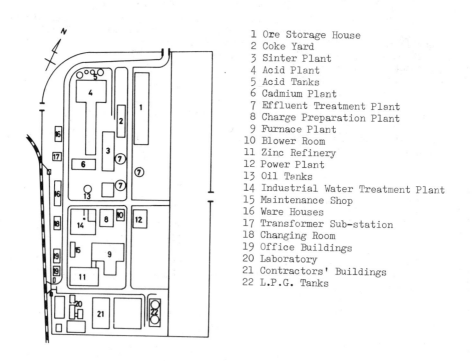

1 Ore Storage House
2 Coke Yard
3 Sinter Plant
4 Acid Plant
5 Acid Tanks
6 Cadmium Plant
7 Effluent Treatment Plant
8 Charge Preparation Plant
9 Furnace Plant
10 Blower Room
11 Zinc Refinery
12 Power Plant
13 Oil Tanks
14 Industrial Water Treatment Plant
15 Maintenance Shop
16 Ware Houses
17 Transformer Sub-station
18 Changing Room
19 Office Buildings
20 Laboratory
21 Contractors' Buildings
22 L.P.G. Tanks

Figure 2. Plant Lay-out of Hachinohe I.S.F. Complex

262

Introduction

Since 1969, Hachinohe Works, a zinc and lead blast furnace complex located close to a 50,000-tonne public berth, has been operating as a toll smelter by Hachinohe Smelting Co., Ltd. This smelter was established in 1967 by six non-ferrous metal companies in Japan of which the major stock holder is Mitsui Mining & Smelting Co., Ltd.

The details of the operation have already been published several times, but the best approach to this smelter is the paper presented at the 1976 AIME-MMIJ Joint Meeting, Denver, Colorado by Kinoshita, et. al. (1)

For the last ten years, the smelter has produced 450×10^3 tonnes of P.W. grade zinc, 190×10^3 tonnes of S.H.G. zinc, 226×10^3 tonnes of lead bullion, 61×10^3 tonnes of copper bearing dross, 2411 tonnes of cadmium, and $1,320 \times 10^3$ tonnes of sulfuric acid (as 98%).

Various attempts have been made to improve production costs. Particularily, the main objective to be dealt with is energy costs which accounts for about 50% of direct production costs. Especially since the 1973 oil crisis, energy savings have become of more and more importance and concern.

In view of resource economy, the smelter handles a great variety of raw materials ranging from sulfides to recycled oxides; most of which are unsuitable for zinc electrolysis. To allow a satisfactory operation using such a wide range of raw materials, a computer control system has played an important role.

Consequently, a higher efficiency is maintained both in energy consumption and in productivity.

This paper reports on the efforts in this field over the ten-year period, mainly in energy savings. As such, the total energy requirement per unit zinc produced have been estimated, and can well be comparable to that of any modern electrolytic zinc plant.

Fig. 1 shows the flowsheet of the smelter and fig. 2 shows the plant layout of the smelter.

Energy Profile of the Smelter for the Past Ten Years

The smelter consumes coke as a reducing agent, butane gas, heavy oil and electricity as the primary energy sources in the ratio shown in Fig. 3. From this, it can been seen that the essential energy source for the I.S.P. is coke.

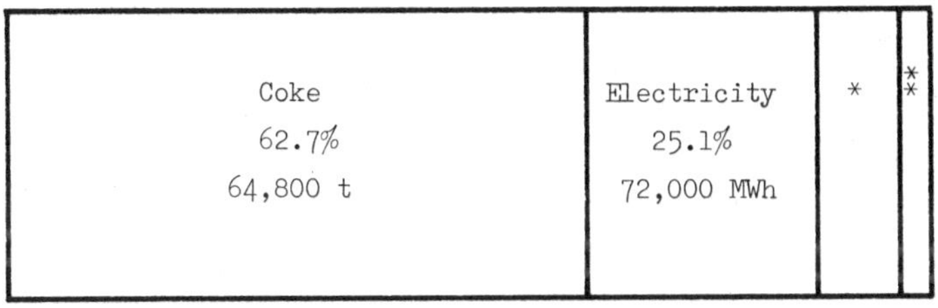

Coke 62.7% 64,800 t	Electricity 25.1% 72,000 MWh	*	**

* Butane   8.8%   5,120 t
* Oil   3.4%   2,380 kl
Total   702,130 x 10⁶kcal/y

Figure 3.
Total Energy Consumption as of 1977

Table I shows the major energy consuming equipment.

Table I. Major Energy Consuming Equipment.

Plant	Energy Source	Equipment
Sintering	Oil	Ignition stoves for sinter machine
Acid	Oil	Preheater (for start-up)
I.S.F.	Coke	Blast furnace
	Butane	Flux bath, Separation bath, Holding bath, 3 Decopperizing kettles, Casting bath
	Oil	Forehearth, Cowper stoves (for start-up)
Zinc Refinery	Butane	3 Refluxers, 3 Holding baths, Liquation bath, S.H.G. bath
Cadmium	Butane	Melting kettle, Baby Column, Refining kettle
Utility	Oil	Power plant (pilot burner)

Fig. 4 shows the trends of annual production and energy use for the past ten years. The zinc and lead productions were increased up to 130% compared with these in the first operating year, but since 1975, they have gone down to 70-80% of the production capacity due to the world-wide recession and its effects on the zinc market.

On the other hand, energy consumption per unit zinc has remarkably been decreased by 20% which includes decrease of 21% in coke, 53% in butane and 59% in oil but an increase of 14% in electricity. The addition of pollution abatement processes meet the national regulation standards tightened several times since 1971 is responsible for this exceptional increase.

The major efforts on energy savings made for the last decade can be summarized as:

1970: – Reductions in coke consumption by applying lower reactivity coke.

1971: – Reductions in coke consumption by increasing blast temperatures.
– Development of new large trays for zinc refluxing columns.

1973: – Utilization of condenser gas (L.C.V. gas) to zinc feed baths at the zinc refinery plant.
– Reduction in oil consumption by improving the forehearth construction.
– Steam savings for the shower-room by utilizing hot water, heat-exchanged with the immersion cooling water of the I.S.F. condenser.

1975: – Application of L.C.V. gas to the existing package boiler.

1976: – Oil savings by modifying the ignition stove for the sinter machine.
– Application of L.C.V. gas to the zinc holding bath at the I.S.F. plant.
– Reduction in butane gas consumption at the refinery plant due to the successful results of a large column operation and other efforts.
– Reductions in coke consumption by blast air dehumidification at the I.S.F. plant.

1977: – Power savings by an efficient operation of the pollution abatement processes, the ventilation systems, etc.

1978: – Construction of the L.C.V. gas burning power plant.

As realized from the above, the effective utilization of L.C.V. gas is one of the features of our energy saving program. A description of these efforts will be presented in the following.

264

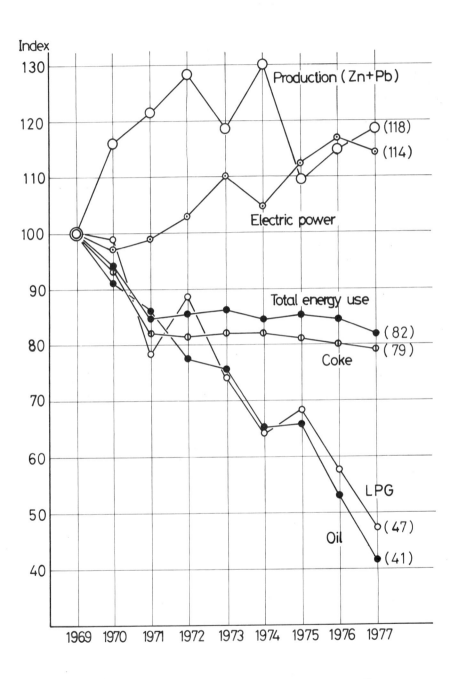

Figure 4. Trends of Annual Production and Energy Use

Developments on Energy Savings

Reduction in Coke Consumption

Low Reactivity Coke In an attempt to minimize the solution loss reaction in the upper region of the shaft within the allowable limits, a test operation with low reactivity coke (medium sized foundry grades) was carried out in comparison with that of standard metallurgical coke. The analyses of the cokes tested are shown in Table II.

The metallurgical coke shows an inferior carbon/zinc ratio and a higher CO/CO_2 ratio as shown in Table III. Based on this, the ratio of the low reactivity coke to the higher one fed to the I.S.F. was gradually increased up to two thirds to get a considerable reduction in coke consumption.

Table II. Analyses of Cokes Tested

Grade	Bulk Density	Rattle Index (% +40mm)	F.C. (%)	V.M. (%)	T-S (%)	Ash (%)
Low reactivity	0.53	83.7	86.8	2.0	0.7	11.2
Metallurgical	0.44	69.5	86.0	1.5	0.7	12.0

Table III. Test Results

	Furnace Feed(t/day) Sinter	coke	Zinc Produced (t/day)	Blast Air Nm³/min	°C	mmH2O	Coke*L.C.V.Gas Ratio	CO2%	CO%
Low reactivity	466	182	199	488	760	3195	0.77	12.0	21.0
Metallurgical	439	197	192	491	758	3307	0.86	10.5	24.0

* Hot coke charged / Zinc in sinter

A typical analysis of the cokes currently used are shown in Table IV. Both the fixed carbon content and the size distribution have been improved further since the above tests were made. At present, four kinds of coke are blended at an appropriate ratio to satisfy the necessary furnace conditions.

Table IV. Analyses of Cokes Currently Used

Brand	Size(%) +75mm	+50mm	Rattle Index %+40mm	Analysis F.C. (%)	V.M. (%)	T-S (%)	Ash (%)
K(Low reactivity)	46.8	40.5	86	92.1	0.68	0.67	7 24
T(")	0	80.3	87	91.3	0.80	0.61	7.95
S(")	67.8	27.2	87	91.5	0.79	0.71	7.74
M(Metallurgical)	61.8	33.0	89	90.3	0.88	0.61	8.88

The size of the blended coke is another very important factor for keeping good furnace conditions, since good draft in the shaft depends upon the coke which occupies more than 50% of the volume of the total burden in the shaft.

High Temperature Blast The I.S.F. is equipped with two cowper stoves with one separate combustion chamber in which L.C.V. gas is fired without any auxiliary fuel. Oil is used only for the start-up period.

At an earlier stage of the operation, the blast temperature ranged from 600 to 850°C, but it has since been increased up to 1,000°C (maximum) as shown in Table V. This was achieved without any modification to the stoves. However the heat balance was carefully re-examined, and the tuyere leads, e.g., the refractory lining, were modified to allow for the 1,000°C blasting.

Table V. Hot Blast Temperature

Campaign No.	1	2	3	4	5	6
Period	Feb./1969 Oct./1969	Nov./'69 Apr./'71	May./'71 Feb./'74	May./'74 Feb./'76	Apr./'76 Mar./'79	Apr./'79
Max. temperature	820	850	950	960	960	1,000(°C)

The relationship of the coke to zinc ratio (C/Zn) and the blast temperature (t°C) was statistically estimated as

$$C/Zn = 1.251 - 0.000573t \qquad (1)$$

This means that the coke ratio is improved by 5.7% per 100°C increase in the blast temperature, though a theoretical calculation suggests only 3.5%. This difference may possibly be due to a more uniform air distribution in the tuyere region and an increase in the hearth efficiency resulting from the higher blast temperatures.

Blast Air Dehumidification 1. Background The past operational data showed a seasonal change in the furnace conditions, for example, the coke consumption per unit zinc (coke ratio) was higher by 2% or more in summer than in winter.

It was known that the absolute moisture content in the atmosphere also changed seasonally from $3g/Nm^3$ in January to $19g/Nm^3$ in August, which corresponded to a change in the hydrogen content in the L.C.V. gas. An assumption was then made that the moisture in the air adversely affected the coke ratio by reacting with coke to form hydrogen according to the following reaction.

$$C(s) + H_2O(g) = CO(g) + H_2(g) \qquad (2)$$

$$\Delta G^o = 32,090 - 34.1(t + 273)$$

Assuming that the hearth temperature is 1,300°C and the PCO in the hearth, equal to 0.3, then

$$\Delta G^o(1,300°C) = -21,550$$

$$K = \frac{P_{H2} \cdot P_{CO}}{P_{H2O}} = \frac{0.3 P_{H2}}{P_{H2O}} = 985$$

$$\therefore \frac{P_{H2}}{P_{H2O}} \doteq 3,283$$

Reaction (2) is endothermic (1,743 kcal/kg-H2O). In the summer, if the excess water, which accounts for $9g/Nm^3$, entirely changes into hydrogen, the heat to be lost in the hearth will be 12.42×10^6 kcal/day (at 33,000 Nm^3/hr blast rate). This value is equivalent to 3.1% in the coke ratio.

Furthermore, moisture adversely affects the furnace operation; i.e. , by deteriorating the condenser efficiency. At the top of the furnace, the temperatures are kept above 1,000°C to avoid reoxidation of zinc vapor by introducing the "top air" which is a portion of the hot blast air.

For these temperatures, the following equilibrium will exist:

$$H_2(g) + CO_2(g) = H_2O(g) + CO(g) \qquad (3)$$

$$\Delta G^O = 8,650 - 7.59(t + 273)$$

Assuming that the temperature (t) is 1040°C, then

$$\Delta G^O (1040°C) = -1,316$$

$$\therefore K = \frac{PH_2O \cdot PCO}{PH_2 \cdot PCO_2} = 1.66$$

Where PCO/PCO2 is equal to about two, then PH2O/PH2 becomes 0.83. This means that about half of the moisture blown into the furnace is still existing as H2O at the furnace top, and it could oxidize the co-existing zinc vapor when entering the condenser.

To prove these hypotheses, a full-scaled test plant was constructed. The dry lithium chloride method, using a rotating asbestos honeycomb drum was selected among several possible ways of dehumidification. Fig. 5 shows the flowsheet of the dehumidifier and the main specifications are listed in Table VI.

Table VI. Main Specifications of Blast Air Dehumidifier

Hot gas fan	: 300 m³/min x 350 mm H2O x 37 kw

Heat exchanger	: Rotor : 1,900mmⱷ x 200 mm thick
	                      Asbestos honeycomb, 1-16 rpm
	  Capacity : 670,000 kcal/hr

Regenerating air fan	: 377m³/min x 220mm H2O x 22kw

Dehumifying drum	: Rotor : 2,400mmⱷ x 400mm thick
	                      Asbestos honeycomb, 8 rpm
	  Capacity : 600Nm³/min

Process fan	: 645m³/min x 150mm H2O x 30kw
After cooler	: Plate fin type, 470,000 kcal/hr

2. Heat balance of the dehumidifier   Fig. 6 shows a typical heat balance of the process at the blast rate of 33,600Nm3/hr with the moisture removal rate of 272.2kg/hr. Waste heat from the cowper stoves is effectively utilized in place of steam or electric heat to re-generate the deliquescent lithium chloride.

3. Effect on the coke ratio   Fig. 7 shows the effect on the coke ratio (hot coke consumed/zinc and lead produced). No comparable data were available in April and August, because of the scheduled shut-down of the I.S.F. for production cut-back during those two months.

The reduction in coke ratio, on the average, from May to December is estimated as 18kg, or 2.6% of the whole. It was made clear that the seasonal

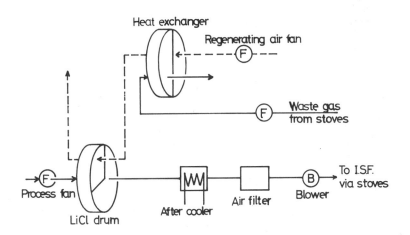

Figure 5.
Flowsheet of Blast Air Dehumidifier

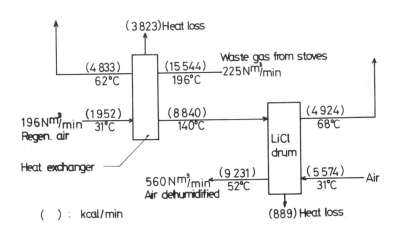

Figure 6.
Typical Heat Balance of Dehumidifier

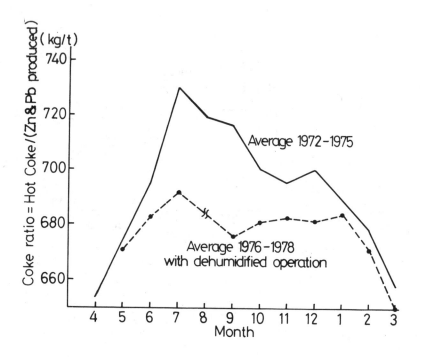

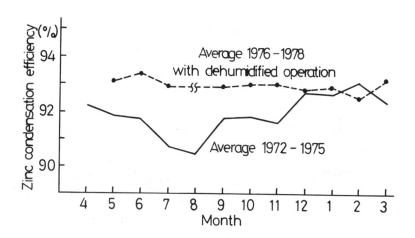

Figure 7-8.
Effect of Blast Air Dehumidification on
Coke Ratio and Condensation Efficiency

changes in coke ratio was lessened to a considerable extent.

   Improvement in the condensation efficiency   Fig. 8 shows the effect on
the condensation efficiency (zinc produced /zinc volatilized), which had
increased by 1.2%, without any clear seasonal changes.

   These phenomena are considered to back-up the hypotheses. Consequently,
approximately 1,000 tonnes of water per annum have been removed from the air
blasted into the I.S.F., giving an annual reduction in coke consumption of
nearly 900 tonnes.

Energy Recovery from L.C.V. Gas

   The L.C.V. gas from the I.S.F. has the following specifications at the
outlet of the gas washing system - a spray tower, a theisen disintegrator
and a T.C.A. (turbulent contact absorber) combined in series - :

   CO : 24 +2 %   Dust loading   : 20   mg/Nm³
   CO2: 11 +1.5 %   Calorific value: 700   kcal/Nm³
   H2 : 1 +0.15%   Moisture   : Saturated at 35-40 °C
   N2 : 64%

   Despite continuous use of L.C.V. gas, up to 50%, for the stoves and the
coke preheaters, the rest had been exhausted from the flare stack because of
the following reasons:
(1) A low calorific value gives low flame temperatures, causing a lack of
combustion stability.
(2) Fluctuations of the gas pressure also disturb stable firing.
(3) Wide limits of imflammability give risks to explosion when the gas misfires.

   However, the engineering developments have finally enabled the complete
utilization of the gas as described below. Fig. 9 shows the progress in the
L.C.V. gas utilization.

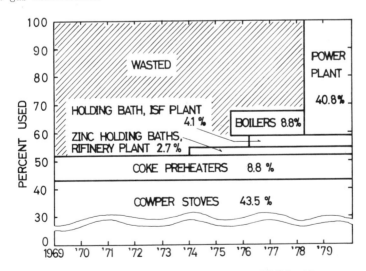

Figure 9. Progress in the L.C.V. gas Utilization

271

<u>Zinc holding baths</u> The butane-air gas burning system for the feed baths and the melting bath at the refinery plant was successfully replaced with L.C.V. gas burners. After that the holding bath at the I.S.F. plant followed.

The main reasons for the success were:
(1) Development of a special L.C.V. gas burner equipped with a butane-air gas pilot burner.
(2) The optimization of each burner's position and angle.
(3) Development of an emergency cut-off valve to be used whenever the gas pressure abnormally fluctuates.
(4) Sufficient capacity of the existing combustion chambers to make L.C.V. gas firing possible.

<u>Package boiler</u> In view of the recent increases both in price and in power consumption, the most promising way of recovering waste energy should be in the form of electricity. There are, however, a lot of problems to be solved regarding firing the L.C.V. gas in water tube boiler with insufficient heat accumulation capacities.

Accordingly, application of the gas to the existing two oil firing package boilers was considered to cope with these problems, in particular with the following:
(1) Load of combustion chamber required is too heavy for the L.C.V. gas.
   (1.6 x 10^6 kcal/m^{3}hr)
(2) Fuel oil should immediately be fired with or without the gas whenever the gas supply is unstable.
(3) Safety measures should be adopted for emergencies.
(4) Waste gas volume increased by 1.8 times will cause difficulties in draft control and increases in heat loss.

After the test work of several kinds of burners, the best performance was achieved by the ring type burner having vanes for air and gas to rotate both of them in the same direction as shown in Fig. 10.

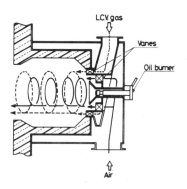

<u>Figure 10.</u>
L.C.V. gas Burner

Operational results by this burner to the boilers are as follows:

- Specification of the boiler
   Kawasaki Heavy Industries : BS-40
   Steam pressure   : 8kg/cm^2
   Fuel rate   : 312kg/hr (Bunker C)

272

Steam production rate   : 4t/hr
Heat conducting area   : 62.68m^2
Size of combustion chamber : 1.9$\mathbf{m^3}$
Burner Conventional   : Pressure atomizing return burner
Burner Modified   : Ring burner
- The maximum steam production by L.C.V. gas firing is 2.6t/hr.
- The maximum mixing ratio of gas to oil to produce 4.0t/hr steam is 2,100Nm^3/hr to 170 l/hr oil.
- The reason for less steam productivity of the gas is an insufficient gas supply due to draft control difficulties in the combustion chamber. This problem can be eliminated by applying a balanced draft system with a larger combustion chamber.
- Fuel oil savings up to 1,500 kl/year resulted from the improvement.

L.C.V. gas burning power plant   Despite the additional applications of zinc holding baths and package boilers, about one third of the L.C.V. gas was still exhausted. In 1976, the time came for utilizing all of the L.C.V. gas. The estimation work showed economic feasibility for the power generation together with the low pressure steam production by extraction.

From a technical point of view, the whole technology established in the package boiler test could be transferred. With a design and construction work period of a little less than two years, the new power plant, specifications of which are shown in Table VII, was put into operation in April 1978.

Table VII. Main Specifications of Power Plant

Power Generation		: 3,150 kw when L.C.V. gas burns at 18,000 Nm^3/hr
Boiler,	type	: 2-drum, natural circulation water tube boiler with an economizer
	efficiency	: 81.5% at L.C.V. gas burning 89.9% at oil burning
	burner	: scroll-type for gas burner, steam atomizing oil burner
	evaporation	: 18.5 t/hr
	steam pressure	: 40 kg/cm^2 x 400 °C
	draft	: balanced
Turbine,	type	: single cylinder, impulse, extration condensing turbine
	output	: 3,150 kw
	inlet steam pressure	: 38 kg/cm^2
	steam extraction	: 2.9 t/h (1.5 t/hr in min.) in summer 6.0 t/h (10.0t/hr in max.) in winter
Generator,	type	: 4-pole, cylindrical field, 3-phase synchronus generator
	output power	: 3,316 KVA
	excitation	: brushless type

The power plant features the following:
(1) To avoid misfires, the boiler is equipped with a burner with high combustion stability, an oil pilot burner and a flame detector. A CO gas detector is also placed in front of the boiler.

(2) To minimize the adverse effects of the boiler operation on the I.S.F.,
a gas pressure control system was developed.
(3) Adoption of an extraction-condensing turbine permits a low pressure
steam supply resulting in a shutdown of the existing package boilers.
(4) Pollution control measures adopted are a 55-m stack and low sulfur oil
burning for SOx control, and a combination of a two-step combustion and
gas recirculation system for NOx reduction.
(5) A wide range of oil mixing ratios is possible to follow any gas conditions
(6) Two kinds of control systems for the turbine-generator are used: a
combination of inlet pressure control and extracted steam pressure control
and extracted steam pressure control when connected with the purchased line,
and a combination of rotation speed control and extracted steam pressure
control in case of independent operations.

Since May 1978, the plant has satisfactorily been in commercial
operation producing 1,960 kw of electricty and 3.9 t/hr of steam available
on the average. A total of about 86 million Nm3 of L.C.V. gas and 944 kl
of oil were burned in 1978.

It is estimated that approximately 13,900 MWH is annually generated at
the power plant under the present operating rate. This amounts to about
20% of the total demand of the smelter. In addition, excess steam is sold
to a nearby factory.

Another benefit of the plant is that it can continue to operate during
possible power stoppages to meet the emergency needs of the plant, i.e. ,
the equipment absolutely required to operate the plant in terms of safety
and pollution control can well be covered with the new 2,000 kw power supply
from the power plant.

Energy Savings at Zinc Refinery

Development of new large trays It was considered that energy savings
at the zinc refining process should be one of the most important objectives
for the economical improvement of the smelter's total energy.

To get a higher thermal efficiency and productivity rate, a newly
designed enlarged tray with a size of 762 x 1,372 mm (30 x 54 inches) was
completed and in 1971 the first enlarged lead column was built in the existing
combustion chamber.

Based on the operational results of the column during the period of
1971-1973, two lead columns and one cadmium column were rebuilt with large
trays. However, this attempt was not completely successful because of the
deterioration in column conditions caused primarily by several earthquakes
affecting the plant. Furthermore, there were some difficulties in the firing
techniques of the narrow combustion chamber.

In the next(4th) campaign, however, a prolonged column life(31 months)
and overall satisfactory plant performance resulted from various improvements
made on the basis of the previous experiences. It has, therefore, been
concluded that the enlarged column in the conventional combustion chambers
should give superior productivity and energy consumption to the conventional
columns.

The necessity of a more accurate combustion control for the enlarged
column mounted in the conventional furnace than needed for the original
column, suggests that the combustion chamber be enlarged when rebuilt. As
such, better results can be expected.

Operating results of campaign no. 4 The zinc refinery plant consists
of 4 refluxers altogether, two lead columns, one cadmium column and one baby
column for cadmium refining. Three main columns have simultaneously been
rebuilt for the past four campaigns.

The operational summary of each campaign is shown in Table VIII.

Table VIII. Operational Summary of the Zinc Refinery

Campgn. No.	Life (months)	Total Production (t)	Energy Consumption* (kcal/t-SHG)	Fume Collected (kg/t-SHG)	(index)	Note
1	25	38,000	2.23×10^6	–	–	std. trays, no fume collection.
2	28	43,200	2.27×10^6	2.60	100	one large column, added electric power for fume collecting.
3	28	46,700	2.68×10^6	4.70	181	all large columns.
4	31	57,100	2.11×10^6	0.80	31	all large columns, 28-month life for one Pb-column

*Fossil fuel basis: 1 kwh = 2,450 kcal, 12,000 kcal/kg-L.P.G.)

Zinc leakage through cracks in the column is one of the major causes
for a shortened column life. The installation of the fume collecting bag
filters for waste gas from the refluxers in the second campaign permitted
the comparison of degree of deterioration in the columns.

From Table VIII, the results obtained in the 4th campaign showed
remarkable improvements not only in the life expectancy but also in general
performance, e.g. energy consumption, productivity throughout the life.

To have enabled this, many efforts had been made since the rebuilding
in 1976. These are (1) making more accurate fittings for each tray including
roof plates, (2) developing a special mortar named "K-mix" to be used for
joint protection and repairing cracks by brushing, (3) strictly controlling
draft in the combustion chamber employing a portable manometer to analyse the
pressure balance throughout the waste gas flue including the bag filter,
and (4) painting radiant mix called "High Emis sion" inside the combustion
chamber in an attempt to increase the thermal efficiency of the refluxers.

Estimation of energy requirement for the zinc refinery equivalent in the
capacity to the I.S.F. At present the smelter produces special high grade
zinc at only one third of the total production from the I.S.F. When more
S.H.G. zinc is needed, some steps to increase the production are possible;
i.e., (1) loading more zinc to each lead column, (2) adding one reboiler to
further reflux the "run-off" metal, (3) adding another set of lead and
cadmium columns, and so forth.

Accordingly, to treat all the zinc from the I.S.F. four lead columns,
two cadmium columns and two reboilers will be necessary.

An estimation of energy requirements has been made for the case shown
above. An addition of one reboiler to the existing columns will improve
fuel consumption by approximately 10% based on the assumption that the
productivity increases by 40 to 45% with only 30% of additional fuel. The
operation of the bag filter for leaking zinc fume collection would be

necessary in the latter half of the column life. The figure thus calculated is at most 1.95×10^6 kcal/t - S.H.G. Zn.

Total Energy Requirement for the Zinc-Lead Blast Furnace Complex

The total energy requirement for the smelter based on the annual data in 1977 plus the above figure for the zinc refinery is summarized in Table IX under the assumptions given by Hopkin, et. al. (2)

This result of the total energy requirement is, if taking the lead bullion into account as a byproduct, quite comparable to that of the most modern zinc electrolysis plant.

Table IX. Total Energy Requirement for the zinc-Lead Blast Furnace Complex

Concentrates - I.S.F. Zinc

	$\times 10^6$ kcal/t Zn
Coke (a)	8.01
Electric power (b)	2.66
L.P.G. (c)	0.22
Oil (d)	0.36
Sub - total	11.25
Zinc Refinery	1.95
Power plant credit	
Power generated	-0.52
Steam extracted	-0.08
Oil pilot burner	0.17
Sub total	-0.43
Grand total	12.77

(a) Including coke oven losses at 15.5%
(b) 2,450 kcal/kwh (c) 12,000 kcal/kg
(d) 9,900 kcal/l

Raw Material Constraints

Zinc production from the zinc electrolysis process exceeds 70% of the world supply. Generally speaking, however, a growing necessity of recycling materials could only be met by pyro-metallurgical processes. Certain kinds of concentrates, in most cases complex type ores, are also rejected, in some cases, from the electrolysis.

Standing on such resource availability, Hachinohe Smelter has been endeavoring to treat all kinds of raw materials consigned.

Table X shows the wide range in assays of the raw materials currently used (some thirty sources per annum). Clearly it can be seen from this table that in the long run it is of great importance to have blended ore satisfying all the constraints within the allowable range of each constituent. The optimization control of production processes are also needed to treat such materials. A computer control system undertakes not only ore blending calculations but also various closed loop controls and data loggings to allow the smelter to be the most efficient and stable.

Table X. Asseys of Raw Materials in Use[**]

Contents	Zinc Sulfides	Zinc Oxides*	Lead Sulfides	Lead Oxides*	Drosses & Hard zinc
Zn	36 - 56	21 - 66	5.4 - 12.5	1 - 28	15 - 90
Pb	0.0 - 7.2	3 - 12	55 - 62	34 - 59	11 - 60
Cu	0.1 - 2.4	–	0.4 - 5.0	0.3 - 1.5	–
S	29 - 35	–	16 - 21	–	–
Fe	1 - 20	3 - 30	3.5 - 8.5	0.3 - 3.7	0.8
Sn	–	0 - 0.2	–	0.05 - 0.5	–
Cl	–	0 - 1.0	–	0.6 - 2.6	–

* Oxides include calcined ore, waelz oxides, zinc leach residue, steel mill dusts, and residues from lead smelters and pyrite roasting plant.
** Bulk concentrates are also treated from time to time.

Conclusions

Energy savings accomplished for the last decade has been described. Roughly 20% of energy consumption has been saved, but the electric power consumption had to be increased due to the growing demand for pollution control.

The present objectives on further savings of energy are,
(1) further reductions in coke consumption by higher temperature blasts and other improvements to raise hearth efficiency,
(2) further reductions in butane gas consumption at the zinc refinery with heavy load operations, and
(3) further steam recovery from the waste gas of the zinc refluxers.

Regarding heat losses, there are still too many, most of which are too low in potential. However, since the total amount of the losses cannot be considered small, the ultimate target should be the encouragement of engineering developments and breakthroughs in waste heat recovery from these sources. The effective utilization of L.C.V. gas is one of the features of the energy conservation program proceeding at the smelter.

Thus the smelter with a zinc-lead blast furnace complex can operate highly efficiently especially from the viewpoint of resource economy.

Acknowledgements

The authors wish to acknowledge the kind suggestions of Dr. D.A. Temple, I.S.P. Limited, Bristol, England on the preparation of the paper. They also wish to thank the smelter's staff for their continued efforts in resource conservation.

References

1. H. Kinoshita, Y. Higashitsuji and D.A. Temple, "Use of the Zinc-Lead Blast Furnace in Japan," paper presented at Joint Meeting of AIME-MMIJ, Denvor, Colorado, Sept. 1976
2. W. Hopkin, and A.W. Richards, "Energy Conservation in the Zinc-Lead Blast Furnace," paper presented at 107th AIME Annual Meeting, Denvor, Colorado, Feb. 1978

TREATMENT OF A DIVERSE RANGE OF LOW GRADE FEED

MATERIALS IN THE ISF

Peter R. Mead
Sulphide Corporation Pty. Limited, Cockle Creek Works,
NSW, Australia

Abstract

The Imperial Smelting Process is ideally suited for the treatment of a diverse range of relatively low grade feed materials containing zinc, lead and copper. Materials of this nature are usually not as attractive to the electrolytic zinc process and conventional lead smelting as their high grade counterparts. Consequently it is within this spectrum of feed materials that the Imperial Smelting Process has found an important role to play.

This could become an expanding role in the future given that an increasing number of new mines will be based on fine grained and complex zinc-lead ore bodies.

The ISF smelter at the Cockle Creek Works of Sulphide Corporation is favourably placed to exploit the low grade zinc-lead feed materials available in Australia. Initially based on single high grade zinc and lead concentrates the smelter feed is now dominated by a wide range of lower grade concentrate materials and residues.

The effect of these materials on the metallurgical performance of the smelter is described and an account is given of the technical developments that have been applied to enable this feed policy to be pursued. In this context the following aspects are dealt with:

- process constraints imposed by low grade feeds
- improved control of sintering process
- metallurgical and economic evaluation of low grade feeds
- binder briquetting of fine materials
- recovery of copper as copper sulphate
- recovery process for selenium
- possible future avenues of research and development.

Introduction

Sulphide Corporation is owned by Australian Mining & Smelting Limited, which represents the lead and zinc interests of Conzinc Riotinto of Australia, and operates a zinc-lead smelter based on the Imperial Smelting Process at Cockle Creek, near Newcastle, New South Wales, Australia.

The Imperial Smelting Process (ISP) involves the simultaneous smelting of zinc and lead in a single blast furnace with recovery of zinc in a lead-splash condenser. At the present time there are twelve ISP plants operating throughout the world.

The smelter at Cockle Creek commenced operations in August 1961 and at that time the Imperial Smelting Process represented a major innovation in non-ferrous metallurgy. It is now the oldest Imperial Smelting Furnace (ISF) in operation. As a result of progressive developments and improvements in operating efficiency and equipment, it continues to be one of the more successful.

The main units of the smelter are:

- an updraught Sinter Machine which produces desulphurised lump feed for the furnace from a variety of zinc, lead and mixed zinc-lead concentrates;
- a Contact Acid Plant producing sulphuric acid from the sinter gas;
- a fluid bed Roasting Plant and associated Acid Plant which produces zinc calcines to supplement sinter production;
- an Imperial Smelting Furnace which smelts a charge of sinter and coke to produce Prime Western grade zinc, and lead bullion of 99.5% purity;
- a Zinc Refinery where furnace zinc is distilled in refluxer columns to produce High Grade zinc.

Copper, cadmium, silver and gold contained in raw materials are recovered in saleable products.

Annual capacity is -

	Tonnes
GOB (or Prime Western) grade zinc	40 000
High Grade and other zinc grades	30 000
Lead bullion	30 000
Copper in products (recovered from lead bullion)	2 000
Refined cadmium	300
Sulphuric acid (as 100% H_2SO_4)	140 000

Background

When the Cockle Creek ISP smelter commenced operation in 1961, it was on the basis of treating only two concentrate raw materials. These were a high grade zinc concentrate from the Zinc Corporation and a high grade lead concentrate from New Broken Hill Consolidated, both of these mines being located at Broken Hill in western New South Wales. This raw materials policy continued unchanged until 1969, and since that time the range of materials processed has become increasingly varied with emphasis on lower grade mixed zinc-lead concentrate feeds.

To understand the reasons for this policy change it is first necessary to

279

appreciate a number of unique features of the ISP smelter that are relevant in this context. These unique features are:

- the ability of the process to simultaneously smelt both zinc and lead. This can be performed with reasonable flexibility and with a good recovery of both metals
- the process is largely untroubled over a wide range of gangue constituents in feed materials and can handle the high slag falls associated with low grade feeds
- the process can handle copper in feed materials up to about 15% of lead production. The copper is recovered as a lead-copper dross product. Ability to handle copper at these levels significantly broadens the scope of feed materials suitable for the process
- a smelter based on the IS Process shows good recoveries for sulphur, silver and cadmium
- the direct smelting of a wide range of secondary zinc and lead bearing materials can be readily handled by the process.

These unique features enable the ISP smelter to successfully treat low grade and mixed zinc-lead feed materials which are usually less attractive to the electrolytic zinc process and the conventional lead smelter than their high grade counterparts.

Thus the most important avenue for maximising the economic position of the ISP is through the use of feed materials which take full advantage of the unique features of the process. This is the correct role of the ISP smelter within the zinc-lead industry.

In the successful pursuance of this policy at Cockle Creek a number of important process developments have taken place in recent years or are in progress.

These are:

- the installation of an ammonia leach process to recover copper from lead-copper dross as copper sulphate
- the development of a process to pilot plant stage for the recovery of selenium
- the establishment of a small binder briquetting process with potential to handle secondary materials
- the setting up of an on-going research programme to optimise Sinter Plant control and develop a better understanding of the desired characteristics of sinters for smelting in the ISF based on low grade feeds
- development of computer based techniques for the metallurgical and financial evaluation of a diverse range of feed materials.

Mixed and Low Grade Materials Currently Processed

Composition

The composition of some of the more important mixed and low grade material is given in Table I.

Table I - Typical Composition of Low Grade Feed Materials

Element		1 Zinc Corporation Low Grade	2 Zinc Corporation Dump Retreatment	3 MMM Dump Retreatment	4 EZ Co. Silver-Lead Residue	5 Cobar Mines Lead Concentrate
Zn	%	46.0	21.7	39.6	9.9	11.0
Pb	%	2.2	30.0	11.1	23.5	41.9
S	%	29.3	18.0	28.0	10.7	22.0
Cd	%	0.15	0.11	0.20	0.08	0.02
Ag	g/t	44	546	955	770	440
Cu	%	0.2	0.3	0.6	0.2	1.8
Fe	%	9.2	5.3	7.8	14.0	13.5
SiO_2	%	5.4	7.7	5.3	6.0	2.1
CaO	%	2.2	1.9	1.2	3.8	1.1
MgO	%	0.1	0.1	0.1	0.2	-
Al_2O_3	%	0.5	0.9	0.6	2.9	0.5

Low Grade Zinc Concentrate

Concentrate 1 in Table I is a low grade zinc concentrate from the Zinc Corporation at Broken Hill. This material provides an interesting example of how active co-operation between a mine and an ISP smelter in the same corporate group made possible improvements in metal recovery and profitability.

The Zinc Corporation concentrator traditionally operated to produce individual high grade zinc and lead concentrates until testwork and plant trials led to an increase in flotation capacity and some rearrangement of the mill circuit. This now gives:

(i)   improved overall recovery of zinc and lead from mined ore
(ii) a higher grade zinc concentrate than was formerly produced, for sale to electrolytic zinc producers
       and -
(iii) a low grade material (concentrate 1) for treatment in the Cockle Creek ISF.

Dump Retreatment Concentrates

Mining at Broken Hill dates from the mid-1880s, well before the flotation process was available. For many years a variety of mineral concentrating methods was tried, and the full development of flotation techniques also took several years. A consequence of this period was the establishment of large residue dumps containing higher metal values than would be expected in the tailings from a modern concentrator.

Recovery of these metal values from the dumps is feasible using today's flotation techniques. However, the zinc and lead contents of some dumps are highly variable, and the situation is at times further complicated by the presence of a motley array of old concentrator reagents. Were it necessary to produce individual high grade zinc and lead concentrates, the recoveries would be significantly less than if a mixed or bulk concentrate were produced. The availability of the Cockle Creek ISF has made such mixed concentrates acceptable.

Concentrates 2 and 3 in Table I are produced by retreatment of residue dumps at Zinc Corporation and Minerals, Mining and Metallurgy (formerly Broken Hill South).

Electrolytic Plant Silver-Lead Residue

Cockle Creek regularly processes the silver-lead leach residue from the Electrolytic Zinc Company's jarosite process at annual rates of up to 25 000 tonnes. Its composition is given in Table I - material 4.

The residue is also of interest because in this case the ISP is successfully consuming a low grade by-product material generated by the other major zinc-producing process.

Cobar Lead Concentrate

This material is of interest because of its very high selenium content, viz about 0.1%. Treatment of this material had to be suspended following the introduction of regulations stipulating a limit on selenium level in liquid effluent from the smelter. The development of a selenium recovery process, described later in the paper, has enabled the treatment of this material to be resumed.

Woodlawn Mines Lead Concentrate

The advent of Woodlawn Mines, which came on stream during 1978, has provided a prospective long term source of lead and zinc concentrates for Cockle Creek.

Woodlawn Mines is a joint venture of St. Joe Minerals, Phelps Dodge and New Broken Hill Consolidated. The Woodlawn deposit is located in New South Wales, about 75 km north east of Canberra. It is a stratiform copper-lead-zinc-silver deposit, which geologically resembles the Rammelsberg in Germany and has considerable similarity to Kuroko in Japan. In common with these, the ore is fine grained and metallurgically "complex". Factors which make selective separation difficult are the presence of talc and the proneness of the complex sulphide ore to oxidation.

Since commencement of this mine Cockle Creek has successfully treated quantities of the lead concentrate which, particularly in the early stages, contained relatively high levels of zinc and copper. The lead concentrate also has a selenium content around 0.03 - 0.04% and this has been satisfactorily handled by the new selenium recovery process.

The optimum concentrate mix to be produced at Woodlawn has not yet been established but it is likely that Cockle Creek will process substantial quantities of the lead concentrate which is comparatively low grade and ideally suited for treatment in the ISF smelter.

Process Constraints Imposed by Low Grade Feeds

The transition at Cockle Creek from a smelter processing only two high grade concentrates to one treating a range of lower grade materials introduced new problems to the smelter and prompted a number of important development projects.

The changes experienced were a consequence of the increased variety and quantity of minor elements in the new feed materials and the substantial increase in the quantity of gangue material handled by the smelter.

In regard to minor elements substantial increases have occurred in the case of copper and selenium and the development projects associated with these metals are discussed later in the paper.

In regard to gangue materials the furnace responded well to the higher slag fall which over the transition period increased by nearly 50%. The ratio of combined zinc and lead production to slag weight fell from 2.2 to 1.4.

However, in attempting to further pursue the treatment of low grade feeds a serious constraint related to the silica level in sinter was encountered. Over the transition period the level of silica in sinter rose from 2.5% to as high as 6% on occasions. At these higher levels severe operating conditions were experienced which at times resulted in extensive build-up of fused charge in the upper shaft region of the ISF. On several occasions this condition was so severe as to cause the furnace to become inoperable.

The cause of this problem was attributed to premature softening of the sinter in the upper part of the furnace shaft.

Less severe conditions of premature softening were characterised by unstable working of the furnace shaft and poor hearth conditions reflected in excessive production of speiss and similar irony "third phase" materials. The failure rate of the copper tapping breasts also increased.

In as much as optimum operating conditions for the ISF require a maximum of zinc reduction to occur in the solid state, premature softening of sinter would be likely to cause the instability observed. Furthermore, since the temperature in the equilibrium zone of the ISF is postulated to be about 1 050°C, any sinter with a softening point lower than this cannot be considered viable for smelting in the ISF.

The constraint this imposed on the treatment of low grade materials is obvious and led to an extensive investigation of the mechanism of early softening. This included the identification of phases in the sinter structure by means of X-ray and electron microprobe examinations and the development of a simple test for the measurement of sinter softening temperature.

This is an ongoing investigation and the findings at the time of writing can be summarised as follows:

- Satisfactory sinter for the ISF must have good strength at elevated temperatures, and should not soften until it passes through the equilibrium zone of the furnace. To achieve this, the sinter must be bonded by a network of refractory crystals which retain its stength up to the required temperatures. Two satisfactory structures have been identified:

    (i) at low gangue levels, bonding by acicular crystals of zincite
    (ii) at high gangue levels, bonding by a network of calcium ferrite.

Unsatisfactory structures which soften early include zincite structures where inadequate fusion has occurred, and higher gangue sinters in which lead silicate glasses constitute the bonding phase.

- A range of compositions has been tentatively identified. In higher silica sinters, it is necessary to have sufficient lime and iron present to form the required calcium ferrite structure. Composition ranges have still to be established in more detail.

- It was found also that physical parameters in the sintering operation can influence sinter structure. In order to obtain the required flame front characteristics and subsequent high temperature zone, it appears that sinter returns sizing must be fine, and that fresh air blowing pattern and sulphur fuel loading have an influence.

- The softening point of sinter, as measured by a laboratory softening test, has been positively linked with a range of serious furnace operating problems.

- If the softening test result is less than $1\ 000^{o}C$, there is a serious risk of the furnace shaft being choked by prematurely softened sinter.

- At results in the range $1\ 000 - 1\ 050^{o}C$, ISF operation is characterise by heavy upper shaft accretion, unstable hearth operation, sudden copper block failures and excessive generation of third phase material

- As results increase above $1\ 050^{o}C$, there is a progressive improvement in the stability and efficiency of ISF operation.

Further work remains to be done to define the theory more precisely, including refinement of the softening test procedure and investigation of a wider range of sinter compositions. In particular, a better understanding is required of operating parameters on the sinter machine which influence sinter structure. When this work is completed it is hoped that significantly higher levels of silica in sinter will be acceptable to the ISF which will enable even more lower grade feeds to be successfully treated.

Recovery of Copper as Copper Sulphate

The ability of the ISF to successfully process quite high levels of copper in concentrates has been established at a number of smelters. Lead bullion containing 10% or more copper has been handled without difficulties using bottom tapping forehearths.

The copper is recovered as a lead-copper dross typically containing 25 - 35% Cu with the balance predominantly lead. At low copper inputs the recovery of copper is at best about 80%. This low recovery is associated with the generally higher slag make of the ISF compared with the lead blast furnace. At higher copper inputs the recovery improves and marginal recoveries of 90% are feasible.

At Cockle Creek the copper intake has increased fourfold in recent times from 500 tpa to about 2 000 tpa. Until recently all of the lead-copper dross was sold to overseas refineries but with the impending increase in production it was realised that the comparatively low return on the contained lead content of the dross would increasingly disadvantage the smelter. Studies were undertaken of suitable processes to extract the copper and thus enable lead to be recycled to the ISF and recovered as lead bullion.

An ammonia leach process was selected and successfully developed to pilot scale by the research arm of Sulphide Corporation's parent company Conzinc Riotinto of Australia (CRA). ISP Research undertook parallel development of the process in the UK.

The process comprises four main stages:

- dross is agitated with an aerated solution of ammonia and ammonium carbonate taking the copper into solution and leaving all other metal values as insoluble metallics or oxide/carbonates in a sludge which can be returned to the Sinter Plant
- the leach solution is contacted in mixer/settlers with a proprietary organic reagent, dissolved in kerosene, which selectively removes the copper
- copper is stripped from the loaded organic solution with sulphuric acid in a second set of mixer/settlers
- copper sulphate is crystallised from acid stripping solution by cooling in a crystalliser.

A commercial plant with capacity of 4 000 tpa of copper sulphate penta-hydrate commenced operation at Cockle Creek in July 1979. This installation is seen as a first step in a strategy which offers the following benefits:

- improved return on sale of lead as bullion compared to lead in dross
- improved return on sale of copper as copper sulphate compared to copper in dross
- increased flexibility for the ISF in the selection of feed materials.

The extent to which this strategy will be further developed will depend on the opportunities that arise in the future with copper bearing feed materials and the technical development of copper in the ISF.

Recovery Process for Selenium

Several of the new concentrates handled by the Cockle Creek smelter contain very high levels of selenium. Input has increased from about 5 tpa to nearly 20 tpa in recent times.

Faced with this situation it became necessary to develop a recovery process to avoid exceeding limits stipulated by the local pollution control authority for the Works liquid effluent. At the same time the quantity of selenium involved made the commercial recovery of selenium an attractive proposition.

The behaviour of selenium is unusual in that it progressively builds up as a massive recycle in the leady fume evolved during sintering. This fume is removed from the SO_2 gas stream in a hot gas precipitator, treated for removal of cadmium, and then recycled back to the sinter machine. The losses to effluent are only a small percentage of the total recycled selenium, but once this recycle builds up even this small percentage becomes a significant quantity. It became clear that the logical point to attack the problem was the accumulation of selenium in sinter fume. Removal of selenium from this stream would have the maximum effect on distribution throughout the smelter and in effluent.

Computer simulation studies on the behaviour of selenium confirmed the above approach and further indicated that treatment of not more than 20% of the sinter fume would be necessary. Under these conditions effluent quality standards could be met at the predicted high levels of selenium input in feed materials and the recovery of selenium would be in excess of 90%.

A suitable process was devised by the CRA Research group and in conjunction with Sulphide Corporation further developed at pilot plant scale to a continuous process.

In brief the process involves the reaction of sinter fume with 98%

sulphuric acid at an elevated temperature. After initial mixing the reaction proceeds in a specially designed rotating vessel which is indirectly heated to approximately 450°C. The selenium rich vapours are collected by a simple scrubbing system operated at controlled acidity. The residue from the reactor is subsequently processed for cadmium recovery.

The selenium is precipitated in collection tanks as "red mud" and typically contains 90% Se with the balance mainly lead. Upgrading of the crude selenium to a marketable product is planned in the near future.

Binder Briquetting of Fine Materials

Normal practice with an ISP smelter is to recycle the drosses produced in the ISF and Zinc Refinery to the Sinter Plant. The observation that this practice had an adverse effect on sinter plant capacity led to the establishment of a small briquetting plant to treat these materials separately and return the dross briquettes direct to the ISF.

The operation was successfully established several years ago at Cockle Creek using petroleum based bitumen as a binder. The plant is relatively simple and it was demonstrated that briquettes of adequate quality could be produced by this means and successfully smelted in the ISF.

The implications of binder briquetting in the future are seen to be of considerable importance for the ISF smelter in a number of ways:

- secondary zinc and lead bearing materials are likely to become more important feed sources in the future and binder briquetting offers a simple means by which fine residues can be exploited by the ISF
- the high cost of lump metallurgical coke could be partly offset by incorporating the fines, generated from the screening of coke, in dross briquettes recycled to the ISF
- a possible further development would be the production of briquettes from 100% coke fines purchased at a substantially lower price than lump coke
- use of briquetting as a means of making special additions to the ISF. In this regard several ISF operators have reported greatly reduced accretion formation and improved condenser efficiency from chloride and alkali metals. These compounds occur in zinc bearing steel plant dusts which are now being treated by some operators. In cases where such materials are not readily available, as at Cockle Creek, the possibility exists of achieving the same result by the incorporation of common salt in the briquetting charge.

These aspects are scheduled for development and trial on the ISF at Cockle Creek and could greatly enhance the merits of binder briquetting as an important adjunct of the process.

Valuation of Raw Materials

The metallurgical assessment and valuation of raw materials is based on a mathematical model of smelter operations which computes mass balances for all revenue contributing elements and slag formers, and provides a complete listing of all raw materials inputs, product outputs and recycles. Recovery factors and metallurgical efficiencies are specified to suit circumstances and new values may be inserted in the computer programme when necessary. The model covers the operation of the Acid Plant, Sinter Plant, Cadmium Plant, ISF and Refinery.

In as much as the operation of the Cockle Creek smelter was originally based solely on high grade zinc and lead concentrates from mines at Broken Hill any new feed material has replaced a proportion of these concentrates in the feed mix. The price of the material being assessed is thus determined in terms of its break-even value with these high grade concentrates.

The valuation also depends on the overriding constraint for the smelter at the time, and this is taken into account in the assessment procedure. This constraint may be one imposed by any one of the plants in the smelter complex constituting a limit to further production increases, or it may result from external market conditions.

The speed and accuracy of the computing technique that has been developed, coupled with its high degree of flexibility, has played a vital role in the acquisition of new feed materials.

Future Avenues of Research and Development

Sinter Structure

Traditionally ISF operators have had to rely on measurements of sinter hardness and residual sulphur in sinter as the principle indices of sinter quality. The deficiency of these measurements for predicting the performance of sinter in the ISF has been acknowledged, but more meaningful indices of sinter quality have not been forthcoming.

The recent interest in the microstructure of sinter has opened up a new approach to sinter quality which promises to elucidate the critically important properties of sinter which affect its performance in the ISF.

More research in this area is required, while the important task remains of identifying the operating parameters in the sintering process that promote the desired sinter properties. Control of the sintering process to continuously achieve this end is a further task.

Higher Slag Fall

The ISF has demonstrated its ability to handle the higher slag fall associated with the treatment of low grade feeds. The carbon efficiency of the furnace as measured by the so called ISP Carbon Rating has not deteriorated. This index is the ratio of actual to theoretical carbon requirement.

This favourable feature of the process offers the prospect of economically treating feed grades that are too low to be acceptable to alternative processes.

To pursue a policy of treating low grade feeds suggests two developments that would be worthwhile:

- Continuous slagging of the furnace in lieu of the present practice of intermittent slagging;
- increased declination of furnace tuyeres to achieve greater activity in the slag pool.

Minor Elements

A diversified feed policy can be constrained by the inability of the process to satisfactorily handle the variety and quantity of minor elements that may ensue from such a policy. Minor elements can have adverse effects on the efficiency of the process, on product quality and on the environment. On the other hand some minor elements can offer prospects for economic recovery.

In the case of the ISF smelter notable examples of minor elements in the above categories are:

Adverse effect on process	Arsenic
Adverse effect on zinc product quality	Tin
Adverse effect on environment	Mercury
Economic recovery	Selenium

Although the behaviour of minor elements on the ISF smelter is reasonably well understood the subject is very complex. More research is needed to gain a more detailed knowledge of their behaviour on the basis that this might lead to the development of new metallurgical practices to enable the process to handle increased quantities of these elements.

Copper

The ability to recover copper enhances the feed flexibility of the ISF smelter. Looking to the future it would appear that an increasing number of new mines will be based on fine grained and complex ore bodies containing zinc, lead and copper.

It is therefore important that the ISF smelter takes steps to improve its metallurgical efficiency with respect to copper and in this regard the following developments are seen to be important:

- Gain a better understanding of the behaviour of copper at the bottom of the furnace with a view to achieving an improvement in recovery.
- Develop a continuous process for copper drossing of ISF lead bullion to replace the present batch operation.
- Explore the maximum copper capacity of the process and seek solutions to the limiting factors.

BIBLIOGRAPHY

1. A. O. Adami, G. R. Firkin and A. W. Robson, "Treatment of Complex Materials and Residues in the Imperial Smelting Process", paper presented at Joint Meeting of IMM and GDMB. "Complex Metallurgy '78", Bad Harzburg, Sept., 1978.

2. R. J. Holliday and P. H. Shoobridge, "A study of the composition, microstructure and softening characteristics of some zinc-lead sinters", paper presented at North Queensland Conference (1978), pp 311 - 321, Australian IMM.

3. D. A. Cain and I. E. Lewis, "Pilot plant studies in the recovery of copper from copper-lead dross", Proc. S. Australia Conf. (1975) Part A, pp 203 - 210, Australian IMM.

4. W. Hopkin, "Processes for treatment of zinc-lead blast furnace copper dross", paper presented at Symposium on Hydrometallurgy, Institution of Chemical Engineers, Manchester UK, April, 1975.

5. "Ammoniacal leaching of copper dross"
   British Patent   1 399 281
   US Patent   3 971 652

6. E. J. Burns, I. E. Lewis and A. E. Parsons, "Recovery of selenium and cadmium from ISF smelter fume" paper presented at North Queensland Conference (1978), pp 307 - 310, Australian IMM.

7. M. G. Taverner and A. S. Buchanan, "Cold briquetting of metallurgical arisings from an ISF", Agglomeration 77, K V S Sastry, ed., Vol. 2, pp 737 - 753, AIME, New York.

ZINC/LEAD SINTERING AND SULPHURIC ACID PRODUCTION

RM Sellwood

Commonwealth Smelting Limited
Avonmouth

Many lead smelter personnel have expressed dissatisfaction regarding quality of sinter produced for lead blast furnaces whilst attempting to satisfy the requirements of a standard or double absorption cold gas acid plant, and in fact have stated that the two operations are not compatible when taking all SO_2 gas produced, for acid production.

Throughout the world, however, plants producing lead/zinc sinter are satisfying the above criteria and, by applying the sintering techniques used to lead sintering, the same criteria can be satisfied. This means the installation of adequate proportioning equipment to control sinter assay and sulphur in feed mix, mixing equipment, and above all an updraught grease seal sinter machine to allow the use of gas recirculation systems. Using gas recirculation the weak ignition and tail end gases can be built up to provide a gas strength suitable for sulphuric acid production, while the sinter is cooled to satisfy handling conditions.

Control over the quality and quantity of sinter returns and the proportioning and return of acid plant sludges, ventilation dusts, etc., to the circuit, are also of paramount importance and, although costly, must be included to satisfy sintering, acid production, and environmental conditions.

For many years sinter plants feeding Imperial Smelting blast furnaces smelting lead and zinc, have successfully used all SO_2 gases produced for sulphuric acid production, at the same time producing a good hard well sized porous homogeneous sinter for the blast furnace.

This has been achieved by adapting the lead sintering practice of up-draughting to zinc/lead sintering, and by upgrading the practices and techniques used **the sintering operation supplies the furnace with good sinter and the acid plant with good SO_2.** There is no reason why lead smelting practice should not in turn develop these techniques for their use.

The ISF uses a sinter containing some 22% lead (at present) but the furnace is primarily a zinc smelter and the hot top furnace shaft operating condition does not at present readily allow use of higher lead charge. As far as the sintering plant is concerned however production of higher lead tenor charge presents no problem, but operating technique does have to be changed as the lead tenor increases.

The following paper outlines proposals which are already adopted in zinc/lead sintering, or would be applicable to lead sintering, and use of all SO_2 gas produced for sulphuric acid production.

For flow sheet see Fig. 1.

Sintering

Proportioning

This can be simple or sophisticated depending upon the number and variability of raw materials. It is essential to ensure that all materials flow freely and feed consistently which is more easily said than done. Design of the bin and discharge boot must be such as to take the weight of the material without creating excess compression, and then allow the material to expand and flow.

Independently driven belt weigh feeders are desirable to give a consistent feed mix, which will in turn allow a homogeneous sinter to be produced.

Mixing and Conditioning

A vigorous mixer to give mixing and conditioning is necessary. Mixing is self explanatory and combines the mixing in of moisture with the mixing of concentrates and returns, but conditioning is that state which gives permeability to the mix. It is partly pelletising, but is really a state which produces large lumpy looking material which falls to pieces on being picked up. A mixing drum with contra-rotating paddle shaft is probably as good as any. One or two of these drums can be used depending on feed rate etc.,and some 3 to 6 min. mixing time is necessary. The last mixer should be placed as near as practicable to the feed boards on the sinter machine to avoid breakdown of mix after it is conditioned. See Fig.2.

Feed Distribution

Most non-ferrous sinter machines built these days are 2½ to 3 metres wide and are updraughted so that feed splitting and distribution to the main igniter layer is important. Slow running oscillating shuttle conveyors will give an even distribution of material across the feed board and are probably as good as anything, although more sophisiticated systems are sometimes used.

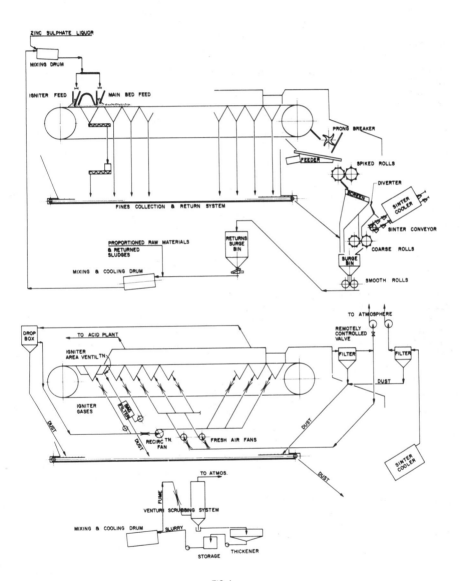

FIG. I

The shape and holding capacity of feed boards is critical. It will take around 7 to 8 min. for the feed mix to come from the proportioning bins to the feed board with a material handling rate of say 3,500 tonnes/day.

Feed Boards

The igniter feed board is of a shape as shown in Fig.3.

This profile is to avoid too great a weight of material on the actual igniter bed, but encourages a certain amount of rilling to fill the gaps between the pallet bars. The depth of the igniter layer is approximately 30 mm, but must be adjustable and is eventually set to suit the ignition conditions of each machine, ie. it depends on the length of the igniter stove and igniter windbox and on the volume pulled through it, machine speed, type of fuel for ignition,etc. It must also be expected that some day a rogue pallet bar will pass through and pull down the feed board. Steps are taken to avoid this by fitting a pallet roller across the feed end of the machine to push bars back into place, and limit switches are fitted so that if the roller is pushed up instead of the pallet bar down, the machine drive will cut out.

The main layer feed board must again be designed to lay down a continuous permeable bed. We have found that a feed board with a water cooled vertical back plate with sloping front plate to take the weight of material is preferred. The throat should be designed as shown; this gives maximum porosity with minimum back pressure. Fig.4.

Material of construction is stainless steel or a similar non-wetting material to reduce build up problems, blockages etc.

The feed board should be mounted on load cells or similar, so that the operator knows the level of his feed board from the control room. The capacity of the feed board should be such as to contain about 1½ min. of feed mix at normal speed (say 4 to 6 tonnes). If it is larger it becomes difficult to clean, if it is smaller there is insufficient capacity to maintain continuity.

Igniter Stove

This needs to be long enough to allow ignition of the igniter layer from top to bottom. This is not critical for zinc/lead sintering where the igniter layer comes back as returns, but can be very important for lead sintering where the ignited layer becomes part of the main sintered cake and can carry significant quantities of sulphur. In general the ignition stove and igniter windbox should be long enough to provide one minute of down-draughting when the machine is running at normal throughput speed.

Final control is exercised by varying the amount of volume pulled through the windbox. The actual control mode can be either direct volume control or pressure control, pressure control using a fixed speed fan and a butterfly damper being the favoured method where the igniter gases have to be recirculated since it gives less sudden variations of volume.

The ignition stove must be mounted as close as practicable to the back of the main layer feed board say 150-200 mm so that the ignited surface can be seen and tested, but as little heat as practicable allowed to escape before the main layer is applied.

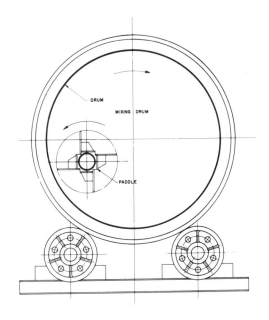

FIG. 2

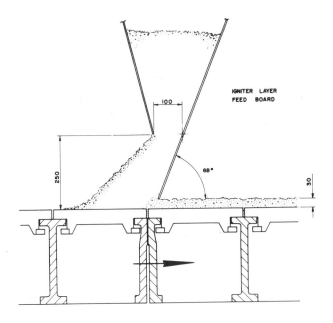

FIG. 3

297

Control of Feed Mix

As mentioned earlier it will take approximately 8 minutes for an alteration in the proportioning system to reach the feed boards, therefore some design criteria and method must be introduced such as:

1. The main feed board to be mounted on load cells and weight indicated and recorded in the control room.

2. The main feed board to hold say 4 to 6 tonnes of material.

3. All feed to be able to be diverted to the main or igniter layer to allow cleaning of the feed boards.

4. The igniter layer to be fitted with conductivity probes to ensure it does not run short of material, or comes to too high a level.

    (The igniter feed board must not hold more than 2-3 minutes feed in general, otherwise a "miss" on the feeds has an overlong reaction, and if material is held too long it dries out.)

5. The machine speed can be linked (to a relative degree) to the weight of material on the main feed board. This gives gradual speeding up or slowing down, but it ensures the feeds can be kept running all the time. Too many stops and starts on feeds and mixing drums means too many burnt out motors, as well as too much sounding of starting sirens, etc.

6. The splitting of the feed mix can be carried out in a variety of ways but it must be controllable.

7. If the sinter is a high lead sinter and there are large plates of semi reduced lead in returns such as at Port Pirie, the igniter layer may need to be screened using such a system as employed at Port Pirie.

Igniter Windbox

The design of the ignition stove and windbox in relation to each other and to the first updraught windbox is also important. The construction of a sinter machine and its pallets is such that dead plates isolating windboxes, and webs dividing and strengthening pallets have to be taken into account. The bottom of the webs must be level with the bottom of the pallet to form a seal.

In most non-ferrous sinter machines the pallets are 1 metre long, and divided to take 2 or 3 pallet bars. The length of the floating dead plate and fixed dead plate providing the maximum seal between the downdraught and the updraught windboxes depends on the number and length of pallet bars, and hence sealing webs, per pallet.

The length of dead plates before the leading edge of the ignition stove and between the other updraught windboxes is determined by the length of the pallet bars i.e.some 50 mm longer than a pallet bar. See Fig.5.

The main feed board back plate, and configuration of igniter windbox dead plate and floating dead plate, is such that any gas escaping past the dead plate is pulled down into the igniter windbox.

As the igniter gas has to be recycled the windbox must be sealed and a system fitted to enable grate spillage to be systematically removed. A possible method is to fit a continuously running screw feeder in the bottom

298

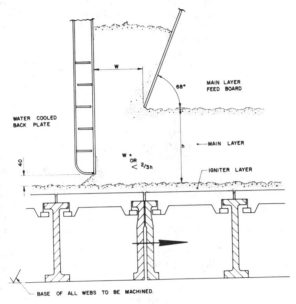

FIG. 4

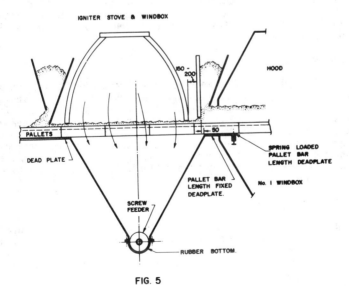

FIG. 5

299

of the windbox with a rubber bottom, and to discharge the material into a seal box. The seal box in turn can be fitted with load cells and a screw feeder. The feeder discharges intermittently into a spillage conveyor below the machine, which returns spillage continuously to the handling system. The windbox must be sealed and false air eliminated to prevent condensation, build-up, and corrosion. See Fig.6.

The whole igniter stove should be enclosed and ventilated.

Sinter Machine

As gas recirculation has to be practised, the machine must be fitted with grease sealed slide rails, and the pallets with replaceable gib plates. The slide rail may be criticised for wear and tear, but provided that the correct materials of construction are used, and a reliable set of grease pumps fitted, this is not a major problem. At Avonmouth (a 3 m wide x 44 m long machine where some 1.3 million tonnes per year is put over the machine), the same slide rails are in use after 12 years. The greaseways have to be positioned correctly for updraughting and the grease pipes do block up and have to be cleaned periodically. For this purpose use is made of a spacer pallet, which is really a frame of pallet dimensions, and this is run through the machine, for inspecting and clearing the grease ways. See Fig.7.

The whole of the underside of the machine is completely enclosed and a spillage conveyor fitted. Rapping rings to keep the pallet bars clean are also used.

Selection of size and the number of windboxes is of course determined by the tonnage of sinter required. For zinc/lead sintering the selection of draught intensities in the windbox and the number of fans used is not too critical, but it can be for lead sintering. Here again use of more fans can be used as a means of accurately controlling draught intensities along the machine. In zinc/lead sintering, draught intensity starts at 17 metres/min. in No 1 windbox, rising to some 22 metres/min.along the machine. The under-side of the machine is completely enclosed and ventilated. For lead sintering draught intensities of 30 metres/min. or more are used to keep the sinter porous and homogeneous

Sinter Machine Hood

Hoods for an updraughting machine have always been large, but because of the increase in draught intensities and the amount of heat evolved, the hoods tend to burn away. The size therefore has tended to become very large. Although water injection and ceramic lining of the hoods have been used with some success the larger hoods are preferred. Water injection under the hood is best avoided as it can cause condensation problems in the recirculation system and settling of dust in mains. The pallet rollers and collars are protected as much as possible by heat deflectors, and kept cool by keeping the hood at slight negative pressure.

Gas Recirculation System

The gas circulation system is suggested as shown in the **Figure 1.** As much use as possible is made of fresh air used on the machine to ventilate those areas of the machine system which carry fine fumes and traces of SO_2, so that environmentally poor discharges are as limited as possible.

Venturis are fitted where possible for the control of fresh air and recirculated gas volumes. A sensitive pressure indicator recorder is also fitted at the discharge end of the hood, and this reading is transmitted back to the control room. The hood must always be run at slight negative or zero pressure, to avoid blow out and contamination of the machine floor, and also to avoid overheating of the pallet rollers as these are in tunnels inside the machine hood. The channels or tunnels should be sealed as much as practicable at the ends of the hood to reduce leakage of false air.

The use of a drop box and the cleaning facilities for gas recirculation mains must receive special consideration, especially of the very hot sections.

Sinter Handling System

Sinter falling from the machine at around 600°C average temperature passes through a pronged breaker and over a vibrating feeder to a spiked roll crusher thus controlling the top sizing of the output sinter. The speed of the pronged breaker and vibrating feeder help to smooth out the flow of sinter through the circuit. If the sinter is too hot, sticking on the spiked rolls could be a problem, and use of conveyor for cooling by water or air instead of a vibrating feeder is an alternative.

The sized sinter is then passed over a vibrating screen which controls the minimum size to the furnace. The system is built and operated so that more lump sinter than required is produced. The excess lump sinter is always crushed and made into returns by passing through a coarse crusher, and rejoins the stream of undersize from the screen. The method of returning excess oversize must be selected to avoid surges of material to the crushers and overloading them.

The returns material now sized down to say < 25 mm is passed to a small surge bin mounted on load cells and fed at a constant rate by vibrating feeder to a smooth roll crusher, from where it passes to the returns bin - still hot. Pan conveyors are used for the handling of hot material.

Returns of basically -6 mm sizing are proportioned from the returns bin mounted on load cells, using a steel belt mounted on a weigher where returns and new materials are fed to the first cooling and mixing drum. At this drum, dewatered ventilation sludges, acid plant sludge and other furnace residues are added, the dewatered sludges being proportioned back into the sinter mix as cooling media as well as wetting media. Some fresh water is also added as a fine control to adjust the right amount of moisture in the mix.

Conductivity probes are probably the simplest and best way of controlling moisture content. If steady feeding of consistent returns can be achieved, it is practicable to achieve steady moisture content of the feed mix. The cooling drum can be positioned earlier in the system, but a surge bin must be positioned before it. Wet returns fed to a proportioning bin causes accretion problems in the bin.

Cooling of Output Sinter

Lead blast furnaces require a cold sinter feed (as distinct from the zinc/lead ISF which takes the sinter as hot as practicable) and the sinter up to the end of the screen is fairly hot. To cool this material, there is the choice of an air cooler such as is used for iron sinter, or a water spray and air cooler, but it is a matter of choice. It will however make a difference to the type of gas cleaning system used ie either a dry filter or a wet scrubber such as a venturi. In either case it will not be a difficult gas to clean.

301

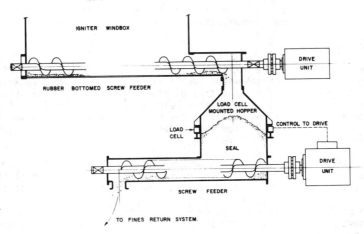

IGNITER WINDBOX SPILLAGE DISPOSAL

IGNITER WINDBOX

DRIVE UNIT

RUBBER BOTTOMED SCREW FEEDER

LOAD CELL MOUNTED HOPPER

LOAD CELL

CONTROL TO DRIVE

SEAL

DRIVE UNIT

SCREW FEEDER

TO FINES RETURN SYSTEM.

FIG. 6

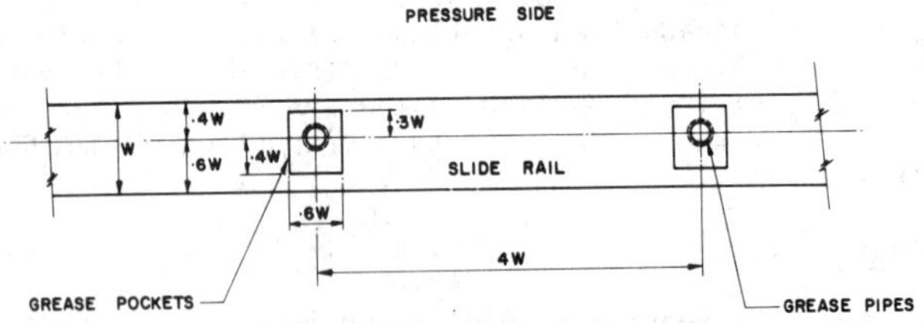

PRESSURE SIDE

.4W .3W

W .6W .4W

SLIDE RAIL

.6W

4W

GREASE POCKETS

GREASE PIPES

COLD SIDE

FIG. 7

Ventilation

Using a hot returns system, the natural choice of gas cleaning equipment is bag filters. The main problem then is how to handle the collected dusts without creating a secondary emissions problem, and how to proportion this material back into the sinter plant.

The most sure way of preventing secondary dust emission is to drown the dust, pump it as a slurry to the point of usage and then dewater it back into the cooling drum. This is also an excellent means of proportioning dusts back into the system.

Dusts from the acid plant, and from the blast furnace, will also need to be stripped of cadmium and such like and returned to the sinter circuit. The effluent treatment precipitates will also need to be pumped back and reabsorbed.

Dewatering equipment is best split into various types, so as to have flexibility to match the water balance, although a large single filter system will work. It has been found that a continuous centrifuge is excellent for dewatering acid plant sludge, a hydrocyclone for sinter plant dusts, and a disc filter for furnace dusts and effluent precipitates.

Acid Plant

General

The Acid Plant is a standard single or double catalysis cold gas plant with indirectly fired preheaters. See Fig.8.

The average gas strength which can be expected at the converter from full gas recirculation is in the region of 5% to 6.5% SO_2. The gas from the sinter machine will be some +300oC, about 12% moisture and carrying 35-40 g/m^3 of dust particulate. Gas strength on start ups especially will vary and the plant should be built to cope with gas strengths as low as 4% SO_2 for lengthy periods.

Gas Cleaning

Hot dirty gases from the sinter machine are ducted through mild steel insulated mains designed to avoid dust deposition, to the hot gas precipitator. These operate at around 300oC with an efficiency of about 99.5+% ie the dust loading is reduced to 0.2 g/m^3.

It is probable that the collected dust will be treated to remove soluble cadmium by leaching with sulphurous/sulphuric acid and the cadmium recovered by standard methods eg. ion exchange and zinc precipitation. The resulting sludge will be returned to the sinter machine circuit. Depending on raw materials etc this circuit may also be used for selenium and other nuisance metals removal.

Gases are cooled to about 65oC by passing through low head venturi scrubbers and separators where about 50% of the residual particulate is removed. As with all venturi scrubbers care must be taken to avoid the formation of a wet/dry line which promotes accretion build up. The venturi scrubber needs to be constructed in stainless steel to stand the temperature and corrosion and if a saturator is fitted ahead of the venturi it will need to be of either stainless steel or mild steel lead lined and acid resistant brick lined. The recirculation liquor, indirectly cooled, requires a bleed off, and this liquor plus sulphuric acid addition is used for leaching the

ACID PLANT FLOW SHEET

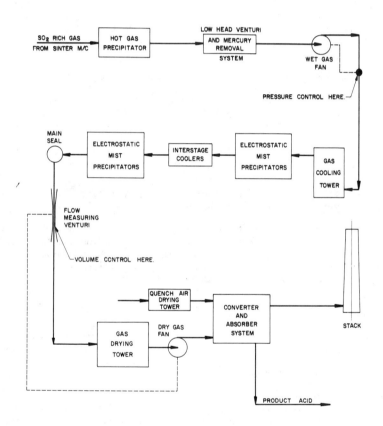

FIG. 8

hot gas precipitator dusts.   If mercury has to be removed, this is probably
the position for doing so, but again this will depend on raw materials supply.

Wet Gas Fan

The fan casing needs to be of stainless steel, with an impeller of
Incalloy 825 or Ferralium, i.e. it needs to be acid resistant and the root of
the impeller blades needs to be resistant to fluoride concentrations.
Ferralium, high chrome, nickel, copper, molybdenum alloy is probably better
for the job.

This fan which has to overcome the pressure drop of the sinter machine
offtake, hot gas precipitator, low head venturi scrubber, mercury removal
system and gas cooling tower, needs a control system compatible with the main
dry gas fan, to provide constant volume to the sinter and acid plants, despite
variation in temperature, moisture content and pressure.

Following the venturi/mercury removal system the gas still has to be
cooled and this is most efficiently done in a packed tower with the recirculated liquor being indirectly cooled.   The gas temperature at this point
should be down to about 45oC, and there will still be some solid particulate
present.

It is therefore suggested that two stages of electrostatic mist precipitators are used with interstage cooling between them to bring the gas temperature down to 34-38oC, but the colder the better.   To produce all acid at
96% H_2SO_4, using 6% SO_2, the temperature of the gas entering the drying tower
will need to be below 38oC.   The constant cleanliness of the gas will
determine the life of the converter catalyst and converter pressue drop.
Fibreglass reinforced plastic construction has been found excellent for these
large diameter mains, but the electrostatic mist precipitators should be in
lead.

Since everything these days tends to be designed to the limit, it is
advisable to be generous with the gas cooling system especially for summer
conditions and design for a lower temperature than 38oC say 35oC.   If there
is a market for 77% acid the temperature is less critical.

Markets for acid will determine the number of drying towers and the
complexity or otherwise of the acid circulating system.   If a market for 77%
acid is available it is likely that two drying towers will be used i.e. 77% and
96%, as this will enable a considerable proportion of the strong acid to be
produced and sold as clean white acid.   If the market is only for 96% acid
the acid will be produced as black acid, and will have to be treated with
hydrogen peroxide to enable it to be sold as white acid.   Depending on gas
strength and gas temperature, it may be possible to produce a small amount
of the acid directly as white acid, but it will need a separate absorption
tower and circulating system to do so.   Should it be necessary to produce
all acid SO_2 free, this can be done bleeding off acid from a quench air
drying tower where the dry air is required for control of the final catalyst
beds of the converter.   As the air is dried it strips any SO_2 from the
drying acid, thus producing a source of SO_2 free acid.

Alternatively treatment of black acid with hydrogen peroxide will oxidise any
SO_2 to SO_3 before it starts to decolourise the acid.   Some 1.5-2.0 kg of
33% H_2O_2 per tonne of acid is used and reaction time is some 4 hours.

The drying tower can be of conventional design, or of the venturi type,
but in both cases it is desirable to fit a mist eliminator on the discharge
to ensure there is no carryover into the dry gas fan.   The venturi drying

tower has some advantages, namely that the changing of the acid supply jet in the venturi can be quickly done, whereas the unpacking and repacking of a tower is expensive and time consuming. Good choice of acid coolers can markedly reduce the likelihood of having to repack the drying tower.

The main fan, of mild steel construction and variable speed drive, probably a fluid coupling, is required to be run at constant volume.

To provide a constant volume control measurement, a venturi can be fitted in the ducting between the outlet of the second stage wet electrostatic precipitators and the drying tower. It does not have to be corrected for temperature and moisture content, as it will always be operating at about 35oC and saturated at that temperature and so temperature and moisture correction is insignificant provided the design of the venturi takes them into account initially. A main safety seal is also necessary in this section to protect lead work. The electrostatic precipitators always need to be operated under suction, but because they are of lead construction, at the minimum practicable.

As mentioned the main fan needs to be run at constant volume, and the wet gas fan to supply at a constant pressure at the outlet of the wet gas fan. This will allow variations in temperature from the sinter plant to be taken care of without upsetting the balance between the fans.

Any new cold gas sulphuric acid plant built these days is almost certain to be a double catalysis plant as acid plant manufacturers say they can build such plants to be self-sustaining on 4-4½% SO_2 as far as the converter is concerned. All that one can recommend is to install adequate good quality catalyst and to bear in mind that a sinter machine is a much more temperamental piece of equipment than a sulphur burner. Therefore the indirect preheaters with which the plant will be equipped must be of adequate size, and never be pushed to their maximum design capacity, otherwise they will not last. The weatherproofing of the converter is also very important as the plant will have longish shutdowns because of stoppages of other sections and keeping the converter hot ready for start-up is of major importance. Good insulation and keeping water out of the insulation will enable stops of 12-15 hours duration to be taken without major start-up problems. On start-up if there is one bed below catalyst bite temperature, a cold slug of gas will pass through the whole converter.

The intermediate and final absorbers can be of the venturi or standard type. Mist eliminators are an essential to keep acid droplets out of the heat exchangers and the stack offtake.

The acid coolers now produced are much better than the CI hairpin coolers which were a source of high iron in acid, and high maintenance, especially on the larger plants. The use of plate coolers and more recently shell and tube coolers electrically protected to prevent corrosion, seems to be the answer, with preference going to the shell and tube type.

Cross bleeding systems are necessary between the absorber and drying systems. With the cleaned gas at 35-38oC it will be carrying a fixed amount of water vapour which will dilute the drying acid. This must be then strengthened with a bleed of absorber acid which is always at constant strength say 98.5% H_2SO_4. The excess drying acid is then bled back to the absorber system as dilution acid to keep the strength at 98.5% or it can be in part bled off as the production acid. Quench air (ie. dry fresh air) is normally used to control the temperature of the last bed of a converter, and the drying acid used is normally from the same circulation system as the drying tower.

As mentioned earlier, if the gas strength is high enough, and the gas cooling system good enough (gas temperatures kept significantly below 38°C), some white acid can be produced directly, but a separate absorber system is necessary for this clean acid.

The acid plant stack for emission of the exit gases needs to be studied with care. The height of the stack is usually set not for normal running conditions, but to take care of accidental discharges when they occur, because occur they will. Most stacks are constructed with concrete outer support and either acid resistant brick lined or with a mild steel liner. Condensation will occur in the stack because of the fall in temperature of the gas giving both moisture and SO_3. If the gas velocity is over 9 metres/second the condensate will form a rising film and be carried out of the top as droplets. If the liner is mild steel sulphation will occur and if velocities are too high, sulphate particulate will be discharged from the stack. This sulphate emission can be prevented by occasional steaming of the stack which dissolves the sulphate and washes it to the bottom. Perhaps the best answer is to design the stack velocity for less than 6 metres/second and reduce heat losses in the stack as much as practicable. Arrangements must be made for removal of condensate from the stack.

Some stacks have been constructed from plastic reinforced with fibreglass, but these have a limited life.

Effluent Treatment

Water has become a resource which cannot be wasted, and where possible recirculation of water should be practised to the full. At Avonmouth, towns water is softened for use in tuyeres and jackets and for full recirculated indirect cooling on the Acid Plant. The bleed-off water is used for washing the electrostatic mist precipitators before passing to the venturi scrubber used for gas cleaning, after which it passes to the ion exchange plant for soluble cadmium recovery. It then passes to the inplant effluent treatment plant.

Industrial water supply is chlorinated treated sewage effluent, and this is used on all ventilation scrubbing equipment. It is recirculated as much as practicable before being bled off to the final effluent treatment plant. All spillages, bleed off water, rain water, wash down water is either put to the inplant effluent treatment or collected in two rhines surrounding the smelter site, which in turn passes the water to the final effluent treatment plant.

In both effluent treatment plants about 5% milk of lime is used to precipitate the metal hydroxides and neutralise any acidity. Final pH is controlled at 9.3-9.5 pH. If necessary, ferrous sulphate solution is added to acidic liquor bleed off in acid condition to aid the precipitation of arsenic.

Final monitoring is by automatic sampler, flow measurement recorded and indicated through a venturi flume, pH recorded and indicated, and indicated temperature measurement. If pH drops below control limits, the outlet valve is automatically shut and alarms are shown in the sinter/acid control room.

The metallic precipitates and any residual $Ca(OH)_2$ are then pumped back for thickening in conjunction with furnace residues, and dewatered back into the sinter plant.

For a detailed description of the Commonwealth Smelting Limited Plant at Avonmouth see pages 581-618 of the 1970 AIME World Symposium on Mining & Metallurgy of Lead and Zinc, Vol. II.   Note should also be taken of the flow sheet of Amax-Homestake Lead Tollers Buick Smelter flow sheet page 739.

DEVELOPMENT OF OPTIMUM FLUXING PROCEDURES FOR ISAMINE LEAD CONCENTRATES

B.C. McLoughlin, J.F. Riley and G.R. McKean

Research Metallurgist, Mineralogist, Smelter Metallurgist
Mount Isa Mines Limited, Mount Isa, Queensland, Australia

Important physical properties are determined by the microstructure of lead sinter. Chemical specifications to ensure the formation of ideal microstructure have been quantified. Verification is provided by reference to physical properties, microstructure and pilot blast furnace performance of sinter. Details are given of the application of findings to optimize flux usage over varying concentrate grades at Mount Isa.

Introduction

Lead concentrates at Mount Isa are fluxed to give a fixed lime and silica level in slag. The lime in slag of 22.5 wt% CaO and silica level of 18.5 wt% SiO_2 give ratios of CaO/Zn, CaO/Fe and CaO/SiO_2 which are among the highest in the industry. This fluxing procedure has been developed to maintain a high standard of productivity over a range of concentrate grades.

High limestone usage however, represents significant cost to the process and attempts to operate at lower lime levels have been tried. Six trials have taken place at irregular intervals from 1960 to 1977 and all but one of these trials were abandoned or resulted in loss of the blast furnace. All the unsuccessful trials displayed certain common characteristics. Viz:

. In the sinter plant, the product became denser and less porous causing crushing and handling difficulties and higher residual sulphur was recorded in sinter.

. In the furnace, feeding was irregular, pressures increased and there was a higher lead level in slag.

Following the trial in 1975 and in preparation for the 1977 trial, a relationship was found between the physical properties and the micro-structure of sinter. A function was developed to specify the chemistry necessary for formation of the desirable microstructure and physical properties. This function predicts different component levels (lime and silica) in slag for different concentrate grades.

Control of the variable flux levels in the 1977 trial posed problems and the trial was abandoned. Subsequent work has involved refinement of the function with examination of local and other plant sinters, operation of a pilot blast furnace, and development of a system of process control which will facilitate the implementation of optimum fluxing procedures.

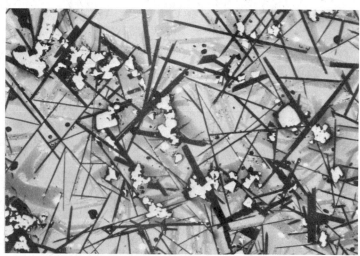

50 μm

FIGURE 1: Standard Sinter Microstructure
Black laths are melilite, white crystals are zinc ferrite

310

The Relationship Between Physical Properties and Microstructure

Physical testing and microscopic examination of trial sinters provided evidence for an explanation of the inferior performance of low lime sinter. Softening of sinter made using the standard fluxing procedure occurred in the range 1080 to 1150°C with a mean of 1120°C. Mean softening temperature for sinter with lime level halved was 980°C. Microscopic investigations revealed major differences in the phase distribution and morphology of the standard and low lime sinters. Figures 1 and 2 demonstrate the different sinter microstructures observed.

The major phases present in sinter are a melilite approaching hardystonite (Ca_2 Zn Si_2O_7) composition, a ferrite approaching franklinite ($ZnO.Fe_2O_3$) composition, lead oxides, lead silicates and glass.

High softening point sinters (Figure 1) show a microstructure of an interlocking network of melilite laths, isolated grains of zinc ferrite and a matrix of predominantly lead oxides. Melilite morphology in the lower softening sinter (Figure 2) however, is quite different being squat, isolated and frequently rimmed. In addition lead is more abundant as a silicate than as an oxide.

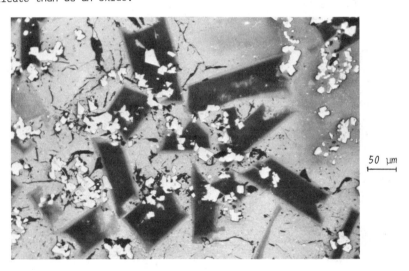

50 μm

FIGURE 2: *Low Lime Sinter Microstructure*
Black rectangular crystals are melilite showing borders where lead is replacing calcium

Changes in porosity pattern were observed between the two types of sinter. The 'low lime' sinter (Figure 4) is denser with a lower volume fraction of pores than the standard sinter (Figure 3). The surface area per unit volume of sinter is higher for 'low lime' than for standard sinter. Areas of poor combustion are frequently observed in the low lime sinter, this feature is rarely observed in standard sinter.

The Variable Fluxing Concept

It was concluded that the desirable physical properties of standard sinter were the result of the formation of the interlocking melilite

FIGURE 3: Standard Sinter Cut Section

FIGURE 4: Low Lime Sinter Cut Section
The dark areas contain a large proportion of galena

network. Conversely, failure of this interlocking structure to form resulted in the inability of the sinter to maintain an open pore structure, and during smelting to soften over a broader height in the shaft giving greater blast pressure and uneven feeding. Electron microprobe analysis of phases in standard and low softening sinters (Table I) reveals that lead replaces some of the calcium in melilite when the lime level of sinter is low. The substitution of lead for calcium in melilite beyond some critical level seems to influence the growth morphology of melilite crystals. It is concluded therefore that there is a threshold lime level for sinter above which calcium will saturate the melilite crystals and ideal morphology will be predominant in the microstructure. Since the ratio of calcium oxide to zinc in calcium saturated melilite is constant (Rm) then:

Threshold lime = Rm x weight of zinc reporting in melilite

Further examination of the electron microprobe analyses of sinter phases indicated that most of the iron is present as ferrite and that the ratio of zinc to iron in the ferrite (R_f) may be assumed to be constant. Further the zinc is present predominantly in the ferrite and melilite. Ignoring the minor occurrences of iron and zinc:

Threshold lime (CaO wt%) = Rm (Zn wt% - R_f x Fe wt%)

which reduces to:

Threshold lime (CaO wt%) = 2.5 x Zn wt% - Fe wt%

		Wt% of Element				Ions on the Basis of 7 Oxygen				
		Pb	Zn	Fe	Ca	Si	Pb	Zn	Fe^{+++}	Ca	Si
Standard	1	3.3	12.9	1.6	25.2	16.4	0.1	0.6	0.1	2.0	1.8
Sinter	2	3.6	15.0	1.5	24.7	16.5	0.1	0.7	0.1	2.0	1.9
Low	1	3.5	13.5	2.0	25.3	17.8	0.1	0.7	0.1	2.0	2.0
Lime	2	5.9	14.5	1.2	24.5	17.5	0.1	0.7	0.1	2.0	2.0
Sinter	3	8.9	10.9	1.8	24.4	16.7	0.1	0.5	0.1	2.0	2.0
	4	30.7	12.0	0.9	12.5	13.7	0.6	0.8	0.1	1.3	2.0
	5	50.4	16.0	1.6	0.9	8.8	1.4	1.4	0.2	0.1	2.0

TABLE 1. *Melilite Composition in Standard and Low Lime Sinter Electron microprobe analysis of major constituents. Analyses 1 to 5 for low lime sinter were taken successively from the centre to the edge of one rimmed melilite crystal.*

This function was applied to interpret results reported from past trials. It was found that, regardless of the total lime level in sinter, if the 'excess lime' (total lime minus the threshold lime value) was greater than three weight percent, then no problems were experienced. When the 'excess lime' was negative, the furnace and sinter plant problems listed previously, were reported, Between these limits variable conditions prevailed. A practical minimum level of lime was therefore defined as the threshold lime level plus a correction factor of three weight percent:

Practical Minimum Lime = Threshold Lime + Correction Factor

Practical Minimum Lime = 2.5 x Zn wt% - Fe wt% + 3

Sinters from Other Plants

On examination of sinter microstructure encompassing samples obtained from fourteen smelters (1), basically similar phase distribution and morphology to that shown in Figures 1 and 2 was found. The sinters were

313

subjected to softening point tests and electron microprobe analyses were performed on the major phases present in sinter from each plant.

The quantity of lead in melilite from microprobe analysis was used as an inverse measure of lime saturation. The softening temperature was plotted against lime saturation of melilite crystals (Figure 5). One sinter with an anomalous microstructure recorded softening temperatures higher than anticipated given the degree of lime saturation of melilite. The sinter was unusual in that it contained an interlocking network of rimmed melilite. The composition of the sinter was also unusual in that it was very high in iron and low in zinc when compared to other sinters. The size of sample was insufficient for detailed study to confirm or resolve this descrepancy.

Ignoring this one sample a good correlation is achieved between softening temperature and lime saturation of melilite.

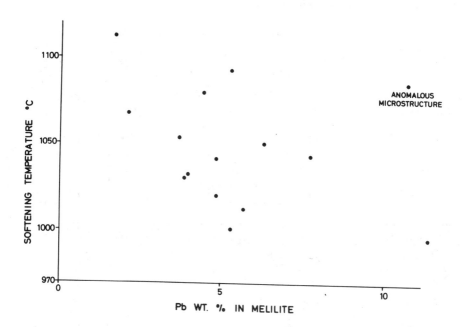

FIGURE 5: *Hot Softening Temperature Variation with Calcium Saturation of melilite phase. Calcium saturated melilite contains a low wt% lead.*

Having found a correlation between lime saturation and softening temperature the derivation of threshold lime level was sought. This was complicated by occurrences of magnesium, manganese and aluminium in higher proportions than in Mount Isa sinter. In general an equation was derived grouping magnesium and manganese with zinc and aluminium with iron to give:

$$\text{Threshold Lime} = f_1 (Zn, Mg, Mn) - f_2 (Fe, Al)$$

The functional relationship was determined for each plant based on electron microprobe analyses of the melilite and ferrite phases.

Figure 6 shows the relationship between softening temperatures and the 'excess lime' (CaO wt% in sinter - Threshold Lime) of all lead sinters examined. Two sinters with high arsenic and antimony levels could not be characterized by this simple approach. Complex phase assemblages and microstructure were observed which made the assumptions necessary to apply the function completely unrealistic. Both these sinters however, displayed properties commensurate with melilite saturation and morphology. Another sinter with massive zinc ferrite occurrence, though not forming a continuous structure, withstood much higher temperatures than predicted. This higher softening temperature was probably determined by the bulk of zinc ferrite present and was independent of the melilite structure. The remaining points show a good correlation between the derived value, 'excess lime' and the softening temperature of sinter.

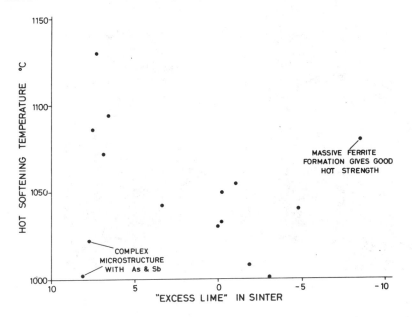

FIGURE 6: Hot Softening Temperature Variation with 'Excess Lime' Level in sinter (calculated from Electron Probe Micro-Analysis).

Pilot Scale Testwork

A pilot blast furnace (Figure 7) was constructed to test batch sinters made to specifications on large (300 kg) laboratory batch sinter pallets. The furnace had a 460 mm wide, 610 mm long and 1520 mm high shaft with seven tuyeres and was continuously tapped during test runs. Six special sinter batches were made. Four were higher in lime than the practical minimum lime composition and two were less than the practical minimum. All six were of lower than standard lime composition. These were tested in the furnace along with standard sinter from plant production. Smelting rates were calculated from the charging rates of sinter and coke when charging frequency became constant. An average pilot furnace run took six hours to reach equilibrium. Figures 8 and 9 show the consistent discrepancy between smelting rates of sinters over practical minimum lime values, and those below.

FIGURE 7: Pilot Blast Furnace
Weighed hoppers of sinter and coke are charged from above. Slag
and metal flow together into a mould.

The sinters performed as had been anticipated. Differences while
consistent, were less than that required for strong statistical
significance because of the scale of the operation. Figure 8 shows the
smelting rates of sinters against the lime level of the slags produced.
All sinters above the practical minimum lime level (this includes
standard sinters) performed better than the two under the minimum. For
the Mount Isa production furnace smelting rate is about 200 kg of sinter
per 100 m³ of air.

Replotting the smelting rates against the percent excess lime in
sinter (the practical minimum level corresponds to 3%) the variation in
excess lime for standard sinter with varying concentrate grades is
highlighted. All plant sinters had lime levels sufficient for 22.5 wt%
lime in slag but the 'excess lime' levels varied from five to eleven.
The high 'excess lime' levels represent material consumed which was not
essential to good plant operation.

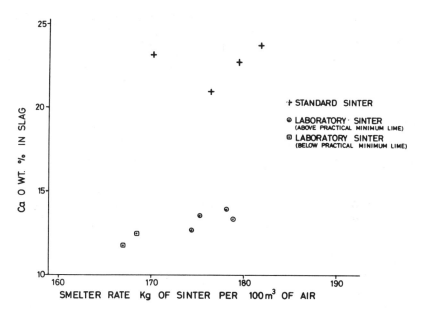

FIGURE 8: *Lime in Slag Versus Smelting Rate of Sinter*

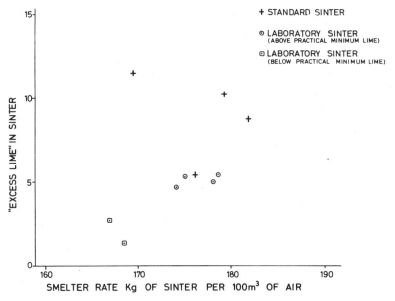

FIGURE 9: *'Excess Lime' in Sinter Versus Smelting Rate of Sinters*

Control of Variable Fluxing Rates at Mount Isa

While it was possible to accurately flux sinter on the static pallet, application of the model to the dynamic plant situation posed some control problems. Concentrate grade may vary significantly over 48 hours, flux grades vary, the feed proportioning system is not highly accurate and different recycled materials are handled from time to time. Failure to control flux rates so that the lime in sinter is always above the practical minimum level for the particular concentrate grade will result in production of inferior quality sinter with subsequent adverse consequences in the blast furnace.

A procedure has been proposed incorporating the flexibility to generate limestone savings without suffering from the consequences of inferior sinter. This procedure involves the use of a fluxing model which consists of a series of fourteen simultaneous equations comprising material balances, component balances and control equations for lime and silica within the smelting process.

The control equations determine the lime and silica levels appropriate for the particular concentrate type. The lime control equation consists of two parts, the practical minimum lime formula and the standard fluxing formula. The practical minimum lime expression is multiplied by an optimizing factor (N) which is set to balance the overall control of lime between practical minimum lime and the standard fluxing lime level.

$$CaO \text{ (sint)} = \frac{N \times \text{Practical Minimum Lime} + \text{Standard Lime}}{N + 1}$$

The optimizing factor may be used to maximize flux savings at any level of plant control sophistication. At N equal to zero the standard fluxing procedure will be applied unmodified. Positive N values will generate savings in material usage approaching that for practical minimum lime fluxing as N becomes larger. Low N values present little risk to the operation but as N increases, depending on the monitoring and control responses available in the plant, the probability of producing inferior material increases and represents cost increases and reduced production. The situation is presented graphically in Figure 10. Variable costs fall as N increases because less fluxing material is being used but as N becomes larger production will be reduced. The reduced production results in increased allocation of fixed costs per tonne of lead produced. The total cost per tonne however will go through a minimum. While the shapes of the curves are known the exact relationship will be a function of the plant. The fixed cost per tonne line will depend on the control features of the plant - more sophisticated control will move the range at which the curve rises sharply more to the right. An optimum level may be obtained for any control configuration at which net savings are maximized.

The silica control equation allows for a lowering of the silica to iron ratio at lower lime levels. The silica to iron ratio is reduced so that lead in sinter will tend to be present as an oxide phase rather than as a silicate. The effect of the lower silica on slag properties must be taken into account. Barlin (2) has proposed that slags which form primary wustite on cooling have higher metal contents than those forming primary dicalcium silicate or primary melilite. Mount Isa slag currently lies in the area of primary dicalcium silicate solidification. The initial silica control equation was derived from the $CaO-FeO-SiO_2$ system (3) phase diagram. Silica is adjusted so that slag composition lies in the dicalcium silicate or olivine phase regions for all lime levels. Because of the complexity of lead blast furnace slags the initial relationship will be varied as knowledge of the operating region increases.

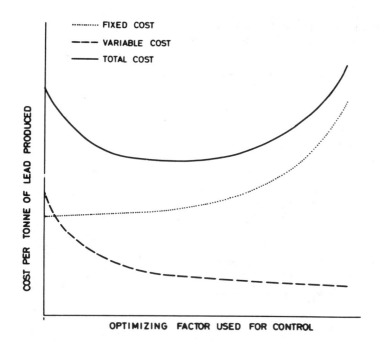

FIGURE 10: *Net Savings Variation with Optimizing Factor Variation*
The position of the curves is not known accurately but their shape
will be of a similar form to that shown.

The initial silica control equation is:

$$SiO_2 \ (sinter) = 0.6 \ Fe \ (sinter) + 0.3 \ CaO \ (sinter)$$

Once a sufficient amount of experience has been achieved with optimum fluxing, it will be possible to determine financial incentives for more sophisticated control features in the sinter plant.

The model is being used off-line with plant data to determine the effect of optimizing factors on plant conditions and economics. A Hewlett Packard 21MX mini-computer is being used to process the data. When the optimum fluxing procedure is implemented, the same machine will be used by sinter plant operators, through a video terminal link, to produce flux feed rates from process assay and rate data.

Conclusion

The high lime levels in Mount Isa sinter ensure production of blast furnace feed with high softening temperatures and open porosity over the full range of concentrates treated. An optimum lime level which varies with concentrate type has been defined which ensures that the desirable physical properties are maintained at lime levels which are frequently below this standard lime level. Adequate control procedures will allow significant limestone savings to be achieved during the treatment of certain concentrate types.

Acknowledgements

The co-operation of the smelters who contributed sinter for the sinter survey is gratefully acknowledged.

Amax-Homestake Lead Tollers, Buick,
ASARCO Inc., East Helena, El Paso, Glover,
Broken Hill Associated Smelters, Port Pirie,
Brunswick Mining and Smelting Corp. Ltd, Belledune,
Bunker Hill Co. Kellogg,
Centromin-Peru, La Oroya,
Cominco Ltd, Trail, Cominco-Mitsubishi Metal Corp., Naoshima,
Metallurgie Hoboken Overpelt, Hoboken,
St. Joe Mineral Corp., Herculaneum,
Soc, Miniére et Métallurgique de Peñarroya, pas De Calais,
Sulphide Corp. Pty Ltd, Cockle Creek

The authors wish to thank the management of Mount Isa Mines Limited for permission to publish this work.

References

1.   B.C. McLoughlin and J.F. Riley, "Examination of Australia and Overseas Sinter Properties, Mineral Chemistry and Microstructure", Mount Isa Mines Limited Technical Report RES SME 29, 1978 - limited distribution.

2.   B. Barlin, "The Evolution of Lead Smelting Practice at Zambia Broken Hill Development Company", AIME World Symposium on Mining and Metallurgy of Lead and Zinc 1970.

3.   E.E. Osborn and Arnulf Muan from "Phase Diagrams for Ceramists", The American Ceramic Society Inc. 1964.

LEAD BLAST FURNACE SMELTING WITH OXYGEN

ENRICHMENT AT THE B.H.A.S. PLANT,

Port Pirie, South Australia

J.H. FERN - BLAST FURNACE SUPERINTENDENT

R.W. JONES - ASSISTANT CHIEF METALLURGIST

SUMMARY

One of the primary objectives of lead blast furnace development
at The Broken Hill Associated Smelters Proprietary Limited,
Port Pirie, has been to achieve the plant production requirements
through a single furnace. In 1964, the first continuous tapping
lead blast furnace at the Port Pirie plant was blown in. Initial
expectations were that this furnace would achieve the necessary
level of production to meet this objective. This was not the case
and it is only since the introduction of oxygen enrichment with
its corresponding 32.7% increase in production, that furnace
performance has attained the desired level.

INTRODUCTION

Prior to 1964, the lead production capacity of the B.H.A.S. plant depended primarily on two furnaces of the type usually referred to as the Port Pirie or Australian lead blast furnace. These furnaces and the support installations were described in detail by White (1) in 1950 with modifications reported by Pelton (2) in 1953 and Green (3) in 1965.

To achieve the maximum plant capacity of 230 000 tonnes of market lead per annum, both of the Port Pirie type furnaces were needed. The loss of either unit and the subsequent "blowing in" of the smaller support furnaces resulted in a marked reduction in production capacity. During 1962, it became apparent that the Port Pirie type furnace had severe limitations with excessive refractory erosion in the lead well area, poor crucible life and irregular production performance. As a consequence of these difficulties, it was decided to build a larger furnace incorporating a continuous tapping arrangement developed by Roy and Stone (4) and used by the ASARCO lead plants.

In constructing this larger continuously tapped furnace, it was hoped that the majority of the lead production requirements could be achieved through the operation of one furnace. The benefits of working such a furnace were anticipated to be:

(1) Increased crucible and furnace life

(2) Reduced material and tool requirements

(3) Improved labour productivity

(4) More reliable and hotter slag for the proposed slag fuming plant

This larger furnace, known as the No. 2 Lead Blast Furnace, was blown in on 17th November, 1964.

NO. 2 LEAD BLAST FURNACE

A description of the No. 2 Lead Blast Furnace and its development was previously published by Fern and Jones (5) in 1975.

The furnace comprises a water jacketed shaft supporting two levels of tuyeres, a central continuous tapping forebay, a replaceable forehearth for separation of lead bullion and slag, and separate product handling facilities. The principal dimensions of the furnace are given in Table I and a general arrangement shown in Fig. 1. The internal dimensions presented in the table are steel to steel dimensions and do not allow for the 150 mm thickness of refractory lining protecting the chair water jacket of the lower part of the shaft. A section through the furnace depicting the lower part of the shaft and the continuous tapping arrangement is shown in Fig. 2.

Following its commissioning, the main benefits of operating such a furnace were gradually realised save for the achievement of all production through one unit. At the maximum capacity of 230 000 tonnes of market lead, the equivalent of 630 tonnes per day, one of the older Port Pirie type furnaces was required to supplement the larger furnace. Only during periods of low world demand and hence production was it possible to run with just No. 2 Blast Furnace. For example, in the period between 23rd July, 1969 and 1st August, 1973, No. 2 Blast Furnace produced all the Company's market lead at a rate of 520 tonnes per day. Over the years of its operation, however, the long term production rate has been close to

480 tonnes per day. It became apparent, therefore, that further furnace development was necessary if the target of 630 tonnes per day was to be achieved.

Table I.   Principal Dimensions of No. 2 Blast Furnace

Dimension		
Operating length	m	10.67
Internal width at bottom tuyeres	m	1.52
Internal width at top tuyeres	m	3.05
Height of shaft above bottom tuyeres	m	4.82
Height of bottom tuyeres above tapper U/F	mm	430
Height of top tuyeres above bottom tuyeres	mm	990
Number of top tuyeres		46
Number of bottom tuyeres		46
Furnace area at bottom tuyeres	m^2	17.28
Chair jacket angle to horizontal	degrees	30
Tuyere insert bore diameter (minimum)	mm	50
Number of tiers of water jackets		3

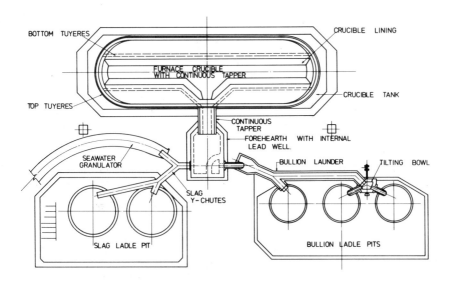

Fig. 1.   General Arrangement of No. 2 Lead Blast Furnace

323

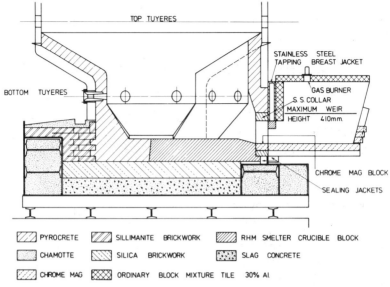

Fig. 2. Section Through Furnace Showing Hearth and Forebay

FURNACE DEVELOPMENT

The investigation into optimising blast furnace performance considered three main alternatives:

(1) Construction of a larger furnace

(2) Use of preheated blast air on No. 2 furnace

(3) Use of oxygen enriched blast air on No. 2 furnace

Based on previous experience, the length of the furnace could be increased without experiencing any difficulty in tapping the lead and slag. However, it was considered that such an increase could magnify the segregation problems associated with the existing charge feeding system and thus be counter-productive. Also, No. 2 furnace had on occasions operated for long periods at high production rates and indications were that the target of improved performance could be achieved by the introduction of some form of blast air conditioning.

Preheating of the air blast has been adopted at some lead plants (6, 7, 8) with apparent mixed success. To introduce preheated blast at B.H.A.S. would involve major redesign of the tuyere-hearth area, interrupt production for a considerable period and pose a possible safety risk in a restricted working area.

Oxygen enrichment of the blast appeared to be the best approach as numerous plants were using oxygen as the basis for most of their recent furnace developments. Reported usage of oxygen on lead blast furnaces (9 - 15) indicated substantial increases in production rate (refer Fig. 3) occurred with concurrent improvement in such factors as coke consumption, continuity of operation, charge top temperatures, fume or dust losses and tuyere problems. Conflicting results were given on the effect of oxygen enrichment on the lead loss in the blast furnace slag but it was anticipated that any increased lead loss could be overcome by a change in slag composition.

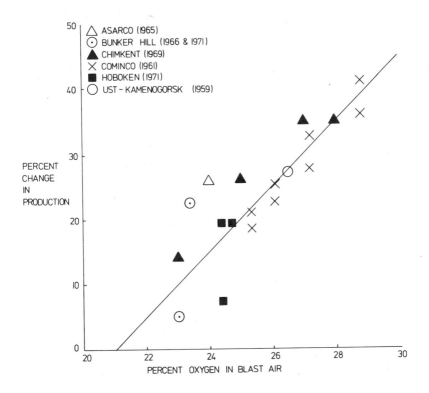

Fig. 3. Effect of Oxygen Enrichment on Furnace Capacity
at Various Lead Plants

A feasibility study into the use of oxygen concluded that providing the claimed benefits were attained, oxygen enrichment could be economically justified. The study subsequently recommended that oxygen enrichment trials be carried out on No. 2 Blast Furnace.

PLANT TRIALS

Arrangements were made with an Australian company, Commonwealth Industrial Gases Ltd., to supply both the equipment and sufficient liquid oxygen for an eight-day plant trial. The main equipment used for the trial consisted of a 70 tonne liquid storage tank, a sea water vaporiser and an automatic oxygen flow controller. Initially the oxygen was injected into the large bustle main at the rear of the furnace but this resulted in uneven distribution of the injected oxygen between the top and bottom tuyeres. The enrichment point was subsequently relocated to allow independent control of oxygen levels at both top and bottom tuyeres.

The initial trial, conducted from 23rd April to 1st May, 1976, proved quite successful. While the duration of the trial was too short to accurately assess the benefits of oxygen, there were significant improvements in such areas as production, coke consumption and top conditions. A comparison of average furnace factors prior to and during the trial is shown in Table II.

Table II.   Comparison of Blast Furnace Operating Performance
Prior To and During the Oxygen Trial

Furnace Factors		Normal Operations	Oxygen Enrichment
Average oxygen in blast air	%	21.0	23.9
Lead content of sinter	%	48.5	50.9
Sulphur content of sinter	%	1.8	1.9
Lead bullion production * tonnes per hour		22.3	28.6
Production increase	%	-	28.3
Slag production * tonnes per hour		26.5	29.7
Lead - slag temperature	°C	1204.4	1201.8
Blast air pressure	kPa	16.5	16.2
Blast air volume	Nm³/s	6.3	6.2
Top gas temperature			
Point a.	°C	75.5	58.0
Point b.	°C	82.5	61.0
Top gas CO_2/CO ratio		1.48	1.54
Lead content of blast furnace slag	%	2.31	2.44

(* Both lead and slag produced were weighed at the furnace.
Lead bullion production differs from market lead production
quoted elsewhere in text.)

The significant improvement in furnace operations associated with
oxygen enrichment led to a long-term trial that commenced on 8th May, 1976.
This trial, while successful, highlighted difficulties with the water-cooled
jackets, a problem not recognised during the shorter eight-day trial. A low
heat transfer rate through the fireface of the water-cooled jackets allowed
high temperature coke to contact the steel fireface, carbonisation of the
metal and finally holing of the jacket. In the first month of the trial
alone, eight leaking jackets were detected. As a result of the continuing
jacket failures and the consequent difficult shaft conditions, the trial was
terminated on 8th November, 1976.

Although the full potential of oxygen on the furnace was not achieved,
the indications were sufficient to justify continuing with this line of
development. It was felt that the problems associated with the water-cooled
jackets could be overcome with a minimum of redesign and modification.

NO. 2 BLAST FURNACE MODIFICATIONS

On 27th February, 1978, the seventh production campaign of No. 2 Blast
Furnace was concluded to enable rebuild of the shaft with the changes
necessary for improved jacket life.

The following major modifications were made during the rebuild.

1.   Redesign of the furnace jackets: The jackets were fabricated
from stress relieved low carbon steel with the fireface increased in

thickness to 12 mm. Additional internal stiffening was incorporated in the middle and top row jackets to prevent buckling and bowing observed during the oxygen trials. Larger diameter pipes and fittings were used inside and between the jackets to reduce the pressure drop over the system and increase water flow through the jackets.

Each jacket was fitted with a central drainage point to enable regular flushing "on the run", and prevent the build-up of fine colloidal clays present in local waters.

Five trial jackets using stainless steel firefaces were located at critical areas of the furnace to assess their ability to withstand the extreme conditions and compare their cost effectiveness against the low carbon steel firefaced units. Previous experience showed the stainless steel was vulnerable to attack from high sulphur - high temperature bullion.

2. Incorporation of tuyere inserts: Inserts made from austenitic stainless steel (containing a maximum of 0.5 per cent nickel) were installed in both top and bottom rows of tuyeres. The inserts reduced the effective internal diameter of the tuyere and increased the blast air velocity. The change in velocity plus the projection of the insert into the furnace pushed the heat zone away from the tuyere reducing the thermal effects on the jacket fireface.

3. Lining of the chair water jacket with castable refractory: A 150 mm lining of high alumina castable refractory was used to cover the most vulnerable water jackets to prevent crazing of the fireface and under-cutting of the jacket by molten lead and slag. The lining uses the bottom tuyere inserts and jacket joints as the sole means of support.

4. Installation of stainless steel collar around tap hole: An area of extreme corrosion and one that has caused premature termination of many a campaign is the stainless steel seal between the furnace shaft and the continuous tapping forebay. In an effort to reduce corrosion in this area, a stainless steel collar covered by a chrome magnesite block was installed to replace the previous stainless steel plate butt joint.

5. Use of filtered, softened and treated cooling water: One of the main reasons for the reduced heat transfer through the jacket fireface was the formation of boiler-like scale on the water side of the jackets. To prevent scale formation, it was decided that all water used for cooling should be filtered to remove the coarse clay particles, softened to less than one part per million of hardness and dosed with a corrosion inhibitor.

Other changes were made to improve the operating conditions around the furnace. The most significant innovation was the installation of a tilting lead bowl between the second and third lead ladles. Tilting the bowl changed the flow of lead from one ladle to the other with a minimum of time and effort. The bowl replaced the arduous practice of diverting the lead flow by the blocking of the launder channel with a fire clay ram mix.

During the period of 101 days required for the major rebuild, an air separation plant was commissioned to supply oxygen for the eighth production campaign.

AIR SEPARATION PLANT

The location of the Port Pirie lead plant presents numerous
difficulties to the supply of oxygen on a continuous basis. A constant
supply could only be guaranteed by the operation of an on-site plant and
thus Linde AG were contracted to erect an air separation plant capable of
meeting both the present and future needs.

The air separation plant would be considered small in comparison with
world standards, having a designed capacity of 1200 normal cubic metres per
hour of 95 per cent pure gaseous oxygen. The plant uses the low-pressure
process with double rectification. The small size of the plant enables
drying of the process air as well as adsorption of the carbon dioxide and
hydrocarbons from the air by means of molecular sieves. The plant is
located outdoors and comprises the basis units common to most separation
operations of this type, viz. air filtering, compression, refrigeration,
purification and rectification. To cope with the changes in demand for
oxygen from the blast furnace, the downturn of the compressor and plant is
sufficient to enable operation at 50 per cent of design capacity. Should
the furnace requirements be lower than 50 per cent of design, the excess
oxygen produced can be blown off to waste at a penalty of increased cost
per unit of oxygen used. Relevant specifications for the air separation
plant are given in Table III.

Table III.   Specifications of the Linde Air Separation Plant
Operating at B.H.A.S.

Factor		Design Load	Minimum Load
Gaseous oxygen	Nm^3/h	1200	600
Purity	%	95	95
Outlet pressure (cold box)	kPa	156	156
Air flow	Nm^3/h	5750	3100
Air pressure after compressor	kPa	679	598
Power consumption			
Air compressor	kW	615	375
Refrigeration unit	kW	45	30
Molecular sieve unit	kW	40	40
Other	kW	30	30
Time required for derime	h	12	
Time required to start-up after derime	h	36	

Furnace oxygen requirements and final oxygen flow control are the
responsibility of the blast furnace operators. A metering system
automatically varies the oxygen flow according to the blast air flow on a
continuous basis. The level of enrichment and the distribution of oxygen
between top and bottom tuyeres may be varied manually.

FURNACE OPERATION WITH OXYGEN ENRICHMENT

The eighth campaign with No. 2 furnace started on 8th June, 1978, and to the 30th May, 1979, a total of 307 days had been completed with oxygen enrichment. Interruptions to the commissioning of the air separation plant and to the operation of No. 2 furnace resulted in the poor time utilisation.

Despite the poor time utilisation, the results of oxygen enrichment have far exceeded project expectations. Typical data for the main operating factors of No. 2 furnace with oxygen are shown in Table IV.

Table IV. Performance of No. 2 Blast Furnace with Oxygen Enrichment

Factor	
Average level of oxygen in blast air %	23.5
Market lead production tonnes per day	637
Percentage increase (based on 480 tonnes per day) %	32.7
Coke -	
Fixed carbon in coke %	82.9
Kilograms of fixed carbon per tonne of market lead	142.5
Lead content of sinter	49.9
Lead content of blast furnace slag	2.6

The average production rate of 637 tonnes per day more than satisfies the plants' requirements with some allowance available for "catch up" of any lost production. The higher level of production and improved continuity of operations have increased the flexibility of the plant in meeting the demands for lead while decreasing unit operating costs.

Experience with oxygen is still limited but the following benefits have resulted so far.

1.   Coke consumption improved by some 5.0 per cent early in the campaign but an increase in the oxidised state of the sinter finally led to a coke consumption close to that achieved prior to oxygen enrichment. The improvement in coke utilisation takes place when the large increase in throughput is not accompanied by a corresponding percentage change in heat losses in the furnace top gases and cooling water. Poor sinter quality and the resultant charge segregation problems have a detrimental effect on the CO_2/CO ratio of the top gases and thus the heat loss in top gases. Under these circumstances, oxygen enrichment maintains production, compensating for the sinter quality at the expense of coke consumption.

2.   The furnace can treat a higher lead tenor sinter at a correspondingly lower level of coke consumption. The limitation to the level of lead in sinter is now the ability of the sintering plant to produce a sinter of acceptable sulphur content.

3.   Operating conditions are more stable with brigher, hotter tuyeres and more consistent top conditions. One difficulty found with the brighter tuyeres is the little warning given by the furnace to a coke deficiency in the hearth area. Previously, a change in the coking rate of the furnace was indicated by a gradual loss of tuyeres but, with

oxygen, the tendency is to lose complete banks of tuyeres at one time.

4.   The initial lead content of blast furnace slag was excessively high with oxygen but changes in slag composition, particularly the calcium level, have improved the situation to such an extent that the lead lost in slag is now similar to that prior to oxygen enrichment.

No effort has been made to determine the change in fume content of the top gases as assessment will necessitate measurement of fume collection over a number of years.

The modifications made to improve furnace shaft life appear successful with no down-time referable to water jacket failure to date. A furnace shut-down of three weeks duration during January 1979 revealed that the refractory lining had eroded to a fraction of its initial thickness but it still afforded protection to the chair jackets. All water jackets appeared to be in good condition.

With the continuous high level of production, weak points in the product handling facilities have been highlighted by an increase in the failure rate of various items of equipment. The operating life of a forehearth has decreased as a result of increased refractory failure and lead chute attack by the hot bullion. A severe failure of the bullion launder culminated in the shut-down of the furnace in January 1979.

FURTHER DEVELOPMENT

At the present stage of experience with oxygen enrichment, the following areas are considered to be in need of further development.

1.   An improvement in the operating life of the handling equipment by way of better refractories, cooling and minor modifications.

2.   Enhancement of the monitoring of furnace operations to ensure earlier warning of changes in coke rates, etc. Currently investigations are underway to continuously measure the CO_2/CO ratio of the top gases and assess if such a method can be used to accurately predict furnace shaft conditions.

3.   Better preparation of the charge materials to reduce the effects of size segregation on the furnace top. Although oxygen enrichment has a beneficial effect, shaft conditions are still heavily dependent on properly sized, good quality coke and sinter. Variation in lead content of sinter with differing lump size combined with the segregation problems found on such a long shaft results in areas of variable coking and uneven top conditions. Typical variations in lead assay with sinter size are shown in Table V.

New crushing facilities have recently been installed at the Sintering Plant in an effort to improve continuity of operations, sulphur elimination and sinter quality.

CONCLUSION

The introduction of oxygen enrichment has finally realised the full potential of No. 2 Blast Furnace in producing the annual production requirements through one unit. The operating life of the furnace between major rebuilds is expected to be around three years with production capacity at risk for the period of the rebuild. Although No. 2 furnace is capable of recovering lost production, some additional furnace capacity is required to supplement production during the rebuild. Consequently, No. 4 Blast Furnace has been converted from a Port Pirie type furnace to a continuous tapping arrangement with oxygen enrichment. Recent operation of the furnace indicated a short-term production capacity close to 500 tonnes of market lead per day, well in excess of its old capacity of 300 to 350 tonnes. With the two furnaces and oxygen enrichment, both the short and long-term production requirements of the plant are guaranteed.

Table V.   Change in Lead Content with Sinter Size

	Percentage Sample	Total Lead %	Total Sulphur %
+ 75 mm	1.5	59.2	1.9
- 75 mm + 50 mm	14.1	57.5	1.6
- 50 mm + 25 mm	32.1	51.2	1.0
- 25 mm + 12 mm	22.4	43.2	0.8
- 12 mm + 6 mm	17.3	42.8	0.7
- 6 mm + 3 mm	8.6	42.1	1.1
- 3 mm	4.0	44.5	1.5

ACKNOWLEDGEMENT

The authors wish to thank the management of The Broken Hill Associated Smelters Pty. Ltd. for permission to publish this paper.

REFERENCES

1.   L.A. White, "The Development of the Lead Blast Furnace at Port Pirie, South Australia", Trans. A.I.M.E., (1950) pp. 1221-1228.

2.   L.A.H. Pelton, "Lead Smelting and Refining Practice at The Broken Hill Associated Smelters Pty. Ltd., Port Pirie, South Australia", Extractive Metallurgy in Australia: Non-Ferrous Metallurgy, IVB (1953) pp. 23-52.

3.   F.A. Green, "Lead Smelting and Refining Operations at Port Pirie", The Australian Mining, Metallurgical and Mineral Industry, Vol. 3 (1965) pp. 83-88.

4.   J.T. Roy and J.R. Stone, "Lead Blast Furnace Continuously Tapped", Journal of Metals, 15 (1963) pp. 827-829.

5.   J.H. Fern and R.W. Jones, "The .evelopment and peration of a ontinuous apping ead last urnace at Port Pirie", paper presented at Annual Aus. I.M.M. Conference, South Australia, June 1975.

6.   J. Leroy, "Non-ferrous Contributions", paper presented at Blast Furnace Injection Conference, Wollongong University, 1972.

7.   M.P. Smirnov, "Trends in the Development of Lead Production Technology in the U.S.A.", Tsvetnye Metally, The Soviet Journal of Non-Ferrous Metals Vol. 12 (1971) pp. 96-97.

8.   A.V. Evdokimeko, "Natural Gas and Oxygen in the Shaft Smelting of Lead", Sb. Nauch. Tr. Gos. Nauch Issled. Inst. Tsvet. Metal,

9.   E.A. Hase, "Oxygen Enriched last at Asarco's Lead Smelter", Journal of Metals 17 (1965) pp. 1334-1337.

10. Private report, G.C. Hancock, "Visit to North American Lead Smelters", The B.H.A.S. Pty. Ltd., November 1966.

11. M.P. Smirnov, op cit. pp. 96-97.

12. A.V. Evdokimenko, op cit. pp. 33.

13. L. Landucci and F.T. Fuller, "Oxygen-enriched Air in Lead and Zinc Smelting", Journal of Metals 13 (1961) pp. 759-763.

14. P.J. Lenoir, J. Thiriar and C. Cockelbergs, "Use of Oxygen-enriched Air at the Metallurgie Hoboken-Overpelt Smelter", paper presented at the Inst. of Mining and Metallurgy Conference, London, October 1971.

15. A.M. Vartanyan and D.S. Koptchenko, "The Experimental Use of Oxygen in the Lead Smelting Shaft Furnaces", Tsvetnye Metally, The Soviet Journal of Non-Ferrous Metals 32 (1959) pp. 46-49.

NEW DETELLURIZING AND DECOPPERIZING PRACTICES AT

OMAHA REFINERY OF ASARCO

Carl DiMartini
General Supt., Metals Research
Central Research Dept.
ASARCO Incorporated

R. F. Lambert
Manager
Omaha Lead Refinery
ASARCO Incorporated

Summary

The bullion received at the Omaha Lead Refinery typically contains 0.02% Cu and 0.02% Te. However, lacking satisfactory methods to extract them, significant amounts of both elements remained in the bullion until desilverizing, at which point they were removed with zinc. Since the Cu and Te in these desilverizing crusts ultimately reported into cupel litharge and were returned to the softener, the plant reached a point where the Cu and Te content of the bullion after softening was about ten times that of the bullion received. In addition, the only bleed for Cu and Te in the circuit, a Cu_2 Te-lead dross skimmed from cooling softened bullion, was difficult to treat in the tellurium plant.

Consequently, two new processes were implemented at Omaha. In the first, a detellurizing operation, softened bullion is treated with a combination of NaOH and Na to remove Te to ≤0.005%. The Na_2Te slag produced is skimmed and is readily amenable to subsequent treatment to recover Te. The second process, decopperizing, consists of treating the detellurized lead with NaOH and pyrite to reduce the Cu content to <0.005% Cu. Aside from eliminating the serious run around of Cu and Te, the use of the new refining processes to remove the Cu and Te to desirably low levels prior to desilverizing has significantly reduced zinc consumption.

INTRODUCTION

As a custom smelter the lead concentrates received by Asarco are varied. Consequently the crude bullion produced for treatment at the Omaha, Nebraska lead refinery has a fairly complex chemistry as shown in Table I.

TABLE I

Typical Bullion Analysis as Received by Omaha

	w/o
Sb	1.2
As	0.3
Sn	0.1
Cu	0.02
Te	0.03
Bi	0.30
Ag	0.80 (250 oz/T)
Au	0.003 (1 oz/T)

Prior to 1975 the lead refining steps used by Omaha, as shown in Figure 1, consisted of reverberatory softening, decopper-izing, Parkes[1] desilverizing, vacuum dezincing, Kroll-Betterton[2] debismuthizing and final refining with NaOH and NaNO$_3$ to remove trace amounts of oxidizable impurities before casting.

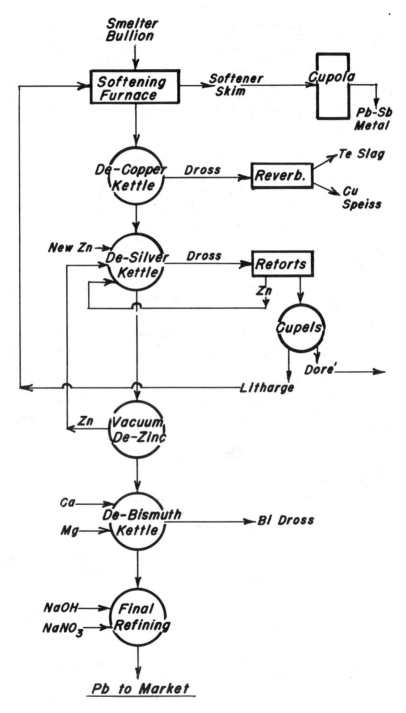

Fig. 1 - Omaha Lead Refinery Flow Sheet.

While ostensibly straightforward, close examination revealed a serious flaw in the Omaha flowsheet. Specifically, lacking methods to effect their removal to low levels, both copper and tellurium remained in the softened bullion until desilverizing at which point they were removed with zinc into the Parkes crust. Upon retorting, both elements reported to the retort metal, then to the cupel litharge. Since litharge is returned to the softener the net result was that bullion exiting the softener contained from 2 to 5 times the copper and tellurium content of the crude bullion received from the smelter.

In order to vent these elements, Omaha employed a "decopper-izing" process which consisted of simply cooling softened bullion in a kettle to about 330°C, to precipitate copper-tellurium compounds, and skimming. It is of note that the copper and tellurium content of the lead after this step was still substantially higher than in the crude bullion. The dross skimmed from the process was subsequently smelted in a reverberatory to produce a low grade tellurium-copper speiss and a tellurium slag. Figure 2 shows the extent of the copper and tellurium circulation.

Predictably numerous problems were experienced as a consequence of the circulating load: zinc consumption in desilverizing was high due to some being expended to remove copper and tellurium; the amount of Parkes crust produced/1000 oz dore increased, effectively reducing Omaha's retorting capacity; cupelling was slowed particularly because of the high copper content of the retort metal; the speiss and slag generated from smelting the copper-tellurium dross could not be readily treated at plants to recover valuable tellurium.

Clearly the need existed for processes, compatible with Omaha's kettle lead refining operation, which would mitigate these problems.

DETELLURIZING

Laboratory Investigation

The first objective was to develop a detellurizing process. According to Harris,[3] tellurium can be extracted from bullion into an alkaline salt melt by reacting with molten sodium hydroxide and niter according to the reactions:

$$3 \text{ Te} + 6\text{NaOH} \rightarrow 2\text{Na}_2\text{Te} + \text{Na}_2\text{TeO}_3 + 3\text{H}_2\text{O} \qquad (1a)$$

$$X \text{ Te} + \text{Na}_2\text{Te} \rightarrow \text{Na}_2\text{Te}_{x+1} \qquad (1b)$$

Emicke[4] reports that tellurium is rapidly removed from lead by pumping it through molten caustic (Harris cylinder) at 460°C at efficiencies ranging from 60-90%.

While tests did indicate tellurium could be removed from Omaha bullion by stirring with NaOH at temperatures up to 500°C, plant experience was not favorable. In fact ratios of 10 NaOH/l Te and reaction times upwards of 8-10 hours were required to reduce the tellurium content of a 250 T kettle of softened bullion from 0.06% to 0.01%. Since Omaha did not have the capacity either to treat the produced

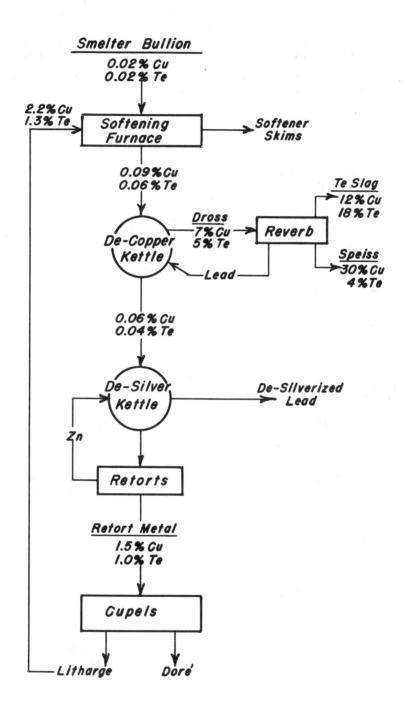

Fig. 2 - Omaha Cu—Te Circulation.

salt melt to recover NaOH for re-use or to improve its tellurium tenor, the process was not commercially feasible.

Since tellurium[5] forms binary tellurides with many elements, this tendency might be exploited to serve as the basis of a detellurizing process. Investigation suggested several reagents which react with and precipitate tellurium from molten lead. However sodium proved preferable in that:

1) it easily alloys with lead at low temperatures,

2) it rapidly reacts to form sodium telluride which has low solubility in lead,

3) it is specific to tellurium (i.e., the other impurities, arsenic, antimony and/or tin in Omaha bullion which could form intermetallics with sodium are removed to low levels in softening),

4) it can be used to detellurize lead at temperatures up to 540°C, although sodium utilization is better at lower operating temperatures,

5) produces a detellurizing slag which is amenable to subsequent treatment to recover tellurium.

Plant Procedure

The plant detellurizing procedure[6] which evolved from this work is described in Figure 3. Skims from the final refining are stirred into a 250 T kettle of softened bullion to remove residual arsenic and antimony. The bath is cooled to 400°C, skimmed clean, and about 200 pounds of NaOH beads added and allowed to melt. The purpose of the NaOH is to serve as a slag to collect Na_2Te. The bath is then vigorously mixed and metallic sodium, as 12 pound bricks, charged directly into the vortex up to a total of 0.45# sodium/# tellurium contained in the lead. (Note - 0.36# sodium/# tellurium is stoichiometrically required for Na_2Te.)

After all the sodium is added, stirring is stopped to prevent the oxidation of the Na_2Te compound to elemental tellurium which will report back into the lead. The fluid telluride slag is skimmed by crane ladle into molds and the detellurized lead, containing <0.005% tellurium and <0.005% sodium, is sent onto desilverizing.

When sufficient slag has been accumulated, i.e, after about 15 kettles have been treated, it is charged into a gas fired steel vessel, melted and held at about 550°C for 15-30 minutes. Molten lead is tapped out from the bottom of the vessel and returned to the refinery and the slag is tapped into 55 gallon drums (about 1000# of slag/drum) for shipment to the Amarillo refinery of Asarco for tellurium recovery. A typical assay of the detellurizing slag is 26% Te, 6% Pb, 30% Na.

DECOPPERIZING

Laboratory Investigation

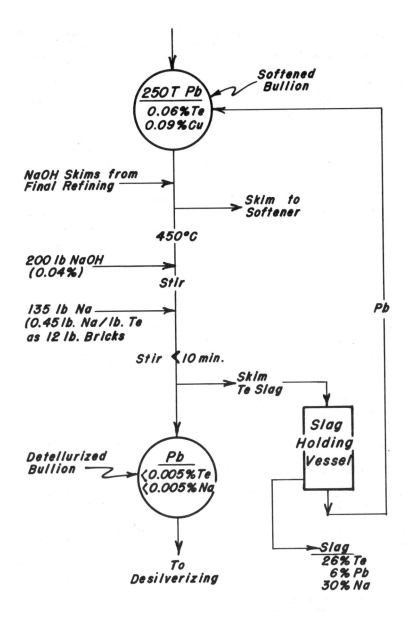

Fig. 3 - Detellurizing Process.

Within several months of the incorporation of the sodium
detellurizing process at Omaha, tellurium circulation was
reduced to low levels. However the problem of copper circulation continued, with the copper content in the softened bullion
remaining at about 0.1% due to the return of coppery cupel
litharge. Attempts to liquate and skim softened bullion only
reduced the copper content to about 0.04%, offering little hope
as a means of disrupting this short circuit.

Lead smelters, including Asarco's, routinely decopperize blast furnace bullion to 0.02% or less with elemental sulfur. Consequently as a first step Omaha examined sulfur decopperizing of their lead. Their results were not encouraging. Low copper contents in the lead were not consistently obtained and the temperature required (<340°C) inconvenienced the refinery flow and resulted in the production of large quantities of wet, leady dross of low copper tenor.

The reason for Omaha's inability to sulfur decopperize lead is not understood. It is held that tin, arsenic, antimony and/or silver[7,8,9] must be present for effective sulfur decopperizing. Although there are only trace amounts of tin, arsenic and antimony in Omaha's detellurized bullion, it does contain up to 1% silver. Nevertheless sulfur alone was not an effective decopperizing reagent.

Decopperizing lead with sulfur can be described by the reaction

$$2 \; Cu + PbS = Cu_2S + Pb \qquad (2)$$

Davey[10] notes that thermodynamic and solubility data suggest the equilibrium Cu content in lead saturated with Cu_2S and PbS is about 0.03% Cu. Consequently in order to decopperize to lower levels the activity of sulfur with respect to copper must be increased or the activity of Cu_2S reduced. Regarding the latter, if a suitable flux, or refining slag, were employed/which the Cu_2S produced could dissolve, better decopperizing might be obtained.

Since Omaha was using NaOH to help collect Na_2Te in their detellurizing, the possibility of using NaOH to collect Cu_2S in decopperizing suggested itself.

Numerous tests were conducted using combinations of NaOH and sulfur to decopperize softened lead. Fixing such variables as sulfur addition and time of mixing, relationships among treatment temperature, quantity of caustic used, and the initial and final copper content of the lead, were established. These data, shown in Figures 4 and 5, indicate that:

1) for a standard addition of sulfur (0.2%) the extent of decopperizing at either 370°C or 400°C is relatable to the amount of NaOH present.

2) decopperizing improved with lower reaction temperature and,

3) with lower initial copper content better decopperizing was experienced for a selected quantity of NaOH.

Analysis of these data suggested that decopperizing lead with sulfur and NaOH might fit a solvent extraction mechanism. That is the solute, in this case Cu, or Cu_2S, partitions itself between lead and the solvent NaOH according to the relationship-

$$w_n = w_o \; [\frac{V}{D(s)+V}]^n \qquad (3)$$

where

V = weight of solution (liquid Phase I - lead)

s = weight of solvent (liquid Phase II - NaOH)

w_O = initial weight of solute in V

n = number of treatments

w_n = weight of solute in V after n treatment

D = distribution coefficient

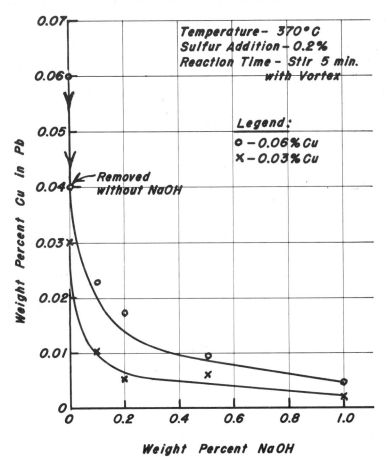

Fig. 4 - Effect of Caustic Additions on the
Removal of Cu from Pb
at 370°C.

Since all tests were conducted in one step the appropriate
values for the variables in equation 3, can be measured
and w_1, the copper concentration in the lead after one
treatment with varying amounts of solvent, s, (i.e., % NaOH),
can be calculated. The correspondence between the values
predicted and experimental are good as shown in Tables II and III.

341

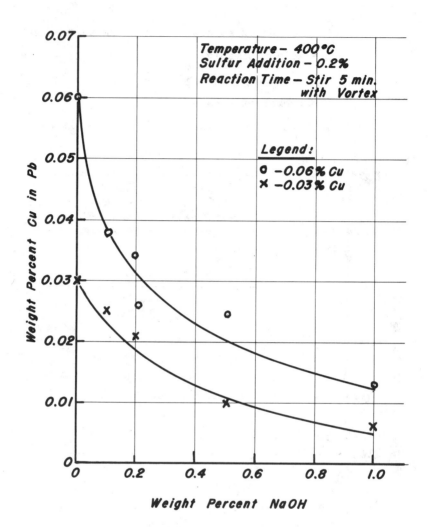

Fig. 5 - Effect of Caustic Additions on the
Removal of Cu from Pb
at 400°C.

A solvent extraction mechanism would also imply that the
copper content in lead could be reduced to extemely low
levels by repeated treatments. Figure 6 supports this
contention. Although large quantities of NaOH were employed,
the data indicates that it is possible to decopperize lead
to <0.001% at temperatures up to 450°C with repeated
treatments of NaOH and sulfur.

TABLE II

Predicted Values, (w_1) From Solvent Extraction Model and Experimental Values (wex) for Decopperizing Lead with 0.2% S and Varying Amounts of NaOH (s) at T = 370° C; D = 1400.

S (w/o NaOH)	w_o = 0.03% Cu		w_o = 0.06% Cu	
	w_1	wex	w_1	wex
0.10	0.012	0.010	0.025	0.022
0.20	0.008	0.008	0.016	0.018
0.50	0.004	0.005	0.008	0.009
1.00	0.002	0.002	0.004	0.004

TABLE III

Predicted Values (w_1), From Solvent Extraction Model and Experimental Values (wex) for Decopperizing Lead with 0.2% S and Varying Amounts of NaOH (s) at T = 400° C, D = 600.

S (w/o NaOH)	w_o = 0.03% Cu		w_o = 0.06% Cu	
	w_1	wex	w_1	wex
0.10	0.019	0.022	0.038	0.038
0.20	0.014	0.020	0.027	0.030
0.50	0.012	0.010	0.025	0.025
1.00	0.005	0.006	0.009	0.012

Plant tests of the NaOH and sulfur decopperizing were initiated with excellent results. However, it was impossible to prevent some of the sulfur charged to the bath at 370°C from burning with the evolution of acrid SO_2 fumes. This was environmentally unacceptable. Adding sulfur at a lower temperature, <330°C, to minimize burning, inconvenienced the refinery flow.

Therefore consideration was given to the possibility of adding sulfur as a sulfide. To do this, it would be necessary that the sulfide have little or no solubility in lead, and that it produce a decopperizing slag compatible to the smelter. Obvious sulfides of interest, Na_2S and PbS, proved suitable. However Na_2S represents a relatively expensive sulfur source while prohibitively large quantities of NaOH were required to effectively decopperize lead with PbS. The search for alternatives led to the mineral pyrite, FeS_2. Tests quickly confirmed that the combination of NaOH and FeS_2 could decopperize lead as effectively as NaOH and sulfur. A

program was established to optimize the decopperizing
parameters, viz., time and temperature of reaction, and
NaOH/FeS$_2$ ratio.

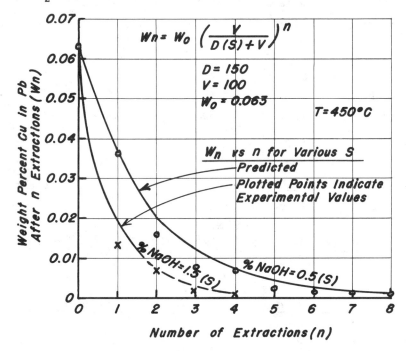

$$W_n = W_0 \left(\frac{V}{D(S) + V} \right)^n$$

$D = 150$
$V = 100$
$W_0 = 0.063$

$T = 450°C$

W_n vs n for Various S
Predicted
Plotted Points Indicate
Experimental Values

Number of Extractions (n)

Fig. 6 - Effect of Repeated Extractions on the Removal of
Cu from Pb. Predicted Curve vs Experimental Data.

Plant Procedure

On the basis of these data a two-stage plant decopperizing
process was developed.[11] Figure 7 is a flow sheet of the
operation and Table IV gives the results of kettle decopperizing 56 charges, each of 245 tons of lead.

In practice detellurized bullion containing about 0.04%
copper, is cooled to 370°C, and the slag from the second
decopperizing stage is added and stirred into the bullion
for about one hour. The purpose of the recycling is to enhance
utilization of the reagents and to produce a dry, readily
removable slag. The slag, about 1 T, and assaying approximately 10% copper is skimmed and ultimately smelted to yield a
copper matte and lead.

The copper content of the bullion after the first stage is
about 0.028%. A total of 700-1000 pounds of NaOH is then
added to the bath and allowed to melt. The mixer is
started and pyrite, FeS$_2$ shoveled into the vortex. Mixing
continues for about 1 hour at which point the decopperizing
slag is skimmed and returned to the first stage. The decopperized lead, assaying $\leq$0.005% Cu, is sent onto desilverizing.

344

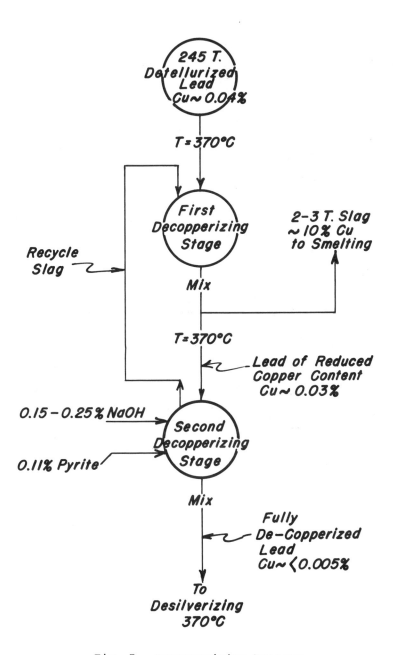

Fig. 7 - Decopperizing Process.

TABLE IV

Results of analyses of 56 kettle charges (each charge 245 T) of
softened bullion decopperized in a two-stage process with NaOH
and pyrite.

	Initial Cu Content in Bullion	Cu Content of Bullion After Reacting With Recycled Slag (w/o)	Final Cu Content of Bullion After Treating With NaOH and Pyrite (w/o)
High	0.082	0.058	0.011
Low	0.025	0.009	0.001
Average	0.042	0.028	0.005

Conclusion

The sodium detellurizing and NaOH and pyrite decopperizing
processes were integrated into Omaha's kettle refining
circuit with a minimum of difficulty. These processes acted
quickly to eliminate the serious circulation of copper and
tellurium in the plant. In addition, substantial benefits
were obtained in improved operating efficiency, due to
increased and more rapid retorting and cupelling, and
improved outcome, due to reduced zinc consumption and
improved tellurium recovery. Consequently, despite the
fact that copper and tellurium content of their softened
bullion now reflects their intake of 0.02% Cu and 0.01% Te,
Omaha continues to use both processes to yield lead containing <0.005% each of copper and tellurium for desilverizing.

References

1) G.K. Williams, "On the Determination of Certain Curves and Their Application to the Desilverization of Lead." Proceedings - Aus. I.M.M., 58 (1925), pp. 88-92.

2) J.O. Betterton and Y.E. Lebedeff, "Debismuthizing Lead with Alkaline Earth Metals Including Magnesium and Antimony," Trans. AIME, 121 (1936), pp. 205-225.

3) Harris, H., British Patents, 142,398,184, 369,189,013.

4) K. Emicke, G. Holzapfel, and E. Kniprath, "Lead Refining and Auxiliary By-Product Recoveries at Norddeutsche Affinerie," pp. 867-889, AIME World Symposium of Lead and Zinc," Vol. II, J. Cotterill and J. Cigan ed.; AIME, New York, NY, 1970.

5) W.A. Dutton, "Inorganic Chemistry of Tellurium," pp. 110-183, Tellurium, W.C. Cooper, ed.; Van Nostrand Rheinhold, New York, NY.

6) Y.E. Lebedeff and W.C. Klein, "Refining Lead," U.S. Patent 3,607,232.

7) T.R.A. Davey and J.V. Happ, "The Decopperizing of Lead, Tin and Bismuth by Stirring with Elemental Sulfur," Proceedings - Aus. I.M.M. No. 237 (1971), pp. 23-31.

8) D. Gallagher, "The Recovery of Copper from Blast Furnace Bullion," pp. 31-58, Proceedings - Aus. I.M.M. 31, (1951).

9) R.F. Blanks and G.M. Willis, "Equilibrium Between Lead, Lead Sulfide and Cuprous Sulfide and the Decopperizing of Lead with Sulfur," pp. 991-1027, The Physical Chemistry of Process Metallurgy, Part 2, G. St. Pierre, ed.; Interscience Publishers, 1961.

10) T.R.A. Davey, "Decopperizing Lead with Sulfur," pp. 121-129, Research in Chemical and Extraction Metallurgy," J.T. Woodcock, A.E. Jenkins, and G.M. Willis, eds.; Aus. I.M.M. Monograph Series No. 2, 1967.

11) Y.E. Lebedeff and W.C. Klein, Process for Decopperizing Lead, U.S. Patent 3,694,191.

METHOD OF TREATING DEBISMUTHIZING

DROSS TO RECOVER BISMUTH

J. E. Casteras
Senior Research Metallurgist
Central Research Dept.
ASARCO Incorporated

Carl DiMartini
General Supt., Metals Research
Central Research Dept.
ASARCO Incorporated

Summary

There are several methods available to upgrade, and ultimately recover bismuth from the dross produced from debismuthizing lead with calcium and magnesium.

A new process developed for this purpose consists of vacuum filtering debismuthizing dross to remove a lead filtrate of low bismuth content. The high bismuth residue cake remaining is then heated to about 475°C in air. At this temperature the cake ignites autogenously, oxidizing much of the contained lead. This oxidic residue is treated with molten lead chloride, producing an alloy of lead, 30-40% bismuth which is, in turn, reacted in the conventional manner with chlorine to yield bismuth and lead chloride.

Advantages to be gained through the use of this process include a low circulation of bismuth, a reduction in chlorine requirement to produce bismuth metal, and a reduction of excess lead chloride.

Introduction

The removal of Bi from Pb constitutes one of the most burdensome operations of Asarco's Omaha Lead Refinery. Omaha employs the Kroll-Betterton process[1] to reduce a typical Bi content of 0.3% to final refined contents as low as 0.01%. The Kroll-Betterton process relies on the formation of a bismuth intermetallic compound with reagent additions of Ca and Mg. The compound formed in debismuthizing, $CaMg_2Bi_2$, collects as a dross with entrained Pb; the dross typically contains about 4-6% Bi. The debismuthizing dross must be upgraded to high Bi contents before subsequent economic recovery of by-product Bi is possible.

Conventionally, upgrading of debismuthizing dross at Omaha, as summarized in Flow Sheet(1), utilizes a coal tar addition to effect flash oxidation of the Ca and Mg components of the Bi intermetallic, freeing entrained Pb. The dross that remains, called dry dross because of the lower Pb content, exhibits a typical Bi composition about three times that of the debismuthizing dross charged, i.e., 12-18% Bi. The dry dross is finally treated in an exchange kettle with $PbCl_2$ for subsequent Bi metal recovery. While the dry drossing operation using coal tar is cheap and effective, the prior history of the debismuthizing dross charged can lead to variable and unpredictable Bi concentration in the dry dross and consequent high and variable reagent use in final Bi metal recovery.

Earlier work at the Central Research Department investigated several alternatives to dry drossing by flash oxidation, including debismuthizing under protective atmospheres, liquation of debismuthizing dross under a chloride flux, pressing of semimolten debismuthizing dross, and hot vacuum filtering of debismuthizing dross.

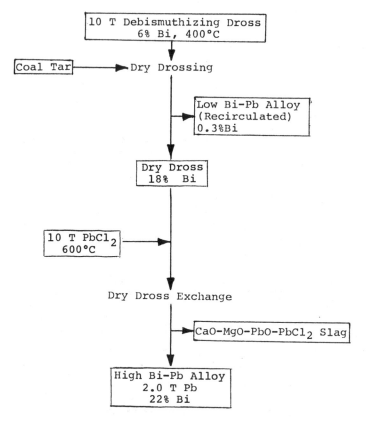

Flow Sheet 1. Pb and Bi Materials Balance for
Conventional Dry Drossing Procedure

Based on the results of those investigations a procedure was
developed (2,3) involving hot vacuum filtering of debismuthizing
dross followed by autoignition of the filter cake; the process
described can produce a higher grade of Bi-Pb alloy, thereby
reducing reagent burdens and costs in subsequent Bi recovery
operations.

Vacuum Filtering

A schematic of the vacuum filter developed is depicted in
Figure 1. The filter unit consisted of a flared pipe to which
an internal flange was attached and upon which a filter vacuum
can be placed. The filter screen consisted of a closely spaced
array of steel bars having triangular cross section, corresponding to about 100 mesh (150μ opening). The smaller end of the
unit is immersed in molten Pb, thereby forming a vacuum seal.

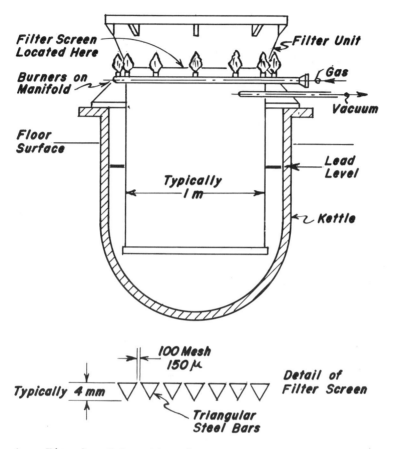

Fig. 1 - Schematic of Vacuum Filter Unit.

Preliminary small scale test work established the parameters
for successfully hot filtering debismuthizing dross, viz.,
temperature of filtration, screen pore size, and filtration
time. The effect of each parameter on filtering is depicted
in Figures 2-4.

In a typical filtering operation, debismuthizing dross, heated
in a small premelter to 450°C and having the consistency of
a wet mud, was charged onto the filter screen under
approximately 95 KPa vacuum (28 inches mercury); Figures 5-10
illustrate the operations performed in a plant test of vacuum
filtering. The porous and gas permeable filter cake is easily
separated from the screen for further treatment by inversion of
the filter unit. Flow Sheet 2a summarizes typical Pb and Bi
element balances observed in vacuum filtration based on a
treatment of 10 T debismuthizing dross. Comparable degrees
of Bi recirculation were obtained in vacuum filtering compared
to conventional dry drossing.

Flow Sheet 2a. Pb and Bi Materials Balance
        for Vacuum Filtering

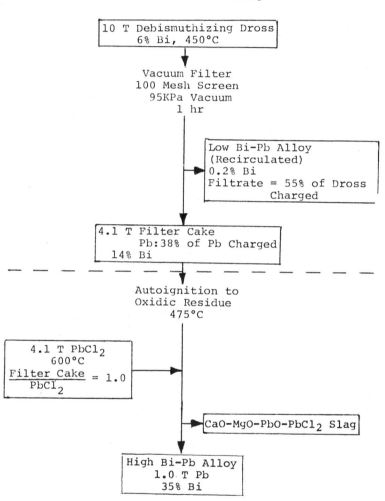

```
            ┌─────────────────────────┐
            │ 10 T Debismuthizing Dross│
            │    6% Bi, 450°C          │
            └─────────────────────────┘
                        │
                        ▼
                 Vacuum Filter
                 100 Mesh Screen
                 95KPa Vacuum
                    1 hr
                              ┌──────────────────────────────┐
                              │ Low Bi-Pb Alloy              │
                              │ (Recirculated)               │
                        ├────▶│ 0.2% Bi                      │
                              │ Filtrate = 55% of Dross      │
                              │             Charged          │
                              └──────────────────────────────┘
            ┌──────────────────────────────┐
            │ 4.1 T Filter Cake            │
            │    Pb:38% of Pb Charged      │
            │ 14% Bi                       │
            └──────────────────────────────┘
                        │
                  Autoignition to
                  Oxidic Residue
                     475°C
  ┌─────────────────────┐
  │  4.1 T PbCl₂        │
  │     600°C           │
  │ Filter Cake  = 1.0  │────────▶
  │ ──────────          │
  │   PbCl₂             │
  └─────────────────────┘
                              ┌──────────────────────────┐
                          ├──▶│ CaO-MgO-PbO-PbCl₂ Slag   │
                              └──────────────────────────┘
                        │
                        ▼
                 ┌──────────────────┐
                 │ High Bi-Pb Alloy │
                 │    1.0 T Pb      │
                 │    35% Bi        │
                 └──────────────────┘
```

Flow Sheet 2b. Pb and Bi Materials Balance for
        Autoignition of Filter Cake

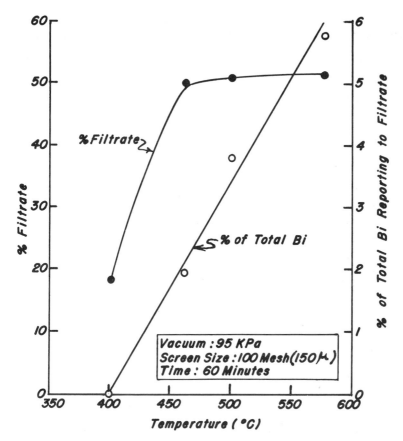

Fig. 2 - Effect of Temperature on Debismuthizing
Dross Filtration.

Early tests, coupled with metallographic and electron probe
analyses, indicated that the efficiency of filtering, i.e.,
as measured by the degree of Bi recirculation, reflected
the composition of the debismuthizing dross charged.
Particularly important were the degrees of Ca, Mg, and Pb
oxidation of the debismuthizing dross and the Sb impurity
content of the bullion prior to debismuthizing. Oxidation of
the Bi intermetallic constituent of the debismuthizing dross
was shown to result in an increased tendency for fracture of
the $CaMg_2Bi_2$ compound with consequent impairment of filtering
as well as increased Bi recirculation. The presence of Sb
above about 0.01% in the bullion to be debismuthized interferes
with the formation and separation of $CaMg_2Bi_2$ intermetallic
by competitive formation of a Sb-Ca-Bi compound. As a result,
procedures were instituted at Omaha to reduce the oxidation of
the $CaMg_2Bi_2$ constituent of debismuthizing dross, e.g., by
reducing the turbulence of stirring during debismuthizing and by
shortening the storage time of debismuthizing dross prior to
upgrading operations. In addition, steps were initiated to
reduce Sb levels to below 0.01% in the bullion prior to the

debismuthizing operation. These steps 1) reduce the variability of the Bi content in the debismuthizing dross experienced
in the past and, 2) improve the efficiency of the filtering
operation.

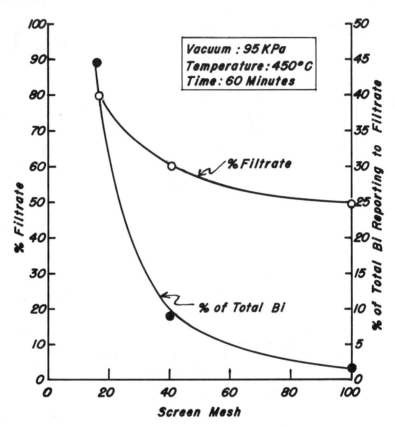

Fig. 3 - Effect of Screen Size on Debismuthizing
Dross Filtration.

Autoignition of Filter Cake

In order to further upgrade the Bi content of the filter cake,
thereby reducing subsequent chlorine use and accumulation of
$PbCl_2$ during Bi metal recovery, a procedure was developed by
which the porous filter cake was autoignited in air at 475°C.
As a result of the substantially flameless combustion, nearly
all of the Ca and Mg and about 60% of the Pb in the filter cake
was oxidized, forming an oxidic residue. The oxidic residue
was then treated with molten $PbCl_2$ in the same fashion as for
dry dross. A Bi-Pb alloy containing about 35% Bi was produced,
along with a PbO-MgO-CaO-$PbCl_2$ slag containing a small amount
of Bi. Flow Sheet 2b (a continuation of Flow Sheet 2a) depicts
the Pb and Bi element balance during autoignition of the filter
cake followed by treatment with molten $PbCl_2$. By comparing
Flow Sheets 1 and 2b, it is apparent that only about 40% as
much $PbCl_2$ is required to treat filter cake as would be needed

for upgrading dry dross. The higher grade Bi-Pb alloy produced by the PbCl$_2$ filter cake reaction compared with conventional Omaha practice (see Flow Sheet 1) results in a 50% decrease in subsequent chlorine reagent costs to recover Bi metal. Accordingly since 50% less PbCl$_2$ is produced during the Bi recovery, one of the attractive consequences of the filter-autoignition process is a significant decrease in PbCl$_2$ accumulations which previously required a separate reduction treatment for Pb recovery.

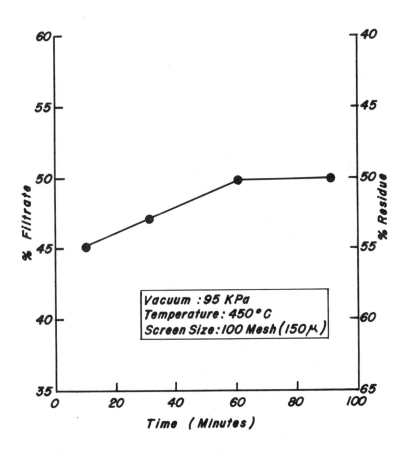

Fig. 4 - Effect of Time on Debismuthizing Dross
Filtration.

Fig. 5 - Charging the Vacuum Filter with
Semimolten Dross.

Fig. 6 - Heating of Dross During Filtration.

Fig. 7 - Residue Cake After Filtering; Cracks in
Cake Result in Loss of Vacuum.

Fig. 8 - General View of Filter Unit During
Removal After Filtration.

Fig. 9 - Inverting Filter Unit to Dislodge Filter Cake.

Fig. 10 - Dry Filtered Residue Ready for Autoignition.

Conclusion

A new process for upgrading debismuthizing dross has been developed, involving hot vacuum filtration followed by autoignition of the filter cake. The new process reduces chlorine use and $PbCl_2$ accumulation in subsequent Bi recovery by virtue of producing a richer Bi-Pb alloy after treatment of the oxidic residue with $PbCl_2$. Bismuth recirculation with the new process is equivalent to that experienced with the conventional operation. Variability in Bi recovery from debismuthizing dross can be minimized by reducing Sb levels in bullion prior to debismuthizing and by instituting measures to reduce oxidation of the $CaMg_2Bi_2$ constituent of debismuthizing dross.

REFERENCES

1. J.O. Betterton, Y.E. Lebedeff, "Debismuthizing Lead with Alkaline Earth Metals, Including Mg, and with Sb," Transactions AIME, 121 (1936) pp. 205-225.

2. U.S. Patent No. 4,039,322. August 2, 1977. C.R. DiMartini, W.L. Scott, Assignee: ASARCO Incorporated. Method for the Concentration of Alkaline Bismuthide in a Material Also Containing Molten Lead.

3. U.S. Patent No. 4,039,323. August 2, 1977. C.R. DiMartini, W.L. Scott, Assignee: ASARCO Incorporated. Process for the Recovery of Bismuth.

THE RECOVERY OF METALLIC ARSENIC IN COMINCO'S LEAD OPERATIONS

H.E. Hirsch
Technical Research
Cominco Ltd.
Trail, B.C., Canada
V1R 4L8

Summary

Historically arsenic oxide and metallic arsenic have been produced as by-products in copper smelters. Lead concentrate and recycle materials (scrap) also contain arsenic which has to be separated in order to obtain commercial grade lead metal. Cominco Ltd. has developed the technology and installed the facilities for commercial recovery of metallic arsenic in its lead smelting and refining operations at Trail. Anode slimes from the Betts electrolytic refining process are the starting material. After drying, the slimes are melted in an oxidizing atmosphere. The volatilized arsenic and antimony oxides are collected and reduced in the presence of suitable lead-bearing materials to yield a lead-arsenic-antimony alloy. Crude metallic arsenic is recovered from this alloy by vacuum distillation. After a chloride treatment to remove traces of cadmium, a second stage of distillation results in commercial grade metallic arsenic suitable for sale or for the production of various alloys with controlled composition.

A significant part of the commercial arsenic capacity is used to provide feed to Cominco's high purity arsenic plant which supplies the electronics and related industries with products grading up to 69+ purity.

Introduction

At Trail Cominco Ltd. produces approximately 160,000 T/yr. of highly refined lead from a variety of raw materials by the sintering - blast furnacing - drossing - electrolytic refining process as outlined in Figure 1.

The operating metallurgy of these plants has to take into account, among others, the following factors:

a) The feed materials contain varying quantities of arsenic and antimony and their inputs can vary with time because of the custom smelter philosophy.

b) The Betts electrolytic refining process, as normally practised, requires a controlled concentration of arsenic plus antimony in the anodes. This requirement of the Betts process constrains the arsenic intake even with complete elimination of all internal recycles.

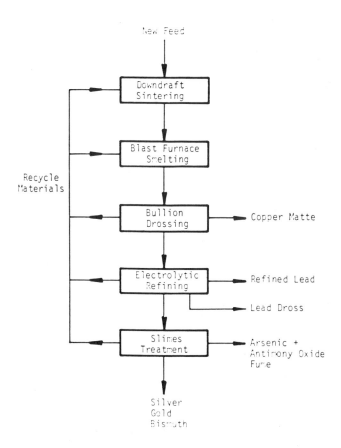

Fig. 1 - Basic Flow Diagram of Cominco's Lead Operations at Trail, B.C.

361

In view of the availability of valuable concentrates with a high arsenic content, it has been logical to examine potential methods for the removal of excess arsenic at some stage prior to casting the bullion into anodes. While it was realized that caustic oxidation of the bullion would be a viable contender, the removal of arsenic by vacuum distillation promised sufficient advantages to justify a closer examination. High temperature large scale vacuum technology had entered the commercial scene, no process chemicals and their attendant cost and disposal requirement would be incurred, and metallic arsenic would be obtained as a potentially saleable product.

Original Investigation

The engineering principles applicable to the vacuum distillation of arsenic from lead appeared to be straightforward. The rate of evaporation W per unit area is described by the Knudsen-Langmuir equation

$$W = \textit{k} \times (\alpha \times N \times P_o - P_e)(\frac{M}{2\pi RT})^{1/2}$$ (1)

where: W = rate of evaporation

$\textit{k}$ = Langmuir coefficient

α = activity coefficient

N = mole fraction

P_o = vapour pressure of the pure evaporating species

P_e = total pressure adjacent to the melt

M = atomic (molecular) weight of the evaporating species

For metallic systems the Langmuir coefficient is often found to be close to one(1).

An early warning signal pointing towards the inadequacy of this approach, appeared when laboratory batch tests indicated distillation rates far below those predicted by the Knudsen-Langmuir equation. Under the experimental conditions the estimated effects of diffusion phenomena in both the liquid and the vapour phase were not limiting the distillation rate and this was confirmed in later work. Formation of an oxide or dross film on the evaporating surface, which effectively reduced the area available for unhindered mass transfer, seemed a logical explanation based on experimental observations. Consequently, the pilot scale cascading tray tower shown in Figure 2 was built and operated. The results, however, remained dismal in spite of the obviously large, clean and constantly renewed surface observed through the sight port.

At this stage it was evident that vacuum processing was not a practical method to significantly reduce the arsenic tenor of Cominco's lead bullion.

In retrospect it became obvious that the low observed distillation rates are linked to the formation of the particular arsenic species present in the vapour phase over the temperature range of practical interest. The evaporation of arsenic from its mono-atomic solution in lead seems to be preceded by the formation of As_4 molecules at the evaporating surface by rapid equilibration

$$4As_{dissolved} \xrightleftharpoons{} As_4 \; surface$$ (2)

$$[As_4]_{surface} = k[As]^4$$ (3)

k = concentration equilibrium constant

362

Based on this postulate the Knudsen-Langmuir equation needs to be expanded for the specific case (and for others where the solute and gaseous species differ from each other). With some simplification the modified equation becomes

$$W_m = \mu \times k \times \left[\alpha N\right]^4 \times P_o \times \left(\frac{M}{2\pi RT}\right)^{1/2} \qquad (4)$$

Although the expanded Knudsen-Langmuir equation expresses a 4th power dependence of the evaporation rate on the concentration of arsenic, it is highly improbable that the underlying reaction is of 4th order. Nevertheless, our observations indicate the exponent to be larger than three. It would naturally be expected to be somewhat less than four because of the likely evaporation of lower arsenic species.

The temperature dependence of the distillation rate, besides the obvious effect on the vapour pressure of arsenic, has not yet been examined with sufficient accuracy to distinguish between the effect on the reaction (3) and that of the inverse square root relationship in the original Knudsen-Langmuir equation. The elucidation of the reaction kinetics and mechanism will require further study.

A cursory examination of the indicated 4th power relationship revealed that at practical temperatures

a) about 2% residual arsenic would be the realistic limit attainable by vacuum treatment, and

b) an arsenic content of more than four percent would result in distillation rates adequate for commercial recovery.

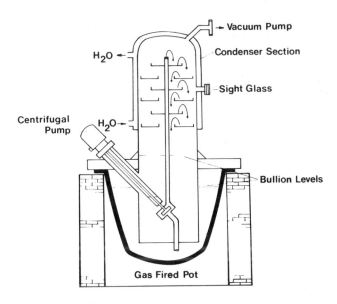

Fig. 2 - Pilot Scale Vacuum Processing Tower for Lead Bullion

The attainable residual arsenic level of 2% is far above the control range for the Betts process, and therefore, marked the end of this specific project.

Recovery of Arsenic from High Arsenic Lead Alloys

Crude Arsenic

Subsequent to the above investigation, changes in the market stimulated renewed interest in vacuum processing for the purpose of recovering metallic arsenic. As shown in Figure 1, electrolytic refining and slimes treatment produce as by-products a lead dross and a fume containing arsenic and anti-mony oxide in approximately equal proportions. For many years these materials have been co-reduced with carbon to yield an arsenic (7-12%)-antimony (18-30%) -lead alloy. The arsenic content of this so-called reduction metal was re-duced to saleable levels of generally less than 6% by caustic oxidation.

Reduction metal was an obvious starting material for arsenic recovery. In addition vacuum de-arsenizing would simultaneously produce marketable antimonial lead without requiring caustic soda.

However, while the indicated 4th power dependence of the distillation rate on the arsenic concentration effectively sets the lower practical limit, this relationship also causes difficulties at the upper end of the arsenic concentration range in reduction metal. To illustrate the case one needs to visualize water evaporating from a one-litre laboratory beaker. At the equivalent of 3% arsenic in lead, the water would calmly evaporate in about four days, while it would have to boil vigorously to dryness in about 20 minutes at the equivalent of 12% arsenic.

Even though a rapid distillation rate appears desirable, the factor that controls throughput is the design of a condenser for easy arsenic removal with high standards of hygiene and with the ability to last in a tough met-allurgical environment. Of the three key variables which determine the dis-tillation rate, namely temperature, residual gas pressure and starting con-centration of arsenic, we chose a combination of temperature and arsenic concentration for designing the condenser and devising the operating strategy. Iterative numerical analysis indicated that a reasonably sized cylindrical water-cooled condenser with a domed top could handle arsenic concentrations up to 7% and meet the production target with a single unit while operating at a temperature which posed no difficulties with regard to materials of construction and equipment life. These estimates were confirmed by operating the full size prototype distillation unit shown in Figure 3.

The equipment is based on a standard Cominco 15-ton lead pot in a stan-dard, automatically controlled, gas-fired setting. A 30-in. steel extension is welded vacuum proof to the top to provide increased volume capacity and access for a side entry agitator and the vacuum connection. The content of the pot provides a clean barometric seal around the agitator shaft and dif-ficulties associated with mechanical feedthrough are avoided.

The vacuum offtake is positioned close to the condenser top. It is fitted with a single plate baffle.

The condenser has a forced spiral water path. It is fitted with trun-nions to support it in the cropping rig. The seal consists of a flat but thick, soft neoprene ring held in a machined groove in the condenser bottom. The sealing surface on the pot extension is machined smooth and is water-cooled. A lip extends from the pot up into the condenser to prevent arsenic from depositing in the contact areas.

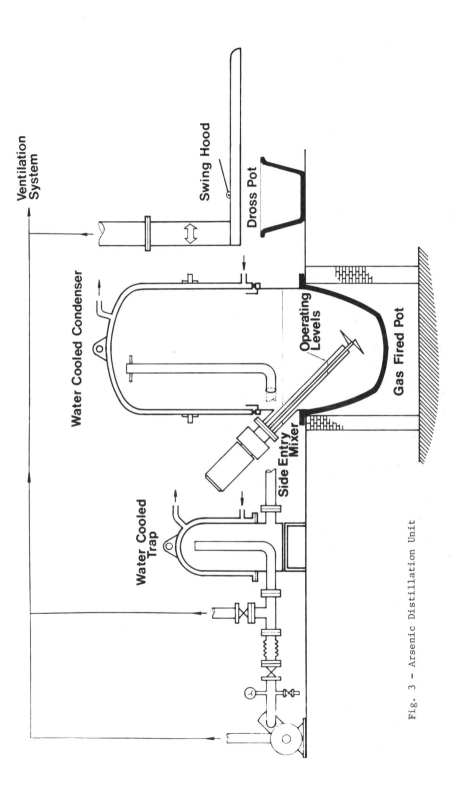

Fig. 3 – Arsenic Distillation Unit

365

· A water-cooled trap prevents arsenic migration to the mechanical vacuum pump. Vibration isolation is achieved by bellows between the trap and the pump.

The vacuum system piping is also used for ventilation purposes. After the vacuum has been broken via the bleed valve at the end of a run, the ventilation system is connected by opening the appropriate valve. The condenser can now be raised with no escape of fumes. The swing hood is then maneuvered into position and the condenser can be moved to the cropping station. The swing hood is fitted with a charge port and additions to the pot do not result in the emission of fumes into the working atmosphere.

Some typical key operating parameters are as follows:

Pumpdown time	20 min.
Cycle time	8 hr.
Temperature	480°C increasing to 540°C
Pressure (at pump inlet)	<100 microns
Residual As Concentration	2-3%

The starting concentration of arsenic is controlled by pumping out a predetermined quantity of the de-arsenized reduction metal and replacing it with fresh reduction metal.

Refining of Crude Arsenic

Vacuum processing of reduction metal yields an impure product consisting of approximately 90% arsenic and 10% antimony. The crude product is directly suitable for some alloying applications where the antimony content can be tolerated. Besides antimony, cadmium at around 0.05% is the only other significant contaminant present in the condensate. All other impurities aggregate to a maximum of 200 ppm and are commonly below 100 ppm.

A second treatment pass comprising re-alloying the crude arsenic with refined lead and re-distillation yields a product containing >99% As, <0.6% Sb, <0.1% Pb and again approximately 0.05% Cd. Contents of all other impurities combined amount to less than 100 ppm. Material of this quality is generally suitable for conventional alloying uses.

However, some high technology applications of arsenic alloys and compounds, for instance gallium arsenide in the opto-electronics field, require a material of much higher purity than that just quoted above. The necessary further reduction of the antimony and lead contents to the fractional ppm level can be attained by physical methods, but cadmium tenaceously remains with the arsenic (similar vapour pressures, compound formation). It was, therefore, necessary to develop a chemical treatment method if Cominco arsenic was to become a suitable feed material for the production of high purity arsenic.

The literature reveals an equilibrium constant for the reaction

$$Cd + PbCl_2 \quad \overset{\rightarrow}{\leftarrow} \quad CdCl_2 + Pb \qquad (5)$$

which is favourable for the removal of cadmium from cadmium lead alloys at temperatures below about 970°C(2). Therefore, cadmium can be extracted from a lead alloy by contacting it with molten lead chloride. The equilibrium constant increases rapidly with decreasing temperature, thereby making the distribution of cadmium between the alloy and the molten salt phase more favourable at lower temperatures. The presence of arsenic was found to have no major influence on the equilibrium constant.

366

With lead chloride melting at 498°C, a practical operating temperature would be around 550°C. However, it was demonstrated in the laboratory that the operating temperature could be lowered to the range of 480 to 500°C by the addition of 8 to 10% potassium chloride to obtain a near-eutectic salt melt(3), without degrading the extraction efficiency significantly. The maximum permissible arsenic content in the alloy to be treated at 480°C is 23%. Higher concentrations will result in partial solidification.

Based on this information a process was devised which involves repeated contacting of a molten lead-arsenic alloy with batches of fresh potassium chloride-lead chloride eutectic in a countercurrent pattern. The alloy is agitated mechanically to facilitate mass transfer for an approach to equilibrium within reasonable time. Separate agitation of the relatively shallow molten salt layer was found unnecessary. The moving metal surface transfers adequate momentum to the interface and into the molten salt layer to overcome the buildup of an overriding constraint on mass transfer.

Clean lead chloride is available in more than adequate quantity as a by-product from local Cominco operations. It is normally reduced with carbon and lime to recover the lead content. The prior use of the compound for cadmium removal is of no significance in the overall metallurgy. The potassium chloride used is fertilizer grade potash obtained in bags from Cominco's Vanscoy, Saskatchewan, Operations.

The equipment for cadmium removal by treatment with lead chloride is shown in Figure 4.

Again use is made of a standard gas-fired 15-ton lead pot with a 3-ft. steel extension welded on. The inside is refractory-lined over the length of the steel extension to protect against lead chloride corrosion. The extension is fitted with a refractory-lined launder with a knock-out refractory weir for the removal of the cadmium-bearing molten salt.

Agitation of the metal phase is provided by a movable rig completely enclosed by a cylindrical vent hood. Two hinged quarter sections allow for easy relocation of the agitator rig to the top and insertion of a metal pump. Temperature control is achieved via graphite-sheathed thermocouples immersed into the salt and metal bodies. The eutectic salt mixture is added through a feed chute attached to the back of the vent hood.

The Arsenic Recovery Process

The three key process steps described before - primary recovery, cadmium removal and secondary distillation - are linked together as shown in the flow diagram of Figure 5.

The arsenic oxide + antimony oxide fume resulting from the treatment of the Betts process anode slimes is reduced together with oxidic lead dross in a reverberatory-type furnace to yield a reduction metal containing 7 to 12% As and 18 to 30% Sb.

As explained before, the starting concentration of arsenic in the primary recovery distillation step must be limited to a maximum of 7% to comply with the condenser design. The same constraint applies to the secondary distillation stage and is readily met by leaving an appropriate heel of 3% arsenic alloy in the pot.

Alloying before secondary distillation can be carried out in either the vacuum or the treatment pot. Operating logistics normally dictate the use of the latter. The alloying procedure is simple and fast. The alloy containing

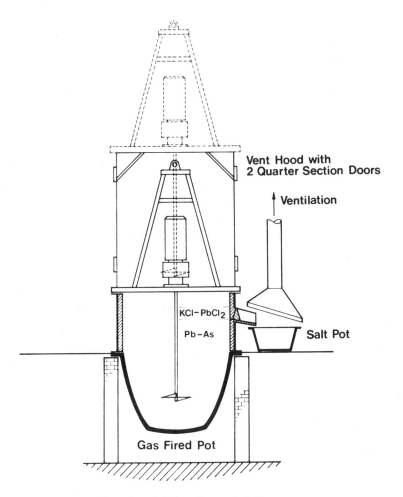

Vent Hood with 2 Quarter Section Doors

↑ **Ventilation**

KCl-PbCl$_2$

Pb-As

Salt Pot

Gas Fired Pot

Fig. 4 - Cadmium Removal Unit

22-23% arsenic is skimmed and the temperature is raised to 530-550°C before the salts are added, one appropriately sized skip of lead chloride plus one bag of potash at a time. Care has to be taken in ascertaining that the alloy level is in the range of the refractory-lined top section of the pot and sufficiently close to the weir level for complete removal of the salt layer. After the salts have melted the temperature is dropped to 480-500°C.

The contact time is usually six hours, mainly for operating convenience - one contact per shift. The number of contacts depends upon the initial cadmium concentration, the alloy to salt ratio and the desired final level. Usually 5 to 6 contacts result in residual cadmium concentrations of less than one part per million. The salt melts from the third contact on usually are low enough in cadmium to warrant recycling to the next lower contact of the following alloy batch. The alloy is held at temperature under the final salt layer until adequate cadmium removal has been confirmed. The agitator is then raised as shown before, a pump is inserted and the alloy is pumped into ventilated button moulds.

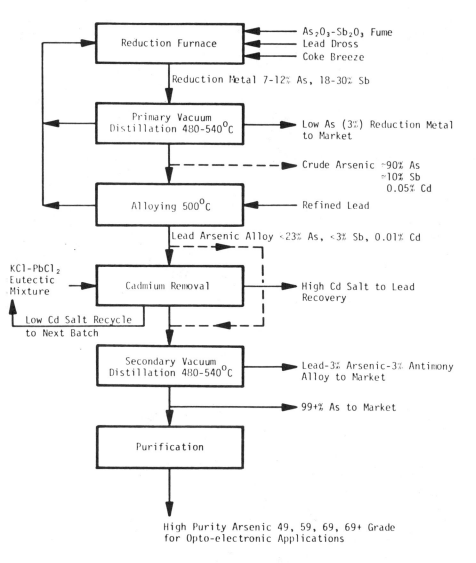

Fig. 5 - Flow Diagram of the Cominco Arsenic Recovery Process

The secondary vacuum distillation is analogous to the primary recovery distillation with the same distillation unit being used. A heel of 3% As alloy is fortified to 6-7% As by melting in 23% As buttons. A 3% As - 3% Sb - lead alloy is the co-product which finds use in adjusting the composition of other alloys to market requirements, or, if necessary, is recycled to the lead refinery or the reduction furnace.

Figure 6 shows a general view of the plant as it exists now. Removal of a crop from the condenser is shown in Figure 7.

Fig. 6 - View of Cominco's Arsenic Recovery Plant

Fig. 7 - Removal of Arsenic from the Condenser

Product Properties

Metallic arsenic exists in at least three allotropic forms.

The preponderant article of commerce is the crystalline α-form, a black, lumpy and hard material. The α-form oxidizes readily in air at ambient temperature until a thick enough coating has formed to protect the underlying material. This thick black coating is soft and readily abraded.

The condensation conditions of the Cominco process result in the deposition of the soft and friable γ-form. This material is of a typical columnar structure as shown in Figure 8.

The dull grey γ-form is amorphous and stable towards oxidation at ambient temperature, and may be stored and shipped in air without a change in appearance. It alloys more rapidly than α-arsenic, a characteristic likely attributable to its amorphous character.

Although the friable nature of the γ-form would indicate a potential dusting problem, experience has shown that this does not occur to the extent originally anticipated. It appears that the columnar material separates along the column walls and breaks transversely to the 0.1 to 0.5 mm ∅ columns. Breakdown essentially stops once the individual column fragments have reached an aspect ratio in the order of unity. As a result the fines do not contain a large fraction of particles of a size which can become and remain airborne. It is needless to say, however, that adequate ventilation must be provided wherever the material is handled, just as with the α-form. For some uses the γ-form has an advantage over the α-form in that crushing is not required, thereby avoiding one source of dust.

Environmental and Hygiene Aspects

Being physical in nature, the process generates no arsenic-containing chemical by-products or wastes which require further treatment and disposal.

At the present time the following TLVs apply to the in-plant atmosphere:

As 0.5 mg/m^3
Pb 0.15 mg/m^3

Cominco is in full compliance with these regulations in the arsenic recovery and refining areas. Campaigns with operators wearing personal samplers have shown that with proper training even short term localized excursions of the airborne concentrations of arsenic can be avoided.

Fig. 8 - γ-Form Metallic Arsenic

371

Ventilation is provided via a 5,000 cfm baghouse with fixed or flexible pickup points and adjustable dampers at the distillation pot-2, the treatment pot-1, the skimmings pot-1, the molten salt receiving pot-1, the casting moulds-1, and the condenser cropping station-2. The baghouse catch was found to be miniscule in quantity and under the resulting light duty baghouse operating cycles, stack losses are close to non-detectable.

Personnel in the arsenic recovery and refining areas are covered by the long established biological monitoring program. Urine and blood are checked when a person begins to work in these areas and thereafter at regular intervals. The normal Cominco procedure is to move personnel out of a hazardous area if significantly elevated arsenic or lead levels are detected. To date no such precautionary moves have been necessary from the arsenic recovery and refining areas.

Summary

Cominco has been a supplier of high purity arsenic to the electronics and related industries for many years. By developing and implementing a process to recover the raw material for high purity arsenic production from indigenous sources, Cominco has gained control over the availability and quality of the final products. The accessible resources and production capacity are more than adequate to satisfy the requirements of the opto-electronics industries and the developing gallium arsenide microwave and integrate circuit applications.

Additional outlets for Cominco's metallic arsenic and arsenic alloys are under active development.

Acknowledgements

Cominco's permission to publish this information is greatly appreciated.

Many people in Trail Operations and Development, in Engineering, as well as in Trail Electronic Materials, and in particular Roland Perri, contributed to the success of this project. The author is deeply indebted to all of them.

References

(1) Hillary W. St. Clair, "Distillation of Metals under Reduced Pressure", Vacuum Metallurgy, pp. 295-305, Rointan F. Bunshah, ed.; Reinhold Publishing Corporation, New York, N.Y., 1958.

(2) Gmelins Handbook of Inorganic Chemistry, 8th ed., Cadmium, suppl. vol., No. 33, pp. 481 ff, Verlag Chemie, Weinheim, Germany, 1959.

(3) Y.A. Ugai and V.A. Shatillo, J. Phys. Chem. U.S.S.R., 23(6) (1949) 745.

A TWO-STEP PROCESS FOR SMELTING COMPLEX Pb-Cu-Zn MATERIALS

L.M. Fontainas and R.H. Maes

Metallurgie Hoboken-Overpelt
Research Department
2710 Hoboken, Belgium

Abstract

Complex Pb-Cu-Zn-based materials which contain various secondary metals may be incorporated in appreciable proportions in the feed of a traditional smelter including the operations of sintering, blast furnace smelting, matte blowing and slag fuming. The requirements for sinter quality and blast furnace charge reducibility make it difficult, however, to increase the proportion of secondary materials in the charge above a certain level.

A new process is therefore being developed according to which complex materials are smelted in two steps. The first step is characterized by a high oxygen potential and is followed by a strong reduction of the slag in the second step. The basic metallurgical principles of the process are explained and it is shown how the separations and recoveries of metals are improved by comparison with treatment in a traditional smelter.

For materials of relatively low sulphur content or for sulphated products, which are normally not suitable for autogeneous smelting, the electric furnace with submerged electrodes, operating continuously, seems to be the most suitable equipment to carry out the first step. For the second step, an electric furnace or a top- or side-blown converter, operating continuously or batch-wise, are possible alternatives.

Pilot scale experiments carried out in an electric furnace at about 200 kW have shown that energy consumption looked favourable and that it seemed possible to extrapolate the process to an industrial scale. A good control of pollution problems and a great flexibility of the process with regard to feeds of varied physical nature and chemical composition should also be emphasized.

Introduction

The present flow-sheet of the Hoboken smelter, which is shown in Fig. 1 and has been described in a previous AIME World Lead-Zinc Symposium (1), is the result of continuous changes involving increasing complexity of feed materials and extraction processes. Starting from a lead smelter, whose construction dates back to the beginning of the century and which included the traditional steps of sinter roasting, smelting and lead refining, the first complication was introduced by the reprocessing of the copper-bearing by-products of the plant in the Pb-circuit. This led to the formation of low grade lead-copper matte in the blast furnace, the processing of which required the development of the siphon converter (2) and of a method of operation allowing the treatment of considerable quantities of lead-containing matte to produce blister copper of acceptable quality from it.

A new complication appeared after the second world war, when the smelter received growing quantities of arsenical lead-copper concentrates, giving rise to the regular production of an arsenical alloy phase in the blast furnace, for which a treatment process had to be developed.

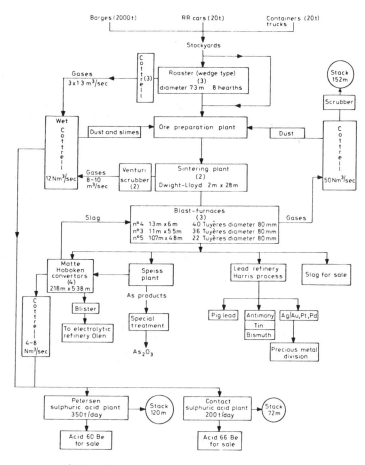

Fig. 1 Flow-sheet of Hoboken plant.

376

The development was accelerated markedly in the nineteen sixties, when the smelter was supplied with increasing quantities of by-products and secondary materials. The flow-sheet remained basically unchanged, but the equipment had to be adapted to materials of increasingly difficult physical nature. The range of metals entering the smelter widened considerably and this was offset by increasing the capacity for the recovery of the secondary metals. More particularly, increasing nickel intakes gave rise to the regular formation of nickel-bearing speiss during the treatment of the arsenical alloy mentioned above and a process for treating this new phase had to be developed.

A thorough investigation of the chemical equilibria in the smelting of complex materials (3) has thus been necessary because the limits of metallurgical complexity of the system had been reached with the production of four distinct liquid phases in the blast furnace. This led to the development of a computer program in order to optimize the operation of the whole smelter as a function of the different feed materials and of the operating conditions of each production unit. This program also allows the introduction of new operations, such as the two-step smelting process described in this paper, in order to estimate their effect on the whole smelter.

In the meantime, the development of ever more complex feed materials continued, particularly at the sinter plant, where increasing quantities of lead sulphate residues were incorporated (4), and at the blast furnaces, where the proportion of concentrates introduced in sinter form at present only represents a small part of the total useful materials charged.

It should be added that all this development took place under the pressure of ever more severe environmental constraints, the implications of which are described in another paper presented at this symposium (5).

Limitations of the Present Process

At this stage of development, it was advisable to review our extraction processes and to enquire whether they remained well integrated for the treatment of feeds whose nature and composition had changed so much. Despite the flexibility of the traditional processing sequence with respect to ever more complex feeds, various limitations must be pointed out :

- the need for a minimum amount of sinter as charge support in the blast furnace, in order to ensure smooth operation, so that the process still depends upon a certain feed of concentrates.

- the necessity to subdivide the feeds of each furnace of the smelter into special charges, in order to optimize the recovery of the different valuable metals, resulting in a certain complexity of dispatching of materials and a considerable internal circulation of intermediate products.

- the impossibility of increasing reduction in the blast furnace much below 1.5 % Pb in the slag, without collecting excessive amounts of iron in the matte and the arsenical alloy, so that the recovery of some less reducible metals, such as Sn and Zn, is unsatisfactory.

- the many problems introduced by the physical nature of complex materials, more particularly with regard to the stability of operation of the furnaces and to the environment.

Considering the autogeneous smelting processes which have become established for several years in copper metallurgy and which are now progressively being introduced in lead metallurgy, a survey was initiated on the direct smelting of our complex materials.

It was hoped that, instead of first desulphurizing materials at the sinter plant and then reducing them in the blast furnace, a direct smelting operation would make better use of the oxygen available in a substantial fraction of the feeds and contribute to the desulphurization of the sulphur-containing materials. Direct smelting should also allow a reduction of the energy consumption to a minimum and to lessen our dependence on metallurgical coke. Finally, a rationalization of the smelting processes would lead to an improvement of the environment of the smelter.

Principles of Direct Smelting of Pb-Cu Materials

Several direct smelting processes are now being commercialized or developed at a pilot scale for treating lead concentrates. The main characteristic of all these processes is that a Pb-rich slag has to be produced in order to desulphurize the charge and to avoid high sulphur contents of the lead bullion (6); a separate reduction step of the slag has then to be considered.

If similar processing conditions are applied to Pb- and Cu-based materials, the first question which arises is that of the desulphurization equilibria between the slag, which is rich in lead, and the matte, concentrating copper. This problem was approached from our previous study (3) of the chemical equilibria in the smelting of complex materials, where we showed that the composition of the mattes could be represented in a triangular diagram Cu_2S-FeS-PbS, in which Cu and Fe iso-activity lines could be drawn corresponding to reactions of the type

$$Cu_2S_{(1)} + Pb_{(1)} \rightleftharpoons PbS_{(1)} + 2 Cu_{(1)}$$
$$FeS_{(1)} + Pb_{(1)} \rightleftharpoons PbS_{(1)} + Fe_{(1)}$$

taking place between the matte and a lead bullion phase which is supposed to be present in equilibrium. Providing some assumptions are made (7), the drawing of Cu and Fe iso-activity lines as represented in Fig. 2 is obtained.

In the same way, the composition of a slag in equilibrium with a lead bullion phase can be represented in a Cu_2O-FeO-PbO diagram (assuming, as a simplification, that all the iron is effectively present as FeO) and the composition of the slag will be defined by reactions of the type

$$Cu_2O_{(1)} + Pb_{(1)} \rightleftharpoons PbO_{(1)} + 2 Cu_{(1)}$$
$$FeO_{(1)} + Pb_{(1)} \rightleftharpoons PbO_{(1)} + Fe_{(1)}$$

As above, Cu and Fe iso-activity lines as represented in the second diagram of Fig. 2 can be drawn.

The figure shows that well defined values of the Cu and Fe activities correspond to each matte and slag composition at any particular temperature and consequently that each matte composition is associated with a well defined composition of the slag. In a system simultaneously containing slag, matte and lead bullion, the vapour pressures of sulphur and oxygen can then be calculated from the matte and slag compositions, by equations such as

$$\log P_{S_2}^{1/2} = -1.397 + \log a_{PbS}$$
$$\log P_{O_2}^{1/2} = -3.048 + \log a_{PbO}$$

which are deduced from the equilibrium constants of the formation reactions of PbS and PbO.

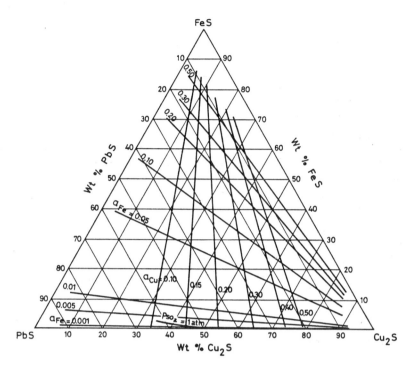

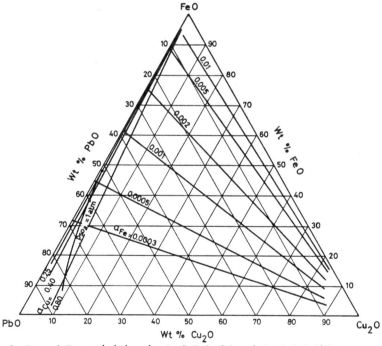

Fig. 2   Cu and Fe activities in Cu$_2$S-FeS-PbS and Cu$_2$O-FeO-PbO
         systems (at 1127°C and for Pb activity of 0.8, i.e. in
         presence of a lead bullion phase).

Finally, the compositions for which the SO_2 pressure is equal to 1 atm, i.e. corresponding to the direct smelting of Pb-Cu materials, can be calculated from the relation

$$\log p_{S_2}^{1/2} + \log p_{O_2} = - 9.676$$

and are represented by the curved lines in both the diagrams of Fig. 2.

It can thus be seen that, in direct Pb-Cu smelting, the composition of mattes is in fact situated along the Cu_2S-PbS line, and that of the slags along the FeO-PbO line, i.e. the mattes will be practically iron-free whereas the slags will be practically copper-free.

The sum of the results can further be summarized for any particular temperature in a single diagram (Fig. 3), where the conditions for SO_2 evolution are shown as a function of the Pb content of the slag and of the Cu content of the matte. The slope of the line calculated for p_{SO_2} = 1 atm shows that, the higher the Cu content of the matte, the higher has to be the Pb content of the slag in order to achieve SO_2 evolution. It can also be shown that the p_{SO_2} line moves downwards as the temperature is raised. The possibilities of direct desulphurization are thus small and, more particularly, smaller than in pure Pb or pure Cu systems. This is explained by the relatively high PbO and Cu_2S stabilities and the rather small tendency for reactions of the type

$$2 \text{ PbO }_{(1)} + Cu_2S_{(1)} \longrightarrow 2 \text{ Pb }_{(1)} + 2 \text{ Cu }_{(1)} + SO_2{}_{(g)} \qquad (K_{1400°K} = 0.9)$$

to take place. On the other hand, the study shows that direct smelting allows the concentration of copper in relatively rich mattes (50-60 % Cu) while collecting lead in Pb-rich slags which are practically copper-free.

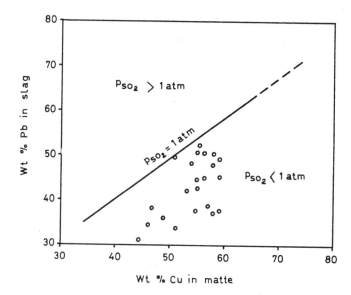

Fig. 3   Direct smelting conditions of Pb-Cu materials (at 1127°C), as a function of Pb content of slag and Cu content of matte; points represent final compositions obtained in furnace trials.

380

This conclusion has been well confirmed by the experimental results of crucible and electric smelting tests, on complex charges such as described below. The experimental results of some smelting tests are represented in Fig. 3 and it can be seen that they are all situated, with some scatter, near the line calculated for $P_{SO_2} = 1$ atm. Deviations can readily be explained by the fact that the SO_2 evolution can take place at a pressure lower than 1 atm and that the reactions are quite sensitive to variations of temperature or of the PbO activity. It is obvious that the latter can easily be influenced by changes in the basiscity of the slag.

Extension to the Direct Smelting of Complex Materials

The behaviour of other metals, associated with lead and copper, in the system which has just been described, is comparatively straightforward, since direct smelting of these materials necessarily gives rise to very high oxygen activities. The metals can thus be classified into three groups :

- metals more readily oxidizable than lead, which will for the greater part concentrate in the slag; the principal examples are Zn, Fe and Sn.

- metals less oxidizable than lead, mainly the precious metals, which will be distributed between the reduced phases.

- if Ni, As and Sb are present, they will concentrate in an arsenical alloy phase, distinct from the other liquid phases of the system and from which it can be separated in the same way as in the traditional smelting process.

This behaviour is well illustrated by the example of Table I, which gives the results of a smelting test of such a complex charge.

Table I : Direct smelting of a complex charge
(compositions expressed as wt %)

	Feed	Productions				
	Charge	Slag	Matte	Speiss	Pb bullion	Dust
Wt, kg	12.000	7.552	1.339	365	743	384
Ag	0.0362	0.0035	0.1187	0.1745	0.2443	0.0094
Pb	36.45	40.00	26.00	23.40	97.00	52.00
Cu	7.71	1.16	53.00	31.63	1.25	0.60
Ni	0.91	0.41	0.72	18.83	0.0	0.0
As	1.98	2.04	0.64	17.82	0.12	2.43
Sb	0.97	1.18	0.27	5.32	0.53	0.18
Sn	1.01	1.59	0.02	0.05	0.0	0.25
Zn	2.13	3.31	0.10	0.0	0.0	1.00
Fe	6.65	10.55	0.10	0.0	0.0	0.20
S	9.11	0.25	15.50	1.80	0.10	7.50
CaO	2.48	3.87	0.0	0.0	0.0	0.0
SiO_2	7.79	12.37	0.0	0.0	0.0	0.0

Reduction of the First Smelting Slag

The slag produced by direct smelting of a complex charge will necessarily have to undergo a reduction operation. Four distinct phases will then normally be produced : depleted slag, lead bullion, an arsenical alloy and zinc-bearing dust.

The depleted slag is a reject product and the objective is thus to extract a maximum of valuable metals from it. Reduction tests in crucibles have shown that a factor of primary importance was the CaO content of the final slag. This is shown in Fig. 4, which gives the results of reduction tests at 1250°C, lasting 6 h and in the presence of an excess of coke, of a given starting slag to which increasing amounts of lime had been added. The figure shows the extraction rate of three metals, Sn, Zn and Fe, as a function of the CaO content of the final slag. In practice, a compromise has to be achieved between the extraction of valuable metals and that of Fe, which has to be kept within reasonable limits. The conclusion of the tests is that we should operate with slags containing 22-24 % CaO, which give a Sn extraction greater than 90 %, while limiting the Fe reduction to about 20 %. In these conditions, the Pb content of the slag is reduced below 0.5 %, which constitutes a marked improvement as compared to traditional extraction processes.

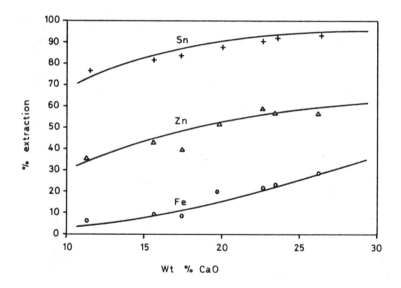

Fig. 4   Extraction of Sn, Zn and Fe from first-step slag as a function of CaO content of end slag.

Lead extracted by reduction is mainly recovered in the lead bullion phase; in this way, it acts as a collector of secondary metals such as Sn and Sb. Iron which is reduced will associate with the small amounts of arsenic present in the initial slag, to form the arsenical alloy. Due to its volatility, zinc will be recovered in the dust, with, however, a contamination by lead, owing to the relatively high vapour pressure of lead oxide at the temperature at which the reduction has to be carried out.

By way of illustration, Table II gives the composition of the products which would be obtained by reduction of the slag whose composition has been given in Table I, providing that sufficient lime had been added to bring the content in the final slag to 24 % CaO.

Table II : Reduction of Pb-rich slag from Table I
(with addition of about 500 kg CaO)

	Feed	Productions			
	Pb-rich slag	Slag	Speiss	Pb bullion	Dust
Wt, kg	7.552	3.231	412	2.901	702
Ag	0.0035	0.0001	0.0017	0.0087	0.0003
Pb	40.00	0.50	5.00	92.44	43.02
Cu	1.16	0.04	10.92	1.42	0.0
Ni	0.41	0.02	6.87	0.07	0.0
As	2.04	0.09	31.61	0.42	1.25
Sb	1.18	0.05	4.85	2.31	0.10
Sn	1.59	0.36	6.64	2.41	1.57
Zn	3.31	0.99	0.10	0.0	30.98
Fe	10.55	21.86	21.61	0.0	0.20
S	0.25	0.16	0.40	0.05	1.50
CaO	3.87	24.00	0.0	0.0	0.0
SiO2	12.37	28.92	0.0	0.0	0.0

Flow-sheet of a Two-Step Smelting Process

Combining the operations of high oxygen potential smelting and reduction of the obtained slag, the general flow-sheet (8) represented in Fig. 5 results.

In a first step, materials are smelted under oxidizing conditions, so as to produce a slag containing about 20 to 40 % Pb, and with a sulphur content which produces a matte containing about 50 to 60 % Cu. Lead should also be present to such extent that in addition to the slag and the matte, a lead bullion phase is produced; this latter must, in particular, collect part of the precious metals of the charge. Arsenic and nickel will further give rise to the formation of an arsenical nickel-bearing alloy.

The achievement of these conditions requires a judicious balancing of the different materials composing the charge and, more particularly, an adequate proportion of oxidized and sulphurized materials. It may possibly be necessary to convert part of the sulphurized materials into oxidized ones, either by roasting or even by autogeneous smelting.

After separation of the phases, the matte is processed by converting; converter slags will normally be recycled to the smelting step, preferably in the liquid state. Lead bullion is sent to the lead refinery. The arsenical alloy is treated for recovery of the contained metals. Smelting dust is recycled with the charge, unless a concentration of valuable elements, such as cadmium, justifies a separate treatment.

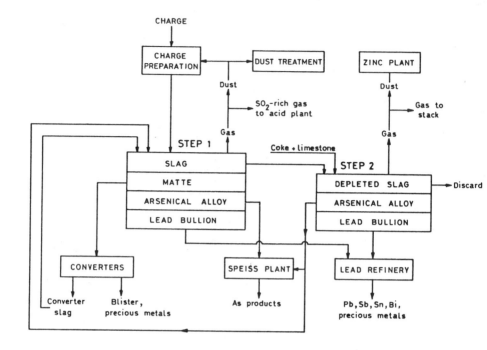

Fig. 5   Flow-sheet of a two-step smelting process (after 8) applied to complex Pb-Cu-Zn materials.

The first stage slag is reduced in a second stage, under such conditions that a depleted slag with about 0.5 % Pb is produced. The resulting lead bullion is sent to the lead refinery. The arsenical alloy is recycled to the first step, unless the concentration of some valuable metals justifies a separate treatment. The dust, concentrating zinc, may be processed in a zinc plant.

Gases produced in the first step will have a high SO_2 content, if the above described desulphurizing conditions (Fig. 3) are achieved; after purification, they will then be sent to the sulphuric acid plant. In the second step, gases have a low SO_2 content, so that they may be released to the atmosphere after purification.

Feasibility of the Two-Step Smelting

The practical realization of the above described flow-sheet must necessarily make use of non-conventional smelting techniques. To define the type of furnace best adapted to each of the two stages, it seems that energy requirements should, first of all, be taken into consideration.

If the feeds are predominantly of sulphurized type, one should logically take a maximum advantage of the energy, which is likely to be released by the roasting reactions of these materials, and the first-step furnace should then preferably be an autogeneous smelting furnace. For example, for feeds consisting of 80 % or more of sulphurized materials, such as concentrates, the first step could take place in autogeneous reactors such as the Kivcet furnace (9) or the QSL converter (10).

For feeds which consist mainly of various secondary materials, partly sulphurized, partly oxidized, and even possibly sulphated, the energy supply of the roasting reactions is insufficient, so that an external heat source has to be used. In this case, the electric furnace with submerged electrodes might be a suitable smelting apparatus, since it presents the advantage, compared to any flame furnace, of producing a minimum volume of gas hence giving rise to a minimum volatilization of the metals, particularly of lead. It would thus also permit pollution problems to be kept under good control.

For the second step, the most important energy aspect is the possibility of feeding the first-stage slags directly in the liquid state. Here, again, the electric furnace with submerged electrodes could be an attractive means, although the feasibility of other techniques, based on the converter, has already been established on a pilot scale (10) and even on an industrial scale (11).

Electric smelting presents the advantage of being a well established technique. It is already largely practiced for two major operations of non-ferrous metallurgy : matte smelting (an operation which is relatively close to the one we consider in our first step) and slag cleaning (relatively close to our second step). It was natural that we first looked in the direction of this technique to carry out the metallurgy we were considering, in order to decrease scaling-up problems.

Small-scale Electric Smelting Tests

The main developments of the process by electric smelting took place in a laboratory furnace, with a nominal power of 60 kW, whose dimensions are given in Fig. 6.

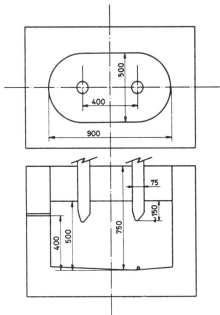

Fig. 6   Dimensions of 60 kW experimental electric furnace (in mm).

The furnace is fitted with two electrodes, dipping into the slag and heating the bath by ohmic resistance. They are connected to a source of alternating current, which can be continuously adjusted between 0 and 160 V, with a maximum of 800 A. The lining of the furnace is made of rammed magnesia. Separate tap holes allow the separate tapping of the slag from the top (400 mm level) and the reduced phases from the bottom. A feeding device permits the continuous charging of more than 200 kg of materials per hour. Gases are filtered in an installation including a cyclone, a bag-filter and a scrubber, before being sent to the stack or to the acid plant if they contain SO_2. A general view of the installation is shown in Fig. 7.

Fig. 7   View of 60 kW furnace, during slag tapping.

Some typical operating conditions are given in Table III. In first-step smelting, the operation normally needs an average power load of about 55 kW and a feed rate of about 140 kg/h. The corresponding energy consumption is of the order of 400 kWh/tonne.

Metallurgical results of these tests have already been illustrated by the examples given in Tables I and II. At the end of a nearly two-year test campaign, the conclusion was that the results which had been foreseen from theoretical studies and by laboratory experiments were well confirmed and that the feasibility of the two steps in electric furnaces was demonstrated, at least in principle. The technological aspects, especially the stability of operation, the energy consumption, the electrode consumption and the production of dust, were promising enough to undertake a pilot scale test campaign.

Feed rate	140 kg/h
Charge composition	40 % crushed converter slag 60 % pelletized materials
Average power load	55 kW
Voltage	80 V
Intensity	687.5 A
Bath temperature	1150°C
Dust production	5 % of charge
SO_2 content of gases	15-25 %
Energy consumption	~ 400 kWh/tonne

Pilot scale Experiments

These tests were performed in the pilot installation of an electric furnace manufacturer, in a 200 kW furnace shown in Fig. 8. It is a single electrode furnace with a bottom contact, fed by a 1500 kVA transformer, whose secondary voltage can be adjusted between 40 and 180 V.

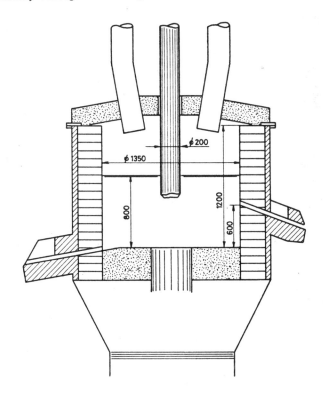

Fig. 8 Sketch of 200 kW pilot furnace (after 12) (dimensions are in mm).

The lining was made of magnesia bricks and separate tap holes were provided for the slag, at the 600 mm level, and for the reduced phases, at the bottom. With an internal diameter of 1350 mm and a bath height of about 800 mm, the corresponding useful volume of the furnace was of the order of 1.1 m³, i.e. about 5 metric tons in the case of the charges we have tested.

First-step Pilot Experiments

The charge was fed continuously, in uninterrupted campaigns of one week. It consisted of about 45 % crushed converter slag and 55 % pelletized fine materials. Two types of charges have, in fact, been tested, one giving rise to the production of slags containing about 30 % Pb and the other producing slags with about 40 % Pb. The corresponding operating conditions are given in Table IV.

Table IV : Typical operating conditions of 200 kW pilot furnace on first-step smelting

	30 % Pb slag	40 % Pb slag
Feed rate, kg/h	425	490
Charge composition	45% converter slag 55% pellets	42% converter slag 58% pellets
Average power load, kW	175	177
Voltage, V	74	75
Intensity, kA	2.4	2.4
Bath temperature, °C	1150	1050
Dust production, % of charge	3.3	3.4
SO_2 content of gases, vol %	20	20
Energy consumption, kWh/tonne	412	361
Electrode consumption, kg per tonne of charge	6.3	6.0

During the campaigns with 30 % Pb slags, the average feed rate was 425 kg/h and the average power load 175 kW. This resulted in an electric energy consumption of 412 kWh/tonne. For the campaigns with 40 % Pb slags, smelting could be performed at a somewhat lower temperature; it was then possible to increase the feed rate for the same power load, so that the energy consumption was only 361 kWh/tonne.

In both cases, the production of dust was low, about 3.3 % of the charge, and the gases contained at least 20 % SO_2. It is assumed that these two factors are more or less connected and that they are strongly influenced by the preparation of the charge. A careful pelletization of the fines contributes to lower dust production, so that the draught on the furnace can be reduced to a minimum and the SO_2 content of the gases is improved.

The consumption of electrodes was 6.0 to 6.3 kg per ton of charge. This is relatively high, compared with traditional electric smelting processes, but remains nevertheless within reasonable limits, considering the strong corrosive effect of Pb-rich slags on carbon.

The general conclusion of the first step experiments was that they confirmed the results of the tests in the 60 kW experimental furnace, and even demonstrated a certain improvement of operating characteristics. The operation was extremely stable and there should be no difficulty in extrapolating it to an industrial scale.

Second-step Pilot Experiments

First-step slags were accumulated for the second-stage reduction experiments. In a first series of runs, tests were performed discontinuously, on batches of previously remelted slag. We intended, however, to extrapolate the results to a continuous operation, with feeding of the starting slag as a liquid.

The reduction of liquid slag in an electric furnace suffers from the difficulty that the reducing agent, normally coke, is spread over the surface of the bath and that contact with the oxides to be reduced is not good, the only motion imparted to the slag being that of the convection at the surface of the electrodes. Furthermore, the Pb-rich slag is denser than the depleted slag so that it tends to stagnate at the bottom of the furnace. These considerations led us to promote convection by injection of natural gas, this gas having the advantage of being non-oxidizing and even of contributing to the reduction reactions. We have been able to show (13) that satisfactory reduction rates could be reached, provided that a relatively small gas flow, not exceeding 2.5 Nm^3/h per ton of treated slag, is injected into the bath. An example of this is given in Fig. 9, which shows the Pb, Sn and Zn contents of a slag as a function of time, during a batch test whose conditions are given in Table V.

Table V : Typical operating conditions of 200 kW pilot furnace on second-step reduction smelting

Feed rate	4 tonne batches
Charge composition	1st step slag 6.5 % pebble lime 4.5 % coke
Power load	100 kW
Voltage	59 V
Intensity	1.8 kA
Bath temperature	1250°C
Reduction time	6 h
Energy consumption	155 kWh/tonne
Electrode consumption	0.7 kg/tonne

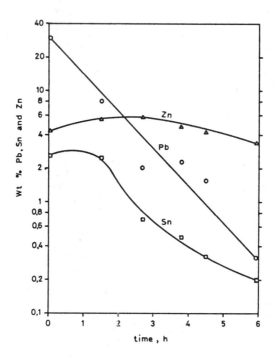

Fig. 9   Reduction of first-step slag as a function of time.

After a 6 hours reduction, the Pb content of the slag had dropped
below 0.5 %. It is interesting to observe that the tin and zinc reduction
only starts when the Pb content of the slag has been brought below about
5 %.

With a power load of about 100 kW maintained during 6 hours for the
reduction of 4 tonnes of slag, a total electric energy consumption of
155 kWh/tonne had been used. For feeds with a higher zinc content,
additional power should be provided in order to take account of the
volatilization of larger amounts of zinc. These figures, together with the
electrode consumption, which is 0.7 kg/tonne, are attractive enough to
justify the development of the process towards continuous operation. Since
then, continuous tests have indeed been performed in the 60 kW experimental
furnace, with satisfactory results. The extrapolation to industrial scale
seems possible but it will require a more thorough investigation of the
convection and reduction mechanisms involved.

Integration of the Two-step Smelting in the Hoboken Smelter

Considering the results of the pilot tests, the combination of the two
smelting steps constitutes an attractive process, both economically and
technically. The operation of the electric furnaces is stable, control of
the operating parameters is easy and the whole offers a great flexibility
with regard to feed materials of varied physical nature and chemical
composition. Furthermore, a good control of pollution problems, by means
of sealing techniques which have now become usual on commercial electric
furnaces, should be emphasized.

The integration of the process in the Hoboken smelter is now considered. A preliminary survey on the effect of two-step smelting on the operation of the smelter and on the circulation of intermediate products was effected by means of the computer program mentioned in the introduction. It led to the conclusion that the progressive integration of the process would allow appreciable improvement in the operation of the existing production units, while ensuring a sufficient return on the capital costs.

In order to approach the scaling-up problems systematically, it is proposed to introduce the process in successive stages, the first being to collect all the secondary materials of complex nature in an electric smelting circuit, where a Pb-rich slag would be produced and where a first separation of the metals would take place according to the principles described above; the Pb-rich slag would then be introduced into the traditional sinter-smelting circuit. In a further stage, the direct reduction of the liquid slag would be considered.

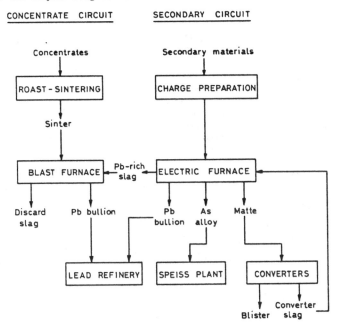

Fig. 10 Flow-sheet of Hoboken smelter after the introduction of 1st step smelting.

The application of the first step to a part of the smelter feed already considerably modifies the flow-sheet of the plant, as shown in Fig. 10. Secondary materials, including the intermediate products of the smelter such as dusts, drosses, converter slags, etc..., are all collected in one single complex circuit, where the main metallic separations take place; it produces, among others, the matte, collecting copper and which is fed to the converters. In parallel, the traditional circuit is mainly fed with concentrates and with the Pb-rich slag from the electric furnace; here lead acts essentially as a collector for secondary metals.

The clearest result of the operation is that the electric furnace will free the traditional circuit from all the secondary materials which have congested it, so that its metallurgical practice will be greatly simplified and that the operation of the furnaces will unquestionably be improved. The traditional circuit will thus be transformed into a pure Pb smelter, for which it was originally designed, eighty years ago.

Conclusion

The prospect of the application of two-step electric smelting at the Hoboken smelter is favourable, so that we hope, in the long run, to adapt the process to all the feed materials of the smelter. If we wish to smelt an important amount of concentrates, this implies that they should be desulphurized by some kind of autogeneous smelting process. The direct combustion of concentrates is not excluded as a part of the first-step electric furnace process.

We expect a simplification of the flow-sheet of the plant, an improvement of the environment and a better recovery of valuable metals. More particularly, when the second smelting step is introduced, a better recovery of some less reducible metals, whose present extraction rate is unsatisfactory, would be achieved.

Acknowledgement

The authors gratefully acknowledge the General Management of Metallurgie Hoboken-Overpelt for their permission to publish this paper. They take this opportunity to express their thanks to the colleagues who have contributed to the carrying out of the present work. They are also grateful for helpful comments of Professor J. De Cuyper, of the University of Louvain, and of Professor F.D. Richardson, Dr. J.H.E. Jeffes and Dr. D.G.C. Robertson, Imperial College, London. The Belgian Institute for the Promotion of Scientific Research in Industry and Agriculture (I.W.O.N.L.) is thanked for financial support.

References

1. Leroy J.L., Lenoir P.J. and Escoyez L.E., "Lead Smelter Operation at N.V. Metallurgie Hoboken S.A.", Extractive Metallurgy of Lead and Zinc, Cotterill C.H. and Cigan J.M. eds (New York, AIME 1970), pp. 824-52 (World Symposium on mining and metallurgy of lead and zinc, vol. 2).

2. Leroy J.L. and Lenoir P.J., "Hoboken Type of Copper Converter and Its Operation", Proceedings of the Symposium on Advances in Extractive Metallurgy, The Institution of Mining and Metallurgy, London, April 1967, pp. 333-43.

3. Fontainas L., Coussement M. and Maes R., "Some Metallurgical Principles in the Smelting of Complex Materials", Proceedings of the Symposium on Complex Metallurgy, The Institution of Mining and Metallurgy, Bad Harzburg, September 1978, pp.13-23.

4. Coekelbergs C. and Delvaux A.L., "Lead Sulfates Processing at Metallurgie Hoboken-Overpelt", 108th AIME meeting, New Orleans, February 1979, AIME-TMS Paper No. A-79-33.

5. Soens M., Van Boven J. and Deckers G., "Results of a Five Years Program in Lighting Pollution by Lead Dust in the Hoboken Smelter", AIME Lead-Zinc World Symposium, Las Vegas, February 1980.

6. Matyas A.G. and Mackey P.J., "Metallurgy of the Direct Smelting of Lead", Journal of Metals, November 1976, pp. 10-15.

7. Fontainas L., Coussement M. and Maes R., "Discussions and Contributions to the Symposium on Complex Metallurgy", Transactions of the Institution of Mining and Metallurgy, Section C, vol. 88, March 1979, p. C67.

8. Metallurgie Hoboken-Overpelt, "Process for Treating Lead-Copper-Sulphur Charges", U.S. Patent applied for.

9. Mueller E., "Die Verhüttung von Bleikonzentraten nach dem Kivcet-Verfahren", Erzmetall, 1976, pp. 322-27.

10. Anon., "Environmentally Clean Lead Bullion Production", Journal of Metals, December 1978, pp. 11-12.

11. Petersson S., Norrö A. and Eriksson S., "Treatment of Lead-Zinc Containing Dust in a TBRC", 106th AIME meeting, Atlanta, March 1977, AIME-TMS Paper No. A 77-12.

12. Gundersen J. and Skretting H., "A Comparison of Electric Smelting on Pilot and Industrial Scale", Journal of the South African Institute of Mining and Metallurgy, August 1978, pp. 18-22.

13. Metallurgie Hoboken-Overpelt, "Process for the Extraction of Non-ferrous Metals from Slags and Other Metallurgical By-products", U.S. Patent applied for.

QSL - A CONTINUOUS PROCESS FOR ENVIRONMENTALLY CLEAN LEAD PRODUCTION

Werner H. Schwartz and Peter Fischer
Lurgi Chemie und Hüttentechnik GmbH
Frankfurt am Main, Germany

Paul E. Queneau
Dartmouth College
Hanover, N.H.

Reinhardt Schuhmann, Jr.
Purdue University
West Lafayette, Indiana

Abstract

A new direct smelting process for the winning of lead from sulfide concentrates is described. Metallurgical considerations based on the system Pb-S-O and pilot plant batch tests with various types of concentrates on a 2 t/h scale, indicate that it is possible to produce continuously and simultaneously lead of low sulfur content and slag of low lead content. In the QSL process green pelletized concentrates are oxidized indirectly by submerged tonnage oxygen injection in the smelting zone and the resulting slag is reduced by submerged injection of low grade powdered coal. The multistage process operates at high rates in a single, sealed reactor. Since heats of reaction are utilized, and equipment is simple and compact, optimum conditions for economic and environmentally clean lead production are ensured.

Introduction

Lead is conventionally produced from galena concentrates via the roast-reduction route. The concentrate is first oxidized on a sintering machine to eliminate sulfur as sulfur dioxide gas that can be converted to sulfuric acid. The lead oxide sinter product is reduced to lead bullion by coke in a second reactor, the shaft furnace. This technology presents difficult problems in the field of energy and environmental conservation.

Many lead compounds, including metallic lead, have high vapor pressures even at rather low temperatures. Aside from sulfur dioxide contamination of the atmosphere, in conventional practice intermediate matter produced is hot and often dusty so that lead may easily become airborne. Thus effective measures must be taken to avoid pollution of the environment: conventional practice in modern chemical plants but extremely costly in old-fashioned lead smelters. In the latter there are so many material conveying and transfer zones where vapors and dusts can be emitted, calling for enormous volumes of exhaust air to be cleaned.

It should be noted here, in view of the difficulties associated with gas and dust control, that wet processes must of course be given careful consideration. Unfortunately, the problems associated with hydrometallurgy on a large scale are also hard to overcome. Lead sulfide is attacked either by roasting or by an oxidizing leach and the metal is extracted by ammoniacal or acid solutions. Due to the usual low solubility of lead compounds, large amounts of these solvents are necessary, some of which can be very corrosive. Lead is then isolated by aqueous electrolysis, by gas reduction, or by fused chloride salt electrolysis. Leach residues need chemical treatment to extract precious metals, copper, antimony, and other elements. Vapors and sprays require control, bleed liquids must be purified, and the large tonnages of unstable solid residues may contaminate ground-water and air when exposed to the weather. It must also be borne in mind that the manufacture and use of solvents can be energy intensive. Electrolysis also calls for some 400 kWh per ton of concentrate, and equivalent amounts of energy are needed in the case of gaseous reduction. It presently appears unlikely that hydrometallurgy can compete successfully with truly modern lead pyrometallurgy, due to the former's inherent burdens in respect to reaction rates, energy efficiency and environmental protection.

It must be emphasized that mineral concentrates of lead contain some zinc, cadmium, copper, nickel, arsenic, antimony, and last but not least, precious metals. Except for zinc and cadmium, these metals are collected in lead bullion and can then be isolated conventionally in the refinery, most of them contributing to the net profit of the smelter. In addition to ability to produce lead from a wide variety of galena concentrates, it would be most desirable if a new plant could treat a spectrum of lead-rich rejects, such as refinery slags, anode slimes from copper scrap recycle, argentiferous slimes from zinc electrolysis, and electric storage battery scrap. An efficient, flexible process is required which will decrease lead recovery costs, improve recovery of by-product metals and of sulfur, save energy by utilizing heats of reaction and - this is a critical necessity - sharply decrease environmental pollution. After three years of test work a process that meets these demands can be introduced now. It is the QSL process, invented by Paul E. Queneau and Reinhardt Schuhmann, Jr., applying injectors invented by Guy Savard and Robert Lee, and developed by Lurgi in Germany.(1)(2)(3)(4)

Metallurgical Considerations

The equilibrium diagram Pb - S - O in a form presented by Schuhmann and co-workers (2) (Fig. 1) indicates that under certain conditions of temperature, oxygen- and sulfur dioxide partial pressures galena concentrates can be smelted to produce lead bullion directly in a single step.

Although, to cite Davey (5), the chemical equation to express this partial oxidation reaction,

$$(1) \quad PbS + O_2 = Pb + SO_2,$$

is deceptively simple to write, the technology to realize it successfully in industrial practice has turned out to be highly sophisticated.

Three main problems of direct lead smelting processes may be pointed out in brief:

i)   The area of stability of metallic lead with a sulfur content low enough to be acceptable to the refinery is restricted to a rather narrow range of oxygen potentials, so that close thermodynamic control of the process is essential to optimize metallurgical performance.

   The question as to whether the invariant equilibrium of the four condensed phases Pb, PbS, $PbSO_4 \cdot 2$ PbO, and $PbSO_4 \cdot 4$ PbO under a SO_2 partial pressure of 1 atm corresponds with a temperature of 900°C or a lower one appears to be more or less of academic interest. In practice much higher temperatures will be necessary to ensure rapid reaction rates and adequate slag fluidity.

ii)   Any lead bullion formed by direct oxidation of PbS will be in equilibrium or, to express it in process terminology, in contact with a slag the PbO-content of which will depend on the prevailing oxygen potential as well as on the partial pressure of SO_2 in the gas atmosphere and the constitution of the slag.

   Low-Pb slags require low oxygen potentials, and the governing equilibrium

$$(II) \quad 2 \text{ PbO} + \text{PbS} = 3 \text{ Pb} + SO_2$$

   indicates that it should be possible in principle to select operating conditions accordingly.

   Slags containing not more than about 4 to 4.5 wt % Pb together with a high-sulfur bullion (about 3 wt % S) are indeed reported produced by the electrothermic Boliden Lead Process (6). Both products, however, need further processing in a slag fuming plant and by blowing in a converter, respectively. Whether or not 4 wt % Pb is now acceptable in a discard slag, the process apparently calls for a rather unusual slag composition (34 wt % CaO), and, consequently a high temperature of around 1350°C.

   According to our experience with more conventional low-melting point slags of the basic type:

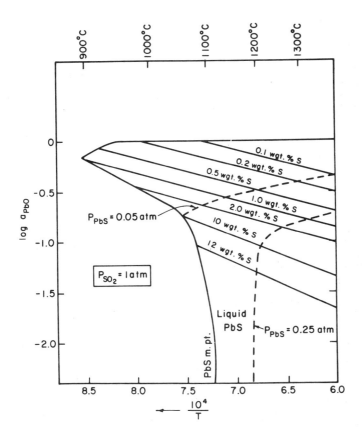

Fig. 1.   Lead converting equilibrium
diagram.

$$(FeO + ZnO + Al_2O_3) \; : \; \text{about 47 wt \%}$$

$$(CaO + MgO + BaO) \; : \; \text{about 16 wt \%}$$

$$SiO_2 \quad\quad\quad\quad : \quad \text{about 37 wt \%}$$

it appears to be impossible to obtain lead contents below ~ 10 wt % at about 3 wt % sulfur in the bullion. Also the Pb content of the slag increases rapidly with decreasing sulfur content of the bullion (Fig. 2).

The curve given here represents a set of quasi-continuous QSL tests on a 2 t/h scale with various types of concentrates. These tests were carried out under comparable conditions with respect to temperature (about 1000 to 1100°C) and SO_2 partial pressure (about 0.6 Bar), and the wide scattering of the individual data is mainly caused by variations of the type of slag, i.e. the activity coefficient of the PbO.

Reaction (II) is governed by the following equilibrium relation, applicable in the presence of a lead bath ($a_{Pb} \sim 1$)

$$(III) \quad K = \frac{P_{SO_2}}{(a_{PbS})(a_{PbO})^2}$$

When temperature, P_{SO_2}, and slag type are substantially fixed in a series of tests, equation (III) then governs the relationship of the sulfur content of the bullion to the lead content of the slag. Using weight percentages to approximate activities, one would expect in such a test series

$$(IV) \quad (\text{Wt \% S in bullion}) \, (\text{Wt \% Pb in slag})^2 \sim \text{Constant}$$

For the test data of Fig. 2, this constant is on the order of 1000. It should be emphasized that such a magnitude applies only to the high SO_2 pressures generated during oxygen converting of PbS. For subsequent stages of slag reduction, the "constant" of equation (IV) will be one or two orders of magnitude lower.

iii) Because of the high vapor pressures (Fig. 3) of the lead compounds involved, considerable volatilization of lead - mainly as PbS - is hard to avoid in direct smelting.

The flue dust, although consisting of PbS, PbO and Pb in statu nascendi, is oxidized to basic lead sulfates during cooling under a SO_2-containing gas atmosphere. It will be kept in mind that, because of endothermic dissociation reactions, the heat balance of the process is considerably improved by minimizing the amount of flue dust recirculated.

Measurements of Yazawa and coworkers (7) at temperatures from 1060° to 1180°C show that the partial pressure of PbS over dilute solutions of sulfur in lead is very sensitive both to sulfur content and to temperature. Their data show that the activity coefficient of PbS in dilute solutions at 1100°C is about 5 - showing a relatively large positive deviation from ideality.

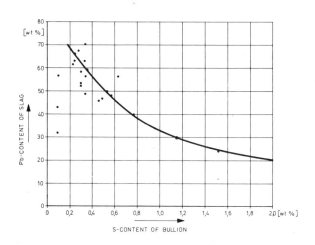

Fig. 2. Lead content of primary slags.

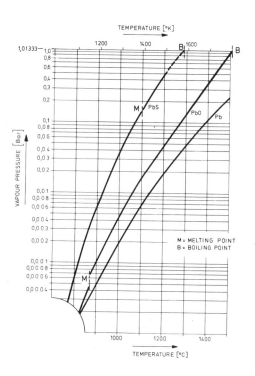

Fig. 3. Vapour pressure of lead and lead compounds .

It appears, therefore, that direct lead smelting should be performed at controlled low sulfur levels in the bullion as well as at lowest possible temperatures. These requirements do not fit together as long as single stage smelting of galena concentrate is considered. Here, high temperatures together with high sulfide potentials, i.e., high sulfur levels in the bullion, are necessary to produce a fluid discard slag at the cost of excessive flue dust formation.

These relations are quantitatively shown in Fig. 4. It summarizes the results of tests that were carried out on a given concentrate over a broad range of sulfur potentials. It should be noted that minimum temperatures were maintained for each individual test, and that temperatures had to be raised at increased sulfur potentials according to the increasing melting point of the slag as well as of the metals phase. These test results demonstrate the advantages of the multistage QSL process. In order to minimize the oxidation temperature a low-melting point primary slag is required. This is easily achieved by closely controlled overblowing of the concentrate so that some PbO is formed together with metallic Pb. To give an example: a 50 to 60 wt % Pb primary slag of the afore mentioned basic type is fluid even at a temperature of about 900°C. The comparatively high oxygen potential prevailing in the oxidation zone of the reactor corresponds to a low sulfide potential. Consequently, the fume make is much lower than is reported for other known direct smelting procedures.

The stoichiometry of the oxidation reaction

$$(V) \quad PbS + \frac{3-n}{2} O_2 = nPb + (1-n) PbO + SO_2,$$

and, consequently, the amount of heat liberated can be adjusted according to thermal requirements so that flexibility is ensured. For certain types of concentrates it is possible to treat oxidic or sulfatic lead materials, in addition to recirculated process flue dust, so as to avoid waste of energy by water cooling or use of lower-grade oxygen. Since the feed is pelletized, intimate contact of all components and, therefore, optimum conditions with respect to slag formation and to the solid-solid reaction

$$(VI) \quad PbS + PbSO_4 = 2 Pb + 2 SO_2$$

are ensured. This strongly endothermic reaction seems to start at rather low temperature (8) thus lowering the PbS-activity at least locally even when heating the pellets.

The PbO-rich primary slag is reduced under a low-SO_2 gas atmosphere by means of powdered coal which is blown into the liquid bath together with a carrier gas. This coal, as well as such additional fuel as may be necessary to satisfy the process heat balance, can be high in sulfur. In step with decrease in PbO content of the slag, its temperature as well as reduction potential are gradually increased so that the thermodynamic conditions at the slag tapping end of the reactor approach those prevailing in the tuyere zone of a lead blast furnace. Excessive co-reduction and consequent volatilization of zinc, is avoided by proper dosing of the reductant and close control of the temperature. The testwork indicated that there is no build-up of zinc in the flue dust circuit as long as slag reduction is controlled so as to produce a final slag containing less than 1 wt % Pb.

Despite the relatively high end-temperature, fume make during slag reduction is quite low and does not exceed about half of the amount of oxidation stage flue dust. According to our calculations, based on the results of several combined oxidation - reduction tests, total volatilization of less than 20% of the lead content of the concentrate feed can be

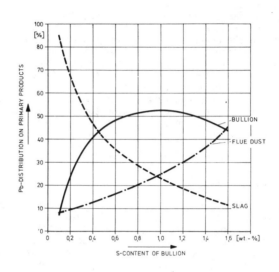

Fig. 4.　Lead distribution on
primary products

expected in a well regulated QSL direct smelting operation.

A typical discard slag of the QSL process is shown in Table I. These data confirm that excellent recoveries of all metal values, except Zn, are feasible. An accompanying lead bullion sulfur content is typically in the 0.2-1.0% range, depending on concentrate copper content. Oxidation of the feed is carried out under conditions in which magnetite forms so that the primary slag is nearly saturated with magnetite. This material provides a thin protective layer on the brick lining of the reactor. Most of the magnetite is decomposed during slag reduction so that local accretion build-up on the bottom of the reactor did not take place. Because excessive bath turbulence is avoided by use of Savard-Lee injectors, good results were obtained in respect to refractory life. Serious attack on the lining was not encountered, either in the injector zone or at the slag line, and material balances indicated that in commercial practice refractory consumption might be below 200 g/t of lead.

The Commercial Direct Smelting Process (Fig. 5)

Lead concentrates are pelletized together with lime, silica, iron oxide if necessary, and the flue dust generated during smelting. The moist pellets are charged continuously into the QSL reactor. The reactor consists of a horizontal brick-lined kiln, about 40 m long and 4 m in diameter. The pellets are fed through openings in the top. They fall into a melt consisting of lead bullion, lead oxide, lead sulfide, fluxes, gangue, and zinc oxide. Submerged injectors are located below the charge openings, and oxygen is blown in from the bottom. The sulfides in the melt are oxidized immediately, SO_2-gas bubbles out, and oxides of lead, iron, and zinc form slag. Metallic lead together with copper and the precious metals pass to the bottom and bullion is tapped continuously at the off-gas end of the reactor. The slag flows steadily to the opposite end for continuous discharge. On its way it passes over a row of additional submerged injectors through which powdered coal is blown. Its lead oxide content is reduced and the resulting lead bullion flows countercurrently along the bottom to join the primary lead mentioned above. The final slag contains less than 1% lead and perhaps 15% zinc, the latter being recovered by any method desired.

Injection of tonnage oxygen through bottom tuyeres into a sulfide or metallic melt would normally destroy both the adjacent brickwork and the tuyere itself, due to reaction temperatures and the formation of aggressive oxides. The idea of Savard and Lee, developed into industrial application by Maxhuette in their steel plant in Bavaria, involves use of a protective shielding fluid surrounding the injected oxygen. Some difficulties with this concept had to be overcome for lead-making but the necessary know-how was developed during the course of pilot plant experiments at Berzelius.

The process can employ two countercurrent movements: lead relative to slag, and slag relative to off-gas. Since the highest temperature is at the slag tapping end, a burner can be located there fired by coal, oil or gas. At the lead discharge end the temperature of both gas and melt is low, so that a minimum of energy is lost with the small volume of SO_2-rich off-gas. Energy losses are confined to the heat contents of the three reactor products and reactor radiation losses. QSL fossil fuel cost is estimated at well less than two-thirds that of the conventional sinter machine-blast furnace tandem, with an electric energy consumption about the same as the latter. Assessment of QSL investment and operating costs also indicates substantial savings - about 30% - in comparison to those of a conventional plant. Since materials transport lines are short, the equipment is compact, and wet concentrates and pelletized materials are handled, escape of contaminants into the atmosphere can be prevented. Furthermore, the volume of air exhausted from

Table I

Analysis of a Typical QSL Discard Slag

Compound	Content in wt %
Pb	0.9
Zn	13.7
Cd	0.003
Cu	0.03
Ag	0.0006
Sn	0.25
Bi	0.002
Sb	0.005
As	< 0.01
FeO	23.6
CaO	13.6
MgO	2.7
Al_2O_3	3.2
SiO_2	36.4

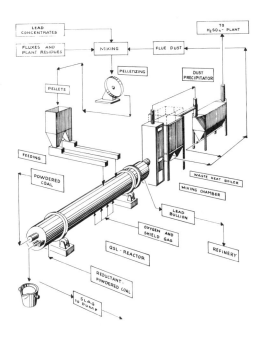

Fig. 5.   Flowsheet of the QSL process.

around equipment and materials transfer points, is only a fraction of that demanded in a conventional smelter. These are the considerations which warrant optimism that an economic process has been developed which will permit environmentally clean lead production - solving a problem now plaguing the industry.

The excellent results achieved in the batch pilot plant at the Berzelius Works in Duisburg, have led to the decision to build a QSL demonstration plant there. It will employ a continuous reactor of about 50,000 tpy bullion capacity when treating high grade galena concentrates. Engineering design is under way and present plans call for initiation of construction during the winter of 1979-1980.

References

1. P. E. Queneau and R. Schuhmann, Jr., "Metallurgical Process Using Oxygen", U.S. Patent 3,941,587, 1976; "Apparatus for a Metallurgical Process Using Oxygen", U.S. Patent 4,085,923, 1978; patents pending.

2. R. Schuhmann, Jr., Pei-Cheh Chen, P. Palanisamy and D. H. R. Sarma, "Thermodynamics of Converting Lead Sulfide", Metallurgical Transactions, Vol. 7B, March 1976, pp. 95-101.

3. P. E. Queneau, "Oxygen Technology and Conservation", Metallurgical Transactions, Vol. 8B, September 1977, pp. 357-369.

4. Anon., "Environmentally Clean Lead Bullion Production", Journal of Metals, December 1978, pp. 11-12.

5. T. R. A. Davey, "Developments in Direct Smelting of Base Metals - A Critical Review", paper presented at International Conference on Advances in Chemical Metallurgy, BARC, Bombay, January 1979.

6. H. I. Elvander, "The Boliden Lead Process", in "Pyrometallurgical Processes in Nonferrous Metallurgy", J. N. Anderson and P. E. Queneau, eds., Gordon and Breach, New York, 1967, pp. 225-245.

7. A. Yazawa, et al: Bull. Res. Inst. Min. Dress. Met., Sendai, Tohoku Univ., 24, 1968, p. 105.

8. A. Melin: "Beitrag zu den physikalisch-chemischen Grundlagen der Röstung von Bleiglanz", Diss. RWTH Aachen, 1965.

PILOT PLANT DEMONSTRATION OF ZINC SULPHIDE PRESSURE LEACHING

E. G. Parker and S. Romanchuk

Senior Research Chemist, Cominco Ltd.
Trail, British Columbia, Canada

Project Manager, Sherritt Gordon Mines Limited
Fort Saskatchewan, Alberta, Canada

SUMMARY

Cominco's modernization program in Trail, British Columbia requires expansion of the current zinc capacity to 272,000 MTPY. A preliminary study indicated that a supplementary zinc pressure leaching operation would be the most flexible and economical means of realizing this increased capacity. Sherritt Gordon Mines Limited was interested in the further development of their direct zinc pressure leaching process with the objective of future licensing of the process to other interested companies. During the summer and fall of 1977 Cominco and Sherritt jointly operated a 3 tonne per day pilot plant in Fort Saskatchewan, Alberta to demonstrate pressure leaching of Sullivan zinc concentrate and the subsequent separation of molten elemental sulphur from the leach slurry.

Following a brief commissioning period, the pilot plant attained or exceeded all objectives. During two, week-long demonstration periods, on-stream times were over 99%, and zinc extraction to solution averaged 97%. A novel method for continuously separating molten elemental sulphur was also successfully demonstrated. During these periods considerable commercial plant design data were collected. Zinc sulphate solutions and iron-lead residues produced in the pilot plant were tested in Trail and were found to be satisfactory for integration with existing Trail operations. A total of 116 tonnes of zinc concentrate was treated through the leach and elemental sulphur separation processes. On-site atmospheric testing found no evidence of potential air pollution or hygienic problems.

The Pilot Plant campaign indicated that pressure leaching of Sullivan zinc concentrate could be conveniently integrated with existing Trail Operations to increase zinc production without necessitating increased sulphuric acid production. As a result of this successful campaign, a commercial pressure leaching facility is scheduled to go into operation in 1981 to supplement Cominco's zinc production in Trail. This plant, which will be the first of its kind, will permit a unique flexibility with respect to sulphuric acid production.

INTRODUCTION

Direct pressure leaching of zinc sulphide concentrates to produce zinc sulphate solutions and elemental sulphur was first successfully carried out by Sherritt in 1959(1). Originally the maximum leach temperature was restricted to below the melting point of sulphur (119°C) because molten elemental sulphur, formed in the leach, coated the partially reacted metal sulphides and limited zinc extraction. In order to obtain high zinc extractions at the low temperature employed, retention times of 6 to 8 hours were required.

Cominco became interested in zinc pressure leaching in the early 1960's. At that time the concept of being able to increase zinc production without increasing sulphuric acid production was attractive. In 1962, Cominco, with co-operation from Sherritt, operated a continuous pilot scale pressure leaching autoclave which treated 2 to 3 tonnes per day of zinc concentrates from Cominco's Sullivan Mine at 110°C. The pilot plant was a technical success, but changing zinc and fertilizer markets did not warrant a commercial installation at that time. In 1971 two new Lurgi fluid bed roasters were installed at Trail, and little further thought was given to pressure leaching.

When Cominco's current modernization plans for expanding and upgrading their zinc plant were formulated, the pressure leaching process was again examined. Since 1963 a number of factors had changed to make pressure leaching more attractive.

1. Further Sherritt research(2) showed that high temperatures, coupled with the use of certain surface active reagents, substantially increased leaching rates, thereby reducing the autoclave volume necessary, and hence, capital cost.

2. Fertilizer markets were unlikely to match precisely the planned zinc expansion. Since elemental sulphur could be stockpiled or burned for acid manufacture, pressure leaching would circumvent the previously mandatory link between zinc and fertilizer production.

3. Techniques developed by Cominco indicated that elemental sulphur produced in the leach could be separated in the molten state, further reducing equipment and handling costs (3).

4. Finally, the availability of the equipment from the Sherritt-Cominco Copper Pilot Plant in Fort Saskatchewan and the willingness of Sherritt and Cominco to co-operate in the venture afforded a unique opportunity to pilot the pressure leach process including all the latest developments, at minimum cost, and in minimum time.

PILOT PLANT OBJECTIVES

The principal objectives of the pilot plant were:

1. To demonstrate continuous pressure leaching of Sullivan zinc concentrate at temperatures above the melting point of sulphur to extract a minimum of 95% of the zinc into solution.

2. To demonstrate the operation of an elemental sulphur-decanting autoclave to separate molten elemental sulphur from a slurry containing zinc sulphate solution and the iron-lead residue.

3. To demonstrate an alternative method of sulphur separation.

4. To collect pilot plant products for further test work to ensure their compatibility with Cominco's existing Trail Operation.

5. To develop commercial plant design data, including corrosion testing of materials of construction.

In addition to the above objectives it was necessary to complete the pilot plant campaign by the end of October, 1977, to comply with Cominco's modernization schedule for the selection of a preferred route for zinc expansion.

ORGANIZATION AND OPERATION

The pilot project was the joint responsibility of Cominco's and Sherritt's Technical Research centres. Each company provided a Co-Manager and technical staff. Shift supervisors, operators and maintenance personnel were recruited from Sherritt's Fort Saskatchewan operation. Sherritt also provided research, engineering, analytical and administrative support. The positions of operating superintendent, technical superintendent, maintenance co-ordinator and roving engineer were filled by engineers. Each shift consisted of a supervisor, a technician and two operators. Two extra operators were required on day shift for concentrate grinding and miscellaneous jobs. The plant was normally operated continuously for five days per week, utilizing the weekend for maintenance and alterations. This schedule routinely demonstrated start-up and shut-down procedures.

Schedule

Pilot plant construction was started mid-April, 1977 and was completed by July 22, 1977. A six-week commissioning period began on July 18, and was followed by an eight-week test and demonstration period ending October 28. Compilation of the results followed immediately.

PROCESS FLOWSHEET

Figure 1 shows the principal operations piloted, and how these operations will fit into Cominco's zinc plant in Trail.

The pilot plant divided naturally into three areas:

1. A grinding section, for concentrate size reduction.

2. A pressure leaching section, where zinc was extracted with dilute H_2SO_4 (return electrolyte) in a 4-compartment autoclave using elevated temperatures and oxygen pressures.

3. A sulphur separation section, in which elemental sulphur was removed from the autoclave leach product leaving a slurry containing the dissolved zinc and an iron-lead residue.

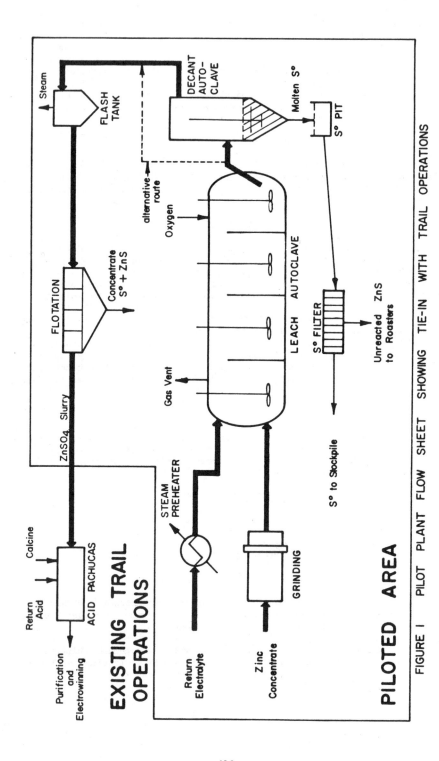

EXISTING TRAIL OPERATIONS

PILOTED AREA

FIGURE I   PILOT PLANT FLOW SHEET SHOWING TIE-IN WITH TRAIL OPERATIONS

410

For incremental zinc production in Trail, the zinc-enriched slurry from pressure leaching will join the existing roast-leach operation in the calcine leaching section. In a "grass roots" pressure leaching plant, the slurry would require further neutralization and separation of the iron-lead residue prior to purification and electrolysis.

FEED MATERIALS

Zinc concentrate from Cominco's Sullivan mine at Kimberley, British Columbia, was treated in the pilot plant. This concentrate was selected for pressure leaching following laboratory tests which indicated that it would leach well and provide satisfactory products for integration with Trail operations. In addition, Sullivan concentrate is less satisfactory for roast-leach processes because its rather high lead content tends to cause problems in fluid bed roasters, and its high iron content lowers first-pass zinc extraction due to the formation of insoluble zinc ferrites.

To increase reaction rates and zinc extraction, a classical closed circuit grinding operation was employed in the pilot plant. Typically, the concentrate size was reduced from 91% minus 44 microns to 98.5% minus 44 microns.

The zinc plant return electrolyte (cell house acid) was shipped from Trail in tanker trucks. Pregnant solution was returned to Trail in the same trucks.

Concentrate and return electrolyte compositions are shown in Table I.

Table I - Concentrate and Return Electrolyte Analysis

	Concentrate %	Electrolyte g/L
Zn	47.6	50.8
Fe	11.3	–
Pb	6.95	–
S_t	30.6	–
SiO_2	1.9	–
Ca	0.15	0.4
Mg	0.13	6.8
H_2SO_4 (Fortified)		169

LEACHING

Chemistry

At elevated temperatures and pressures, zinc sulphide, lead sulphide and some iron sulphide minerals react with oxygen and sulphuric acid to form simple sulphates, elemental sulphur and water.

The initial reactions may be represented as follows:

$$ZnS + H_2SO_4 + 1/2\ O_2 \longrightarrow ZnSO_4 + S° + H_2O \qquad (1)$$

$$PbS + H_2SO_4 + 1/2\ O_2 \longrightarrow PbSO_4 + S° + H_2O \qquad (2)$$

$$FeS + H_2SO_4 + 1/2\ O_2 \longrightarrow FeSO_4 + S° + H_2O \qquad (3)$$

Subsequently, iron is further oxidized from the ferrous to the ferric state:

$$2FeSO_4 + H_2SO_4 + 1/2\ O_2 \longrightarrow Fe_2\ (SO_4)_3 + H_2O \qquad (4)$$

It has been observed that in the absence of iron, pure zinc sulphide is only slowly decomposed by oxidative pressure leaching. In the presence of iron, the leaching mechanism is probably represented by:

$$Fe_2\ (SO_4)_3 + ZnS \longrightarrow 2FeSO_4 + ZnSO_4 + S° \qquad (5)$$

followed by regeneration of ferric iron as shown in reaction (4).

At elevated temperatures (150°C) and diminishing acid levels, ferric iron precipitates in the form of complex basic sulphates. The nature of the compounds formed is variable. Two typical reactions are shown below:

$$3Fe_2\ (SO_4)_3 + PbSO_4 + 12H_2O \longrightarrow PbFe_6(SO_4)_4(OH)_{12} + 6H_2SO_4 \qquad (6)$$

(Plumbojarosite)

$$3Fe_2\ (SO_4)_3 + 14\ H_2O \longrightarrow (H_3O)_2Fe_6(SO_4)_4(OH)_{12} + 5H_2SO_4 \qquad (7)$$

(Oxonium jarosite)

Plumbojarosites, which tend to become highly complex, retain approximately 2% Zn in their crystal structure.

In the pilot plant, approximately 80-85% of the total reaction took place in about 27 minutes in the first compartment of the 4-compartment autoclave.

Under standard pilot plant leach conditions, concentrate surface area was rate limiting in the first compartment. When concentrate was fed at 150% of the standard rate, the Fe^{+++} level fell to zero, and the rate of oxidation of ferrous iron, reaction (5), became the practical rate limiting factor. In the presence of a molten sulphur dispersant (lignin sulphonate) to prevent encapsulation of the sulphide minerals, concentrate surface area was probably the chief rate limiting factor in the latter three compartments of the autoclave.

Circuit and Process Description

Figure 2 shows a flowsheet for the leaching section.

The leach autoclave was a lead and brick-lined carbon steel vessel. The operating volume approximated 1.27 m³. Agitation was provided in each of the four compartments. Return electrolyte was continuously pumped at 12 litres/minute to the first compartment of the leach autoclave via a shell-and-tube heat exchanger, which raised its temperature to 70°C. Concentrate slurry at 50 to 70% solids was pumped at 2.3 litres/minute into the autoclave using an oxygen-operated pressure egg system. The pressure egg pump was constructed on-site to enable a small and precise flow of concentrate to be delivered to the leach autoclave. The free acid concentration in the first and last compartments of the autoclave was constantly monitored, and final acidity, so important to the process, was controlled by maintaining a constant electrolyte flow, and adjusting the timed cycle of the concentrate pump. Oxygen was continuously added to the autoclave and a small bleed was required to remove the inerts. The leached slurry was continuously discharged from the fourth compartment.

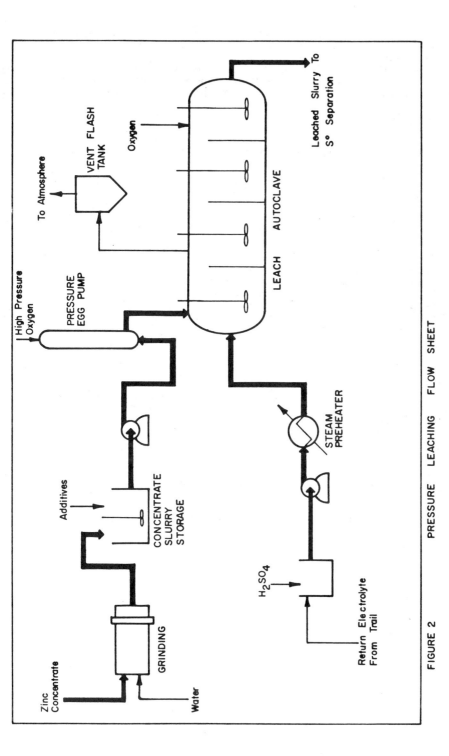

FIGURE 2   PRESSURE   LEACHING   FLOW   SHEET

413

Eighty to eighty-five percent of the total reaction took place in the first autoclave compartment, raising the temperature to about 146°C. The highest temperature (average 155°C) occurred in the second compartment, then declined to 152°C and 146°C in the last two compartments. With the exception of start-up, this operating temperature profile was maintained without the need for auxiliary heating or cooling.

Leach Conditions

Table II shows the standard conditions employed in the leach section, and imposed variations. Under these conditions elemental sulphur remains in the molten state, near its minimum viscosity range.

Table II - Leach Conditions

	Standard	Variation
Temperature	154°C	138-160
O_2 Partial Pressure	620 kPa (90 psi)	480-900 (70-130 psi)
Total Pressure (absolute)	1140 kPa (150 psig)	1000-1410(130-190 psig)
Solids/Liquid ratio	145g/1.0 litres	small
Retention time	90 min.	60-120
Final Free Acid	25 g/L	18-35
Additives-Lignin Sulphonate	0.1 g/L	0.05-0.2

Results

1. Ninety-seven percent of the concentrate zinc reported to the leach solution, 1% remained as unreacted sulphide, and 2% was trapped in the plumbojarosite crystal lattice. Approximately 96% of the original sulphide was oxidized to elemental sulphur, 3% to sulphate, and 1% remained unattacked. Figure 3 shows a typical zinc extraction profile in the leach autoclave.

2. Final free acid tenors could be maintained to ± 3 g/L of target (usually 20 to 25 g/L) by simple manipulation of the slurry feed cycle timer.

3. On-stream time was high, averaging over 99% during two, week-long demonstration periods.

FIGURE 3   Zn EXTRACTION vs RETENTION TIME, AND BY COMPARTMENT

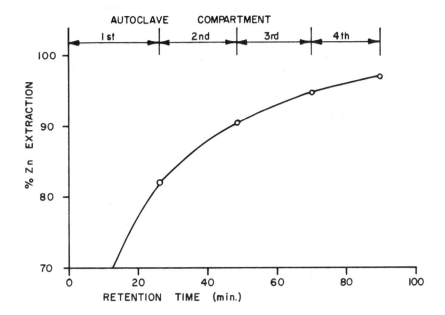

4. Following the separation of S° and the Fe-Pb residue as described in the next section, the pregnant liquor was returned to Cominco's zinc plant in Trail. Typical pregnant liquor composition is shown in Table III. Standard neutralization, purification and electrolysis tests indicated that the pregnant liquor from pressure leaching was suitable for integration with Cominco's zinc operations.

Overall, the leaching process was judged highly successful.

Table III - Typical Liquor Composition

	g/L		mg/L
Zn	120	Sb	8
Fe_t	3.5	As	15
Fe^{+++}	2.8	Sn	0.3
Mg	6.4	Se	0.8
Mn	2.3	Hg	0.002
Free H_2SO_4	25-30	Cl	48
Cu	0.25	F	3
Cd	0.21		

SULPHUR SEPARATION

Circuit and Process Description

The block flowsheet for the sulphur separation section is shown in Figure 4. The slurry from the leach autoclave consisted of an aqueous phase containing acidic zinc sulphate, a molten sulphur phase, and a solid phase, essentially a lead-iron residue (plumbojarosite) and gangue. Table IV shows typical proportions and compositions of these phases.

Two methods of separating elemental sulphur from the leach slurry were piloted. The preferred method (3) utilized a decantation autoclave, in which the heavier, molten sulphur phase was coalesced and collected as an underflow. The alternative procedure directly flashed the leach slurry to atmospheric conditions and collected the solidified elemental sulphur by froth flotation.

Table IV - Leach Autoclave Slurry Composition

		Aqueous Phase	Solid Phase	Liquid S° Phase
Wt/litre of slurry		–	52 g	41 g
Composition:	Zn	110-120 g/L	3.7 %	–
	Pb	–	17.4	–
	Fe	3	24.2	–
	S_t	–	11.8	100 %
	Free Acid	25	–	–

Sulphur Recovery by Decantation

The decant autoclave consisted of a conically bottomed, vertical, stainless steel tube, 39 cm in diameter and 3 metres high. A central shaft, driven by a variable speed motor, was used to mount a variety of agitators. This vessel operated at the same temperature and pressure as the leach autoclave i.e., 150°C, 1140 kPa. Under these conditions elemental sulphur remains a molten, immiscible liquid. · Unreacted sulphides (mainly marmatite) are preferentially wetted by molten sulphur, and therefore tend to remain in the sulphur.

Slurry from the leach autoclave was fed directly and continuously into the decant vessel. Rise velocity within the vessel was nominally 13 cm per minute. A little further zinc dissolution and iron precipitation took place in the decant autoclave. The elemental sulphur phase which accumulated in the lower, conical section was discharged into a drum at hourly intervals and the liquid sulphur was poured into molds, or allowed to freeze in the drums.

The aqueous phase, including the finely divided lead-iron residue, was continuously discharged from the top of the decant autoclave, and let down to atmospheric pressure via a choke in a modified flash tank.

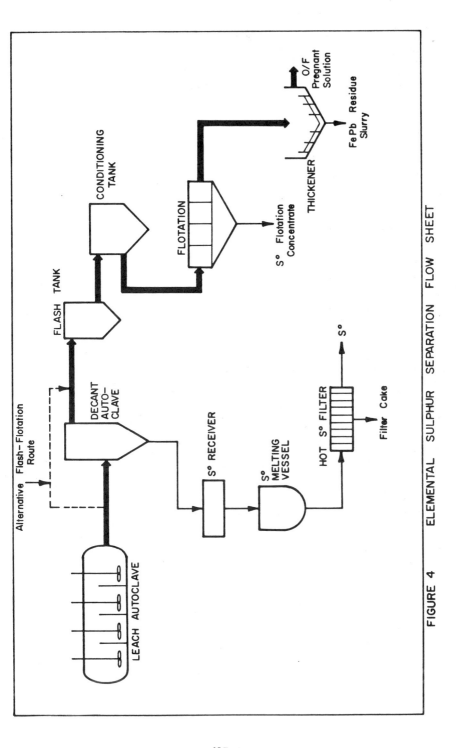

FIGURE 4   ELEMENTAL   SULPHUR   SEPARATION   FLOW   SHEET

Following let-down, the slurry was retained in a conditioning tank to allow any residual elemental sulphur to be converted from the sticky amorphous state to monoclinic crystals. Conditioned slurry was then subjected to froth flotation in a battery of four flotation cells to remove residual elemental sulphur and sulphides. The first cell was used as a rougher, and subsequent cells as scavengers, with the scavenger concentrate being returned to the rougher feed. Total residence time in flotation was about nine minutes. No reagents were required.

Finally, the flotation tailings were thickened and the zinc-enriched overflow was returned to Cominco's zinc operation in Trail.

Decantation Results and Discussion

1.   During the demonstration period, approximately 91% of the elemental sulphur formed during leaching reported to the decant autoclave underflow, yielding a product assaying 94 to 98% S°. The underflow discharge retained a small amount of aqueous phase, estimated at a maximum of 3% of the underflow during normal operation. Mild agitation in the sulphur phase was essential to effect coalescence of the sulphur droplets.

2.   Almost all of the elemental sulphur remaining in the decant overflow was collected by flotation, which produced a concentrate assaying 80 to 90% S°. This material could be recycled to a number of places in the circuit.

Typical product compositions and element deportments through the decantation-flotation process are shown in Table V.

3.   Hot sulphur filtration was carried out on decant underflow in batch campaigns. Over 98% of the elemental sulphur reported to a filtrate assaying 99.9% S°. The filter cake consisted largely of unreacted sulphides, principally zinc concentrate. Details are shown in Table VI. Polish filtration of decant underflow using diatomaceous earth as a filter medium further reduced impurities to the following level (parts per million) Zn 3, Fe 4, Pb 2, Cu 1, Te <2, Hg <1. Se was unchanged at 14 ppm.

During pilot plant start-up, numerous operating problems were encountered with the decant autoclave, principally due to underflow blockages with frozen sulphur. The most troublesome plugging occurred when a significant proportion of aqueous phase was allowed to accompany the decant underflow, causing sulphur to freeze as flashing lowered the temperature to below about 120°C. Eventually, all pipes, valves, etc. that carried elemental sulphur were steam jacketed at a controlled steam pressure, and high pressure steam purge points were installed at critical locations. Subsequently, on-stream time was nearly 100%, and sulphur decantation was demonstrated to be a viable process step.

Table V – Composition and Deportment of Solids Through Sulphur Decantation and Flotation

	Throughput		Solids Assay (%)				Distribution (%) Based On Solids in Decant Feed			
Decantation Balance	Slurry (L/min.)	Solids (kg/min.)	Zn	Fe	Pb	S°	Zn	Fe	Pb	S°
Decant Feed	14.0	1.30	2.1	13.7	9.8	43.8	100	100	100	100
Decant Underflow	0.27	0.54	1.6	0.5	0.1	95.6	32	1.5	0.4	91
Decant Overflow	13.7	0.76	2.15	23.9	16.6	7.0	60	102	99	9
Flotation Balance										
Flotation Feed	12.9	0.76	2.15	23.9	16.6	7.0	60	102	99	9
Flotation Conct.	0.17	0.063	0.6	1.7	1.5	85.0	1.4	0.6	0.7	9
Flotation Tails	12.4	0.70	2.3	25.8	17.9	0.1	59	101	98	0.1

Notes:
1. Decant overflow is flotation feed. Volume losses are due to flash discharge and evaporation during flotation.

2. Distributions do not total 100% because a little further zinc dissolution and iron precipitation occur during decantation.

3. Final solution assayed: Zn 119 g/L, Fe 3.2 g/L, free acid 26 g/L

Table VI - Filtration of Molten Sulphur from Decant Underflow

Element	Filter Feed (398.2 kg) Calculated Composition	Filter Cake (23.2 kg) Assay%	Distribution%	Filtrate (375 kg) Assay%	Distribution%
S°	95.8	29.2	1.8	99.9	98.2
S⁻⁻	0.79	13.0	96.4	0.03	3.6
Zn	1.46	25.0	99.6	0.0062(0.0003*)	0.4
Fe	0.50	8.5	98.1	0.010 (0.0004*)	1.9
Pb	0.08	1.3	97.9	0.0017(0.0002*)	2.1
Cu	0.08	1.43	99.4	0.0005(0.0001*)	0.6
Se	0.0014	0.0005	2	0.0014	98
Te	0.0014	0.0005	2	0.0014(0.0002*)	98
As	0.005	0.08	94	0.0003	6
Hg	0.0048	0.047	57	0.0022(0.0001*)	43

*Analysis when using diatomaceous earth as a filter medium.

Sulphur Recovery by Flash-Flotation

This process by-passed the decant autoclave and flashed the leach slurry directly. Subsequently, the slurry was conditioned, floated, and thickened as described under the decantation process. Originally it had been planned to remove the bulk of the elemental sulphur by screening following conditioning, but the pellets were disappointingly small, and this approach was abandoned.

Direct Flash-Flotation Results

1.   Direct flash let-down of the leached slurry produced sulphur pellets assaying approximately 87% S°. Pellet size was much smaller than had been anticipated from laboratory tests, some 30% passing through a 150 mesh screen.

2.   Flotation of flashed slurry removed virtually all of the elemental sulphur as a concentrate assaying 88% S°. Despite the fact that there was ten times as much sulphur in the flotation feed, the products were similar to those obtained from flotation of decant overflow. Average results are shown in Table VII.

Advantages of direct flash-flotation include elimination of the decant autoclave, and the ability to treat slurries containing a significant amount of unreacted sulphides, such as pyrite. However, if the flotation concentrate is to be treated to recover clean sulphur or metal values, additional equipment and energy would be required.

Table VII - Composition and Deportment of Solids in Flotation
Following Direct Flash Let-Down

	Weight g/L Solids*	Assay (%)				Distribution %			
		Zn	Fe	Pb	S°	Zn	Fe	Pb	S°
Flot. Conc.	53	1.0	2.3	1.5	88.0	34	9	8	100
Flot. Tails	46	2.2	25.8	19.2	0.07	66	90	92	0.1
Solids Feed	99	1.6	13.2	9.7	47.1	100	100	100	100

* grams of solids per litre of flotation feed slurry.

MATERIALS OF CONSTRUCTION

Prior to the design of the pilot plant, a considerable amount of test work on materials of construction had been carried out on a laboratory scale by both Sherritt and Cominco. Added to this was the in-house experience of both companies. Consequently, the selected materials of construction stood-up very well. At the start of the pilot plant campaign, metallic corrosion coupons and non-metallic materials were installed throughout the circuit for testing. Materials that behaved satisfactorily in various parts of the circuit included lead and brick linings, Alloy 20, Incoloy 825, 316 SS and FRP.

APPLICATION

Cominco's Trail Operation

When expansion plans were being formulated in 1976, Cominco's zinc production capacity was approximately 227,000 tonnes/year. This capacity was achieved with the conventional roast-leach process using two modern Lurgi roasters and an old suspension roaster. Expansion to 272,000 tonnes/year with existing technology would have required extensive modifications to the roasters, acid and fertilizer plants. Economic comparisons of expansion by pressure leaching versus roast leaching are difficult in an operation as complex as Cominco's, where the effects on zinc, lead and fertilizer plants are inter-related. Nevertheless, it was determined that expansion via pressure leaching would have a clear capital cost advantage. In addition, pressure leaching would permit a flexibility with respect to fertilizer production, and allow the old, inefficient suspension roaster to be retired.

Currently, one autoclave treating 64,000 tonnes/year concentrate (31,000 tonnes zinc) is being installed, and should be in operation in 1981. A second autoclave is planned for 1982.

The process will be almost identical to that piloted, i.e., grinding, leaching, sulphur decantation and flotation. Cleaned flotation concentrates will be combined with molten sulphur from the decant underflow and then filtered. Clean sulphur will be sold, burned for acid locally, or stockpiled, and the filter cake recycled to the zinc roasters. Integration with the existing zinc plant will occur in the acid leaching section. At this point the pressure leaching slurry (flotation tailings, consisting of acidic zinc sulphate solution and lead-iron residues) will join the calcine and return acid from the roast-leach process. Following neutralization and separation of the lead-iron residue, the pregnant solution will be purified and the zinc electrowon. The residues will be treated in the lead smelter to recover lead, zinc and precious metal values.

INTEGRATED PLANTS

Over the past year, Sherritt has had discussions with several zinc refiners who are interested in the integration of zinc pressure leaching with their existing operation. Some, such as Cominco are interested in expanding their facility, some are interested in treating lower grade zinc materials and others are interested in modernizing their facilities for better efficiency and flexibility.

"GRASS ROOTS REFINERY"

A simplified flowsheet for a "grass roots refinery" utilizing a one-stage leach is presented in Figure 5. Basically the flowsheet is similar to that piloted. Ground concentrate is treated in a pressure leach to extract at least 97% of the contained zinc. The sulphur in the zinc sulphide feed material reports to the leach residue as elemental sulphur which can be recovered or impounded with the tailings for future recovery. Typically, the zinc rich solution contains 15 to 20 g/L free acid and 3 to 5 g/L iron. During the first stage iron removal limestone is added at a controlled pH and ferric iron concentration to give a low zinc loss to the iron oxide-gypsum solids. Roasted zinc dross with the chloride removed is added to the second stage of iron removal. The solution is then processed for the normal removal of copper, cadmium and other minor components prior to electrowinning. The zinc is recovered by electrowinning and the acid solution is returned to the pressure leach.

The one-stage leach process can be modified to a two stage leach (4) for more efficient utilization of the acid as shown in Figure 6. The product solution from the first stage or neutral leach contains only 0.2 to 1.5 g/L iron and 0.2 g/L H_2SO_4 which is much lower than the 3 to 5 g/L iron and the 15 to 20 g/L free H_2SO_4 in solutions from the single stage leach. Zinc extractions of 97% to 99% are obtained with a 45 minute retention time (batch) in each leaching stage. The elemental sulphur reports to the second-stage leach residue and can be recovered or impounded with the tailings for future recovery.

An independent evaluation of the Sherritt Zinc Process, as compared to the well-known roast leach process, has shown a significant capital cost advantage in favor of the Sherritt process (5). Operating costs are essentially the same.

Extensive test work has been carried out to show that the optimum process conditions developed for high grade zinc concentrates can also be applied to bulk concentrates with good success (6).

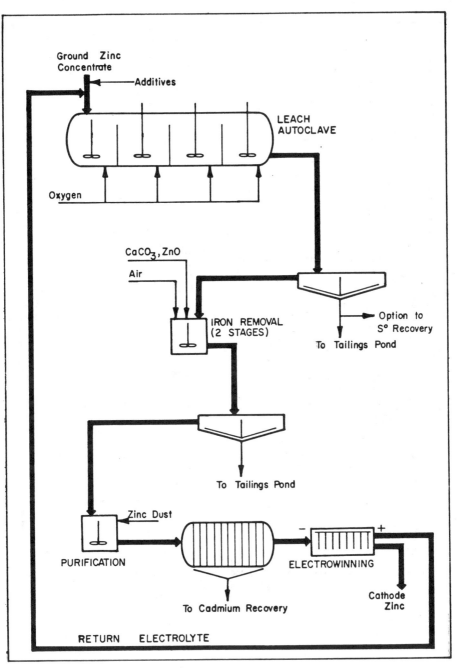

FIGURE 5   SINGLE – STAGE LEACH ZINC PROCESS

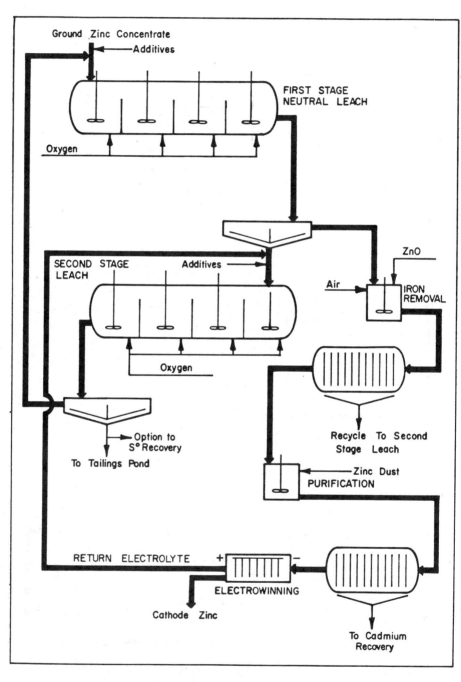

Ground Zinc Concentrate
Additives

FIRST STAGE
NEUTRAL LEACH

Oxygen

ZnO

SECOND STAGE
LEACH
Additives

Air

IRON
REMOVAL

Oxygen

Recycle To Second
Stage Leach

Option to
S° Recovery

To Tailings Pond

Zinc Dust
PURIFICATION

RETURN ELECTROLYTE

ELECTROWINNING

Cathode Zinc

To Cadmium
Recovery

FIGURE 6   TWO-STAGE LEACH ZINC PROCESS

CONCLUSIONS

The pilot plant project demonstrated that direct pressure leaching of zinc concentrates to produce zinc sulphate solution and elemental sulphur is a viable process. Two methods of separating the elemental sulphur from the main process stream were successfully demonstrated. Zinc extractions of 97% and elemental sulphur recoveries of 96% were routinely obtained. During sustained periods of steady operation considerable data were collected to facilitate commercial plant design. Cominco showed its confidence in the process by immediately using these data to design the world's first commercial zinc pressure leach plant for treating zinc concentrates . This plant, which is scheduled to commence operation by 1981, will eventually treat about 25% of the zinc concentrates processed in Trail.

Two flowsheets are presented to illustrate the operation of a grass roots zinc pressure leaching plant. Capital cost advantages over a roast-leach process are obtained because roasters, acid plants, smoke stacks, etc., are not required, and high zinc extractions remove the need for residue treatment unless lead and silver values are worth recovering. In addition, atmospheric emissions of sulphur dioxide and particulates are eliminated, and workplace hygiene greatly improved. Finally, independence from sulphuric acid production would allow zinc production to continue unimpaired regardless of acid and fertilizer markets.

ACKNOWLEDGEMENTS

The authors wish to thank the Management of Sherritt Gordon Mines Limited and Cominco Ltd. for permission to publish this paper.

Acknowledgement is also made of the financial assistance received for this work from the Government of Canada through its Enterprise Development Program, Department of Industry, Trade and Commerce.

REFERENCES

1. Forward, F. A. and Veltman, H., "Direct Leaching Zinc Sulphide Concentrates by Sherritt Gordon," Journal of Metals, 12, (1959), pp. 836-840.

2. Kawulka, P., Haffenden, W. J. and Mackiw, V.N., "Recovery of Zinc from Zinc Sulphides by Direct Pressure Leaching," U.S. Patent 3,867,268, Feb. 18, 1975.

3. Hirsch, H.E., Higginson, J. F., Parker, E.G., Swinkels, G.M., "Leaching of Metal Sulphides," U.S. and other patent applications pending.

4. Veltman, H., Mould, G. J. J. and Kawulka, P., "Two-Stage Pressure Leaching Process for Zinc and Iron Bearing Mineral Sulphides," U.S. Patent 4,004,991, Jan. 25, 1977.

5. Doyle, B.N., Masters, I. M., Webster, I. C. and Veltman, H., "Acid Pressure Leaching Zinc Concentrates Produces By-Product Elemental Sulphur," paper presented at The Eleventh Commonwealth Mining and Metallurgical Congress, Hong Kong, 6-12 May, 1978.

6. Bolton, G. L., Zubryckyj, N., Veltman, H., "Pressure Leaching Process for Complex Zinc-Lead Concentrates," paper presented at the Thirteenth International Mineral Processing Congress, Warsaw, Poland, June 1979.

HYDROMETALLURGICAL TREATMENT OF LEAD CONCENTRATES*

J.M. Demarthe
& A. Georgeaux
MINEMET RECHERCHE, Trappes, FRANCE

MINEMET RECHERCHE have devised a new hydrometallurgical process for the electrolytic production of lead from galena.

The involved operations are :
a) solubilization of lead in a brine by means of an oxidizing agent (Fe^{+++})
b) purification of the resultant solution
c) direct electrolysis of the clear solution, which allows the deposit of lead powder with simultaneous leach regeneration.

This process, which has been developed on the laboratory and pilot scales, enables the exclusively hydrometallurgical recovery of lead with a very high degree of purity and a 99 % yield. The minor elements included in the galena (Ag, Zn, Cu) are similarly reclaimed and beneficiated.

* This work was carried out by MINEMET RECHERCHE and supported by PENARROYA, both companies jointly belonging to the IMETAL Group.

I. INTRODUCTION

Despite the weak but steady growth (2 to 2.5 %) of the lead market over the past 20 years, new investments have remained scarce in the field of lead metallurgy.

There is a steady fall-off in the availability of high-grade lead concentrates and the smelters are accordingly bound to rely to a greater extent on products originating from complex ores. In addition to their low lead content, the latter are contaminated by other metals (Fe, Bi, Sn, Sb, As) and rich in sulphur, which makes them ill-adapted to treatment facililites that have been conceived for diverse ores in other times.

Thirdly, the increasing apprehension of public opinion as well as state authorities regarding the environmental effects of lead processing should not be underestimated. In a number of industrialized countries, stringent rules are about to be applied, especially to monitor the amount of lead contained in fumes and surroundings.

These various constraints incited PENARROYA to evolve a new process. The four-year investigation that was performed in the MINEMET RECHERCHE laboratories resulted in the development of a wholly hydrometallurgical process capable of directly producing commercial lead. The process is very flexible as to the nature of the input product. Part of the minor elements associated with the galena are reclaimed and no gases are released.

II. DESCRIPTION OF THE PROCESS

The new hydrometallurgical process evolved by MINEMET RECHERCHE
to recover metallic lead from galena comprises the following successive
stages:

Dissolution of Lead in a Brine of Sodium Chloride (or another alkaline/alkaline-earth chloride) by Means of Ferric Chloride

The following reaction occurs :

$$2 FeCl_3 + PbS \longrightarrow 2 FeCl_2 + PbCl_2 + S \qquad (1)$$

The high chloride concentration in the leach medium is aimed at increasing
the solubility of lead chloride. In the course of the leach, a number of
metallic sulphides associated with lead, such as Zn, Cu, Ag, Bi, are
likely to pass into solution. These minerals also react with ferric
chloride and dissolve :

$$2 FeCl_3 + Zn S \longrightarrow 2 FeCl_2 + ZnCl_2 + S \qquad (2)$$
$$4 FeCl_3 + CuFeS_2 \longrightarrow 5 FeCl_2 + CuCl_2 + 2 S \qquad (3)$$
$$2 FeCl_3 + Ag_2S \longrightarrow 8 FeCl_2 + 2 AgCl + S \qquad (4)$$
$$6 FeCl_3 + Bi_2S_3 \longrightarrow 6 FeCl_2 + 2 BiCl_3 + 3 S \qquad (5)$$

There appears to be no concomitance in the occurrence of these reactions.
A certain selectivity in the dissolution of lead is observed. However,
the direct production by selective leaching of a solution free of impurities proved impractical, even by lowering the recovery yield of lead.

Coarse Purification

After leaching it is essential to withdraw from the solution the metals nobler than lead (Cu, Bi, Ag etc.). To this effect a two-stage purification is undertaken; the first stage consists in a cementation with electrolytic lead powder :

$$2 \text{ CuCl} + \text{Pb} \longrightarrow \text{PbCl}_2 + 2 \text{ Cu} \quad (6)$$
$$2 \text{ AgCl} + \text{Pb} \longrightarrow \text{PbCl}_2 + 2 \text{ Ag} \quad (7)$$
$$2 \text{ BiCl}_3 + 3 \text{ Pb} \longrightarrow 3 \text{ PbCl}_2 + 2 \text{ Bi} \quad (8)$$

The impurities in question are thus removed to a great extent, yet it derived from experience that the resultant solution is not suited for the direct electrowinning of commercial lead.

Fine Purification

This stage is necessary to produce lead metal via a fully hydro-metallurgical process. The solution resulting from cementation is treated by an ion-exchange resin that quantitatively fixes the copper and silver still remaining in the liquor.

The applied resin -designated IMACTI GT73- is inactive with respect to lead and has the particular feature of inducing quite high rate distributions relevant to the Cu^+ and Ag^+ ions, in the medium under consideration. Once the solution has flowed through a column lined with such a resin, the concentrations of both elements are brought down to values below the conventional threshold for analytical detection.

Electrowinning of Lead Powder from Clear Lead
Solution and Coincident Regeneration of Leach Reagent

This operation is performed in a cell of special design. The cathodic and anodic compartments are separated by a porous diaphragm. The pure lead chloride solution is fed into the cathodic compartment, where the lead collects in the form of a powder that slightly adheres to the supporting cathode.

The solution wherefrom part of the lead has been removed then enters
the anodic compartment in which the oxidation of ferrous to ferric
ions takes place according to the equations :

$$\text{Cathode} : Pb^{++} + 2\ e \longrightarrow Pb \qquad (9)$$
$$\text{Anode} \quad : 2\ Fe^{++} - 2\ e \longrightarrow 2\ Fe^{+++} \quad (10)$$

A parasitic reaction occurs in that chlorine is released at the anode:

$$2\ Cl^- - 2\ e \longrightarrow Cl_2 \qquad (11)$$

This unwanted chlorine emission is avoided by maintaining in the
solution about 10 g excess of ferrous irons over the theoretical
quantity likely to be oxidized by the anodic reaction. The base of
the cell comprises a special device for the recovery of lead powder.
The latter is subsequently melted in the presence of soda and cast into
ingots. The use of soda is intended to protect the metal from air
oxidation and to remove the possibly included traces of arsenic and
antimony.

Zinc Removal

Zinc is the major solubilized element associated with lead.
Its presence in the solution does not pollute the electrolytic lead
powder but on increasing, it lowers the solubility of lead and hence
a bleed is required.

Depending on the importance of the latter, three solutions may
be envisaged :

a) take part of the catholyte with a low lead content and subject
it to a two-stage treatment by means of sodium sulphide. At first the
reagent is added until a pH value between 2 and 3.5 is achieved, which
causes a total and selective precipitation of lead sulphide.
The filtered resultant solution is next exposed to a second stage using
the same reagent up to a pH value ranging from 4.5 to 5. Zinc sulphide
precipitates selectively with respect to the ferrous ions remaining in
solution. Then the lead sulphide is recirculated to leaching, whereas
the zinc sulphide is marketed.

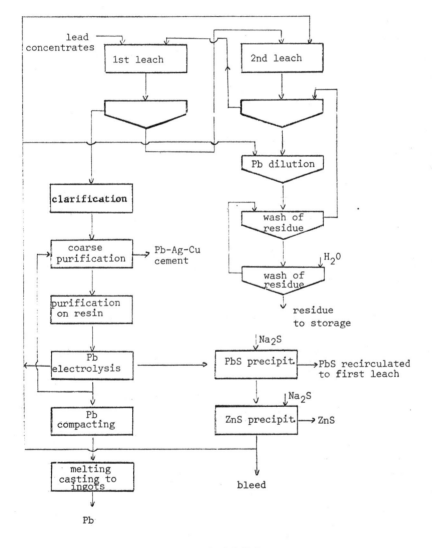

F L O W S H E E T

<u>Hydrometallurgical process for the production of electrolytic lead</u>

431

b) Take advantage of the anionic complex condition of zinc in the solutions. The $ZnCl_4^{--}$ complex may be fixed on an anionic exchanger, i.e. a secondary amine as a resin or solvent. The elution with water of the loaded exchanger produces a pure zinc chloride solution wherefrom zinc hydroxide may be precipitated by means of an alkaline agent.

c) Proceed as for b) with this difference that a neutral solvent such as tributyl phosphate is used to extract zinc chloride.

The solutions issuing from the zinc recovery stage are subsequently blended with the anolyte and the resultant stream is directed to leaching.

This unique process as outlined above is depicted on the attached flow-sheet.

Possible Alternatives

Under some particular local circumstances, slight modifications to the process will permit to bring down the investment costs that are required for the hydrometallurgical unit.

a) The lead leaching stage may be limited to a recovery yield between 80 and 90 %. In this case, the quantity of zinc and other minor elements that will be solubilized is very low. Therefore the solution can be directly treated with an ion exchange resin, which obviates the necessity of cementation.

The leach residue is then suited for a pyrometallurgical treatment; such as for instance the Imperial Smelting Process, with high recovery rates for zinc and silver.

b) It is also possible, after selective leaching or a complete leach followed by cementation, to avoid the operation of fine purification. One proceeds then directly with electrolysis and the resultant lead powder is subjected to a pyrometallurgical purification by melting and zinc addition to remove the residual traces of copper and silver.

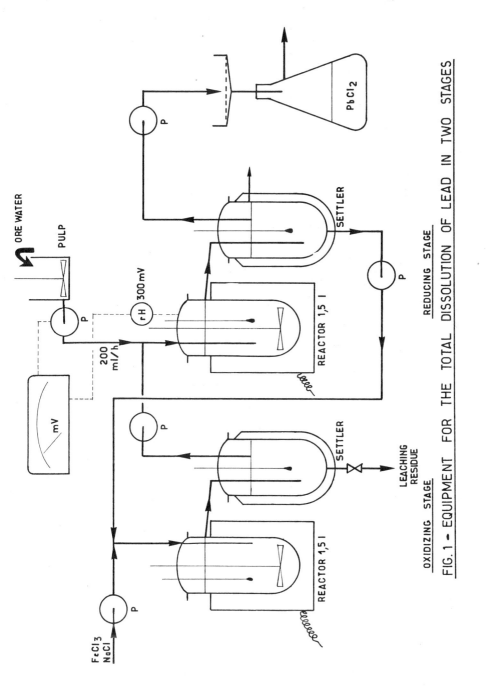

OXIDIZING STAGE   REDUCING STAGE

FIG. 1 – EQUIPMENT FOR THE TOTAL DISSOLUTION OF LEAD IN TWO STAGES

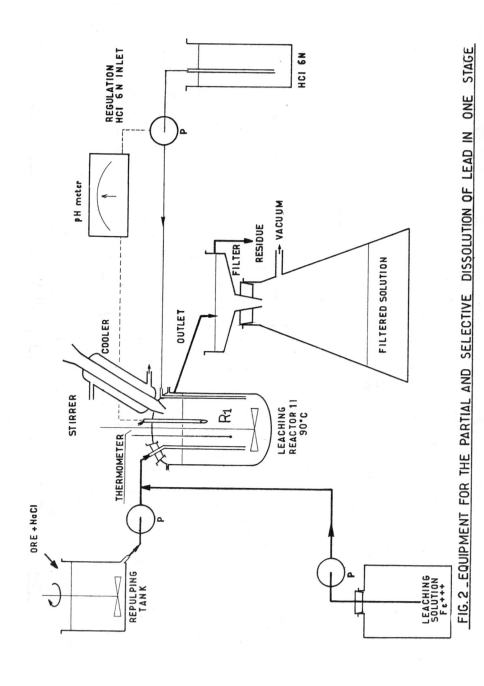

FIG. 2 _ EQUIPMENT FOR THE PARTIAL AND SELECTIVE DISSOLUTION OF LEAD IN ONE STAGE

III. LABORATORY AND PILOT SCALE TESTS

Laboratory Tests

a) Equipment : The whole process has been experimented on the laboratory scale using the equipment described in the attached diagrams :

i) Figure 1
Complete dissolution of lead and most of the associated elements in two stages, each comprising a leaching tank and a decanter, fitted so to ensure a counter-current circulation of leachant and ore between the two stages.
Volume of reactor : 1.5 litre.

ii) Figure 2
One-stage selective but partial lead leaching.

iii) Figure 3
Equipment for the thorough purification of the solution, comprising:
. a cementation reactor of special design, at the bottom of which a turbulent zone is created where lead powder is kept suspended in the solution to be purified, whereas the solution is clarified in a quiet zone before being injected at the base of the resin bearing ion exchange column.
. an ion exhange column containing 85 cm^3 of resin.

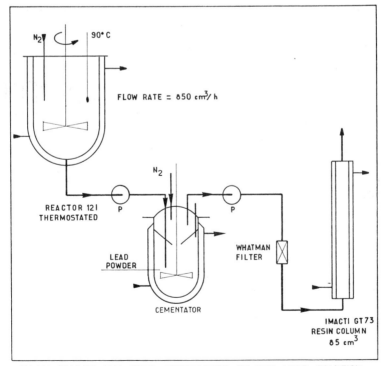

FIG. 3   COARSE AND FINE   PURIFICATION   OF THE LEAD SOLUTION

435

iv) Figure 4

Electrolysis cell, with two compartments separated by a diaphragm designated 7330 D manufactured by Fyltis. The anode is made of graphite and the cathode of titanium. A pump makes the catholyte circulate between the storage vessel and the cathodic compartment. This operation affects the thickness of the diffusion layer in the vicinity of the cathode and improves the quality of the deposit. The lead powder that is produced in contact with the cathode is collected in a storage vessel fitted at the bottom of the electrolysis cell.

The spent catholyte passes by **gravity** through the diaphragm and reaches the anodic compartment where the ferrous chloride is oxidized into ferric chloride. The oxidized solution then leaves the cell.

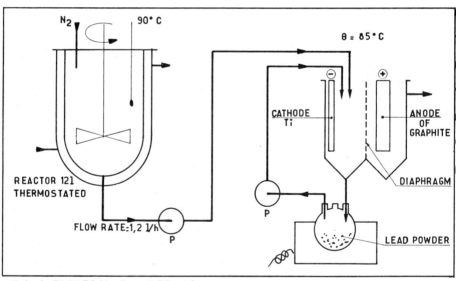

FIG.4_ELECTROLYTIC FITTINGS

b) <u>Results</u> : The results attained on treating two lead **concentrates** of different nature and subjected to equally different processes are outlined below :

The first concentrate is a galena with few impurities supplied by St. Joe Lead Company. The second product is a complex lead/zinc/copper ore originating from Portugal and subjected to selective lead recovery. The analysis of both products is:

	Pb %	Zn %	Cu %	Fe %	Ag ppm	S %
St. Joe galena	71.1	1.34	1.07	3.32	41	15.4
Pb/Zn/Cu rough concentrate from complex ores	3.2	9.5	2.5	33	0.07	43.9

The first concentrate underwent a thorough leaching process **using** the **equipment** illustrated in Figure 1, whereas the second was selectively leached in one stage in the **equipment in Figure 2.**
The **resultant** leach solutions were then successively subjected to cementation, treatment on resin (Fig. 3) and electrolysis (Fig. 4).

The results are :

	Dissolution yields in %			
	Pb	Cu	Ag	Zn
St. Joe concentrate	97.2	13	63.4	44.8
Portuguese concentrate	92.3	0.3	19.1	2.77

Both solutions were purified in two stages :

		Pb g/l	Cu mg/1	Ag mg/1	Bi mg/1
St. Joe concentrate	Leach solution	27.6	60	0.9	not detected
	Purified solution	"	< 0.2	< 0.1	"
Portuguese concentrate	Leach solution	25.2	38	7	47
	Purified solution	"	< 0.2	< 0.1	< 15

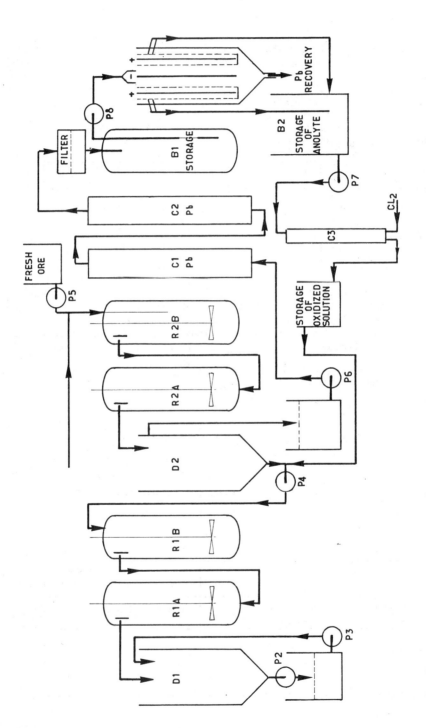

FIG.5_SMALL PILOT PLANT FOR THE PRODUCTION OF ELECTROLYTIC LEAD POWDER FROM A SULPHIDE ORE

The electrolysis of the clear solutions allowed for the recovery of lead powder with the following analysis:

ppm \ Product	Zn	Cu	Ag	Bi	As	Sb	Sn
Lead powder from St. Joe concentrate	2	11	< 5	< 5	20	20	
Portuguese concentrate lead powder	10	< 10	< 10	< 10	40	100	< 10

Small Pilot Tests

a) Equipment : The assembly as per Figure 5 is set up using :

- A counter-current leach system, including a reducing stage with the R2A and R2B reactors and the D2 settler. The ore concentrate is introduced into the R2B reactor together with a partly reduced leach solution and the mixture flows successively through R2A, R2B and D2. The overflow of the D2 settler is filtered and directed to the subsequent purification stage. The underflow of D2 is combined with a fresh ferric chloride solution and sent to the oxydizing stage. The solution and solid pass in succession through R1B, R1A and D1. The overflow of the D1 settler is reintroduced in the R2B reactor and the D1 underflow constitutes the final leach residue.
- 2 columns loaded with lead turnings designed to cement the impurities contained in the solution.
- An electrolysis cell comprising 1 cathode and 2 anodes, which receives the clear solution issuing from cementation.
- A chlorine oxidation column to offset the slight shortage of oxidizing agent due to parasitic ferric chloride consumption by other elements than lead.

This installation has been operated for 300 hours with the parameters listed below.

439

```
Volume of R1A, R1B, R2A and R2B      20 litres
Total flux of solution               40 L/h
NaCl concentration                   250 g/L
Fe++ excess in leach solution        10 g/L
Ore introduced per hour              1162 g/h
Electrolysis current density         600 A/m²
Volume of one lead column            5 litres
```

b) <u>Results</u> : The materials flowing through the installation during the pilot trial showed the following chemical analysis :

	Pb g/l or %	Zn g/l or %	Cu g/l or %	Fe g/l or %	Ag g/l or %	S g/l or %	Bi g/l or %	As g/l or %	Sb g/l or %	Cl⁻ g/l or %
Initial Feed	65.5	2.8	0.59	7.5	0.13	19.6	0.098	0.46	0.18	
Final residue	2.60	3.96	1.65	22.5	0.060	58.5	0.014	1.25	0.23	<0.05
D 1 solution	25.3	8.79	0.011	25.4	0.005		0.012	0.004	0.028	
D 2 solution	39.7	8.54	0.016	24.1	0.025		0.018	0.006	0.042	
Purified solution	39.7	8.70	0.003	25.1	<0.002		0.009	0.002	0.010	
Electrolytic Pb powder	97.0	0.0012	0.013	0.023	0.0061		0.0077	0.0045	0.0082	0.37
Cu – Ag cement	82.0	0.19	1.45	0.56	5.4	1.3	1.9	0.59	2.04	0.3

It is observed that the obtained lead powder contains a certain number of impurities as no purification on resin was carried out.

The following percentages of elements were recovered from the initial feed:

Pb	Zn	Cu	Fe	Ag	Bi	As	Sb
98.7	52.6	6.3	< 1	84.5	95.2	8.9	57.1

Pilot Plant Trial

After the small pilot experiments described above, a pilot plant
operation was conducted, which included :
- dissolution of lead by ferric chloride
- cementation
- electrolysis.
Figures 6 and 6A depict the assembly of the pilot installation:
- a leach reactor of 1 m^3 followed by a settler
- a cementation reactor which is similar in layout to that used for
  the laboratory tests comprising two compartments (see Fig. 3),
  with a total capacity of 1.5 m^3
- an electrolysis cell, with electrodes of 0.8 m^2 and a cone-shaped
  bottom, completed by a set of sieves for the continuous recovery of
  lead powder
- a furnace designed to melt the lead powder.

The plant was run continuously for two periods of three and two
months respectively and the achieved results enabled :
(i) corroboration of the chemical results as to the dissolution yields
     of the elements, the effectiveness of purification by cementation,
     and the quality of the electrolytic lead powder obtained.
(ii) the development of a specific technology for the electrolytic for-
     mation of lead powder.

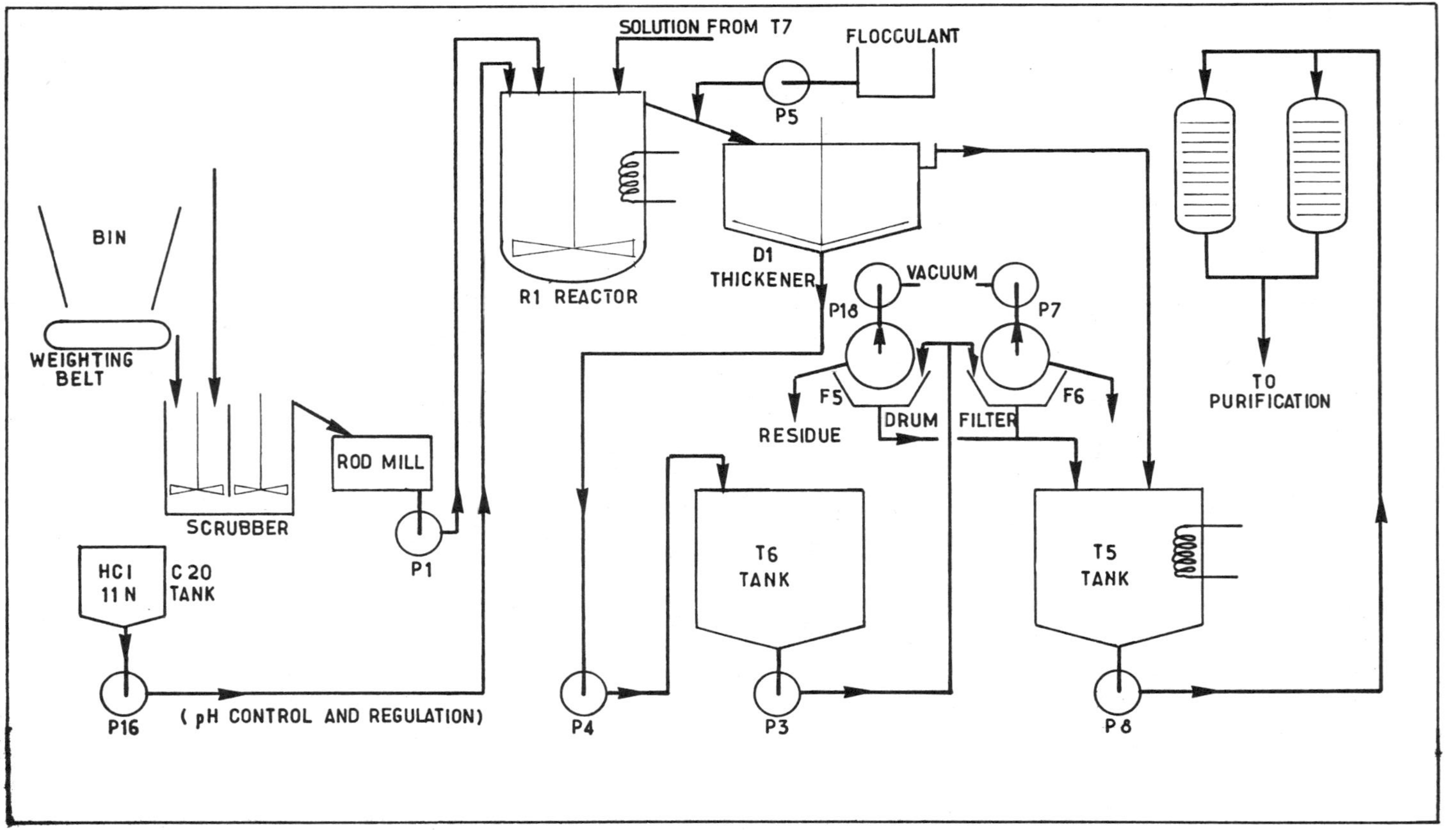

FIG. 6 _ PILOT SCALE TREATMENT OF A LEAD CONCENTRATE: LEACHING, FILTRATION

442

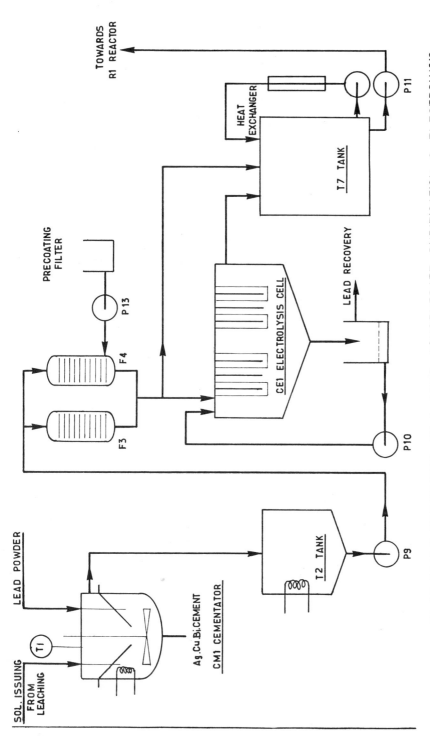

FIG. 6A - PILOT SCALE TREATMENT OF A LEAD CONCENTRATE: PURIFICATION & ELECTROLYSIS

443

IV. CONCLUSIONS

The previously described process combines the advantages listed below.

1) A narrow range of operating costs

The overall cost elements boil down to the following:

- Energy : (i) current supply for the electrolysis and various apparatus (stirrer, pumps, etc.),
  (ii) the necessary heat to maintain all solutions at 90° C. This moderate caloric consumption ought to be stressed, considering that the process is wholly isothermal.
- The reagents used to treat the zinc bleed : Na_2S, resin or solvent as the case may be.
  It is noteworthy to state that the expenditure on account of zinc reclamation is offset by the sale of the produced concentrate.
- The recovery of copper and silver is effected by means of the recirculated electrolytic lead powder. Thus the electric energy circumstantially acts as a reagent. Other expenses at this stage pertain to the replacement of the ion-exchange resin.
- The amount of soda consumed at the lead melting stage is also very low, since this reagent can be recycled.

2) A conventional technology

All operations, except for the melting and casting of lead ingots, are conducted at a temperature below the boiling point and under atmospheric air pressure.

3) Limited investment costs

Owing to the low solubility of lead chloride in the medium under consideration, considerable solution flow rates are required. Yet, allowing for the elementary technology, the price of the leaching installation remains reasonable; the cost of the electrolytic equipment is of the usual order of magnitude whenever diaphragms are involved. The fittings are of small size, considering the high current density (400 to 600 A/m^2) and the great molecular weight of lead.

4) Conformity with the legal provisions for industrial environments

Since the system dispenses with roasting, neither lead-polluted nor sulfurous gases are released.

FERRIC CHLORIDE LEACH-ELECTROLYSIS

PROCESS FOR PRODUCTION OF LEAD

M. M. Wong, F. P. Haver, and R. G. Sandberg
U. S. Bureau of Mines
Reno, Nevada

The U. S. Department of the Interior, Bureau of Mines, under a cooperative research agreement with lead producers, is conducting a feasibility study of a process to minimize pollution caused by sulfur oxide and lead. The process involves ferric chloride leaching of galena to produce lead chloride and fused-salt electrolysis of the lead chloride to produce lead and chlorine. The chlorine produced by electrolysis is used directly to regenerate ferric chloride in the leach solution.

The study is being conducted in a process development unit designed to produce about 500 pounds per day of lead metal, operating on an integrated basis. This report describes the equipment and operating procedures, and presents a preliminary evaluation of the results with respect to process performance, equipment corrosion, impurity buildup, and air monitoring.

Introduction

Smelting provides a low-cost, energy-efficient method for producing lead metal from sulfide ores. Unfortunately, it prodces sulfur oxide and lead emissions. As part of its mission to minimize undesirable environmental impacts and workplace health hazards that are caused by processing minerals and metals, the Bureau of Mines is investigating an alternative method for the production of lead (1). This method consists of leaching galena concentrate with a ferric chloride-sodium chloride solution to produce lead chloride (2), electrolyzing the lead chloride in a fused-salt bath to produce lead metal and chlorine (3), and using the chlorine to regenerate ferric chloride in the leach solution:

$$PbS + 2FeCl_3 \xrightarrow{\text{NaCl}} PbCl_2 + 2FeCl_2 + S^0$$

$$PbCl_2 \xrightarrow{\text{Electrolysis}} Pb + Cl_2,$$

and
$$2FeCl_2 + Cl_2 \rightarrow 2FeCl_3.$$

As early as 1923, ferric chloride leaching of galena was investigated by Christiansen (4); later by Agrecheva et al. (5,6); and recently by Cottam et al. (7); Baker et al. (8); and Milner et al (9). Similarly, fused-salt electrolysis of lead chloride was studied by Ashcroft in 1925

(10) and later by Starliper et al. (11). The process as a whole, however, has never been developed to the industrial production stage, probably for economic and material corrosion reasons. As promulgation of environmental protection regulations adds to the smelter cost for pollution control and new corrosion-resistant materials become available, a reexamination of the process appears warranted.

The Bureau's laboratory-scale investigations (1,2,3) and preliminary cost evaluations (12) indicate that this process appears to offer a promising alternative to smelting in terms of meeting environmental and health protection regulations and helping small producers to overcome the increasing cost of sending their concentrates to custom smelters (13). In July 1978 four lead producers joined the Bureau of Mines in a cost-sharing program to study this process further on an expanded scale. (The authors gratefully acknowledge the financial support from St. Joe Minerals Corp., ASARCO, AMAX, and COMINCO, and also the technical advice from their representatives during periodic meetings.)

The objective of the present study is to obtain information on integrated operation of the process, corrosion of materials of construction, impurity buildup in the system, long-term electrolytic cell operation, identification of wastes generated, and monitoring lead levels of the workplace and of the workers. A flow diagram for the integrated operation is shown in Figure 1. The unit is sized to produce 500 lb of lead metal per day. The investigation is not yet complete; this paper reports only the progress of the project as of March 1979.

Equipment and Operating Procedure

A batch of 125 lb of lead concentrate (120 lb on dry basis) is leached at about 95° C in 300 gallons of solution, which was initially made up to contain 73 g/l $FeCl_3$, 254 g/l NaCl, and enough HCl to give a pH of about 0.3. The leaching vessel is a 400-gallon steel tank lined with Kynar (polyvinyldene fluoride) and equipped with a titanium heating coil and a titanium stirrer. The concentrate is added to the tank from a feed hopper by a screw feeder. After a 15-minute leach, the slurry is filtered in a polypropylene plate and frame filter press dressed with a polypropylene felt cloth. The feed pump is a spring-activated, fluorinated rubber-lined Oliver diaphragm. Air is blown through the filter cake following each filtration. The residues accumulated after three to four batches are washed with a steam-water mixture before they are removed from the filter press. About 20 to 30 gallons of water at the rate of 1 gal/min are used to wash each load of residue.

The hot pregnant solution from the filter is transferred through a chlorinated polyvinyl chloride (CPVC) pipe to a crystallizer where it is cooled to about 20° C to precipitate the lead chloride. The crystallizer is a 400-gallon polypropylene-lined steel tank equipped with a titanium cooling coil and a titanium stirrer. The slurry from the crystallizer is filtered through a polypropylene cloth in a fiber reinforced plastic (FRP) vacuum pan filter. Twenty gallons of water are used to wash each batch of lead chloride.

The lead chloride is loaded in trays and dried in an electric oven at 150° C for 6 hours. The dry lead chloride is crushed and then placed in a portable, closed feed hopper inside a fume hood that has an air velocity of 125 ft/min at its face.

446

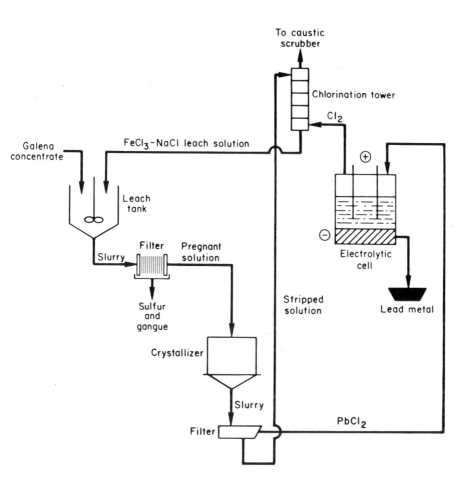

FIGURE 1. - Flow Diagram of Ferric Chloride Leach/Fused-Salt
Electrolysis Process.

The barren solution from the pan filter is transferred by a polypropylene centrifugal pump to a 400-gallon polypropylene-lined steel
tank (prep tank). The prep tank is equipped with a titanium heating coil
and a titanium stirrer. The line connecting the crystallizer and the prep
tank is made of CPVC. When the solution in the prep tank has reached
70° C, it is discharged into a 600-gallon polypropylene-lined steel tank
(surge tank). The surge tank is equipped with a titanium heating coil and
a titanium stirrer. Some water, besides the wash water, is added at this
point to the circuit to compensate for the water losses such as entrain-
ments in the leach residue and lead chloride and vapors from the tanks
and chlorination tower.

The solution in the surge tank is maintained at 75° C; a stream from
this tank, at about 8 gal/min, is constantly circulated through a
chlorination tower to convert ferrous chloride to ferric chloride. Four
hundred and fifty gallons of solution are kept in the surge tank to

provide at least a 150-gallon heal for chlorination when 300 gallons of the regenerated leach solution are transferred to the leach tank before another batch of barren solution is treated.

The dry lead chloride is added to the electrolytic cell from a closed feed hopper by a screw feeder. Figure 2 is a diagram of the electrolytic cell. The cell exterior dimensions are 53" long, 44" wide, and 30" high; the cavity is 34" long, 25" wide, and 18" deep with the inside walls constructed of silica bricks. Two graphite plates--30" long, 24" wide, and 3" thick, are separated by 3/4" silicon oxynitride spacers to give a constant electrode gap. Two 6" diameter graphite rods are threaded into the top graphite plate, which is used as the anode. The bottom graphite plate serving as the cathode is supported by four graphite blocks partially immersed in a pool of molten lead metal. A steel bar through the bottom of the cell connects the lead metal electrically to a 5,000-ampere rectifier. The respective surfaces of the anode and cathode plates are grooved 3/8" deep and 1/4" wide to guide the flow of chlorine and lead metal to opposite sides of the cell.

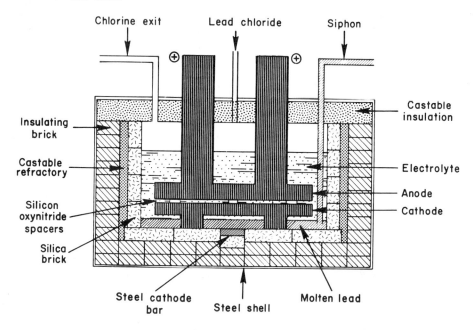

FIGURE 2. - Electrolytic Cell.

The initial bath was composed of 834 lb of salt mixture with a composition of 25 wt-pct LiCl, 32 wt-pct KCl, and 43 wt-pct $PbCl_2$. Heat from electrolysis maintains the electrolyte temperature at about 450° C and the lead metal temperature at about 400° C. The lead metal produced is periodically siphoned from the cell and discharged into a mold in a vacuum chamber.

Chlorine produced in the cell is drawn through an FRP pipe to the bottom of the chlorination tower where the spent leach solution is fed to the top, converting ferrous chloride to ferric chloride. The chlorinator is a 12" diameter, 10' long FRP tower loaded with 5' of 1" saddles. Chlorine absorption is usually over 98 pct. To ensure that the exhaust is free of chlorine, the gas from the chlorinator is scrubbed with a caustic solution in a second tower identical to the chlorination tower in construction.

All tanks in the leach section are covered to prevent vapors from escaping. A separate water scrubbing tower is used to collect fumes from tanks in the leach section.

Results and Discussion

An overall view of the process development unit (PDU) is shown in Figure 3. The leach section occupies two levels on the right, and the electrolytic cell is shown at the lower left. As of the end of March 1979, 182 leaching cycles have been completed. Presently, there are insufficient data to make a complete evaluation of the process, but some trends in the operation have become apparent.

FIGURE 3. - Overall View of Process Development Unit.

The major impurities in the leach liquor (shown in Table I) are copper, zinc, and sulfate. Copper and zinc come from chalcopyrite and sphalerite in the concentrate, and sulfate comes from dissolving $PbSO_4$ formed by oxidation of PbS during storage of the lead concentrate. Magnesium, cadmium, and silver also showed a moderate increase. Calcium and iron did not show any trend of buildup. The lead content varied with the sodium content as would be expected from the solubility of PbCl$_2$ in a NaCl solution. Substantial portions of the silver and cadmium in the concentrate report to the leach liquor.

TABLE I. - Analysis of Leach Liquor, g/l

Cycle	Ca	Cu	Mg	Na	Pb	Fe	Zn	Ag	Cd	SO$_4$
0	0.59	0.3	0.2	100	4.2	25	0.3	tr	tr	1.6
19	1.60	2.6	.9	96	14.0	32	3.8	.01	.03	8.9
37	1.30	7.2	1.6	91	11.0	26	2.6	.02	.05	14.5
64	.59	13.0	2.5	86	10.0	31	6.0	.04	.07	19.0
81[1]	.76	13.0	2.3	66	6.2	27	5.4	.03	-	16.7
98	.72	10.0	2.2	90	9.1	19	4.5	.03	-	16.2
119	.53	13.0	2.5	110	14.0	21	7.2	.05	.12	21.3
135	.40	15.0	3.0	110	15.0	24	9.3	.06	.15	26.5
161	.53	19.0	3.4	110	17.0	21	12.0	.06	.19	25.6
182	.57	18.0	4.3	110	12.0	19	12.0	(2)	(2)	27.4

[1]A 300-gal bleed was made and NaCl concentration adjusted thereafter.
[2]Analytical results of cycle 178: Ag, 0.04 g/l and Cd, 0.21 g/l.

Table II shows the analysis of the leach residue along with the analysis of lead concentrate for comparison. About 30 lb (dry basis) of residue are produced from leaching 120 lb (dry basis) of concentrate. The residue contains about 50 pct elemental sulfur. The remainder of the total sulfur is present as insoluble sulfates and unreacted sulfides of iron, copper, zinc, and lead. Most of the arsenic was probably present in the residue as insoluble ferric arsenate.

TABLE II. - Analysis of leach residue,[1] pct

Cycle	Cu	Pb	Fe	Zn	Ag	As	Cd	Sb	SO$_4$	Tot. S
Conc.[2] residue	1.5	73.5	3.7	1.6	0.006	0.01	0.02	0.002	2.4	16.7
1-4	3.0	5.6	13	2.7	.010	.05	.04	.01		68.7
18-21	3.5	4.6	15	2.4	.007	.08	.03	.03		57.6
37-40	2.2	6.6	13	2.1	.007	.04	.03	.02		-
58-61	1.2	8.0	14	2.6	-	-	.04	-	4.3	61.0
80-82	1.6	5.2	14	3.1	.006	-	.04	-		-
99-101[3]	3.3	4.2	15	1.8	.006	-	.03	-		-
117-119	2.0	2.5	12	5.0	.009	-	.08	-		-
132-134	1.1	6.2	9	1.6	.005	-	.02	-		-
159-164	1.8	7.6	10	2.5	.005	-	-	-	5.7	59.0
176-181	1.7	7.3	-	2.4	.005	-	.04	-	6.5	59 8

[1]Sample taken from filter press and rewashed in laboratory.
[2]Also Ca, 0.35 and Mg, 0.20.
[3]After a 300-gal leach liquor bleed.

Metal extractions, based on the analysis of leach residue, are shown in Table III. The lead extraction was about 98 pct. Extractions of copper, zinc, silver, and cadmium are higher than those obtained in our earlier bench-scale work (2) in which better control of leaching time was possible. Ferric chloride leaching is selective (5), and over 99 pct lead extraction can be obtained in 15 minutes (2). Additional leaching time increases solubilization of chalcopyrite and sphalerite. Because leaching and filtration are done batchwise in the PDU operation, it is difficult to control leaching time precisely. Generally, the actual contact time between solution and concentrate is longer than the prescribed leaching time, thereby increasing the extraction of Cu, Zn, and Cd.

450

Table III. - Extraction,[1] pct

Cycle	Cu	Pb	Zn	Ag	Cd
1-4	50	98	59	48	54
18-21	32	98	57	63	60
37-40	68	98	71	73	67
58-61	80	97	59	-	54
80-82	64	98	34	59	23
99-101[2]	44	99	71	67	68
117-119	70	99	30	62	14
132-134	80	98	73	74	68
159-164	73	98	65	80	-
176-181	71	98	61	76	60

[1]Based on leach residue.
[2]After a 300-gal leach liquor bleed.

The major impurities in the lead chloride (Table IV) are Na, Fe, Cu, Zn, and SO_4. These impurities, possibly with the exception of copper, do not show a consistent buildup in the leach liquor. This lack of a consistent trend could indicate, at least in part, the variability in washing the leach chloride from batch to batch.

TABLE IV. - Analysis of lead chloride,[1] ppm

Cycle	Cu	Fe	Na	Zn	SO_4
1-2	<20	81	79	<20	-
17-20	<20	120	180	60	-
37-41	<20	80	390	55	6,900
63-66	<20	60	110	410	3,200
79-81	<20	130	120	<20	2,800
98-99[2]	<20	71	210	<20	310
117-119	<20	38	150	24	470
136-140	24	94	1,400	31	800
154-160	63	92	370	34	800
176-177	43	41	190	45	1,000

[1]Also Ag, <10; and Ca, <60.
[2]After a 300-gal leach liquor bleed.

Table V shows the overall operating data of the electrolytic cell since its startup in November 1978. Operation at about 3,000 amperes maintains the cell at the desired temperatures of about 450° C for the electrolyte and about 400° C for the lead metal. When the cell is not running, as during the period of intermittent operations, the electrolyte and the lead metal are kept molten by passing alternating current between two graphite electrodes. The lower cell voltage of 4.3 at a higher current of 3,030 amperes and the lower energy requirement of 0.53 kWh/lb of lead obtained during March 13-22 were caused by the increased conductivity of the electrolyte with an addition of LiCl and KCl to the bath. The average cathode current efficiency of about 92 pct is compared with the value of 98 pct that was obtained consistently in our previous small-scale work (3).

Analysis of the electrolyte is shown in Table VI. As expected, the major impurities consisted of metals that are higher than lead in the electromotive series such as Ca, Na, and Fe. Zinc, Mg, and Cu are also present in the electrolyte, but at a much lower level. Since copper is more electronegative than lead, copper would be expected to contaminate the lead metal product by codeposition. Sulfate was introduced into the electrolyte by the lead chloride. The present levels of sulfate contamination have had no noticeable effect on cell operation.

TABLE V. - Electrolytic cell operating data, Nov. 1978 - March 1979

	Nov. 6-Jan. 31[1]	Feb. 6-15[2]	March 13-22[3]
Average current, amp · · · · · · ·	3,300	2,860	3,570
Average voltage, v · · · · · · · ·	5.1	4.8	4.3
Average current density, amp/in[2]. .	4.0	3.5	4.3
Amp-hr · · · · · · · · · · · · ·	483,000	603,500	532,500
PbCl2 added, lb · · · · · · · · · ·	6,837	6,584	5,375
Lead metal produced, lb · · · · · ·	3,760	4,707	4,245
Cathode current efficiency, pct · ·	91	92	94
Energy requirement, kWh/lb · · · ·	0.64	0.60	0.53

[1]Intermittent operation.
[2]Continuous operation.
[3]Continuous operation with LiCl and KCl concentrations in electrolyte adjusted.

TABLE VI. - Analysis of electrolyte[1]

Amp-hr (x 10[3])	Percent				Ppm		
	Ca	Fe	Na	SO4	Cu	Mg	Zn
139	0.01	0.03	0.04	–	<10	8	75
556	.17	.03	.09	0.56	13	6	33
694	.16	.07	.10	.74	16	9	41
979	.15	.11	.15	.62	34	15	59
1,269	.11	.11	.17	.36	42	23	80
1,556	.22	.10	.52	–	76	27	120
1,881	.34	.21	1.4	–	90	48	280

[1]Silver is below determination limit of 10 ppm.

As expected, the more electronegative metals such as copper and silver were present in the lead metal product (Table VII). The steady increase of copper in the lead metal product corresponded to a similar increase of copper in the lead chloride and electrolyte. The introduction of copper and silver with the lead chloride must be controlled to maintain a high level of purity in the lead product. This, too, is scheduled for future investigation.

Table VII. - Analysis of lead metal,[1] ppm

Amp-hr (x 10[3])	Ag	Cu
139	1	5
556	2	11
694	2	14
979	2	22
1,269	2	24
1,556	13	44
1,881	14	49

[1]Also Bi, <15; Cd, <0.7; Co, <2; Fe, <4; Mo, <25; Ni, <2; Sb, 25; Sn, <50; and Zn, <1.

We have found the equipment and materials described in this paper worked reasonably well. In general, polymeric materials and titanium are best suited to withstand the corrosion of the chloride solutions, and silica bricks are compatible with the molten salt mixture and lead metal.

Air samples were taken with samplers worn by operating personnel to determine lead exposure, with samplers placed at various parts of the workplace to determine the lead level in the ambient air, and with samplers placed near the equipment of the individual unit operations to determine the lead emission of various sources. During normal operation of the PDU, personnel air samples range from 3.9 to 34 μg Pb/m^3 air measured over 8-hour periods. Air samples taken near the covered tanks, filter press, electric oven, electrolytic cell, and fume hood in which dry lead chloride is handled show that the lead concentration is not significantly higher than that of the ambient air. The highest lead concentration of 800 μg/m^3 was found in the vapors inside the leach tank, demonstrating that all tanks should be covered, and vapors from the hot, lead-containing solutions should be vented and scrubbed.

Only limited data have been obtained thus far on air monitoring and biological monitoring for lead exposure associated with the operation of the PDU. The results are within the new permissible exposure limits according to Section 1919.1025 of Title 29 of the U.S. Code of Federal Regulations.

Conclusions

From the initial period of operation of the PDU, it appears that the process is simple and straightforward. No major operating problems have been encountered. The corrosive chloride solutions can be handled adequately by a variety of commercially available polymeric materials or titanium; therefore, materials selection for process equipment should not be an insurmountable problem. Operation of the PDU will be continued to demonstrate the reliability of equipment and to monitor the lead concentration in air. The process eliminates sulfur oxide generation and has the potential for reducing the difficulties in controlling lead emissions because all the unit operation equipment can be closed.

Investigation of methods for controlling impurities in the leach circuit and the electrolyte will be our major objective for future work. Additional research will be needed to develop methods to dispose of the solid and liquid wastes in an acceptable manner. Data available are still insufficient to make a complete evaluation of the process, but it appears at this time that the ferric chloride leach/fused-salt electrolysis approach offers a promising alternative to smelting for producing lead metal in compliance with the newly promulgated Federal regulations governing lead concentration in air.

References

1. Murphy, J. E., F. P. Haver, and M. M. Wong. Recovery of Lead From Galena by a Leach-Electrolysis Procedure. BuMines RI 7913 (1974) 8 pp.

2. Haver, F. P., and M. M. Wong. Ferric Chloride-Brine Leaching of Galena Concentrate. BuMines RI 8105 (1976) 17 pp.

3. Haver, F. P., C. H. Elges, D. L. Bixby, and M. M. Wong. Recovery of Lead From Lead Chloride by Fused-Salt Electrolysis. BuMines RI 8166 (1976) 18 pp.

4. Christiansen, Niels C. Process of Treating Ores Containing Galena. U. S. Patent 1,456,784. May 29, 1923.

5. Agracheva, R. A., A. N. Volskii, and A. M. Egorov. Treatment of Lead Sulfide Concentrates by the Application of Ferric Chloride Solutions. Izv. Akad. Nauk. SSSR, OTD. Tekhn. Nauk. Met. i. Toplivo. 3 (1959) pp. 37-46.

6. Agracheva, R. A., and A. N. Volskii. Processing of Lead Sulfide Concentrates by Treatment With Ferric Chloride. Sb. Nauchn. Tr. Mack. Inst. Tsvetn. Metal. i. Zolota, 33 (1960) pp. 26-33.

7. Cottam, Stephen M., Howard E. Day, and William A. Griffith. Production of Lead and Silver From Their Sulfides. U.S. Patent 3,929,597. Dec. 30, 1975.

8. Baker, Richard D., Stephen M. Cottam, Howard E. Day, and William A. Griffith. Method of Producing Metallic Lead and Silver From Their Sulfides. U.S. Patent 3,961,941. June 8, 1976.

9. Milner, Edward F. G., Ernest G. Parker, and Godefridus M. Swinkels. Hydrometallurgical process for Treating Metal Sulfides Containing Lead Sulfide. U.S. Patent 4,082,629. April 4, 1978.

10. Ashcroft, Edgar Arthur. Process for Electrolyzing Fused Metallic Salts. U.S. Patent 1,545,385. July 7, 1925.

11. Starliper, A. G., and H. Kenworthy. Recovery of Lead and Sulfur by Combined Chlorination and Electrolysis of Galena. BuMines RI 6554 (1964) 20 pp.

12. Phillips, T. A. Economic Evaluation of a Process for Ferric Chloride Leaching of Chalcopyrite Concentrate. BuMines TC 8699 (1976) 22 pp.

13. Rovirosa, Nicolas Yris. The Hydrothermal Processing of Lead Concentrates in Small Scale Mining. Adapted by John D. Wiebmer. Mining Engineering, February 1979, pp. 147-149.

454

THE PHYSICAL CHEMISTRY OF LEAD EXTRACTION

G.M. Willis,

University of Melbourne.

Phase diagrams are given to illustrate the relations between the lead-sulfur melt and the oxide-sulfate melt which are formed by the oxidation of lead sulfide at high temperatures. The thermodynamic properties of the former are calculated, and from these, the equilibria with slags of varying lead oxide activity are obtained. In direct smelting, low sulfur in lead can only be obtained with high lead in slag.

The rate of oxidation of lead concentrates in direct smelting can be very high. The phase relations in sinter, particularly the presence of refractory solids (which generally contain zinc) are important for its smelting behaviour. Some of the consequences of the exothermic reduction of lead oxide in the blast furnace are given.

The activity coefficients of lead oxide in slags are reviewed. Lead in slag and hence the oxygen potential for bullion-slag equilibrium can be estimated from Fe^{3+}/Fe^{2+} in slag, its magnetic susceptibility, and the iron activity or the sulfur content. These can be related to oxygen or sulfur partial pressures, and it is shown how these also determine the bullion-matte equilibrium and the distribution of metals such as copper between metal and matte.

Introduction

The physical chemistry of lead extraction has attracted less interest than that of the other common nonferrous metals. This has probably been due to the feeling that the metallurgy of lead is easy. Primary lead is derived almost entirely from sulfide concentrates, which are sintered and smelted; there are only minor variations in practice from plant to plant.

With one of the lowest energy requirements per ton of metal produced (1) lead has been comparatively free from the pressure of energy considerations. In the Imperial Smelting Furnace, the production of lead makes little or no demand on the carbon requirements of the furnace (2). Conventional smelting is not free of technical (3,4) and environmental criticisms, and these have given rise to a number of direct smelting processes. Matyas and Mackey (5) have reviewed these, but without any special attention to their physical chemistry.

Improvements in established processes and the development of new ones must be subject to restraints derived from the basic physical chemistry of the various reactions used. This paper will thus be concerned with these reactions, rather than with the technology needed for their practical realization. The heterogeneous equilibria between the various phases encountered in lead smelting are described, rather than the thermodynamic properties of appropriate lead compounds. It is believed that reliable thermodynamic data have been used but detailed assessment has been included only where it has been considered necessary.

Oxidation of PbS and the Pb-S-O System

Both conventional and direct smelting processes involve the oxidation of lead sulfide, according to the greatly over-simplified reactions :

$$PbS + 3/2 \ O_2 \ = \ PbO + SO_2, \tag{1}$$

$$PbS + O_2 \ = \ Pb + SO_2. \tag{2}$$

Reaction (1) is a text book version of the roasting of galena with the elimination of sulfur as SO_2, while (2) is the hoped-for reaction in direct smelting processes.

Unfortunately the Pb-S-O system is considerably more complicated than most, as was first shown clearly by the work of Kellogg and Basu (6), which has been confirmed with only minor changes by later workers (7), (8) and (9). The phase relations for 1100K are shown in Fig. 1.

The stability of the sulfates of lead is obvious. The problem in direct smelting of pure galena to metallic lead at this temperature is shown to be that of carrying out an oxidation while maintaining an oxygen partial pressure of ca. 10^{-11} with P_{SO_2} of 0.1 atmosphere say. Higher P_{O_2} leads to the production of oxide sulfate instead of metallic lead. The P_{CO_2}/P_{CO} scale shows that oxidation say by pure CO_2 is thermodynamically unfavourable. What is required is an oxidant which has an adequate oxygen capacity and is buffered at oxygen partial pressures below the boundary separating lead from the oxidized phases.

Since no such material appears to be available, the obvious solution is to go to higher temperatures, which has the qualitative effect of moving the SO_2 isobars down and to the left relative to the fields of stability of the oxidized phases. Pb-S-O diagrams for higher temperatures have been given by Yazawa and Gubčová (10), by Rosenqvist (11), and by Schuhmann et al. (12).

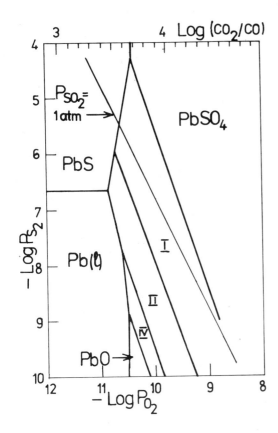

Fig. 1

The Pb-S-O system at 1100K.
I, II and IV indicate

PbO · PbSO₄

2PbO · PbSO₄

and 4PbO · PbSO₄

respectively.

The higher temperature phase relations are less well-defined because of lack of data on the oxidized liquid, and to a lesser extent, on Pb-S melts.

A more detailed description can be derived from the measurements of Jacob and Toguri (13) and of Derriche and Perrot (8). From the measured activities in PbO-PbSO₄ melts, the equilibrium P_{O_2} for equilibrium with pure lead can be obtained. P_{O_2} at PbS/PbSO₄ equilibrium can be extrapolated from the results of Fredriksson et al. (9). Finally, the PbS-melt equilibria can be calculated from the activities of PbO and PbSO₄ in the melt. The results for 1253K are shown in Fig. 2. The lines for the binary equilibria PbS(c) + oxidized melt, Pb-S melt + oxidized melt, and PbS(c)-Pb-S melt, should meet at a point. The sulfur partial pressure for this can be obtained approximately from Scanlon (14). If the solubility of oxygen in the lead-sulfur melt is neglected, this line is straight. A second three (condensed) phase equilibrium involves PbS, PbSO₄ and the oxidized melt. In spite of uncertainties due to lack of knowledge of the mutual solubility of all the phases, Fig. 2 gives more insight than does Fig. 1. It can be seen that lead in equilibrium with oxidized melt at 1 atm. P_{SO_2} contains low sulfur (≈2.5%).

Fig. 3 shows the relations between the lead-sulfur melt, the PbO-PbSO₄ melt and SO₂ at 1 atm. The compositions of both melts are fixed, since the system at constant temperature and pressure is invariant. What can be altered is the relative proportion of the two liquid phases, as PbS or O₂, or both, are added to the system. The system, with provision for SO₂ to be removed at constant pressure, is self-regulating. An excess of PbS will

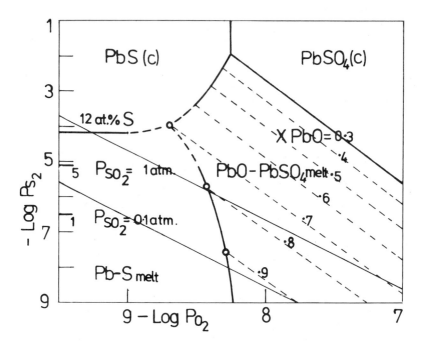

Fig. 2 — The Pb–S–O system at 1253K, from Jacob and Toguri (13). XPbO is the mole fraction of PbO in PbO-PbSO₄ melts. The atomic % sulfur scale is approximate, as it is extrapolated from low temperature solubilities (28).

increase the proportion of the metal–sulfur melt, while excess O_2 will increase the relative amount of the oxidized melt.

Of necessity, any mutual solubility of the metallic and oxidized phases in one another has to be neglected, but the general relations are qualitatively correct. Approximate phase relations at a higher temperature (1150°C) are given by Byerley, Rempel and Takebe (15); these are incidental to their study of the kinetics of the reaction between PbS and PbO. They confirm the presence of two immiscible liquid phases :– one a lead–sulfur melt and the other an oxide–sulfate melt. Fig. 4 shows that there is considerable solubility between the phases, and in fact the miscibility gap closes near the Pb–S side of the system, as might be expected from the flat liquidus of the binary. The trend of the tie-lines showing that sulfur is concentrated in the oxide melt, rather than the metal, agrees with that in Fig. 3.

These phase relations should correspond more closely to those of the initial oxidation products of an isolated grain of galena at actual sintering temperature than do those of the solid phases at low temperatures. The product of complete oxidation will be a PbO-PbSO₄ liquid with perhaps some dissolved PbS. Figs. 2 and 3 show that the higher the partial pressure of SO_2, the greater the sulfur content of both the lead and the oxide phases. Final elimination of sulfur in sintering may depend on flushing out SO_2 with fresh air passing through the hot zone just after combustion is completed. This of course neglects for example the probable role of CaSO₄ in retaining sulfate sulfur because of its high stability.

The occurrence of lead spheres surrounded by PbS crystals in a lead

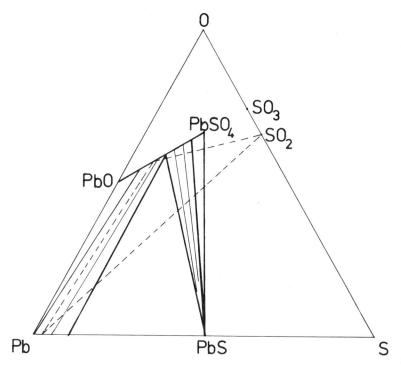

Fig. 3 — The Pb-S-O system at 1253K from Fig. 2. The triangle formed
by the broken lines shows the equilibrium between Pb-S melt,
PbO-PbSO₄ melt and SO₂ at 1 atmosphere. The solubilities
of the two liquids in one another is not known, and have not
been given.

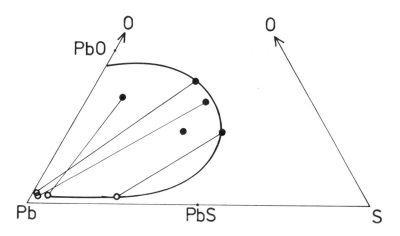

Fig. 4 — The Pb-S-O system at 1150°C from Byerley, Rempel and
Takebe (15).

461

sinter as observed by O'Keefe, Bennett and Cole (16) (their C and D phases respectively) is an indication that local conditions in sintering may permit the production of a Pb-S melt, which rejects PbS on cooling. In sinter, lead sulfate or oxide-sulfates do not appear as a result of the solidification of the oxide-sulfate liquid, as it can react with e.g. CaO and SiO_2, with the removal of sulfate and the formation of lead silicates.

Rate of Oxidation of Lead Concentrates

Low temperature oxidation studies of PbS are numerous and have been briefly reviewed (17) and will not be considered further here.

In many of the direct smelting processes (5) which have been proposed, oxidation of fine particles of galena in air or oxygen takes place. Fig. 5 shows some of the changes which take place in 2 metres in the shaft of a KIVCET furnace (18). Comparable temperatures have been reported for the Outokumpu flash smelter (19) (where some fuel is burnt) and for the Cominco pilot plant (20).

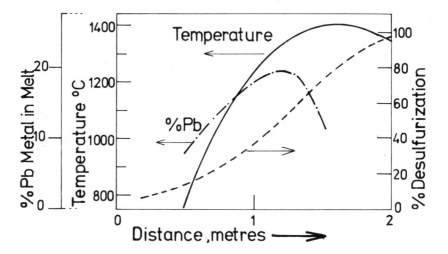

Fig. 5 — Reactions in the shaft of a KIVCET furnace (18); distance is from the top of the shaft.

Although the products of these vary because of different process conditions it is clear that rapid rates are possible under intensive conditions. Over a range of such conditions, there appears to be little limitation from inherent reaction kinetics.

Sychev et al. (21) observed ignition temperatures around 400°C for a range of concentrates, the exact temperature depending very much on particle size. The early stages of oxidation involve solids, but liquids appear at temperatures depending on composition of the concentrates, and enable intensive oxidation to occur. This later declines as sulfides are consumed and large amounts of lead silicate are formed. Reaction between "sulfide" lead and lead oxide in slag is strongly temperature dependent, being small at 1200°C and requiring temperatures of over 1300°C (22). Satisfactory oxidation of a concentrate thus seems to require a particle size, composition, and mineralogy, which will permit initial rapid oxidation in order to achieve high

462

temperatures so that the final stages can proceed fast enough to achieve adequate sulfur removal.

Lead-Sulfur Melts in Equilibrium with Slags Containing PbO

Because of impurities in the concentrates, in direct smelting processes the product of oxidation will be a slag containing PbO, not the relatively simple PbO-PbSO₄ melts discussed earlier. An ideal process should produce lead with very low sulfur concentrations, to avoid the danger of precipitating PbS or other sulfide phases on cooling. For a proper understanding of the oxidation of lead or sulfur or both, the thermodynamic properties of the lead-sulfur system are important.

The only direct measurements over a wide range of composition are those of Gubčová, Yazawa and Igouye (23). Activities of PbS were read off their graph for 1140°C. There is no obvious effect of temperature in the range 1060 to 1160°C. Activities of lead from a Gibbs-Duhem integration are shown in Fig. 6. Combining the equation of Schuhmann et al. (12) for

$$Pb_{(1)} + \frac{1}{2} S_2 = PbS_{(c)} \qquad (3)$$

$$\Delta G° = -38,200 + 20.15T$$

with the heat of fusion of PbS of Blachnik and Igel (24) (8,700 cal) gives

$$Pb_{(1)} + \frac{1}{2} S_{2(g)} = PbS_{(1)} \qquad (4)$$

$$\Delta G° = -29,500 + 13.88T$$

and so P_{S_2} can be calculated from the activities of Pb and PbS. The flat curve for P_{S_2} vs. N_{PbS} shown in Fig. 6 implies that, unless P_{S_2} is carefully controlled in a direct smelting process, the sulfur content of the lead is likely to vary widely.

From these results, the behaviour of lead and sulfur can be established for oxidation under equilibrium conditions. Oxygen partial pressures can be calculated either from lead activities at given PbO activities, or from P_{S_2} at chosen P_{SO_2} values. Fig. 7 shows some results for $a_{PbO} = 1$ and 0.1 and $P_{SO_2} = 1$ atmosphere. A low sulfur content can be obtained at $a_{PbO} = 1$, but if a_{PbO} is lowered to 0.1, high sulfur in lead is to be expected. Only at very low partial pressures of sulfur dioxide can low sulfur in lead and low PbO in slag be obtained under equilibrium conditions. Matyas and Mackey (5) and Davey (25, 26) also conclude this from a survey of operating results of a number of direct smelting processes.

The experimental results (23) do not extend to dilute solutions of sulfur in lead. Grant and Russell (27) note the disparity between PbS solubility calculated from H₂S/H₂ ratios and that measured directly (28) and that only the latter are consistent with the high-temperature liquidus of Miller and Komarek (29). The direct binary system measurements are confirmed by Davey's measurements in the Pb-Cu-S ternary (30). Schuhmann et al. (12) give a regular solution model which relies heavily on the high-temperature liquidus (29) but which does not reproduce the low temperature solubilities. The regular solution parameter α is a weak function of temperature (31, 32) but for dilute solutions, where the liquidus is very steep, it must change rapidly with concentration (33). A model with constant α may thus fail for low sulfur in lead, and it is clear that more experimental work is needed here.

463

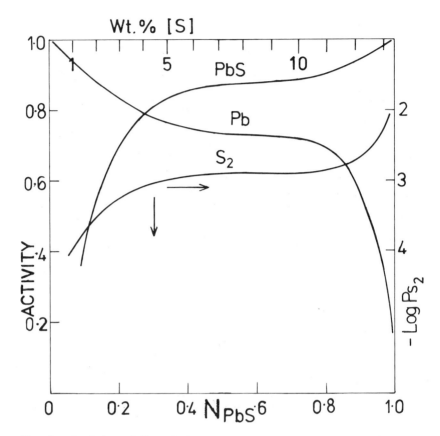

Fig. 6 — Activity of Pb and PbS; and P_{S_2} in Pb-S melts at 1140°C.

Davey (25, 26) has approached the equilibrium oxidation of Pb-S melts in a different way. Starting with the reaction

$$2PbO + [S] \text{ in lead } = 2Pb + SO_2 \qquad (5)$$

$$K = \frac{a^2_{Pb} \cdot P_{SO_2}}{a^2_{PbO} \cdot a_{[S]}}$$

Taking $a_{Pb} = 1$, he takes a_{PbO} as proportional to (wt. % Pb in slag) and a_{PbS} as proportional to [wt. % S in lead] and obtains

$$\text{wt. \% [S](in lead)} = \frac{610 \cdot P_{SO_2}}{(\text{wt. \% Pb in slag})^2}$$

where the constant is obtained empirically. This of course neglects variations in the activity coefficient of PbO in the slag, and in that of sulfur in lead e.g. with concentration, or from interaction with other solutes such as Cu (27, 34). Nevertheless, this result explains the incompatibility of low S in lead with low PbO in slag which Matyas and Mackey noted (5). Fig. 8 shows Davey's relation, and some plant results. The need for two

464

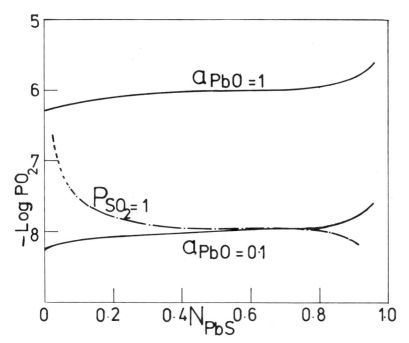

Fig. 7 — Oxygen partial pressures at 1140°C for Pb-S melts in
equilibrium with PbO at activities of 1 and 0.1, and
with one atmosphere of SO_2.

stages in a direct smelting process is clear.

Oxidation of Lead Concentrates in Sintering

In conventional smelting, the product of oxidation is a complex solid
mixture, sinter, in contrast to the liquid slags needed in direct smelting
processes. Reaction (1) is a greatly over-simplified account of the oxidation of actual concentrates. Oxides and silicates of lead are found as
expected (16, 35) but rather surprisingly no oxide sulfates are present,
according to O'Keefe et al. (16). Sulfate sulfur is presumably present as
calcium sulfate.

Two zinc-bearing phases are of interest, namely zinc ferrite and the
melilite hardystonite $Ca_2ZnSi_2O_7$, in which Fe apparently can substitute for
some of the zinc. According to O'Keefe (16) hardystonite may contain up to
several percent lead. Riley (36) assumes that lead can substitute for
calcium in hardystonite (in lead B.F. slags, at least). Apart from the oxide
and silicate, hardystonite appears to be the only major oxidized phase which
contains lead. Very roughly, sinter contains low-melting lead oxide and
silicate phases, and refractory zinc-bearing phases such as $Ca_2ZnSi_2O_7$
(melting point 1425°C) and $ZnFe_2O_4$. In the blast-furnace, the lead-rich
phases are presumably readily reduced, and the others will eventually melt
to form the slag. It is therefore to be expected that the zinc-bearing and
related phases should also predominate in solidified lead blast-furnace slag.
Although there are not exact correspondences, the predominance of $Ca_2ZnSi_2O_7$,
zinc ferrite and magnetite (and solid solutions between these and $ZnAl_2O_4$)
is noteworthy (35, 36, 37).

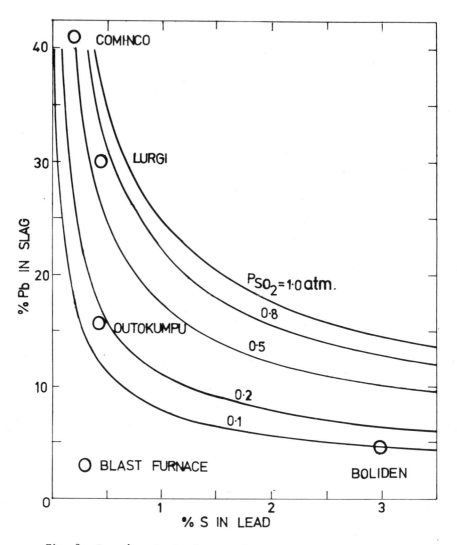

Fig. 8 — Davey's relation between lead in slag and sulfur in lead
for some lead smelting processes; from (25).

Holliday and Shoobridge (38) have recently established very well-defined
relations between the nature and distribution of the various phases in lead-
zinc sinter and its softening behaviour as measured in a simple laboratory
heating test on the one hand, and on the other hand, its actual smelting
behaviour in an Imperial Smelting Furnace. Good resistance to premature
softening is obtained with interlocking needles of a high melting phase, such
as zinc oxide. By analogy, a satisfactory lead sinter should have some
refractory framework which would provide strength for the lumps even when
lead silicates and oxide have melted or been reduced to liquid lead.
Although lead-rich phases may provide the bond at low temperatures, it must
be the more refractory zinc-bearing phases which survive to higher temperatures, and if properly distributed may provide good high temperature strength.

466

In lead-zinc sinters the amount and distribution of the various phases are quite sensitive to the bulk composition (38) and again much the same can be expected for lead sinters.

The reaction $ZnO + 3FeO = Zn_{(g)} + Fe_3O_4$ means that phases with high ZnO and FeO activities cannot co-exist at high temperatures. Although the oxygen partial pressures are important for the state of oxidation of iron oxide and for the stability of zinc oxide it appears that the important phases in the $CaO-ZnO-iron$ oxide-SiO_2 system are $ZnFe_2O_4-Fe_3O_4$ solid solutions, and solid solutions based on $Ca_2ZnSi_2O_7$, Zn_2SiO_4 and ZnO. The first three of these co-exist in the $CaO-ZnO-SiO_2$ system (39). Phases such as wüstite and fayalite may be expected at lower oxygen potentials, which appear to characterize the rather complex results in the $CaFeSiO_4-Zn_2SiO_4$ system studied by Dobrotsvetov et al. (40).

Reactions of Sinter in Smelting

At first sight, the reduction of sinter to lead looks simple. There is no problem with gas-solid equilibrium, and reduction with $CO-CO_2$ mixtures is inherently fast, so that for lead oxide compacts at 538-630°C, the rate is controlled by pore diffusion (41). However, sinter is not a compact of lead oxide and it is generally believed that the reduction of the silicates and other components of lead is slower. The literature contains very little on the behaviour of sinter on heating under reducing conditions, although many smelters must have attempted some such work in the hope that small scale experiments will provide some insight into the behaviour in the blast-furnace. Softening tests in air, or under reducing conditions, require interpretation in terms of structural changes rather than merely bulk changes in composition.

One feature of the reduction of lead oxide is unusual, and of practical importance for the behaviour of sinter in the blast-furnace. This is the fact that the reduction by carbon monoxide is exothermic.

$$PbO + CO = Pb(1) + CO_2, \qquad \Delta H_{1000} = -15.4 \text{kcal.} \qquad (6)$$

Reduction by zinc vapour is strongly exothermic.

$$PbO + Zn_{(g)} = Pb(1) + ZnO, \qquad \Delta H_{1000} = -60.3 \text{kcal.} \qquad (7)$$

This cannot be overlooked if conditions in the lower part of the furnace are sufficiently reducing to allow a modest partial pressure of zinc to survive cooling as the gas ascends the shaft. Lumsden (42, 43) calculates that with high-lead sinter the temperature rise from these exothermic reactions could be high enough to allow a substantial portion of the sinter constituent to melt. The heat of these reactions causes the charge temperature to rise and exceed that of the gas. This temperature may pass through a maximum because when reduction (i.e. heat evolution) is complete, the charge is cooled by the gas. Partly molten charge resolidifies and so causes bridging and hanging-up. Lumsden's proposed remedy for this problem is the use of steam and preheated blast, since steam should produce less zinc vapour, and the reduction of PbO by H_2 is much less exothermic than with CO ($\Delta H_{1000} = -7.1 \text{kcal}$). NL Industries Inc.'s patent for the secondary lead blast-furnace is to use steam with O_2 enrichment (44).

A more general approach to this problem may be provided by the theory of countercurrent solid-gas reactors with an exothermic reaction. Gupta and Geiger (45) show that such reactions can lead to maxima in gas and solid temperatures, as Lumsden found. Also, two stable regimes are possible under certain conditions, in which bed temperatures are perhaps hundreds of degrees different, and with different temperature profiles down the shaft. It is tempting to identify the high temperature steady state with a hot-top in a blast furnace. A quantitative test of the theory would be of great interest; the physical and chemical history of sinter from top to bottom of the furnace is still largely unknown. Surveys of gas temperature and composition in the blast-furnace, such as that of Chao et al. (46), unfortunately do not provide much detailed information on the course of reduction reactions. This is partly due to the reduction of Fe_2O_3 compounds to FeO overlapping with that of PbO in the upper parts of the furnace.

Slag-Metal Reactions

Equilibria between lead bullion and slags are important for both the lead blast-furnace and for direct smelting processes. It is desirable to know the proportions of lead in solution and in suspension in the slag at tapping temperatures. The activities of both lead and zinc oxides in conventional blast-furnace slags are needed for equilibrium models of slag fuming (47, 48) and for the reduction stage of direct smelting processes. The activity of zinc oxide in blast-furnace slags is also important for an understanding of its reduction to zinc vapour in the lower part of the furnace, and hence of its over-all circulation in the furnace.

Many workers have considered that lead enters the slag as oxide

$$[Pb] + \tfrac{1}{2} O_2 = (PbO) \text{ in slag.} \qquad (8)$$

It has also been suggested that lead can dissolve as sulfide; although Wiese (49) found 6-19% Pb in slag after melting PbS with $FeO-SiO_2$ and $FeO-CaO-SiO_2$ slags, he found difficulty in separating the resultant phases, which included metallic lead and a lead-iron matte. These observations do not give definite evidence for the existence of a homogeneous solution of PbS in slag. Nolan (50) found that partial replacement of FeO by FeS in $FeO-SiO_2$ slags (saturated with iron) lowers γ_{PbO}. This cannot be interpreted as showing that PbS enters the slag. As Richardson and Pillay (51) point out, although a slag must have a lead sulfide activity (since it contains lead and sulfur) "it is the same dissolved lead ions which give rise to both activities" :

$$[Pb] + [S] = (Pb^{2+}) + (S^{2-})$$
$$\text{and} \quad [Pb] + [O] = (Pb^{2+}) + (O^{2-})$$

It is therefore misleading to distinguish between "sulfide" and "oxide" solubility as has sometimes been attempted for copper in slags (52, 53).

For (8) above, the activity coefficient of PbO in the slag is needed. Most laboratory work has obtained γ_{PbO} from measured lead activity and oxygen partial pressure. The second method assumes equilibrium in slag fuming and works back from observed elimination of lead (and zinc) to obtain the activity coefficients of the oxides in slags (47, 48).

Some results for γ_{PbO} are given in Table I. The effect of increasing CaO/SiO_2 at constant iron oxide is shown by slags 1 and 2, and by 8 and 9, and is confirmed by the negative correlation between CaO and Pb observed by Riley (36). A similar correlation with CaO/SiO_2 is found (59) for I.S.F. slags; their higher CaO content is a contributing factor (apart from more strongly reducing conditions) in their lower lead content. Riley finds Al_2O_3 to reduce lead in slag, in contradiction to the conclusion of Matyas (57). The high activity coefficient of slag 5 with its high ZnO content is to be noted; it should be possible to check the role of ZnO in more realistic FeO-bearing slags by working at higher oxygen pressures, since γ_{PbO} appears to be

Table I — Activity Coefficient of PbO in Slags 1200–1300°C

No	FeO	Fe$_2$O$_3$	CaO	SiO$_2$	Al$_2$O$_3$	MnO	ZnO	γ_{PbO}	Remarks	Ref.
1	53	5	13	29	0	0	0	0.81		
2	53	5	20	22	0	0	0	1.21	Slags in iron	
3	49	5	13	21	7	5	0	0.64	crucibles;	(54)
4	75	7	0	18	0	0	0	0.31	Au–Pb alloys.	
5	0	0	18	32	0	0	50	2.0		
6	56.5	3.5	0	25	10	0	0	0.07	Cu–Pb alloys.	(55)
7	FeO$_x$	~ 67	0	33	0	0	0	0.25	Cu–Pb alloy, Cu$_2$S, SO$_2$	(56)
8	51	3	15	20	10	0	0	0.10	Pb and CO/CO$_2$	(57)
9	45	2	19	14	20	0	0	0.40	gas mixtures.	
10	28.5	3.5	14	23	0	0	16	0.07	Slag fuming.	(47)
11	"FeO" = 27		15	21	5	4	22	0.45	Slag fuming.	(58)

unrelated to those P_{O_2} values which can be calculated from the results in Table I, and from Nolan's work (50).

In agreement with equation (8) laboratory work (55, 56, 57) shows that the lead content of slags is proportional to the square root of the oxygen partial pressure. It is not so obvious what partial pressures should be used in practice. Matyas and Mackey (5) give from log P_{O_2} = -6.6 for the strongly oxidizing Cominco process to -11.3 for a lead blast furnace. For direct smelting processes oxygen partial pressures can be calculated from those of sulfur dioxide and of sulfur. For the lead blast-furnace, the equilibrium between zinc in bullion and zinc oxide in slag, the Fe^{3+}/Fe^{2+} ratio in slag, and measured or calculated CO_2/CO ratio for the gas in the tuyere zone have been used to obtain oxygen pressures (43, 51).

Freni et al. (60) and Riley (36) find the lead content in practice to correlate well with Fe^{3+}/Fe^{2+} of slags. Matyas and Street (61) and Riley (36) both find that magnetic susceptibility and lead content of slags increase together. High magnetite contents result from more strongly oxidizing conditions and can cause lead losses by entrainment in slag (61).

Unfortunately the oxygen partial pressures for saturation with magnetite are not known for CaO-FeO$_2$-SiO$_2$ slags. The usual phase diagram is for slags saturated with iron; otherwise the system has been studied only in air (62) or at well above lead smelting temperatures (1450 and 1550°C) (63). Zinc oxide is a further complication with practical slags, because of the possibility of saturation with ZnFe$_2$O$_4$-Fe$_3$O$_4$ solid solutions.

Yazawa and Azakami (64) and more recently Fontainas, Coussement and Maes (65) suggest the activity of iron as a measure of reducing conditions. They deduce the activities of iron from the Cu$_2$S-FeS-PbS system, and the latter (66) give a diagram from which iron activities can be read. These are related to lead in slags by the empirical relation for Hoboken practice :

$$a_{Fe} \simeq \frac{0.13}{(\% \text{ Pb}) \text{ in slag} - 0.4} \ .$$

Riley's observation of a negative correlation between lead and sulfur in slags suggests another measure of the oxidation-reduction potential in the blast-furnace.

Writing $$[S] + (O^{2-}) = [O] + (S^{2-})$$ (9)

if sulfur forms dilute solutions in both phases, and the slag composition is assumed constant, so that $a_{(O^{2-})}$ is constant, then:

$$K = \frac{a[O] \ N_{(S^{2-})}}{N_{[S]}}$$

or $$\frac{N_{(S^{2-})}}{N_{[S]}} = K/a_{[O]} = K'/P_{O_2}^{1/2}$$

or since $a_{Pb} = 1$:

$$\frac{N_{(S^{2-})}}{N_{[S]}} = K''/a_{PbO}$$ (10)

Equation (10) shows that the distribution of sulfur between metal and slag is controlled by the oxygen partial pressure; high P_{O_2} favours sulfur in metal, while reducing conditions favour sulfur in slag. Analogous relations are well-known in iron and steelmaking. Alternatively, the reactions can be formulated electrochemically :

$$[Pb] = Pb^{2+} + 2e^-$$
$$2e^- + [S] = S^{2-}$$

That reducing conditions favour sulfide in slag is shown by the fact that ISF slags are higher in sulfur than lead blast-furnace slags, while ISF lead is lower in sulfur than blast furnace lead (67).

Relations between Bullion and Matte

The distribution of metals between bullion and matte is basically controlled by reactions such as

$$2Cu + PbS(matte) = Pb + Cu_2S(matte)$$ (11)

The Cu-Pb-S system has recently been evaluated by Choudhary, Lee and Chang (69) who give the sulfur partial pressure — composition diagram of Fig. 9. Although the lead-rich side is not known exactly, small amounts of copper split the Pb-S binary into metal and matte. At low P_{S_2}, a Pb-Cu alloy is stable; at high P_{S_2}, only matte, and for intermediate P_{S_2}, bullion and matte co-exist with copper relatively concentrated in the latter.

The phase diagram of the Fe-Pb-S system in Fig. 10 is based on the work of Chaskar (69) and of Gontarev (70). Earlier diagrams tend to overlook the field of stability of solid iron. From the course of the tie-lines, it follows that P_{S_2} increases, and a_{Fe} decreases, as the Pb-S side is approached on moving across the miscibility gap. Thus in the bullion + matte region, high sulfur potentials favour high lead in the matte.

Yazawa and Azakami (64), and Fontainas, Coussement and Maes (65) point out that the distribution of copper could be controlled by

$$Cu_2S(in \ matte) + Fe = 2Cu(in \ lead) + FeS(in \ matte).$$ (12)

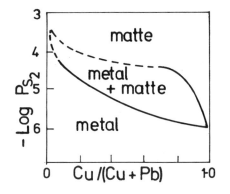

Fig. 9 — Sulfur pressures at 1200°C for the Cu-Pb-S system (68).

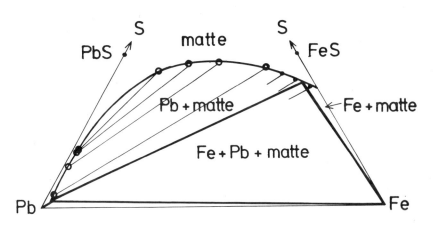

Fig. 10 — Fe-Pb-S system; open circles and Fe-Pb-matte boundaries from Gontarev (70) 1167°C; filled circles, Chaskar (69) 1200°C.

Similarly

$$Fe + PbS(matte) = FeS(matte) + Pb. \qquad (13)$$

The activity of iron enters into both of these. For the second, at 1127°C :

$$N_{FeS}/N_{PbS} = 30.2 \ (a_{Fe}/a_{Pb}) \ (67)$$

However iron activities are generally not known for slags, although they can be obtained if a_{FeO} and P_{O_2} are known. Their use can be avoided, e.g. by writing (13) as :

$$(FeO) + PbS(matte) = (PbO) + FeS(matte) \ ,$$

which then requires a knowledge of a_{FeO} and a_{PbO}.

Yazawa and Azakami assumed constant a_{FeO}, in a slag, and then a_{Fe} can be varied through the CO_2/CO ratio of the gas :

$$CO_2 + Fe_{(\gamma)} = (FeO) + CO \ . \qquad (14)$$

Although all the necessary restrictions are not met experimentally, they find varying CO_2/CO leads to changes in copper and sulfur in lead, and in lead sulfide in matte as predicted from (12) and (13).

The partial pressure of oxygen is a more familiar method of expressing the intensity of reduction than is the activity of iron. The relation between oxygen and sulfur pressures however is not obvious, since they are not determined by equilibria such as (5). In these reactions, sulfur is behaving as a typical nonmetal, and is an oxidant similar to oxygen, so that changes in oxygen potential should be accompanied by qualitatively similar changes for sulfur :

$$1/2 \ O_2 + 2e \ = \ O^{2-} \ \text{(in matte or slag)}$$
$$1/2 \ S_2 + 2e \ = \ S^{2-} \ \text{(in matte or slag)}$$

Electrons from the metal are involved in both these reactions.

Fig. 11 shows how these equilibria can be re-written in terms of P_{O_2}. This is based on the Fe-S-O system at 1200°C, but has been simplified by neglecting the solubility of oxygen in Fe-S melts at higher P_{O_2}. Approximate activities for the Fe-S melt are from Stofko et al. (71). Although the actual boundaries between iron and melt, and wüstite and melt will differ from those shown, the qualitative relation between P_{S_2} and P_{O_2} at constant a_{FeO} will still hold.

Lines of constant iron activity are also shown. At low P_{S_2} and P_{O_2}, iron is the stable phase, and addition of metallic lead will make no difference, because of their mutual insolubility.

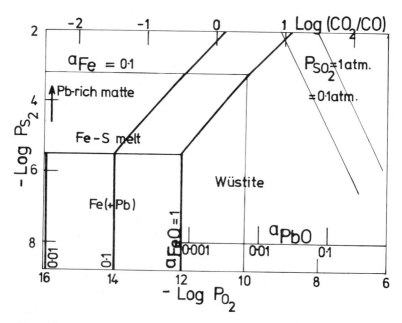

Fig. 11 — The Fe-S-O system at 1200°C; the wüstite + Fe-S melt boundary is simplified (see text).

Along a line of constant a_{FeO}, Fig.11 shows P_{O_2} and P_{S_2} to change together, so either can be used to give the intensity of reducing conditions. If the P_{S_2} scale of Fig. 9 is transferred to Fig. 11, the relation between copper in matte and P_{S_2} or P_{O_2} can be seen. In agreement with Yazawa and Azakami (64) strongly reducing conditions i.e. low P_{O_2}, low P_{S_2} and high a_{Fe}, as in the Imperial Smelting Furnace, lead to the production of copper in bullion, not copper in matte. The lower limit of the Fe-S melt is not necessarily pure iron; it could very well be a speiss with $a_{Fe} < 1$. Another conclusion is that oxidation of Fe-Pb-S mattes under a slag of reasonable a_{FeO} will remove iron, leaving lead and sulfur in the matte which is therefore enriched in lead. The value of P_{S_2} again determines whether the matte will be high or low in lead; at low enough P_{S_2} it will be a high iron matte and even metallic iron or a speiss could be produced.

For the lead blast-furnace, the system Pb-Cu-Fe-S should be considered. Again a high P_{S_2} is to be expected for lead-rich mattes, and these can be correlated over a wide range of practical operations (65). Metals with a lower affinity for sulfur than that of lead will naturally tend to report in bullion rather than in matte.

Conclusion

It is not possible to cover all the topics which are of real or potential interest in a paper such as this. Thus, the activity of zinc oxide in slags has not been discussed; but the recent measurements of Filipovska and Bell (72) should be mentioned. An examination of the factors controlling the lead content of the gas is desirable, particularly in view of the high local temperatures in some of the direct smelting processes. It is disappointing to find practically nothing helpful in the literature on the fundamental properties of even simple speisses.

Both conventional and direct smelting methods still present many intriguing problems : not enough is known about the role of sinter constitution, and its behaviour in the blast furnace. The combustion of coke and the reactions of the gas in the furnace and the equilibria between slag, bullion and other phases need more quantitative work.

Finally, the possibilities of alternatives such as hydrometallurgical or chloride routes for lead production should be kept in mind.

Acknowledgments

The writer is indebted to Prof. A.E.Morris for copies of theses by Chaskar and by Gontarev; to Mr G.D.J.Smith for Figs. 2 and 3, and to staff of Broken Hill Associated Smelters Pty Ltd, Sulphide Corporation Pty Ltd, and of Mt Isa Mines Ltd for provision of material, and to Prof. T.R.A.Davey for helpful discussion.

References

1. H.H.Kellogg, "The Role of Recycling in Conservation of Metals and Energy," J. Metals, 28 (12) (1976) pp. 29-32.
2. S.W.K.Morgan, and D.A.Greenwood, "The Metallurgical and Economic Behaviour of Lead in the Imperial Smelting Process," J. Metals, 20 (12) (1968) pp. 31-35.
3. P.E.Queneau, "Oxygen Technology and Conservation," Metall. Trans. B, 8B (1977) pp. 357-369.

4. W.Schwartz, "The Future of Lead Recovery," Metallgesellschaft : Review of the Activities, Edition 20 (1970) pp. 46-55.
5. A.G.Matyas, and P.J.Mackey, "Metallurgy of the Direct Smelting of Lead," J. Metals, 28 (11) (1976) pp. 10-15.
6. H.H.Kellogg, and S.K.Basu, "Thermodynamic Properties of the System Pb-S-O to 1100°K," Trans. Met. Soc. AIME, 218 (1960) pp. 70-81.
7. J.Esdaile, "Equilibria in the Lead-Sulfur-Oxygen System," pp. 65-80 in Research in Chemical and Extraction Metallurgy, J.T.Woodcock, A.E. Jenkins, and G.M.Willis, eds.; Aust. Inst. Min. Metall., Melbourne, 1967.
8. Z.Derriche, and P.Perrot, "Thermodynamic Study of Solid and Liquid Phases in the System PbO-PbSO₄," Rev. Chim. Minér. 13 (1976) pp.310-323.
9. M.Fredriksson, E.Rosén, and L.Wittung, "Thermodynamic Studies of High Temperature Equilibria XVI. Solid State emf Determinations of Equilibrium Oxygen Pressures in the Systems PbS-PbSO₄, PbS-PbO·PbSO₄-SO₂ and PbO·PbSO₄-PbSO₄-SO₂," Chem. Scr. 11 (1977) pp. 32-36.
10. A.Yazawa, and A.Gubčová, "Diagrammatic Representation of Equilibrium Relations in the Lead-Sulfur-Oxygen System," Trans. Met. Soc. AIME, 239 (1967) pp. 2004-5.
11. T.Rosenqvist, "Phase Equilibria in the Pyrometallurgy of Sulfide Ores," Metall. Trans. B, 9B (1978) pp. 337-351.
12. R.Schumann,Jr., P.C.Chen, P.Palanisamy, and D.H.R.Sarma, "Thermodynamics of Converting Lead Sulfide," Metall. Trans. B, 7B (1976) pp. 95-101.
13. K.T.Jacob, and J.M.Toguri, "Thermodynamics of Oxide-Sulfate Melts. The System PbO-PbSO₄," Metall. Trans. B, 9B (1978) pp. 301-306.
14. W.W.Scanlon, "The Physical Properties of Semi-conducting Sulfides, Selenides and Tellurides," Mineralogical Society of America Special Paper No. 1 (1963) pp. 135-143.
15. J.J.Byerley, G.L.Rempel, and N.Takebe, "Reactions of Lead Sulfide with Lead Oxide in the Molten State," Scand. J. Metall., 4 (1957) pp. 9-16.
16. T.J.O'Keefe, C.Bennett, and E.R.Cole, "A Microscopy Study of Lead Sinter," Metall. Trans. 5, (1974) pp. 427-432.
17. N.B.Gray, M.R.Harvey, and G.M.Willis, "Roasting of Sulfides in Theory and Practice," pp. 19-32 in Physical Chemistry of Process Metallurgy : the Richardson Conference, J.H.E.Jeffes, and R.J.Tait, eds.; Inst. Min. and Metall., London, 1974.
18. K.B.Chandhuri, and G.Melcher, "Comparative View on the Metallurgy of the KIVCET-CS and Other Direct Lead Smelting Processes," Can. Min. Metall. Bull., 71 (799) (1978) pp. 126-130.
19. P.Bryk, R.Malmstrom, and E.Nyholm, "Flash Smelting of Lead Concentrates," J. Metals, 18 (12) (1966) pp. 1298-1302.
20. S.C.Liang, E.F.G.Milner, G.W.Toop, and R.W.Anderson, "Lead Smelting Process," Canadian Patent No 934968 (1973).
21. A.P.Sychev, N.I.Kopylov, V.N.Novoselova, I.P.Polyakov, and N.P. Pestunova, "Behaviour of Sulfidic Lead-Zinc Concentrates during Roasting-Smelting," Sborn. Trud. VNIIT Tsvet Met., 25 (1975) pp. 204-209.
22. A.P.Sychev, V.F.Larin, and I.P.Polyakov, "A Study of the Interaction between Sulfide Lead and Oxide Lead Dissolved in a Slag Melt," Ibid., pp. 196-203.
23. A.Gubčová, A.Yazawa, and H.Igouye, "Activity of PbS in Pb-PbS Melt," Bull. Res. Inst. Miner. Dressing and Metallurgy, Sendai, Tohoku Univ., 24 (1968) pp. 105-106.
24. R.Blachnik, and R.Igel, "Thermodynamic Properties of IV-VI Compounds : Lead Chalcogenides," Z.Naturforschung, 29(b) (1974) pp. 625-629.
25. T.R.A.Davey, "Developments in Direct Smelting of Base Metals — A Critical Review," paper presented at International Symposium on Chemical Metallurgy, Bombay, Jan. 1979.
26. T.R.A.Davey, "Advances in Lead Zinc and Tin Technology — Projections for the 1980's," This Volume, pp.

27. R.M. Grant, and B. Russell, "A Thermodynamic Study of Dilute Sulfur and Cu-S Solutions in Lead," Metall. Trans.1, (1970) pp. 75-83.
28. R.F. Blanks, and G.M. Willis, "Equilibria between Lead, Lead Sulfide and Cuprous Sulfide and the Decopperizing of Lead with Sulfur," pp. 991-1027 in Physical Chemistry of Process Metallurgy, G.R.St. Pierre ed.; Interscience Publishers, New York, 1961.
29. E.Miller, and K.L.Komarek, "Retrograde Solubilities in Semiconducting Intermetallic Compounds, Liquidus Curves in the Pb-S, Pb-Se and Pb-Te Systems," Trans. Met. Soc. AIME, 236 (1966) pp. 832-840.
30. T.R.A. Davey, "Phase Systems concerned with the Copper Drossing of Lead," Trans. Inst. Min. Metall 72 (8) (1962-3) pp. 553-620.
31. J.Esdaile, "Thermodynamics and Phase Diagram of the Lead-Lead Sulfide Binary System," Proc. A'asian Inst. Min. Met., No 241 (1972) pp. 63-71.
32. J.Lumsden, Thermodynamics of Molten Salt Mixtures,pp. 311-312; Academic Press, London and New York, 1966.
33. G.M.Willis, "Thermodynamics of Metal-Sulfur Melts," paper presented at Extractive Metallurgy Symposium, University of New South Wales, Sydney Nov. 1977.
34. L.G.Twidwell, and A.H.Larson, "The Influence of Additive Elements on the Activity Coefficient of Sulfur in Liquid Lead at 600^0C," Trans. Met. Soc., AIME, 236 (1966) pp. 1414-1420.
35. W.McA.Manson, "Laboratory Studies of the Updraught Lead Sintering Process," pp. 143-172 in Sintering Symposium, A'asian Inst. Min. Metall., Melbourne, 1958.
36. J.F. Riley, "Phase Relations and Lead Losses in Mount Isa Blast-Furnace Slags," Trans. Inst. Min. Met., 88 (1979) pp. C19-24.
37. W.McA.Manson, and E.R. Segnit, "Fundamental Research in the Port Pirie Lead Blast Furnace Slags," Proc. A'asian Inst. Min. Met. No 180 (1956), pp. 1-29.
38. R.J.Holliday, and P.H.Shoobridge, "A Study of the Composition, Microstructure and Softening Characteristics of Some Lead-Zinc Sinters," pp. 311-321 in A'asian Inst. Min. Metall., North Queensland Conference, 1978.
39. E.R.Segnit, "The System $CaO-ZnO-SiO_2$," J. Am. Ceram. Soc., 37 (1954) pp. 273-277.
40. B.L.Dobrotsvetov, E.I.Bogoslovskaya, and V.E.Rudnichenko, "The $CaFeSiO_4$-Zn_2SiO_4 system," Russian J. Physical Chemistry, 12 (8) (1967) pp. 1153-1158.
41. Y.K.Rao, and I.J.Lin, "The Kinetics of Reduction of Lead Oxide by CO + CO_2 Gas Mixtures," Can. Met. Quart. 12 (2) (1973) pp. 125-130.
42. J.Lumsden, "Lead Blast Furnace Process," United States Patent 3,243,283 (1966).
43. The Physical Chemistry of the Lead Blast Furnace, B.H.A.S. Report No. 915, (1960).
44. K.D.Libsch, and M.V.Rao, "Secondary Lead Smelting Process," United States Patent 4,115,109 (1978).
45. D.Gupta, and G.H.Geiger, "Stability of Countercurrent Packed-bed Reactors with Internal Heat Generation," Ironmaking Steelmaking, 6 (4) (1979) pp. 196-203.
46. J.T.Chao, P.J.Dugdale, D.R.Morris, and F.R.Steward, "Gas Composition, Temperature and Pressure Measurements in a Lead Blast Furnace," Metall. Trans. B, 9B (1978) pp. 293-300.
47. H.H.Kellogg, "A Computer Model of the Slag-Fuming Process for the Recovery of Zinc Oxide," Trans. Met. Soc., AIME, 239 (1967) pp. 1439-1449.
48. R.M.Grant, and L.J.Barnett, "Development and application of the Computer Model of the Slag Fuming Process at Port Pirie," pp. 247-265 in South Australia Conference 1975 Part A, A'asian Inst. Min. Metall., Melbourne, 1975.

49. W.Wiese, "Solubility of Sulfides in Slags," Erzmetall. 16 (8,9) (1963) pp. 377-386, 452-458.

50. J.B.Nolan, "Oxide and Sulfide Activities in Molten Slags," Ph.D. Thesis, London, 1964.

51. F.D.Richardson, and T.C.M.Pillay, "Lead Oxide in Molten Slags," Trans. Inst. Min. Metall., 66 (1957) pp. 309-330.

52. M.Nagamori, "Metal Loss to Slag Part I. Sulfidic and Oxidic Dissolution of Copper in Fayalite Slag from Low Grade Matte", Metall. Trans. 5 (1974) pp. 531-538.

53. F.Sehnálek, and I.Imris, "Slags from Continuous Copper Production," pp. 39-62 in Advances in Extractive Metallurgy and Refining, M.J.Jones, ed.; Inst. Min. Metall., London, 1972

54. H.W.Meyer, J.B.Nolan, and F.D.Richardson, "On the Activity of Lead Oxide in Blast-Furnace Slags," Trans. Inst. Min. Metall., 75 (1966) pp. C121-122.

55. M.Nagamori, P.J.Mackey, and P.Tarassoff, "Copper Solubility in $FeO-Fe_2O_3$-SiO_2-Al_2O_3 Slags and Distribution Equilibria of Pb, Bi, Sb and As between Slag and Metallic Copper," Metall. Trans. B, 6B (1975) pp. 295-301.

56. M.Kashima, M.Eguchi, and A.Yazawa, "Distribution of Impurities between Crude Copper, White Metal and Silica-Saturated Slag," Trans. Jap. Inst. Met. 19 (1978) pp. 152-158.

57. A.G.Matyas, "Solubility of Lead in Lead Blast Furnace Slags," pp. 999-1011 in Metal-Slag-Gas Reactions and Processes, Z.A.Foroulis and W.W.Smeltzer, eds.; The Electrochemical Society, Princeton, 1975.

58. R.M.Grant, Personal Communication.

59. J.W.Sangster, "The Solubility of Lead in ISF Slag," Private Communication, Sulphide Corporation Pty Ltd, 1975.

60. E.Freni, F.Massazza, and P.Virdis, "Relation between Coke Sizing, Mineralogical Composition and Lead Content of Blast Furnace Slag," Metall. Ital. 57 (1965) pp. 407-414.

61. A.G.Matyas, and M.D.Street, "Factors Influencing the Lead Content of Brunswick Blast-Furnace Slag," Can. Min. Met. Bull., 70 (786) (1977) pp. 132-136.

62. B.Phillips, and A.Muan, "Phase Equilibria in the System CaO-Iron Oxide-SiO_2 in Air," J. Am. Ceram. Soc., 42 (1959) pp. 413-423.

63. M.Timucin, and A.E.Morris, "Phase Equilibria and Thermodynamic Studies in the System $CaO-FeO-Fe_2O_3-SiO_2$," Metall. Trans. 1 (1970) pp. 3193-3201.

64. A.Yazawa, and T.Azakami, "Thermodynamic Considerations of Zinc Blast Furnace Smelting," Can. Metall. Quart. 8 (1969) pp. 313-318.

65. L.Fontainas, M.Coussement, and R.Maes, "Some Metallurgical Principles in the Smelting of Complex Materials," pp. 13-23 in Complex Metallurgy, M.J.Jones, ed.; Inst. Min. Metall., London, 1978.

66. Trans. Inst. Min. Metall., Reply to Discussion on Above, 88 (1979) pp.C67.

67. T.R.A.Davey, Personal Communication.

68. U.V.Choudhary, Y.E.Lee, and Y.A.Chang, "A Thermodynamic Analysis of the Copper-Lead-Sulfur System at 1473K," Metall. Trans. B, 8B (1977) pp. 541-546.

69. V.D.Chaskar, "The Solubility of Lead in Iron Sulfide and Oxysulfide Mattes at 1100°C and 1200°C," M.Sc. Thesis, University of Missouri-Rolla, 1976.

70. V.Gontarev, "Thermodynamical Activity of PbS in the Fe-Pb-S Ternary System," Master's Thesis, University of Ljubljana, 1976.

71. M.Stofko, J.Schmiedl, and T.Rosenqvist, "Thermodynamics of Iron-Sulfur-Oxygen Melts at 1200°C," Scand.J.Metall.3 (1974) pp. 113-118.

72. N.J.Filipovska, and H.B.Bell, "Activity Measurements in $FeO-CaO-SiO_2$ and $FeO-CaO-Al_2O_3-SiO_2$ Slags Containing Zinc Oxide Saturated with Iron at 1250°C," Trans. Inst. Min. Metall., 87 (1978) pp. C94-98.

THE PHYSICAL CHEMISTRY OF LEAD REFINING

T.R.A. Davey

Metallurgy Department, University of Melbourne, Australia.

Lead refining includes examples of all techniques used for refining metals generally: electrolytic, fractional crystallisation, fractional distillation, and precipitation by reagents.

Of particular concern for optimising operations is the question of whether each process is governed essentially by equilibrium or kinetic considerations.

Lead refining processes which proceed under equilibrium conditions include copper drossing (precipitation of copper compounds out of solution by cooling), softening (preferential oxidation of antimony, arsenic and tin to form a slag or dross), desilverising (precipitation of Ag-Zn crystals by addition of zinc as a reagent), and debismuthising (preferential precipitation of bismuthides by addition of alkaline-earth metals).

Kinetic considerations govern the operations of decoppering (by stirring sulfur into lead at low temperature), Harris softening (oxidation of antimony, arsenic and tin and adsorption of the oxides into a caustic melt), vacuum dezincing, and final refining by stirring with caustic soda.

An understanding of the physical chemistry of these processes permits mathematical modelling in most cases, as a prerequisite to optimisation and automation.

Introduction

Lead is normally produced by a smelting process at a temperature of about 1150°C, in equilibrium with a silicate slag. At this temperature it is an almost universal solvent, and so contains numerous impurities derived from the original ore or concentrate, fluxing materials, or refractories with which it has come into contact. Nearly all of these impurities have value, and warrant recovery as well as elimination from the lead. They include Cu, Fe, Ni, Co, Zn, In, As, Sb, Sn, Ag, Au, Bi, S, Se, Te and O.

Copper normally ranges from a few tenths of a percent to several percent, with extreme maximum values of 10-12%. Fe and Zn are of the order of a tenth of a percent. Ni and Co are normally less than a tenth of a percent. As, Sb, Sn and Bi vary from ppm to several percent, being normally in the range 0.1-1% for As and Sb, and from nil to a few tenths of a percent for Sn and Bi. Au is usually a small fraction of the Ag, which may vary from a few ppm to nearly 1%, but most frequently is in the range 0.1-0.5% (about 30-150 ozs/ton). Se and Te are normally vanishingly small. S and O depend upon the degree of reduction in the smelting furnace; in a lead blast furnace S will normally be of the order of 0.2-0.3% in the lead, and O of the order of 0.1%; whereas in the more highly reducing Imperial Smelting furnace lead, S will be less than 0.1%, and O probably an order of magnitude less. S is drawn into the slag where Fe has an activity approaching unity.

The impurities are invariably removed in groups: Cu, Fe, Ni, Co and Zn in the first (copper drossing) operation, which entails cooling the lead nearly to its freezing point, to throw out of solution all those elements which can be removed by cooling. In the second (softening) operation, elements more readily oxidised than lead are removed by oxidation, and these include As, Sb, Sn and In. Then only metals nobler than lead remain as impurities: Ag, Au and Bi. These may be removed by partial crystallisation (as in the long since defunct Pattinson process, or its modern revival termed "reflux refining") or by precipitation from lead as intermetallic compounds after special reagent additions: Zn for Ag and Au; a mixture of alkaline-earth metals for Bi. Residual reagents must also be removed from lead, by chlorination or oxidation. In the case of zinc much of the residual reagent in the lead may be recovered by a vacuum distillation process. The final traces of all elements less noble than lead are invariably removed by a final refining operation comprising stirring with an oxidising agent (nitre or air) and caustic soda.

In special cases noble metals and bismuth are removed by electrolytic refining, not further mentioned here, recently reviewed by Kraus[1].

General

Figure 1 shows the course of the production of lead from the ore to the refined product. All stages of production may be considered as refining operations, in that impurities are being separated at all stages from the desired product. However, the present discussion will be restricted to those operations normally understood by the term "refining" - carried out after the smelting operation, as shown in Figure 2. The impurities remaining in lead only in minor amounts at each stage are in square brackets.

An understanding of the physical chemistry of each operation is necessary in order to be able to optimise it, and we may distinguish between operations in which equilibrium is substantially achieved (governed essentially by consideration of the equilibria involved) and those far from equilibrium, in which the degree of refining achieved is governed by kinetic factors. In the

478

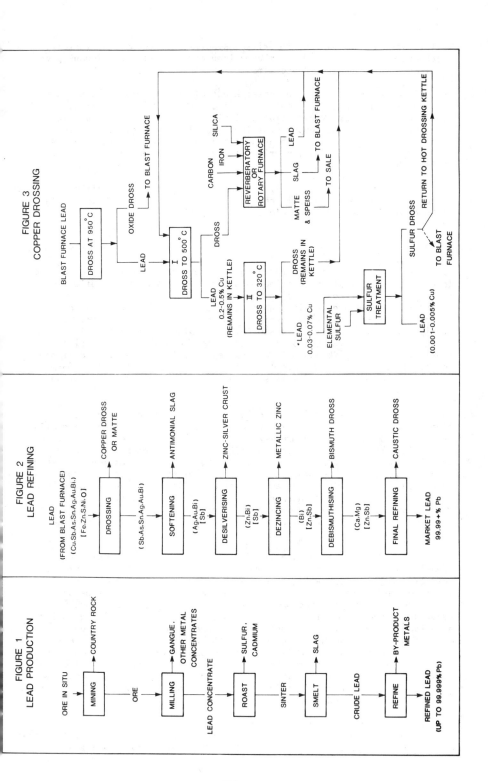

FIGURE 1
LEAD PRODUCTION

ORE IN SITU → MINING → COUNTRY ROCK
MINING → ORE → MILLING → GANGUE,
OTHER METAL
CONCENTRATES
MILLING → LEAD CONCENTRATE → ROAST → SULFUR,
CADMIUM
ROAST → SINTER → SMELT → SLAG
SMELT → CRUDE LEAD → REFINE → BY-PRODUCT
METALS
REFINE → REFINED LEAD
(UP TO 99.999%Pb)

FIGURE 2
LEAD REFINING

LEAD
(FROM BLAST FURNACE)
$(Cu.Sb.As.Sn.Ag.Au.Bi.)$
$[Fe.Zn.S.Ni.O]$
→ DROSSING → COPPER DROSS
OR MATTE
$(Sb.As.Sn.Ag.Au.Bi)$
→ SOFTENING → ANTIMONIAL SLAG
$(Ag.Au.Bi)$
$[Sb]$
→ DESILVERISING → ZINC-SILVER CRUST
$(Zn.Bi)$
$[Sb]$
→ DEZINCING → METALLIC ZINC
(Bi)
$[Zn.Sb]$
→ DEBISMUTHISING → BISMUTH DROSS
$(Ca.Mg)$
$[Zn.Sb]$
→ FINAL REFINING → CAUSTIC DROSS
→ MARKET LEAD
99.99 + % Pb

FIGURE 3
COPPER DROSSING

BLAST FURNACE LEAD → DROSS AT 950°C
OXIDE DROSS → TO BLAST FURNACE
LEAD → I DROSS TO 500°C
DROSS
LEAD
0.2-0.5% Cu
(REMAINS IN KETTLE) → II DROSS TO 320°C
DROSS
(REMAINS IN KETTLE)
CARBON IRON SILICA
→ REVERBERATORY
OR
ROTARY FURNACE
LEAD → TO BLAST FURNACE
MATTE SLAG
MATTE
& SPEISS → TO SALE
*LEAD
0.03-0.07% Cu
ELEMENTAL
SULFUR → SULFUR
TREATMENT
SULFUR DROSS → RETURN TO HOT DROSSING KETTLE
LEAD
(0.001-0.005% Cu)
→ TO BLAST
FURNACE

latter case, the removal of impurities sometimes proceeds indefinitely, if a batch process is continued for a very long time, or if a continuous process provides a very long residence time.

Although in a few special cases electrolytic refining is used in conjunction with pyrometallurgical, in most of the world's lead refineries the operations are entirely pyrometallurgical. Those governed by equilibrium considerations include copper drossing, normal softening, desilvering, debismuthising and possibly chlorine dezincing. Those governed by kinetics include sulfur decoppering, Harris softening, vacuum dezincing, and final refining by caustic treatment.

Most refining operations are capable of being performed either batchwise or continuously, but the basic physical chemistry is not affected by this, although of course process engineering aspects are.

Copper Drossing

The first stage of lead refining is invariably drossing, whereby the lead is cooled almost to freezing in order to throw out of solution any elements (or their compounds) which are much less soluble at the lower temperature. The process is usually termed "*copper drossing*", since copper is the main by-product which becomes concentrated into the dross. Copper is precipitated as the sulfide, arsenide, antimonide or stannide. These compounds precipitate in that order, with of course a degree of overlap as each constituent (S, As, Sb and Sn) becomes depleted in the lead. Nickel and cobalt, if present, tend to follow the copper. Lead sulfide also forms at the lower temperatures and enters the dross. Selenium tends to follow the sulfur.

It is not generally mentioned in the literature, although known to some plant operators, that the first compounds to separate from lead on cooling from blast furnace temperatures are not sulfides but oxides. The oxygen solubility in blast furnace lead at 1000-1200°C (in the presence of the normal impurities) is not known, but the amounts and proportions of zinc, iron and oxygen are such that they separate between tapping temperatures and about 950°C as spinels or magnetite, with some litharge, and a "black slag" or "chocolate slag" forms on the surface of the cooled lead, and may be removed separately from the subsequently precipitated sulfides, etc., which can then be separated from the lead without contamination by significant amounts of oxide.

Figure 3 shows the flowsheet of copper drossing as it could be practised ideally, although no plants are known to follow this scheme completely. The refractory oxides separating from the lead down to about 950°C, containing little except oxides of iron and zinc, and entangled lead, may be separated first so that their presence does not complicate subsequent recovery of by-product copper. Next, the main dross constituents, comprising sulfides, arsenides, antimonides and stannides of copper (and nickel and cobalt, if present), are skimmed off down to about 500°C, at which temperature the physically entangled lead metal amounts to about 60-65%. The dross separating from the lead at temperatures below 500°C contains a great deal of entangled lead, and is allowed to remain in the kettle when the lead is pumped away, close to its freezing point (310-327°C, depending upon impurity content) and is liquated by the next charge of hot lead.

The *continuous copper drossing* processes(2-5) are arranged so that drosses separating at lower temperatures rise through incoming hot lead and are completely melted, to produce mattes (mixtures of sulfides) and speisses (mixtures of arsenides and antimonides) and of course slags if oxides are still present.

Figure 4 shows the solubility of copper in lead as a function of temperature, and as dependent upon S, As, Sb or Sn contents, which modify it considerably.

Figures 5-8 show the phase diagrams of relevance in understanding the course of the copper drossing operation, and the thermodynamic data from which the solubility of copper in lead may be calculated was given in detail by Davey(6), the source of all these data. A more recent Pb-Cu-S diagram(7) is in reasonable agreement.

Due to the form of the Pb-Cu-S phase diagram, shown in Figure 5, it is not possible to avoid the formation of PbS as well as Cu_2S on cooling lead(6). Thus the copper dross inevitably contains lead in combined form, as well as mechanically entrained metallic lead adhering to the crystalline dross particles. It is therefore not sufficient merely to liquate the dross in order to remove its lead content: a chemical removal of the combined lead fraction is required - as, for example, displacement of Pb from PbS by Fe.

Copper Dross Treatment

Space does not permit a full discussion of this topic.

The drosses produced by precipitation from the cooling lead contain a great variety of crystals, which must be completely melted in order to free them from the entangled metallic lead.

There will be produced in any case a *matte* phase (predominantly copper-lead sulfides) and a metal phase. If oxides are present there will also be a slag phase, and to ensure that this is fluid, silica flux is added. The slag phase will then contain essentially iron, zinc and lead silicates. It will also contain much of the indium and tin present in the dross. If a *speiss* phase forms, this will be basically copper arsenide (and antimonide, if the antimony/arsenic ratio in the bullion was very high). All elements, in fact, distribute themselves between all four phases, but have certain preferences for concentrating in one or another phase: copper tends to enrich in the matte, nickel in the speiss, silver in lead and matte, gold in lead and speiss, cobalt and iron in slag and speiss.

The lead content of the matte depends upon its main constituents (Cu, Fe and S), as shown in Figure 10, which shows the data of Figure 9 in a more usable form. In Figure 10 all constituents except Cu, Fe, S and Pb are considered simply as diluents. Points on the diagram represent adjusted compositions, calculated to make Cu + Fe + S = 100%. The lead content is calculated with Cu + Fe + S + Pb = 100%, and this represents the saturation lead content of mattes of these compositions.

Figure 10 has been found to be accurate to ±2% of lead content for many industrial and experimental mattes. According to recent work by Fontainas *et al.*(8), Zn can be regarded as equivalent to Fe, so possibly the adjusted composition should be calculated taking account of this.

Figure 10 holds only for mattes with little or no content of alkali or alkaline-earth metals. Mentzel(9) and Hesse(10) showed many years ago that sodium or calcium introduced into mattes lowers their solubility for lead, but no use appears to have been made of this until the development of the soda matte smelting process at El Paso in the 1940's.

481

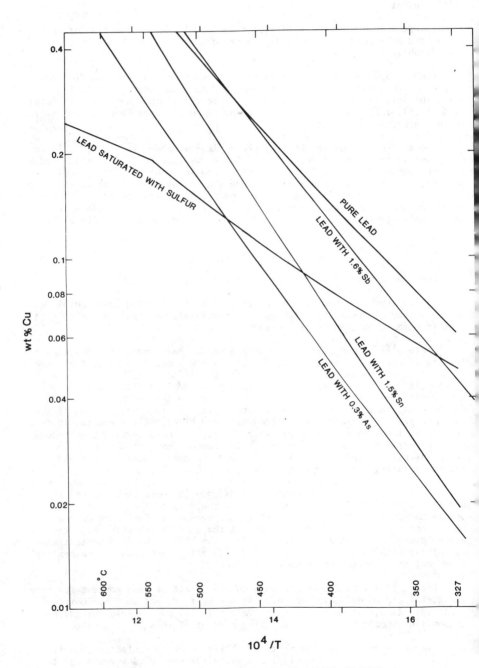

FIGURE 4 - SOLUBILITY OF COPPER IN LEAD

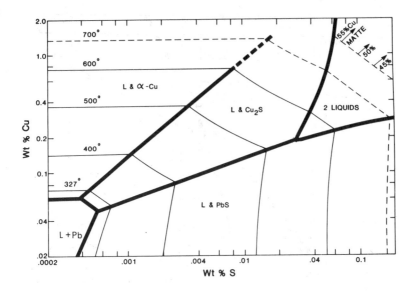

FIGURE 5 - Pb-Cu-S SYSTEM, LEAD CORNER

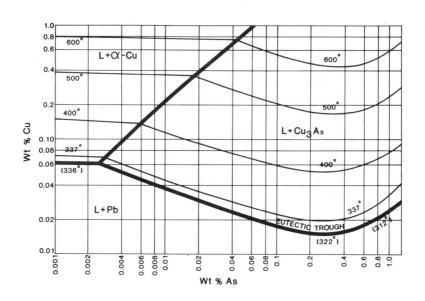

FIGURE 6 - Pb-Cu-As SYSTEM, LEAD CORNER

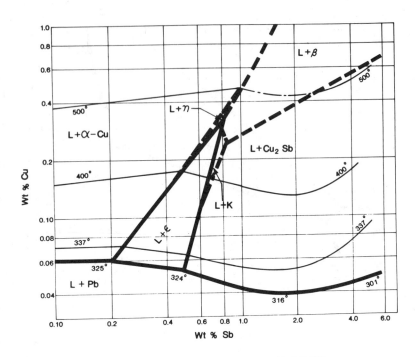

FIGURE 7 - Pb-Cu-Sb SYSTEM, LEAD CORNER

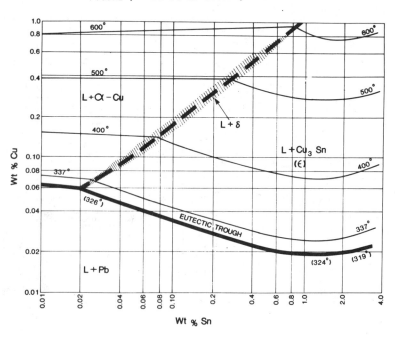

FIGURE 8 - Pb-Cu-Sn SYSTEM, LEAD CORNER

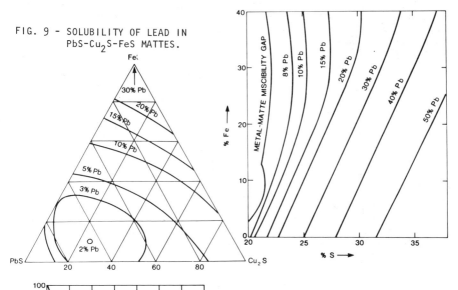

FIG. 9 - SOLUBILITY OF LEAD IN PbS-Cu₂S-FeS MATTES.

FIG. 10 - SOLUBILITY OF LEAD IN Cu-Fe-S MATTES.

The data of Fig. 9 are here re-plotted, as "adjusted" compositions:

$$Adj. \% Fe = \frac{Fe \times 100}{Cu+Fe+S} \; ;$$

$$Adj. \% S = \frac{S \times 100}{Cu+Fe+S} \; ;$$

$$Adj. \% Pb = \frac{Pb \times 100}{Cu+Fe+S+Pb} \; .$$

Example: What is the lead content of a matte containing 7% Fe, 37% Cu, 18% S?

Adjusted compositions are: 11.3% Fe, 59.7% Cu, 29.0% S. Figure 10 shows 35% Pb expected in such a matte. Actual lead content would be:

$$35 \times 62 \div 65 = 33\% \ Pb, \pm 2\% \ Pb.$$

There are thus 5% of other materials in the matte.

Note: All solubilities are those applying at temperatures just above the respective melting points.

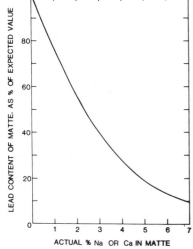

FIG. 11 - EFFECT OF Na OR Ca IN LOWERING LEAD CONTENT OF MATTE.

Example: What is the lead content of a matte containing 7% Fe, 37% Cu, 18% S, 3% Na?

In the absence of Na, 33% Pb would be soluble in the matte, as calculated above. Fig. 11 shows that 3% Na lowers the lead to 40% of expected value, i.e. 0.40 × 33 = 13.2% Pb. The lead solubility is thus 13% Pb ± 2% Pb.

485

Figure 11 shows how the saturation value of lead in mattes is lowered by Ca or Na, below the value expected from Figure 10. Once again, the diagram is usually accurate to within about ±2% of lead, but the relative error is larger in this case, as the lead solubilities are generally much lower in *soda mattes*.

The *iron-silica or soda-matte processes* for smelting copper dross(11-14) may be understood in the light of these principles.

Normally there is a speiss phase present as well as a matte phase, when copper drosses are smelted. The lead loss in speiss is also a matter of serious concern, and it was long ago shown that Pb content is a function of Cu/As ratio in the speiss(15). This was thought to be largely due to the temperature effect - the lead minimum solubility occurring near the eutectic composition, but the work of Matyas(16) shows the lead solubility to be composition-dependent even at constant temperature.

Recent developments in copper dross treatment include the development of hydrometallurgical processing - a sulfuric acid leach followed by solution purification and electrowinning (suitable for sulfide material) or an ammonia leach, followed by solvent extraction and electrowinning (suitable for metallic copper or oxidic material, which results from Imperial Smelting furnace lead)(17).

Decoppering Lead with Sulfur

Although the minimum value of Cu in lead that may be produced by precipitation with sulfur is 0.05%, as already discussed, it is possible to remove copper to vanishingly small levels by stirring sulfur into the lead at temperatures near the freezing point, and although the literature abounds with supposed "explanations" of the phenomenon, on examination these invariably reveal that the facts of the situation are unknown to the investigators.

In the presence of more than about 0.02% Ag, if 0.1% of S is stirred into lead at about 330°C, then the copper content is lowered to around 0.001% Cu within about 15-20 minutes(18,19). The sulfur particles must be not too large nor too small, must be added gradually, a shovel-full at a time, and the stirring must be adequate to provide a good vortex, preferably with an impellor producing considerable shear in the lead, to promote reaction. The stirring must be concluded about 5 minutes after the sulfur addition is completed, and the dross must be removed without raising the temperature, otherwise copper reverts from the dross into the lead, and the final figure may be 0.01-0.02% Cu instead of 0.001-0.002% Cu.

Figure 12 shows the course of decoppering under several sets of conditions:

Curve 1 represents copper content of silver-free lead as a function of time while stirring in 0.1% S at 330°C, and curve 4 similar conditions for a lead containing 0.04% Ag. Curve 2 resulted when only 0.05% S was added, and curve 3 when a total of 0.1% S was added as ten equal additions of 0.01% S each, at 1-minute intervals, approximating plant practice whereby the sulfur is added gradually over ten minutes. The tests were performed on a 16 kg kettle of lead, exposed to atmosphere(19), and the conditions approximate to plant practice.

Figures 13 and 14 show the influence of the silver or the tin contents of lead on the degree of decoppering achieved by a 5 minute stir (adding five

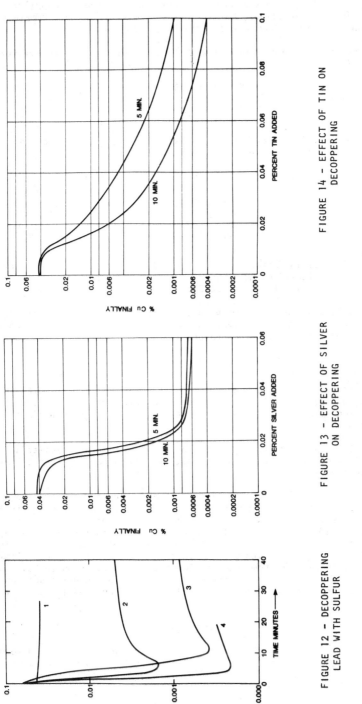

FIGURE 14 – EFFECT OF TIN ON
DECOPPERING

FIGURE 13 – EFFECT OF SILVER
ON DECOPPERING

FIGURE 12 – DECOPPERING
LEAD WITH SULFUR

Note: All experiments conducted at 330 – 335°C

487

equal amounts of 0.02% S every minute) or a 10-minute stir (with further additions of 0.02% S every minute) as determined by Davey & Happ[19]. Tin behaves similarly to silver, but is more effective at higher temperatures of operation up to 100° above the freezing point of lead.

Elemental sulfur can combine with lead, or with copper in solution in lead, as follows:

$$S + Pb = PbS \qquad\qquad (1)$$

$$S + Cu = CuS \qquad\qquad (2)$$

In the presence of Ag or Sn, reaction (1) is greatly slowed, but even in their absence it is considerably slower than (2), so that some decoppering can occur below the equilibrium value of 0.05% Cu.

CuS can further react with lead or dissolved copper:

$$CuS + Pb = Cu \;\; + PbS \qquad\qquad (3)$$

$$2CuS + Pb = Cu_2S + PbS \qquad\qquad (4)$$

$$CuS + Cu = Cu_2S \qquad\qquad (5)$$

Reaction (5) is faster than (3) and (4) - and none of these reactions appears to be affected by Ag or Sn - so that further decoppering to low values can occur by this means (20,21).

Finally, Cu_2S can react with Pb and allow copper to revert from the dross to the lead, and this reversion occurs if stirring is continued too long, or especially if the temperature is raised:

$$Cu_2S + Pb = 2Cu + PbS \qquad\qquad (6)$$

The dross contains Cu_2S and PbS crystals, with entangled metallic lead, and some CuS may also still be in evidence at the conclusion of the operation, but reaction (5) is rather fast, and very little CuS is present in the final dross as a rule.

No elements other than Ag or Sn are known to have any similar "catalytic" effect - certainly As, Sb and Bi do not. Au and Pt metals are not known to have been investigated.

In the absence of Ag or Sn, it is not possible to decopper lead to less than 0.01% Cu unless repeated sulfur stirs are given, when prohibitively large amounts of PbS are drossed off. However within the last decade it has been found that decoppering to low values is still possible if FeS_2 (pyrite) and elemental sulfur are stirred into the lead simultaneously. In this case, bornite and chalcopyrite predominate in the dross, and the copper elimination is much faster and more complete at higher temperatures - up to 450°C at least[22]. The simultaneous stirring in of sulfur and caustic soda has also been found effective in decoppering lead in the absence of silver or tin[23].

Softening

The removal of the elements chiefly responsible for hardening lead (antimony, tin and arsenic) is termed "softening". This may be accomplished by preferential oxidation due to differences in the free energies of formation of the oxides of these metals and lead.

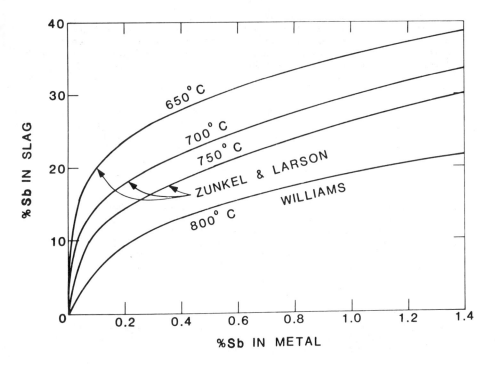

FIGURE 15 - DISTRIBUTION OF ANTIMONY BETWEEN
SLAG AND METAL PHASES.

In a low-temperature variant of this procedure - the Harris Process -
the oxidation is performed in kettles, and the oxides are absorbed into a
caustic soda melt.   The classical procedure, due to Parkes, involves
oxidation in a reverberatory furnace at elevated temperature, so that most of
the impurities are removed as a liquid slag, although in the last stages of
removal the major product may be PbO, giving a solid dross requiring to be
skimmed.

At temperatures of over 600°C, the order of formation of the oxides is
SnO_2, As_2O_3, Sb_2O_3, PbO.   By rapid stirring of air into lead at red-heat
(600-650°C) it is possible to form a solid dross in which SnO predominates,
and which may be worked up for tin.

In reverberatory softening practice, it is customary to reduce the Sb
content of lead to about 0.03%, while still producing a liquid slag (and
leave the removal of the last of the antimony until after desilverising and
dezincing, so that the last traces of all oxidisable elements are removed in
a "final refining" stage) when As and Sn will have been reduced to traces
only.

Figure 15 shows the antimony content in a lead-antimony-oxide slag as a
function of the lead content of lead in equilibrium with it, according to
data from Williams(24) and Zunkel & Larson(25), which are in reasonable
harmony.   These are preferred to much different results by Pelzel(26) and
several other authors(27).

There is evidence that the oxidation of antimony occurs faster than that of lead, so that the slag formed has an antimony content higher than that corresponding to equilibrium, and on standing the antimony content would partially revert to the lead. Although most publications refer to the slag as being composed of PbO and Sb_2O_3 (and minor amounts of As_2O_3) at Port Pirie the *softener slags* contain more oxygen than this, and the composition analysis comes close to 100% if the antimony is considered as Sb_2O_4.

The physical chemistry of softening has not been fully examined, largely because there is small possibility of significant improvement by optimising operations. The thermodynamic data on the metal oxides are uncertain, and sound activity data on the oxide systems do not exist.

Figure 15 may be used to estimate the slag composition in equilibrium with lead of a given antimony content, and so give an approximation to the softening (oxidation) stage, and also to the metal alloy that may be produced by subsequent reduction of the slag. Since most antimony has been disposed of to date as antimonial-lead alloys of 2-11% Sb content, it has not been necessary to produce pure antimony as a by-product. Any arsenic content of the alloy may be reduced if required by preferential removal by a modified Harris process - stirring with caustic soda and/or nitre - as later described in the final refining operation.

Harris Softening

Until recently, the only correct description of the fundamentals of softening by the *Harris process* was that of Lauterbach(28). Now an excellent account of modern practice is also available from Emicke *et al.*(29).

When tellurium is present in excess of 0.01% Te, it is removed first, then As and Sn are removed together, and finally Sb. Essentially the process consists in oxidising the impurities, which are then converted to sodium salts, and these are suspended in a caustic melt of NaOH or NaOH-NaCl, so that their separation from lead is a simple matter without requiring drossing off which would result in considerable entangled metallic lead.

Tellurium is reacted with NaOH in the absence of oxidation, as far as possible:

$$3Te + 6NaOH = 2\ Na_2Te + Na_2TeO_2 + 3H_2O \tag{7}$$

$$xTe + Na_2Te = Na_2Te_{x+1} \tag{8}$$

Unavoidably some oxidation of As, Sn and Sb also occurs, and this contaminates the melt with arsenate, stannate and antimonate, which must be dealt with in the subsequent wet-way treatment.

The melt is granulated into water, when metallic Te-sludge precipitates:

$$2Na_2Te_{x+1} + Na_2TeO_3 + 3H_2O = (2x+1)Te + 6NaOH \tag{9}$$

Any excess sodium telluride in solution shows a reddish coloration, and may be precipitated by air-oxidation:

$$Na_2Te_x + \tfrac{1}{2}O_2 + H_2O = xTe + 2NaOH. \tag{10}$$

The Te-sludge, contaminated by As, Sb, Sn and Pb, is worked up for Te.

An alternative procedure for Te removal is due to Asarco(30), and does not require the use of the special Harris apparatus. Metallic sodium is stirred into the lead, together with NaOH, and the resulting sodium telluride is dissolved in the molten NaOH, and skimmed off the lead. If the stirring is continued too long, oxidation causes Te to revert to the lead.

Although the order of oxidation is As, Sn, Sb and then Pb, in practice As and Sn are removed together, as they are more difficult to separate from each other than from Sb, which is· scarcely oxidised at all until As and Sn have been removed to very low values. As and Sn can react with NaOH even without oxidation:

$$2As + 6NaOH = 2Na_3AsO_3 + 3H_2, \qquad (11)$$

$$Sn + 2NaOH = Na_2SnO_2 + H_2 . \qquad (12)$$

In this case the lower valency forms (arsenite and stannite) of the sodium salts are formed.

However, the main reactions occur by oxidation, using sodium nitrate as a strong oxidising agent, when the higher valency salts are formed in each case:

$$2As + 2NaNO_3 + 4NaOH = 2Na_3AsO_4 + N_2 + 2H_2O \qquad (13)$$

$$5Sn + 4NaNO_3 + 6NaOH = 5Na_2SnO_3 + 2N_2 + 3H_2O \qquad (14)$$

$$2Sb + 2NaNO_3 + 4NaOH = 2Na_3SbO_4 + N_2 + 2H_2O . \qquad (15)$$

Any lower valency salts formed by displacement of hydrogen are also oxidised by sodium nitrate, so that the products are sodium arsenate, stannate and antimonate. These are separated, and the caustic soda regenerated, by wet-way processing(29). Sodium antimonate is practically insoluble at all NaOH concentrations, and can be freed from As and Sn by repeated washing and decantation. Addition of lime precipitates calcium stannate from the concentrated NaOH solutions (around 300 g/l NaOH) whereas calcium arsenate is precipitated only after dilution to <80 g/l NaOH. The final solution, almost free from Sn and Sb, and low in As (about 3 g/l) is evaporated to produce "regenerated Harris salt" for re-use.

As well as entering into chemical combination with the impurity oxides, NaOH also acts as a suspension medium for them, facilitating their separation from the lead. It can hold about 20% of either As or Sn in suspension, or 30% of Sb, before becoming too thick to flow. In earlier times a near-eutectic mixture of NaOH and NaCl was always used as the suspension medium, but this has fallen into disuse at some plants.

The calcium stannate precipitated is smelted to produce lead-tin alloys. The calcium arsenate precipitate is heavily contaminated with calcium carbon-ate, and is discarded. CO_2 from the atmosphere tends to be absorbed by the molten caustic soda or the solutions, and so when lime is added the reaction occurs:

$$Na_2CO_3 + Ca(OH)_2 = CaCO_3 + 2NaOH \qquad (16)$$

The pure (white) sodium antimonate recovered from the latter stage of the process (after As and Sn have been largely removed) is sold as such, after drying, crushing and sizing. The impure (grey) sodium antimonate recovered from the earlier stage of the process, while chiefly As and Sn are being removed from the lead, is recycled to the lead blast furnace.

The NaOH regenerated from the spent solutions generally contains several percent of chloride, sulfate and carbonate, as much as 5% As, and minor amounts of Pb, Sb, Sn and Cu(16).

Desilverising

After removal of impurities less noble than lead by the softening (oxidation) process, the only notable impurities remaining are those more noble than lead - silver and gold, and bismuth. These cannot be removed from lead by reaction with oxygen, chlorine, sulfur, etc., but rather lead can be removed from them by these reagents - and this fact is utilised in working up the by-products.

Silver can be removed by precipitation with zinc, forming silver-zinc crystals of ε- or η-phase, depending upon the relative concentrations and temperature. As the solubilities are strongly temperature-dependent, the finishing temperature is the lowest practicable - just above the freezing point of the lead - about 320°C, since it contains almost the eutectic concentration of zinc still in solution.

Provided that sufficient time is allowed for stirring in of reagent to achieve equilibrium, and to settle out precipitates, kinetics plays no part in the operation, which is then governed entirely by equilibrium consider-ations. In order to optimise the operation, it is necessary to operate in two counter-current stages, whereby the zinc consumption is appreciably less than for a 1-stage operation. Although multi-stage counter-current operation has been proposed in the patent literature, this has not proved to be economical in practice, and two-stage operation is universally used for the batch procedure. A continuous desilverising operation has been practised at Port Pirie since the early 1930's, but has not found wide application else-where; although it was a great advance on batch desilverising when introduced, the modern batch operation is now more economical in terms of zinc and fuel consumption, by-product recovery, maintenance, and even of labour. A new continuous desilverising process to overcome these disabilities has been demonstrated in a pilot plant, but not yet developed(31,32).

The reaction

$$Ag + rZn \rightleftarrows (Ag\text{-}Zn_r)_{\varepsilon\text{-} or \eta\text{-}phase} \qquad (17)$$

is difficult to treat in a simple thermodynamic fashion, because the ratio r varies with both solution composition and temperature.

Graphical solutions to the course of the process for several variants of the 2-stage process were presented over 20 years ago by Davey(34), and al-though the fundamental data have been refined since, the overall picture has not sensibly changed.

Figure 16 shows the lead corner of the Pb-Ag-Zn phase diagram, as recently re-assessed by the writer. In most practical cases the entire desilverising operation is conducted in the phase region liquid + ε-phase, until the last stages of cooling in the second stage, when the solution enters the phase region of liquid + η-phase.

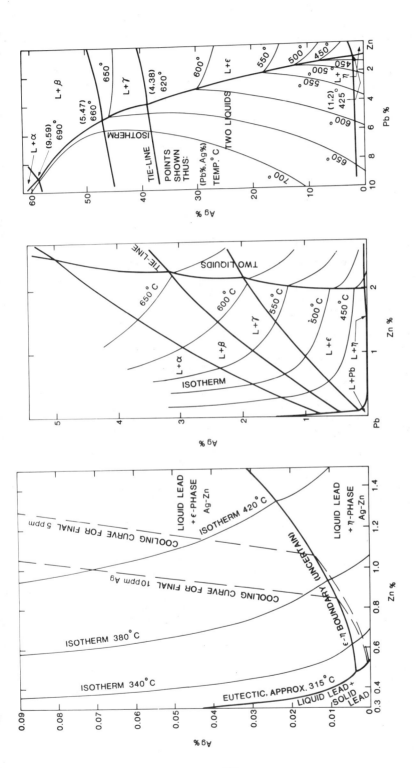

FIGURE 18 – Pb-Ag-Zn SYSTEM,
ZINC SIDE OF MISCIBILITY GAP

FIGURE 17 – Pb-Ag-Zn SYSTEM,
LEAD SIDE OF MISCIBILITY GAP

FIGURE 16 – Pb-Ag-Zn SYSTEM,
LEAD CORNER.

493

The best method of conducting the 2-stage desilverising process is to stir the second-stage precipitate (or "crust", consisting of Ag-Zn crystals and entangled metallic lead) from the previous charge into softened lead at a temperature of 460-470°C. With an efficient stirrer, this takes 20-30 minutes. Zinc (and to a lesser extent silver) is dissolved out of the Ag-Zn crystals, and the leached crystals come close to equilibrium with the lead solution. The "rich crust" is then skimmed off for further treatment, and fresh zinc is added to the lead bath, when once again 20-30 minutes of stirring at 460-470°C suffices to dissolve the zinc completely and bring any Ag-Zn crystals formed into equilibrium with the lead solution. The lead bath is then cooled to nearly freezing, when ε-phase Ag-Zn crystals separate out, until a temperature of about 370°C is reached, after which η-phase Ag-Zn crystals (with a much lower Ag/Zn ratio) separate out. If the zinc additions have been calculated correctly, the final lead, just before freezing, contains 5-10 ppm Ag and about 0.55% Zn.

The eutectic trough shown in Figure 16 slopes downward from a maximum of about 320°C to the lead-zinc eutectic temperature of 318°C, and there is no practical limit to the degree of desilverising that can be achieved using this process: the greater the zinc addition, the lower the final Ag-value in desilverised lead. If very low Ag-values are required, it is possible to re-heat the lead, add further zinc, and cool again, and this operation may be repeated indefinitely, giving ever lower final Ag-values. It is not economical to use more than two counter-current stages for normal lead refining purposes, and it is doubtful whether the additional silver recovered below about 10 ppm pays for the added costs of zinc required.

The Williams *continuous desilverising* process(34) is a one-stage operation, whereby softened lead (after deoxidation by a preliminary addition of zinc, forming an oxide dross which is removed before desilverising) is heated to about 600-650°C and flowed into the top of a deep kettle, where there is a layer of molten zinc, in the initial stages, or of molten silver-zinc-lead alloy in the later stages of a cycle. The lead dissolves zinc (and later silver as well) from the top layer, and descends in the kettle, where it is progressively cooled to freezing, throwing out Ag-Zn crystals which rise up through the descending lead and dissolve in the zincy layer at the kettle top.

The lead composition changes along a "cooling curve" to the right of those shown in Figure 16, and reaches a composition of about 3 ppm Ag, 0.57% Zn, at about 318°C at the kettle bottom. The composition of the zinc alloy layer at the top of the kettle gradually changes from zero-silver, about 2% Pb, to about 15% Ag, 20% Pb, 65% Zn over a period of about 16 hours (depending upon the rate of supply of silver in the incoming lead) by which time it begins to freeze at 600°C. The mushy alloy is then dipped off and sent for by-product treatment, and the zincy layer is replenished with fresh zinc. The lead content of this alloy is about ten times the theoretical, due to imperfect liquation.

The Williams (Port Pirie) procedure has been adopted by the Met-mex Penoles plant in Mexico, and (with slight modifications) by the Ust-Kamenogorsk refinery in Russia. It is about to be adopted by Trepca, in Yugoslavia.

An alternative proposal for continuous desilverising, utilising mixer-settler units, was made by Davey in 1954(34), but not actually tested until 1960(31,32). This would enable continuous desilverising to be performed with a minimum of fuel and zinc consumption, and without serious corrosion problems, if performed as a 2-stage counter-current process. The proposal of Castle & Richards(31) to perform this low-temperature continuous desilverising procedure as a 1-stage process is less attractive, as it practically doubles the zinc-silver ratio in the crystals as compared with the 2-stage counter-current procedure.

Since Cu-Zn and Au-Zn intermetallic compounds have a higher negative free energy of formation than Ag-Zn, decoppering and degolding by zinc occur preferentially to desilverising. Copper and gold are therefore removed to traces during desilverising. The amount of gold present is normally too small to have any appreciable effect on the amount of zinc consumed. Copper is thought to require four times its weight of zinc.

Silver Crust Treatment

The rich silver crust removed in the first stage of desilverising comprises ε-phase Ag-Zn crystals plus a good deal of entangled lead. This lead may be partially removed by physical pressure, and the Howard press is still widely used for this purpose, yielding a content of about 50-60% Pb in the pressed crust.

An alternative to pressing is liquation, in which the entire crust is melted, whereby it separates into two conjugate liquid layers, the upper one zinc-rich, and the lower lead-rich. A flux of zinc ammonium chloride helps to prevent atmospheric oxidation, which would hinder the liquation. In the absence of oxide, a temperature of 600-650°C serves to melt the crust completely, yielding a zinc-silver alloy of 2-4% Pb theoretically (see Figure 18), but usually more than this in practice; the liquated lead, containing about 2% Zn and usually 1-2% Ag (see Figure 17) is returned to the desilverising kettle.

The partially deleaded silver-zinc alloy is next subjected to distillation, to recover most of its zinc content for re-use. In older plants, Faber du Faur furnaces are used, in which an oil- or gas-fired plumbago crucible is heated to a temperature above 1000°C, and the zinc is thereby evaporated until only a few percent are left in the alloy, the zinc vapour being condensed in a brick-lined steel-cased condenser luted to the plumbago crucible by clay. The distilland residue comprises mainly silver, with up to several percent each of Pb, Zn and Cu. These are removed by cupellation - air oxidation forming a slag which is tapped off and returned to the smelter - followed by a further refining operation in which nitre is added for further oxidation of the base metals and the incorporation of their oxides in a caustic slag - also returned to the smelter. The refined silver bullion is then parted from gold (if warranted) by electrolysis, and the electrodeposited silver is given a final clean-up by bubbling Cl_2 through it to remove the last traces of base metals, which are chloridised in preference to silver.

More modern plants employ the Penarroya-Leferrer vacuum distillation furnace(35) rather than Faber du Faur furnaces, to distil off zinc from the rich crust. The same principles apply to vacuum dezincing of silver alloy and of desilverised lead, as described subsequently, although the concentrations, temperatures and pressures are considerably different. Since the furnace operates in the range 800-1000°C (below the melting point of silver) it is desirable to retain considerable lead in the alloy so that it does not freeze when most of the zinc has been removed. The liquation of the crust is still carried out, mainly with the object of removing oxides, so that the free metal surface in the vacuum still is not obstructed, and lead is added as necessary to the still, to keep the melting point of the final Ag-Pb alloy low enough to ensure that it is liquid.

Vacuum Dezincing

The lead after desilverising contains 0.55-0.60% Zn, and this can be removed by preferential oxidation or chloridation. The latter is capable of producing a saleable $ZnCl_2$, which was once in demand for tanning, but this market disappeared some decades ago and so the Betterton chlorine dezincing process has fallen into disuse. The principles are clearly understood, and the process has been quantitatively evaluated(36).

Removal of zinc by oxidation produces a dross which is a mixture of zinc and lead oxides, and entangled metallic lead, which must be returned to the smelter for recovery of its lead content, and loss of its zinc to the slag (for possible recovery by slag fuming). The Harris process for zinc removal operates on much the same principles as those for softening, already described above, except that zinc does not require an additional oxidising agent, as it can replace hydrogen in NaOH:

$$Zn + 2\ NaOH \rightarrow Na_2ZnO_2 + H_2. \qquad (18)$$

The slurry of sodium zincate in molten NaOH can be separated by wet-way procedures, to produce products for sale and re-use. However Harris dezincing, like Betterton chlorine dezincing and drossing or slagging removal of zinc, has given way almost universally to vacuum dezincing, which is capable of recovering 90-95% of the zinc content of desilverised lead, followed by a caustic soda stir (sometimes termed "modified Harris process") in a final refining stage to remove the last traces of all base metal impurities by oxidation and incorporation into a caustic dross.

The vacuum dezincing operation is kinetically controlled, as there is no question of achieving an equilibrium between the liquid and vapour phases, because a condensing surface is placed close to the evaporating surface and so volatile constituents are removed from the vapour phase as fast as they enter it. Under these conditions, zinc continues to leave the metal phase in a proportion greater than its mole fraction in the metal, even in a single stage process. If molecular conditions apply (i.e. there is only a negligible chance of an evaporated atom or molecule suffering a collision in the vapour phase before reaching the condensing surface) this can readily be seen to be so, as in the following calculation. (It is also true for vacuum distillation in a higher pressure range, but the calculation is more complicated).

Taking the data for the vapour pressures of pure zinc and lead as $P_{Zn} = 11.5$ torr, $P_{Pb} = 0.38$ torr, and the activity coefficient as $\gamma_{Zn} = 10$ in dilute solution in liquid lead, all at 600°C, and $M_{Pb} = 65.4$, $M_{Pb} = 207.2$, we have the relative rates of evaporation of zinc and lead as:

$$\frac{E_{Zn}}{E_{Pb}} = \frac{\gamma_{Zn} \cdot P_{Zn} \cdot N_{Zn}}{P_{Pb} \cdot N_{Pb}} \cdot \sqrt{\frac{M_{Pb}}{M_{Zn}}} \quad \text{molar ratio,}$$

$$= 540\ N_{Zn}, \text{ since } N_{Pb} \simeq 1. \qquad (19)$$

The symbols are defined in Table 1. Zinc removal relative to lead is therefore proportional to 540 times its mole fraction at all mole fractions, right down to zero. The refining action would cease only if the zinc removal relative to lead fell to the value of its mole fraction in the remaining lead. The refining to arbitrarily low values of impurity content is possible obviously because the impurity does not accumulate in a phase adjacent to the metal, but is continuously removed, by condensation.

496

TABLE I

Nomenclature for Vacuum Distillation Equations

$a \quad = \quad \alpha \sqrt[]{\dfrac{M}{2\pi RT_1}}$

$\alpha \quad = \quad$ accommodation coefficient (unity for clean metal surfaces)

$b \quad = \quad 0.796 \ DP/TL$

$C \quad = \quad$ condensation rate, $gm/cm^2 sec$

$c \quad = \quad P_c \sqrt{T_1/T_c}$

$\bar{c} \quad = \quad \dfrac{2}{\sqrt{3}} \sqrt{\dfrac{2RT}{M}}$, mean projected molecular velocity, cm/sec

$D \quad = \quad$ Maxwell's diffusivity, cm^2/sec

$E \quad = \quad$ evaporation rate, or overall distillation rate, $gm/cm^2 sec$

$e \quad = \quad$ exponential function, 2.71828

$f \quad = \quad$ ratio $p/$(bulk concentration of volatile species, in gm/cm^3)

$k \quad = \quad$ mass transfer coefficient, cm/sec

$L \quad = \quad$ distance between evaporating and condensing surfaces, cm

$M \quad = \quad$ molecular weight of volatile species, gm

$P \quad = \quad$ total pressure in distillation space, torr

$p \quad = \quad$ partial pressure of volatile species in bulk of distilland, torr

$p' \quad = \quad$ partial pressure of volatile species in evaporating surface, torr

$P_1 \quad = \quad$ partial pressure of volatile species above evaporating surface, torr

$P_2 \quad = \quad$ partial pressure of volatile species above condensing surface, torr

$P_c \quad = \quad$ partial pressure of volatile species in condensing surface, torr

$R \quad = \quad$ gas constant, $82.07 cm^3$ atm./deg. mole

$r \quad = \quad$ ratio $v/\bar{c}$

$T \quad = \quad$ temperature of vapour between evaporating and condensing surfaces

$T_1 \quad = \quad$ temperature of evaporating surface

$T_2 \quad = \quad$ temperature of vapour above condensing surface

$T_c \quad = \quad$ temperature of condensing surface

$V \quad = \quad$ vacuum, or residual gas pressure, torr

$v \quad = \quad$ nett velocity of vapour distilling, cm/sec

The process has been extensively studied, both theoretically and experimentally, and may be considered as comprising a series of steps:

(1) transfer of zinc from the bulk of the liquid to the liquid surface;

(2) evaporation (less back-condensation) of zinc from the liquid to the gaseous phase at the surface (interface);

(3) diffusion of zinc across the gaseous phase between evaporating and condensing surfaces;

(4) condensation (less back-evaporation) of zinc from the gaseous to the solid or liquid phase of the condenser.

Expressions may be derived for the rates of each of these steps, and at the steady state they are all equal. Following this approach, Davey(37) derived:

$$p' = V+C+2E/a - Ve^{-E/b}, \qquad (19)$$

where a, b, c are constants for a given metal system and distillation plant:

$$a = \alpha \sqrt{\frac{M}{2\pi RT_1}} \qquad (20)$$

$$b = 0.796DP/TL \qquad (21)$$

$$c = p_c \sqrt{T_1/T_c}. \qquad (22)$$

This equation gives the rate of distillation *in vacuo* only if conditions are such that the nett rate of vapour flow is much less than the free molecular velocity, so that the pressure of the vapour is almost the same in all directions. If the nett rate of vapour flow v, is related to the average *projected* velocity of zinc atoms $\bar{c}$, by the relation:

$$r = v/\bar{c}, \qquad (23)$$

then the corrected rate of distillation, allowing for the fact that the pressure of zinc atoms in the direction of distillation will be increased by the factor $(1 + r)^2$, and the back-pressure will be decreased by the factor $(1 - r)^2$, is given by the expression:

$$E = \frac{a}{2+2r^2} [p' (1+r)^2 - (1-r^2)^2 V(1-e^{-E/b}) - (1-r)^2 c] \qquad (24)$$

(The factors $(1+r)^2$ and $(1-r)^2$ arise by consideration of the rates of evaporation or condensation using coordinates moving with the average projected velocity, $\bar{c}$, of zinc atoms, and the relationship $p \propto v^2$. It is not just an intuitive argument, as has been suggested(38).

Warner(39) showed that allowance must be made for the rate of transfer of zinc from the bulk of the liquid to the surface, when the evaporation is occurring from the surface of lead flowing in a channel:

$$E = \frac{k}{f} (p-p'), \qquad (25)$$

and estimated k = 0.05 cm/sec, by analogy with an empirically determined rate of absorption of CO_2 in water(39). He later estimated k = 0.02 cm/sec.(40),

498

but does not give sufficient details to enable a check to be made on his calculations. One must have some reserve in accepting this value, especially as there was a conceptual error in his earlier work(39), which led to the erroneous conclusion that α, the accommodation coefficient, is much less than unity(41). Adding the diffusion equation to the three equations for evaporation, condensation, and vapour transfer, we obtain:

$$E = \frac{a}{2+2r^2 + \frac{af}{k}(1+r)^2}[p(1+r)^2 - c(1-r)^2 - V(1-e^{-E/b})(1-r^2)^2]$$

$$(26)$$

This equation calculates the distillation rate only for evaporating and condensing surfaces of equal area, and parallel to each other. Modification of the equation when these conditions do not apply has been discussed(37).

Since the numerical values of a, b and c may all be determined from fundamental data, the only empirical constant in this equation is k - apart from molecular weights, atomic diameters, vapour pressures, activity coefficients, etc..

In dilute solution in lead, the partial pressure of zinc in solution is approximately proportional to the weight percent:

$$p = x \, w, \qquad\qquad (27)$$

where x is a temperature-dependent constant.

Thus the elimination curves for zinc from lead as related to area, mass of lead and time may be calculated for either batch or continuous processes, as shown by Davey(37,42). Figure 19 shows the results of such calculations when the lead is at a temperature of 600°C, the condenser is water cooled, about 30 cm away from the lead surface, the surface concentration of zinc equals the bulk concentration, and evaporating and condensing surfaces are equal in area. Then, from ref.(27):

$$af/k \to 0, \text{ and } a = 0.016, \, b = 2.83.10^{-5}, \, c = 0, \, x = 3.46 \, .$$

The figure shows that failure to make the abovementioned correction for the high velocity of zinc atom across the distillation space over-estimates the time required by a factor of about 2. In the higher concentration ranges the rate, with a vacuum of 100 torr, is almost the same as the rate with a complete vacuum - i.e. molecular distillation.

Debismuthising

Bismuth, like silver, is more noble than lead, and thus cannot be removed from it by combination with sulfur, oxygen or the halides. Lead can in fact be removed from bismuth by the use of these reagents, and this property is used in bismuth refining.

Bismuth (and silver) can be separated from lead by fractionally oxidising or fractionally crystallising the metal, to yield a liquid slag or solid crystals respectively, purer in lead and lower in impurity than the incoming metal.

Preferential oxidation of lead into slag, followed by reduction of slag to produce a low-bismuth lead, results in about a 20:1 enrichment of bismuth between metal and slag at equilibrium. A process involving a number of stages

499

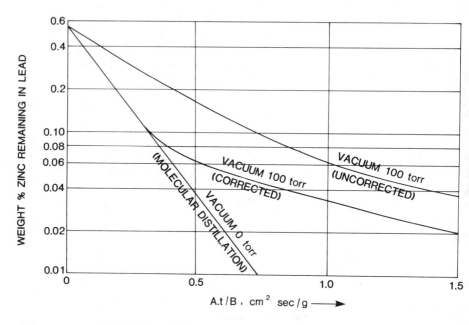

FIGURE 19 : ZINC ELIMINATION BY VACUUM DEZINCING

of cupellation and reduction, with recycling of intermediate products, was at one time practised, and indeed it was once the only method available for debismuthising and for desilverising. In some circumstances a single cycle of oxidation and reduction, to produce a proportion of the output with a specified bismuth content, might still be economic today. The principles involved are identical with those discussed previously for antimony removal, and the quantitative evaluation of the process has been adequately described previousl (43,44).

Bismuth (and silver) can also be separated from lead by fractional crystallisation, the primary lead crystals being purer than the mother liquor. The Pattinson process utilised this phenomenon, using a many-cycled freezing and remelting procedure which entailed high fuel and labour consumption. It became obsolete before the turn of the century with the advent of the Parkes process for desilverising, and left lead refineries without an economic possibility of debismuthising. A recent revival of essentially the same process using more sophisticated equipment is currently under development(45), and the principles of such phase separations were thoroughly explored at the time of zone refining development techniques(46).

Isothermal squeezing during solidification or melting has also been used recently to purify metals in the laboratory, although it has not been suggested that this will be developed for industrial use(47,48).

Bismuth may be precipitated from lead by addition of alkali or alkaline-earth metal reagents, just as silver is precipitated by addition of zinc - but with the important difference that the precipitate is of fixed composition, so that it is not possible to reduce reagent consumption by reacting in two counter-current stages. The reagent addition is determined by the amount required to combine with the bismuth, plus that required to enter into solution in the lead and drive down the concentration of bismuth remaining in the lead:

$$Bi + xA = A_x Bi \qquad (28)$$

or

$$Bi + xA + yB = A_x B_y Bi \qquad (29)$$

where x and y are stoichiometric constants, and A and B represent reagent metals. In the latter case a double bismuthide is postulated.

Kroll found that either Ca or Mg could debismuthise lead, and for a time they were used commercially, but the degree of debismuthis·ing achievable was insufficient for many purposes.

According to the most recent data(49), the solubility relations near the freezing point of lead can be represented by:

$$\log ([Mg]^3[Bi]^2) = 5.36 - 4.280/T \qquad (30)$$

or

$$\log ([Ca]^3[Bi]^2) = 12.46 - 11,160/T \qquad (31)$$

where T is in Kelvin, [X] represents wt. %X in liquid lead.

More complete debismuthising may be achieved by using simultaneous additions of Ca and Mg, as found by Betterton and Lebedeff, the process being generally referred to as the Kroll-Betterton process.

The Pb-Bi-Ca-Mg diagram given by Davey(33) should be revised in the light of more recent data(49,50) as shown in Figure 20. At the lead liquidus temperature, in the lead corner of the quaternary system the relation between Mg, Ca and Bi in solution in the lead is given approximately by:

$$\log ([Mg]^2[Ca][Bi]^2) = -7.37 . \qquad (32)$$

This is certainly more accurate than Davey's value of -7.74, based on experimental work of 25-30 years ago, and probably to be preferred to the figure of -7.1 found by Moffatt and Iley(50). Figure 20 shows the "isobis" based on this formula, together with the approximate positions of phase boundaries in the lead corner of the system. Its interpretation and use in optimising reagent additions have been previously described by Davey(33).

Reagents (magnesium as the metal, calcium as a 5% Ca-alloy in lead) are stirred into the lead to be debismuthised at about 420°C, and the lead is cooled back almost to the freezing point, while $CaMg_2Bi_2$ crystals separate, leaving lead debismuthised to the extent shown in Figure 20, dependent upon the residual Ca and Mg contents. The crusts separating are very diluted with entangled metallic lead, and general practice is to stir them into the next batch of lead to be debismuthised, to liquate out portion of this lead at the higher temperature. This operation does *not* reduce reagent consumption (as in the analogous desilverising operation) because the $CaMg_2Bi_2$ crystals do not change in composition. The liquation simply increases the yield of treated lead per batch and reduces the amount of crust for by-product treatment.

Further reduction of bismuth content is possible by stirring in antimony (normally in the form of antimonial lead) after removal of the $CaMg_2Bi_2$ crystals as completely as possible. Only recently was it shown(34) that the action of the antimony is to form $CaMg_2Sb_2$ crystals, isomorphous with $CaMg_2Bi_2$, and capable of substituting Bi for Sb in the lattice. Thus bismuth can be removed from lead to any desired limit by stirring in antimony, as

501

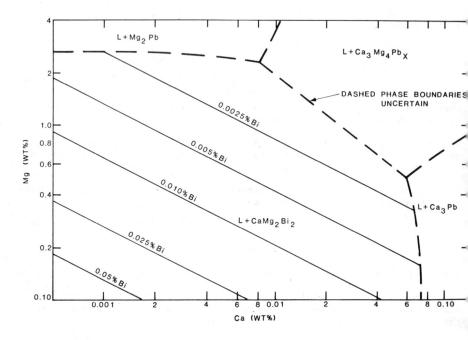

FIGURE 20 - Pb-Bi-Ca-Mg SYSTEM, LIQUIDUS SURFACE OF LEAD CORNER

long as Ca and Mg remain in the lead; of course, Ca and Mg in the lead may be replenished by further additions if necessary.

Both continuous and batch versions of Kroll-Betterton debismuthising are practised, as also of the subsequent fine debismuthising by stirring with antimony.

The Jollivet-Penarroya debismuthising process, using K and Mg additions (33) is no longer practised. Its principles are identical with those of the Kroll-Betterton process.

Residual Ca and Mg may be removed by oxidation or chloridation, or both, and the final traces of these and antimony and arsenic are removed in the final refining operation discussed below.

The by-product crust treatment consists of first oxidising or chloridising the Ca and Mg away, and then removal of lead from the resultant Bi-Pb alloy by preferential oxidation or sulfidation, to produce fine bismuth metal. A recent Asarco development(51) involves partial preferential oxidation of lead, as well as Ca and Mg, leaving an enriched Bi-Pb alloy after separation of the oxides.

Final refining

The last stage of treatment of lead before casting is final refining to remove all but traces of all the elements more easily oxidised than lead. The almost universally used process nowadays is the modified Harris process, whereby caustic soda and/or nitre is stirred into the lead at 450-500°C, and

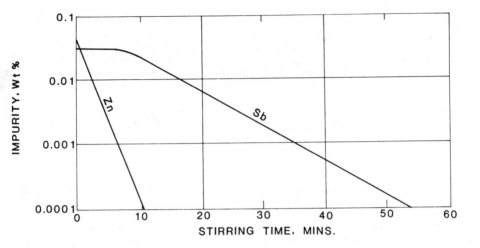

FIGURE 21 - FINAL REFINING BY MODIFIED HARRIS PROCESS

oxidation proceeds by reaction with oxygen from the atmosphere or from nitre decomposition, the latter being more intensive, and less selective: it refines rather more quickly, but produces more dross because more lead is oxidised in the process.

About 0.1% of NaOH or NaNO$_3$ are stirred into the lead, and these remove the 0.02-0.05% Zn and 0.03-0.04% Sb to less than 1 ppm each, the product being a granular or balled dross, which contains appreciable PbO. There is insufficient NaOH to act as a suspension medium for the impurity metal oxides, and thus permit a clean separation of these from lead. The dross is returned to the blast furnace.

Zinc is removed prior to antimony, which begins to be eliminated only when the zinc concentration has fallen to a few ppm, as shown in Figure 21. As, Sn, Ca, Mg, Fe and any other elements more electropositive than antimony are likewise eliminated to very low values early in the final refining operation. The reaction is first order with respect to antimony, as indicated by the straight line relationship between logarithmic concentration of anti- mony and time of stirring. The slope of the line (proportional to the rate of elimination) depends upon a number of factors: the rate is increased by faster impeller speed, increased temperature (in the range 400-600°C at least) and increased flux additions. Generally speaking, an increased total dross make (i.e. increased oxidation of lead) accompanies faster reaction rates. The relative refining rates of caustic soda and nitre depend upon stirrer impellor design, as discussed by Davey *et al*.(52) who also consider power requirements of the stirrer. Paschen(53) extensively investigated all the reactions occurring in final refining.

Although as far as is known all final refining operations throughout the world are batch processes, Davey's proposals for a continuous operation were pilot-planted on a limited scale with considerable success(31) and warrant further attention.

Conclusions

Although the physical chemistry of the various refining stages has not been investigated as thoroughly for some as for others, sufficient is known to enable both batch and continuous operations to be optimised fairly completely in relation to local conditions. It is obvious from many practices of long standing that this optimisation has not always been done, and that most refineries could with profit examine their procedures in the light of the principles set out here. Savings can be made without capital expenditure, and a contribution made to conservation of energy and other resources, as well as reducing labor - often of an unpleasant nature.

References

1. I.C.J. Kraus, "Cathode Deposit Control in Lead Electrorefining", *Journal of Metals*, 28 (11) (1976) pp. 4-9.

2. Y.Z. Malkin *et al.*, "Plant Investigation of a New Continuous Process for the Decoppering of Crude Lead", *Tsvetnye Metalli*, 34 (3) (1961) pp. 15-22. (In Russian).

3. S.S. Novoselov *et al.*, "Pilot Plant Study of a Lead Decoppering Process, Producing a Copper Matte", *Tsvetnye Metalli* 35 (5) (1962) pp. 25-31. (In Russian).

4. T.R.A. Davey & H.T. Webster, assgd. to B.H.A.S., "The Copper Drossing of Lead Bullion", *Aust. Pat.* No. 256,553 (1965).

5. W.H. Peck & J.H. McNicol, "An Improved Process for Continuous Copper Drossing of Lead Bullion", *Journal of Metals*, 18 (1966) pp. 1027-1032.

6. T.R.A. Davey, "Phase Systems Concerned with the Copper Drossing of Lead", *Trans. (Brit.) Instn. Min. & Met.* 72 Pt. 8 (1962-3) pp. 553-620; pp. 770-781; 73 (1963-4) pp. 614-5.

7. R.J. Moffatt & G.M. Willis, "The Copper-Lead-Sulphur System at High Temperatures, and its Application to Continuous Copper Drossing", *Aus. I.M.M. South Australian Conference*, 1975, Part A, pp. 167-185.

8. L. Fontainas, M. Coussement & R. Maes, "Some Metallurgical Principles in the Smelting of Complex Materials", pp. 13-23 in *Complex Metallurgy '78*, M.J. Jones, ed.; (Brit.) I.M.M., London, 1978.

9. W. Mentzel, "Treatment of Lead-Containing Copper Mattes", *Metall und Erz*, 10 (1913) pp. 193, 230. (In German).

10. R. Hesse, "The Smelting of High Lead Content Copper Mattes", *Metallurgie*, 8 (1911) pp.333, 336. (In German).

11. Fleming & McIntosh, assgd. to A.S.A.R.C.O.; *U.S. Patent* 2,343,760, 1944.

12. R.A. Perry, "Treatment of Speisses and Drosses in Lead Smelting", *Trans. AIME*, 159 (1944) P. 75.

13. A.A. Collins, assgd. to A.S.A.R.C.O., "Smelting Copper-Lead Drosses and Removal of Iron as the Sulfide"; *U.S. Patent* 2,381,970, Aug.14, 1945.

14. A.A. Collins, "Soda Treatment of Blast Furnace Drosses at El Paso Smelter", *Metals Technology* (Feb. 1977) T.P. 2193.

15. R.H. Emery & L.A.H. Pelton, "The Soda Treatment of Copper Dross Produced in Lead Refining", Aust. I.M.M. Regional Conference, Pt. Pirie, 1951.

16. A.G. Matyas, "Metallurgy of Lead-Copper Dross Smelting", *Trans. (Brit.) I.M.M.*, 86 (1977) pp. C190-194.

17. W. Hopkin, "Processes for treatment of zinc-lead and lead blast furnace copper drosses," TMS-AIME Paper Selection A76-99 (1976).

18. T.R.A. Davey, "Decoppering Lead with Sulphur", pp. 121-129 in *Research in Chemical and Extraction Metallurgy*, J.T. Woodcock, A.E. Jenkins & G.M. Willis ed.; *Aust. I.M.M.*, Melbourne, 1967.

19. T.R.A. Davey & J.V. Happ, "The Decoppering of Lead, Tin and Bismuth by Stirring with Elemental Sulphur", *Proc. Aust. I.M.M.*, No. 237 (1971) pp. 63-70.

20. I.S.R. Clark, L.A. Baker & A.E. Jenkins, "Decoppering of Lead Alloys with Sulphur: 1 - Reactions in the Binary Pb+Cu System", *Trans. (Brit.) I.M.M.*, 81 (1972) pp. C195-203.

21. Idem, "2- Reactions in the Pb+Cu+Ag and Pb+Cu+Sn Systems", *ibid*, 82 (1973) pp. 61-69.

22. T.R.A. Davey, G. Jensen & E.R. Segnit, "Decoppering Hard Lead with Sulfur and Pyrites", to be published in *Australia-Japan Extractive Metallurgy Symposium;* Aust. I.M.M., July 1980.

23. Y.E. Lebedeff & W.C. Klein, assgd. to A.S.A.R.C.O., "Process for Decopperizing Lead"; *U.S. Patent* No. 3,694,191, Sept. 26, 1972.

24. G.K. Williams, "Continuous Lead Refining", *Trans. AIME*, 121 (1936) pp.226-277.

25. A.D. Zunkel & A.H. Larson", "Slag-Metal Equilibria in the $PbO-Sb_2O_3$ System", *Trans. AIME*, 239 (1967) pp. 473-477.

26. E. Pelzel, "Reaction Equilibria Between Liquid Lead and Metal Oxides", *Erzmetall*, 11 (1958) pp. 56-63. (In German).

27. J. Gerlach & G. Herrmann, "Investigations into the Oxidation of Liquid Lead-Antimony Alloys", *Erzmetall*, 15 (1962) pp. 132-138. (In German).

28. H. La erbach, "Lead Refining by the Harris Process", *Metall und Erz*, 28 (1931) p. 317. (In German).

29. K. Emicke, G. Holzapfel & E. Kniprath, "Lead Refining at the Norddeutsche Affinerie", *Erzmetall* 24 (1971) pp. 205-215. (In German).

30. C. DiMartini & R.F. Lambert, "New Detellurizing and Decopperizing Practice at the Omaha Refinery of Asarco"; this Symposium.

31. J.F. Castle & J.H. Richards, "Lead Refining: Current Technology and a New Continuous Process", pp. 217-234 in *Advances in Extractive Metallurgy 1977*, M.J. Jones ed.; Instn. of Min. & Met., London, 1977.

32. T.R.A. Davey, discussion to ref. 31, *ibid.*, D16-18.

33. T.R.A. Davey, "Desilverising and Debismuthising of Lead", *Erzmetall*, 10° (1957) pp. 53-60. (In German).

34. T.R.A. Davey, "Desilverizing of Lead Bullion", *Trans. AIME*, 200 (1954) pp. 838-848.

35. V.F. Leferrer, "Vacuum Dezincing of Parkes Process Zinc Crusts", *Trans. AIME*, 209 (1957) pp. 1459-1460.

36. T.R.A. Davey, "The Physico-Chemical Principles of Refining Metals", pp. 29-50 in *International Symposium on Fifty Years of Metallurgy*; Banaras Hindu University, Varanasi, Dec. 1973.

37. T.R.A. Davey, "Distillation Under Moderately High Vacuum", *Vacuum*, 12 (1962) pp. 83-95; 14 (1964) pp. 227-30.

38. J.J. Poveromo & J. Szekely, "Analysis of Vapor Deposition Processes", *Met. Trans.*, 5 (1974) pp. 289-297.

39. N.A. Warner, "Kinetics of Continuous Vacuum Dezincing", pp. 317-332 in *Advances in Extractive Metallurgy*, M.J. Jones ed.; I.M.M. London, 1968.

40. J.G. Herberton & N.A. Warner, "Liquid Metal Mass Transfer in Vacuum Distillation of the Circulating Lead of a Zinc Blast Furnace", *Trans. (Brit.) I.M.M.*, 82 (1973) pp. C16-20.

41. T.R.A. Davey, Discussion to ref. 39, *ibid*, pp. 417-420.

42. T.R.A. Davey, "Vacuum Dezincing of Desilverised Lead Bullion", *Trans. AIME*, 197 (1953) pp. 991-996.

43. T.R.A. Davey, "Debismuthising of Lead", *Trans. AIME*, 206 (1956) pp. 341-350.

44. G.M. Willis, "Principles of Refining by Slag-Metal Reactions", pp. 107-137 in *Advances in Extractive Metallurgy and Refining*, M.J. Jones ed., I.M.M., London, 1971.

45. T.R.A. Davey & G.M. Willis, "Lead-Zinc-Tin Extractive Metallurgy Review", *Journal of Metals*, 29 (1977) pp. 24-30.

46. W.G. Pfann, *Zone Melting*; John Wiley & Sons, Inc., New York, 1958.

47. A.L. Lux & M.C. Flemings, "Refining by Fractional Solidification", *Met. Trans.*, 10B (1979) pp. 71-78.

48. A.L. Lux & M.C. Flemings, "Refining by Fractional Melting", *Met Trans.*, 10B (1979) pp. 79-84.

49. J.M. Moodie, "Debismuthising of Lead", *M.App.Sc. Thesis*; Univ. of Melbourne, 1976.

50. R.J. Moffatt & J.D. Iley, "A New Look at the Kroll-Betterton Process", Paper VII-4 in *Extractive Metallurgy Symposium*; Univ. of Melbourne, 1975.

51. J.E. Casteras & C. DiMartini, "Method of Treating Debismuthizing Dross to Recover Bismuth", this Symposium (1980).

52. T.R.A. Davey *et al.*, "Flow Characteristics of Molten Lead", pp. 105-118 in *Symposium on Chemical Engineering in the Metallurgical Industries;* Instn. Chem. Eng., London, 1963.

53. P. Paschen, "The Refining of Lead with Sodium Nitrate", Thesis; Technical University, Aachen, 1965. (In German).

THE PHYSICAL CHEMISTRY OF TIN SMELTING

John M. Floyd

C.S.I.R.O. Division of Mineral Engineering
Clayton, Victoria,
Australia

The tin extractive industry is at present undergoing rapid changes brought about by pressures from high tin prices and a greater proportion of lower-grade concentrates in feed. New smelting techniques, new slag treatment procedures and fuming as a means of up-grading ores and concentrates are showing promise of improving recoveries and lowering energy requirements for tin production. This review compares the physical chemistry of the old and emerging tin smelting processes and points to areas where more or better data are required.

Basic thermodynamic data for pure materials and solutions involved in the tin smelting and fuming processes are presented and factors affecting the distribution of constituents between phases in the slag-metal-gas and slag-matte-gas systems are discussed. The sparse kinetic data on tin smelting systems are evaluated and compared with data from other systems. The significance is shown of reaction mechanisms and of possible and established rate limitations on process efficiencies. Some recent work on process dynamics is discussed in relation to process design.

INTRODUCTION

In the decade since Wright's book on the Extractive Metallurgy of Tin(1) was published tin smelting has undergone great changes. The classical reverberatory smelting of high grade concentrates in a concentrate smelting stage and a slag smelting stage, which was the predominant smelting technique is still well entrenched but in Russia, Bolivia and Britain the development of tin slag fuming(2) has allowed smelters to overcome the iron reduction problem associated with slag smelting which limited the recovery of tin, especially from lower-grade concentrates. Established smelting furnaces such as slag resistance electric furnaces(3) and rotary furnaces(4) have been employed to improve on the concentrate smelting in reverberatory furnaces, but in general these did not result in very great improvements. At present there are a number of concentrate smelting techniques which have been proposed to improve on the reverberatory smelting operation, but at the time of writing none have become fully established. These include smelting and fuming operations on ores, preconcentrates, and concentrates in cyclone (5), submerged combustion (Sirosmelt)(6) and top blown rotary furnaces (TBRC)(7).

It is very unlikely that any new smelter would now be built employing the conventional two stage reverberatory smelting process and existing plants using this process will probably be modified to include a process for recovering tin from liquid slag.

The process routes now available for tin smelting are shown in the flow-sheets of Figure 1. Table I lists the important features of the various furnace types now available for smelting concentrates (smelting operation A in Figure 1) while Table II lists the important features of the furnace types available for recovering tin from liquid slags (smelting operation B and C in Figure 1) and fuming from preconcentrates and ores (smelting operation D in Figure 1).

The Sirosmelt and TBRC furnaces are capable of carrying out all four of the smelting and slag treatment procedures shown in Figure 1 and are also capable of carrying out a smelting and slag treatment cycle without tapping slag from the furnace.

With the large variety of process and furnace options available, there is a need for a review of the physical chemistry of tin smelting which includes data relevant to the new and emerging technology.

The newer technology, such as fuming in suspension or in matte or slag baths and smelting and slag treatment in the TBRC or Sirosmelt furnaces, involves highly turbulent reaction systems which result in high specific smelting rates in contrast to the relatively slow smelting in the reverberatory bath. In the turbulent systems there is a potential for further increasing throughput and efficiency by determining reaction mechanisms and rates so that the cause of rate limitations can be recognised and overcome when possible. Furthermore, in the long smelting time in reverberatory furnaces, reactions could be expected to achieve equilibrium so that an understanding of equilibrium properties was sufficient for an understanding of the process. In contrast the turbulent systems provide conditions where equilibrium may not be achieved due to transport or reaction rate limitations. If equilibrium is not achieved this must be recognised so that processes can be designed to take advantage of this where possible, or to reduce its effect if it is undesirable.

EQUILIBRIUM THERMODYNAMICS AND PHASE RELATIONS

Figures 2 and 3 are free energy diagrams for oxides and sulphides which are relevant to the processes used for tin smelting and fuming.

509

Table I - Concentrate Smelting Furnaces

Furnace Type	Features
1. Reverberatory	Heat recuperation or regeneration. Batch operation. Mixture charged fixes product composition. Low specific smelting rate. Laborious rabbling required. Products tapped Slag equilibrates with metal but not with gas
2. Electric	Low flue gas volumes if sealing adequate. Batch operation. Electric energy involves low fuel efficiency. Reductants with low volatiles required. Limitations on slag properties. Mixture charged fixes product composition. Low specific smelting rate (> reverb.) Products tapped. Sophisticated electrode controls. Slag equilibrates with metal but not with gas.
3. Rotary	Maximum temperature not greater than 1300°C. Batch operation. Some bath modification possible - but slow. Low specific smelting rate (> reverb.) Products tapped by furnace rotation. Slag equilibrates with both metal and gas. High refractory wear due to rotation.
4. Sirosmelt	Efficient heat transfer to bath but only limited recuperation. Semi-batch or continuous operation. Products controlled by changing variables. Very high specific smelting rate. Sophisticated lance controls. Products tapped or poured. Any reductant suitable (except coke) Slag equilibrates with gas but can operate at will to equilibrate or not with the metal. Well stirred .˙. refractory wear significant. Simple, cheap furnace unit. Head room for lance removal required.
5. TBRC (RTSV)	Efficient heat transfer to bath but no recuperation. Oxygen combustion, .˙. small gas volumes. Batch operation. Products controlled by changing variables. Very high specific smelting rate. Sophisticated lance and furnace control. Products poured. Slag equilibrates with both metal and gas (no control) High refractory wear due to rotation.

Table II - Liquid Slag Treatment and Fuming Furnaces

Note: Furnaces 4 and 5 are also suitable for liquid
slag treatment or fuming .˙. can operate as
a complete smelter with one furnace.

Furnace Type	Features
6. Slag Fuming	Batch operation. Both oil and pyrite used as fuel. Water jacketed .˙. high heat losses and high fuel requirement. Turbulent bath gives high treatment rates. Attention to tuyeres and accretions required. Flue gases contain SO_2 Slag equilibrates with gases? Fuel-rich injection .˙. low combustion efficiency in bath but can use air preheat or O_2 No refractories.
7. Matte Fuming	Efficient heat transfer to bath but limited recuperation Semi-Batch or continuous operation. Simple, cheap furnace unit. Matte bath maintains low oxygen potential and collects valuable constituents. Flue gases contain SO_2 Very high specific smelting rate. Products can be tapped or poured. Slag equilibrates with matte and gas. Well stirred so that refractory attack significant. Sulphides can be used as fuel. Sophisticated lance control.
8. Cyclone Fuming	Efficient heat transfer to feed but water cooling and electric furnace separator required. Continuous operation with intermittant settler tapping. Can operate to maintain matte bath which collects valuable constituents. Fine dry feed required. Slag requires further tin extraction. Suspension smelting will cause over oxidation so that excess sulphides are required. Flue gases contain SO_2 Sulphide combustion supplies part of fuel requirement. Sophisticated electrode controls.
9. Top Jetting	High level of O_2 enrichment possible. Continuous operation. Can be used for fuming or reduction. Products tapped. Sophisticated lance control and many lances required. Well stirred so that significant refractory attack. Can operate with a matte or metal bath.

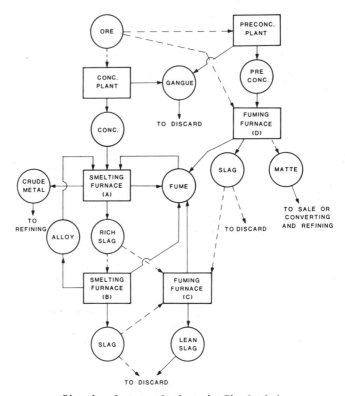

Fig. 1 - Process Options in Tin Smelting

Most of the data have been obtained either from the most recent critical compilations of data by Stull and Prophet(8) or Barin and Knacke(9). Other data are shown algebraically in Appendix 1. The data for SnO is from Carbo-Nover and Richardson(10) and that for SnO_2 is from Belford and Alcock(11) corrected for the Ni/NiO data of Steele(12). The SnO_2 data have been confirmed by two recent investigations(13,14).

The shaded areas in Figure 2 show the regions where the various smelting and fuming operations are carried out.

Figure 4 is a vapour pressure diagram for the oxides, sulphides and chlorides relevant to tin smelting operations. Most of the data are from the compilation of Barin and Knacke(9), but that for SnO was derived from Colin and Drowart(15) and the data for SnS are from Davey and Joffré(16). (Appendix 1). It can be seen from Figure 4 that both SnO and SnS are quite volatile at tin smelting temperatures. It is this volatility of tin compounds which causes the major physico-chemical differences between the various smelting and slag treatment processes. It results in advantages such as the effective separation of tin from iron in oxide or sulphide slag fuming processes but also causes the recycle of much tin fume in all smelting processes.

Table 111 lists activity coefficient data for the main metal, oxide and sulphide solutions involved in tin smelting operations. There have been a number of investigations of activities in the Sn-Fe system but other data are in reasonable accord with the equations of Shiraishi and Bell(17) shown here.

512

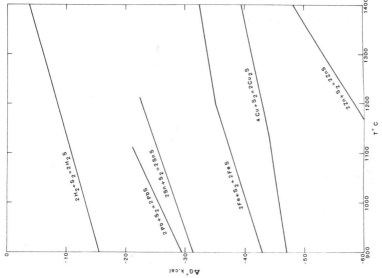
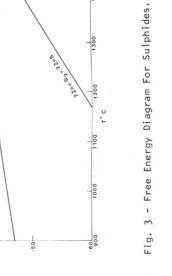

Fig. 3 - Free Energy Diagram For Sulphides.

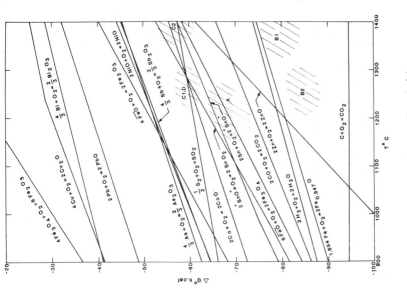

Fig. 2 - Free Energy Diagram For Oxides.
A = Concentrate smelting; B1 = Conventional slag reduction;
B2 = Sirosmelt slag reduction (final slag);
C1,D = Sulphide fuming; C2 = Oxide fuming.

Table III. Activity Coefficients in Tin Smelting Phases

System	X	γ_X	T°C	Ref.
Fe-Sn	Sn	$\log\gamma_{Sn}=0.6125N_{Fe}^2+0.371N_{Fe}^3$	1165	17
	Fe	$\log\gamma_{Fe}=1.169N_{Fe}^2-0.371N_{Fe}^3$	1165	17
$SnO-SiO_2(+SnO_2,+Sn)$	SnO	0.9 to 1.1	1160 and 1250	
$SnO_2-SnO-SiO_2(+Sn)$	SnO_2	20 - 32	1320	18
Indus. tin slag (+Sn) (18.3%SnO,41.6%SiO$_2$,18%FeO, 9.1%CaO,5.0%Na$_2$O)	SnO	1.8	1160	18
$FeO-CaO-SiO_2(+Fe_2O_3+Fe)$ $CaO/SiO_2=1.0$	FeO	0.7(25 mol.%FeO),1.0(32 mol.%FeO),1.4(55 mol.% FeO)	1305	19
$CaO/SiO_2=0.5$	FeO	0.7(30mol.%FeO),1.0(45 mol.%FeO),1.3(63 mol.%FeO)	1300	19
SnS-FeS	SnS	$\log\gamma_{SnS}=\dfrac{283}{T}N_{FeS}^2$	800 to 940	16

Figures 5 and 6 show the oxygen and sulphur pressures for stability of the condensed phases in the Sn-O-S and Fe-O-S systems at 1200°C calculated using the data in Figures 2 and 3. Also shown in these diagrams are the stability regions for high- and low-tin slags, which were calculated using the data in Figures 2 and 3, and Table III.

The diagrams ignore the solution of oxygen and the presence of higher valent tin and iron in the matte phase. It is of interest that the diagrams indicate that SnO_2 is formed at a lower oxygen pressure than Fe_3O_4, and it is possible that oxidation of a tin-containing slag will result in solid SnO_2 precipitation before solid Fe_3O_4 separates out. Tin slags could be more susceptible to rapid viscosity increases on oxidation than other non-ferrous smelting slags.

The vapour pressure of SnO is shown in Figure 7 as a function of oxygen partial pressure for the Sn-O system and for high-and low-tin slags at 1400°C. This differs in detail from Kellogg's(20) relationship for the Sn-O system in that SnO is now known to be stable at high temperatures so that the vapour pressure is constant over a range of oxygen potentials. SnO is stabilized in the slag solution so the maximum occurs over a larger range of oxygen partial pressures, but the maximum vapour pressure of SnO is decreased due to its lower activity in the slag. The decreasing vapour pressures at low oxygen partial pressures are due to metal formation while at high oxygen partial pressures they are due to formation of SnO_2.

The vapour pressure of SnS is shown in Figure 8 as a function of sulphur partial pressure for the Sn-S-O system and for high- and low-tin slags at 1200°C. The decreased vapour pressure of SnS at low sulphur partial pressures is caused by formation of tin metal and SnO respectively at the two oxygen partial pressures of $10^{-13.3}$ and $10^{-9.6}$.

514

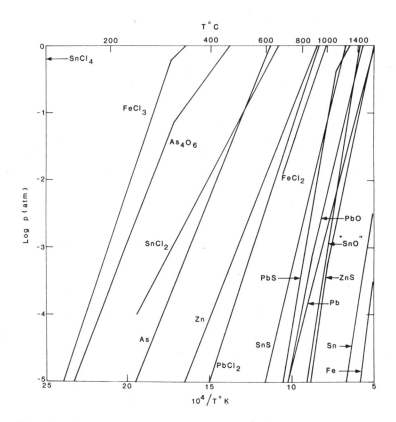

Fig. 4 - Vapour Pressure Diagram for Relevant Metals and Compounds.

Tin smelting involves the formation of slags containing mainly lime, silica and ferrous oxide with some stannous oxide and in order to achieve clean separation of metal from slag it is necessary that the slag be liquid at the temperature of operation. Figure 9 shows the areas of liquid in the CaO-SiO$_2$-FeO system at 1200°C. Because SiO$_2$ is always present in tin smelting slags, only the central area is of interest.

The effect of the presence of SnO is to increase the area of liquid, especially at high silica levels, so that the compositional requirements are less stringent for the smelting stage with high SnO levels than for the slag treatment stage with low SnO levels (if this is carried out at low temperatures). Other constituents such as Al$_2$O$_3$, MgO, TiO$_2$ and ZnO will generally lower the liquidus temperature at small concentrations but will increase it at higher levels. The viscosity of slags should also be low to ensure clean separations and high processing rates, and this is generally favoured by high ratios of basic components to acid components.

Figure 10 shows the Sn-Fe phase diagram, which contains a liquid miscibility gap at smelting temperatures. The liquidus temperature of metal is quite low until the iron content of metal approaches 20 wt%, when temperatures of 1130°C or higher would be required to ensure liquid metal.

The equilibrium data are generally well established and can be used to indicate the likely distribution of smelting components between metal, slag and fume if equilibrium conditions prevail for any given smelting process.

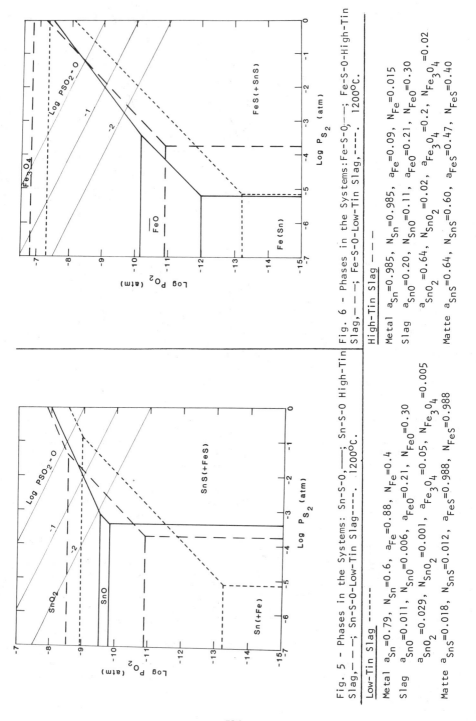

Fig. 5 – Phases in the Systems: Sn-S-O, ——; Sn-S-O High-Tin Slag, — — ; Sn-S-O-Low-Tin Slag----. 1200°C.

Fig. 6 – Phases in the Systems:Fe-S-O,——; Fe-S-O-High-Tin Slag,— — ; Fe-S-O-Low-Tin Slag,----. 1200°C.

Low-Tin Slag ------

Metal a_{Sn}=0.79, N_{Sn}=0.6, a_{Fe}=0.88, N_{Fe}=0.4

Slag a_{SnO}=0.011, N_{SnO}=0.006, a_{FeO}=0.21, N_{FeO}=0.30
a_{SnO_2}=0.029, N_{SnO_2}=0.001, $a_{Fe_3O_4}$=0.05, $N_{Fe_3O_4}$=0.005

Matte a_{SnS}=0.018, N_{SnS}=0.012, a_{FeS}=0.988, N_{FeS}=0.988

High-Tin Slag — — —

Metal a_{Sn}=0.985, N_{Sn}=0.985, a_{Fe}=0.09, N_{Fe}=0.015

Slag a_{SnO}=0.20, N_{SnO}=0.11, a_{FeO}=0.21, N_{FeO}=0.30
a_{SnO_2}=0.64, N_{SnO_2}=0.02, $a_{Fe_3O_4}$=0.2, $N_{Fe_3O_4}$=0.02

Matte a_{SnS}=0.64, N_{SnS}=0.60, a_{FeS}=0.47, N_{FeS}=0.40

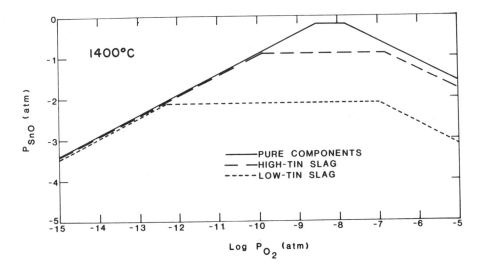

Fig. 7 Vapour pressures of SnO over the Sn-O system
and over tin-containing slags. 1400°C.

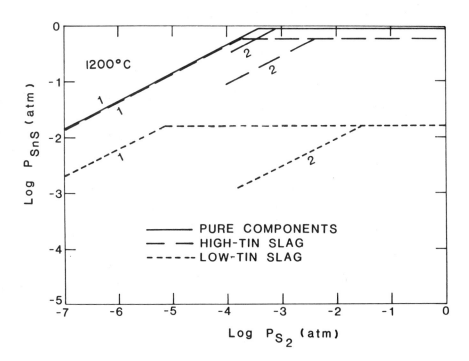

Fig. 8 Vapour pressures of SnS over the Sn-S-O system and over
tin-containing slags in equilibrium with matte. 1200°C.

1. $pO_2 < 10^{-13.3}$ 2. $pO_2 = 10^{-9.6}$

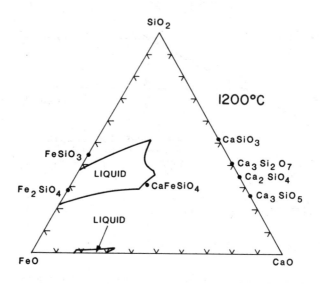

Fig. 9 Liquid regions in the FeO-CaO-SiO$_2$ system at 1200°C.

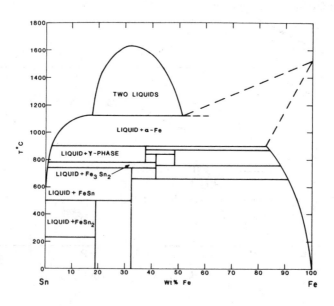

Fig.10 The Sn-Fe phase diagram.

The activity data for tin-containing slag systems do not cover a wide range of compositions and have only been determined for equilibrium with tin metal. The effect of SnO_2 and Fe_2O_3 dissolved in slag on the activity coefficient of SnO at high oxygen potentials requires more study and the use of Mössbauer spectroscopy for determining Sn^{2+}, Sn^{4+}, Fe^{2+} and Fe^{3+} in slags would be of considerable assistance in this work[21]

KINETICS AND MECHANISMS OF REACTIONS AND PROCESS RATE LIMITATIONS

Tin Concentrate Smelting

In conventional smelting, concentrate is mixed with flux, recycled fume, and excess coal to ensure reduction of most of the cassiterite to metal and iron minerals to ferrous oxide. The mixture is dropped into the furnace and melted over a period of 8 to 24 hours. Padilla and Sohn[22] found that solid state reduction of SnO_2 with carbon proceeded very rapidly above 850°C according to the reactions:

$$SnO_{2(c)} + 2CO = Sn_{(\ell)} + 2CO_2 \qquad (1)$$

$$\text{and} \qquad C + CO_2 = 2CO \qquad (2)$$

They found that reaction (2) was rate controlling and was catalyzed by tin. It is well established that the reduction of other solid oxides, including Fe_2O_3, also occurs through gaseous CO rather than solid carbon and for fine material is also very rapid at temperatures above 850°C. The long smelting time in reverberatory furnaces is not required for reduction reactions, but for the slow heat transfer through the finely ground concentrate in the static bed[1].

Other reactions taking place are the reactions between solid oxides to form the liquid slag phase such as:

$$2CaO_{(c)} + SiO_{2(c)} = Ca_2SiO_{4(slag)} \quad (\text{or} = 2Ca^{2+} + SiO_4^{4-}) \qquad (3)$$

and the reaction of iron in recycled alloy with tin oxide in slag:

$$Fe_{(metal)} + SnO_{(slag)} = FeO_{(slag)} + Sn_{(metal)} \qquad (4)$$

The quiescent bath also gives unfavourable conditions for these reactions to proceed rapidly and the early formation of liquid tin probably causes dissolution of much of the alloy before the iron removal reaction occurs. Much of the tin produced can be tapped off before slag melting and Wright has discussed the advantages of this practice in terms of decreased iron in metal[1].

Similar smelting reactions and mechanisms will apply to smelting concentrate in the TBRC but the charge is stirred by the rapid rotation of the furnace and the flame and hot gases are in intimate contact with the charge, so that the smelting process time is much shorter. The tumbling of the charge and the contacting of gases with the charge would cause more dusting and fuming of SnO than with the reverberatory furnace. Iron removal from alloy would be rapid but the speed of the process does not favour early pouring of metal.

In the Sirosmelt process concept the smelting mixture is either injected with fuel and air down the lance or pelletized and dropped into the agitated slag bath. Similar smelting reactions will occur but it is also likely that some dissolution of tin oxide in slag and reduction from slag occurs.

In this case melting and slag forming reactions occur in the liquid slag bath almost instantaneously so that the smelting rate is only limited by the rate of supply of fuel and air. The rate of reaction between the metal bath and the slag can be fast or slow depending on whether the lance is operated in a low or high position. Thus the removal of iron from recycled alloy is effectively carried out by the exchange reaction (Equation 4) by initially operating with a low lance position while smelting is carried out with a high lance position. By an increase in coal addition rate a little before the final metal tapping a relatively low tin slag can be achieved without any significant increase in iron content of metal. The slag bath is effective in preventing significant dust evolution but the gases will equilibrate with the slag so that fuming of SnO will be greater than reverberatory smelting and also greater than TBRC smelting which uses oxygen for combustion.

Tin Reduction and Tin Oxide Fuming from Liquid Slags

In the presence of coal the reduction of liquid tin slags may occur according to the overall reactions:

$$SnO_{(slag)} + C = Sn_{(metal)} + CO \tag{5}$$

$$SnO_{(slag)} + CO = Sn_{(metal)} + CO_2 \tag{6}$$

with similar reactions for reduction of other slag components SnO_2 to SnO, $FeO_{1.5}$ to FeO and FeO to Fe.

The Boudouard reaction causes regeneration of CO according to:

$$C + CO_2 = 2CO \tag{7}$$

The reaction involves liquid slag, solid carbon and a nucleating metal phase whose composition becomes richer in iron as tin is reduced from the slag, and a nucleating gas phase consisting of CO, CO_2 and SnO vapour. Actual reduction paths probably involve charge transfer between ferrous and ferric ions in the slag. The means by which the reduction is carried out will be an important factor in the relative importance of possible reaction rate limitations. In a reverberatory furnace, electric furnace, rotary furnace, Sirosmelt furnace and top-blown rotary furnace the contacting and turbulence between phases will be different and it is therefore possible that different rate controlling factors and mechanisms may apply.

Katkov(23) attempted to simulate the electric furnace reduction of slag by melting slags in a graphite crucible and found that temperatures in excess of 1400°C were needed to achieve rapid reactions. Floyd and Conochie(24) attempted to simulate the reactions in the Sirosmelt furnace by adding various forms of solid carbonaceous reductants to liquid slag held in a refractory crucible and agitated by injected gases. Rapid reactions were found as low as 1150°C and the rate appeared to be limited by mass transfer in the slag. When carbon monoxide was injected into the slag reduction reactions were much too slow to achieve equilibrium during bubble rise, but the rate was improved by allowing contact above the slag surface. However hydrogen reduction proceeded rapidly during bubble rise and for a fluid slag the rate conformed to the predictions of an equilibrium model. For a less basic, more viscous slag the rate was slower than the predictions of the equilibrium model. The results of reduction with C and CO suggested that reaction 5 was more rapid than 6, indicating that direct reduction of liquid slag by solid carbon was important in the Sirosmelt reduction process. This is contrary to the presently accepted mechanism for slag reduction and more work is required to prove this point.

When reducing tin slags with a stationary graphite sphere Sokolov *et al.* (26) found that reactions 6 and 7 were the main reduction reactions. The rate limitation was the Boudouard reaction at high tin levels and diffusion in the slag at low tin levels.

In the case of iron making direct carbon reduction of slag is also not considered to be a significant contributor to reduction of FeO, but the reductant in that case is less reactive, the system is less violently agitated, and the gas phase is continuous. However, much work has been done on this reaction with quite conflicting conclusions(25).

Chandrashekar and Robertson(27) also carried out a study of the reduction of liquid tin slags by injected hydrogen but concluded that the reactions did not achieve equilibrium and results conformed to a kinetic model. The different interpretation of the quite similar results hinges on the value accepted for the tin oxide activity coefficient in the slag. Floyd and Conochie used a value of between 1 and 2 based on the experimental data of Carbo-Nover and Richardson while Chandrashekar and Robertson used a value of 3-4 based on the distribution of tin and iron between slag and metal and thermodynamic data for hydrogen reduction reactions. Chandrashekar and Robertson assumed that the pool of reduced metal was in equilibrium with slag but did not allow for re-equilibration in their program. Floyd and Conochie showed that the distribution coefficients obtained from their work were consistent with the metal pool not equilibrating with the slag.

Floyd and Conochie also found that the injection of nitrogen into slags resulted in fuming of SnO at a rate similar to an equilibrium model and the equilibrium fuming rate was similar to that found during reduction experiments and pilot plant trials. Deb Roy *et al.*(28) found similar fuming rates with injected argon but on the basis of weight losses they concluded that during reduction the fuming rate was much higher. They could offer no reason for this anomaly but speculated about possible enhanced fuming during oxidation of metal or formation of a volatile hydroxide.

The tin removal from slag in pilot plant Sirosmelt reduction(29) in the range 15% to 3% Sn conformed to a first order plot (i.e. a straight line plot of log %Sn in slag vs. time) characterised by a reduction half time (the time to halve the tin level in slag) of 20 to 30 minutes. At tin levels below 2% Sn the half time increased to 40-50 minutes. Lower reduction half times than these were achievable (15 to 20 minutes) in both crucible and pilot plant work and it is considered that the rate was generally limited by the supply of reductant.

Temperature had no significant effect on the reduction rate between 1150 and 1350°C. SnO fuming has not been studied as a function of temperature but if it does occur with equilibrium attained between gas and slag there will be a rapid increase in fuming rate with increasing temperature.

The rates of tin removal in 50kg pilot plant Sirosmelt trials and the commercial 4-tonne pilot plant were identical and are shown in Figure 11. The equilibrium SnO fuming rate for the 4 tonne plant at 1400°C is also shown, calculated from equilibrium data and the known gas volumes evolved from the plant. It can be seen that the tin removal rate to tin levels down to about 1% is much faster for Sirosmelt reduction than the other processes. For instance the time to reduce from 10% to 1% Sn is less than 2 hours compared with about 4½ hours for Vinto slag fuming. Below 1% Sn the reduction still proceeds at a similar rate to the fuming processes, but the final tin level achievable is limited ultimately by the formation of solid alloy which causes an increase in tin in slag due to entrainment of reduced metal. For a slag containing 30% Fe initially this limit is about 0.4% Sn at 1250°C.

Fuming as SnS

There has been little work published on the kinetics and mechanism of reactions in fuming processes but the generally accepted reactions for slag fuming are:

$$2SnO + \frac{3}{2} S_2 = 2SnS + SO_2 \qquad (8)$$

$$\text{and} \quad SnO + FeS = SnS + FeO \qquad (9)$$

Wright(30) reported crucible experiments in which pyrite and SO_2 in a reducing gas mixture were added to slags to generate the S_2 and FeS required for fuming and found high efficiencies of utilisation of the sulphidising agents, if it were assumed that the labile sulphur produced from pyrite by the reaction: (ignoring non-stoichiometry)

$$FeS_2 = FeS + \frac{1}{2} S_2 \qquad (10)$$

did not take part in fuming reactions. This loss of sulphur was caused by adding pyrite to the surface but if it had been injected into the slag the labile sulphur would have been used in reactions and greater utilisation of sulphur would result.

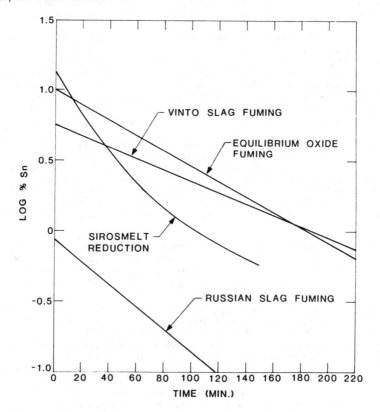

Fig. 11 - Tin Removal Rates from Liquid Slags by Reduction and Fuming Processes.

The rate of tin removal in the fuming process follows first order kinetics as shown in Figure 11 and the rate reported from Russian plants appears to be higher than in the Bolivian operations. The first-order kinetics could be caused either by slag diffusional control or from equilibrium being achieved between gas, matte and slag with a constant activity coefficient of SnO in slag. In either case the difference in rates could be explained by different slag compositions.

Wright(30) suggested that an excessive use of pyrite in the slag fuming process was due to inefficient combustion of injected oil within the bath. Combustion of oil in Sirosmelt plants is only slightly fuel-rich and leads to some bath oxidation due to slow combustion(31). Oxygen enrichment or the use of pulverized coal as fuel have been found to prevent this because of improved combustion rates. A Bulgarian zinc fuming plant uses oil-rich combustion to provide fuel as well as reductant and high efficiencies are obtained with high oil atomizing pressures and careful design of the tuyeres(32). The matte fuming process overcomes the problem by injecting air-rich combustion mixtures into the matte bath and taking advantage of the high rate of the matte-air reaction to rapidly remove oxygen from the gas. The matte acts as an oxygen and sulphur potential buffer and ensures operation in the desired SnS + FeS region of Figure 5. In the absence of the matte phase the oxygen potential may increase and the sulphur potential may decrease to cause operation in the SnO region where lower vapour pressures of SnS will decrease tin removal rates. Continued oxidation may cause operation in the even less favourable SnO_2 and Fe_3O_4 regions where precipitation of solids will cause high viscosities with resultant lower combustion and reaction rates. These factors may be the cause of the lower tin removal rate for Bolivian slag fuming practice than the calculated equilibrium rate of oxide fuming in the 4 tonne Sirosmelt plant. The higher maximum vapour pressure of SnS shown by comparing Figures 7 and 8 and the higher flue gas volumes for slag fuming furnaces should lead to higher tin removal rates for slag fuming.

The material being smelted in matte fuming is added as lumps or agglomerates to the bath surface and this will mean that labile sulphur will not take a part in fuming reactions in the bath and excess air must be blown into the furnace space above the bath to utilize its heat of combustion. However for ores, pre-concentrates, or copper concentrates containing tin, this labile sulphur ensures good sulphidizing conditions during melting so that much of the tin can be fumed as SnS before it enters the slag or matte phases. The FeS formed by melting joins the matte phase to replenish that oxidized by the injection and maintains low oxygen potentials in the matte and slag.

In cyclone fuming fine concentrate is smelted in suspension in oxygen (+ air). Reactions are very rapid and the rate is probably controlled by gas diffusion to the solid particles(33). Sulphide particles achieve very high temperatures during their oxidation and over-oxidation of finer particles would occur. For this reason tin volatilization is not complete and a further slag treatment procedure is required to remove the remaining tin(34). In order to separate a matte phase high sulphur ratios in feed are also required(5) because of this tendency to over-oxidize.

PROCESS DYNAMICS

In evaluating processes it is not sufficient to understand the equilibrium properties and reaction kinetics and mechanisms. The way phases physically interact in a given reactor will also have a bearing on the material and thermal efficiency of a process.

Water-air modelling was used to determine the flow patterns in a copper converter and this suggested substantial horizontal penetration by the gas jet cone(35). More recently Oryall and Brimacombe(36) carried out mercury-air modelling studies, which should more closely approach the copper converting situation but found little horizontal penetration of the gas and in fact there was a significant backwards component to the gas bubble stream.

It is questionable whether the tuyere injection of combustion mixtures into slag in the sulphide slag fuming furnace is more closely represented by an air-water or air-mercury system, but the consequences are not as significant since refractories are not used and the furnaces are quite small with tuyeres on opposite walls so that great horizontal penetration is not required. The requirement for punching of tuyeres is also not excessive, with punching required every half hour in Bolivian practice. The formation of solid magnetite in the slag by reaction of air with FeO in slag is the obvious cause of the build up across the tuyere mouth which is probably aided by the pulsing nature of the air flow.

With Sirosmelt lancing into a liquid bath it might be expected that channelling of gases(37) would limit the rates of reaction and heat transfer achievable(24). This would suggest a minimum effective lance depth, but in small scale and commercial pilot plant operations no significant effect of lance depth on these parameters was observed.

In the top-jetting reactor of Wuth(38) impingement of gas on liquid is physically similar to the BOS or LD Process for steelmaking. Many studies of the BOS process have been carried out (e.g. 39), and the flows in the case of Wuth's reactor are similar, but it is unlikely that there will be an emulsion of metal in slag because of the small amount of metal present in slag reduction compared with BOS steelmaking.

Plant investigations can be used to determine flow and mixing patterns, phase separation limitations and process inefficiencies, but no work has been published on tin smelters. Tracer studies together with detailed sampling and analysis have provided a good insight into the operation and design requirements for copper smelting reverberatory furnaces, and a lead blast furnace settler(40,41,42) and these techniques, together with in-situ sensors(43), would be invaluable in studies of tin smelter efficiencies.

INTERFACIAL PROPERTIES

The interfacial properties of systems can have a bearing on process design and optimization. For instance it has been shown(44) that flotation of tin metal in slags occurs in injection stirred systems so that suitable settling conditions are required before tapping metal from the furnace to avoid undue recycle of metal suspended in slag. This is not so much of a problem in slag reduction because dissolution of iron in tin results in an increase in the interfacial tension with slag and improved conditions for phase separation. Minto and Davenport(45) studied the problem of matte prill suspension in copper slags and Conochie and Robertson(46,47) are carrying out hot model studies of the behaviour of a third phase at gas bubble-liquid interfaces. Interfacial turbulence(48) can also have a considerable bearing on reactions at interfaces and can produce a much higher reaction rate than predicted.

Interfacial phenomena can also be a limitation on reaction rates(25) and it has been suggested that this is the cause of slow reduction of tin slags by CO gas(24).

524

DISTRIBUTION OF IRON AND TIN BETWEEN PHASES DURING THE TIN REDUCTION AND FUMING PROCESSES

Tin smelting is mainly concerned with the separation of tin from iron and laboratory investigations have concentrated on the distribution of tin and iron between metal and slag phases. The low vapour pressures of iron and its oxides and sulphide at tin smelting temperatures results in iron reporting to the metal and slag phases, but not to the gas phase.

The classical means of separating iron from tin was to smelt concentrates under mildly reducing conditions in the oxygen potential and temperature region marked as A in Figure 2. This produced a metal containing only a small amount of iron in the range .05 to 1% Fe, together with a slag of high SnO content. The metal and slag were tapped and the metal was slowly cooled to remove iron as a dross which was recycled to the smelting operation. The slag was granulated and resmelted under more strongly reducing conditions, in the oxygen potential and temperature region marked as B1 in Figure 2. This recovered most of the tin from slag, which could then either be discarded if the tin level was acceptably low or smelted yet again. The metal produced was an alloy of tin and iron, containing 20 to 60% Fe, which was granulated and recycled to the concentrate smelting stage.

The distribution of iron and tin between metal and slag in tin smelting operations has been discussed in a number of papers and this work was recently summarized(24). The iron content of the metal produced in tin smelting is dependent on the exchange reaction (Eq.4) which can be characterised by an exchange coefficient,

$$ k = \frac{\text{wt.% Fe in slag}}{\text{wt.% Fe in metal}} \times \frac{\text{wt.%Sn in metal}}{\text{wt.%Sn in slag}} = K_1 \times \frac{\gamma_{Fe}}{\gamma_{Sn}} \times \frac{\gamma_{SnO}}{\gamma_{FeO}} $$

where K_1 is the equilibrium constant for the exchange reaction and γ_i is the activity coefficient of component i.

The variation of k with iron content of metal has been studied by a number of workers with a very wide scatter of results. The effects of temperature and slag composition on k have been reported to be minor but the reverse reaction of Equation 4 (i.e. reaction of Sn with FeO) was shown to be too slow to achieve equilibrium in a reasonable time and it appears that this is the main reason for the large scatter. Hallett's work(49) is considered to be the most reliable because equilibrium was only approached with the faster reaction of Fe with SnO and the system was sealed so that loss of SnO as vapour did not require the slow back reaction for equilibration. The following equations represent the results of Hallett:

0% Fe to Miscibility Gap : $\log k = 2.23 - 5.36\ N_{Fe} - 13.69\ N_{Fe}^2 - 12.84\ N_{Fe}^3$

Within the Miscibility Gap : $\log k = 1.26 + \log\ (1-N_{Fe})/N_{Fe}$

Miscibility Gap to 90% Fe : $\log k = 2.08 - 2.19\ N_{Fe} + 0.86\ N_{Fe}^2$

The laboratory work in which k was determined in open systems and the results of commercial operations give higher k values at a given metal composition. Higher k values are economically desirable because they imply a higher tin recovery for a given iron recycle. Thus factors that avoid equilibrium by not allowing time for the slow back reaction to take place are advantageous in tin reduction smelting.

In conventional reverberatory or electric furnaces the concentrate and slag smelting stages are carried out in relatively quiescent baths. Smelting reactions occur before and during melting of the charge and the oxygen potential cannot be adjusted significantly in a reasonable time after the charge has been prepared. Furthermore only minor fume generation occurs, unless the charge contains high sulphur levels, so that the exchange coefficient is close to the equilibrium value. The main detractions of the process are its need to cool and remelt slag, its inability to handle more than a small quantity of alloy because of slow reactions between the metal and slag baths and high tin in discard slag resulting from limits to the iron recycle. The use of rotary furnaces improved the removal of iron from hardhead, but did not succeed in the reduction of liquid slag(50).

The Sirosmelt procedure involves rapid smelting reactions in an agitated slag bath and the oxygen potential can be changed rapidly. This allows improved conditions for separating iron from tin, as explained earlier and results in production of slags with about 1/3rd of the tin content of reverberatory furnaces when producing metal of the same iron content. Thus the tin recycled with alloy is decreased substantially and the tin content of discard slag is reduced to about a third of that achievable with reverberatory smelting of the same concentrate. Thus a two stage Sirosmelt reduction process is much more attractive than two stage reverberatory reduction smelting and there is less incentive for a final fuming stage.

The TBRC process also involves rapid reactions but the metal-slag interface is very turbulent because of the furnace rotation and any departure from equilibrium conditions will be small. The slag will also equilibrate with the gas phase but combustion of fuel with oxygen will produce less flue gas and therefore less fume will be generated for recycle than with Sirosmelt. These factors will make the TBRC less attractive than Sirosmelt for two stage reduction smelting and a final fuming stage would be essential for high tin recovery.

Fuming as SnO entails operation at the very high temperature of about 1400°C. In the turbulent refractory-lined Sirosmelt and TBRC furnaces this will result in high refractory wear. This would be more serious in the TBRC because the turbulence is generated by the furnace walls so that very high shear rates could be expected. Fuming as SnS in either the matte fuming or slag fuming mode of operation could be used in these furnaces to avoid the use of high temperatures, but this would require an environmentally acceptable system for collection and disposal of SO_2 in the flue gases.

Slag fuming furnaces treat slags from other smelting furnaces but cannot be used for smelting, reduction or matte fuming because the water-jackets cannot be operated with a metal or matte bath. In principle a refractory-lined hearth could be used to hold this bath and then the relative merits of the processes would mainly depend on factors such as the distribution coefficients achieved and the balance of refractory maintenance against higher fuel requirements.

In the presence of a matte bath the distribution of tin and iron between matte and slag can be represented by an exchange coefficient:

$$C = \left(\frac{\%Sn}{\%Fe}\right)_{matte} \times \left(\frac{\%Fe}{\%Sn}\right)_{slag}$$

Davey and Flossbach(51) reported C to be about 2 for mattes containing 1 to 15% Sn but at the lower levels of tin associated with the matte fuming process (0.05 to 0.4% Sn) C had a value of 0.8 to 0.9 for iron sulphide mattes and 2 to 3 for copper-iron sulphide mattes(52,53).

DISTRIBUTION OF MINOR CONSTITUENTS BETWEEN PHASES

There has been no systematic study of the deportment of minor constituents in tin smelting and fuming operations and there is a need for work
such as that carried out for copper smelting by Nagamori and Mackey(54,55).

The general trend of distributions can be deduced from the standard free
energies of formation of oxides and sulphides in relation to the smelting
and slag treatment regions shown in Figure 2 and from the vapour pressures
of metals, oxides, and sulphides, but more detailed analysis requires a
knowledge of activities in metals, mattes and slags and more data are
required in this area. Since relative reaction rates may also affect the
deportments the predictions would require confirmation by laboratory or
plant work.

Injection processes give conditions where gases can more nearly approach
equilibrium than in reverberatory furnaces, so that volatile constituents
will tend to report to fume to a greater extent and this is an important consideration in the physical chemistry of tin smelting and slag treatment
processes. Sequential oxidation and reduction and temperature changes can
be employed in the Sirosmelt and TBRC furnaces to affect the distribution
of components in different phases.

Oxides which have free energies of formation much higher than SnO and
FeO, such as SiO_2, Al_2O_3, MgO, Ta_2O_5 and Cr_2O_3 report almost completely to
the slag phases in all smelting and fuming operations. Oxides more easily
reduced than these will report to the various phases in relative amounts
controlled by their level of concentration, their vapour pressures and the
stability of their oxides and sulphides. Table IV shows the expected
deportment of the more easily reduced impurities present in concentrates at
levels of 0.5% or greater.

It can be seen that in the fuming operations most of the As, Sb, Bi and
Pb present at high levels will report to the fume and this places a limitation on the fuming of concentrate or ore. Treatment of feed or fume for
removal of these constituents would then need to be considered as an
alternative to removal during refining.

Table IV. Deportment of Minor Constituents in Tin Smelting Operations

($\geq$0.5% impurity in feed; Minor deportments shown in brackets)

Constituent	Concentrate Smelting	Slag Reduction	Oxide Fuming	Sulphide Fuming
Zn	Slag	Fume (Slag)	Slag	Fume,Slag(Matte)
Co	Slag	Metal (Slag)	Slag	Slag,(Matte)
Ni, Cu	Metal (Slag)	Metal	Slag	Matte(Slag)[*1]
As, Sb	Metal, Fume(Slag)	Metal, Fume	Fume	Fume
Bi, Pb	Metal, Fume(Slag)	Metal,(Fume)	Fume	Fume
Au, Ag	Metal			Matte(slag)[*1]

(*1 : If no matte present deportment to slag)

CONCLUSIONS

Physico-chemical data at present available allow a satisfactory comparison of the many process and equipment options available for tin smelting but there is still a need for considerable effort in studies of reaction rates and mechanisms, process dynamics, interfacial properties and activities of minor constituents in phases.

REFERENCES

1. P.A. Wright, "Extractive Metallurgy of Tin", Elsevier, 1966.

2. Jorge Lema Patiño, "The fuming of Tin Slags in the Extractive Metallurgy of Tin", in Fourth World Conference on Tin, Kuala Lumpur, 1974, N.L. Phelps, ed.; Vol. 3, pp.113-145.

3. K. Niwa and S. Kudo, "Current Operations and Future Possible Improvements at the Ikuno Tin Smelter", in A Second World Tin Conference, Bangkok, 1969, Vol. 3, pp.1133-1142, W. Fox, ed; International Tin Council, London, 1970.

4. Kasmir Batubara, "Tin Smelting in Indonesia", paper presented at International Tin Symposium, La Paz, Bolivia, Nov. 1977, 13p.

5. Erich A. Muller, "Verflüchtigung von Zinn aus Armen Schwefelhaltigen Konzentraten im Zyklonofen", Erzmetall 30(2) (1977), 54-60.

6. J.M. Floyd, "The Submerged Smelting of Tin Slags - A New Approach to Lower-Grade Concentrate Smelting", in Fourth World Conference on Tin, Kuala Lumpur, 1974, N.L. Phelps, ed.; Vol. 3, pp.179-190.

7. E.B. King and L.W. Pommier, "The Future of the Texas City Tin Smelter", paper to International Tin Symposium, La Paz, Bolivia, Nov.1977, 9p.

8. D.R. Stull and H. Prophet, "JANAF Thermochemical Tables", Second Edition, U.S. Dept. of Commerce, Nat. Bur. of Standards, Washington, 1971. Supplements in J.Phys.Chem. Ref. Data., 3(1974), 311, 4(1975) 1, 7(1978) 793.

9. I. Barin and O. Knacke, "Thermochemical Properties of Inorganic Substances", Springer-Verlag, Berlin etc., 1973. Supplement with O. Kubaschewski, 1977.

10. J. Carbo-Nover and F.D. Richardson, "Stannous Oxide and the Solubility of Oxygen in Tin", Trans. I.M.M., 81 (1972), C63-68.

11. T.N. Belford and C.B. Alcock, "Thermodynamics and Solubility of Oxygen in Liquid Metal from E.M.F. Measurements Involving Solid Electrolytes Part 2 - Tin", Trans. Farad. Soc., 61 (1965), 443-53.

12. B.C.H. Steele, "High Temperature Thermodynamic Measurements Involving Solid Oxide Electrolyte Systems", pp.3-27 in Electromotive Force Measurements in High-Temperature Systems, C.B. Alcock, ed.; I.M.M. London, 1968.

13. S. Seetharamon and L. -I. Staffanson, "On the Standard Gibbs Energy of Formation of SnO_2", Scand. J. Met. 6(1977), 143-144.

14. T.A. Ramanarayanan and A.K. Bar, "Electrochemical Determination of the Free Energy of Formation of SnO_2", Met. Trans.,9B(1978), 485-486.

15. R. Colin, J. Drowart and G. Verhaegen, "Mass Spectrometric Study of the Vaporization of Tin Oxides", Trans. Farad. Soc. 61(1963), 1364-71.

16. T.R.A. Davey and J.E. Joffré, "Vapour Pressures and Activities of SnS in Tin-Iron Mattes", Trans. I.M.M.,82(1973), C145-150.

17. S.Y. Shiraishi and H.B. Bell, "Thermodynamic Study of Tin Smelting.: 1. Iron-Tin and Iron-Tin-Oxygen Alloys", Trans. I.M.M. 79(1970), C120-127.

18. J. Carbo-Nover and F.D. Richardson, "Activities in $SnO-SiO_2$ Melts", Trans. I.M.M., 81(1972), C131-6.

19. C. Bodsworth, "The Activity of Ferrous Oxide in Silicate Melts", J. Iron and Steel Inst., 193(1959), 13-24.

20. H.H. Kellogg, "Vaporization Chemistry in Extractive Metallurgy", Trans. AIME, 236(1966), 602-15.

21. K.V. Alen'kinn, V.A. Varnek and V.S. Filatkina, "Use of Thermo-Emf Mössbauer Spectroscopy, and X-ray Diffraction Analysis to Investigate Disproportionation of Stannous Oxide", Inorganic Materials, 13(1977), 1474-8.

22. D. Rafaél Padilla and Hong Yong Sohn, "The Reduction of Stannic Oxide with Carbon", Met. Trans. B., 10B(1979), 109-15.

23. O.M. Katkov, "Mechanism of the Reduction Process for Tin Concentrate and Slag", Tsvetnye Metally, 8(1967), 45-51.

24. J.M. Floyd and D.S. Conochie, "Reduction of Liquid Tin Smelting Slags Part I: Laboratory Investigations", Trans. I.M.M., 88(1979), C114-122.

25. M.W. Davies, G.S.F. Hazeldean and P.N. Smith, "Kinetics of Reaction Between Liquid Iron Oxide Slags and Carbon", Physical Chemistry of Process Metallurgy: The Richardson Conference, J.H.E. Jeffes and R.J. Tait, eds., I.M.M., London (1974), 95-107.

26. A.E. Sokolov, V.I. Deev and A.T. Tikhanov, "Kinetics of Reduction of SnO from Melts with Solid Carbon", Izvest V.U.Z. Tsvetnaya Metallurgiya, No. 1(1972), 75-8.

27. S.R. Chandrashekar and D.G.C. Robertson, "Recovery of Tin from Slags" Presented to CIM Annual Conference, August, 1978, 41 p.

28. T. Deb Roy, S.R. Chandrashekar and D.G.C. Robertson, "Fuming of Stannous Oxide from Slags", Trans. I.M.M., 87(1978), C225-230.

29. J.M. Floyd and D.S. Conochie, "Reduction of Liquid Tin Smelting Slags, Part II: Development of a Submerged Combustion Process", Trans. I.M.M., 88(1979), C123-7.

30. P.A. Wright, "Possible Developments in the Tin Sulphide Fuming Process", paper presented at International Tin Symposium, La Paz, Bolivia, Nov. 1977, 19p.

31. J.M. Floyd, G.J. Leahy, R.L. Player and D.J. Wright, "Submerged Combustion Technology Applied to Copper Slag Treatment", in The Aus. I.M.M. North Queensland Conference, 1978, The Aus. I.M.M., Parkville, 1978, pp.323-337.

32. G. Abrashev, "Slag Fuming by the Use of Liquid Fuel", in Advances in Extractive Metallurgy and Refining, I.M.M., London(1972), pp.317-325.

33. Frank R.A. Jorgensen and E. Ralph Segnit, "Copper Flash Smelting Simulation Experiments", Proc. Aus. I.M.M., No. 261(1977), 39-46.

34. Jorge Lema Patino, "Fundicion de Minerales de Estaño de Baja Ley en Bolivia", paper presented at International Tin Symposium, La Paz, Bolivia, Nov. 1977, 32p.

35. N.J. Themelis, P. Tarassoff and J. Szekely, "Gas-Liquid Momentum Transfer in a Copper Converter", Trans. Met. Soc. AIME, 245(1969), 2425-2433.

36. G.N. Oryall and J.K. Brimacombe, "The Physical Behaviour of a Submerged Gas Jet Injected Horizontally into Liquid Metal", Met. Trans. 7B(1976), 391-403.

37. A.E. Wraith, "Gas Lancing in Metal Refining: An Air-Water Model Study", in Advances in Extractive Metallurgy and Refining, I.M.M., London(1972) pp.303-316.

38. L. Sevila and W. Wuth, "Zinnverflüchtigung durch Aufblasen", Erzmetall, 31(1978), 57-61.

39. A. Chatterjee, N. -O. Lindfors and J.A. Wester "Process Metallurgy of L.D. Steelmaking", Ironmaking and Steelmaking, 3(1976), 21-32.

40. N.J. Themelis, "Techniques of Process Analysis in Extractive Metallurgy", Met. Trans. 3(1972), 2021-2029.

41. D.J. Milne, G.E. Casley, and G.S. Stacey, "The Efficiency of Reverberatory Furnace Smelting", J. Aust. Inst. Metals, 16(1971), 49-62.

42. A.G. Matyas and M.D. Street, "Factors Influencing the Lead Content of Brunswick Blast Furnace Slag", paper presented at 106th AIME Annual Meeting, Atlanta, 1977, 17p.

43. J.M. Floyd, D.S. Conochie and N.C. Grave, "Measurement of Oxygen Potentials in a Nickel Smelter Using Disposable-Tip EMF Cells", Proc. Aus. I.M.M., No. 270(1979), 15-23.

44. J.M. Floyd, "Flotation of Metal During Injection of Gases into Liquid Slags", Trans. I.M.M., 82(1973), C51-2.

45. R. Minto and W.G. Davenport, "Entrapment and Flotation of Matte in Molten Slags", Trans. I.M.M., 81(1972), C36-42.

46. D.S. Conochie and D.G.C. Robertson, "The Behaviour of the Third Phase Produced in Gas Bubble-Liquid Reactions", in Gas Injection into Liquid Metals, A.E. Wraith, compiler, Dept. of Met. and Eng. Mat., University of Newcastle-Upon-Tyne, U.K., 1979, 21p.

47. D.S. Conochie and D.G.C. Robertson, "A Ternary Interfacial Energy Diagram", Ibid., 8p.

48. J.K. Brimacombe, "Interfacial Turbulence in Liquid Metal Systems" in The Physical Chemistry of Process Metallurgy: The Richardson Conference, J.H.E. Jeffes and R.J. Tait eds.; I.M.M., London (1974), 175-185.

49. G.D. Hallett, Consolidated Tin Smelters (1967). Private communication.

50. T.R.A. Davey, "Smelting of Low-Grade Tin Concentrates", <u>Aust. Mining</u>, <u>61</u>(1969), 62-65.

51. T.R.A. Davey and F.J. Flossbach, "Tin Smelting in Rotary Furnaces", <u>J. Metals</u>, <u>24</u>(1972), 26-30.

52. K.A. Foo and W.T. Denholm, "Matte Fuming of Tin from Sulphidic Ores", paper presented at <u>International Tin Symposium</u>, La Paz, Bolivia, 1977.

53. K.A. Foo and J.M. Floyd, "Development of the Matte Fuming Process for Tin Recovery from Sulphide Materials", paper to this Symposium.

54. M. Nagamori and P.J. Mackey, "Distribution of Sn, Se and Te Between FeO-Fe$_2$O$_3$-SiO$_2$-Al$_2$O$_3$-CuO$_{0.5}$ Slag and Metallic Copper", <u>Met. Trans.</u>, <u>8B</u>(1977), 39-46.

55. M. Nagamori and P.J. Mackey, "Thermodynamics of Copper Matte Converting: Part II. Distribution of Au, Ag, Pb, Zn, Ni, Se, Te, Bi, Sb and As Between Copper, Matte and Slag in the Noranda Process", <u>Met. Trans.</u>, <u>9B</u>(1978), 567-79.

Appendix 1

Equilibrium Data for Tin Compounds
(Data not available in critical compilations)

Reaction	Equation (cal. or atm.)	Temp. Range ($^{\circ}$K)	Ref.
$2Sn_{(1)}+O_2=2''SnO''_{(c)}$	$\Delta G^{\circ}=-138,500+50.70T$	1170-1230	(10)
$2Sn_{(1)}+O_2=2''SnO''_{(1)}$	$\Delta G^{\circ}=-128,800+42.75T$	1250-1500	(10)
$Sn_{(1)}+O_2=SnO_{2(c)}$	$\Delta G^{\circ}=-139,650+50.61T$	770-980	(10)
$''SnO''_{(c)}=Sn_xO_{x(v)}$	$\log \overset{\circ}{p}Sn_xO_x = -\dfrac{16,340}{T}+9.902$	1070-1250	(15)&(20)
$''SnO''_{(1)}=Sn_xO_{x(v)}$	$\log \overset{\circ}{p}Sn_xO_x = -\dfrac{15,247}{T}+9.028$	1250-1500	*1
$SnS_{(c)}=SnS_{(v)}$	$\log \overset{\circ}{p}SnS = -\dfrac{10,963}{T}+7.64$	1170-1143	(16)
$SnS_{(1)}=SnS_{(v)}$	$\log \overset{\circ}{p}SnS = -\dfrac{8,877}{T}+5.81$	1143-1500	(16)

*1 Calculated from vapour pressure data of $''SnO''_{(c)}$ using melting data for $''SnO''$ from(10).

THE PHYSICAL CHEMISTRY OF IRON PRECIPITATION IN THE ZINC INDUSTRY

J.E. Dutrizac*
CANMET
Department of Energy, Mines and
Resources
Ottawa, Ontario
Canada K1A 0G1

Iron may be conveniently precipitated from zinc hydrometallurgical solutions as jarosite, goethite, hematite or magnetite, and the general advantages and disadvantages of each precipitation method are surveyed. The mechanism of iron precipitation is discussed in terms of Fe^{3+} hydrolysis, followed by the dimerization and subsequent polymerization of the iron hydroxyl complexes, that leads to the eventual precipitation of some crystalline iron compound. The phase diagram relationships governing the iron compounds precipitated are presented, and the physical chemical factors such as solution concentrations, oxidation potential, ionic strength, seeding and temperature that affect iron precipitation are reviewed and discussed. The stability of goethite with respect to Fe_2O_3 and β and γ FeO.OH is discussed with emphasis on particle size, pH and temperature effects. Special attention is paid to jarosite formation because of its more general utility in conventional zinc processing and its potential importance in the oxygen pressure leaching of zinc. In this regard the stabilities and solubilities of the various jarosite – type compounds are presented and compared to similar values for goethite and hematite.

* Head, Metallurgical Chemistry Section, Mineral Sciences Laboratories, CANMET, Department of Energy, Mines and Resources, Ottawa, Canada K1A 0G1

Introduction

Iron is commonly associated with zinc concentrates, both as a replacement for zinc in sphalerite and as separate minerals such as pyrite or chalcopyrite. During roasting at temperatures above 900°C, some of the iron reports as hematite but most combines with zinc oxide to form ferrite: $ZnO.Fe_2O_3$. During a conventional sulphuric acid neutral leach, free zinc oxide is dissolved as $ZnSO_4$; zinc ferrite is not, however, significantly leached and its contained zinc values would, therefore, be lost. It has long been known that the zinc in the ferrite can be liberated by leaching with excess sulphuric acid at temperatures near 100°C; the problem with such a process is that the associated iron values are also largely dissolved and must subsequently be precipitated in an easily filtered form prior to zinc electrolysis. Simple neutralization of the resulting zinc-iron sulphate solution is not advised since an iron gel is formed which settles with extreme difficulty and includes significant zinc. It was only with the development of effective iron precipitation techniques, such as the jarosite or goethite processes, that the hydrometallurgical recovery of zinc from the ferrite became commercially feasible.

In the mid 1960's patent applications on the jarosite process were filed independently by Asturiana de Zinc S.A. of Spain (1), Electrolytic Zinc Company of Australasia Ltd. (2) and Det Norske Zinkkompani A/S (3). The jarosite process is being used by at least twelve companies and the commercial operation of the process has been well documented (4,5,6). The goethite process was developed somewhat later in the sixties by Societe de la Vieille Montagne (7) and the Electrolytic Zinc Company of Australasia, and its commercial application has also been discussed (8). The hematite process is a more recent innovation in the zinc industry although the precipitation of iron as Fe_2O_3 at elevated temperatures and pressures has been known for some time. The commercial application of the hematite process as practised by the Akita Zinc Company has also been presented (9). The magnetite process has yet to find commercial application in the zinc industry.

In the above processes the precipitation of iron is controlled by a complex set of thermodynamic and kinetic factors operative both in the iron solution and in the iron precipitates. It is important that these factors and their interrelations be identified so that the processes can be controlled to yield crystalline iron precipitates. The purpose of this paper is to outline the physico-chemical properties of typical solutions and to show how these ultimately influence the iron compounds formed. A mechanistic pathway of iron precipitation is presented and, lastly, the conditions leading to the precipitation of jarosite, goethite, hematite and magnetite are enumerated.

Iron Species in Solution

Although a number of valence states are known for iron (10), Figure 1 shows that only the +2 and +3 oxidation states are stable for the oxidation potentials and pH's encountered during zinc processing where strongly complexing ligands are absent (11). In the presence of air or oxygen, ferrous ion will thermodynamically be oxidized to the ferric state although the rate of oxidation is relatively slow in acidic media. Ferrous ion shows only a slight tendency to hydrolyze before precipitation as $Fe(OH)_2$ or Fe_3O_4 at pH's above 7. The following 25°C formation constants indicate that Fe^{2+} and Zn^{2+} would likely co-precipitate under such conditions (12).

$$Fe^{2+} + H_2O \rightarrow FeOH^+ + H^+ \qquad K_{11} \approx 10^{-9} \qquad (1)$$

$$Fe^{2+} + 2H_2O \rightarrow Fe(OH)_2 + 2H^+ \qquad K_{12} \approx 10^{-20} \qquad (2)$$

$$Fe^{2+} + 3H_2O \rightarrow Fe(OH)_3^- + 3H^+ \qquad K_{13} \approx 10^{-29} \qquad (3)$$

$$Zn^{2+} + H_2O \rightarrow Zn(OH)^+ + H^+ \qquad K_{11} \approx 10^{-9} \qquad (4)$$

$$Zn^{2+} + 2H_2O \rightarrow Zn(OH)_2 + 2H^+ \qquad K_{12} \approx 10^{-17} \qquad (5)$$

Because of the dual problem of the high solubility of ferrous ion in acid zinc-sulphate solutions and the difficulty of preventing zinc coprecipitation in alkaline media, iron removal processes in the zinc industry have been based on the precipitation of ferric ion.

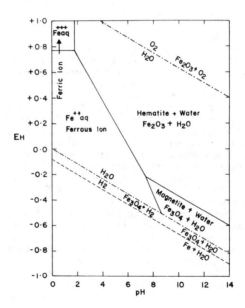

Fig. 1 - Eh-pH diagram for the iron-water system at 25°C and 1 atm; ion fields are for activities $>10^{-6}$ (Ref. 11).

Ferric sulphate itself is extensively soluble in zinc sulphate solutions as the 50°C data of Figure 2 indicate (13), and removal of Fe^{3+} by crystallization of $Fe_2(SO_4)_3 \cdot 9H_2O$ or $ZnSO_4 \cdot 2Fe_2(SO_4)_3 \cdot 22H_2O$ is most unlikely for the iron concentrations encountered in zinc processing (0-50 g/L Fe). Hence, iron removal must be effected by hydrolysis and subsequent precipitation of the hydrolyzed ferric species.

If it can be assumed that the compound precipitated by hydrolysis is closely related to the ferric complexes present at the instant of precipitation, then it becomes very important to know which ferric species exist in solution. As can be seen from the thermodynamic data presented in Table 1

for the $Fe^{3+}-SO_4-H_2O$ system at 25°C (14), the simple hydrated Fe^{3+} ion is
stable at low iron concentrations, acidic conditions and in the absence of
strong complexing anions such as sulphate. Small concentrations of
$Fe(H_2O)_6^{3+}$ probably exist in most ferric solutions although the light pur-
ple colour of this species is generally masked by the more strongly absorbing hydroxyl and sulphate complexes.

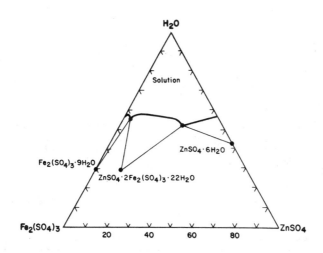

Fig. 2 - 50°C solubility isotherm in the $Fe_2(SO_4)_3-ZnSO_4-H_2O$
system (Ref. 13).

As hydroxyl ion is added to the system (i.e., as the pH is raised),
a series of Fe^{3+} - hydroxyl complexes is formed. For ferric concentrations
less than 10^{-3} M simple complexes exist:

$$Fe^{3+} + H_2O \rightarrow Fe(OH)^{2+} + H^+ \qquad \log K_{11} = -3.05 \qquad (6)$$

$$Fe^{3+} + 2H_2O \rightarrow Fe(OH)_2^+ + 2H^+ \qquad \log K_{12} = -6.31 \qquad (7)$$

$$Fe(OH)^{2+} + H_2O \rightarrow Fe(OH)_2^+ + H^+ \qquad (8)$$

In more concentrated iron media a dimerized species predominates:

$$2Fe^{3+} + 2H_2O \rightarrow Fe_2(OH)_2^{4+} + 2H^+ \qquad \log K_{22} = -2.91 \qquad (9)$$

The dimer is assumed to be linked via hydroxyl bridges:

$$(H_2O)_4 \; Fe \underset{OH}{\overset{OH}{<}} Fe \; (OH_2)_4$$

There is general agreement concerning the above major species (12,15,16) although other polynuclear complexes such as $Fe_3(OH)_4^{5+}$ and $Fe_2(OH)^{5+}$ have also been advanced as minor species. Regardless of the complex, the extent of hydrolysis increases with increasing temperature; for example, the formation constant for the dimer, K_{22}, varies as follows (17):

T (°C)	$10^3 \times K_{22}$
15	4.9
25	7.3
35	10.2
45	16.3
51	25.1

Figure 3 illustrates the approximate stability regions of the various ferric hydroxyl complexes as a function of the ferric ion concentration and the pH (14). In the region of concern to zinc processors (Fe^{3+} = 0.01 – 1.0 M; pH = 1–4), the predominant solution species would be the dimer, $Fe_2(OH)_2^{4+}$, and Fe^{3+}.

Table 1. Equilibrium Data for the Iron(III)-Sulphate-Water System at 25°C

Reaction		Medium		log K
$\alpha FeOOH + H_2O$	$= Fe^{3+} + 3OH^-$	3M	$NaClO_4$	-38.7
am. $Fe(OH)_3$ (active)	$= Fe^{3+} + 3OH^-$	3M	$NaClO_4$	-38
am. $Fe(OH)_3$ (inactive)	$= Fe^{3+} + 3OH^-$	3M	$NaClO_4$	-39.1
$0.5 \; \alpha Fe_2O_3 + 1.5H_2O$	$= Fe^{3+} + 3OH^-$	$\mu = 0$		-42.7
$\alpha FeOOH + 3H^+$	$= Fe^{3+} + 2H_2O$	3M	$NaClO_4$	3.96
$\alpha FeOOH + H_2O$	$= FeOH^{2+} + 2OH^-$	3M	$NaClO_4$	-27.5
$\alpha FeOOH + H_2O$	$= Fe(OH)_2^+ + OH^-$	3M	$NaClO_4$	-16.6
$\alpha FeOOH + H_2O$	$= Fe(OH)_3$	dil.	NaOH	- 6.53
$\alpha FeOOH + H_2O + OH^-$	$= Fe(OH)_4^-$	3M	$NaClO_4$	- 4.5
$2 \; \alpha FeOOH + 2H_2O$	$= Fe_2(OH)_2^{4+} + 4OH^-$	3M	$NaClO_4$	-51.9
$Fe^{3+} + H_2O$	$= FeOH^{2+} + H^+$	3M	$NaClO_4$	- 3.05
$Fe^{3+} + 2H_2O$	$= Fe(OH)_2^+ + 2H^+$	3M	$NaClO_4$	- 6.31
$2Fe^{3+} + 2H_2O$	$= Fe_2(OH)_2^{4+} + 2H^+$	3M	$NaClO_4$	- 2.91
$Fe^{3+} + HSO_4^-$	$= FeHSO_4^{2+}$	1.2M	$NaClO_4$	0.78
$Fe^{3+} + HSO_4^- + SO_4^{2-}$	$= FeHSO_4 \cdot SO_4$	1.2M	$NaClO_4$	2.58
$Fe^{3+} + SO_4^{2-}$	$= FeSO_4^+$	1M	$HClO_4$	2.03
$FeSO_4^+ + SO_4^{2-}$	$= Fe(SO_4)_2^-$	1M	$HClO_4$	0.97
HSO_4^-	$= H^+ + SO_4^{2-}$	1M	$NaClO_4$	- 1.02
H_2O	$= H^+ + OH^-$	3M	$NaClO_4$	-14.22

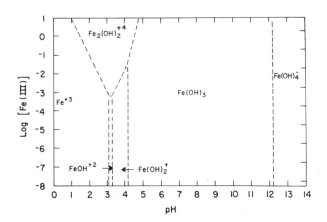

Fig. 3 - Solution species present in the $Fe^{3+}-H_2O$ system at
25°C as a function of iron concentration and pH (Ref. 14).

In strongly acidic sulphate solutions (i.e., in the comparative absence of OH), simple sulphate-iron(III) complexes are formed by reactions of the type:

$$Fe^{3+} + SO_4^{2-} \rightarrow FeSO_4^+ \qquad\qquad \log K_{11} = 2.03 \qquad (10)$$

$$FeSO_4^+ + SO_4^{2-} \rightarrow Fe(SO_4)_2^- \qquad\qquad \log K = 0.97 \qquad (11)$$

According to Ashurst and Hancock (18) only the Fe^{3+} and $FeSO_4^+$ species exist in moderately dilute solutions of iron(III) perchlorate in Na_2SO_4 – $HClO_4$ media. Evidence for the formation of $Fe(SO_4)_2^-$ at SO_4 concentrations greater than 0.3 M was presented, and it would appear that for the range of SO_4 concentrations of relevance to zinc processing (0.2 – 2.0 M SO_4) the predominant species would likely be $Fe(SO_4)^+$ and $Fe(SO_4)_2^-$. X-ray diffraction studies of concentrated ferric sulphate solutions (19) reported average ligand numbers, $Fe(SO_4)_x$, between x = 1 and x = 1.2, apparently confirming the presence of both mono- and di-sulphate complexes in such solutions. The Fe-S bond length in the solutions (3.24 Å) is similar to the value reported for solid iron sulphates (3.23 – 3.35 Å), thereby supporting the view that the structure of the complex exercises an important influence on the precipitated iron compound and illustrating the stability of the $Fe-SO_4$ bond in aqueous solution. It is also known (20) that the stability constant for reaction 10 increases with increasing temperature:

$$\ln K_f^o = 21.7130 - 8895.59/T + 115508/T^2 \qquad (12)$$

That is, the iron-sulphate complexes also become more stable as the temperature is increased.

The role played by bisulphate ion (HSO_4^-) in the precipitation of iron

compounds is far from clear. Figure 4 indicates that bisulphate ion is prevalent at low pH values whereas SO_4 is the major species for pH values > 2.
Bisulphate stability increases with increasing temperature by virtue of the
reaction (14):

$$SO_4^{2-} + H^+ \rightarrow HSO_4^- \tag{13}$$

and it might be expected to play a major role during the hydrolytic precipitation of iron. Calculations by McAndrew et al. (14) have shown bisulphate-
sulphate complexes to be of major importance under iron precipitation conditions, with the bisulphate regions of stability increasing with increasing
temperatures. Lister and Rivington (21) were able to identify a series of
Fe(III)-bisulphate and bisulphate-sulphate complexes in acidic media using
spectrophotometric methods, and their calculated formation constants are
listed in Table 1. Significantly, the formation constant for the sulphate
complex (log K_{11} = 2.03) is at least an order of magnitude greater than the
corresponding value for the bisulphate complex (log K_{11} = 0.78). Nikolaeva
and Tsvelodub (20) have observed that the high concentration of bisulphate
ion in acidic solutions would likely override the lower stability of the
Fe(III)-HSO_4 complexes with respect to their Fe(III)-SO_4 analogues. The net
effect would be the presence of SO_4^{2-}, HSO_4^- and $HSO_4.SO_4^{3-}$ complexes of
Fe(III) in solution. Direct determination of the presence and concentration
of bisulphate-iron(III) complexes in typical zinc processing solutions is an
area definitely requiring more work.

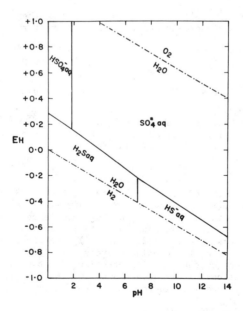

Fig. 4 - Eh-pH diagram for the sulphur - water system at 25°C
and 1 atm (Ref. 11).

538

Once the various equilibrium constants are known, it is possible to calculate the predominant sulphate, bisulphate and hydroxyl solution spec- ies at 25°C as a function of pH and the concentrations of Fe^{3+} and SO_4^{2-}. These results can then be displayed as a three dimensional predominance area diagram. Similar diagrams at elevated temperatures can also be estim- ated using the Entropy Correspondence Principle of Criss and Cobble. This has been done (14) at 25°C and 140°C and the results are summarized in Fig- ures 5 and 6, respectively. At either temperature, sulphate and sulphate- bisulphate species predominate in acidic media (pH < 3) containing > 0.1 M total SO_4 as $Fe_2(SO_4)_3$, H_2SO_4, $ZnSO_4$, etc. The region of stability of the sulphate-bisulphate complex expands significantly with increasing tempera- ture. Various hydroxyl complexes are noted, but for the iron concentrat- ions encountered during iron removal from hot acid leach liquor in the zinc industry (Fe_3^+ > 0.01 M), the dimer $Fe_2(OH)_2^{4+}$ and the simple hydrated Fe^{3+} ion are prevalent.

Figures 5 and 6 were drawn assuming the total absence of mixed hydrox- yl-sulphate complexes (e.g., $[Fe_2(OH)_2SO_4]^{2+}$) and this may well be a major oversimplification. When a solution containing predominately iron(III)- sulphate species is heated or neutralized, the sulphato complexes themselv- es probably undergo hydrolysis to form mixed sulphato-hydroxyl iron(III) species.

$$2FeSO_4^+ + 2H_2O \rightarrow Fe_2(OH)_2SO_4^{2+} + H_2SO_4 \qquad (14)$$

Direct experimental evidence for mixed species is not extensive, however, partly because most work has been done in regions where only one type of complex would be formed and partly because of problems in interpreting the more complex system. Yakovlev et al. (22) showed the hydrolysis behaviour of Fe(III) sulphate solutions to be approximately the same as uncomplexed Fe(III) solutions and concluded that the sulphate-hydroxyl dimer, $Fe_2(OH)_2(SO_4)_2$, was the prevalent species. This or similar species would appear necessary to explain the precipitation of iron hydroxysulphates, the incorporation of anions in FeO.OH precipitates and the effect of the anion on the compound precipitated (e.g., α, β or γ FeO.OH). Later, the same authors (23) confirmed the presence of mixed species and noted that the ex- tent of iron hydrolysis was slightly less in sulphate than in perchlorate or nitrate media. Kiyama and Takada (24) explained the hydrolysis behavi- our of sulphate and chloride solutions of Fe^{3+} by the formation of the fol- lowing mixed complexes:

$$Fe_2(SO_4)_3 + 3H_2O \rightarrow Fe_2(OH)_3(SO_4)_{\frac{3}{2}} + \frac{3}{2}H_2SO_4 \qquad (15)$$

$$\frac{3}{2} Fe_2(SO_4)_3 + 2H_2O \rightarrow Fe_3(OH)_2(SO_4)_{\frac{7}{2}} + H_2SO_4 \qquad (16)$$

$$2FeCl_3 + 3H_2O \rightarrow Fe_2(OH)_2Cl_2O + 4HCl \qquad (17)$$

Although the complexes differ substantially from those reported in the ab- sence of complexing anions, they do illustrate the greater complexity of the mixed systems. Although the future elucidation of the mixed complexes is obviously very difficult, it is an area where the physical chemist can play an important role in advancing our knowledge of iron precipitation in in the zinc industry.

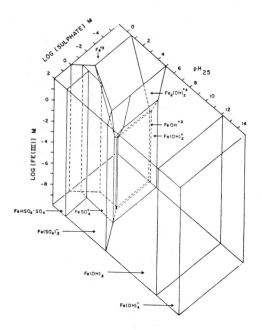

Fig. 5 - The Fe^{3+}-sulphate-water system at 25°C and 1 atm (Ref. 14).

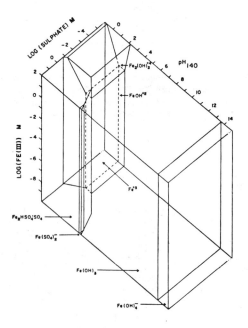

Fig. 6 - The Fe^{3+}-sulphate-water system at 140°C and 1 atm (Ref. 14).

Precipitation Pathways

The iron complexes described above are true solution species which attain equilibrium with their surroundings in a matter of seconds (12); such species do not themselves appear to precipitate although they form polymers which eventually lead to iron precipitation. The mechanism by which a soluble entity is transformed into a solid precipitate is very complex and not all steps are well understood. As was noted above, the addition of ferric ion to an aqueous sulphate system can result in the rapid formation of hydroxyl complexes:

$$Fe^{3+} + H_2O \rightarrow Fe(OH)^{2+} + H^+ \tag{18}$$

$$2Fe^{3+} + 2H_2O \rightarrow Fe_2(OH)_2^{4+} + 2H^+ \tag{19}$$

where it is understood that the hydroxyl species can also contain sulphate, bisulphate or sulphate-bisulphate ligands. Such iron solutions undergo slow reactions with the aqueous environment that can continue for several days (25,26). During this period the pH of the solution falls gradually and the colour changes from yellow to intense reddish-brown; the solution begins to scatter light, indicating the presence of small colloidal particles. The rate at which such changes occur depends on temperature, Fe^{3+}/OH, pH, iron concentration, ionic strength as well as the specific anions and cations present. Reactions which take days at 25°C can occur in a few hours at elevated temperatures and a suitable pH. The above changes seem to be associated with the formation of small iron polymers:

$$2\left[\begin{array}{c} \text{OH} \\ Fe \quad Fe \\ \text{OH} \end{array}\right]^{4+} \rightarrow \left[\begin{array}{c} \text{OH} \quad\text{OH} \quad\text{OH} \\ Fe \quad Fe \quad Fe \quad Fe \\ \text{OH} \quad\text{OH} \quad\text{OH} \end{array}\right]^{6+} + 2H^+ \tag{20}$$

$$\underline{A} \qquad\qquad\qquad \underline{B}$$

$$nB \rightarrow \left[\begin{array}{c} \text{OH} \quad\text{OH} \\ Fe \quad Fe \\ \text{OH} \quad\text{OH} \end{array}\right]_n + nH^+ \tag{21}$$

$$\underline{C}$$

The reactions are reversible although the rates of acid dissolution of polymers are also very slow, and evidence suggests the larger polymers are progressively more inert. Although the polymers are shown as Fe-OH species, they probably include significant amounts of the anions present in the solution. For example, a water soluble amorphous polymer was isolated (27) from ferric nitrate solutions and was found to have the composition $[Fe_4O_3(OH)_5 \cdot NO_3 \cdot 4H_2O]_n$. Margulis et al. (28) coagulated a colloidal basic iron sulphate $[Fe_4O_5(SO_4) \cdot xH_2O]_n$ from iron sulphate solution and this suggests the sulphate is bonded to the iron in the polymer before coagulation.

The value of n in polymer structure C is probably variable depending on the solution parameters, although the value, at least initially, appears to be fairly small. Davydov et al. (29) reported a polymer with an average of seven irons and Ciavatta and Grimaldi (30) noted one of average composition $Fe_{12}(OH)_{34}^{2+}$ in partially neutralized ferric perchlorate solutions. At this state the polymers are probably linear chains although cross linking of chains via oxo-bridges in SO_4, or OH-bridges is a possibility.

The critical step in the precipitation of crystalline iron compounds seems to be the growth of the small polymers (C) into larger polymer units which become the immediate precursors of precipitation. For iron sulphate media the polymer growth process can occur by at least three pathways. If the pH of the solution is rapidly increased, thereby raising the OH/Fe^{3+} ratio, continued growth of the Fe-OH chains occurs because the hydroxyl ions neutralize the acid produced during polymerization.

$$nB \rightarrow \left[\begin{array}{c} OH \quad OH \\ Fe \qquad Fe \\ OH \quad OH \end{array} \right]_n + nH^+ \qquad (22)$$

$$H^+ + OH^- \rightarrow H_2O \qquad (23)$$

The lengthy polymer chains become cross linked via hydroxyl bridges and large iron hydroxide gel colloids are formed that still contain some coordinated sulphate and appreciable bound water.

$$2 \left[\begin{array}{c} OH \quad OH \\ Fe \qquad Fe \\ OH \quad OH \end{array} \right]_n + 2H_2O \rightarrow \begin{array}{c} \left[\begin{array}{c} OH \quad OH \\ Fe \qquad Fe \\ OH \quad OH \end{array} \right]_n \\ OH \quad OH \\ \left[\begin{array}{c} OH \quad OH \\ Fe \qquad Fe \\ OH \quad OH \end{array} \right]_n \end{array} + 2H^+ \qquad (24)$$

Simple coagulation of the large gel colloids by OH^- or other anions leads to massive gel "precipitation"; such a process is nearly temperature independent in contrast to the other precipitation processes which are strongly temperature dependent. This is the situation existing when a ferric sulphate solution is simply neutralized with excess base and is clearly undesireable given the poor filtering and settling properties of the iron gel. Gel formaton is promoted by the presence of excess base, the rapid addition of base, poor agitation and low temperatures which prevent the formation of oxygen, as opposed to hydroxyl, bridges.

If the quantity of base is insufficient to cause immediate gel formation but is still moderately high, oxolation (formation of -O- bridges) can occur.

$$\left[\begin{array}{c} OH \quad OH \\ Fe \qquad Fe \\ OH \quad OH \end{array} \right]_n \rightarrow \left[\begin{array}{c} OH \quad OH \\ Fe \qquad Fe \\ O \quad O \end{array} \right]_{2n} + 2H^+ \qquad (25)$$

The presence of moderate base concentrations is required to neutralize the acid produced in Equation (25) and to encourage further polymer growth. The tolerance of the oxolation reaction to acid increases with increasing temperature, and Fe_2O_3 can be precipitated at 200°C in the presence of large amounts of acid (Fig. 12). The oxolated structure is the precursor to goethite or hematite precipitation with growth likely occurring by the addition of low molecular weight species such as $Fe(OH)_2^+$ or $Fe_2(OH)_2^{4+}$ to the polymer unit. The large polymers formed by the oxolation process are of colloidal size and appear to be crystalline or partly so. Little is known of the crystallization mechanism although some of the kinetic and thermodynamic factors determining the crystallization of FeO.OH or Fe_2O_3 are known and

will be discussed later. Spiro et al. (31) found the polymers to be spherical with a diameter of 70-90 Å and containing about 900 iron atoms; they had
the approximate formula $[FeO_{x/2} (H_2O)_z]_n$ and were closely associated with
the nitrate counterions. The particles formed quickly but dissolved very
slowly in acids once formed. Murphy et al. (32) also observed spherical
large polymer species that ranged in diameter from 15 to 30 Å; the polymer
formation reaction was the same in chloride, nitrate and perchlorate media.
Dousma and DeBruyn (33) reported spherical polymer particles about 40 Å in
diameter and containing both Fe-O and Fe-OH bonds (Equation 25).

The growth of the large polymers into crystallites and the eventual agglomeration of the crystallites into coarse precipitates has also been studied, at least for FeO.OH precipitation. The spherical polymer particles
tend to stick end-to-end to form rod-like clusters (32,25,26) which appear
to be about 30 x 150 Å in size (32), or slightly larger (26). The individual rods mat together to form relatively coarse particles which settle rapidly, but the conditions affecting the agglomeration are ill defined. Hematite apparently is precipitated by a different mechanism since it tends to
form distinct crystals as opposed to agglomerated particles.

The precipitation of jarosite or other basic iron sulphates such as
$FeSO_4$.OH must occur by yet another reaction pathway which is not at all well
defined. Jarosite possesses the R3m structure of alunite and consists of
sheets of hydroxyl- and sulphate-bridged Fe^{3+} distorted octahedra wherein
each Fe^{3+} is bonded to $4OH^-$ groups and $2SO_4^{2-}$ groups and each distorted oct-
ahedron is linked to its neighbours by $4OH^-$ groups. Each SO_4^{2-} group bonds
three $Fe(OH)_4$ units, with one oxygen in the SO_4^{2-} being "free" (34,35). The
structure of alunite, which is isostructural with jarosite, is given in Figure 7 (36). If associated sulphates are considered, the small polymer
chains (Equation 21) can be considered to consist of distorted Fe^{3+} octahedra shared at two edges. Such a structure could give rise to that of solid
jarosite by the opening of OH bonds such that all octahedra would be shared
at a corner instead of by edges. Since jarosite tends to be an acidic species, it might be assumed that protons play an important role in the opening
of the polymer chains. Sulphate or similar anions such as CrO_4 (36) help
stabilize the structure in some manner which is not totally clear. Much
work remains to be done on the crystallization of jarosite.

The Jarosite Process

The jarosite process in its most simplified form is shown in Figure 8
(6). The neutral leach residue is subjected to leaching for several hours
at 85-95°C in solutions containing > 100 g/L H_2SO_4 to dissolve zinc ferrites.
Provided the terminal acid concentration is greater than 20-25 g/L H_2SO_4,
almost complete (> 98%) ferrite leaching occurs. A jarosite forming cation
(NH_4^+, Na^+, K^+, etc.) is added to the hot solution and the pH is regulated
to 1.1 - 1.5 by calcine addition. Under these conditions jarosite is readily precipitated leaving a zinc-rich solution containing only 1-3 g/L Fe
that is recycled. The jarosite settles rapidly and can be thickened and
filtered to give relatively low zinc losses. The principal zinc losses into
the residue are in the form of ferrite originating with the neutralizing
calcine and as water soluble zinc caused by process water balance restraints.
The jarosite itself contains only a small amount of zinc substituted in the
jarosite lattice likely for Fe^{3+}. Figure 8 is the simplest possible jarosite flowsheet and other processes have been evolved that include multiple
stage leaching, preneutralization, jarosite leaching, silver-lead residue
recovery etc. (6,37,38).

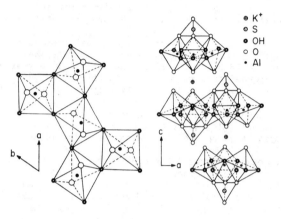

Fig. 7 - Structure of $KAl_3(SO_4)_2(OH)_6$ as viewed along the a and c axes (Ref. 34).

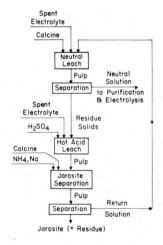

Fig. 8 - Simplified flowsheet for jarosite precipitation (Ref. 6).

The strengths of the jarosite process include:

- The process is fairly simple and can be integrated readily into new or existing plants

- the precipitate is readily thickened and filtered; low zinc losses are associated with jarosite precipitation. Filtration rates of 5-15 tons residue/m^2.day are being achieved.

- can achieve 90-95% iron rejection if the input solution contains 35g/L Fe and the final solution contains 2-3 g/L Fe

- can control both sulphate and alkali in the circuit

- relatively small amounts of alkali are needed and the theoretical amount is further reduced by the substitution of H_3O^+ for alkali. In practice (37) only 0.05-0.08 ton NH_3/ton Fe is required

- relatively low neutralizing requirements. Less calcine is needed than if the iron were precipitated as $Fe(OH)_3$ and because the process operates at pH $\sim$ 1.5, there is less free acid to be neutralized

- operates at ambient pressure using standard hydrometallurgical process equipment

- gives high recoveries of Zn, Cd, Cu, etc. (i.e., is relatively selective for iron) and can produce a low iron residue suitable for Ag-Pb recovery

Weaknesses associated with the process include:

- the need to purchase alkali

- the low iron content of the residue and the high residue volume with its attendant disposal costs

- Fe^{2+} produced by reaction of the hot solution with residual sulphides is not removed. Because the solution is acidic, the Fe^{2+} ion is only slowly oxidized by air

- potential environmental problem caused by basic sulphate disposal; ammonia leaching could be especially troublesome under certain conditions

- close process control, especially of pH, is needed. The development of self cleaning pH electrodes has tended to minimize this problem.

The precipitation of jarosite is governed by phase relationships in the $Fe_2(SO_4)_3-H_2SO_4-H_2O$-alkali sulphate-$ZnSO_4$ system at the operating temperature. Since $ZnSO_4$ does not appear to alter the equilibrium phases or the solubilities "significantly" in the range of compositions of interest during iron precipitation, this complex system can be reduced to the quaternary: $Fe_2(SO_4)_3-H_2SO_4-H_2O$-alkali sulphate where the "alkali" can be NH_4, Na, K, 1/2 Pb, etc. (39). Such quaternary systems have not been extensively studied at 100°C. The $Fe_2(SO_4)_3-H_2SO_4-H_2O$ system has, however, been investigated over the temperature range 50-200°C, and there is fair agreement among the various studies (40,41,42). Figure 9 shows the 110°C isotherm as determined by Posnjak and Merwin (40). The acidic ferric sulphate solution is in equilibrium with goethite (α FeO.OH) at low iron concentrations and with jarosite [$(H_3O)Fe_3(SO_4)_2(OH)_6$] at the intermediate concentrations of concern during zinc processing. Very high iron sulphate levels are required to form

Fe(SO$_4$)(OH). Essentially similar diagrams were also obtained at 140 and 75°C and none of these reported the compound Fe$_4$(SO$_4$)(OH)$_{10}$. Although iron can theoretically be precipitated at about 100°C as hydronium jarosite without the addition of alkali, the reaction is slow and incomplete except at higher temperatures.

$$3Fe_2(SO_4)_3 + 14H_2O \rightarrow 2(H_3O)Fe_3(SO_4)_2(OH)_6 + 5H_2SO_4 \qquad (26)$$

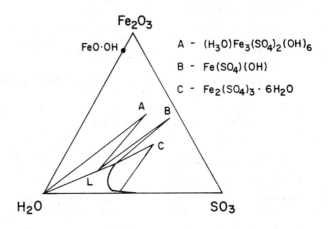

Fig. 9 - The Fe$_2$O$_3$-H$_2$O-SO$_3$ phase system at 110°C (Ref. 41).

Figure 10 shows part of the 100°C isotherm for the Fe$_2$O$_3$-H$_2$O-SO$_3$ system as determined by Walter-Levy and Quemeneur (41). The solubility measurements are in agreement with (40) but the compound Fe$_4$(SO$_4$)(OH)$_{10}$ is prevalent between jarosite and goethite. Experience in our laboratory confirms the presence of Fe$_4$(SO$_4$)(OH)$_{10}$ and its omission by Posnjak and Merwin (40) is somewhat inexplicable given the care with which they carried out their studies. The compound Fe$_4$(SO$_4$)(OH)$_{10}$ is formed at relatively low iron concentrations and could be produced during the terminal stages of jarosite formation when the iron concentration is just a few grams per litre; FeO.OH is precipitated only at the lowest iron and acid concentrations.

Addition of an alkali sulphate decreases the iron solubility somewhat at a given pH and seems to promote the rate of jarosite formation as well. In the absence of alkali, a jarosite is formed that appears to contain hydronium ion, H$_3$O$^+$, although the spectroscopic evidence for H$_3$O$^+$ is not overwhelming (43). A complete solid solution series appears to exist among the jarosites H$_3$O$^+$-K$^+$-Na$^+$-NH$_4^+$-1/2Pb^{2+}, etc. (44). The addition of even small quantities of alkali produces an alkali jarosite containing only limited hydronium ion substitution. The extent of hydronium ion substitution

546

increases with increasing temperature for Na-H$_3$O jarosite (45) and 1/2Pb-H$_3$O jarosite (46); for example, the molar percentage of Na in the Na-H$_3$O jarosite is: 64% at 50°C, 57% at 70°C and 50% at 90°C. The molar percentage of NH$_4$ in NH$_4$-H$_3$O jarosite also decreases with increasing temperature (47): 83% at 50°C, 71% at 70°C and 62% at 90°C. This behaviour is presumably tied to the increased stability of hydronium jarosite at elevated temperatures. Potassium appears to be selectively incorporated into jarosite at the expense of, say, sodium (44) and this accounts for the wider natural distribution of the potassium mineral. Further phase studies in the presence of alkali sulphate and zinc sulphate are required to elucidate fully the system relevant to iron removal in the zinc industry.

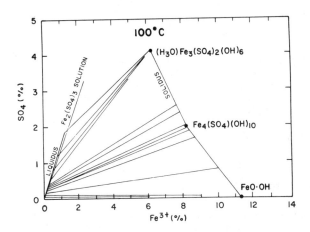

Fig. 10 - Part of the Fe$_2$O$_3$-H$_2$O-SO$_3$ phase system at 100°C (Ref. 41).

The standard free energies of formation of various jarosites have been calculated from solubility data (48) and the free energy values are listed below together with the standard free energies of the corresponding alkali ions (49).

Jarosite	ΔG°_{298} (kcal/mole)	M$^+$	ΔG°_{f298} (kcal/mole)
KFe$_3$(SO$_4$)$_2$(OH)$_6$	- 788.6	K$^+$	- 67.5
NaFe$_3$(SO$_4$)$_2$(OH)$_6$	- 778.4	Na$^+$	- 62.6
H$_3$O Fe$_3$(SO$_4$)$_2$(OH)$_6$	- 772.5	H$_3$O$^+$	0.0
NH$_4$Fe$_3$(SO$_4$)$_2$(OH)$_6$	- 736.2	NH$_4^+$	- 19.0
Ag Fe$_3$(SO$_4$)$_2$(OH)$_6$	- 701.3	Ag$^+$	18.4
Pb$_{0.5}$Fe$_3$(SO$_4$)$_2$(OH)$_6$	- 722.5	1/2 Pb^{2+}	- 2.9

Some idea of the accuracy of the available data can be inferred from the reported values for the potassium compound: -788.6 (48), -790.1 (50) and -763 kcal/mole (51). The data (48) confirm that potassium jarosite is the preferred species (44):

$$NaFe_3(SO_4)_2(OH)_6 + K^+ \rightarrow KFe_3(SO_4)_2(OH)_6 + Na^+ \qquad (27)$$

$$\Delta G^\circ_{Reaction} = -5.3 \text{ kcal/mole}$$

The data also illustrate the low stability of hydronium jarosite at low temperatures in the presence of alkali ions:

$$H_3OFe_3(SO_4)_2(OH)_6 + Na^+ \rightarrow NaFe_3(SO_4)_2(OH)_6 + H_3O^+ \qquad (28)$$

$$\Delta G^\circ_{Reaction} = -56.7 \text{ kcal/mole}$$

The use of thermodynamics to predict behaviour in such systems is complicated by the high and variable ionic strength and by the extensive formation of solid solution compounds. The data are useful, however, for predicting trends and for estimating solubilities under various conditions. The total dissolution of jarosite:

$$NH_4Fe_3(SO_4)_2(OH)_6 \rightarrow NH_4^+ + 3Fe^{3+} + 2SO_4^{2-} + 6OH^- \qquad (29)$$

is controlled by:

$$K = [NH_4^+][Fe^{3+}]^3[SO_4^{2-}]^2[OH^-]^6 \qquad (30)$$

Hence, the solubility of a potential pollutant such as ammonium ion is related to the iron and sulphate concentrations of the medium but, especially, to the pH. At pH = 2.1 and 25°C in a solution containing 3.5 g/L SO_4 and 0.74 g/L Fe^{3+}, the solubility of ammonium ion is about 60 mg/L (48). At pH = 2.3 with a solution containing 0.8 g/L SO_4^{2-} and 0.07 g/L Fe^{3+}, the ammonium ion solubility is only 19 mg/L. For a given pH, iron and sulphate concentration, potassium jarosite is least soluble, followed by the sodium and ammonium compounds.

The conversion of jarosite to goethite is relatively unfavourable (52):

$$KFe_3(SO_4)_2(OH)_6 \rightarrow 3FeO.OH + K^+ + 2SO_4^{2-} + 3H^+ \qquad (31)$$

$$\Delta G^\circ_{25°C} = +19.7 \text{ kcal/mole}$$

and this decomposition route is unlikely, at least for K, Na and NH4 jarosites.

The precipitation of jarosite is controlled by both thermodynamic and kinetic factors. For the reaction:

$$M^+ + 3Fe^{3+} + 2SO_4^{2-} + 6OH^- \rightarrow MFe_3(SO_4)_2(OH)_6 \qquad (32)$$

the equilibrium concentration of iron is approximated by:

$$[Fe^{3+}]^3 = K^{-1}[M^+]^{-1}[SO_4^{2-}]^{-2}[OH^-]^{-6} \qquad (33)$$

$$[Fe^{3+}] = K^{-1/3} [M^+]^{-1/3} [SO_4^{2-}]^{-2/3} [OH^-]^{-2} \qquad (34)$$

where K is the equilibrium constant for reaction (32) and it is assumed that activity may be approximated by concentration. Iron removal may be facilitated by a larger value of K, and this can be achieved by using potassium in preference to ammonium or sodium which in turn are superior to hydronium jarosite. Elevated temperatures also produce larger values of K and such temperatures encourage iron removal. Alkali concentrations above stoichiometric requirements are unlikely to have much effect and sulphate concentration tends to be relatively fixed by the process. Increasing $ZnSO_4$ concentration promotes the thermodynamics of iron removal but retards the settling of the precipitate (53). Hydroxyl concentration (i.e., pH) exercises a major effect on iron removal as jarosite, a property noted early in the development of the process (54). Temperature, and especially, pH are probably the two most important control variables with the restriction that the pH must remain below about 1.8 - 2.0 to prevent precipitation of other iron phases. High temperatures can be used to counteract high acidities; for example, the same iron level can be achieved at 180°C and 40 g/L H_2SO_4 as at 95°C and 5 g/L H_2SO_4 (pH $\sim$ 1.5).

The kinetics of jarosite precipitation are very complex and their elucidation has been complicated by the fact that the iron concentration and pH both vary during the reaction. There is a need for studies carried out under conditions of constant pH. When a ferric sulphate-alkali sulphate solution is heated rapidly to 90-100°C, there is often an induction period before appreciable jarosite formation occurs. This effect is observed in laboratory studies and has been reported for industrial batch precipitations where induction times of about an hour were noted (55). Such periods are presumably related to the formation of the "right" iron polymer species since they are observed commercially in the presence of seed. Once jarosite precipitation commences, rates are fairly rapid with essentially complete reaction being noted after 4-6 h at 95°C; complete reaction is achieved in less than 1 h at 140°C (46). The activation energy for jarosite precipitation is, therefore, fairly high with the following values being noted (56).

Compound	ΔH^* (kcal/mole)
$KFe_3(SO_4)_2(OH)_6$	17.7
$NaFe_3(SO_4)_2(OH)_6$	21.7
$NH_4Fe_3(SO_4)_2(OH)_6$	20.1

The rate of iron precipitation increases as the initial iron concentration increases; there is some evidence (56) that the rate varies as $[Fe^{3+}]^{0.5}$ but this is not well established. The rate tends to be independent of the alkali concentration provided that above-stoichiometric amounts of alkali ion are present. Getskin et al. (53) found potassium ion concentration to be without effect whereas ammonium ion exercised a greater influence. The concentrations of $ZnSO_4$ likely to be encountered in practice were without effect on the rate of jarosite formation (53) although high $ZnSO_4$ levels retarded jarosite sedimentation. It appears that the rate (as well as the extent) of jarosite formation increases as the pH increases (54,57), although this conclusion is open to discussion given that most work has not been carried out at constant pH. The role played by jarosite seeding is not clear. It seems to be necessary in practice, but laboratory work has been less convincing. It appears that seeding under optimum conditions in small·vessels (46) does not significantly alter the rate of precipitation. Under less than ideal conditions (46,53), the presence of seed is advantageous. Since jarosite precipitation appears to occur by a growth process, the presence of nuclei is advantageous. In large vessels or under conditions where natural seeding is slow, the addition of seed is recommended.

Lead jarosite forms solid solutions with alkali metal jarosites and this can affect lead recoveries in processes using jarosite precipitation. Lead jarosite can be formed by reaction of $PbSO_4$ with ferric sulphate media:

$$PbSO_4 + 3Fe_2(SO_4)_3 + 12H_2O \rightarrow 2Pb_{0.5}Fe_3(SO_4)_2(OH)_6 + 6H_2SO_4 \qquad (35)$$

Although this reaction does not occur significantly at $< 100°C$ (46), it is fairly rapid above 130°C. Since processes are now being advanced for oxygen pressure leaching of zinc concentrates at 140-160°C, the danger exists that the lead will report as jarosite ($\sim 18\%$ Pb) and not as $PbSO_4$ (68% Pb). Formation of lead jarosite gives rise to substantial copper and zinc losses into the precipitate and this is a further reason to avoid Pb-jarosite precipitation. Jarosite effectively removes other elements from solution as well. Arsenic and antimony will substitute to a limited degree for sulphate and can, therefore, be precipitated; fluoride ion will substitute to a limited extent for OH ion in jarosite and will be removed. The presence of fluoride seems to lead to amorphous deposits which are difficult to filter and this is probably related to the fact that F^- does not "bridge" two iron octahedra as does OH^-. Unfortunately, chloride and magnesium ions are not incorporated into the jarosite lattice; base metals such as Cu^{2+}, Zn^{2+}, Co^{2+}, Mn^{2+}, Ni^{2+} etc. are slightly dissolved in the jarosite but the amounts, in the absence of lead, are fairly low (58).

The Goethite Process

The goethite (α FeO.OH) process is something of a misnomer since goethite residues may consist of α FeO.OH, β FeO.OH and α Fe_2O_3 as well as amorphous phases (59,60); γ FeO.OH is normally not encountered since this phase is produced at low temperatures ($< 45°C$) and at elevated pH (61,62). The characterization of goethite residues is not a straightforward task since some of the phases can be amorphous to X-rays. The use of a variety of examination techniques, especially X-ray diffraction, DTA-TGA and spectroscopic methods, is adviseable (59). Goethite precipitation can be effected by one of two processes. In the Vieille Montagne technique the iron-bearing solution is reduced to the ferrous state with concentrate and is then oxidized at 80-90°C with concomitant neutralization to pH 2-3.5. In the Electrolytic Zinc process the concentrated ferric solution is added with a neutralizing agent to a heated precipitation tank at a rate equal to the goethite precipitation rate so that the ferric concentration remains low. For either approach it is important that the Fe^{3+} concentration be < 1 g/L at the time of precipitation and that the pH be controlled at a moderately high value.

Advantages claimed for the goethite process include:

- good iron precipitation, $Fe^{3+}_{Final} < 1$ g/L
- excellent filterability with filtration rates as high as 500 kg/m^2.h or 12 t/m^2.day
- no need for added alkalis
- apparently stable residues·
- able to eliminate fluoride from solution

Disadvantages seen for this process include:

- the need for iron reduction - iron oxidation in the Vieille Montagne modification
- a high neutralizer requirement in the EZ approach
- limited control capabilities for alkalis and sulphate
- pH control appears more stringent than for jarosite processes
- goethite residues include some cations and anions (e.g. to 12% SO_4 and 6% Cl) and these can "leak" during residue storage (60).

Figure 10 (redrawn from Ref. 41) shows the iron compounds in equilibrium with ferric sulphate solutions at 100°C; presumably similar diagrams exist at 80-90°C in the presence of small amounts of acid and significant $ZnSO_4$ concentrations. One percent Fe^{3+} corresponds to about 10 g/L. Iron solutions containing > 4 g Fe^{3+}/L produce jarosite during hydrolysis; solutions containing from about 2-4 g Fe^{3+}/L yield $Fe_4(SO_4)(OH)_{10}$ or mixtures of this compound and jarosite. Since the basic ferric sulphate is frequently amorphous and difficult to filter, its presence is to be avoided. Goethite formation occurs only for ferric ion concentrations less than 1-2 g/L. If the advantageous aspects of goethite precipitation are to be realized, the precipitation reaction must be carried out from relatively dilute Fe^{3+} media. This desideratum has been realized with ingenuity by both the Vieille Montagne and EZ processes.

Although Figure 10 shows FeO.OH as the stable phase, hematite (α Fe_2O_3) may, in fact, be slightly more thermodynamically stable although there are apparently kinetic barriers to its formation at low temperatures. Such a situation could explain the common precipitation of both α FeO.OH and α Fe_2O_3 (59,63) as well as the observation that hematite frequently forms from goethite but not vice versa (63,64). Recent free energy data (65) for the reaction:

$$2FeO.OH \rightarrow Fe_2O_3 + H_2O_{liquid} \qquad (36)$$

indicate that $\Delta G° = -1.18$ kcal at 27°C and -2.15 kcal at 127°C; production of water vapour further enhances the decomposition of goethite and hematite. For such small free energy changes, the course of the reaction can be reversed by factors not normally operative, such as activity differences caused by incorporation of foreign ions in the goethite and, especially, variations in surface area. The free energy equation assumes that the surface energy terms of the solids are negligible but this is true only for "massive" material (> 1 μ). The initial particles formed during iron precipitation are very much smaller, and their surface energies are significant. Using measured surface energy values, Langmuir (64) deduced the following equation for the free energy of Reaction 36.

$$\Delta G_T° \text{ (cal/mole)} = 1328 - 3.54 \text{ T} - 14.92 \log\frac{T}{343} -$$

$$\frac{1}{x}(144.6 - 0.211 \text{ T}) + \frac{1}{y}(68.6 - 0.0997 \text{ T})$$

where x and y are the edge lengths in microns of hypothetical cubes of goethite and hematite, respectively. Figure 11 is a plot of $\Delta G°$ versus particle size for the reaction at 25°C. Although the absolute free energy values differ slightly from those of Ref. 64, the following trends are significant. Goethite stability increases with decreasing size only if the goethite is coarsely crystalline and the hematite is finely divided. If the goethite is also finely divided, then hematite increases in stability as the particle sizes decrease. Since the initial particles which form are very fine (< 1000 Å ∿ 0.1 μ), it may well be that goethite represents a metastable phase formed because of kinetic barriers to the nucleation and growth of Fe_2O_3.

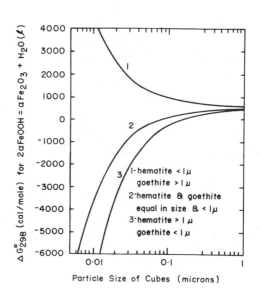

Fig. 11 - Particle size effect on the free energy of the goethite dehydration reaction at 25°C (Ref. 64).

Goethite is precipitated by oxolation of iron polymers in solution (Equation 25). The mechanism by which this reaction proceeds is obscure although several of the factors affecting the reaction have been delineated. pH control is of obvious importance since a high pH must be used to neutralize the protons generated, but not too high to drive the polymerization reactions to the iron gel stage. Commercial precipitations are conducted in the pH range 2-3.5 although it seems that part of the reason for the high pH is the need to oxidize Fe^{2+} that is facilitated by high pH. All hydrolysis reactions are favoured by increases in temperature and this is certainly true of goethite precipitation which is carried out at 75-95°C. The higher temperatures favour the formation of hematite at the expense of goethite:

$$2FeO.OH \rightarrow Fe_2O_3 + H_2O \qquad (37)$$

For example, $\Delta G°$ for the above reaction at 60, 80 and 100°C is -0.96, -1.10 and -1.36 kcal/mole, respectively (66).

The iron polymer which is the precursor of goethite is certainly an anion-hydroxyl polymer, and the anion plays a significant, if ill understood, role in determining the actual compound precipitated. The following data illustrate the products precipitated from hydrolyzed ferric nitrate solutions at 25°C (17).

Anion	Conditions	Product
Fluoride	------	β-FeOOH
Chloride	------	β-FeOOH
Bromide	------	α-FeOOH
Nitrate	------	α-FeOOH
Perchlorate	Low base addition	γ-FeOOH
Perchlorate	High base addition	α-FeOOH
Sulphate	As ammonium alum	$(NH_4)Fe_3(OH)_6(SO_4)_2$
Sulphate	In presence of Cl^- or F^-	β-FeOOH

It appears that anions with weak complexing tendencies for iron produce α or γ FeO.OH (e.g., ClO_4, NO_3, Br); anions such as F or Cl which are more strongly attached seem to give β FeO.OH. This argument is simplistic in that sulphate (as $MgSO_4$) also leads to α FeO.OH formation and sulphate is a stronger ligand than Cl. Also the compound precipitated sometimes depends on other factors such as temperature as can be seen from the following data obtained for the oxidation of 0.33 M $FeBr_2$ solutions at various temperatures (67):

Temp°C	Compound
10	β FeO.OH
20	β FeO.OH > α FeO.OH
40	γ FeO.OH $\overset{\sim}{\sim}$ α FeO.OH
65	α FeO.OH > γ FeO.OH > α Fe_2O_3
80	α Fe_2O_3 $\overset{\sim}{\sim}$ α FeO.OH > γ FeO.OH

The increasing importance of Fe_2O_3 as the temperature is raised is notable.

That anion-hydroxyl iron polymers are precursors of goethite is reflected by the fact that FeO.OH precipitated from sulphate or chloride media contains several percent of these anions. Sulphate concentrations ranging from 5-14 wt% and chloride concentrations from 0.5-6 wt% have been reported (60) and such anions cannot be readily removed by washing; i.e., they are incorporated into the structure of the precipitate. Transition metal cations such as Zn, Cu, Co and Ni are also readily incorporated into the precipitate (60, 63), presumably by substitution for Fe in the FeO.OH structure, and this suggests that the polymer precursor itself contains such cation substitution. The precipitate has a high surface area and both anions and cations can adsorb on the FeO.OH particles and this is yet another source of "contamination" that does not affect the iron removal capabilities of the process but would likely complicate any attempt to utilize the product as an iron blast furnace feed.

The Hematite Process

The hematite process has been in operation since 1972 at the Iijima Electrolytic Zinc Plant (9,68), its only site of commercial application for zinc processing. In this process the neutral leach residue is subjected to attack with return electrolyte and SO_2 at 95-100°C; this results in the reduction of iron to the Fe^{2+} state:

$$2Fe^{3+} + SO_2 + 2H_2O \rightarrow 2Fe^{2+} + SO_4^{2-} + 4H^+ \qquad (38)$$

After copper removal with H_2S, the solution is neutralized in two stages to pH = 4.5 with limestone to give a marketable gypsum precipitate. Since the iron is present as $FeSO_4$, it remains in solution during neutralization. Finally the iron is precipitated as $\alpha\ Fe_2O_3$ by heating to 180-200°C for 3 h under 18 atm O_2 pressure.

Advantages seen for the hematite process include:

- potentially marketable product although reported hematite residues (9) indicate 0.5% Zn and up to 3% S

- environmentally stable product should storage be required

- good iron elimination properties (1 g/L) although iron dissolution during filtration etc. raises this value to 3-4 g/L

- excellent filtration properties

- low volume product for storage

- process can recover Ga and In from residues

The limited application of this process is related to some inherent problems which include:

- an expensive process because of the need for pressure vessels, SO_2 liquifaction plant, etc. (42)

- potential residue marketing problems especially for the gypsum

- requires a separate iron reduction step with SO_2

Figure 1 indicates that the oxidation of a ferrous sulphate solution at pH = 4.5 will, thermodynamically, lead to the production of $\alpha\ Fe_2O_3$. Magnetite formation will be entirely avoided provided the pH remains below 7-8. Although Fe_2O_3 appears to be slightly more stable than $FeO.OH$, goethite formation would occur (probably for kinetic reasons) under these conditions at temperatures below 100°C (see below). Above 130°C, however (40), goethite is not formed and Fe_2O_3 is both kinetically and thermodynamically favoured. Figure 12 presents the relevant phase diagram at the 200°C temperature of the process (40). Although later work (42) has shown these iron solubilities to be too high, the essential features of this diagram remain correct. Solutions low in H_2SO_4 (< 6% SO_3) yield only Fe_2O_3 on hydrolysis; solutions containing more acid will produce various mixtures of Fe_2O_3 and $Fe(SO_4)(OH)$, depending on the iron level. The following data for the hydrolysis of 20 g/L $FeSO_4$ solution indicate this point (42).

H_2SO_4 (g/L)	% S in precipitate	% $FeSO_4 \cdot OH$ in precipitate
0	4.6	24.2
10	9.7	51.1
15	11.3	59.5
20	17.2	90.6

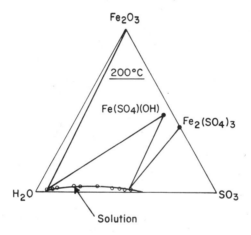

Fig. 12 - The Fe_2O_3-H_2O-SO_3 phase system at 200°C (Ref. 40).

The co-precipitation of the basic sulphate is the likely cause of sulphur contamination of the hematite product produced commercially; the degree of contamination can be somewhat controlled by regulating the iron concentration since the acid produced comes from the hydrolysis of ferric sulphate:

$$2\,FeSO_4 + 1/2\ O_2 + H_2SO_4 \rightarrow Fe_2(SO_4)_3 + H_2O \qquad (39)$$

$$Fe_2(SO_4)_3 + 3H_2O \rightarrow Fe_2O_3 + 3H_2SO_4 \qquad (40)$$

$$2\,FeSO_4 + 1/2\ O_2 + 2H_2O \rightarrow Fe_2O_3 + 2H_2SO_4 \qquad (41)$$

The presence of $ZnSO_4$ in solution reduces the amount of sulphate precipitation by enlarging the region of hematite stability as seen from Figure 13 (42). The need for low sulphur levels depends on whether the Fe_2O_3 will be marketed

555

or impounded; low sulphurs are desireable in the former case but higher concentrations can be tolerated in the latter instance, and may be advantageous in maintaining the plant sulphate balance.

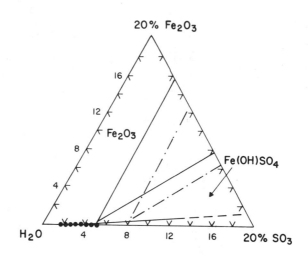

Fig. 13 - The Fe_2O_3-H_2O-SO_3 phase system at 200°C; solid lines are for 0 g/L Zn and dashed lines are 75 g/L Zn (Ref. 42).

Increasing the temperature increases the degree of iron rejection for a given acid level or permits higher tolerable acid concentrations for a given final iron concentration. For example, an Fe^{3+} concentration of 5 g/L can be produced in the presence of 90 g/L H_2SO_4 provided the temperature is 200°C. The iron precipitation reaction:

$$Fe_2(SO_4)_3 + 3H_2O \rightarrow Fe_2O_3 + 3H_2SO_4 \qquad (42)$$

suggests:

$$\log [Fe^{3+}] = a + b \log [H_2SO_4] \qquad (43)$$

and the measured equations (42) are:

$$\log [Fe^{3+}] = 3.8 \log [H_2SO_4] - 6.05 \qquad 150°C \quad (44)$$

$$\log [Fe^{3+}] = 3.7 \log [H_2SO_4] - 6.10 \qquad 170°C \quad (45)$$

$$\log [Fe^{3+}] = 3.9 \log [H_2SO_4] - 6.65 \qquad 185°C \quad (46)$$

$$\log [Fe^{3+}] = 3.9 \log [H_2SO_4] - 7.00 \qquad 200°C \quad (47)$$

Increasing the temperature from 150°C to 200°C results in an order of magnitude reduction in the dissolved iron.

The Magnetite Process

Although the magnetite process has yet to find application in the zinc industry, variations of it have long been used both for the preparation of brown-black paint pigments (69) and for the synthesis of ferrites for various electronic applications (70). Its use for the removal of metallic impurities from process waste streams was also suggested by Okuda et al. (70). The application of the process to metallurgical process streams such as iron-containing zinc solutions from the hydrometallurgical treatment of sphalerite was suggested by Sherritt Gordon Mines Limited (71).

Examination of the Eh-pH diagram for the iron-water system (Fig. 1) indicates that magnetite is stable under weakly reducing and neutral-to-alkaline conditions. The solubility of $Fe(OH)_2$ is known (12):

$$Fe(OH)_2 \rightarrow Fe^{2+} + 2OH^- \qquad (48)$$

$$\log K_s = 12.85$$

The corresponding solubility of magnetite in the presence of hydrogen is:

$$1/3 \ Fe_3O_4 + 2H^+ + 1/3 \ H_2 \rightarrow Fe^{2+} + 4/3 \ H_2O \qquad (49)$$

$$\log K_s = 12.02 \ (1 \ atm \ H_2)$$

Hence under the reducing conditions imposed by hydrogen the magnetite is slightly more stable with respect to $Fe(OH)_2$. Moderate variations in temperature, pressure, ionic strength, etc. do not appear to alter the position of the magnetite field significantly. Any process designed to precipitate Fe_3O_4 must take into account the need for both the weakly reducing conditions and the elevated pH. Complete oxidation of the solution followed by neutralization is likely to give rise to iron hydroxide, hematite or goethite. Based on traditional chemical techniques outlined by Mellor (72), the Sherritt workers identified three methods capable of removing iron from metallurgical streams as magnetite. These three techniques were:

1 - Partial oxidation of iron to the ferric state with subsequent neutralization with ammonia

2 - Simultaneous oxidation and neutralization of ferrous sulphate solution with ammonia

3 - Separate precipitation of $Fe(OH)_2$ and $Fe(OH)_3$ followed by their mixing and reaction in alkaline media.

The second method was the preferred technique. The process solution would be completely reduced to Fe^{2+} and would then be heated to temperatures > 50 < 100°C. A slight oxygen overpressure (0.05 MPa) was applied and the solution was continuously neutralized with injected ammonia. Rates were fast with complete iron removal being achieved in 5-15 min. at 100°C for an NH_3/S ratio of 2.0 and a final pH > 8. Tests done using separately oxidized solutions showed that the reaction was relatively independent of the total iron content but was most effective at a Fe^{2+}/Fe^{3+} ratio of 1.0. This suggests that the oxidation potential is more important in determining the form of the precipitated oxide than the stoichiometric ratio of

Fe^{2+}/Fe^{3+} (in magnetite itself Fe^{2+}/Fe^{3+} = 1:2). Because ferric hydroxide is insoluble at the pHs used but $Fe(OH)_2$ is soluble in excess alkali, it was tentatively postulated (72) that the reaction involved the penetration of an Fe^{2+} species into the $Fe(OH)_3$ particles:

$$Fe^{2+} + 2Fe(OH)_3 \rightarrow FeO.Fe_2O_3 + 2H_2O + 2H^+ \qquad (50)$$

There appears to be some evidence that the precipitation of Fe_3O_4 passes through an hydroxide intermediary stage and some support for the mechanism comes from the rapid rates observed when the $Fe(OH)_2$ and $Fe(OH)_3$ precipitates are mixed in spite of the obvious physical problems involved in their intermingling.

Advantages claimed for the magnetite process include:

- high iron content in the precipitate ($\sim$ 70% Fe)
- good settling and filtration rates
- low water content in the filter cake
- low sulphur content of the precipitate
- fast precipitation reactions
- nearly complete iron removal under the preferred conditions
- environmentally stable precipitates
- possibility of using the precipitate for iron making and for magnetic devices

There are also some disadvantages of the process, especially as applied to zinc solutions; some of these limitations are:

- need for a low pressure autoclave to oxidize the solution
- control of Fe^{2+}/Fe^{3+} ratio is fairly critical
- need for prior reduction of the solution
- reported sulphur levels (0.13 - 0.89%) may be too high for iron making
- ammonium sulphate byproduct is produced which must be recovered from solution by crystallization
- precipitate also removes zinc, cadmium, copper, etc.

The last item listed is certainly the most serious problem associated with the application of the process to conventional zinc solutions. There is a great tendency for the zinc to precipitate with iron to form $(ZnFeO).Fe_2O_3$ ferrites; in essence the ferrites destroyed during the hot leaching stage are reconstituted by magnetite precipitation. The incorporation of zinc in the magnetite precipitate is a consequence of similar free energies of formation of magnetite and zinc ferrite; for example, at 300 K the free energy for the reaction:

$$FeO.Fe_2O_3 + ZnO \rightarrow ZnO.Fe_2O_3 + \tfrac{1}{4}Fe_3O_4 + \tfrac{1}{4}Fe \qquad (51)$$

is + 2.8 kcal (65), suggesting there will be considerable zinc substitution in the magnetite lattice. Sherritt found that from 30 to 90% of the zinc in solution was precipitated with the iron, with the amount increasing as the dissolved zinc concentration increased . from 1.3 to 3.2 g/L. The ferrite precipitate formed in the presence of 3.2 g/L Zn contained 2.8% Zn; it is doubtful if such losses could be sustained by most zinc producers, especially since the proportion of zinc ferrite will increase with increasing zinc concentrations in solution. The magnetite process could be seriously considered for a zinc operation only where Zn, Cd, Cu etc. were removed .from solution by a solvent extraction process, leaving an iron rich, zinc-deficient solution for treatment.

Conclusions

The physico-chemical factors affecting iron precipitation from zinc hydrometallurgical solutions have been discussed. Although our understanding of iron hydrolysis-precipitation reactions has increased considerably, many areas still remain obscure. Data are needed on mixed sulphate-hydroxide solution species and on the role of bisulphate ion. The nature of the polymers requires clarification and their growth into crystallites should be studied in detail. Phase diagrams for the $Fe_2(SO_4)_3-H_2SO_4-H_2O$ system in the presence of alkali sulphates and $ZnSO_4$ should be evaluated; lastly, very precise thermodynamic data are required for the various iron compounds and for the jarosite solid solutions at iron precipitation temperatures. The resolution of these complex problems will surely occupy physical chemists for years to come.

References

1. Asturiana de Zinc S.A., Spanish Patent No. 304,601; application date October 12, 1964; cf. U.S. Patent 3,434,798; March 25, 1969.

2. Electrolytic Zinc Company of Australasia Ltd., Aust. Patent No. 401,724; application date March 31, 1965; cf. U.S. Patent 3,493,365; Feb. 3, 1970.

3. Det Norske Zinkkompani A/S, Norwegian Patent No. 108,047; application date April 30, 1965; cf. U.S. Patent 3,434,947; March 25, 1969.

4. G. Steintveit, "Electrolytic Zinc Plant and Residue Recovery Det Norske Zinkkompani A/S," AIME World Symposium on Lead and Zinc, vol. 2, pp. 223-246 (1970), AIME, New York.

5. C.J. Haigh and R.W. Pickering, "The Treatment of Zinc Plant Residue at the Risdon Works of the Electrolytic Zinc Company of Australasia Limited," AIME World Symposium on Lead and Zinc, vol. 2, pp. 423-448 (1970), AIME, New York.

6. A.R. Gordon and R.W. Pickering, "Improved Leaching Technologies in the Electrolytic Zinc Industry," Metal. Trans. 6B, 43-53 (1975).

7. Societe de la Vieille Montagne, Belgian Patent No. 724,214; application date November 20, 1968; cf. U.S. Patent 3,652,264; March 28, 1972.

8. R. Delvaux, "Lixiviation de Minerais de Zinc Grilles Selon le Procede Goethite," Metallurgie 16 (3), 154-163 (1976).

9. S. Tsunoda, I. Maeshiro, M. Ewi and K. Sekine, "The Construction and Operation of the Iijima Electrolytic Zinc Plant," AIME T.M.S. Paper No. A73-65, Chicago 1973.

10. F.A. Cotton and G. Wilkinson, Advanced Inorganic Chemistry, Inter-science Publishers, New York, 1962.

11. R.M. Garrels and C.L. Christ, Solutions, Minerals and Equilibria, Harper & Row Publishers, New York, 1965.

12. C.F. Baes and R.E. Mesmer, The Hydrolysis of Cations, Wiley-Interscience Publishers, New York, 1976.

13. V.I. Emelyanov, N.N. Sevryukov and V.P. Dolganev, "Ferric Sulphate-Zinc Sulphate-Water System at 50°," Zh. Neorg. Khim. 21 (1), 255-257 (1976).

14. R.T. McAndrew, S.S. Wang and W.R. Brown, "Precipitation of Iron Compounds from Sulphuric Acid Leach Solutions," CIM Bull. 68 (753) 101-110 (1975).

15. J.C. Landry, J. Buffle, W. Haerdi, M. Levental and G. Nembrini, "Contribution to the Study of Hydroxocomplexes of Fe(III) in Aqueous Solution," Chimia 29 (6) 253-256 (1975).

16. R.J. Knight and R.N. Sylva, "Spectrophotometric Investigation of Iron(III) Hydrolysis in Light and Heavy Water at 25°C," J. Inorg. Nucl. Chem. 37 (3) 779-783 (1975).

17. R.N. Sylva, "The Hydrolysis of Iron(III)," Rev. Pure Appl. Chem. 22, 115-132 (1972).

18. K.G. Ashurst and R.D. Hancock, "The Thermodynamics of the Formation of Sulphate Complexes of Iron(III), Cobalt(II), Iron(II), Manganese(II) and Copper(II) in Perchlorate Medium," NIM Report No. 1914, Randberg, South Africa (1977).

19. M. Magini, "Solute Structuring in Aqueous Iron(III) Sulphate Solutions. Evidence for the Formation of Iron(III)-Sulphate Complexes," J. Chem. Phys. 70 (1) 317-324 (1979).

20. N.M. Nikolaeva and L.D. Tsvelodub, "Complex Formation by Iron(III) in Sulphate Solutions at Elevated Temperatures," Russ. J. Inorg. Chem. 20 (1) 1677-1680 (1975).

21. M.W. Lister and D.E. Rivington, "Ferric Sulphate Complexes and Ternary Complexes with Thiocyanate Ions," Can. J. Chem. 33 1591-1602 (1955).

22. Y.B. Yakovlev, F.Y. Kul'ba, A.G. Pus'ko and M.N. Gerchikova, "Hydrolysis of Iron(III) Sulphate in Zinc Sulphate Solutions at 25, 50 and 80°C," Russ. J. Inorg. Chem. 22 (1) 27-29 (1977).

23. Y.B. Yakovlev, F.Y. Kul'ba, A.G. Pus'ko and N.A. Titova, "Hydrolysis of Iron(III) Sulphate in Solutions of Mixtures of Zinc, Ammonium and Copper Sulphates at 25°C," Russ. J. Inorg. Chem. 23 (2) 229-232 (1978).

24. M. Kiyama and T. Takada, "The Hydrolysis of Ferric Complexes. Magnetic and Spectrophotometric Studies of Aqueous Solutions of Ferric Salts," Bull. Chem. Soc. Jap. 46 (6) 1680-1686 (1973).

25. J. Dousma and P.L. DeBruyn, "Hydrolysis - Precipitation Studies of Iron Solutions, II-Aging Studies and the Model for Precipitation from Fe(III) Nitrate Solutions," J. Colloid Interface Sc. 64 (1) 154-170 (1978).

26. J.R. Fryer, A.M. Gildawie and R. Paterson, "Polymeric Film Precursors in the Homogeneous Crystallization of Some Metal Oxides," Nature 252 574-576 (1974).

27. O.E. Zvyagintsev and Y.S. Lopatto, "Tetranuclear Oxohydroxo-Complexes of Iron(III)," Russ. J. Inorg. Chem. 6 (4) 439-442 (1961).

28. E.V. Margulis, L.S. Getskin, N.A. Zapuskalova and L.I. Beisekeeva, "Hydrolytic Precipitation of Iron in the $Fe_2(SO_4)_3$-KOH-H_2O System," Russ. J. Inorg. Chem. 21 (7) 996-999 (1976).

29. Y.P Davydov, M.A. Grachok and L.A. Molchun, "Hydrolysis of Trivalent Iron. II-Formation of Polynuclear Hydroxy Complexes," Vestsi Akad. Nauk. Belarus. SSR, Ser. Fiz. - Energ. Nauk 1 42-44 (1974).

30. L. Ciavatta and M. Grimaldi, "On the Hydrolysis of the Iron(III) Ion, Fe^{3+}, in Perchlorate Media," J. Inorg. Nucl. Chem 37 163-169 (1975).

31. T.G. Spiro, S.E. Allerton, J. Renner, A. Terzis, R. Bils and P. Saltman, "The Hydrolytic Polymerization of Iron(III)," J. Am. Chem. Soc. 88 2721-2726 (1966).

32. P.J. Murphy, A.M. Posner and J.P. Quirk, "Characterization of Hydrolyzed Ferric Ion Solutions - A Comparison of the Effects of Various Anions on the Solutions," J. Colloid. Interface Sc. 56 (2) 312-319 (1976).

33. J. Dousma and P.L. DeBruyn, "Hydrolysis - Precipitation Studies of Iron Solutions. I - Model for Hydrolysis and Precipitation from Fe(III) Nitrate Solutions," J. Colloid Interface Sc. 56 (3) 527-539 (1976).

34. R. Wang, W.F. Bradley and H. Steinfink, "The Crystal Structure of Alunite," Acta Cryst. 18 249-252 (1965).

35. S. Menchetti and C. Sabelli, "Crystal Chemistry of the Alunite Series: Crystal Structure Refinement of Alunite and Synthetic Jarosite," Neues Jahrb. Mineral., Monatsh. 1976 (9) 406-417 (1976).

36. D.A. Powers, G.R. Rossman, H.J. Schugar and H.B. Gray, "Magnetic Behaviour and Infrared Spectra of Jarosite, Basic Iron Sulfate and Their Chromate Analogs," J. Solid State Chem. 13 1-13 (1975).

37. J.T. Wood, "Treatment of Electrolytic Zinc Plant Residues by the Jarosite Process," Aust. Min. 65 (1) 23-27 (1973).

38. C.J. Haigh, I. Oakes and J.S. Garrigan, "Residue Treatment Plant Practice at Risdon, Tasmania," Proc. Aus. I.M.M. Conference, Tasmania, May 1977.

39. J.E. Dutrizac and S.Kaiman," Synthesis and Properties of Jarosite - Type Compounds," Canadian Mineral. 14 151-158 (1976).

40. E. Posnjak and H.E. Merwin, "The System $Fe_2O_3-SO_3-H_2O$," J. Am. Chem. Soc. 44 1965-1994 (1922).

41. L. Walter-Levy and E. Quemeneur, "On the Hydrolysis of Ferric Sulphate at 100°C," C.R. Acad. Sc. Paris 258 3028-3031 (1964).

42. Y. Umetsu, K. Tozawa and K. Sasaki, "The Hydrolysis of Ferric Sulphate Solutions at Elevated Temperatures," Canada. Metal. Quart. 16 111-117 (1977).

43. J. Kubisz, "Studies on Synthetic Alkali - Hydronium Jarosites. III. Infrared Absorption Study," Mineral. Polonica 3, 23-36 (1972).

44. G.P. Brophy and M.F. Sheridan, "Sulfate Studies IV: The Jarosite - Natrojarosite - Hydronium Jarosite Solid Solution Series," Am. Mineral. 50 1595-1607 (1965).

45. E.V. Margulis, L.S. Getskin, N.A. Zapuskalova and L.I. Beisekeeva, "Hydrolytic Precipitation of Iron in the $Fe_2(SO_4)_3-NaOH-H_2O$ System," Russ. J. Inorg. Chem. 22 (4), 558-561 (1977).

46. J.E. Dutrizac, O. Dinardo and S. Kaiman, "Factors Affecting Lead Jarosite Formation," Hydrometallurgy, in press.

47. E.V. Margulis, L.S. Getskin and N.A. Zapuskalova, "Hydrolytic Precipitation of Iron in the $Fe_2(SO_4)_3-NH_3-H_2O$ System," Russ. J. Inorg. Chem. 22 (5), 741-744 (1977).

48. C.M. Kashkay, Y.B. Borovskaya and M.A. Babazade, "Determination of ΔG°_{f298} of Synthetic Jarosite and Its Sulphate Analogues," Geokhim. 5 778-784 (1975).

49. "Handbook of Chemistry and Physics" 54th edition, The Chemical Rubber Publishing Company, Cleveland (1973).

50. A.V. Zotov, G.D. Mironova and V.L. Rusinov, "Determination of ΔG°_{f298} of Jarosite Synthesized from a Natural Solution," Geokhim. 5 739-745 (1973).

51. J.B. Brown, "A Chemical Study of Some Synthetic Potassium - Hydronium Jarosites," Canadian Mineral. 10 (4) 696-703 (1970).

52. J.B. Brown, "Jarosite - Goethite Stabilities at 25°C, 1 Atm," Mineral. Deposita 6 245-252 (1971).

53. L.S. Getskin, E.V. Margulis, L.I. Beisekeeva and A.S. Yaroslavtsev, "Investigation of the Hydrolytic Deposition of Iron in the Form of Jarosites from Sulphate Solutions," Izv. Vyssh. Uchebn. Zaved., Tsvetn. Metall. 6 40-44 (1975).

54. C.J. Haigh, "The Hydrolysis of Iron in Acid Solutions," Proc. Aust. Inst. Min. Metal. No. 223 49-56 (1967).

55. G. Steintveit, "Treatment of Zinc Leach Plant Residues by the Jarosite Process," Advan. Extr. Met. Refining, Proc. Int. Symp. 521-528 (1972).

56. E.V. Margulis, L.S. Getskin, N.A. Zapuskalova and F.I. Vershinina, "Concentration Conditions for Formation of Primary Crystalline and Amorphous Phases During Hydrolytic Precipitation of Iron(III) from Sulphate Solutions," Z. Prikl. Khim. 49 (II), 2382-2386 (1976).

57. E.V. Margulis, L.S. Getskin, N.A. Zapuskalova and M.V. Kravets, "The Mechanism and Kinetics of the Hydrolytic Precipitation of Fe(III) from Zinc Sulphate Solution," Izv. Yvssh. Uchebn. Zaved., Tsvetn. Metall. 77 (5) 49-55 (1977).

58. A.S. Yaroslavtsev. L.S. Getskin, A.U. Usenov and E.V. Margulis, "Behaviour of Impurities when Precipitating Iron from Sulphate Zinc Solutions," Tsvetn. Metall. 16 (4) 41-42 (1975).

59. P.T. Davey and T.R. Scott, "Formation of β-goethite (β-FeOOH) and α-iron Oxide (α-Fe$_2$O$_3$) in the Goethite Process," Trans. Inst. Min. Metall. 84 C83-C86 (1975).

60. P.T. Davey and T.R. Scott, "Removal of Iron from Leach Liquors by the Goethite Process," Hydrometallurgy 2 25-33 (1976).

61. S. Hamada and K. Kuma, "Preparation of γ-FeO.OH by Aerial Oxidation of Iron(II) Chloride Solution," Bull. Chem. Soc. Jap. 49 (12) 3695-3696 (1976).

62. W. Wolski and A. Burewicz, "Formation and Several Physical-Chemical Characteristics of γFeO.OD," Z. anorg. allg. Chem. 383 90-95 (1971).

63. R. Derie, M. Ghodsi and R.J. Hourez, "A New Method for the Preparation of Acicular Goethite Crystals," J. Appl. Chem. Biotechnol. 25, 509-513 (1975).

64. D. Langmuir, "Particle Size Effect on the Reaction Goethite = Hematite + Water," Am. J. Sc. 271 147-156 (1971).

65. I. Barin, O. Knacke and O. Kubaschewski, "Thermochemical Properties of Inorganic Substances - Plus Supplement," Springer-Verlag, Berlin, 1977.

66. M. Collepardi, L. Massidda and G. Rossi, "Aging of Iron Oxide Gels," Trans IMM C43-C46 (1972).

67. M. Kiyama and T. Takada, "Iron Compounds Formed by the Aerial Oxidation of Ferrous Salt Solutions," Bull. Chem. Soc. Jap. 45 1923-1924 (1972).

68. T. Ohtsuka, T. Yamada, H. Abe and K. Aoki, "Progress of Zinc Residue Treatment in the Iijima Refinery," TMS Paper A78-7 presented at AIME Annual Meeting, Denver, 1978.

69. C.C. DeWitt, M.D. Livingood and K.G. Miller, "Pigment Grade Iron Oxides," Ind. Eng. Chem. 44 (3), 673-678 (1952).

70. T. Okuda, I. Sugano and T. Tsuji, "Removal of Heavy Metals from Wastewater by Ferrite Co-Precipitation," Filtration and Separation Sept-Oct 472-478 (1975).

71. W. Kunda and R. Hitesman, "Recovery of Iron as Magnetite from Aqueous Ferrous Sulphate Solution," Canada. Metal. Quart. 16 (1-4), 118-125 (1977).

72. J.W. Mellor, "A Comprehensive Treatise on Inorganic and Theoretical Chemistry," Longman Green and Co., Vol. XIII 739-740 (1934).

73. W. Kunda and H. Veltman, "Decomposition of Jarosite," Metal. Trans. 10B 439-446 (1979).

THE JAROSITE PROCESS - PHASE EQUILIBRIA

Mervyn G. Kershaw and Ralph W. Pickering
Electrolytic Zinc Company of Australasia Limited
Melbourne, Victoria, Australia.

Extensive phase equilibrium studies in the system ferric sulfate, sulfuric acid, ammonium sulfate and water were carried out at 75 and 100°C. Phase diagrams illustrate the results obtained, which provide an improved understanding of the Jarosite Process.

It was shown that the establishment of quite low concentrations of ammonium ions in the system resulted in a marked decrease in ferric iron solubility. Carphosiderite (hydronium jarosite) was replaced as a stable solid phase by the less soluble solid solution of ammonium hydronium jarosite. As the ammonium concentration in solution was increased, the composition of the solid solution changed towards that of stoichiometric ammonium jarosite, and the ferric iron solubility decreased still further.

Increasing the ammonium ion concentration in the system also decreased the concentrations of ferric iron and sulfuric acid in solutions with which goethite and solid solutions of ammonium hydronium jarosite can co-exist as stable solid phases.

Introduction

The Jarosite Process is now used widely in the electrolytic zinc industry for the recovery of zinc and other metals (principally lead, silver, gold, copper and cadmium) that would otherwise remain undissolved in residues.

The zinc contained in zinc plant residues is mostly in the form of zinc ferrite which is formed during the roasting processes that convert zinc sulfide into zinc oxide. Part of the zinc combines with the iron that is present as an impurity in the zinc sulphide concentrates. It is not uncommon for as much as 10 per cent of the contained zinc to react with iron to form zinc ferrite. Lead in zinc plant residues is present as the insoluble sulfate derived from galena that is also a common impurity in zinc sulfide concentrates.

With the advent of the Jarosite Process it became feasible to adjust the cut-off concentration for iron in calculating ore reserves, and it was also possible to adjust the selectivities sought during beneficiation. It has to be remembered, however, that the costs of recovering zinc from zinc oxide are lower than the costs of recovering zinc (and other metals) from zinc ferrite residue. The main benefit of the Jarosite Process is in the provision of an economically viable method of recovery of metals that might otherwise have little or no value.

Development of Hydrometallurgical Residue Treatment Processes at E.Z.

From the time of commencement of production of electrolytic zinc at Risdon (and elsewhere) attention was concentrated on possible methods of recovery of the zinc lost in zinc ferrite. It was known quite early that zinc ferrite could be dissolved in sulfuric acid (spent electrolyte) if of sufficient strength and at a temperature of about 100°C. The problems arose in trying to remove the large amounts of dissolved iron from the resulting solution.

In 1955 E.Z. embarked on a programme of research into leaching of zinc ferrite and high temperature (pressure) hydrolysis of the dissolved iron. It was already well known that partial hydrolysis (precipitation) of dissolved iron could be achieved during the leaching of zinc ferrite at 100°C, and it was hoped that the use of higher temperatures would increase the extent of such hydrolysis and precipitate the iron in a form that could effectively be removed from solution.

The classic phase equilibrium studies of Posnjak and Merwin[1] in the system $Fe_2O_3 - SO_3 - H_2O$ indicated that it should be possible to precipitate iron from solution at (say) 200°C either as the oxide Fe_2O_3 or the hydroxysulfate $Fe_2O_3 . 2SO_3 . H_2O$ or as a mixture of the two compounds.

Polytherms derived from the data of Posnjak and Merwin are shown in Figures 1 and 2 for temperatures from 50 to 200°C. Values for Fe_2O_3 and SO_3 concentrations have been converted to the equivalent Fe and SO_4. Figure 1 shows the data from 0 to about 40 weight per cent sulfate, and Figure 2 the data from about 30 to 70 weight per cent sulfate. The polytherms are necessarily shown from different angles because of the high solubility of ferric iron at sulfate concentrations of 30 to 40 per cent.

From Figure 1 it is evident that quite concentrated solutions of ferric sulfate can be stable at temperatures up to 140°C.

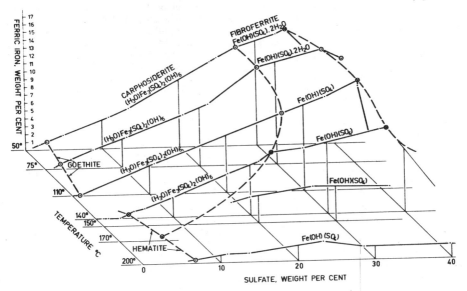

Figure 1 - System $Fe_2(SO_4)_3$ - H_2SO_4 - H_2O :
Polytherm 50°C to 200°C, 0 to 40% SO_4 :
from data of Posnjak and Merwin   (1).

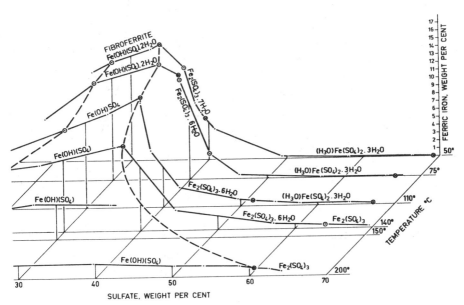

Figure 2 - System $Fe_2(SO_4)_3$ - H_2SO_4 - H_2O :
Polytherm 50°C to 200°C, 30 to 70% SO_4 :
from data of Posnjak and Merwin   (1).

At higher temperatures, towards 200°C, carphosiderite is replaced by the hydroxysulfate* as a stable solid phase, and the solubility of iron is decreased quite markedly. At lower sulfate (or sulfuric acid) concentrations, goethite is replaced by hematite as the stable solid phase. Hematite also becomes stable to somewhat higher sulfate concentrations.

It is evident that a solution of ferric sulfate when heated to 200°C will hydrolyze according to one or other (or both) of the following reactions:

$$Fe_2(SO_4)_3 + 2 H_2O = 2 Fe(OH)(SO_4) + H_2SO_4 \qquad (1)$$

$$Fe_2(SO_4)_3 + 3 H_2O = Fe_2O_3 + 3 H_2SO_4 \qquad (2)$$

The completeness of the precipitation of iron could be limited by the concentration of sulfuric acid (if any) that was present initially or was generated by the reactions unless a neutralizing agent was present or was added.

The difference in solubility of ferric iron at about 100°C and at 200°C is shown more clearly in Figure 3, in which the two isotherms are projected on to the $Fe-SO_4-H_2O$ plane. The dashed line represents mixtures of ferric sulfate and water, and its position (in relation to lines representing solutions in equilibrium with carphosiderite or fibroferrite) indicates that some sulphuric acid (i.e., additional sulfate for these coordinates) is necessary to stabilize solutions of ferric sulfate even at 100°C.

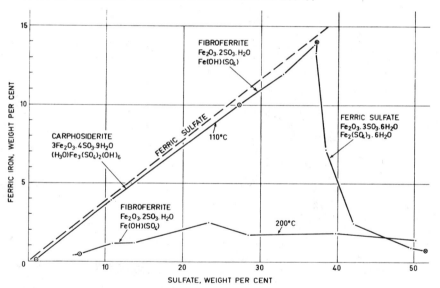

Figure 3 - System $Fe_2(SO_4)_3 - H_2SO_4 - H_2O$
110°C and 200°C Isotherms from
data of Posnjak and Merwin. (1).

*The dihydrate, $Fe(OH)(SO_4) \cdot 2H_2O$ or $Fe_2O_3 \cdot 2SO_3 \cdot 5H_2O$, found at lower temperatures is identical with the mineral Butlerite; there does not appear to be a mineralogical counterpart for the compound $Fe(OH)(SO_4)$ or $Fe_2O_3 \cdot 2SO_3 \cdot H_2O$ that replaces the dihydrate at about 100°C. The use of the generic name fibro-ferrite for these compounds may not be justified.

Results obtained by E.Z. in the leaching of zinc ferrite and hydrolysis of iron were qualitatively as indicated by these equilibrium studies of Posnjak and Merwin.   Quantitative differences were attributed initially to the effect of non-equilibrium conditions.

Patents were filed (2,3) for particular process designs that were developed.

Development of the Jarosite Process

Surprising features of the earlier work had been that carphosiderite appeared to be the solid phase being precipitated, instead of the expected hematite and/or iron hydroxysulfate;   also that the iron concentration of 1 to 4 grams per litre remaining in solution was much lower, at 50 to 70 grams of sulfuric acid per litre, than would have been expected from the data of Posnjak and Merwin.

Haigh (4), in an early review of the development of the Jarosite Process, reported, how when pure solutions of ferric sulfate in sulfuric acid were substituted for the plant solutions that had been used previously, hematite was in fact precipitated at $200^{\circ}C$ at low acidities, and iron hydroxysulfate at higher acidities, and the residual iron concentrations were not as low as those obtained when using leach solutions derived from plant spent electrolyte.

Haigh prepared a sample of carphosiderite from a pure solution of ferric sulfate in sulfuric acid by heating it to $170^{\circ}C$, and showed that the X-ray diffraction pattern, although similar, was different from those of the solids precipitated from solutions derived from plant spent electrolyte.   He had also confirmed the presence of potassium and sodium in "carphosiderite" precipitated from plant solutions.

It was thus shown that the unexpected extent of the hydrolysis of ferric iron had been due to its precipitation as potassium and sodium jarosites.   It was confirmed that potassium, sodium and/or ammonium could be added to otherwise pure solutions of ferric sulfate in sulfuric acid to precipitate the respective jarosites or mixtures or solid solutions of them.   It was also confirmed that the low solubility of ammonium jarosite (and the other jarosites) made it possible for the process to be conducted at temperatures of about $100^{\circ}C$ and at atmospheric pressure.

A plant, which has been described previously (5), was built to use this new process (6) for the recovery of zinc and other metals from currently produced zinc plant residue and from a large quantity of residue that had been stock-piled.

Aqueous ammonia was selected for use at Risdon, because ammonia and ammonium sulfate were already being produced on the site.   Aqueous ammonia had the added advantage that it provided a neutralizing capability compared with the sulfate:

$$3 \ Fe_2(SO_4)_3 + 2 \ NH_4OH + 10 \ H_2O = 2 \ (NH_4)Fe_3(SO_4)_2(OH)_6 + 5 \ H_2SO_4 \qquad (3)$$

$$3 \ Fe_2(SO_4)_3 + (NH_4)_2SO_4 + 12 \ H_2O = 2 \ (NH_4)Fe_3(SO_4)_2(OH)_6 + 6 \ H_2SO_4 \qquad (4)$$

Phase Equilibria - System $Fe_2(SO_4)_3-H_2SO_4-(NH_4)_2SO_4-H_2O$

Although sufficient information was available for design and construction of a plant to use the Jarosite Process for the treatment of zinc ferrite residues at Risdon, an investigation was initiated of the phase equilibria in the system $Fe_2(SO_4)_3 - H_2SO_4 - (NH_4)_2SO_4 - H_2O$. It was considered to be necessary to determine whether there were any other process designs that should be used.

Experimental Techniques

Mixtures to be equilibrated, consisting of 100 mls of a ferric sulfate solution containing varying amounts of ammonium sulfate and sulfuric acid, were placed in sealed pyrex phials and suspended in an oil bath for at least 30 days. The mixtures were agitated by rocking the phials through $120°$ once per second. After equilibrium had been established, the phials were removed shock cooled and the phases separated by filtration and analysed. The temperature variation of the oil bath was less than $\pm 0.1°C$.

Reagent grade chemicals were used wherever possible. The ferric sulfate solutions were usually prepared by oxidizing A.R. ferrous sulfate with A.R. hydrogen peroxide; but, in some experiments, commercial grade basic ferric sulfate (Monsel's salt) was used.

Some starting mixtures also contained one or other of the following solid phases - goethite, butlerite, carphosiderite, or ammonium jarosite. These solids were prepared by precipitation from ferric iron sulfate solutions of the appropriate acidity.

Phase separation by shock cooling, followed by filtration, could not be used for solutions containing more than 10% NH_4, because of changes in the composition of the phases which occurred on cooling. For these mixtures, filtration was carried out at temperature by means of a sintered glass disc sealed into the phial.

The equilibration times used for the mixtures varied from 30 to 50 days. Generally, as no changes in the composition of the phases occurred after the first 30 days, the equilibration time was standardised to 30 to 40 days; but up to 50 days were used for viscous solutions in the region of stability of the compounds $Fe(OH)SO_4$, $Fe(OH)SO_4.2H_2O$ and $Fe_2(SO_4)_3.6H_2O$.

Standard procedures were used to analyse the solution and solid phases for ferric iron, sulfate and ammonium.

Wet, unwashed solid phases were first examined by X-ray diffraction (XRD), then washed with water, alcohol and ether, and air dried. Water could not be used to wash the acid sulfates $(H_3O)Fe(SO_4)_2.3H_2O$ and $(H_3O)Fe(SO_4)_2$ because they dissolved too readily. The XRD analyses were then repeated and, if unchanged, the dried solids were analysed. In most cases the XRD patterns remained unchanged; but, upon ageing, the acid sulfate, $(H_3O)Fe(SO_4)_2.3H_2O$, and the normal sulfate, $Fe_2(SO_4)_3.6H_2O$, tended to dehydrate.

The experimental results that were obtained are shown in the various figures that follow. The co-ordinates sulfate, ferric iron and ammonium are used. Some authors have used "sulfuric acid" or "free sulfuric acid", determined by titration procedures or by deduction of sulfate equivalent to ferric and other cations present in the solution. It is known that ferric iron is present in such solutions in the form of complex ions, so that titration procedures are subject to misinterpretation. Calculation may be more appropriate, particularly when large amounts of other sulfates, such as those of copper or zinc are also present in the solutions.

No attempt was made to explore regions of the system containing free ammonia, because these would be irrelevant in the context of the Jarosite Process.

Isotherms for 100°C

The results obtained in the main regions of interest in consideration of any possibility of modification of the Jarosite Process are illustrated in Figures 4, 5 and 6. Figure 4 shows equilibrium compositions of saturated solutions (and the stable solid phases) for the range of concentrations - SO_4 0 to 50, NH_4 0 to 10 and Fe 0 to 15 weight per cent. Figure 6 shows equilibrium compositions of saturated solutions (and the stable solid phases) for the more limited range of solution compositions - SO_4 0 to 5, NH_4 0 to 0.4 and Fe 0 to 2 weight per cent. Assuming an approximate specific gravity of the order of 1.2 for the solutions shown in Figure 6, the range of concentrations is equivalent to - SO_4 0 to 60, NH_4 0 to 5 and Fe 0 to 20 grams per litre. This range includes the concentrations of greatest interest in present installations of the Jarosite Process.

Referring first to Figure 4, the 100°C isotherm in the ternary system $Fe_2(SO_4)_3$ - H_2SO_4 - H_2O is shown on the Fe - SO_4 - H_2O plane. Glockerite, considered to be a non-equilibrium solid phase, was encountered where shown, separating goethite from carphosiderite. The phase boundary between carphosiderite and the hydroxysulfate (fibro-ferrite) was not found experimentally, but is considered to be close to the position indicated. Similarly, the boundary between the hydroxysulfate and ferric sulfate was not found experimentally, but is considered to be close to the position shown for it.

For saturated solutions containing ammonium (i.e. the quaternary system), the ammonium and sulfate concentrations are shown as points on the NH_4 - SO_4 - H_2O plane, and their iron concentrations are shown as vertical lines of the appropriate lengths. Ideally, there would be sufficient experimental results to allow the planes of stability of the equilibrium solid phases, and the lines and points of intersections of these planes, to be defined exactly. Although this has not been so in all cases, it is possible to obtain a reasonably clear understanding of the planes and their curvature, and of the intersections of the planes.

Figure 5 has been adapted from Figure 4. Ferric iron solubilities have been determined by interpolation for selected ammonium and sulphate concentrations. Lines have been drawn in two directions along the surfaces of the planes of stability of the solid phases. The resulting figure gives a clearer definition of the large changes of solubility that occur when ammonium is introduced into the system:

(a) From (close to) 0 to 30 per cent sulfate, ammonium jarosite, or rather the solid solution $(NH_4,H_3O)Fe_3(SO_4)_2(OH)_6$, has replaced carphosiderite (hydronium jarosite) as the stable solid phase, and there is a marked reduction in the solubility of ferric iron.

571

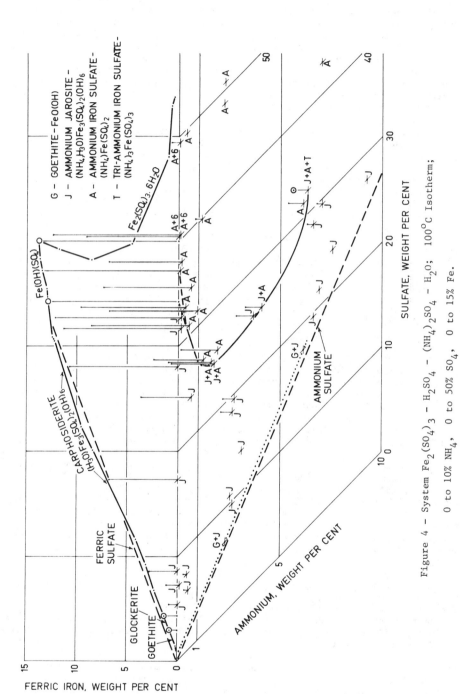

Figure 4 – System $Fe_2(SO_4)_3$ – H_2SO_4 – $(NH_4)_2SO_4$ – H_2O; $100^\circ C$ Isotherm;

0 to 10% NH_4,   0 to 50% SO_4,   0 to 15% Fe.

G – GOETHITE – FeO(OH)
J – AMMONIUM JAROSITE –
    $(NH_4,H_3O)Fe_3(SO_4)_2(OH)_6$
A – AMMONIUM IRON SULFATE –
    $(NH_4)Fe(SO_4)_2$
T – TRI-AMMONIUM IRON SULFATE –
    $(NH_4)_3Fe(SO_4)_3$

FERRIC IRON, WEIGHT PER CENT

AMMONIUM, WEIGHT PER CENT

SULFATE, WEIGHT PER CENT

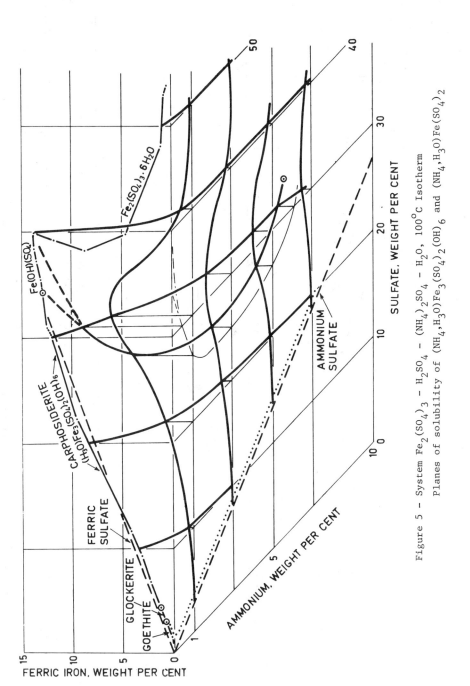

Figure 5 – System $Fe_2(SO_4)_3 - H_2SO_4 - (NH_4)_2SO_4 - H_2O$, $100°C$ Isotherm

Planes of solubility of $(NH_4,H_3O)Fe_3(SO_4)_2(OH)_6$ and $(NH_4,H_3O)Fe(SO_4)_2$

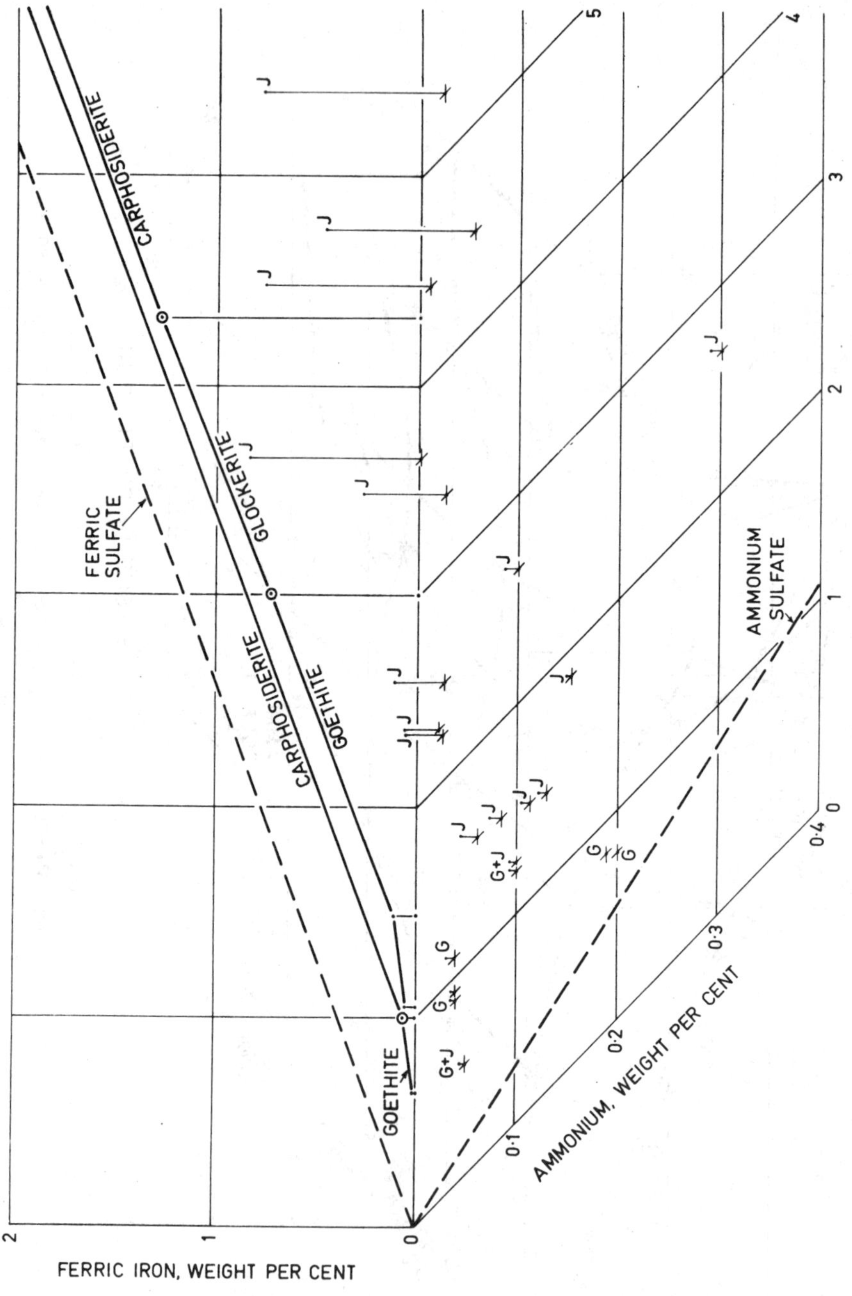

Figure 6 – System $Fe_2(SO_4)_3$ – H_2SO_4 – $(NH_4)_2SO_4$ – H_2O; 100°C Isotherm;   0 to 0.4% NH_4, 0 to 5% SO_4, 0 to 2% Fe.

(b)   From (close to) 0 to 15 per cent sulfate, the reduction in solubility of
       ferric iron is very pronounced, even for ammonium concentrations less
       than 1 per cent.

(c)   From 15 to 30 per cent sulfate, the reduction in solubility of ferric
       iron is less pronounced in absolute terms, although it is still very
       significant in relative terms.

(d)   From (close to) 0 to 30 per cent sulfate, for solutions containing sulfate
       only slightly in excess of that equivalent to the ammonium concentration,
       the solubility of iron is very low.

(e)   Between about 35 and 40 per cent sulfate, introduction of a small concen-
       tration of ammonium results in the replacement of the hydroxysulfate
       (fibro-ferrite) by ammonium iron sulfate or the solid solution
       $(NH_4,H_3O)Fe(SO_4)_2$.   As the ammonium concentration is increased, the
       region of stability of ammonium iron sulfate extends to cover a wide
       range from about 25 per cent sulfate upwards, and the solubility of iron
       decreases quite markedly.   The solubility of iron as ammonium iron sul-
       fate is very low for sulfate concentrations above 50 per cent and ammon-
       ium concentrations above 0.5 per cent.

       In Figure 6 also, part of the $100^{o}C$ isotherm for the ternary system
$Fe_2(SO_4)_3$ - H_2SO_4 - H_2O is shown on the Fe - SO_4 - H_2O plane.   The upper line
represents the results obtained by Posnjak and Merwin, with the phase boundary
between goethite and carphosiderite appearing at 1 per cent sulfate.   The
lower line represents the results of the present authors, with goethite per-
sisting as the stable solid phase up to 3 per cent sulfate, where an apparent
phase boundary with glockerite was found.   Glockerite then appeared to persist as the stable solid phase up to about 4.2 per cent sulfate where an apparent phase boundary with carphosiderite was found, and carphosiderite remained
the stable solid phase as sulfate concentration was increased further.

       As in Figure 4, the compositions of saturated solutions containing ammon-
ium in the quaternary system are shown as points on the NH_4 - SO_4 - H_2O plane,
and their iron concentrations are shown as vertical lines of the appropriate
lengths.   For the sulfate and ammonium concentrations that are of particular
interest in operation of the Jarosite Process, the equilibrium solubility of
ferric iron as ammonium jarosite is shown to be about 0.1 weight per cent, or
1.2 grams per litre.

       The results illustrated in Figure 6 are not included in Figure 4, but
the relationship between the two figures can be seen by reference to the position of the phase boundary between goethite and glockerite on the Fe - SO_4 -
H_2O plane in the two figures.

       The results illustrated in Figures 4, 5 and 6 do not show any evidence of
equilibrium solubilities of ferric iron compounds that indicate that the Jarosite Process would benefit from any major variation of the operating conditions now in use.   The results do, however, indicate the possibility of process
designs for other purposes, based on the precipitation of iron as ammonium iron
sulfate.

Projections for $100^{o}C$ and $75^{o}C$

       Some of the data shown in Figures 4, 5 and 6 are reproduced as double
projections in Figures 7 and 8.   Figure 7 shows the data for $100^{o}C$, and 20
to 60 per cent SO_4, projected orthogonally on to the Fe - SO_4 - H_2O plane

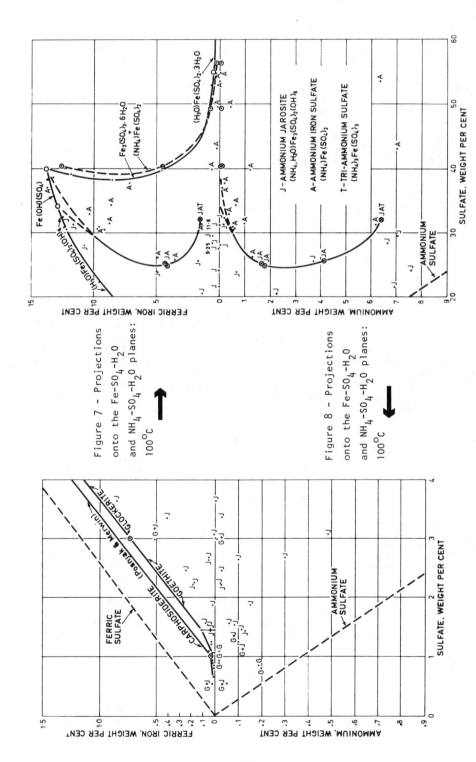

Figure 7 - Projections onto the Fe-SO₄-H₂O and NH₄-SO₄-H₂O planes: 100°C

Figure 8 - Projections onto the Fe-SO₄-H₂O and NH₄-SO₄-H₂O planes: 100°C

J - AMMONIUM JAROSITE
(NH₄, H₃O)Fe₃(SO₄)₂(OH)₆

A - AMMONIUM IRON SULFATE
(NH₄)Fe(SO₄)₂

T - TRI-AMMONIUM SULFATE
(NH₄)₃Fe(SO₄)₃

(upper part of the figure) and on to the NH_4 - SO_4 - H_2O plane (lower part of the figure). The figure shows (in particular) the phase boundary between ammonium jarosite and ammonium iron sulfate. The phase boundary between ferric sulfate and ammonium iron sulfate is not well defined.

The appearance of tri-ammonium iron sulfate is indicated, but the extension of the region of stability of this phase is not defined. It is evident, however, that ammonium jarosite remains the stable phase at high ammonium concentrations at sulfate concentrations up to 30 per cent; and that ammonium iron sulfate remains the stable phase at high ammonium concentrations at sulfate concentrations above 40 per cent. It is not clear whether ammonium iron sulfate appears first at the fibroferrite - ferric sulfate phase boundary or at the fibroferrite - carphosiderite phase boundary (see Figure 9).

Figure 8 shows the projections of the data from Figure 6, at low concentrations of NH_4, Fe and SO_4. The phase boundary between goethite and ammonium jarosite is not well defined, because the two phases were encountered together in equilibrium with a solution containing 3 per cent SO_4 and a low concentration of ammonium. This appears to add weight to the argument for goethite and glockerite to co-exist (in the absence of ammonium) at 3 per cent SO_4.

There is, however, ample evidence that quite low concentrations of ammonium result in a drastic reduction in the solubility of iron as ammonium jarosite.

Some of the data for the system at $75^{\circ}C$ is shown in the projections in Figures 9 and 10. These projections are for similar concentration ranges to those in Figures 7 and 8. It is evident that the field of existence of ammonium iron sulfate has not extended as far in the region of low sulfate concentrations, but has extended into the regions of high sulfate concentrations. Here again, the phase boundary between ferric sulfate and ammonium iron sulfate is poorly defined. There is (at $75^{\circ}C$) evidence of ammonium jarosite and tri-ammonium iron sulfate co-existing; but a phase boundary between tri-ammonium iron sulfate and ammonium iron sulfate was not encountered. There is also (at $75^{\circ}C$) evidence of a field of continued existence of fibroferrite at low ammonium concentrations, and there is some evidence for assuming that ammonium iron sulfate appears first at (or near) the phase boundary between fibroferrite and ferric sulfate.

As was the case for $100^{\circ}C$ (Figure 8) the data for low concentrations of Fe, SO_4 and NH_4 at $75^{\circ}C$ in Figure 10 is not explicit. Goethite and ammonium jarosite were encountered together over a range of concentrations of sulfate (at low concentrations). It can be seen, however, that the solubility of iron as ammonium jarosite is appreciably higher at $75^{\circ}C$ than at $100^{\circ}C$ for comparable ammonium concentrations.

Composition of Ammonium Jarosite

The projections in Figure 8 (which are based on the same information that is shown in Figure 6) show that ammonium concentrations of at least 0.1 per cent are required if iron solubilities less than 0.1 per cent as ammonium jarosite are to be obtained.

The relationship between concentrations of ammonium in solutions and the ammonium contents of ammonium hydronium jarosite in equilibrium with the solutions is shown (for $100^{\circ}C$) in Figure 11. Solutions containing as little as 0.02 per cent ammonium are in equilibrium with ammonium hydronium jarosite containing about 3 per cent ammonium. Solutions containing 0.1 to 0.2 per

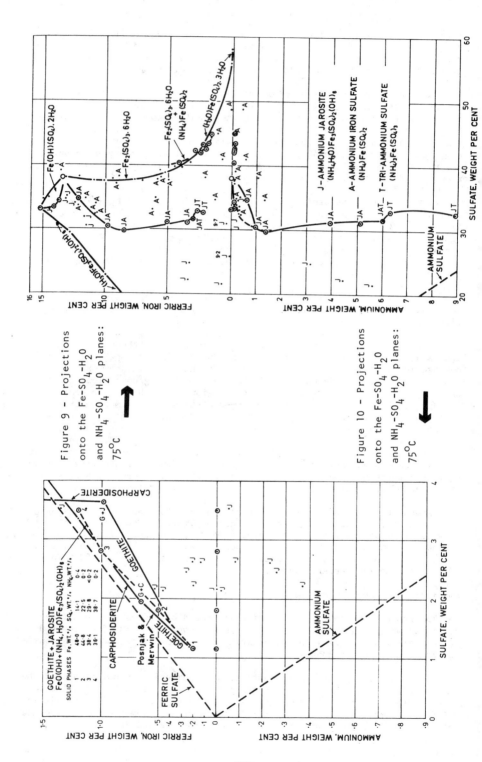

Figure 9 - Projections onto the Fe-SO$_4$-H$_2$O and NH$_4$-SO$_4$-H$_2$O planes: 75°C

Figure 10 - Projections onto the Fe-SO$_4$-H$_2$O and NH$_4$-SO$_4$-H$_2$O planes: 75°C

cent ammonium are in equilibrium with jarosite containing 3.5 to 3.6 per cent
ammonium.

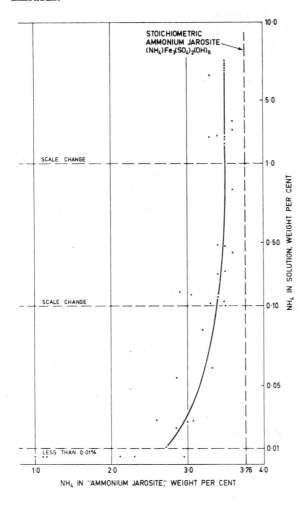

Figure 11 - Relationship
NH_4 content of
$(NH_4,H_3O)Fe_3(SO_4)_2(OH)_6$
and NH_4 concentrations
in solution.

For effective operation of the Jarosite Process when using ammonium as
the reactant, the value of x in the equation below, therefore, has to be
approximately 1, to replace the ammonium that is precipitated as ammonium
jarosite, and to maintain the ammonium concentration in solution at a concentration that will result in sufficiently complete precipitation of ferric iron.

$$3\ Fe_2(SO_4)_3 + 2x\ NH_4OH + (12-2x)\ H_2O =$$
$$2\ (NH_4)_x(H_3O)_{1-x}Fe_3(SO_4)_2(OH)_6 + (6-x)\ H_2SO_4 \qquad (5)$$

Acknowledgements

The authors wish to thank the Electrolytic Zinc Company of Australasia Limited for permission to publish this paper.

It is also wished to acknowledge that part of the work was carried out with financial assistance from the Australian Research Grants Committee.

References

(1)   E. Posnjak and H.E. Merwin, "The System, $Fe_2O_3 - SO_3 - H_2O$", Journal American Chemical Society, 44(2)(1922) pp. 1965-1994.

(2)   R.W. Pickering and E. Whayman, "Recovery of Zinc from Zinc Plant Residue", U.S. Patent No. 3,143,486; Aug.4, 1964.

(3)   H.O.K. Veltman, E. Whayman, C.J. Haigh and R.W. Pickering, "Process for the Recovery of Zinc from Zinc Plant Residues", U.S. Patent No. 3,193,382; July 6, 1965.

(4)   C.J. Haigh, "The Hydrolysis of Iron in Acid Solutions", Proceedings Australasian Institute of Mining and Metallurgy, No. 223 (1967) pp. 49-56.

(5)   C.J. Haigh and R.W. Pickering, "The Treatment of Zinc Plant Residue at the Risdon Works of the Electrolytic Zinc Company of Australasia Limited", World Symposium on Mining and Metallurgy of Lead and Zinc, Vol. 2, pp. 423-448; AIME, New York, N.Y., 1970.

(6)   R.W. Pickering and C.J. Haigh, "Treatment of Zinc Plant Residue", U.S. Patent No. 3,493,365; Feb. 3, 1970.

U.S. ENVIRONMENTAL LAWS AND REGULATIONS

AS APPLIED TO THE LEAD AND ZINC INDUSTRIES

Jerome F. Cole

International Lead Zinc Research Organization, Inc.

Lead Industries Association, Inc.

New York, New York

The environmental movement in tbe U.S. has had a major impact on the lead and zinc industries. Environmental regulations were developed as a result of a mood of public distrust of most institutions which developed during the 1960's. The lead and zinc industries have been particularly affected by the Clean Air Act of 1970 and the Occupational Safety and Health Act of 1970. These laws have resulted in regulations which have impacted both production and marketing in the lead and zinc industries. While the future does not look bright, there are some signs that the public and legislators are realizing that in order to avoid stagnation of the industry, remedial legislation may be in order.

Introduction

The environmental movement in the U.S. is a product of the turbulent 1960's. That era of marching, chanting, public disobedience, violence and near anarchy has left us with a legacy of laws and regulations, but, more important, it has left us with an ingrained attitude of public distrust of institutions of any kind. That attitude and its political ramifications have made it increasingly difficult for the United States to solve problems involving public policy toward the private sector. Nowhere has this been more obvious than in its inability to develop an effective energy policy. No one trusts the oil companies; no one trusts the U.S. Department of Energy; no one has any faith in the politicians. Therefore, no one has an energy policy that can gain enough of a consensus to make any significant progress toward solving our energy problems.

The environmental movement, which was embraced by the news media and politicians as an issue with broad popular appeal, has been characterized by Eisenbud [1] as being "led by the economically elite". As such, there has been little appreciation for the economic needs of the not-so-elite or for the industrial machinery which provides the real wealth of society. Today, after a decade of legislative reaction to one real or imagined pollution episode after another, industry now finds itself bound up so tightly in regulations and standards that there is real worry about survival. Generally, most responsible people now recognize that the pendulum has swung too far and that sound public policy demands reconsideration of much of the environmental legislation and regulations that are now constricting industry. Yet that same public distrust of institutions that has left us so impotent in the fact of our energy shortage prevents any serious attempt at redressing the errors of environmental over-regulation.

Until this attitude of the public changes and until industry and government regain credibility, industry is in the position of trying to cope as best it can with a regulatory situation which in the long run can only be described as disastrous. While this description varies in severity from industry to industry the long-term outlook for the lead industry is particularly bleak.

Impact of Environmental Regulations on the Lead and Zinc Industries

The Clean Air Act

The Clean Air Act of 1970, passed by Congress with great public support because of the interest generated by Earth Day in 1969, was, in effect, an expression of public impatience with the progress of air pollution efforts up to that time. The U.S. Environmental Protection Agency was formed on January 1, 1971, to put into action the requirements of the new law.

The impact of the Clean Air Act and resulting EPA regulations have had a profound impact on the lead and zinc industries. These effects have either been direct, such as those requiring expensive controls on sulfur oxide and particulate emissions from smelters, or indirect, through the loss of important and lucrative markets.

Arguably rigid sulfur dioxide and particulate regulations contributed to the demise of nearly one-fourth of the nation's slab zinc production capacity

[1]

Merril Eisenbud, Environment, Technology, and Health, Human Ecology in Historical Perspective, New York University Press, New York, N.Y. 1978.

in the early 1970's. However, as the smelters which closed were considered somewhat obsolete and only marginally economic, it is difficult to place the blame entirely on environmental regulation. On the other hand, it is clear that environmental regulations have played a critical role in the decline of zinc diecastings in automobiles and, of course, in the decline of lead anti-knock compound usage in gasoline.

Automobile fuel economy took a nose dive in the early 1970's as a direct result of the automotive emissions standards included in the 1970 Clean Air Act. The legislation passed by Congress mandated a 90% reduction in emissions of unburned hydrocarbons, carbon monoxide and oxides of nitrogen by 1975. While there have been several postponements of various deadlines over the years, the near state of panic in which the surprised automobile companies found themselves led to measures which cost dearly in fuel economy. These included the general lowering of compression ratios so as to allow engines to burn lower octane unleaded fuel in order to pave the way for the introduction of catalytic converters on automobiles. Then, along came the 1973 Arab oil embargo and federally mandated fuel economy improvements. Hemmed in on one side with inherently fuel inefficient cars because of governmental environmental regulations and on the other by governmental energy regulations, the auto makers began a massive effort to reduce the weight of automobiles, and hence, to effect an improvement in fuel economy. Zinc diecastings, heavy and largely decorative were an obvious target for elimination. Figure 1 shows just how dramatic the decline in zinc usage has been since 1973. Almost this entire loss can be attributable to losses in automotive diecastings. Only lighter weight thin-wall diecastings developed through ILZRO research has managed to keep zinc in automotive diecastings at all. The loss of approximately 200,000 tons of annual zinc consumption in this market, however, certainly helps explain a large part of the zinc industry's current problems.

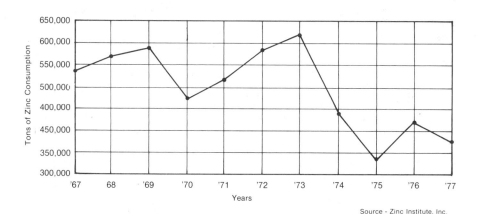

Source - Zinc Institute, Inc.

Fig. 1 - Zinc Consumption for Diecastings - U.S.

Catalytic converters, which were the answer of automobile companies to the stringent automotive emissions controls mandated by the 1970 Clean Air Act, have also resulted in the loss of approximately 50,000 tons per year of lead for the production of lead antiknock additives (Figure 2).

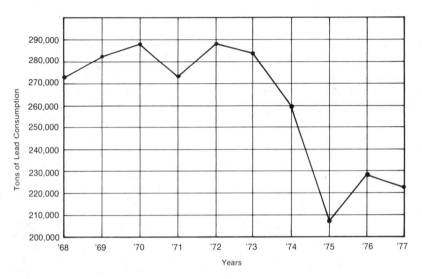

Fig. 2 - Lead Consumption for Production of Antiknock Compounds- U.S.

While the Clean Air Act led to catalytic converters, and hence, to the requirement of generally lead-free gasoline to avoid inactivating the catalyst, the Clean Air Act contained another provision, Section 211, which gave EPA the authority to regulate fuel composition and fuel additives. For alleged health reasons, EPA required that beginning in 1975, lead levels in gasoline be reduced stepwise from pre-control levels of approximately 2.0 - 2.5 grams per gallon to 0.5 grams per gallon by 1979. These schedules have been postponed and delayed for energy reasons but, unless there is rethinking of automotive emissions control strategies by the automobile industry moving them away from catalysts, and a moderation of the lead-in-gasoline phase-down schedule by EPA, lead antiknock consumption in the U.S. will steadily decline.

A new ambient air quality standard for lead promulgated by EPA in late 1978 may have a particularly severe impact on the lead industry. Studies carried out by Charles River Associates and The Research Corporation of New England for Lead Industries Association concluded that compliance with the standard by 1982, as required by law, would be impossible for virtually the entire primary and secondary lead industries. Table 1 indicates the expected economic impact that compliance with this standard of 1.5 ug/m^3 (90-day average) will have on the U.S. lead industry. These impacts would result in the loss of about 23,250 jobs.

(See Table 1 next page)

Table 1

Impact of Attempting to Meet
EPA Ambient Air Lead Standard

Industry Sector	Costs		Plant Closings
	Capital	Annual	
Primary Smelting	$190 million	$35 million	All Seven
Secondary Smelting	$285 million	$60 million	40 of 50 (90% loss of capacity)
Mining	Not Applicable	Not Applicable	All non-Missouri Missouri (?)
Battery Production	No Data	No Data	6 of 200

Source: Lead Industries Association, Inc.

The development of the EPA ambient air lead standard is particularly
interesting because its development spans the entire history of the "environmental movement". EPA made an attempt to set an ambient air standard
for lead in its earliest days. Copies of draft criteria documents from
1971 still remain in our files. EPA's failure to put together information
which could be used to justify an ambient air lead standard resulted in their
calling upon the U.S. National Academy of Sciences to prepare a "criteria
document." This effort resulted in the publication of the book Airborne
Lead in Perspective in 1972. However, there still was no scientific basis
for an air lead standard. Therefore, EPA made an internal decision to control air lead by utilizing Section 211 of the Clean Air Act which, as previously noted, pertains to automotive fuel additives. This section did not
contain the bothersome requirement that specific criteria for a standard be
identified.

However, the environmentalists, organized under the Natural Resources
Defense Council, won a 1976 court action against EPA that required it to set
an ambient air quality standard for lead. EPA tried again to prepare a criteria document. While this was going on EPA itself was becoming more and
more oriented toward the extreme environmentalist viewpoints. While EPA
was working on the new draft, Mr. David Hawkins was appointed as the new
Deputy Administrator for Air and Hazardous Waste Programs for EPA. Prior
to being appointed to this position by President Carter, Mr. Hawkins was a
Staff Attorney for the Natural Resources Defense Council, the environmentalist
group which forced EPA to set the standard and which screamed so loudly at
the first criteria document.

EPA prepared two more draft criteria documents before receiving environmentalist approval. Industry complaints that the documents were becoming more
and more politicized were ignored. Finally, a standard of 1.5 ug/m^3 was proposed in December of 1977. Hearings were held in February of 1978 and the
lead industry through LIA presented a strong case, buttressed by scientific
testimony both written and oral from some of the very experts EPA had earlier
hired to assist them with evaluation of their criteria document. However, as
the record shows, again their testimony was ignored by EPA and in October
1978, a final standard of 1.5 ug/m^3 was announced by EPA Administrator Costle,
noting that it was probably impossible for the primary lead and copper smelters

to meet. He promised a study to determine the real facts as to whether or not industry could comply but noted that Congress in passing the Clean Air Act had tied his hands. He maintained that he could not consider economics in setting ambient air quality standards -- only health effects.

The Lead Industries Association, Inc. appealed the new standard in the U.S. Court of Appeals and is now awaiting a decision from the court as to whether or not the standard will remain.

The Occupational Safety and Health Act

Another important law, the Occupational Safety and Health Act, was being considered by Congress concurrently with the Clean Air Act and it too was passed in 1970. Its purpose, "to assure so far as possible that every working man and woman in the nation has healthful working conditions," could only be applauded by those of us in the occupational health professions. Certainly, there were a multitude of examples of poor health practices and inattention to health protection of workers.

The single most important impact of the Occupational Safety and Health Act in the lead industry has been the development and promulgation of the new occupational health standard for lead. This standard, which was promulgated in November of 1978, is currently being challenged by the Lead Industries Association, Inc. and others in the U.S. Court of Appeals for the District of Columbia Circuit, as is LIA's appeal of the EPA ambient air standard for lead.

This standard established a Permissible Exposure Limit (PEL) for occupational exposure to lead in air of 50 ug/m^3. Under the standard, the PEL must be met by engineering controls and work practices alone -- not through the use of respirators. The timetable for compliance with the standard varies from one industry sector to another as shown on Table 2.

Table 2

Compliance Dates for OSHA PEL for Lead

Industry Sector	Number of Years to Reach Compliance		
	200 ug/m^3	100 ug/m^3	50 ug/m^3
Primary Pb Production	Immediate	3	10
Secondary Pb Production	Immediate	3	5
Pb-Acid Battery Mfg.	Immediate	2	5
Non-Ferrous Foundries	Immediate	1	5
Pb-Pigment Production	Immediate	3	5
All Other Industries	Immediate	0	1

The standard also requires extensive biological monitoring and specifies that workers with blood lead concentration above certain "trigger" levels be removed from exposure with full pay for up to 180 days. There is a stepwise reduction in the "trigger" blood lead level requiring removal as shown in Table 3.

(See Table 3 next page)

Table 3

Criteria for Removal and Return Under
OSHA Medical Removal Protection Provisions

Years After Effective Date	Removal Criteria		Return Criteria
	Air Lead	Blood Lead	Blood Lead
1st	>100 ug/m^3	>80 ug/100 g	≤ 60 ug/100 g
2nd	> 50 ug/m^3	> 70 ug/100 g	≤ 50 ug/100 g
3rd and 4th	> 30 ug/m^3	> 60 ug/100 g	≤ 40 ug/100 g
5th and thereafter	> 30 ug/m^3	> 50 ug/100 g *	≤ 40 ug/100 g

* Average of last 3 blood samples

In addition to these very costly requirements, the standard also contains provisions that would possibly require three different medical opinions to be rendered on a single contested medical issue, all at company expense, as well as extensive training and record keeping requirements.

The Lead Industries Association, Inc. carried out a survey to determine if first-year costs would be sufficiently great to justify a stay of the standard while an appeal was being considered. We learned that a conservative estimate of the first-year costs alone on an industry-wide basis would be in excess of one-half billion dollars. Fortunately for the lead industry, most of the costliest provisions of the standard were stayed by a court order on March 1, 1979.

As with the EPA air lead standard, the lead industry simply cannot meet the requirements of the new OSHA lead standard. It testified, quite genuinely, that it could not meet OSHA's original proposed standard and presented alternative approaches to protecting employee health. OSHA rejected the industry's position and made the final standard far more stringent than it had originally proposed.

Why did all this happen? The responsibility lies in all quarters. Industry still creates and tolerates exposure situations which its own professionals condemn. When these incidents are discovered, they destroy the credibility of the entire industry. Despite the fact that for nearly two decades LIA has publicly urged that blood lead testing be utilized throughout the industry, we are getting calls for advice from some members who are only now instituting blood-lead testing programs. Obviously, it took an OSHA standard to force these companies to do what they should have been doing all along.

Organized labor sees the Occupational Safety and Health Act as a labor law. They have been able to use it to gain far more power than they have previously held, and they have been able to gain, through regulatory fiat, objectives which they were not able to obtain at the bargaining table. Perhaps it is an unfair assessment, but power, not health protection, is their primary goal.

Occupational health professionals, including physicians, industrial hygienists and toxicologists have not been able to convince the public and the legislators that occupational health protection results from a combination of skills and judgment which cannot be codified into numerical regulations.

589

Government occupational health regulators, including professionals, have been too concerned about political pressure and internal power struggles to make sound judgments and decisions.

The result of all this has been a lead standard and other standards that are too extreme, which cannot be supported by scientific fact, and which must be challenged in the courts. This approach has led to ridiculously long periods for the preparation of standards by OSHA. The National Institute for Occupational Safety and Health issued its criteria document for an occupational health standard for lead in 1972. OSHA proposed a standard in 1975. Today, the OSHA lead standard is still not in effect.

Other Environmental Regulations

While the Clean Air Act and the Occupational Safety and Health Act are currently the most troublesome to the lead and zinc industry, there are other environmental regulations that will be felt in the future.

The Toxic Substances Control Act, the Resource Recovery Act, the Clean Water Act and various regulations related to environment in the Department of Health, Education and Welfare, the Consumer Product Safety Commission, and the Department of Transportation all affect the lead and zinc industry to greater or lesser extents. Though not discussed here, regulations pertaining to sulfur oxides, particulates, noise, heat stress, non-deterioration, and by-products, such as cadmium and arsenic, must be understood and obeyed.

The Future

Already heavily impacted by environmental regulation, the lead and zinc industry can only look to the future with trepidation. However, there do appear to be a few hopeful signs. Environmental concerns now have the attention of the top echelon officers of the industry. This should lead to a commitment toward sound environmental planning and practices. Perhaps the mistakes of the past, which have brought the wrath of the regulators down so strongly on the industry, will not be repeated.

Congress seems to be getting the message that we are in an era of environmental overkill that could do lasting damage to industry. Interestingly, in its 1979 Committee Report, the House Appropriations Committee made the following comments and regulations about the EPA air lead standard:

> "In October 1978, the Environmental Protection Agency
> adopted a National Ambient Air Quality Standard for
> lead of 1.5 micrograms per cubic meter of air. The
> present lead content in air is estimated to be two
> to four micrograms per cubic meter. Although the new
> standard applied to lead in the air regardless of
> source, its practical impact is limited to approxi-
> mately five percent of airborne lead. That five per-
> cent is generated primarily by lead and copper smelters.
> The source of 95 percent of airborne lead is automo-
> bile exhausts, which is addressed by separate regula-
> tions.
>
> The EPA lead standard is the subject of intense con-
> troversy. It is being challenged in a pending law-
> suit in the U.S. Circuit Court of Appeals for the Dis-
> trict of Columbia. The Agency has indicated that the
> entire standard may have to be eased or altered on a
> case by case basis because of potential significant

disruption in the smelting industry. The EPA
estimates the standard could cost the industry
up to $650,000,000 by 1982.

The EPA Administrator has stated the Agency's
figures on lead pollution and the cost to control
it are inadequate and that further study is needed in view of the potentially severe impact on the
smelting industry. As a result of these concerns,
EPA is undertaking a two-year study of the technical and economic problems that would confront the
lead industry in attempting to achieve compliance
with this controversial standard. In addition,
the scientific and medical bases for the standard
will be reviewed by the General Accounting Office
and by the independent scientific review committee
created by Congress in 1977 to periodically review
the scientific adequacy of air quality standards.

The Committee urges, therefore, that the Administrator not approve or disapprove any State plan
submitted pursuant to section 110 of the Clean Air
Act, as amended, for implementing the National Ambient Air Quality Standard for lead until the above
studies have been completed."

Although only a straw in the wind, such a statement is, perhaps, signal-
ling a new conservatism toward environmental legislation. Possibly we can
enter a new period when environmental and health protection can be achieved
without the bludgeon of over-regulation.

If that happy circumstance should evolve, it is more important than
ever that the industry have available sound information on the health and
environmental impacts of its operations and products. We can ask no more
than a chance to be heard by really objective regulatory agencies.

In many ways, I think we are at the nadir of environmental over-regulation. If the lead and zinc industries can hold on and cope through the legal
and political means available to it, a better day will dawn. However, there
will be no return to the "old days," and there shouldn't be. Although the
impact of environmental over-regulation has been a disaster for the lead and
zinc industries, truly sound environmental regulation should be welcomed and
encouraged.

591

EUROPEAN ECONOMIC COMMUNITY LAWS AND REGULATIONS

AS APPLIED TO THE LEAD AND ZINC INDUSTRIES

A. K. Barbour

RTZ Services Limited, Bristol, England

United Kingdom systems for environmental control of the lead-zinc industry are outlined and an assessment made of the possible impact of relevant European Economic Community Directives in the areas of emissions to atmosphere, liquid effluents, solid waste disposal and in-plant hygiene. Regulations on products are mentioned briefly.

Attention is drawn to significant differences in thinking underlying the long-established UK systems and that of EEC proposals, most of which have not yet been implemented in the member states. Thus, atmospheric pollution in the UK is based on control of individual emissions from chimneys with, as yet, no ambient air standards whilst the Commission has published Directives on ambient air control but has not yet issued stack emission standards. The UK is firmly opposed to uniform standards for the concentrations of pollutants in liquid effluents, preferring the Environmental Quality Objective (EQO) approach. The latest UK and EEC in-plant hygiene proposals also differ, both between themselves and in comparison with US proposals, due probably to differences in view relating to the necessary degree of protection against biochemical change in exposed personnel.

A plea is made for detailed discussion between the regulatory agencies, and industry to ensure that environmental standards are both justified scientifically and consistently achievable on well equipped, well managed plants.

I. Introduction

The fact that two entire sessions of this Symposium are devoted to environmental matters indicates its current and future importance to the industry. At the corresponding meeting in St. Louis ten years ago, the conference proceedings made only one reference to environmental control and that related to the disposal of tailings from mineral dressing. In the descriptions of sulphuric acid plants 99% refers to the concentration of the absorber acid, not to the percentage conversion of sulphur dioxide! Going further back to the 1936 volume, protection of the environment is not even mentioned. The manufacture of sulphuric acid from the sintering of lead concentrates was mentioned as a possibility which had not then been seriously attempted in the USA. Only two zinc smelters are mentioned as manufacturers of sulphuric acid (Bartlesville and Magdeburg) although in the UK both Imperial Smelting Corporation HD plants were making sulphuric acid by the Contact Process at that time.

In the decade since the St. Louis Lead-Zinc Conference the process industries in general, and ours in particular, have seen both processes and products investigated and challenged to a degree hitherto unknown. Whilst there are sociological and political reasons for this, there are also some scientific aspects to these challenges to which industry must respond scientifically if it is to avoid serious and unjustified harm. Perhaps the most important of the scientific factors is the immense progress which has recently been made in analytical chemistry, particularly spectroscopy in several forms, and data processing. This permits the analysis of metals in many matrices in part-per-billion concentrations which were almost unheard-of ten to fifteen years ago. Science has become much more specialized in recent years and dozens of new disciplines have emerged. Perhaps to compensate for this, research has become much more inter-disciplinary in nature, not always, in my view, between partners of comparable weight and knowledge. Finally, an increasing number of scientists wish to direct their researches to "real" problems, often problems having a social orientation - something which in other contexts many of us would support and commend.

We have also to recognize the inescapable fact that the ordinary person's expectations are demanding even higher performance standards from the chemical and metallurgical process industries as a result of generally increasing material and educational standards. So it is the duty of all concerned with process operations, and the marketing of industry's products, to sift thoroughly, logically and, above all, scientifically, through the avalanche of information currently arising to make a judgement on the performance standards which minimize adverse effects on health and environment. Having made professionally credible judgements on these matters, we have then to ensure, as far as is possible, that the regulatory agencies, the media and the general public fully understand the data and the thinking on which our judgements are based. The Lead-Zinc industry is indeed fortunate in receiving scientific advice, analysis and judgement of the highest quality from ILZRO/LIA and other industry-linked organizations to assist in accurately assessing these complex matters.

All parties, not least the general public, have to accept the limitations of scientific knowledge at any given time and to accept that increasing knowledge will almost inevitably require changes in standards accepted at that time.

We need to be equally professional in attempting to determine the
equally difficult problem of how standards are to be achieved. It is
clearly in no one's interest, and is utterly wasteful of time and resources,
for the regulatory agencies to propose standards far beyond the capability
of current technical knowledge. If operations or products are so clearly
devastating to health or environment as to justify such a judgement, it is
probable that they should be closed down. Both standards and their
achievement must be evolutionary wherever possible and should avoid massive,
order-of-magnitude change. Equally, industry must avoid "Crying Wolf" too
often. Given reasonable time scales for implementation, it is rare that
major industries will be bankrupted by the application of more efficient
environmental technology. Credibility is too easily lost by over-emphasis
on dubious economic arguments. But industry, the regulatory agencies and
the general public should be seriously concerned at the real prospect of
"running out of technology" to meet some of the new proposed standards
where the health or environmental justification is marginal.

In practice, the precise title of this paper has caused a few problems
in authorship simply because health and environmental activity by the
European Economic Community as such is relatively recent. Only a very
small amount of the apparently immense Commission regulatory activity (1)
has yet penetrated through into implemented national legislation. The
degree to which EEC directives will be applied in each member nation or the
influence which EEC thinking will have on national legislation or regulation
is not known now, and will probably form a good subject for the next Lead/
Zinc conference a decade hence. The paper therefore outlines the current
and immediate-future situation in the United Kingdom and includes a
strictly personal assessment of the possible impact of EEC proposals on
this legislation.

II. Administration and Enforcement in the United Kingdom and EEC

In the context of this conference UK environmental regulations as
applied to the zinc/lead industries are concerned with the major products
together with sulphur dioxide, cadmium and minor impurities including
arsenic and mercury. In the United Kingdom, emissions to atmosphere,
liquid effluents and solid waste disposal are dealt with as parts of
general environmental legislation which is applied to all metallurgical and
chemical process industries. In addition, however, legislation specific
to the lead smelting industry and the manufacture of red lead and litharge
relating to in-plant or workplace conditions has been operational in the
UK since 1911 (2). It was introduced by no less a person than the late Sir
Winston Churchill.

The main parts of the UK system are now as follows:

Table I

UK Regulatory Bodies

Environmental Parameter	Enforcing Officer	Controlling Body	Control Level	Parliamentary Responsibility
Emissions to Atmosphere				
(i) Works registered under the Alkali and Clean Air Act	Alkali and Clean Air Inspector	Health and Safety Executive	Central Government	Secretary of State for Employment
(ii) Non-registered Works	Environmental Health Officer	Local Authority	Local - major city or county	Varies
In-plant Conditions	Factory Inspector	Health and Safety Executive	Central Government	Secretary of State for Employment
Liquid Effluents	Technical Officer	Regional Water Authority (RWA)	Autonomous Regional Authority	Secretary of State for Environment
Solid Waste Disposal	Technical Officer	Major Local Authorities designated as Waste Disposal Authorities in conjunction with RWA's	Local - county or major city	Secretary of State for Environment

Although this organisation may seem cumbersome, and from an industrial standpoint necessitates most companies dealing with several regulatory Inspectors for each plant, this situation is not in fact much different from systems outside the UK. In practice, it works effectively because by and large industry respects the professionalism of the inspectors and the systems permit rational technical interchange before standards are defined. The other characteristic of the British system is that it deals with individual plants or point sources rather than imposing national or state-wide fixed standards. The philosophy of deciding upon individual criteria differs, however, between atmospheric emissions and aqueous effluents, for in the latter the basic UK criterion is the receiving body's capability to absorb contaminants without damage to so-called target organisms depending on the use to which the receiving body is to be put. The Alkali and Clean Air Inspectorate's "best practicable means" approach only indirectly concerns itself with the national ambient atmosphere; in the UK we have to date avoided the pitfalls of some recent overseas systems of setting ambient air standard targets first and then deriving point source emission standards.

1. Emissions to Atmosphere

The Alkali and Clean Air Inspectorate system is based on the following:

Table II

Best Practicable Means

Practicable - "Reasonably practicable having regard, amongst other things, to local conditions and circumstances, to the financial implications and to the current state of technical knowledge."

Means   - "Design, installation, maintenance, manner and periods of operation of plant and machinery, and the design, construction and maintenance of buildings."

In general, emission standards are not specified in UK legislation; "Best Practicable Means" for each individual emission are ultimately expressed by the Alkali Inspectorate as so-called "Presumptive Limits" which are based upon the following criteria:

Table III

Alkali Inspectorate Criteria for Presumptive Limits

(i) No emission can be tolerated which constitutes a demonstrable health hazard, either short- or long-term.

(ii) Emissions, in terms of both concentration and mass, must be reduced to the lowest practicable amount.

(iii) Having secured the minimum practicable emission, the height of discharge must be arranged so that the residual emission is rendered harmless and inoffensive. (For highly toxic metals, the concentration of each source to the existing background concentration shall not exceed one-fortieth of the Threshold Limit Value for a factory atmosphere on a three-minute mean basis. In deciding on the most important parameter, the effects on vegetation, animals and amenity are also considered.)

It will be seen, therefore, that implementation of the above criteria provides a fully satisfactory level of health protection for persons living in the proximity of major point sources of pollution, consistent with the best current medical knowledge at the time. Two points need to be emphasised. One is that TLV's or Threshold Limit Values (that concentration which can be tolerated by an average person exposed eight hours per day forty hours per week for prolonged periods) are currently set by an independent group of professional specialists (the American Conference of Governmental Industrial Hygiene). In the past, the ACGIH list of TLV's was adopted in toto by the UK Factory Inspectorate, but there are now a few modifications for application in the UK not, however, in heavy metals (3). The ACGIH is, of course, totally independent of either the Alkali Inspectorate or the Factory Inspectorate. The second point concerns definition of the phrase "demonstrable health hazard, either short- or long-term". It must be emphasised that this is a moving target, reflected usually in lower (or more stringent) TLV's adopted by the ACGIH in the light of increasing toxicological and biochemical knowledge. Scientists can usually accept without difficulty that increasing knowledge can result in changing standards; the general public often finds difficulty in grasping the idea that current knowledge is far from complete at any given time and hence that standards are not fixed for all time.

   i. Emissions to Atmosphere - Lead. It is sometimes felt outside the UK (and by some within the UK) that the best practicable means system is vague and non-quantitative. The following table relating to lead emissions should help to dispel this mistaken view (4). It is typical of the "presumptive limit" values which the Inspectorate issues for all major pollutants.

Table IV

Presumptive Limits for Lead Works

Works Category	Gas Volume		Individual Stack Concentration*		Total Mass Emission*	
	cu.ft./min.	m³/min.	grains/ cu.foot	g/m³	lbs/hour	kg/hour
Class I	7000	200	0.05	0.115	0.6	0.27
Class II	7000-140000	200-4000	0.01	0.023	6.0	2.7
Class III	> 140000	> 4000	0.005	0.0115	12.0	5.4

*Measured as elemental lead

Total Particulates: < 25,000 cu.ft./min. (715 m³/min.)
                    < 0.2 grains/cu.ft. (0.46 g/m³)
                    reducing progressively to 50,000 cu.ft./min.
                    and above 0.1 grains/cu.ft. (0.23 g/m³)

   Although the underlying philosophy of EEC environmental legislation is to reduce emissions at source, there are, as yet, no comparable EEC figures for stack emissions of lead. Under the EEC "Action Programme" of November 1973, lead and its compounds were designated as pollutants for top-priority consideration.

   To date, this has resulted in the following Directives:

Table V

EEC Directives on Lead

Title	Date	Status	Main Provisions
Air Quality Standards	April 1975	Proposed	max 2 μg Pb/m^3 annual mean max 8 μg Pb/m^3 monthly median heavy traffic areas
Biological Standards and Population Screening	March 1977	Adopted	20% : Pb_B ≥ 20 μg/100ml 90% : Pb_B ≥ 30 μg/100ml 98% : Pb_B ≥ 35 μg/100ml
Lead Content of Petrol	June 1978	Adopted	Pb 0.40g/l max. (1 January 1981)

The two last Directives were subjected to lengthy periods of discussion; the Lead content of Petrol Directive, in particular, was proposed by the Commission in 1973 and then contained a long-term guideline value of 0.15 g/litre which has now been dropped. The UK Government has put both adopted Directives into effect but neither have any direct bearing on lead smelting operations. As indicated earlier, there are no UK ambient air regulations but the 5th Report of the Royal Commission on Environmental Pollution under Sir Brian Flowers proposed in 1976 that consideration should be given to air quality guidelines for lead (and sulphur oxides, smoke and nitrogen oxides) in consultation with the Alkali Inspectorate whose "best practicable means" approach to stack emission control was strongly commended.

ii. Emissions to Atmosphere - Cadmium. The current UK presumptive limits for cadmium emissions to atmosphere are summarised in Table VI.

Table VI

Presumptive Limits for Cadmium Works

Works Category	Gas Volume		Individual Stack Concentration*		Total Mass Emission*	
	cfm x 10^3	m^3/min	gr/ft^3	g/m^3	lbs/hr	kg/hr
Class I	1 ↓ 7	30 ↓ 200	0.03 ↓ 0.005	0.067 ↓ 0.012	0.26 ↓ 0.30	0.10 ↓ 0.10
Class II	10 ↓ 140	280 ↓ 4000	0.005 ↓ 0.0013	0.012 ↓ 0.003	0.43 ↓ 1.50	0.20 ↓ 0.70
Class III	> 140	> 4000	Judged individually			

*measured as elemental cadmium

iii. **Emissions to Atmosphere - Sulphur Dioxide.** In the United Kingdom capture as sulphuric acid of sulphur dioxide derived from the roasting of zinc and mixed zinc/lead concentrates has been practised since the early 1930's when the former Imperial Smelting Corporation devised procedures for dedusting sinter-plant gases to a level where the gas was acceptable for conversion in a vanadium contact sulphuric acid plant without rapidly destroying catalyst activity.

The current Alkali and Clean Air Inspectorate requirements for emissions from acid plants fed by sinter-derived sulphur dioxide are similar to those for acid plants based on the burning of sulphur (8, 9). They require that the sulphur loss to air, as acid gases, shall be not greater than 0.5% of the sulphur burned, or, in the case of sinter-based plants, sulphur dioxide fed to the acid plant. In practice, this is usually achieved by the installation of double-absorption acid plants, but it would also be permitted to achieve the same level of final emission by tail-gas scrubbing. There is also the requirement that the waste gases shall be substantially free from persistent mist. Sampling procedures and analytical methods are also laid down; it is agreed that continuous monitoring will be applied when industry and the Inspectorate accept that suitably reliable instruments are available and these are now being used. Chimney heights are calculated individually but for acid plants of 400-1000 tonnes per day capacity are in the range 200-275 ft (60-85m) before additions for local topography.

In line with historical practice, there are no UK Ambient Air Standards for sulphur dioxide although national surveys of smoke and sulphur dioxide have been conducted and published continuously for the last twenty years by the Department of Industry Warren Spring Laboratory as the National Survey of Air Pollution. These surveys consolidate less comprehensive monitoring first started in 1914 (10). Through a combination of the "tall-stack" policy at power stations and other stationary sources and a "Clean Air" policy of permitting only smokeless solid fuels to be burned domestically in most city areas, overall annual average smoke concentrations in urban areas fell from 180 $\mu g/m^3$ to 52 $\mu g/m^3$ between 1958 and 1973, while sulphur dioxide concentrations fell from 177 $\mu g/m^3$ to 95 $\mu g/m^3$ during the same period.

EEC legislation on sulphur dioxide and smoke is in a position somewhat similar to that for lead. It was early decided that these were "first category" pollutants and the first proposed Directive appeared in February 1976. This followed the normal EEC philosophy of proposing ambient air standards, to be met mainly by the burning of fuels of reduced sulphur content. Discussions on modified versions of this Directive continue and at this time, it is not clear what values will be adopted finally. As with lead, no proposals have yet been published in relation to stack emissions, so it is again difficult to judge precisely how the UK and eventual EEC systems will come into balance. In the meantime, however, extensive monitoring continues in the industrialized areas of mainland Europe, as in the UK.

2. Liquid Effluents

In the United Kingdom, liquid effluent standards are set by one of the ten Regional Water Authorities which are public autonomous bodies charged with the management of the total water resources (i.e. supply and disposal) for each region of the country. Standards are implemented by "consents" or licences to discharge which relate usually to each individual discharge from a site. A "consent", which can be reviewed every two years, usually specifies daily volume, temperature of discharge, pH, oxygen demand, suspended solids content, heavy metals content and any other pollutants of specific concern either to the site or to the receiving body of water which may be a sewer, a river or an estuary.

The philosophy underlying the UK system is to categorise receiving bodies into end-uses (e.g. potable, estuarial, recreational, sewerage, etc.) and then to set pollutant concentrations in the receiving bodies at levels consistent with the natural survival of "critical organisms" appropriate for that end-use. Individual effluent standards are then derived so as to be consistent with these receiving body criteria. Under the Control of Pollution Act, 1974, when fully implemented, each "consent" will be available for public inspection, a significant change from historical British practice.

The UK water industry is justly proud of the progress which it is making in regenerating the quality of previously polluted rivers and the vastly improved quality of the River Thames outside the Houses of Parliament in Westminster is reflected in the larger number of fish species found there. The Water Authorities also conduct extensive and sophisticated monitoring of the various estuaries in the UK (e.g. the Severn Estuary (11)) and is convinced that any deterioration in overall quality would be detected quickly by such monitoring programmes.

Accordingly, the UK Government, strongly supported by the Water Authorities and Industry generally, sees little need or justification for systems based on fixed concentrations of pollutants in effluents, regardless of the nature and use of the receiving body into which the effluent flows. This fixed-standard in effluent principle is embodied in the EEC "Dangerous Substances in the Aquatic Environment" Directive adopted in May 1976 - at least for the so-called List I ("Black List") pollutants which include mercury and cadmium. Definite values for permitted concentrations of cadmium in effluent have not been finalized but may well be in the 0.1 to 0.5 mg/litre range.

The UK regards a fixed emission policy of this kind as uneconomic and wasteful of resources; for many UK receiving bodies further reduction in cadmium or mercury concentration is a relatively low priority compared with other pollutants in that receiving body. The UK authorities are convinced that the EQO approach can provide an equal level of environmental protection and will carry out appropriate monitoring to demonstrate the point.

There is, however, concern in some UK Governmental circles regarding the long-term accumulation of cadmium from the general environment by non-occupationally exposed persons. One manifestation of this concern is to place low limits (5kg/ha) for permitted cadmium in sewage sludge to be applied to agricultural land (12). This, in turn, has caused some Water Authorities to restrict sharply the amount of cadmium in industrial effluents (e.g. from plating shops) flowing into sewers from which the sludge produced in biological treatment plants may retain as much as 80% of the supplied cadmium and other heavy metals. It is clear that some of the more dramatic analyses of the health effects of general environmental cadmium will, if passed uncorrected into legislation, sharply affect the economics of cadmium to the zinc smelter by restricting market outlets in some parts of the world, e.g. Sweden and Japan.

3. Solid Wastes

The potential hazards, particularly to children, of uncontrolled dumping of toxic wastes, especially cyanides, led to the hasty passage of the

600

Deposit of Poisonous Wastes Act in 1972. The provisions of this Act currently apply, though they will be superseded when the Control of Pollution Act, 1974 becomes fully operative.

By the Deposit of Poisonous Waste Act, all wastes other than a nominated list have to be notified to the relevant local authority who direct where and how the waste is to be disposed, depending on an analysis of its composition provided by the supplier of the waste. Subsequent disposals of similar wastes are controlled by renewable licences.

Discussions between central Government, the major local Authorities designated as Waste Disposal Authorities and Industry to streamline the documentation relating to waste disposal under the Control of Pollution Act, 1974 and also to provide a basis for identifying seriously toxic wastes which require special treatment. Such treatment could either be by disposal into nominated, hydrologically secure sites or, for heavy metal containing sludges (e.g. from plating or metal treatment, etc.) by chemical fixation by a proprietary process such as "Sealosafe" or "Chemfix". Under current discussion in the UK is what I believe to be a novel attempt to categorize wastes on the basis of their likely toxicity if ingested by children, rather than on the basis of chemical analysis alone. If such an approach proves acceptable, and this is uncertain at present, it could have many benefits in assessing the multi-element wastes which are frequently encountered in metal processing.

A valuable example of collaboration between the UK Department of the Environment, the Waste Disposal Authorities, the Waste Contractors and Industry exists in the DOE Waste Management Series of Technical Memoranda on Reclamation and Disposal, which often include a Code of Practice. This series is colloquially known as the DOE "Red Books". They are compiled by groups of specialists from all of the above sectors and represent agreed, practicable and workable guidelines for the disposal of wastes which present unusual difficulty in disposal. Examples relevant to this Conference include Metal Finishing Wastes, Mercury and a forthcoming one on Arsenic.

The main piece of EEC legislation on solid wastes is the Directive on Toxic and Dangerous Wastes which was submitted in July 1976 and approved by the Council of Ministers in December 1977. A total of twenty-seven categories of waste are designated as requiring priority consideration and these include elements and compounds of lead, cadmium, mercury, arsenic, antimony and soluble copper compounds as items of interest to the Lead/Zinc industry.

Differences in the approaches of the UK and the EEC on solid waste disposal are much smaller than in other environmental areas. The major problem for the lead industry was to ensure that supplies for the secondary lead industry, existing as it does in part to recover values from slags, drosses and the like regarded as "waste" by other processors, was adequately safeguarded. Strong industry representation via national Governments eventually resolved this problem.

4. In-Plant Hygiene

i. Lead. The UK has a long history of legislation to protect plant operators, for the lead smelting industry dating back to the Lead Smelting Regulations of 1911 (2). It is noteworthy that these regulations require monthly medical examinations of lead-exposed workpeople, they prohibit women and persons under 16 from employment in lead smelters and they refer to the necessity of adequate ventilation

and respiratory protection. They also provide for the regular launder-
ing of protective clothing, the banning of food and drink in the
workplace (except non-alcoholic drink approved by the Surgeon) and the
prohibition of smoking (except where the person's hands are free of
lead). Whilst techniques have progressed significantly in the intervening sixty-eight years, the basic principles continue to underlie
many current discussions on in-plant hygiene. Other lead-using
industries including the manufacture of lead compounds, paints and
colours, electric batteries, vehicle painting, shipbuilding and
repairing, were also regulated by legislation spanning the period 1903
to 1964.

This piecemeal legislation will be replaced by the new "Control of
Lead at Work" regulations (13) which are now in an advanced state of
discussion and consultation between all interested parties before
Government eventually decides the precise form of the regulations.

In the UK it appears likely that the present Threshold Limit
Value (TLV) will remain for the time being at its present value of
150 $\mu g/m^3$ and the normal excursion provisions will apply. However,
the regulations place more emphasis on the measurement of air-lead
values by personal samplers rather than the traditionally more frequently
used static samplers. Whilst the thinking behind this change of
emphasis is fully accepted, viz. that the important parameter is the
amount of lead in the breathing zone of the operator, actual experience
shows that personal sampling using currently-available equipment tends
to give both higher and more variable air-lead values than those
obtained from static samplers. No doubt, experience of operating both
systems in parallel will provide the necessary correlations as guides
to interpretation. The present UK objective is to measure total air-
lead values, not respirable lead values, and, in the future, it may be
that more convenient badge-type samplers will be shown to provide
reliable air-lead values. Samplers of this type would be much more
acceptable to the operators. At present, however, it seems imprudent
to dispense with static samplers and rely solely upon personal samplers
for the determination of air-lead concentrations.

The UK proposes to retain statutory medical examination of persons
whose normal work exposes them to more than a specified level of lead.
The protocol for such medical monitoring of exposed workpeople is
difficult to state briefly without loss of accuracy and emphasis but
broadly it requires:

Table VII

Proposed UK Guidelines for
Biological Monitoring of Lead-Exposed Workpeople

- three-monthly medical examination with the provision of
  samples for the determination of blood lead and urinary
  coproporphyrin;

- withdrawal from exposure to lead if blood-lead values
  exceed 80 microgrammes/100 ml. of whole blood;

- more frequent medical examination (usually monthly) if the
  Appointed Doctor thinks appropriate (e.g. if blood leads
  vary appreciably or are consistently in the range 70-80
  microgrammes per 100 ml. whole blood).

The above proposals will correctly apply to industries engaged in the chemical processing and fabrication of lead, as well as to smelters, but at present we are unclear regarding the status of women. Modern medical evidence tends to support the 1911 Regulations' ban on the employment of women in lead works, particularly in that the exposure of pregnant women may possibly affect the child during gestation and subsequently. However, another official view suggests that any ban is depriving women (particularly those who are not pregnant) of traditional and well-paid employment and this constitutes a deprivation of a basic right. I do not propose to attempt to weigh the evidence any further!

EEC activity in the area of in-plant hygiene is concentrated on a March 1979 proposed framework Directive entitled "The Protection of Workers from Harmful Exposure to Chemical, Physical and Biological Agents at the Workplace", together with a specific derived Directive on Lead, "The Protection of Workers from Harmful Exposure to Inorganic Lead at the Workplace". Other specific Directives on Cadmium and Mercury will follow.

These proposed Directives are under detailed discussion at the present time and so their full impact on national legislation is difficult to judge while such discussions are in progress. In principle, the proposals follow the general lines of the UK Consultative document, but significant differences exist in the proposals in relation to both permitted air-lead values and worker-absorption values. Also, measurements of delta-aminolaevulinic acid levels in urine are specified as absorption indicators in addition to measurements of blood-lead values.

In addition to the above EEC Directives, draft proposals from the World Health Organisation - "Programme of Internationally Recommended Health-Based Permissible Levels for Occupational Exposure to Chemical Agents - Inorganic Lead" will, when finalised, undoubtedly influence regulatory and other thinking.

As summarised in the next Table (No. 8) significant differences now exist in the separate proposals for the control of in-plant lead within the UK, the EEC and the USA.

Whilst these differences doubtless reflect several factors - scientific, social, economic and political - I believe there are some basic differences in the scientific and medical thinking underlying the proposals which, in my view, have been insufficiently emphasised to date, and are certainly not fully appreciated by the lay public who often find it quite inexplicable that different Governmental authorities propose significantly different standards for worker protection against lead and in many other situations.

The Health and Safety Executive proposals in the UK are based on medical experience derived from nearly seventy years of statutory medical monitoring of occupationally-exposed lead workers which has virtually eliminated clinical lead poisoning from such workers. I believe that the current variation in proposals for lead exposure standards promulgated by UK, EEC and US governmental specialists reflects differences in view regarding the acceptance of any measurable change in a biochemical parameter, regardless of its reversibility or non-reversibility and also without regard to clinically-observable health effects.

Table VIII

Proposed Blood and Air Lead Standards in the UK,
EEC and USA (July 1979)

	Blood Lead (μg/100ml)	Air Lead (μg/m^3)
UK (workplace)	80	150 (8-hour average)
EEC (workplace)	70 (individuals) 50 (group average)	100 (40-hour average)
EEC (ambient)	50% $\leqslant$ 20 90% $\leqslant$ 30 98% $\leqslant$ 35	2 (annual average)
US (workplace)	50 (if exceeded for 6 months) (return level $\leqslant$ 40)	50
US (ambient)		1.5 (90-day average)
WHO (workplace)	40 (men, and women in reproductive age range)	10-80 (depending on level of non-occupational exposure)

Where clearly defined adverse health effects are directly related
to workplace exposure, it is correct to set standards of exposure at
levels which will prevent these effects whether or not technology
exists to meet them. In effect, this would mean that many plants
responsible for these effects would be closed. But where such clearly
defined adverse health effects are absent - and bear in mind that the
lead industry, at least in the UK, has a record of medical surveillance
unparalleled in effectiveness by any process industry other than perhaps
the nuclear power industry - it is in my view quite unnecessary, even
irresponsible, for regulatory agencies to promulgate standards which
are unnecessarily stringent for the protection of worker health,
particularly where they are unattainable by any known engineering
systems.

Responsible people in the regulatory agencies, management and the
unions can and should reach practicable and uniform standards for
worker protection based on a consensus of underlying philosophy.
Judgement on medico-scientific matters should not be made on the basis
of prolonged legalistic arguments in the courts. The present confused
picture does nothing to improve worker protection and reduces the
credibility of all concerned when it reaches the lay public with little
or no commentary or interpretation of underlying principles.

ii. Cadmium. We must all hope that a similarly confused picture
does not emerge in relation to cadmium. In 1977, the British
Occupational Hygiene Society, a professional but non-regulatory body,
proposed the following (14):

Table IX

Workplace Cadmium Exposure - British Occupational Hygiene
Society Proposals (1977)

(i) Hygiene Standard

   (a) the respirable faction of the total dust of cadmium or
       its compounds shall not exceed 0.05 mg cadmium/m^3
       (50 μg Cd/m^3);

   (b) the total weight of any cadmium or cadmium compound dust
       that is soluble in 0.1 N hydrochloric acid shall not
       exceed 0.20 mg cadmium/m^3 (200 μg/m^3);

   (c) the total weight of acid insoluble compounds shall not
       exceed the "nuisance" dust level of 10 mg/m^3 (10000 μg/m^3);

   (d) a Special Short Exposure Limit (SSEL) of 2 mg cadmium/m^3
       (2000 μg/m^3) for a maximum of ten minutes is acceptable,
       so long as the worker is not exposed again for the rest
       of the shift.

(ii) Hygiene Monitoring

   Personal monitoring on each exposed worker for at least four
   hours should be carried out at least four times per annum.

(iii) Medical Monitoring

   (a) a urine sample should be tested for LM proteinurea prior
       to initial employment and preferably at six-monthly
       intervals thereafter;

   (b) a pre-employment questionnaire on respiratory symptoms
       should be completed and respiratory function tests should
       be carried out at six-monthly intervals while exposure
       continues.

(iv) Review

   In three years' time.

However, it must be emphasised that these proposals have not been
accepted by the Health and Safety Executive which is, of course, the
UK regulatory body. In line with normal UK practice, the Health and
Safety Executive has produced proposals in the form of a discussion
document. The H&SE proposals provide a detailed medical monitoring
protocol including action levels of biochemical parameters for the
withdrawal of workers from exposure and for their return to work. This
document is currently under discussion by the Executive in conjunction
with representatives of management and the unions.

Much discussion on cadmium exposure is also current in both
scientific and regulatory circles in the European Economic Community
(15). As noted earlier a proposed directive on in-plant cadmium is
expected as a specific directive under the framework Directive on "The
Protection of Workers from Harmful Exposure to Chemical, Physical and
Biological Agents at the Workplace". WHO proposals are also under review.

III. The Future - Problems and Solutions

The introduction to this paper stated that the lead/zinc industry faced many environmental pressures. The allocation of space to the different topics discussed so far reflects the belief that the significance of the problems and the difficulty of solution is in the order:

Table X

Summary of Process Environmental Problems Facing the Lead/Zinc Industry

. Future in-plant hygiene standards and medical monitoring of lead and cadmium.

. Emissions to atmosphere, including ambient air standards, in relation to lead, cadmium and - for lead smelters - sulphur dioxide.

. Liquid effluent standards, particularly cadmium.

. Solid waste disposal.

In addition, there are a number of product problems facing the industry in the market place for which time permits only a brief summary:

Table XI

Environmental Pressures on the Products of the Lead/Zinc Industry

. The reduction or banning of tetra-alkyl lead additives to gasoline.

. The reduction in permitted levels of cadmium in zinc used for galvanising, particularly potable water pipes.

. Pressures against the use of cadmium-containing products in some countries, particularly Sweden and Japan.

Fortunately the terms of reference for this paper do not require the proposal of detailed solutions to these intractable questions. Perhaps other papers at the Conference will do this. But, at least, some of the technical and engineering matters may be stated which are causing the industry serious concern at present. These concerns are stated not in any sense of "Crying Wolf" - Industry has engaged in this fruitless occupation too often in the past - but rather in the expectation that rational discussion between professionals from both industry and the regulatory agencies will continue in the future as it has in the UK for many years and lead to practicable working solutions.

1. In-plant hygiene

The law of diminishing returns is beginning to apply in the area of ventilation volumes - particularly in the unavoidable context of rising energy costs. Everyone here will know how difficult it is to maintain lead

$< 150 \ \mu g/m^3$ in all sections of a lead smelter at all times of the day and night: we all understand the natural worker resistance to the wearing of respiratory protection for long periods of arduous work.

Mechanical and other engineers must develop the science (and art) of applying ventilation volumes selectively and devising workable and effective balancing systems. I believe that considerable scope exists for designing both equipment and handling systems which are much less prone to leakage, particularly after long periods of operation in hot and abrasive atmospheres. An essential pre-requisite to this is to develop and use equipment which will measure leakage rates in the actual round-the-clock working situation.

Such equipment would also assist materially in designing operator work-schedules which minimise actual exposure to lead and other in-plant toxicants. Fortunately, there has been a great improvement in worker understanding of the dangers of lead absorption and the vital need for operator-cooperation if it is to be minimised. We must hope that more comfortable and efficient respirators continue to be developed and accepted by the men.

In the context of union (and some medical) pressures in the direction of personal samplers for the measurement of absorption, we must hope that operators will use such equipment properly and with integrity. It may be that dietary modifications can be developed which will assist the excretion of absorbed lead - we need all the help that is available if proposed EEC or US standards become accepted generally.

2. Emissions to atmosphere

i. Sulphur Dioxide. The widespread adoption of double-absorption sulphuric acid plants by zinc smelters, both electrolytic and ISF, means that such operations are now immune from the pressures to reduce SO_2 emissions which have been encountered by, for example, reverberatory copper smelters. In some areas, however, pure lead smelters may find difficulty in meeting modern mass emission criteria and dispersion through tall stacks may not always be acceptable on a mass basis.

ii. Heavy Metals. The emission of heavy metals to atmosphere has been ameliorated greatly in recent years by the more effective use of conventional arrestment equipment. In particular the advent of modern mechanical bag-plants has been very beneficial in reducing emissions of fine fume. However, recent American proposals to adopt an ambient air standard of $1.5 \ \mu g/m^3$ will undoubtedly pose severe problems for the industry. There seems little medical or other justification for this standard and it is sincerely to be hoped that wiser counsels will prevail in those countries which choose to use ambient standards as the ultimate basis for emission criteria.

In the UK, "fall-out" monitoring is now a part of the "best practicable means" approach to Alkali and Clean Air Inspectorate control for lead emissions. I believe that the not inconsiderable expense of such monitoring is well justified in the context of reassuring the general public, particularly near-neighbours, of the safety of the operations of our industry. In addition, the long-term results and trends can give very significant pointers to management regarding the efficacy or otherwise of plant control equipment. For many years at one UK site dairy farming on a closely managed basis was practised "up to the fence" and this gave additional reassurance to the local farming community.

3. Liquid Effluent Standards

Correctly-designed and operated lime-treatment plants provided with adequate settling capacity meet most current regulatory standards. Where such standards are particularly stringent, multi-stage precipitation can be employed. Accurate pH control is essential, the actual control value in single-stage treatment plants being something of a compromise to ensure adequate precipitation of cadmium without redissolution of zinc.

The main worry facing European industry is that exceedingly low standards for cadmium (and mercury) may be imposed, since current "exotic" technology such as ion-exchange or reverse osmosis is unproven and would be impossibly expensive to apply to the large volumes of water to be treated at most smelters. As noted earlier, the UK Government does not support the EEC "fixed standards" approach to the control of these metals in aqueous effluents and, from all standpoints, it is to be hoped that the UK view gains wider acceptance.

4. Solid Wastes

In general, disposal of slags from the pyrometallurgical processes for zinc/lead production causes few problems since their siliceous matrix largely prevents solubilisation of the metals. In most cases, sludges from water treatment and the wet scrubbing of gases can be recycled to process.

However, chemical precipitation from the purification of solutions for electrolytic zinc processing do require careful storage in sealed compounds with recirculation of run-off water to the effluent treatment plant. Whilst the recent innovation of Jarosite technology improves overall zinc recovery quite significantly, Jarosite is not, in general, a recyclable form of iron and expensive processing to Haematite is being resorted to in areas where severe restrictions on solid deposition exist.

Increasing attention is now being paid by regulatory authorities to the minimisation of windage losses from both waste dumps and feed stock-piles. The latter should now be enclosed and the former either dampened when necessary or alternatively covered by one of the proprietary polymeric treatments.

5. Products

Although this symposium is primarily concerned with processing, we must always be on guard against unjustifiable restrictions on the sale of the products of our industry. Restrictions on the use of lead in petrol and in paints are too recent to require repetition here, while we are all aware of the intense study being made on cadmium in products by regulatory agencies worldwide. Such studies sometimes have unforeseen consequences, for example, reduction in the cadmium content of zinc used for the galvanizing of potable water supply pipes, even for the galvanizing of general structural steel where leaching to agricultural land can occur. As with process environmental problems, our industry must continue to take a professional stance in evaluating these problems scientifically and then to make every possible effort to ensure that regulations have a credibly scientific basis.

References

(1) S. P. Johnson, "The Pollution Control Policy of the European Communities", Graham and Trotman Ltd., London, 1979.

(2) The Lead Smelting and Manufacture Regulations, 1911.

(3) Guidance Note EH15/75, "Threshold Limit Values for 1978", Health and Safety Executive, Her Majesty's Stationery Office, 1979.

(4) "Industrial Air Pollution 1975", Health and Safety Executive, p.58, Her Majesty's Stationery Office, London, 1977.

(5) S. A. Hiscock, "Production, Consumption and Uses of Cadmium in the European Community (1965-1976)" EEC Study Contract on Cadmium, Contract No. ENV 223/74-E, Zinc Development Association/Cadmium Association, 1978.

(6) A. Rauhut, "Survey of Industrial Emission of Cadmium in the European Economic Community, EEC Contract No. ENV 223/74-E, Landesgerverbeanstalt Bayern, 1978.

(7) J. S. Alabaster, "Ecotoxicity of Cadmium", EEC Contract No. ENV 223/74-E, Water Research Centre, Stevenage, 1978.

(8) 110th Annual Report on Alkali, etc. Works, 1973, p.70, Her Majesty's Stationery Office, London, 1974.

(9) See reference (4), p.56.

(10) A Parker, "Industrial Air Pollution Handbook" (ed. A. Parker), p.9, McGraw-Hill Book Company (UK) Ltd., London, 1978.

(11) Severn Estuary Survey and Systems Panel, First Report to the Technical Working Party of the Severn Estuary Joint Committee, Severn-Trent Water Authority, 1977.

(12) Report of the Working Party on the Disposal of Sewage Sludge to Land, Department of the Environment/National Water Council, London, 1977.

(13) Control of Lead at Work Draft Regulations, Health and Safety Commission, London, Her Majesty's Stationery Office, 1978.

(14) Hygiene Standard for Cadmium, British Occupational Hygiene Society, Pergamon Press, London, 1977.

(15) Proceedings, 1st International Cadmium Conference, San Francisco, 1977, Metal Bulletin, London, 1978.
Proceedings, 2nd International Cadmium Conference, Cannes, 1979, in the press.

DEVELOPMENTS IN LEAD SMELTER HYGIENE AND

ENVIRONMENTAL PRACTICES AT MOUNT ISA

L.D. White and R.H. Marston
Metallurgical Works Manager, Occupational Health Superintendent
Mount Isa Mines Limited, Mount Isa, Queensland, Australia

By keeping abreast of world wide developments in environmental and occupational hygiene practices, Mount Isa Mines Limited has reached outputs of 150 000 tonnes of crude lead per annum, while at the same time diminishing the impact of the processes on the working and ambient environment.

The progressive upgrading of the plant, associated hygiene monitoring and control equipment, and employee biomonitoring programmes which have produced an improvement in in-plant working conditions are described, as are the parallel developments which have substantially diminished the impact of the Company's operations on the ambient, urban environment.

Introduction

Mount Isa is one of Australia's most geographically isolated industrial communities. It is located at latitude 21° south, longtitude 140° east, 960 km west of Townsville and 1560 km north-west of the state capital, Brisbane. With a population of 27 000, Mount Isa is the major centre of the tropical north-west Queensland region which is otherwise very sparsely settled.

Mount Isa Mines Limited is a major producer of silver-lead, zinc as well as copper, treating 2.6 million tonnes of silver-lead and zinc ore and 4.7 million tonnes of copper ore annually. Although both ores are extracted from the same mine, the orebodies are completely separate, and the ores are treated in separate plants. The Mount Isa silver-lead, zinc orebodies lie in the Urquhart Shale formation which is approximately 1070 metres thick. The shale extends several kilometres north and south of the existing mine.

History of Lead Smelting Operations

Lead Smelting operations have been conducted on the existing site since 1931. There was a short disruption to lead production from 1943 to 1946, when copper was produced in the Lead Smelter for national strategic needs during World War 2. The smelting operations up to 1969 have been described previously (1, 2, 3) but are summarised as follows.

Changes in lead-zinc-silver concentrating techniques (4, 5) have also occurred to achieve the current production outputs of 150 000 tonnes crude lead and 220 000 tonnes zinc concentrates. Concentrates for lead smelting were originally produced from predominantly oxidized carbonate ore which progressively changed to lead-zinc sulphide ore at increasing depth. The smelting operations are based on a Sinter Plant - Blast Furnace duplex.

Sinter Plant operations commenced with three 1.07 m wide down-draught machines. The capacity was subsequently increased to eight machines which operated until 1966. In order to increase capacity from 66 000 tonnes to 142 000 tonnes, the eight down-draught machines were replaced in 1966 by a single up-draught unit (6) with a grate area of 93 m^2. This change in technology also laid the foundation for significant improvements in plant hygiene conditions.

To match the improvement in Sinter Plant capacity, a programme of Blast Furnace developments was undertaken to achieve the target lead production. This aim was achieved in 1969 with the development of a 'Mount Isa' type design.

Ventilation of the process also changed over this period. Initially, process gases from both the Sinter Plants and Blast Furnaces were cooled and treated in a single dust collection system, an electrostatic precipitator. This was supplemented by, and finally replaced by, a shaking baghouse. The original baghouse was extended in two stages to contain 14 sections in 1974. The woollen bags were progressively replaced by wool/terylene and eventually terylene bags.

Hygiene ventilation of the up-draught Sinter Plant, as commissioned, was covered by low energy wet scrubbers (6). As the hygiene requirements for both the working environment and scrubber stack emissions were made more rigid (7), the multi-point scrubber concept was replaced by a centralized baghouse system using the more economical reverse pulse principle (8). The first stage of this baghouse was commissioned in 1978 and stage 2 will be completed in 1980.

With the smelter operating at maximum capacity, production of 149 200 tonnes was achieved in 1970. Owing to a combination of depressed lead prices in the early 1970's and the effect of a Company-initiated SO_2 control programme commenced in 1974, the effective production was restricted to 130 000 tonnes per annum. A 270 m stack (8) was commissioned in November 1978 to reduce the smelter downtime caused by the air quality control prog-ramme while still maintaining acceptable SO_2 ground level concentrations in the adjacent community. The standards are equivalent to the National Sulphur Air Quality Standards for Sulphide Dioxide in the United States. Production at a rate of 150 000 tonnes of crude lead per annum has been restored.

In conjunction with the 270 m stack construction, a new vertical gas cooler and associated ductwork were built. In addition to allowing for the handling of the Sinter Plant hygiene gases in stage 2, the ductwork enhanced plant hygiene conditions, especially in the Blast Furnace area.

A summary of the improvements and modifications to the Lead Smelter from 1955 to 1979 are shown in Figure I. The aim of the changes has been to increase production levels while improving working conditions and protecting employee health. This has been achieved by:

. Elimination or reduction of manual jobs associated with high lead exposure e.g. flue cleaning, and constant attention to molten flows at the furnace taphole, forehearth, settler and lead pots.

. Improved operating reliability of equipment to reduce cleaning by operators and engineering maintenance requirements e.g. scheduled shutdowns and planned maintenance.

. Upgrading process and hygiene ventilation e.g. shaking and reverse pulse baghouses.

. Updating process measurement and instrumentation techniques to allow remote and automatic operation of equipment e.g. moisture control, baghouse operations and gas flow monitoring.

The current operations are shown in an aerial view of the plant in Figure II. A schematic layout of the significant hygiene and ventilation features is shown in Figure III.

Duration	1955-1960	1960-1965
Yearly Crude Lead Production (tonnes) Average	48 840	51 560
Range	37 280 - 58 180	40 000 - 60 000
Average Number of Personnel	185	178
Process Activity Filtering	Two 2.6 m Morse disc filters for concentrates (1956)	Continuous metering of concentrates (1963)
Mixing	Single shaft mixer	
Sintering	Eight down draught Dwight-Lloyd machines 114 m^2. Improved grate bars, seals, spillage control	Rotoclones at machine discharge ends. One machine converted to Updraught for testing 14 m^2 (1962) - Trial Hydrofilters (1961)
Smelting	Roy continuous tappers on blast furnaces (1957) Single furnace operation with standby. No. 2, 1.52 m x 6.25 m No. 3, 1.22 m x 5.49 m at the tuyeres.	No. 3 Blast Furnace widened at tuyeres - 1.75 m x 5.49 m at the tuyeres (1965)
Crude Lead Handling	Intermittent tapping 5 tonne lead pots.	Cooling bays.
Drossing	Four x 40 tonne cast iron, wood fired kettles in operation (1953).	Kettles fired by oil (1960).
Gas Handling	12 section shaking bag-house; 8 @ 230 mm diameter, 4 @ 305 mm diameter. Individual sinter and blast furnace spray chambers. 65 m stack	

Figure I - Summary of Major Changes in Lead Smelter Practice 1955-1979

1965-1970	1970-75	1975-79
96 000	126 400	132 250
62 600 - 149 200	112 500 - 146 700	131 400 - 137 500
134	139	141
Two 6.1 m Stockdale string discharge filters for concentrates (1966) Double shaft mixer and pelletiser.		
New Updraught sinter plant 93 m^2 grate area (1966). Twenty eight hydrofilters.	Automatic moisture control. Missing grate bar detector.	Ventilation Baghouse Stage 1 (1978) Conveyor Spillage Control (1977-78)
No. 2 Blast Furnace rebuilt to 1.98 m x 6.25 m at tuyeres (1967)	No. 3 Blast Furnace rebuilt fully water jacketted to 1.78 m x 7.01 m at tuyeres (1970) Shortened to 1.78 m x 6.25 m at tuyeres (1971)	One furnace only 1.78 m x 6.25 m at tuyeres.
Lead well inside fore-hearth (1966). Two pot tapping - continuous flow (1968). 10 tonne pots.	Cooling bays extended. Tilt and Swivel launder for molten lead handling (1975)	Pot Pouring Vent hood (1976) Forehearth Ventilation Upgrading (1978)
Five x 100 tonne steel oil fired kettles in operation (1969). Improved ventilation.	Mechanical dish dross removal (1971) Mechanical scoop dross removal (1973).	Ventilation upgrading (1976)
New 70 m steel stack (1969)	Extension of shaking baghouse to 14 sections. Maximum capacity 210 m^3/sec. at 105° C Semi automatic dust pulping.	270 m stack (1978) Gas Cooling Tower and Ducts (1978). Videospec instrumentation control system (1978)

Figure II - Aerial View of Lead Smelter and 270 m Stack

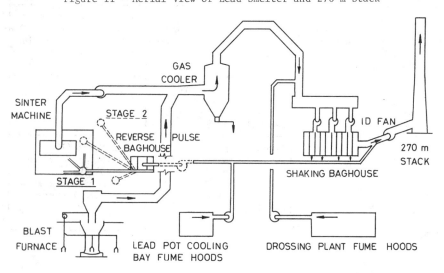

Figure III - Schematic Diagram Lead Smelter Gas Handling Layout

The effect of the recent changes described has been to decrease operator exposure to areas of the process which are necessarily high lead areas and to reduce the general lead-in-air levels in the plant. An example of reduced operator exposure is illustrated at the blast furnace forehearth location in Figure IV.

Figure IV - Blast Furnace Forehearth Area

Significant features here are the Roy continuous tapper, the forehearth with access steps, the tilt and swivel launder on the right of the forehearth, the slag settler to the left and the compact multipoint ventilation system to a single dirty gas fan.

A major contribution to reduced lead-in-air values was the installation of source flues and plenum coupled to a reverse pulse baghouse to ventilate selected, high exposure areas in the sinter plant. This unit is shown in Figure V. The improvement achieved by this equipment is shown in Figure VI and has been the basis of plans to expand the principle of single collection flues to plenums connected to a central baghouse system in stage 2.

The future hygiene improvement programme in the Lead Smelter is centred around the extension of the reverse pulse baghouse in 1980 to ventilate other areas in the Sinter Plant as well as possible extension to the blast furnace and drossing areas.

By this approach and the vigilant application of supervisory, personal hygiene and biomonitoring techniques the improvements in employee health, as discussed in the following sections, will attain regulatory standards.

Figure V - Sinter Plant Reverse Pulse Baghouse

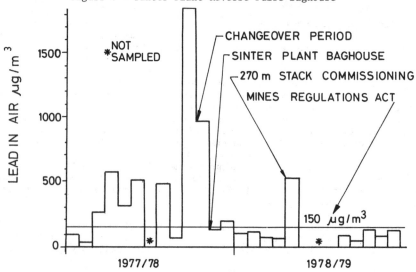

Figure VI - Lead in Air Values - Sinter Plant Spike Rolls

Pre-employment Medical Examination and Induction

Medical

Mount Isa Mines has always undertaken routine medical examinations on prospective employees to ensure the workforce is fit and safe to undertake various activities in the differing environments of its large mining, concentrating and smelting complex. In the late 1960's, the medical examination for prospective lead smelter workers was extended to include an initial haemoglobin level and a measurement of blood urea using a sensitised paper technique: 'Azpstick'. Employees with hypertension, haematological, renal and neurological diseases have always been excluded from lead work both in the smelter and in relation to carbonate ores. The present biochemical standards for new lead workers are a haemoglobin of 14 g/dl, or better, and a blood urea level, as indicated using the above technique, of less than 35 mg/dl. These standards together with a medical questionnaire and examination, ensure that prospective employees with a known or occult medical problem are screened from Lead Smelter work. Women of child-bearing age are regarded as medically unsuitable for lead work.

Induction and Training

All Lead Smelter employees are required to undergo a hygiene and safety induction programme before working in the plant. The programme covers general industrial safety and awareness lectures, and a special hygiene lecture highlighting both the Company's current hygiene programme and the employee's responsibilities for his own personal health while working with a toxic material. These lectures are supported by locally produced video tapes to illustrate these topics and by a questionnaire to reinforce the issues discussed.

The initial induction is supported by on-the-job training which involves a progression through a series of job responsibilities under the guidance of an instructor. The hygiene programme is strengthened by regular respirator inspection and testing, daily respirator washing, maintenance of clean, air-conditioned crib room facilities, and feedback to operators of their personal blood lead results.

Employees are issued with two sets of lockers, one for clean clothes and the other for safety and hygiene gear. Clean clothes are provided daily and are washed on site in a laundry attached to the changehouse.

Biomonitoring in Lead Workers

Historical Background

In the 1940's, Mount Isa Mines began routine stipple cell counts on its lead workers two or three times a year. Initially, these were undertaken by the local hospital, but by the mid-1950's, the Company had developed its own industrial medical section and was undertaking haemoglobin measurements and stipple cell counts on all lead workers on a monthly basis. By 1958, urinary coproporphyrin analysis was also regularly conducted.

Concurrently, medical staff investigated the feasibility of routine urinary lead analysis. The initial work in this area was performed by the Queensland State Health Department, but gallon lots of urine with chloroform preservatives were also forwarded to ASARCO's laboratories in the U.S.A. Urinary lead analysis was undertaken on a random selection of employees, plus those in whom excessive lead absorption was suspected. By the early 1960's however, the Occupational Health Section of the Company had developed its own laboratory techniques, and urinary lead tests in conjunction with stipple cell counts and haemoglobins were taken every three weeks on shift workers and every four weeks on day workers. The results were presented to management as urinary averages by shifts, gangs and areas, and the individual results were returned to supervisors for individual counselling on high variances. As an indicator of the levels identified, Figure VII shows yearly averages by the two main areas of the Lead Smelter together with the range of group averages over the years. Urinary lead gave a good indication of lead load, both for individuals and groups, so urinary coproporphyrin analysis was discontinued.

YEAR	SINTER PLANT WORKERS		BLAST FURNACE WORKERS	
	Urinary Pb Average μg/1	Group Ranges μg/1	Urinary Pb Average μg/1	Group Ranges μg/1
1966/67	280	170 - 370	210	150 - 280
1967/68	195	120 - 310	180	110 - 290
1968/69	205	120 - 300	200	140 - 310
1969/70	180	100 - 260	200	120 - 340
1970/71	180	120 - 230	250	140 - 390
1971/72	168	130 - 220	200	140 - 300
1972/73	150	140 - 170	182	170 - 220
1973/74	130	100 - 150	190	160 - 220
1974/75	120	100 - 150	165	140 - 195
1975/76	126	110 - 147	155	123 - 208
1976/77	127	103 - 153	136	111 - 157
1977/78	121	102 - 137	128	109 - 147
YEAR	Blood Pb Average μg/100 ml	Standard Deviation	Blood Pb Average μg/100 ml	Standard Deviation
1978/79	* 47	5	44	5

* Half Year Only

Figure VII - Group Urinary Lead and Blood Lead Averages

By 1970, techniques for the accurate analysis of blood lead were available and employees having suggestively high levels at routine screening were subjected to blood lead analysis from venous sample. The aim at that time was a group average of 150 μg/1 Pb in urine. Individual levels of 300 μg/1 Pb in urine were carefully monitored and if blood lead was shown to exceed 80 μg/dl or there were symptoms suggestive of intoxication, the employee was transferred from the Lead Smelter for three months or until his blood lead returned to normal (<35 μg/dl) - whichever was the longer. At this time, the use of urinary ALA was also considered but routine urinary lead was regarded as the better approach.

Present Biomonitoring Programme

The difficulties of obtaining a sample of urine and the problems of contamination from the employee or from his clothing were major problems in the urinary lead programme. Confusion also arose where employees were low lead excretors, dehydrated from the heat or ardent beer drinkers. To accurately measure the level of absorption of lead in a lead worker, blood

lead was obviously the pathognomic test. By the mid-1970's the Occupational
Health Section at Isa Mine had perfected the 50 μl blood sampling and analytical technique for lead. The finger prick sample overcame the problem of
venous puncture but still allowed both blood lead and haemoglobin to be
accurately analysed. Contamination was countered by pricking the finger of
the clean worker as he arrived at work through a petroleum jelly film over
the finger. For four years, blood lead, urinary lead, and haemoglobin were
run as a biomonitoring package. With the confirmation of analytical
techniques in relation to international standards, the urinary lead has been
discontinued and blood lead is used for group, activity and area averages.
The inclusion of 1 standard deviation in these results gives management and
supervision an indication of the scatter of results and the potential for
people to reach unacceptable levels.

The 80 μg/dl level is still used as the redeployment level, but any
employee with a microlead over 70 μg/dl has his venous blood lead and urinary
lead checked as well as being counselled at some length. Average blood lead
levels are shown in Figure VII.

Z.P.P.

For over a year zinc protoporphyrin, Z.P.P., as measured by the AVIV
Haematofluorometer, was undertaken routinely in parallel with other biomonitoring controls. The delay in Z.P.P. movement in new workers and the
scatter of results indicated that this procedure would not adequately
identify employees with a considerable lead load on an individual basis.
If all blood leads over 60 μg/dl must be identified, then use of Z.P.P. as
a screening procedure would only reduce blood lead analysis work load by 7%.
The results are illustrated by the scatter as seen in Figure VIII.

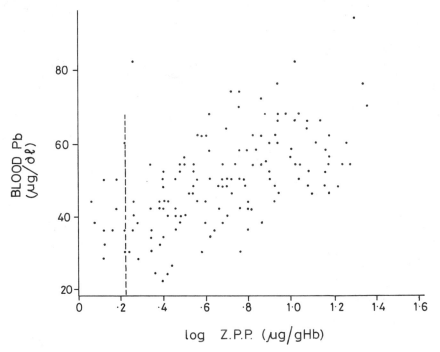

Figure VIII - Relationship between Z.P.P. and Blood Lead

Regulations

The setting and enforcement of environmental standards is a Queensland Government function. In the Lead Smelter the in-plant lead in air standards are as specified in the Mines Regulations Act. These are currently 200 $\mu g/m^3$ but are being revised to 150 $\mu g/m^3$ based on the levels recommended by the National Health and Medical Research Council of Australia.

The Queensland Department of Health accepts a blood lead limit of 80 $\mu g/$ 100 ml, but there is no specific regulations, either State or Federal. Using this level as a transfer control, the Lead Board, as constituted under the Workers' Compensation (Lead Poisoning, Mount Isa) Act, became redundant in 1974. Up to this time a Company nominated medical practitioner and a Trade Union nominated medical practitioner would assess cases of suspected lead poisoning together with representatives of the Queensland State Workers' Compensation Board.

In 1978, the National Health and Medical Research Council, a Federal body, began promulgation of 'Approved Occupational Health Guides'. The latest draft, at time of writing, recommends area monitoring and biomonitoring with blood lead as the criteria for action - 3 $\mu mol/l$, approximately 60 $\mu g/dl$: and for removal from the hazard - 4 $\mu mol/l$, approximately 80 $\mu g/dl$. There is no recommended levels for females.

Environmental Monitoring

Lead in Air Sampling in Plant

Lead in air monitoring and performance checks on the ventilation equipment have been a routine procedure in the Lead Smelter since the plant was commissioned.

Lead in air monitoring has been done by tape samplers at a number of fixed locations in the smelter. Typically, the sampler is run for 19 shifts continuously set at 2 hours per sample spot. Tapes are then removed and individual sample spots analysed for lead. The use of 2 hour samples in this way has enabled identification of plant upset conditions which produce high lead-in-air levels.

The current tape sampling programme involves 14 locations in the smelter with samples running for 9 shifts at each location per month. A major problem with this programme has been a high failure rate of tape samplers; this has been largely overcome by making the sampler cases airtight except for a small ventilation area and fitting the case with a Vortex tube to provide a positive pressure and cool atmosphere inside the sampler.

In addition to this routine programme, special extended sampling programmes at a particular location are carried out as nominated by Smelter or Hygiene personnel to help identify particular operating problems. Personal sampling using open-faced samplers at 2 1/m is done as required for particularly dusty jobs or operations.

Performance checks on the hydrofilters are undertaken routinely and include measurement of the exit volume, dust loading and pressure drop across the marble bed. Where necessary, more exhaustive performance tests are conducted for particular operating problems. The maintenance of these systems has been a very important aspect of controlling lead-in-air in the working environment.

The Urban Environment

Introduction

Major developments at Mount Isa from the years 1965 to 1979 coincided with a period of increasing interest and concern for environmental matters. Legislation was introduced in the Queensland Parliament to regulate and control emissions to the air, liquid effluents. and noise levels.

The effect of the air quality requirements on the operations were subject to an internal investigation in 1977 (7) and was reported to the Queensland Air Pollution Council.

Air Quality Control Programme

The Mount Isa lead and copper smelters are within 200 m of the closest residential areas and emit sulphur dioxide directly into the atmosphere. Evaluations of various means of reducing these emissions have indicated there is no practicable means which the Company could adopt. In order to minimise the impact of sulphur dioxide on urban areas, a combination of intermittent control and tall stacks was adopted.

The Mount Isa intermittent control system, called the Air Quality Control Programme, consists of two concentric rings of Philips sulphur dioxide analysers placed generally towards the nearer and further perimeters of the residential area. The location of the analysers in relation to the operations and the control centre is illustrated in Figure IX.

In addition, there is a comprehensive meteorological system involving peripheral wind stations for velocity and direction, acoustic sounding, and radiosonde flights to provide information for control decisions based on meteorological factors. Continuous information is provided by the National Bureau of Meteorology from Melbourne, and evaluation is made by a Company meteorologigist or by trained air quality control operators.

The operation of the system requires reduction in smelter emissions to ensure that sulphur dioxide values do not exceed the standards which are equivalent to the USA National Secondary Air Pollution Standards for sulphur dioxide. In the case of the Lead Smelter the sinter plant is shut down to reduce sulphur dioxide emission. Different tactics are adopted in the Copper Smelter involving the fluosolids roaster and converters.

In order to minimise production losses resulting from the Lead Smelter environmental shutdowns, the Company constructed a 270 m lead smelter stack in 1978 and has thus been able to achieve production targets while maintaining satisfactory air quality as shown in Figure X.

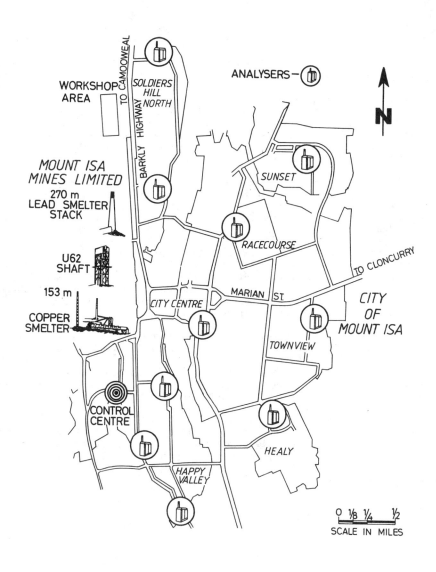

Figure IX – Schematic Layout of SO$_2$ Analyser Locations

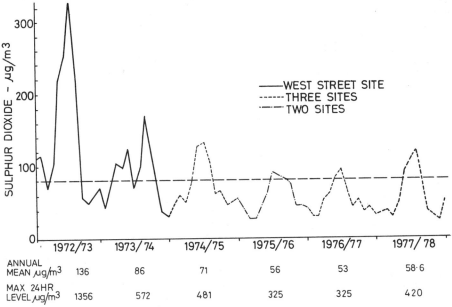

	1972/73	1973/74	1974/75	1975/76	1976/77	1977/78
ANNUAL MEAN $\mu g/m^3$	136	86	71	56	53	58·6
MAX 24HR LEVEL $\mu g/m^3$	1356	572	481	325	325	420

Figure X - Ground Level Sulphur Dioxide Concentration 1972-78

  The data for Figure X is provided by the Annual Reports of the Queensland Air Pollution Council. The significant reduction between 1974 and 1975 was caused by a repositioning of the analysers by the Air Pollution Council. The target level is an annual average of 80 $\mu g/m^3$.

  Due to the close relationship between emissions of heavy metals and sulphur dioxide, it is apparent that by reducing emissions when SO_2 levels in the urban area tend to rise, the Company is providing a very positive control over heavy metal impact on the population. This control is in addition to that achieved by the application of efficient dust and fume control technology applied to the Lead Smelter and Copper Smelter equipment. The Lead Smelter baghouse contains nearly 4500 bags providing an emission which is at one-third the lead level allowed by the Queensland Air Pollution Council. As a consequence of these two factors, measurements made by the Company, in conjunction with the Air Pollution Council, have shown that the air quality in the residential areas of Mount Isa is most satisfactory as regards heavy metal levels.

Minor Heavy Metals

  Small amounts of the elements cadmium, thallium and mercury are associated with the silver-lead-zinc orebodies at Mount Isa. Fortunately, the mercury levels of Mount Isa ore are so low that no adverse environmental effect occurs on processing. As far as the Mount Isa Lead Smelter is concerned, cadmium and thallium do not present any significant occupational or general environmental problem in processing.

Evaluation of Environmental Factors

  Mount Isa Mines Limited has been able to establish excellent relationships with the Queensland Government Departments responsible for administering the Clean Air Act and the Clean Waters Act. They have been achieved in part by the Company's willingness to enter into joint ventures with Government Departments for monitoring and evaluation of the environmental impact of the operation.

Almost all the measurements carried out on environmental factors surrounding the smelter and township are made by Company employees working under a sampling schedule agreed with the relevant Director of Department concerned.

In this way the Queensland Government has been able to make environmental decisions based on an in-depth knowledge of the true situation at Mount Isa rather than on hypothetical situations or alarmist predictions.

Ecology

The Company's concern for the disturbance of ecological systems in the vicinity of the operation has been reflected in the employment of an Ecologist to investigate these matters. Studies have been conducted into the impact of sulphur dioxide on native vegetation and have established that many local species have a considerable tolerance to sulphur dioxide.

Demonstrable effects are not apparent at distances of more than 5 km from the smelter stacks in the prevailing wind direction. However, areas within 1 to 2 km from the smelter stacks in the prevailing wind direction show significant vegetation damage from SO_2, and an investigation to identify resistant native species suitable for revegetating these areas is underway.

Another area of concern in smelting operations is the possible impact on local water bodies. In the case of Mount Isa, the main water storage is downstream from the smelter, so that special care must be taken in minimising or eliminating heavy metal losses into the streams. A major study in conjunction with the James Cook University of North Queensland on the hydrobiology of Lake Moondarra has established that the water of Lake Moondarra is of an extremely high standard, and that the impact of the Company's operation on water quality is negligible as shown in Figure XI.

	Heavy Metal Concentration (μg/l)			
	Cu	Pb	Zn	Cd
*	6.2 (1-56)	2.4 (<1-11)	(<1-85)	(<1-10)
N.H. & M.R.C. Recommended Drinking Water Standards	1000	50	5000	10
Recommended Level at which Biological Investigation should commence	10	20	100	4

* Each set of values represents the mean and the range (in brackets) from 14 sets of samples taken at approximately 4 weekly intervals over the period 7/4/76 to 7/3/77.

Figure XI - Mean Concentration of Copper, Lead, Zinc and Cadmium at Six Sampling Stations within Lake Moondarra

Acknowledgements

The permission of the Management of Mount Isa Mines Limited to publish this paper is acknowledged by the authors along with the recognition of the efforts of members of the Company staff who have contributed ideas and implemented technical developments summarised in the paper.

References

(1) D.T. Buchanan, B.V. Borgelt, A.J. Perry, H.K. Wellington, "Current Lead Smelting Practice at Mount Isa Mines Limited, Mount Isa, Queensland." Fifth Empire Mining and Metallurgical Congress Australia and New Zealand, Proceedings, Australasian IMM 1953.

(2) Lead Smelter Staff - Mount Isa Mines Limited, "Review of Sintering Operations at Mount Isa Mines Limited", Australasian IMM Sintering Symposium, September 1958.

(3) D.T. Buchanan, N.L. Nelson, "Lead Sintering and Blast Furnace Developments at Mount Isa", Paper A69-9, Presented at IMS - AIME Annual Meeting, February 1969.

(4) J.M. Davey, P.J. Slaughter, "Changes in Lead-Zinc Flotation Practice at Mount Isa Mines Limited 1955 - 1970", AIME World Symposium on Mining and Metallurgy of Lead and Zinc, Proceedings, 1970.

(5) J. Bartrum, H.J. Dobrowolski, I.S. Schacke, "Developments in Milling of Silver-Lead-Zinc Ores in the Mount Isa Area since 1970", Lead-Zinc Update, Society of Mining Engineers - AIME 1977.

(6) A.W. Cameron, S.A. Eyers and N.L. Nelson, "The New Updraught Sinter Plant at Mount Isa Mines Limited", Australasian IMM, Proceedings No. 226, Part 2, June 1968.

(7) Mount Isa Mines Limited Staff Committee, R.W. Greenelsh, Chairman; "Environmental Task Force Report on Air Quality Management at Mount Isa", Mount Isa Mines Limited Report (unpublished), January 1977.

(8) K. Ramus, N. Whitworth and P. Anderson, "Developments in Gas Handling and Associated Environment Activities at Smelting Plants of Mount Isa Mines Limited", Australasian IMM, Proceedings, September, 1978.

ENVIRONMENTAL COMPLIANCE BY THE BUNKER HILL COMPANY

IN NORTHERN IDAHO

Ralph N. Gilges and James H. Boyd
The Bunker Hill Company
Kellogg, Idaho

The Bunker Hill Company has been involved in extensive modification to, and new installation of, plant and equipment in order to comply with environmental regulations. These efforts have been costly, burdensome and have consumed a major portion of management's time, but significant advancement in pollution control has been made. Where non-compliance suits have been filed against the company, they have involved only minor excursions.

Bunker Hill initiated litigation and ultimately settled a long standing SO_2 dispute with the Environmental Protection Agency which resulted in the issuance of the first Nonferrous Smelter Order.

The history and impact of the regulations and the methods utilized to reduce emissions from The Bunker Hill Company's lead and zinc operation in Kellogg, Idaho are discussed for water, sulfur dioxide and particulates. Mention is made of the reforestation program undertaken by the company by use of an underground greenhouse.

The Bunker Hill Company, located in the famous Coeur d'Alene Mining District of the North Idaho panhandle, is an important supplier of the nation's requirements for lead, zinc, silver, gold and cadmium. A wholly-owned subsidiary of Gulf Resources & Chemical Corporation since 1968, Bunker Hill employs about 2,000 people in Kellogg, center of the "Silver Valley".

Mining operations commenced at Bunker Hill in 1885 with the discovery of the ore body and construction of a small concentrator. By 1891 there were some 26 mines and 13 concentrators operating in the Silver Valley along the south fork of the Coeur d'Alene river. Effluents, both mine and domestic, were discharged directly into the Coeur d'Alene river which, with justification, became known as "lead creek." The Coeur d'Alene river flows through the Silver Valley, then along farm lands and eventually discharges into a major lake.

In 1908, the mining companies found themselves in court for allegedly polluting farmland with tailings. The suit was settled when rights to use the river for tailings disposal were purchased from property owners.

In 1918, The Bunker Hill Company lead smelter commenced operations and 10 years later the zinc plant came on line. Sulfur dioxide, which is generated in the lead smelting and zinc roasting operations, was discharged from short stacks to the atmosphere. A tailings pond was constructed in 1928 to remove solids from mill effluent before discharge into the river.

Activity on the environmental control front through the depression, war and post-war years was minimal as environmental concern was minimum and society's concern was with jobs and the national requirement for metal. Emissions from major industries in the Silver Valley were similar to those evident in other major industrial areas and were, at worst, considered by the local population to be at times irritating and less than desirable, but a necessary by-product of the mining industry.

As industry expanded and the quality of life improved, with physical and material needs and desires becoming satisfied, a growing unrest in America developed in the fifties and sixties. This unrest manifested itself in many ways, some of which are remembered as dramatic. Less dramatic, but equally important was increased concern for protecting the environment before it was destroyed by the growing nation and its industry. The mid-fifties and sixties saw the introduction of the Federal Water Pollution Control Act (1956), the Clean Water Restoration Act (1966) and the Water Quality Improvement Act (1970). In 1970 the mood in America was set and the the Environmental Protection Agency (EPA) emerged. Lauded in infancy, it rapidly flexed its muscle and soon emerged as the largest regulator of all time.

In January of 1971 the interpretation of the existing Refuse Act was broadened from one intended to protect boats from debris in navigable waters to control of "virtually anything that flows." This definition permitted every stream, no matter how small, to be protected under the Act. In 1970 Congress enacted the Clean Air Act and the nation launched into a massive regulatory program aimed at cleanup and control of the environment which extended into almost every phase of America's life style.

It is within this framework that Bunker Hill found itself attempting to comply with staggering regulatory requirements while still remaining a viable mining and smelting company. It soon became, and still is, the policy of Bunker Hill to comply with all reasonable regulations which promote the health and welfare of its employees and the community, and which improve and/or maintain the environmental quality of the surrounding countryside. Out of necessity, it is also the policy of the company to vigorously oppose those proposed or promulgated regulations which are unreasonable and offer little or no benefit in relation to their cost.

A separate environmental affairs function was established in the early seventies with ever increasing emphasis being placed on these activities. It became a department in its own right in 1974 and now plays a significant and prominent part in the daily and long-term affairs of the company.

The present environmental affairs division, headed by a vice president, maintains as staff a manager, secretary, three meteorologists, two control specialists, two senior environmental engineers, an environmental analyst and a compliance administrator.

Water

Prior to passage of the Federal Water Pollution Control Act, Bunker Hill began to formulate plans for the treatment and/or recycle of all contaminated liquid effluents. This plan was initially developed to reduce mercury discharges, but later expanded to include all significant quantities of heavy metals.

The existing 160 acre tailings pond was rebuilt to become a Central Impoundment Area (CIA) and waste water from the zinc plant was routed together with discharges from the phosphoric acid plant, mill and Bunker Hill mine to this location where the large reservoir provided a suitable settling area for suspended solids. In addition, an effluent treatment plant was constructed at the smelter in 1969 with a neutralizing and thickener/filter circuit which recovered some 90 percent of the solids for recycle. Overflow from this plant was directed to a 300 foot x 40 foot pond which settled the final 10 percent of suspended solids.

Effluent from the CIA discharged into the south fork of the Coeur d'Alene river and, although visibly clear, still contained dissolved zinc, lead, cadmium and mercury. In conjunction with the University of Idaho and Envirotech Systems, a private consulting firm, a study of the discharges of each plant was undertaken in 1971 to determine the best method of treatment.

In addition to maximum recycle of all plant waters, the final plan consisted of directing all flows, not capable of being recycled to the CIA for mixing and clarification. The discharge from the CIA would be processed through one central treatment plant.

The treatment method developed was unique to nonferrous mining and smelting because, although the treatment system was well known, the system for sludge densification had only been laboratory tested. In practice the full scale plant turned out to be outstandingly successful and its operation is described below.

The clarified effluent, at a pH of about 2.5, containing zinc plant, lead smelter, mill wastes and mine drainage is decanted from the CIA to an aeration basin feed launder where it is comingled with recycled sludge from a 236-foot diameter thickener. The pH is raised by milk of lime addition to 9.3. The aeration basin oxidizes the iron contained in mine water from ferrous to ferric which on precipitating assists in co-precipitation of other metallics. The slurry then enters a 32-foot square conditioning tank in which American Cyanamid 212 flocculant is added. Settling of the floc takes place in a 236 foot diameter thickener (see Figure 1). Clarified overflow from the thickener enters the concentrator reservoir and is used as feed

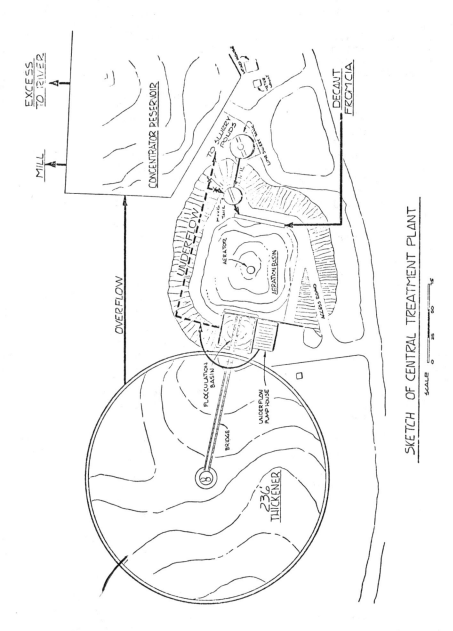

EXCESS TO RIVER

MILL

CONCENTRATOR RESERVOIR

OVERFLOW

UNDERFLOW

TO SLURRY PONDS

DECANT FROM CIA

AERATOR

AERATOR BASIN

FLOCCULATION BASIN

BRIDGE

UNDERFLOW PUMPHOUSE

236' THICKENER

SCALE 0 25 50 75

SKETCH OF CENTRAL TREATMENT PLANT

water to the mill or discharged to the river.

To increase thickener underflow sludge density to about 12% solids, a major portion of the underflow is recycled and mixed with the incoming feed in the aeration basin. Sludge recycle rate is controlled by a regulated bleed off of sludge which is pumped to storage ponds adjacent to the CIA. At present, consideration is being give to processing of these sludges by solvent extraction.

The chart (Figures 2 & 3) shows the success of the operation since it was commissioned in May of 1974 at a cost of $1.3 million. (Only the reduction in lead and mercury in the final effluent are shown, but significant reductions were also obtained in cadmium and zinc concentrations.)

In 1973 before construction of the treatment plant was completed, Bunker Hill received its first National Pollutant Discharge Elimination System (NPDES) permit, which placed limits on discharge of cadmium, zinc, lead, mercury and suspended solids. Although the treatment plant itself proved very successful and generally met all permit requirements, accidental and upset discharges from other areas continued to occur through broken lines, plant upsets and a myriad of minor problems associated either with old plant and equipment or new unforeseen problems arising from the installation of new equipment and the use of water recycle. Considerable expense, time and effort was expended in attempting to minimize or eliminate these occurrences and major progress was made as experience was gained, but the problems were not completely eliminated. A regulatory battle over these issues ensued with the EPA and in 1975 a civil action was filed against Bunker Hill alleging numerous illegal discharges.

In January 1979 the Court found Bunker Hill to have been in violation as alleged by EPA, but noted "Bunker Hill has shown a willingness to invest heavily in environmental protection projects." and that "Discharges that issued did not result in any water quality standards being violated under the proof offered. The record contains no evidence showing that the alleged violations resulted in any harm to persons, fish, wildlife or property, and there is evidence to the contrary." The assessed penalty was $114,640. EPA had requested a penalty of $5,000,000.

In 1977 the initial NPDES permit expired and a second permit was issued which further reduced the allowable discharge of suspended solids, zinc, cadmium and mercury. The reduction required the installation of sand filters to meet heavy metal treatment requirements. These filters are now being installed and should enable Bunker Hill to meet the new NPDES permit requirements which become effective in 1980 and are outlined below.

TABLE I. NPDES Effluent Limitations Effective January 1, 1980 for Discharge from the Water Treatment Plant

	Daily Average	
	Lbs/Day	ug/L
Total suspended solids	N/A	20
Total zinc	163	5.0
Total cadmium	N/A	0.2
Total mercury	0.5	N/A
Total lead	N/A	N/A

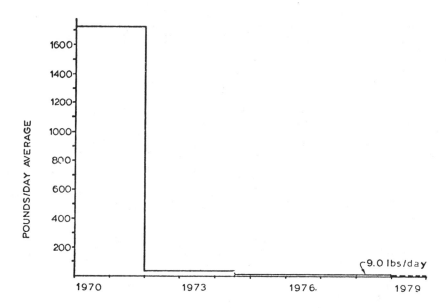

FIG.2   TOTAL LEAD DISCHARGED WITH PROCESS AND
        MINE WASTE WATERS

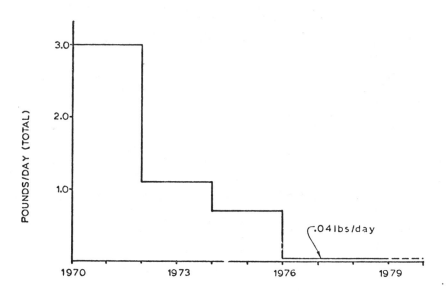

FIG.3 TOTAL MERCURY DISCHARGED WITH PROCESS AND
      MINE WASTE WATERS

There is a permit provision allowing for increased quantity of metals discharge should mine water flow and/or precipitation increase above a base level.

Air

Bunker Hill began control of sulfur dioxide in 1954 with the conversion from wedge roasting to flash roasting at its zinc plant and the construction of the first sulfuric acid plant. When zinc production was increased, a second plant was installed in 1966 to handle the additional SO_2 being generated. Both plants are in use today. In 1968 the company began work to replace the 10 existing Dwight/Lloyd downdraft sintering machines at the lead smelter with a Lurgi updraft facility. This replacement was completed in 1970 with its companion sulfuric acid plant. Particulates had always been controlled as a necessary element of minimizing metallurgical losses.

In March of 1971 the National Primary and Secondary Ambient Air Quality Standards for sulfur dioxide were adopted. A year later Regulations A through S were promulgated by the State of Idaho together with a state implementation plan for meeting the national standards for SO_2 in Kellogg by July 1975. The plan included both 85 percent permanent control and intermittant curtailment, referred to as SCS, or Supplementary Control System. The State Implementation Plan was disapproved by the EPA in May of 1972; thus commenced a period of confrontation which was not settled until June 1979.

While an endless series of meetings, reviews, hearings, changes in regulations, etc. occurred, the significant events are summarized:

In January 1975 the State of Idaho revised their implementation plan extending the compliance date for meeting national primary standards to 1977 and requiring Bunker Hill to capture 72 percent of its SO_2 emissions because it was concluded, after lengthy discourse, that this was the maximum control feasible under currently available technology. In addition to this permanent control, Bunker Hill was required to use a supplemental control system as necessary to assure compliance with the ambient air standards by mid-1977. Seventy-two percent permanent control with emission's through the 200 foot stacks would have permitted Bunker Hill to meet the present and any foreseeable future standards for control of sulfur dioxide, were it not for the fact that the company's operations are located in a valley which creates unique adverse meteorological problems.

The Silver Valley historically records approximately 200 temperature inversions each year. This condition is caused when cold air near the valley floor is trapped under a canopy of warmer air. Smelter and other man-made emissions are trapped under this canopy which normally forms in the early morning hours and dissipates by mid-day. Unless emissions are reduced during these periods, sulfur dioxide concentrations could build to a point where the standards might be exceeded. The SCS system predicts buildup of SO_2 concentrations and provides information necessary to curtail operaton of SO_2 generating equipment.

Bunker Hill realized that curtailment on 200 days per year would destroy the economic viability of the plant. Therefore, in August 1975, Bunker Hill made a decision to build taller stacks at the zinc plant and the lead smelter. The taller stacks would either remove curtailment requirements altogether or, at the minimum, reduce curtailment to an acceptable level.

In November 1975 EPA rejected Idaho's plan and substituted standards that required SCS and 85 percent control. EPA proposed that Bunker Hill could achieve 85 percent control by installing a sulfur burning facility to "even out" the fluctuations in SO_2 gas strengths from the sinter machine. This proposal was rejected by Bunker Hill as being totally unworkable.

In December 1975 Bunker Hill filed a petition for judicial review of the EPA decision.

In July 1977 the Ninth District Court filed a decision that the EPA administrator had not "exercised a reasoned discretion" in issuing the 85 percent control regulation and, furthermore, stated that to require Bunker Hill's plants to meet the limitation would be "arbitrary and capricious". The case was remanded to the EPA for further proceedings.

In August of 1977 the tall stacks were completed together with new flue and fan systems.

The zinc plant stack is 610 feet high. The outside diameter at the chimney base is 45 feet 3 inches and 20 feet 3 inches at the top. A six foot diameter plastic-reinforced fiberglass liner carries the gases up the column at a velocity of 175 feet per second with a capacity of 70,000 cubic feet per minute driven by a 250 horsepower fan. The stack is connected to the two acid plant exhaust systems by a 461 foot long fiberglass duct supported on concrete pillars.

The smelter stack is 715 feet high with an outside diameter of the chimney of 53 feet 3 inches at the base and 26 feet at the top. The stack contains a 13 foot 6 inch diameter liner, which is connected by a fiberglass flue system to the acid plant and the exhaust plenum of the baghouse. Particulate dust from the blast furnace is captured in a new 14 foot square uptake and transferred to the baghouse through a 13 foot 6 inch steel flue. Four 1,000 horsepower fans push and pull the gas stream through the baghouse maintaining a neutral point in the baghouse before being sent up the stack. The stack has a capacity of 600,000 cubic feet per minute and an air-rise velocity of 75 foot per second at the exit point.

The price tag for both stacks, including all ancillary facilities, totaled $11,420,000, one of the largest single expenditures for an environmental control project in Idaho's history.

From the time the stacks were completed in August 1977 to the time of writing, the National Ambient Air Quality Standards have not been exceeded and this appears to negate the need for SCS. The SCS system, however, has been retained to protect against upsets and to monitor fugitive emissions (see Figures 4 and 5).

Negotiations on the remanded SO_2 case continued through 1978 and in June 1979 both parties signed a "settlement agreement."

The agreement brought to an end the litigation which had been pending over the years and resulted in the issuance of the first Nonferrous Smelter Order (NSO) in the United States. This NSO will expire on January 1, 1983 and until such time Bunker Hill will be required to operate within the settlement agreement which forms the basis for the NSO. The key provisions of the agreement are:

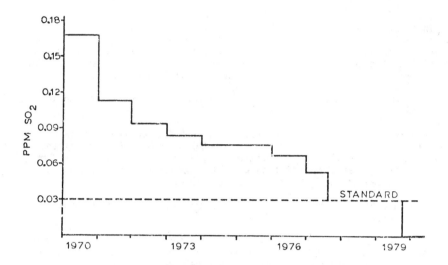

FIG.4 AMBIENT AIR IMPROVEMENT SULFUR DIOXIDE
ANNUAL AVERAGE

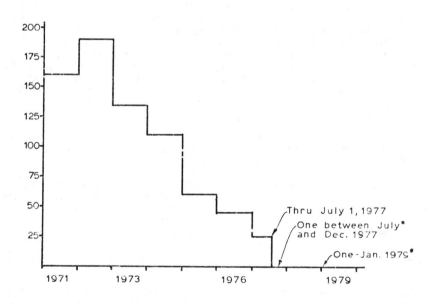

FIG.5 AMBIENT AIR IMPROVEMENT SULFUR DIOXIDE
NUMBER OF DAYS EXCEEDING .14 PPM

ONE EXCURSION ALLOWABLE

636

1) SO_2 emissions shall be limited to 2600 ppm by volume averaged over any running 6-hour averaging period and a maximum of 625 tons of SO_2 shall be allowed over a daily (midnight to midnight) running 7-day period. Violations of the 7-day limitation are determined when any 7-day cumulative tonnage exceeds the limitation tonnage more than once per quarter. The agreement states that no two violations shall contain any common daily data points.

2) All SO_2 gas streams from the zinc roasters and the strong gas exit point of the lead sinter machine shall at all times be processed through the acid plants.

3) Bunker Hill shall continuously monitor all gas flows for SO_2 and with rates of flow indication in practically all ducts.

4) Monitoring of specific parameters of the various processes to ensure compliance with startup/shutdown/malfuction provisions are required. The provisions do permit the bypass around SO_2 treatment facilities under specific conditions and the monitoring is to ensure compliance with the provisions. (Although Bunker Hill would have preferred greater flexibility, the startup/shutdown/malfunction regulations were agreed by Bunker Hill as being stringent, but attainable.)

5) The usual provision of best engineering techniques to maintain the operations in a "leak free" condition is part of the agreement. One additional aspect is the installation of manual or automatically controlled tuyere air flow system on the blast furnaces to minimize upset conditions in the furnace. This installation has been completed, but at time of writing it is too early to judge the effectiveness of the system.

6) Aside from excess emissions caused by startup/shutdown/malfuctions which are presumptively excused, excess emissions are also excused if the administrator is satisfied that bypass of gas was necessary to prevent loss of life, personal injury or severe property damage or sudden and unavoidable excess acid plant tailgas SO_2 emissions which are beyond the control of the operators.

7) In addition, the Bunker Hill complex must employ a supplementary control system (SCS) to the extent necessary to meet National Ambient Air Quality Standards and the SCS program must meet with the approval of the administrator.

8) And finally, Bunker Hill is required to undertake one or other of the following:

8.a) An SO_2 removal facility (flue gas desulfurization system) to capture the weak gas stream exhausted from the sinter machine, or

8.b) Substantially complete recirculation of the sinter machine weak stream and treatment of the resultant gas stream in the acid plant, or

8.c) Conduct a reduced scale research program to treat a portion of the weak gas streams with a minimum capacity of 5,000 SCFM from the sinter machine or the blast furnace.

9) Subject to certain compliance provisions with the above research and development projects, Bunker Hill would be permitted to bypass the acid plant during annual acid plant overhaul periods provided this shutdown for annual maintenance does not exceed 14 days in any one calendar year and that the National Ambient Air Quality Standards are not exceeded.

Particulates

Bunker Hill, over the years, has installed dozens of control devices to reduce particulate emissions from point sources by over 99% and is in compliance with applicable emission regulations.

The remaining emissions come either from numerous tiny contributive sources, or from specific exhaust points where the cost of controlling the emissions is prohibitive in relation to the reduction to be gained (see Figure 6).

The EPA, however, recently filed suit against the company citing the zinc fuming plant as a source of "excessive particulate emission," and asked the court to fine Bunker Hill $25,000 per day for each day the company failed to take corrective action. It would cost over $500,000 to further reduce emissions from the zinc fuming plant at the lead smelter. This source accounts for less than one percent of the smelter's lead particulate emissions. The company contends this minor reduction does not warrant such a capital outlay.

Bunker Hill contested the action on the grounds it has already taken "all reasonable precautionary measures" as required by the regulation through the installation of a baghouse on the furnace and a scrubber on the plant's granulator. A settlement agreement and consent decree was entered into in October 1979 which resulted in no cost penalties to Bunker Hill but required Bunker Hill to operate the plant in such a manner as to assure that fugitive particulates do not exceed present emission levels.

Non-Attainment Area

The 1977 amendments to the National Clean Air Act require each state to monitor the air quality within its boundaries. Those regions failing to meet the national air quality standards set for gaseous and particulate contaminants were to be classified as "non-attainment" areas.

Each state was required to design and submit a plan to the EPA detailing how each non-attainment area would be brought into compliance. The EPA had authority to approve, reject, or revise such plans.

In December 1977 the Idaho Department of Health and Welfare proposed that Kellogg be declared a non-attainment area for failing to meet the standards for both sulfur dioxide and particulates.

The non-attainment status due to sulfur dioxide levels may only be a technicality. State and federal environmental regulatory agencies agree that the standards are now being met in monitored areas. However, the EPA requires one year of monitoring data to confirm compliance. That requirement was completed in August 1978, but as yet no action has been taken on removing the non-attainment status from the area.

Determination of attainment status for particulates is based on a measurement of the total amount of particulates in the air from industrial,

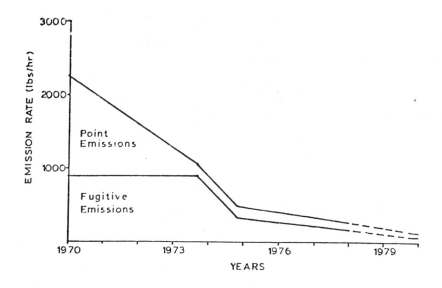

FIG.6 PARTICULATE CONTROL LEAD SMELTER

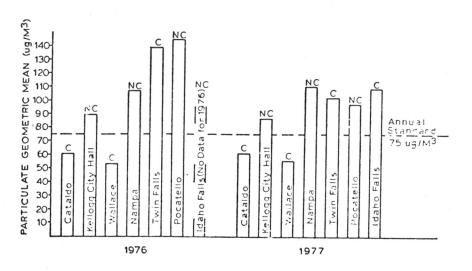

FIG.7 TOTAL PARTICULATES FROM IDAHO STATE MONITORS
C - Compliance   NC Noncompliance

domestic, vehicular and natural sources. Data supports the State's contention that certain areas of the Silver Valley exceed total particulate standards. However, there is some controversy over the sources of particulates which should be included in the measurement.

In several other Idaho communities (see Figure 7), the State has authorized a deduction for background sources of particulates such as wind-blown dust as has EPA in other areas. No such deduction has been considered in evaluation of the monitoring data for the Silver Valley. If such a deduction were made for rural dust blown into Kellogg from rural sources, the area might well be reclassified to attainment status.

If the State and EPA fail to treat this situation as it has others, and the area continues to carry its non-attainment status, there will be severe restrictions on industrial and commercial growth in the area.

In addition to the total particulate standards now in effect, the EPA has promulgated a national ambient air standard for lead of 1.5 micrograms per cubic meter of air, based on a ninety day average. The nonferrous mining and smelting industry is opposing this unrealistic standard and has filed for judicial review. Testimony before a recent EPA hearing on the matter indicated such a standard could not be met in the Kellogg area, nor in the vicinity of most of the primary and secondary smelters in the United States. In addition, many American cities would find themselves unable to comply with such a standard due to lead contained in automobile emissions. Exactly how the Court and EPA will ultimately deal with the issue is difficult to imagine.

Resource, Conservation and Recovery Act (RCRA)

The company will soon be faced with the, as yet, undefined regulations to be promulgated on December 31, 1979 under RCRA. With what has been seen so far, there is little doubt that the burden of these regulations will be felt in the not too distant future. Coupled with these will be the increased attention that the EPA will pay to Section 208 of the Clean Water Act which requires the containment of slurry runoff from stockpiled materials. Of concern will be the classification of mill tailings pond, or mine waste dumps, because that classification will determine what action, if any, will be required under the regulations.

Revegetation

In 1972, Bunker Hill embarked on a long-term program to revegetate 18,000 acres of land in the Silver Valley disturbed by forest fires and early mining and smelting activities.

The first two years of the project involved research to determine tree species and planting methods best suited for the varied soil and climatic conditions found in the Valley.

Actual large scale planting began in the spring of 1974, with over 325,000 seedlings established on 690 acres of the Valley since that time. Plans now call for up to 400,000 trees and shrubs to be planted annually during the years ahead.

To help provide trees for the revegetation program, a unique greenhouse was established in late 1975. It is located 3,000 feet underground in the sunless depths of Bunker Hill mine.

The underground nursery was constructed in an abandoned ventilation drift on the mine's number five level. This area provides a constant temperature, as well as ideal levels of humidity, oxygen and carbon dioxide levels. The drift also provides an environment free of disease and insects. The only missing element--sunlight--is supplied artificially by high intensity lamps.

The greenhouse presently has the capacity to produce over 100,000 seedlings per year. Ultimately, it may be expanded to permit propagation of 350,000 trees annually.

The underground greenhouse has saved the construction costs of a conventional surface facility and also eliminated the expense of winter heating and summer air conditioning. Trees are being grown for one-fourth the cost of surface grown trees, and at a growth rate which far exceeds that achieved in conventional greenhouses. Underground seedlings are grown in individual containers and in six months attain the size required for successful transplanting in the field.

Bunker Hill's pioneering achievements in subterranean horticulture recently earned the company the Industrial Forester of the Year award from The Idaho Tree Farmer's Association.

Reforestation efforts and natural revegetation is also making a mark on the hills surrounding Kellogg. The Silver Valley is again green and the local citizens are verbally happy with the progress.

OSHA & MSHA

1. Safety Standards

Inasmuch as the plants were not constructed to OSHA standards 60 years ago, the task of compliance is both onerous and costly. Here again a major impact is the time spent by maintenance, operational and staff people on projects which contribute little or nothing to improving the safety of personnel. While the principles of worker protection are endorsed without reservation by Bunker Hill, the application has bordered on the bizarre as regulation after regulation has issued forth from the OSHA paper mill and applied across the board. While hundreds of thousands of dollars have been spent on inane projects, the frequency rate of accidents has not shown any improvement at Bunker Hill attributable to OSHA. At the national level, the frequency rate has indeed risen.

It is Bunker Hill's belief that Union/Management/Safety Department committees meeting and acting on safety issues are much more effective and sensible than a plethora of regulations.

2. Health

Great strides have been made in improving the in-plant environment to more effectively protect worker health and this has been supplemented by an enforced respirator program for all workers in "lead exposure" areas. Furthermore, employees, whose routine tests indicate a blood lead count of over 80 ug/m^3*, are removed from an exposure area until their levels have dropped to below 70 ug/m^3. The latest blood tests indicated that less than 4 percent of smelter

* The 80 ug/m^3 is the generally accepted level below which no medical impairment occurs.

workers show blood lead levels above 80 ug/m^3 and most of those were only marginally higher. What Bunker Hill now views with concern is the changing regulations as standards continue to be arbitrarily lowered beyond the point at which it is either practically or economically feasible to comply.

OSHA has recently promulgated standards requiring primary lead producers to implement engineering controls which would reduce all employees' exposure, without the use of respirators, to below 100 ug/m^3 by 1982 and to below 50 ug /m^3 by 1989. Immediately, workers would have to be provided with respirators if an eight hour exposure exceeded 50 ug/m^3.

The standard would impose other onerous requirements, including extensive monitoring of the workplace; biological monitoring and medical examinations of employees; training programs; construction and operation of special changing rooms, lunchrooms, and other facilities; preparation of an elaborate written compliance program, including engineering and other construction details and purchase orders for materials and equipment; and various additional housekeeping practices requiring extensive and intrusive monitoring of employee work, eating, smoking and personal hygiene habits.

An extensive "medical surveillance program" would be required, including not just regular medical examinations, but examinations by a second physician at the employer's expense and by a third consulting physician if the first two physicians disagreed. A more onerous medical removal program would be required, similar to the one Bunker Hill now manages, but involving significantly lower blood lead contents of employees. If an employee is removed from his normal job, and perhaps from work altogether, the employer would be required to maintain his earnings, seniority, and other employment rights and benefits for up to eighteen months. Extensive record-keeping would be required with some records having to be maintained for forty years.

In short, the standard is as burdensome as any that OSHA has sought to impose.

These standards have currently been stayed pending judicial review.

Of almost greater concern is the recently promulgated arsenic standard which at 10 ug/m^3 is as close to zero as can be imagined. Since it is designated a carcinogen, OSHA need only demonstrate the presence of arsenic, without offering any proof of a dose/response relationship, to deem any operation to be in noncompliance. The arbitrary setting of this standard has also been challenged in court.

3. Mine Safety and Health Act (MSHA)

The volumes of regulations issuing forth from MSHA are yet another example of an agency's penchant for overkill, stifling the mining industry with some 1,012 pages of regulations, not including the long list of standards incorporated by reference. The saga of MSHA is too voluminous to even document here in summary; suffice to say the impact places as great a strain on the miners as the environmental regulations place on metallurgical operations.

Impact

This paper has discussed only some of the major issues that Bunker Hill has faced since 1970 and, generally, how they have been dealt with to date. It is not suggested that even those issues have been solved because, as standards are revised, the old issues will be remobilized under different situations.

Compliance has been costly. From 1969 through 1978, The Bunker Hill Company spent 56 percent of its available capital on pollution control, 32 percent on maintenance of existing plant and equipment, and only 12 percent on modernization and profit improvement.

At a time when capital was sorely needed for the updating and modernization of the operations, the disproportionate amount of money allocated to pollution control has placed a severe strain on Bunker Hill.

More difficult to quantify, but equally important, is the increased expense necessary to operate and maintain the additional pollution abatement facilities and the excessive nonproductive hours required by personnel on paperwork to satisfy the requirements of the regulatory agencies. We have estimated that over 20 percent of all administrative time at Bunker Hill is consumed on regulatory compliance. This, of course, does not include the brain drain on key people where innovativeness and full devotion is sorely needed to plan the company's future.

In Conclusion

While the outcries from remote critics continue with their "rape, ruin and run" rhetoric, the general populace within the Silver Valley now look favorably upon Bunker Hill and the mining community for their efforts and applaud the improvement in the environment. The hills are once again turning green, the air is cleaner and the river is clear and capable of supporting aquatic life as evidenced by fishing and the return of the kokanee salmon to spawn.

All operating and maintenance staff at Bunker Hill are acutely aware of, and spend a great deal of time on, pollution related activities. They do share concern as to whether or not the ever tightening regulatory requirements will permit them to seek their livelihood in this area, or even in the mining industry. The community at large shares this concern because their future is inextricably tied to the viability of mining and smelting in the Silver Valley.

The Bunker Hill Company and, we believe, the citizens of the Silver Valley are proud of the significant progress made in the environmental arena. Bunker Hill feels strongly that some recognition and public acknowledgment by the agencies of this fact is in order in place of the apparent unceasing efforts to produce the impossible dream regardless of the economic consequences to the community and nation. While there appears to be an element of reason growing in this country, the radical environmentalists still command a substantial voice. We hope that reason will prevail.

RESULTS OF A FIVE YEAR EFFORT TO REDUCE LEAD POLLUTION

AT THE HOBOKEN PLANT

by MARCEL SOENS, Director Metallurgie Hoboken-Overpelt (M.H.O.)
and Plant Manager
JAN VAN BOVEN, Assistant to the Plant Manager
GILBERT DECKERS, Engineer Environmental Labs

Abstract

The Hoboken plant was started in 1887. Eighty five years later, the smelter was confronted with the problem of cattle perishing due to the absorption of lead present on the grass in the neighbourhood of the plant.

The company had to set up a program for measuring the fall-out on the grass, and the immission of dust and lead in the air outside and inside the plant. Different methods were applied to obtain rapid and reliable results.

On the basis of these results a program of actions was worked out, divided in 3 major groups :

1. immediate actions without long research work such as water spraying in various ways;

2. limitation of the existing emission from sources, such as the 152 m stack and construction of supplementary baghouses;

3. conversion of flue dust into compacted filter cake.

It is not really possible to distinguish the influence of each of these actions separately, but the total results were very rapidly noticeable regarding fall-out, which fell to 20 % of its former value. The immission values measured at the perimeter of the plant dropped by more than 5 % per year.

To obtain in 1983 the yearly mean immission value of 2 μg/m3 Pb in the air close to the plant, investments will exceed the amount of 30 M $.

Introduction

The authors wish to express their thanks to all those of the M.H.O. staff, who in one way or another had to deal with the complex problems of fighting pollution by lead dust, more especially to Robert Nicaise, deputy general manager, Martin Bomans, head of the central quality control lab, and André De Troy, head of the environmental dept.

Since 1970, an important number of studies have been published concerning

644

the impact of lead on the environment. Practically every lead smelter, wherever in this world, has to deal with the important problem of how to reduce lead dust and fume emissions to comply with ever more stringent standards.

The purpose of this paper is to explain how M.H.O. dealt with the lead problem, how the outside effect of lead operations was measured, how M.H.O. proceeded to identify the various sources of pollution, and finally what is being accomplished to solve these problems in a large lead plant with a lead activity initiated more than 90 years ago.

The Hoboken plant has been fully described at the AIME world symposium in 1970 in Saint Louis published in : Lead and Zinc Extractive Metallurgy - volume II, Chapter 28.

O

O   O

1. Location of the Hoboken smelter

Hoboken is an industrial suburb situated SW of the harbour of Antwerp. The industrial area is located on the right bank of the river Schelde   . The population of Hoboken reached 30.000 inhabitants in 1950 of which 8.000 are living close to the plant downwind from the prevailing S-SW wind direction within a distance of 25 m to 750 m.

On the east and south east side of the plant, there is still local but limited farm activity with the presence of several hundred heads of cattle.

In the spring of 1973, first mention was made of cattle dying due to the presence of lead dust on grass. Traces of lead poisoning were found in the liver and kidneys of cows and horses.

To explain more clearly the problem, it is necessary to recall that the Hoboken works, because of their location near densely populated centers have, ever since their foundation in 1887, been equipped progressively with an important number of pollution limiting systems   :

a)   a 125 m high stack (1900) for the process gases from the smelter, and a 120 m high stack for tail gases.

b)   high voltage dry and wet electrostatic precipitors for all gases from blast furnaces and Hoboken copper converters (1923)

c)   two sulfuric acid plants for SO_2 gases from roasting, sintering and matte converting (1930 and 1970).

d)   an important number of baghouses, for process gases as well as for the protection of the workers, totalising 35 Nm^3/sec

e)   several wet scrubbers.

Consequently the pollution around the plant was in 1973 certainly not worse than it had ever been before. The apparently sudden cattle mortality attributed to lead was more the improvement of veterinary diagnosis, than the result of any kind of change in the plant activity.

2. <u>Survey of dust deposit on grass</u> (fall-out)

Since 1973, M.H.O. started a vast program of grass analyses, at 14 different points on the east side of the plant at distances between 700m and 2.000 m from the foot of the 152 m stack. Analyses were made twice a month, in total more than 1.000, spread over a 4 year period.

Grass covering a surface of 1 m^2 is cut off at 3 cm from the ground and dried at 105°C during 12 h. The dry matter is weighed, then chemically leached with a mixture of HCl-HNO3. After sample preparation, the lead content is measured by atomic absorption.

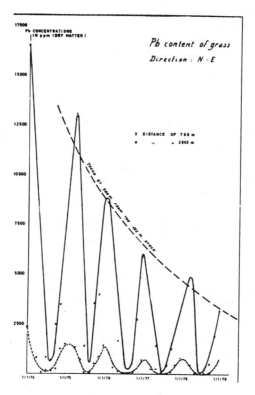

Fig. 1 shows the evolution of the lead content of grass since 1974, in the NE direction at distances of 700 and 2.050 m from the main 152 m-stack.

It is clear that lead concentrations have been seriously decreasing, because of technical improvements since 1974, this in all directions from the plant and at all distances. In late 1978, they dropped to about 20 % of those found 5 years earlier when the analyses were initiated.

There are noteworthy seasonal variations, due to the long exposure time in the winter (no grass growth) in comparison with normal grass production in summer. The lead content in the grass was measured in both the NE and the SE directions. SE values are about 3 times lower than those of the NE, because of the influence of prevailing winds.

Other investigations show that up to 95 % of the Pb in the grass is due to fall-out and 5 % due to absorption from the soil.

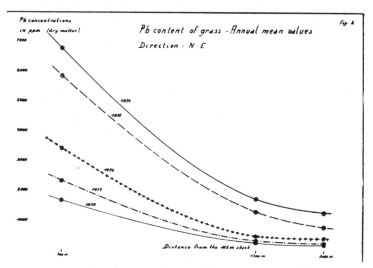

Fig. 2 shows the evolution of lead deposit on grass, after elimination of seasonal variations. This figure gives the annual mean value of Pb at various distances from the 152 m stack in the NE direction **from the smelter.** One can directly see that, from 1974, Pb concentrations on grass have significantly decreased, especially at short distances from the plant. Please note also the sharply decreasing lead concentrations in function of distance from the plant.

Due to the difficulty in obtaining representative samples of grass at lower levels of contamination, fall-out has been measured by means of dust collectors. since 1978 (see section 5 hereunder).

3.   Immission measurements (Ambient air quality measurements)

Immission is measured by the amount of lead in the air. This may be done by low-volume samplers, which have been officially recognized as giving valid results in connection with public health.

Most of the work, however, is done with high volume samplers which are better suited for identification purposes : they provide the best means for detecting sources of emission and their relative importance.

3.1   Immission measurements with a high-volume sampler (HVS)
      at 700 m NE of the 152 m stack

Since 1975 measurements of air suspended particles are made by the local Centrum voor Lucht en Water (CLW Institute Antwerp) on the roof of a nearby school. These particles are measured north of the factory at a distance of about 200 m from the factory walls on top of a flat roof, 8 m from the ground.

Employed in the study is a high-volume sampler, with a Whatman 41 – Ø 15 cm filter. About 450 m3 per day of air are sucked and filtered, gas velocity at the entrance of the filtercover is

+ 9 cm/sec. Every 24 hours the filter is replaced. Analyses are made by atomic absorption and X-ray fluorescence.

Fig. 3 gives in a graphical form the results since 1975 (monthly mean values).

Fig. 4 shows the evolution of the monthly mean values since 1975, by grouping for each month the values for successive years.

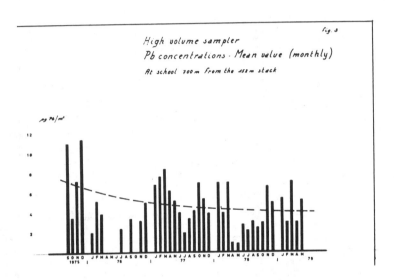

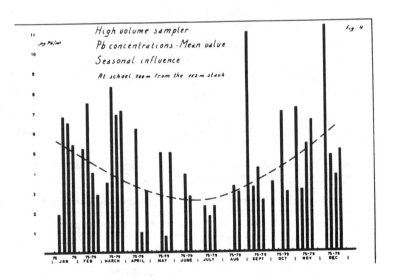

3.2 Immission measurements with a low volume sampler (LVS)

Since 1973, the Institut d'Hygiène et d'Epidémiologie (IHE) measures suspended dust in the air by means of a low-volume sampler. This instrument is placed at the same point as the high volume sampler. It draws in + 15 m3/day through a covered millipore filter. The porosity of the filter is 0,45 m. Once a day, at 0 h the filters are changed by an automatic system containing 8 filters permitting a weekly collection. The velocity of the air at the filter surface is + 13 cm/sec. Analyses of the collected samples are made by X-ray fluorescence.

Fig. 5 gives the measured monthly mean values since 1973 and slide 7 the peak values.
Fig. 6 gives the evolution of the monthly mean values since 1973.

The low-volume sampler is of the type accepted by the Belgian Ministry of Health.

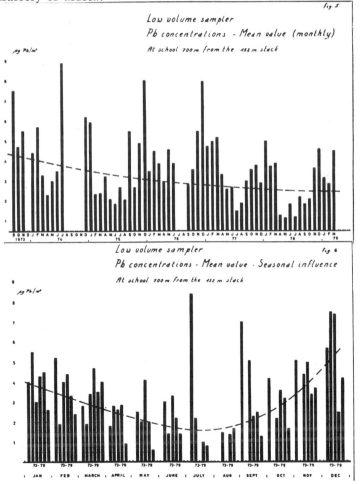

Low volume sampler
Pb concentrations - Mean value (monthly)
At school 700m from the 152m stack

Low volume sampler
Pb concentrations - Mean value - Seasonal influence
At school 700m from the 152m stack

3.3 Correlation between high-volume and low-volume samplers

From the monthly mean values we can calculate the equation

HVS = 0.736 + 1.46 LVS

Based on 30 comparisons, the correlation factor is r = 0.901

In July 1978, a Belgian Arrêté Royal (Royal Decree) was published especially for Hoboken. It requires the smelter to limit its total emissions so as to reach by 1983 a maximum yearly mean value of 2μg lead/m3 at 200 m from the plant down from prevailing winds. The yearly mean values must decrease by $0,2 \mu$g Pb/m3 per year, according to the following schedule :

2,7 (1979); 2,5 (1980); 2,3 (1981); 2,1 (1982) and finally 2,0 (1983).

4.   Technical results obtained since 1974

Soon after death of cattle became known in 1973, serious efforts were initiated with a view to a better identification of the nature and importance of all possible sources of lead dust within the plant. This gave rise to a three phase five-year program, each phase being defined by priorities and the type of investments required.

First phase projects were characterized by the possibility of rapid implementation, requiring no long term study and for which equipment was readily available. Results expected were rather limited but very quickly perceptible. This led to the installation of sprinkling systems to prevent escape of dust from stockpiles and roads.

Second phase projects required some study of the nature and importance of the sources of pollution, the choice of the best solutions and the time necessary for procurement or construction of equipment. The time for implementation was about 2 years. Installation of additional baghouses and scrubbers were typical at this stage.

Third phase projects, which were completed in 1978, were those requiring greater R & D efforts and more sophisticated solutions. They included the control of very small particles (sub-micron size) which are responsible for the greater part of the problems. Typical of this phase were new installations, for which own flue dust was to be transformed into filter cakes, after pneumatic transportation and airtight handling. Solutions had to be found which would not create any waste water problems.

More details of the implementation of the 3 successive phases are given below :

4.1   Control of fugitive dusts from open stockpiles and roads

Since 1960, in order to facilitate the handling of 2 to 3 million tons of concentrates and by-products per year, the system of concrete roads was extended throughout the entire plant area. Moreover as most of the raw materials for the plant are trucked, the handling or moving of the materials within the plant is done by motorized loading and unloading equipment. This transportation method, altough quite efficient, presented problems which had to be corrected without waste of time.

Vehicles using service roads within the plant area lifted, during dry weather, noticeable amounts of dust from the road surfaces. On windy days some of the finer particles of raw materials may have been carried by winds beyond the limits of the plant.

To fight this particular dust problem, three large watertank-trucks, equipped with fantype sprays, regularly sprinkle the surface of all the roads in all areas of the plant used for stockpiling raw materials.

It soon became necessary to use new and more sophisticated sprinkling methods on part of the roads. The new installation should not have any projecting spray heads that would interfere with traffic on the roads. For this reason, a sunk flat nozzle with wide spray and a 15° angle of deflection was chosen.

The level of sprinkling is high enough to wet the surface of the road, but at the same time low enough not to cause any inconvenience to vehicles moving in the center of the roads.

The quantity of water used to spray the roads is very small in order to avoid the formation of mud on the road surface. The circuit is automatically programmed to operate approximately 20 seconds every 12 minutes. It has been determined that this is adequate to maintain the required moisture to keep the dust on the roads, even in dry weather.

The small amount of dust that settles on the surface of the roads, mostly carried by the tires of the trucks is recovered by two large mechanical sweepers.

On the west side of the plant there are two cranes to unload ore from river barges. The ores are dumped in large drop-bottom buckets that are used to convey them to the storage areas where they are stockpiled on the ground. When these buckets are emptied, fine particles would cause dust, were it not for the fact that a fine water spray or mist is continuously applied all around the buckets as the ore is discharged. The amount of water used in the spraying operation is controled by the crane operators.

The above mentioned water screen not only prevents dust formation from being dispersed over the plant area, but at the same time materials are adequately moistened to avoid any further loss by dust.

The top layers of the powdery ores stored in exposed areas over a long period of dry weather would be blown away by high winds, were it not for constant attention given to these particular stockpiles.

For the preservation of these materials under all climate conditions, a system of fixed rotary sprinklers has been installed in and around the stocking areas. The radial range of these sprinklers is sufficient to cover the entire designated areas. They are operated automatically from a central control room. This equipment easures that the storage areas have an adequate supply of spray water to maintain a moisture level that will prevent any metal losses through wind erosion.

It has been proved that, even during extremely dry periods such

as the summer of 1976, the outer layers or "skin" of stockpiled
materials develop a "crust" as a result of systematic sprinkling of
materials at intervals of 4 minutes per hour, 24 hours per day.

The same spraying technique is also used to moisten the furnace
charges in the charge preparation areas of the blast furnaces.

In addition to the larger stockpiles of materials, which
undergo constant and repeated treatment, attention is also paid to
treatment of the smaller stockpiles which are normally retained over
a longer period of time before being processed.

The treatment of the small piles include atomization under high
pressure of a resinous material. The application of this material
provides a water resisting film which hardens and prevents dust formation during stockpiling operations. At the same time, it stabilizes the surface of the piles, preventing loss of metal through heavy
rain. This resinous material is sprayed by portable equipment.

4.2 Control of emission from gas cleaning filters

As far as gas cleaning is concerned, the 3 blast furnaces have
been equipped with two supplementary baghouses controlling and
cleaning an increased volume of ambiant air. The existing Cottrell
Electrostatic installations for the treatment of the process gases
have been improved and modernized for maximum efficiency of dust
collection. After the blast furnace gases have passed through the
electrostatic separators, they are once more cleaned in an additional installation of scrubbers. The overall efficiency is 99.50 %,
on precleaned gases before release in the atmosphere through the new
152 m concrete stack. The blast furnace gases are under permanent
control before and after cleaning by a new, fully automatic
installation. It has a warning system that immediately informs the
operators of any malfunction of the equipment .

At present, the baghouses have a total capacity of 70 Nm3/sec.,
twice that of 1973. The total losses in the atmosphere through the
152 m stack are less than 3 kg Pb/h, 50 % of the 1974 figures.

4.3 Transformation of flue-dusts into aggregates or stable cakes

The particular physical composition of recovered dusts,
sometimes even incandescent, has led to the choice of a method of
direct pulp formation.

Dust collected from electrostatic precipitators and baghouses
are transferred in closed individual conveying systems, such as
pneumatic transport or scraper haulage.

Both systems deliver the fine dust in bins, filled with water
or filtrate and provided with stirring arms producing sludge of
about 300 g dry material per liter. This sludge is pumped into
several variable chamber filters where a final pressure of 15 bars
produces a hard cake 1,5 cm thick with a water content of 18/20 %.

The cake may be transported and stockpiled in the open air
without further dust problems. This method can be used even for the
most difficult dusts and by-products.

Hoboken now handles about 60 tonnes dust a day from 14 different places spread over the plant. Four filters are located in a convenient place of the plant.

Other dusts which are collected in batches can be sucked by an especially built giant aspirator carried on a truck and having a capacity of 6 tons. The truck is driven near the closed bins and can transfer its dust content into one of the bins, leading to the variable filters.

Most of the sludgeforming liquid is obtained from filtrate from the filters, allowing a closed water circuit of great importance for waste water treatment in the plant.

5. Locating sources of emission

Future improvements and the Investment Programme which is to last until 1983, will depend on the interpretation of results obtained from two systems of measurement, quite separate of each other. The first system measures fall-out, or the amount of lead in dust which settles on a given surface. The second consists of immission measurements within the plant, which, with proper interpretation, provides information on the location and the importance of the sources of emission.

5.1 Fall-out

We have seen that the lead content in grass has been considerably reduced. However, since such content is seriously influenced by the rate of growth of grass, it does not provide a direct measurement of lead-dust fall-out.

For this reason, a system of some 50 dust collectors, located both inside and outside the plant, was installed in 1978.

Each collector, located in an open area at about 1.7 m above groundlevel, contains 3 liters of water. It receives all the dust which settles on it. The water and dust content are removed once a month and chemically prepared for analysis by atomic absorption.

In Hoboken, the method used corresponds to the Belgian norm NBN T94-101. A corresponding international ISO standard is under preparation.

It has been observed during about 12 months that there is very little direct fall-out beyond 100 to 200 m from the source of dust. Knowing this, it may be deduced that lead dust which has been found in the populated area adjacent to the north wall of the plant, must originate from the cumulated effect of the Harris lead refining operations and wind turbulance around high buildings. On the other hand, the storage areas are located too far away from the limits of the plant to have any significant direct influence on fall-out outside the plant.

5.2 Internal Immission Measurements

To determine with greater precision the origins of lead in the air, a system of 5 high-volume samplers was placed within the plant area. The samplers are located at the points which should logically show the highest concentration of lead in the air, taking due account of prevailing winds (S - S.W.), height above ground at which emissions occur, influence of surrounding buildings. Such criteria have naturally led us to place the samplers at or near the northern limits of the plant, and also N - N.E. of the major sources of dust and fumes : ore stockyards and handling facilities, smelter, sinterplant and lead refinery.

The samplers contain a filter, which is replaced every 24 hours. The amount of lead in the air is known from the quantity of dust collected and the volume of air which has gone through the filter. Values are recorded daily in μg/m3.

The interpretation of figures recorded in the samplers requires correlation with wind direction. Our wind recorder is placed in an open non-turbulent area 12 m high. Its degree of precision allows us to divide wind recordings in 16 sectors, each having an angular width of 22,5°. Wind direction is averaged over one-hour intervals, so that 24 wind readings are recorded each day.

Computations correlating wind and content of lead in the air are always made monthly and for 6 month periods, respectively october to march (winter) and april to september (summer). They are computed for other periods when there are reasons to do so : for example, when a part of the plant has not been in operation for a sufficiently long period, computations are made for that period. This provides additional information regarding the influence of that particular installation on lead emissions.

Two principal types of figures are determined for each sampling station and for wind blowing from each of the 16 wind sectors :

1°) average lead concentration in the air for wind blowing from a given sector. This provides 16 sets of figures for each month and for each of the above-mentioned 6 month periods, or any other selected period

2°) for the same periods, the effective % contribution of each of the 16 wind sectors to the total amount of lead measured.

On Figs. 7 and 8, the sampling stations are located at points A, B, C, H and G.

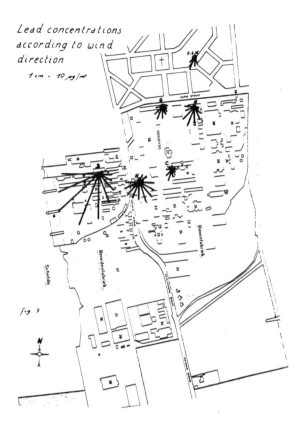

Lead concentrations
according to wind
direction

1 cm . 10 µg/m³

fig 7

Fig. 7 shows at each of these points the computed average
concentration of lead dust in µg/m3 for each windsector, when the
wind comes from that sector, during the 6 month period october 1978-
march 1979. On the original diagram, the scale is 1 cm = 10 µg/m3.
Peak figures appear for S - S.W. and S.W. winds, as could be
expected considering the location of the samplers.

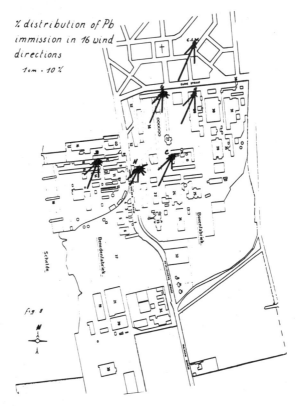

% distribution of Pb
immission in 16 wind
directions

1cm = 10 %

fig 8

Fig. 8 shows the % influence of each wind sector on the total amount of Pb dust measured during the same period. The dominant winds are those which carry the highest leadcontent in the air, the directional nature of the phenomenon is even more striking than on Fig. 7.

Both types of figures mentioned above provide very useful information in setting priorities for dealing with problem areas.

Both figures show that the highest lead in the air concentrations are found downwind from the blast furnaces area. It shows that they constitute a major source of lead in the air in the residential area north of the plant, whereas, as it was shown this was not so in the case of fall-out.

6.   Cost of environmental protection

During the 5 financial years up to september 1978, the value of investments carried out at Hoboken for protecting the environment may be summarized as follows (in million US $, at the exchange rate of 30 BF = 1 US $) :

1973/74	1974/75	1975/76	1976/77	1977/78	Total
1.7	1.8	1.4	1.5	4.3	10.7

Every year the amount was greater than US $ 1,5 million equivalent, with a striking increase to more than $ 4 million in 1977/78. The total for the 5-year period was more than 30 % of total investments at the Hoboken plant. These amounts cover new investment made for the purpose of protecting the environment and, to a lesser extent, expenditure on existing equipment to ensure cleaner operation.

We have also reviewed operating costs to determine to what extent they have been increased by anti-pollution measures. The additional cost, excluding the price of raw materials and equipment depreciation, was 7,5 % in 1977/78.

7. Conclusions

To meet the 1983 goal of yearly mean maximum 2 μg/m3 lead in the air in its close vicinity, the Hoboken plant is moving ahead with an increased program of investment, in parallel with continued training of people, improved monitoring and accelerated response to any occasional malfunction.

The unknown are the weather conditions. If, as in the majority of recent years, winter season winds from the critical SSE to SW directions occur only about 45 % of the time, success is very probable.

However, probability of success decreases as the frequency of these winds increases, particularly should they exceed 60 % of total time, a figure almost reached in 1976.

The Management of M.H.O. wishes to substantiate further its contribution to the fight against pollution, by expressing its willingness to discuss the matter with other companies.

DEMONSTRATION OF FUGITIVE EMISSION CONTROLS

AT A SECONDARY LEAD SMELTER

Richard T. Coleman Jr., P.E.
Texas Metals
Austin, Texas 78756

Robert Vandervort, P.E.
Radian Corporation
Salt Lake City, Utah 84115

Test results from sampling studies performed at a secondary lead smelter are reported. The studies were performed before fugitive emission controls were installed. Some preliminary tests of the installed control system have been completed. Additional work is scheduled when the control installation is complete.

In-duct air flows and hood capture characteristics were measured for both the new and old ventilation systems. Order-of-magnitude increases in some hood face velocities were achieved. A list of engineering controls installed is presented. Work practice changes and other suggested improvements are also discussed.

Stack sampling tests were conducted to determine lead, antimony, sulfur, and chlorine emissions. Personal breathing zone and in-plant area samples were collected and analyzed for lead, antimony, and sulfuric acid. High volume ambient air monitoring for total particulate and lead completed the sampling effort. Blood lead data are also presented.

Introduction

The Environmental Protection Agency (EPA) and the National Institute for Occupational Safety and Heatlh (NIOSH) are engaged in a demonstration of the Bergsøe agglomeration furnace and best management practices. The demonstration project is a test of these techniques for controlling fugitive and workplace lead emissions. The purpose of this study is to gather data regarding the reduction of both employee exposure to lead and fugitive lead emissions to the ambient environment which can be accomplished by retrofitting controls to an existing secondary lead smelter.

Like almost all secondary lead smelting facilities in the United States, the test smelter has had difficulty controlling both fugitive emissions and the resultant employee exposures to lead. The company is currently in the process of implementing a major exposure abatement plan which involves improved process enclosure and local exhaust ventilation, yard paving, modified materials handling, and agglomeration of flue dust.

The demonstration of the agglomeration furnace and best management practices involves before and after characterizations of the smelter. Additionally, the demonstration study provides for limited technical assistance during the final design and installation of engineering controls. This paper contains the data developed in the "before" study and through subsequent contacts with the smelter.

Background

Flue dust handling has historically been a serious problem to secondary lead smelters in the United States and other parts of the world. Flue dusts generated by secondary lead blast and reverberatory furnaces contain appreciable amounts of lead. The collection, handling, storage, and reintroduction of this dust to the smelting process involves opportunities for the dust to enter the ambient, as well as, the workplace environment. The Bergsøe flash agglomeration furnace is a means of helping to control emissions from the handling of flue dust beyond the point of its collection in a baghouse. The agglomeration furnace melts the flue dust into a slag-like material which can be handled in bulk form and reintroduced to the process without generating an appreciable amount of dust.

Flue dust handling, however, is not the only source of employee exposure at a secondary smelter. Emissions from furnaces and other smelting equipment also can contribute to workplace and ambient pollution if they are not properly controlled. The test smelter is the first U.S. secondary lead smelting firm to utilize the Bergsøe agglomeration furnace to help control fugitive dust. This smelter is also actively engaged in the installation of exhaust ventilation and improved management practices which should provide substantial reduction in fugitive emissions and workplace exposures. The amount of improvement will be gauged by the work performed during this study and also by comparing the new conditions with those measured by the company during a two year period preceding this study.

Smelter Description

Equipment and Controls

At the test smelter, scrap batteries are broken using slow-moving shears. Battery plates, mud, plant scrap from an adjacent battery manufacturing facility, drosses, and a variety of other lead scrap are fed to a vertical shaft blast furnace. This blast furnace is charged using a skip hoist and traditional slag-tapping and metal-tapping equipment is employed.

The smelter employs approximately 12 people per shift. Much of the work is manual. Charging the furnace and handling of crude and refined lead in the smelting building involve simple materials handling equipment. A small Bobcat loader is utilized to handle charge materials within the smelting building.

An open-air building houses the blast furnace, refining kettles and other smelting equipment. Several of the walls are open to allow movement of fresh air in and out of the building. The floor of the building is paved but rough. It is heavily contaminated with muds formed by water and paste from battery plates, as well as by other materials tracked in from the yard surrounding the smelting building. Housekeeping, at this particular smelter, is a difficult problem due to the rough surface of the floor and the lack of appropriate drains to flush accumulated lead-containing materials.

Local exhaust ventilation is provided for various emission points associated with the blast furnace and to some extent for the refining kettles. Figure 1 presents an overview of the original local exhaust ventilation system for the smelter building. Most of the system was serviced by a 4.7 m^3/s (10,000 cfm) baghouse and fan system. The slag tapping hood is connected to a separate 2.4 m^3/s (5,000 cfm) baghouse and fan. Not included in this overview sketch is the blast furnace flue gas control system. Flue gases are ducted through knockout chambers and cooling tubes to a separate baghouse and fan duct collecting system. A new baghouse and fan system now collects exhaust gases from the local exhaust ventilation hoods. The slag tapping hood is still served by the separate baghouse and fans.

Figure 2 presents a sketch of the skip hoist ground level loading station. Local exhaust pickups are provided at each side of the hoist enclosure. As indicated in Figure 2 the skip hoist dumping station at the top of the hoist shaft is exhausted via the Skip Hoist Furnace Charging Hood.

Figure 3 presents a sketch of the local exhaust hood associated with slag tapping. As stated earlier, this hood is served by a separate dust collector and fan. This exhaust system is operated during slag tapping. The front door to the hood is raised by means of a cable and pulleys. The sides to the hood are hinged and swing away to allow access to the slag tapping port. Slag containers are moved via forklift trucks.

Figure 4 illustrates the relationship of the three local exhaust hoods which serve the molten lead tapping and molding operation. The Tapping Hood is stationary and is provided with fold up side curtains to facilitate access to the tapping port. The Launder Hood moves with the launder which is supported from beneath by a pivoting mechanism. The sides to the Launder Hood fold up to allow cleaning of the launder. The Mold Filling Hood is portable and is manually lifted into position over the mold being filled. Molds are contained in a tub filled with water. Finished ingots are removed from the molds through use of an overhead monorail hoist.

Following the initial characterization, the Smelter Building exhaust ventilation system was modified. These modifications are shown in part in Figure 5. The slag tapping hood is still served by the 2.4 m^3/s (5,000 cfm) baghouse and fan. However, the remainder of the system is now served by a new baghouse and fan system with capacity in excess of 19.8 m^3/s (42,000 cfm). An additional local exhaust ventilation hood has been

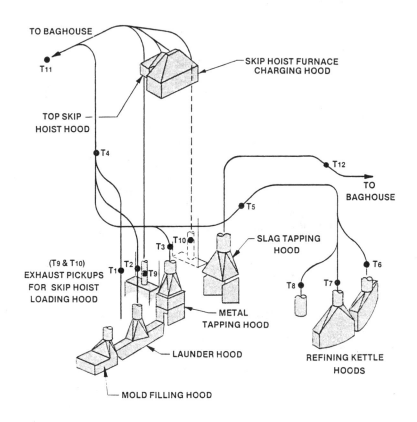

TO BAGHOUSE

T_{11}

SKIP HOIST FURNACE
CHARGING HOOD

TOP SKIP —
HOIST HOOD

T_4

T_{12}

TO
BAGHOUSE

T_5

T_{10}

T_3

SLAG TAPPING
HOOD

T_6

(T9 & T10)
EXHAUST PICKUPS
FOR SKIP HOIST
LOADING HOOD

T_1 T_2 T_9

T_8 T_7

METAL
TAPPING HOOD

LAUNDER HOOD

REFINING KETTLE
HOODS

MOLD FILLING HOOD

FIG. 1 - OVERVIEW OF ORGINAL LOCAL
EXHAUST VENTILATION SYSTEM

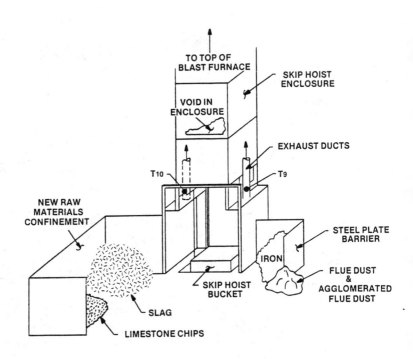

FIG. 2 · SKIP HOIST GROUND LEVEL LOADING STATION

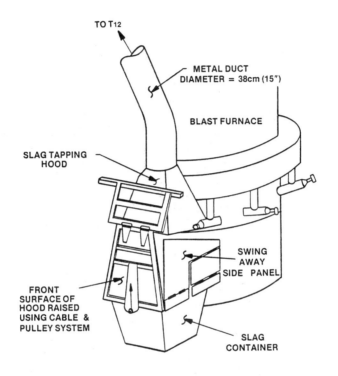

TO T12

METAL DUCT
DIAMETER = 38cm (15″)

BLAST FURNACE

SLAG TAPPING
HOOD

SWING
AWAY
SIDE PANEL

FRONT
SURFACE OF
HOOD RAISED
USING CABLE &
PULLEY SYSTEM

SLAG
CONTAINER

FIG. 3 - BLAST FURNACE SLAG TAPPING HOOD

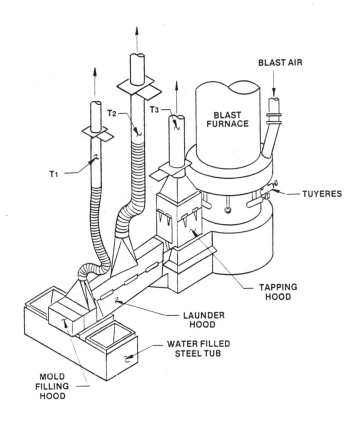

BLAST AIR

BLAST
FURNACE

T_1

T_2

T_3

TUYERES

TAPPING
HOOD

LAUNDER
HOOD

WATER FILLED
STEEL TUB

MOLD
FILLING
HOOD

FIG. 4 · LOCAL EXHAUST VENTILATION CONTROLS FOR
MOLTEN LEAD TAPPING AND MOLDING

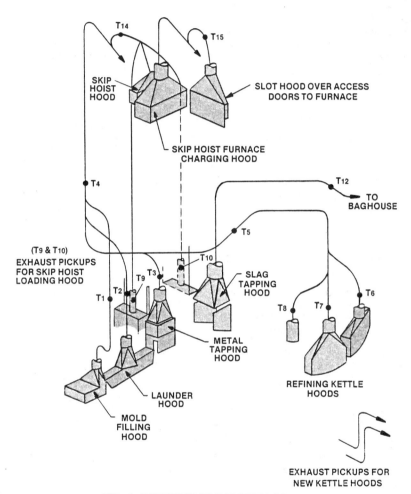

T14

T15

SKIP
HOIST
HOOD

SLOT HOOD OVER ACCESS
DOORS TO FURNACE

SKIP HOIST FURNACE
CHARGING HOOD

T4

T12

TO
BAGHOUSE

T5

(T9 & T10)
EXHAUST PICKUPS
FOR SKIP HOIST
LOADING HOOD

T10

T3

SLAG
TAPPING
HOOD

T9

T1

T2

T8

T7

T6

METAL
TAPPING
HOOD

REFINING KETTLE
HOODS

LAUNDER
HOOD

MOLD
FILLING
HOOD

EXHAUST PICKUPS FOR
NEW KETTLE HOODS

FIG. 5 - OVERVIEW OF MODIFIED LOCAL
EXHAUST VENTILATION SYSTEM

provided at the top of the blast furnace to capture fugitive emissions leaking from side access doors to the furnace. Other local exhaust hoods have not been modified.

In addition to the local exhaust hoods and ductwork shown in Figure 5, the new ventilation system also serves the flue dust agglomeration furnace and associated flue dust handling equipment. As indicated in Figure 5, ductwork has been provided to serve the new refining kettle hoods which were in the final design stage at the time of the initial visit. Eventually both refining kettles will be provided with near total enclosure. The existing backdraft hoods will be removed and refractory chimneys installed to handle kettle burner exhaust gases.

Operating Conditions

The smelter operated normally during the entire initial test period. Battery plates were the major lead-bearing feed material charged to the blast furnace. Small amounts of flue dust and dross were also being recycled to the furnace. The plate material was generated by two men operating a hydraulic battery shear.

Rubber-cased automotive (SLI) and industrial batteries comprised a majority of the batteries processed. Some polypropylene-cased batteries were also present. All battery cases were removed and discarded prior to smelting. Only fragments of the rubber or plastic cases were included in the furnace charge.

Materials Handling and Housekeeping

The smelter began a major effort to clean up yard areas and improve housekeeping after the initial visit. Much of the large flue dust storage pile located adjacent to the process control baghouse system was removed. Unpaved yard areas were being graded to begin removal of lead bearing dirt and rubble. Flue dust being collected in knockout chambers and baghouses was being fed to the agglomeration furnaces. However, since the agglomeration furnace was not completely online, some flue dust was being stockpiled. Tote boxes used to handle flue dusts collected in knockout boxes and small baghouses had been fitted with lids to prevent dispersal of lead bearing dust during transportation of the tote boxes.

During an interim visit, it was observed that the floors of the smelter building were visibly cleaner and wetter than during the initial visit. Major accumulations of dust had been removed from floor surfaces and additional physical barriers (low walls, partitions, etc.) had been provided to contain the spread of charge materials. Although simple in design, the raw materials confinement structure provided at the base of the skip hoist did appear to improve housekeeping in this work area.

The implemented improvements in housekeeping and materials handling were discussed with smelter management. Although major strides have been made to improve housekeeping, much work remains to be accomplished. Plans to improve enclosure of the smelter building and isolate raw materials handling were reviewed. A major housekeeping cleaning effort was completed during the summer shutdown which concentrated on removing accumulations of lead particulates from ceilings, walls, girders, etc.

A major problem at this smelter involves the removal of flue dust from yard areas and floor surfaces. Lead bearing flue dust is difficult to wet. It has a slippery almost greasy property which causes water to bead up and enhances its ability to adhere to surfaces. At present, large unpaved areas of the yard are coated with compacted flue dust. Much of this lead contaminated earth will be excavated and replaced with clean fill and paving. Future contamination by flue dust should be greatly minimized by improved flue dust handling methods and agglomeration of flue dust into a slag-like material. However, paved areas beneath baghouses, U-tube coolers, and knockout chambers will be periodically contaminated by spills of flue dust. These spills could result in widespread yard area contamination if not promptly and completely cleaned up. Improved methods for cleaning flue dust from concrete and asphalt surfaces must be developed. Use of wetting agents and vacuum systems will be explored.

Housekeeping within the Smelter Building will be made easier with the elimination of flue dust handling associated with furnace charging. However, lead bearing particulate from battery plates, drosses, etc. will continue to contaminate the charge materials handling area. Isolation of this area from other work areas by means of a floor to ceiling partition may help isolate this problem. Maintenance of wet floors and prohibition of dry sweeping will also help if strictly implemented. Floor surfaces within the smelter building are rough and should be smoothed to facilitate cleaning.

Flue Dust Agglomeration

At the time the interim visit was made, the agglomeration furnace had been operating for several days. Several startup problems had been encountered which mainly dealt with furnace burner adjustment and furnace temperature control. These problems were nearly solved during the interim visit. A suitable furnace operating temperature must be found once the flue dust composition and dust generation rate stabilizes.

A more difficult problem was encountered in providing a steady feed of flue dust to the furnace. After several days of operation it was determined that individual baghouse compartments released collected flue dust in highly irregular quantities following a bag shaking cycle. Since flue dust is fed directly from the baghouses to the agglomeration furnace, inconsistent feeding of the furnace was being experienced. Surges of flue dust would cause excessive charging of the furnace requiring operation at high heat to clear the furnace. These surges also created plugging at the slag tap which required almost continuous attention by an operator. Plans were being implemented to provide better control of furnace charging to eliminate these problems.

When uniformly fed, the agglomeration furnace did produce a steady stream of agglomerated material which was in appearance very similar to the material produced at the Bergsøe smelter in Glostrup, Denmark. However, the problem of emptying tote boxes of dust cleaned from the numerous dropout points in the system into the agglomeration has not been solved. This is one problem which could be avoided in a new smelter design. In this case, the retrofit of the agglomeration system did not include eliminating the dropout points. Thus, the tote box problem remains. Vacuum systems are being investigated as a means of emptying the tote boxes and charging the cleanout material to the agglomerator.

Control Evaluation

One major objective of the test work performed was to determine what contribution stack emissions make to ambient and workplace lead-in-air concentrations and what contribution fugitive emission sources make. To accomplish this, the following tests were performed:

- ventilation measurements,

- stack sampling,

- ambient air testing, and

- workplace monitoring.

In addition, the smelter provided blood lead data which had been collected over the previous three years. Similar data will be collected during a final smelter characteization once all the fugitive emission controls are installed.

The importance of these measurements to this particular smelter is evident when a plot plan of the smelter and the surrounding property is examined. As can be seen in Figure 6, the smelter actually extends beyond the company property line. This creates a significant problem when inter-preting the new EPA ambient air quality standard for lead of 1.5 μg Pb/m^3. As a result, it is of particular interest to this smelter and EPA to deter-mine what the ambient lead levels are and where they come from. The re-sults of the testing are presented below.

Ventilation Measurements

Table I presents a comparison of exhaust air flows at several loca-tions in the smelter building exhaust ventilation system. Several of the local exhaust hoods are now collecting two to three times the volume of contaminated air previously collected. Greatly increased quantities of air are being collected by the hoods serving the skip hoist and blast furnace charging operation. The slag tapping emission control system was measured to be collecting somewhat more air than during the initial visit. As indicated by measurements made in the new open ended ducts which will serve the refining kettles, approximately 4.0 m^3/s (8,500 cfm) of exhaust capacity will be available for the new refining kettle hoods.

The increased air volume being collected by each of the local exhaust hoods has increased hood face velocities. Face velocity measurements were made for several of the hoods using an Alnor® Velometer. As can be seen from the data in Table I, substantial improvement has been realized and improved control was apparent by observation of fume capture. However, the degree of emission source enclosure offered by many of these hoods could be improved. Hood modifications are planned once more major modifications are complete.

Stack Sampling

Stack sampling was performed in the exhaust duct leading to the main stack. Sampling personnel worked from a platform located approximately 6 meters from ground level. The duct was rectangular and was inclined at approximately a 30 degree angle from horizontal. The interior dimensions of the duct were 813 mm in height and 711 mm in depth.

At the platform level, the duct was equipped with three access ports on the vertical surface of the duct. These were located approximately 135 mm from the edge of the duct on 270 mm centers. The place in which the sample ports were located was approximately two equivalent diameters up-stream and eight equivalent diameters downstream of any flow disturbances. Velocity and sampling traverses were made at four points in each of these three ports, for a total of twelve points.

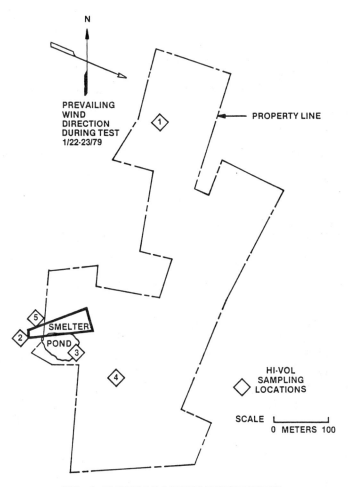

N

PREVAILING
WIND
DIRECTION
DURING TEST
1/22-23/79

PROPERTY LINE

1

5

SMELTER

2

POND

3

4

HI-VOL
SAMPLING
LOCATIONS

SCALE
0 METERS 100

FIG. 6 - PLOT PLAN OF SMELTER PROPERTY

02-4403-1

669

Table I. Comparison of Exhaust Air Flows

Flow Measurement Location	Volumetric Flow Rate, Q			
	m^3/s (1)	cfm (1)	m^3/s (2)	cfm (2)
Mold Filling Hood (T_1)	0.30	630	0.78	1650
Launder Hood (T_2)	0.39	820	1.2	2460
Metal Tapping Hood (T_3)	0.65	1380	1.7	3660
Refining Kettle Hoods (T_5)	0.70	1480	2.2	4700
Skip Hoist Hoods + Skip Hoist Furnace Charging Hood	2.8	5990	7.6	16030
Skip Hoist Furnace Charging Hood	---	----	5.7	12100
Skip Hoist Hoods (Top and Bottom) (T_{14})	---	----	1.9	3930
Slot Hood Over Furnace Access Doors (T_{15})	---	----	0.91	1940
Slag Tapping Hood (T_{12})	1.8	3850	2.1	4540
Exhaust Pickups for New Kettle Hoods	---	----	1.4 2.6	2940 5470
Discharge From Slag Tapping Baghouse	---	----	2.4	5000
Inlet To Main Exhaust Fan For Smelter Exhaust Ventilation System	---	----	20	41700

(1) Measurements made 1/23-24/79 with old exhaust system in operation. (Refer to Fig. 1.)
(2) Measurements made 6/19-20/79 with new exhaust system in operation. (Refer to Fig. 5.)

The EPA Method 5 sampling train was used to determine average duct temperature, velocity, and moisture content on January 21, 1979. These measurements were used to select the appropriate nozzle and isokinetic sampling rates for subsequent tests. Isokinetic sampling was performed at each point for the remaining tests conducted on January 22 and 23.

On January 22 one Method 5 experiment was conducted. Isokinetic sampling was conducted at each point within each port for a period of five minutes. When all the points in all the ports had been traversed, sampling was repeated. This continued for approximately six hours so that sufficient sample for analysis could be obtained.

On January 23, the experiment was repeated, however, each point was sampled over a fifteen-minute period to minimize sampling time. Sampling took approximately three and one-half hours.

Sampling was performed for SO_x emissions using the Method 5 equipment as well. One thirty-minute sample was obtained at the center of the duct. Proportional sampling was conducted to obtain a 0.217 Nm^3 sample.

Table II presents the results of this experiment for both test days. It is important to note that the average grain loadings for both days were nearly identical. This indicates that the baghouse was operating normally and was allowing a constant amount of fine particulate to pass through with the gas streams.

Sulfur and chlorine compounds were collected in a series of three impingers which were placed behind the heated filter in the Method 5 sampling train. Two impingers in series containing 1 percent NaOH followed by an impinger containing 6 percent H_2O_2 were used for the experiment. A Dionex® ion chromatograph was used for the sulfur determinations. A colorimetric determination using mercuric thiocyanate was used for the chlorine determinations.

An analysis of the matter collected as particulate and vapor was made for lead, antimony, sulfur, and chlorine. The results are presented in Table III. These results indicate that both particulate and vaporous species are being emitted from the baghouse.

Lead is emitted mostly as particulate matter (99.99 percent). Antimony, however, is present mostly as a vapor (57 to 66 percent). It is likely that lead oxide (PbO), lead chloride ($PbCl_2$), and antimony pentachloride ($SbCl_5$) are the compounds being emitted. The boiling point of $SbCl_5$, 79°C, would account for the antimony being at least partially present as vapor.

Chlorine emissions were significantly higher on 1/23/79 than on 1/22/79. This might be due to a larger recycle of flue dust from the knockout chambers or because of a larger percentage of polyvinyl chloride separators in the feed materials that day.

Sulfur was emitted mostly as sulfur dioxide (SO_2). Some sulfur was detected in the collected particulate matter, perhaps present as calcium sulfate ($CaSO_4$). This compound is formed by the reaction of SO_3 or sulfuric acid mist with the lime injected into the flue gas. The sulfur emission rates of 33500 and 30400 g/hr were checked on 1/23/79 using the EPA Method 5 equipment with isopropyl alcohol in one impinger and hydrogen peroxide in another. The emission rate using this technique was 29600 gS/hr which agrees closely with the Method 5 results.

Table II. EPA Method 5 Sampling Results

	Average velocity determination	Particulate, chlorine, and sulfur determinations	
Date	21 Jan 79	22 Jan 79	23 Jan 79
Time	1744-1803	1313-1905	1012-1344
Sample duration	19 min	3 hr 40 min	3 hr
Sample volume at meter	0.286 m^3	3.64 m^3	2.89 m^3
Avg. meter temperature	18°C	19°C	19°C
Meter pressure	747 mm Hg	773 mm Hg	779 mm Hg
Avg. stack temperature	70°C	83°C	85°C
Stack pressure	746 mm Hg	772 mm Hg	778 mm Hg
Avg. gas velocity	21.45 m/s	20.56 m/s	20.33 m/s
Total gas flow	$3.36 \times 10^4 \text{ Nm}^3/\text{hr}$	$3.22 \times 10^4 \text{ Nm}^3/\text{hr}$	$3.18 \times 10^4 \text{ Nm}^3/\text{hr}$
Moisture content of gas (Volume %)	3.8	3.5	3.6
Gas molecular weight	28.63	28.6	28.8
Mass collected, filter		43.14 mg	35.48 mg
Mass collected, probe		1.16 mg	0.49 mg
Mass collected, total		44.30 mg	35.97 mg
Nozzle diameter		0.442 cm	0.442 cm
Nozzle area		$1.53 \times 10^{-5} \text{ m}^2$	$1.53 \times 10^{-5} \text{ m}^2$
Stack dimensions		8.13 mm x 711 mm	8.13 mm x 711 mm
Stack area		0.578 m^2	0.578 m^2
Avg. sample velocity at nozzle		22.7 m/s	22.2 m/s
Total particulate concentration		12.8 mg/Nm^3	13.0 mg/Nm^3
Total particulate mass rate by area		0.456 kg/hr	0.453 kg/hr
Total particulate mass rate by concentration		0.412 kg/hr	0.414 kg/hr
Total particulate mass rate		0.434 kg/hr	0.434 kg/hr
Percent isokinetic		110.7%	109.4%

Table III. EPA Method 5 Analytical Results

Element/Date	Stack gas flow (10⁴ Nm³/hr)	Particulate (filter) (g/hr)	(% of total filter catch)	Particulate (probe wash + cyclone) (g/hr)	(% of total probe wash + cyclone)	Condensibles* (g/hr)
Lead (as Pb)						
1/22/79	3.22	10.6	2.6	1.15	10.6	0.05
1/23/79	3.18	12.3	3.0	4.65	82.6	0.17
Antimony (as Sb)						
1/22/79	3.22	0.06	0.01	0.006	0.06	0.13
1/23/79	3.18	0.09	0.02	0.025	0.45	0.15
Chlorine (as Cl)						
1/22/79	3.22	0.62	0.15	1.23	†	4.6
1/23/79	3.18	1.72	0.42	12.64	†	4.7
Sulfur (as S)						
1/22/79	3.22	13.5	3.4	3.93	36.4	33500.
1/23/79	3.18	13.2	3.2	0.46	8.21	30400.
Total particulate (by concentration)†						
1/22/79	3.22	402.	100.0	10.8	100.0	unknown
1/23/79	3.18	408.	100.0	5.6	100.0	unknown

*Material collected in impinger solutions.

†Some HCl is believed to have evaporated when probe and cyclone wash sample was evaporated.

Ambient Air Testing

Four high volume air samples were operating during the sampling study. Figure 6 shows the location of these four hi-vols (locations #5 through #4) and one other location measured by smelter personnel (location #5). Location #4 was also a smelter hi-vol. Two samples were collected from locations #1, #3, and #4 and #5 were provided by the smelter. Table IV presents the hi-vol test results.

The weather on January 22, 1979, was cold, clear, and windy. Winds from the W-NW ranged from 9 to 16 meters per second (20 to 35 miles/hr) during January 22. On January 23, 1979, it was clear, dry, and relatively calm. The ground was snow-covered during the entire test period, however, the snow did not cover the flue dust pile behind the smelter.

The results presented in Table IV are not conclusive, but they do show some trends. For example, location #3 shows a significantly higher lead-in-air concentration on 1/22/79, a windy day, than on 1/23/79, a calm day. The calm conditions apparently reduced fugitive windblown dust from the smelter yard. Also, lead-in-air levels at location #4 remained fairly constant while location #3 varied by an order of magnitude. This indicates that the fugitive lead particulate from the smelter yard may be settling within a relatively short distance (location #3) while some constant background level, perhaps generated by the stack, accounts for conditions at location #4. The windy conditions at 1/22 could thus cause an order-of-magnitude increase in lead concentrations at location #3 without seriously affecting location #4. Certainly more data are required before these trends can be substantiated.

Two Gaussian dispersion computer models were used to estimate the dispersion of particulates from the lead smelter. These two models are the Climatological Gaussian plume assumption. These models assume that the pollutant concentration in a smokestack plume advects downwind with the ambient air while diffusing in both the horizontal and vertical directions. The pollutant distributions in the plume are assumed to obey Gaussian, or normal, distributions.

Table IV. Hi-Vol Sampling Results

Location	Date	Total particulate collected (μg)	Total lead collected (μg)	Ambient lead concentration (μg Pb/m^3)
1	1/22/79	54,000	1,750	0.88
	1/23/79	83,300	2,220	1.3
2	1/22/79	101,000	6,740	3.0
3	1/22/79	106,000	29,500	16.
	1/23/79	72,800	2,830	1.6
4	1/22/79	–	–	9.3
	1/23/79	–	–	7.4
	8/24/78	–	–	1.1
	8/30/78	–	–	0.8
	9/1/78	–	–	1.0
	9/5/78	–	–	3.3
5	5/22/78	–	–	370.

The emission data used in the two models are as follows:

emission rate	0.126 g/sec
stack height	16.8 m
stack diameter	8.6 m
stack gas velocity	20.5 m/sec
stack gas temperature	84°C

Only total suspended particulate (TSP) was modelled. No attempt was made to separate the lead fraction of TSP from the total emission.

One significant assumption involved with Gaussian models is no loss (deposition) of mass from the plume. No pollutant settling or depositional losses are allowed from the plume. This is a conservative assumption for this analysis. The lead fraction of the particulate matter emitted by the smelter is dense and may settle rapidly. The result is that ambient concentrations predicted by the models will be higher than actual concentrations since no settling is allowed in the models.

The CDM model predicts annual or seasonal concentrations due to pollutant emissions. It uses meterological information in the form of a joint frequency distribution of wind speed, wind direction, and atmospheric stability. For this analysis, ten years of meterological observations at a nearby local airport were averaged together to yield the joint frequency distribution.

The CRSTER model predicts both short term and seasonal or annual concentrations based upon actual observed meterological data. The CRSTER model calculates the hourly concentration at a grid of receptors for each hour of a year's worth of data. It then averages these concentrations together to yield average concentrations for averaging times of up to a year in length. Local meterological data for 1964 were used and the model averaged concentrations over the winter months January, February, and December to yield a seasonal average concentration.

Both the short term 24 hour average concentrations predicted by CRSTER and the seasonal average concentrations predicted by CDM are very small. Using a 200m by 200m grid, the highest concentration predicted by CDM for the winter season is 0.5 $\mu g/m^3$, 200m east of the smelter (Figure 7). This is one half the concentration value used for PSD analysis to determine area of impact of a source. All other concentrations produced by the CDM model fall off rapidly with distance from the smelter and are at or below 0.1 $\mu g/m^3$ 600 m away from the smelter.

The 24 hour maximum concentration predicted by CRSTER is 0.9 $\mu g/m^3$ and is located 600m west-southwest of the smelter. This is well below the 5 $\mu g/m^3$ level of significance used in PSD analysis for TSP.

Based upon these dispersion modelling results the conclusion can be drawn that particulate emissions from the test smelter stack have a negligible impact on ambient particulate concentrations in the vicinity (<1 km) of the smelter.

Workplace Lead-in-Air Contamination

The test smelter has monitored workplace lead-in-air contamination for a number of years. To assist this demonstration study, monitoring data covering approximately the last two years were submitted for review and

675

.007	.008	.008	.009	.009	.025	.043	.039	.036	.065	.057
.008	.009	.010	.010	.011	.031	.053	.048	.085	.074	.065
.016	.010	.012	.013	.015	.042	.069	.061	.100	.085	.085
.017	.021	.026	.017	.020	.061	.096	.152	.144	.113	.092
.019	.023	.029	.040	.032	.114	.296	.229	.162	.124	.100
.038	.048	.063	.092	.170		.528	.268	.178	.131	.103
.082	.101	.129	.179	.204	.141	.183	.172	.121	.092	.073
.076	.092	.116	.109	.056	.077	.078	.092	.107	.083	.067
.071	.064	.075	.037	.041	.054	.055	.049	.060	.051	.062
.050	.056	.064	.030	.033	.041	.043	.039	.051	.044	.038
.044	.050	.023	.025	.027	.033	.034	.032	.029	.038	.034

* stack |———— 1 kilometer ————|

Fig. 7.   Modelled map of TSP concentration (CDM model, no deposition, winter conditions) around test smelter stack in μg TSP/m^3.

evaluation. The results of this monitoring are summarized in Table V. These data indicate that severe lead-in-air concentrations have accompanied operation of this smelting complex.

As part of the "before" study of this smelter a limited amount of breathing zone and work area air sampling was conducted. The purpose of this sampling was to 1) corroborate the sampling performed previously by the smelter, and 2) define employee exposures during a known set of smelter operating and ambient weather conditions.

Study resource limitations confined air sampling to two consecutive operating days. The selection of an optimum time for sampling was constrained by the fact that the smelter was actively engaged in a major modification program. To obtain air contaminant information before the agglomeration furnace and other controls were operational required sampling at the site during midwinter.

Breathing zone and work area sampling for lead and antimony were performed on January 23 and 24, 1979. January 23, 1979 was a relatively normal operating day at the smelting complex. The weather was sunny and cold with light winds blowing from several compass directions. Employees at the smelter performed normal tasks and rotated jobs at the halfway point in the work shift as is indicated by the change in employee initials associated with each of the jobs or operations shown in Table VI. January 24, 1979 was a relatively unusual day at the smelter complex. Heavy rains and strong winds were present throughout the period monitored. The strong gusty winds caused vigorous drafts through the smelting building which disturbed the capture characteristics of local exhaust ventilation hoods and resulted in the introduction of dust into the atmosphere from building surfaces. By the afternoon of January 24, large volumes of runoff water were entering the smelting complex and leaving on the downhill side. The smelting building itself was very wet with pools of water in most of the work areas. Several of the employees did not show up for work on this day due to flooded roads in the countryside. The sampling conducted on these two days indicates that exposures were generally higher on January 24, 1979. This result is possibly due to the strong air currents which entrained contaminants from the capture zone of the local exhaust ventilation hoods. As stated earlier, the smelting work area was very wet, however, the production of fume from slag tapping, metal tapping and refining operations was still present. On both days, the amount of antimony in the air was very low by comparison to the current permissible exposure limit of 0.5 mg/m^3 (29 CFR 1910.1000).

Several employees were found to have very high exposures (>500 μg Pb/ m^3) during portions of work shifts on each day sampled. Battery breakers on the first day sampled showed high exposures to lead and the hoist operator encountered very high lead exposure during the afternoon of the second day sampled. It is possible that relatively large particles of lead entered the filter monitoring cassette during the sampling period in each of these operations. In the case of battery breaking, a splash of lead laden solution may have caused excessive contamination. In the case of the hoist operator, the handling of flue dust as it is charged to the hoist could have caused the deposition of a lead particle onto the filter. However, in each of these cases, the measured exposure may not have been adversely influenced by artifact particulate and may represent the actual lead exposure.

The general conditions described at this smelter are similar to many other secondary smelters. The employee exposures which result from relatively unsophisticated materials handling procedures and marginally

677

Table V. Historic Employee Exposures to Lead-in-Air (mg/m^3)

Sampling date	Hoist/coke	Payloader/furnace	Clean-up	Lead pot	Furnace	Payloader	Hoist	Coke	Foreman	OMC loader	n	Mean
9-11-78	0.30	0.51										
9-11-78	0.28	0.38									4	0.37
8-17-78	0.77	0.31										
8-17-78	0.22	0.18	0.10								5	0.32
7-18-78	0.31	0.35										
7-18-78	0.41	0.30									4	0.34
6-21-78	0.47	0.35		0.19								
6-21-78		0.17									4	0.30
5-12-78		0.91			0.53		1.10				3	0.85
4-26-78	1.90	1.85										
4-26-78		2.55									3	2.10
3-21-78	0.58	0.78					0.88					
3-21-78		1.02									4	0.82
2-22-78						0.58	0.89	0.85			3	0.77
1-04-78					0.73	1.04	0.09	0.13			4	0.50
12-16-77						0.30	0.38		0.42		3	0.37
12-26-77					0.42						1	0.42
11-16-77		0.85					0.82					
11-16-77		0.39									3	0.69

678

Table V. (Continued)

Sampling date	Hoist/ coke	Payloader/ furnace	Clean-up	Lead pot	Furnace	Payloader	Hoist	Coke	Foreman	OMC loader	n	Mean
					Function or position							
10-14-77		0.8					1.0				3	0.9
10-14-77		0.8										
9-15-77					0.13	0.43	0.37				3	0.31
8-12-77					1.46	0.35	1.00	1.49			4	1.08
8-30-77										0.19	1	0.19
8-31-77										0.29	1	0.29
7-27-77					0.29	0.36	0.18	0.55			5	0.39
7-27-77						0.55						
6-20-77		0.63					0.64				3	0.57
6-20-77		0.45										
5-23-77					0.48	0.46	0.52				3	0.49
4-20-77					0.64	0.90	1.12	0.41			4	0.77
2-23-77			1.49	0.56	0.61	0.33	0.79	2.04			6	0.97
n	9	19	2	2	9	10	14	6	1	2		
Mean	0.58	0.71	0.80	0.38	0.59	0.53	0.70	0.91	0.42	0.24		

Table VI. Results of Breathing Zone Sampling for Lead and Antimony

Job/Operation	1-23-79			1-24-79		
	Sampling interval	Pb* mg/m^3	Sb** mg/m^3	Sampling interval	Pb* mg/m^3	Sb** mg/m^3
Furnace						
(LW)/(DH)	0758-1232	0.51	0.051	0803-1212	0.62	<0.030
(NH)/(LW)	1300-1547	0.38	<0.044	1234-1559	0.87	0.047
Noise						
(DH)/(LW)	0803-1205	0.37	<0.031	0801-1233	0.87	<0.027
(RH)/(JM)	1300-1547	0.20	<0.044	1238-1558	7.8	<0.037
Payloader						
(NH)/(RH)	0807-1300	0.78	<0.025	0807-1300	1.2	<0.025
(LW)/(NH)	1258-1547	0.48	<0.044	1343-1557	0.83	<0.025
Coke						
(RH)/(NH)	0828-1233	0.18	<0.030	0818-1342	0.66	<0.023
(DH)/-	1205-1547	0.25	<0.033	-	-	-
Lead Pot						
(WE)/-	0807-1206	0.21	<0.031	-	-	-
(WE)/-	1206-1425	0.37	<0.064	-	-	-
Foreman						
(CS)/(CS)	0825-1207	0.27	<0.033	0821-1205	0.66	<0.033
(CS)/(DH)	1207-1547	0.28	0.037	1214-1603	1.2	<0.032
Industrial Battery Breaker						
(JM)/(JM)	0805-1300	0.43	<0.025	0815-1237	0.73	<0.028
(JM)/-	1300-1615	0.35	<0.029	-	-	-
Yardmen						
(DS)/-	0817-1235	0.69	<0.029			
(DS)/-	1235-1615	0.25	<0.034			
(RM)/-	0818-1236	0.37	<0.029			
(RM)/-	1236-1615	0.23	<0.034			
Battery Breaking						
(KA)/(KA)	0833-1209	2.5	<0.034	0823-1200	0.26	<0.034
(WD)/-	0826-1211	3.5	<0.029	-	-	-
(KA)/-	1210-1515	0.18	<0.040	-	-	-
- /(WD)	-	-	-	1203-1520	0.20	<0.037

* Time-weighted-average lead-in-air concentration for the period samples.

**Time-weighted-average antimony-in-air concentration for the period sampled. OSHA permissible exposure limit 0.5 mg/m^3 (29 CFR 1910.1000).

controlled emission sources are high by comparison to the recommended limits for employee exposure. Exposures monitored during the initial part of this survey confirm smelter sampling results and indicate that the smelter does operate with workroom lead-in-air concentrations well above, or many times, currently accepted exposure limits. There is no apparent reason to discount the values measured by the company and they are assumed to characterize exposures (without regard to the use of respirators) during the period monitored.

The work area sampling results shown in Table VII indicate the same trend from the first to second day of sampling. Sampling results of the second day are generally higher than those of the first. Work area concentrations tend to be much lower than breathing zone concentrations and reflect the fact that employees are much closer to the sources of emission within the smelting building than were the stationary area samples.

Area sampling results do indicate that the employee break room or lunchroom is significantly contaminated with airborne lead. This condition offers the distinct possibility for employees to eat food and use smoking and chewing materials which are contaminated by lead. The contamination of the lunchroom area may result from lead from an industrial battery department which may give rise to infiltration of lead contaminated air. A more obvious source of contamination is the traffic of employees dressed in work clothing to and from the lunchroom. Lead particulate lodged on clothing, hardhats, etc., may be dislodged while these materials are removed or during normal movement within the lunchroom.

Table VIII indicates the results of air sampling for sulfuric acid mist which was performed in the battery breaking operation. Exposures were measured to be well below the permissible exposure limit for sulfuric acid mist of 1.0 mg/m (29 CFR 1910.1000). Observation of work performed in the battery area indicated the very distinct possibility of employee eye and skin contact with battery electrolyte (sulfuric acid).

Biological Monitoring Data

The development of biologic monitoring data was outside the scope of this demonstration study. However, the test smelter did supply blood lead monitoring data covering a period of approximately two years directly preceding this study. Tables IX and X contain summaries of these data.

The majority of blood lead concentrations fall in the range of 0.06 to 0.07 or 60 to 70 μg of lead per 100 grams of whole blood. This is true for both employees working in the smelter building and in the battery breaking area. A significant number of measured blood lead concentrations have indicated even more serious lead absorption.

The elevated blood lead concentrations measured by the company indicate that existing employee exposures to lead are significant despite use of respiratory protection devices. There are many possible sources of insufficiently controlled lead exposure. The contaminated lunchroom facility and improper use of respirators by employees are perhaps the major sources of exposure. Improved hygiene facilities and practices would be expected to help reduce continued excessive contact with lead.

Table VII. Results of Work Area Monitoring for Lead and Antimony

	1-23-79			1-24-79		
	Sampling interval	Pb* mg/m³	Sb** mg/m³	Sampling interval	mg/m³	mg/m³
Employee Breakroom/ Lunchroom (Refer to Figure 6)	0838-1214	0.15	<0.034	0739-1145	0.11	0.030
	1215-1559	0.17	<0.033	-	-	-
Location No. 1 (Refer to Figure 1)	0852-1219	0.050	<0.034	0745-1144	0.37	0.031
Location No. 2 (Refer to Figure 1)	0854-1222	0.089	<0.036	0755-1142	0.33	0.033
Location No. 3 (Refer to Figure 1)	0910-1224	0.18	<0.038	0749-1144	0.25	0.031
Location No. 4 (Refer to Figure 1)	0901-1222	0.12	<0.037	0758-1158	0.30	0.030

* Time-weighted-average lead-in-air concentration for the period sampled.

**Time-weighted-average antimony-in-air concentration for the period sampled.
   OSHA permissible exposure limit 0.5 mg/m³ (29 CFR 1910.1000)

Table VIII. Results of Breathing Zone Sampling for Sulfuric Acid Mist

Job/Operation	1-23-79		1-24-79	
	Sampling interval	H_2SO_4* mg/m^3	Sampling interval	H_2SO_4* mg/m^3
Battery Breaking				
--/(WD)	–	–	0821-1200	0.18
--/(KA)	–	–	1203-1520	0.067
(WD)/-	1211-1515	0.10		

*Time-weighted-average sulfuric acid mist-in-air concentration for the period sampled. OSHA permissible exposure limit 1.0 mg/m^3 (29 CFR 1910.1000)

Control Critique and Recommendations

Planned Improvements

The test smelter has recognized the serious lead exposure problem associated with their secondary lead smelting complex. The company is in the process of making major modifications to its smelting complex to better control emissions to the ambient environment and workplace exposure of employees. An important element in this modernization program is the Bergsøe flash agglomeration furnace. The furnace will work in concert with a new baghouse facility. The new baghouse system is of much greater capacity than the system which has served the smelter historically. The new baghouse exhaust ventilation system will provide the capability to exhaust between 68,000 and 93,500 m^3/hr of air from the smelter building. This exhausted air volume will be gathered from strategically located exhaust ventilation hoods which will serve major sources of emission associated with the smelting facility. The following list of major emission sources will be served by the new ventilation system:

1) the charging location at the top of the blast furnace,

2) the skip hoist charging elevator,

3) the lead tapping and pouring station.

4) the refining kettles, and

5) the flue dust tote box dumping station associated with the agglomeration furnace feed.

The existing slag tapping exhaust ventilation system will not be modified since it is quite effective in capturing emissions. It will still be serviced by a separate 8640 m^3/hr (5000 acfm) baghouse.

The final hood designs for local exhaust ventilation of these emission sources are in the development stage at present. The first step in implementing the new exhaust ventilation control program will be to attach existing exhaust ventilation hood structures to the new system. The next stage in implementation of the total control program will be to

Table IX. Blood Lead Concentrations – Smelter Employees

Sampling date	Number of persons with blood lead concentrations* of:										n	mean*
	.03	.04	.05	.06	.07	.08	.09	.10	.11	.12		
12-04-78		1	5	8	5	1					20	.06
10-30-78			5	7	3	3	1				19	.06
10-02-78		1	3	3	4	1					12	.06
09-05-78				3	9						12	.07
06-31-78				2	6	4	1				13	.07
06-27-78			1	1	3						5	.06
05-30-78			3	6	12	2					23	.07
04-24-78			1		2						3	.06
04-04-78			1	7	10	1					19	.07
02-07-78			4	10	5	5					24	.06
01-03-78				1	2	1					4	.07
12-06-77			4	8	7	1					20	.06
11-07-77			2		2	3					7	.07
10-04-77		1		11	3	2	1				18	.06
09-06-77			1		2	4		1			8	.08
08-01-77			6	6	3	4	2				21	.06
06-29-77				5		1	1				7	.07
05-31-77			2	6	7	7					22	.07
05-03-77				2	2	3					7	.07
03-29-77			4	7	6	2		1			20	.06
02-01-77			4	4	4	3	1	2			18	.07
01-04-77					2		2	1			5	.08
12-07-76	2		2	7	6	3		1		1	22	.07
n	2	3	48	104	105	51	9	6		1		
% of Total	0.6	0.9	14.5	31.6	31.9	15.5	2.7	1.8		0.3		

*Milligrams lead per 100 grams whole blood

Table X.   Blood Lead Concentrations - Battery Breakers

Sampling date	Number of persons with blood lead concentrations* of:						n	mean*
	.03	.04	.05	.06	.07	.08		
10-30-78		1	1	1	2		5	.06
09-05-78			1	2	1		4	.06
05-30-78	1		1	1	2		5	.06
03-13-78						4	4	.08
02-07-78			1	1	1		3	.06
12-06-77		1			2		3	.06
10-04-77		2			2	1	5	.06
09-06-77					1		1	.07
08-01-77			1				1	.05
06-29-77		1		1			2	.05
05-31-77				2	1		3	.06
12-07-76		1	1	1			3	.05
n		7	6	9	12	5		
% of Total		17.9	15.4	23.1	30.8	12.8		

*Milligrams lead per 100 grams whole blood.

complete the installation of the flash agglomeration furnace. Once the flash agglomeration furnace is operating in concert with the dust collecting system, attention will be given to the final design of local exhaust ventilation hoods for refining kettles, etc. Completion of this control program will consume several months. The data collected in this "before" study describe employee exposures and engineering controls at a point in time when the plant was being serviced by its historical exhaust ventilation system and local exhaust hoods located within the smelting facility.

Minimizing employee exposure and fugitive emissions will require more than simply increasing the amount of air exhausted from the smelting building and installing better local exhaust hoods. The smelter has plans to improve paving of outside yard surfaces; to enclose portions of the smelting building (cutting down cross drafts which would interfere with contaminant capture); and to make materials handling modifications which should help to control escape of contaminants into the work environment.

Recommended Control Considerations

Following the field work associated with this "before" study, a list of recommendations was developed to assist the test smelter in controlling their lead exposure problem. These recommendations were made within the framework of retrofitting controls to existing process equipment. This constraint ruled out some more radical changes that could be incorporated in a new plant construction. Examples of changes which cannot be accomplished at this time are:

- complete removal of existing smelter building floors and replacement with elaborate washdown and water collection sumps and drains, and

- complete separation of all raw materials handling from the furnace operating and refining areas of the smelter.

The recommendations which have been forwarded to the test smelter are described below. They are organized by smelting operation and other industrial hygienic control considerations. During the interim, while engineering controls are being designed and installed, it is imperative that employees be afforded maximum protection from exposure through use of respiratory protective devices and rigorous personal hygiene. Several recommendations are included which pertain to respiratory protection, hygiene facilities and practices, employee training, etc.

Battery Breaking.

- A suitable eye fountain and emergency shower station should be provided in close proximity to battery shearing work station.

- Lead mud deposits should be cleaned from all surfaces in the battery breaking area (walls, roller conveyor, shear, floor, etc.)

- Plexiglas or other transparent enclosures should be installed at the shear to prevent splashing of battery acid and mud onto employees.

- A means of periodically washing down the battery breaking work area to remove accumulations of caked-on mud, etc. should be provided.

- Depending on the reduction in employee exposures afforded by the above recommendations, an exhaust ventilation system should be designed and installed to serve the battery shears.

- Employees should be instructed to clean accumulations of battery mud from their protective aprons at periodic intervals. Water and a sponge should remove most of this material and prevent it from becoming dry and entering the air in the breathing zone of the worker.

Skip Hoist Charging and Materials Handling.

- Sweeping and shoveling of dry lead bearing materials should be prohibited. Shoveling of wetted materials or vacuuming of dry materials should be substituted.

- Improved materials handling should be instituted in association with skip hoist charging.

  a) Raw materials stored (piled) in the charging area should be kept damp. Sprinklers or hoses with spray nozzles could be utilized.
  b) Deliveries of plant scrap which can consist of reject battery plates, etc. should be brought to the smelter in covered tote boxes or other covered containers. Before dumping into a storage pile, dry materials should be thoroughly wetted.
  c) The floor area near the skip hoist should be kept as clean as possible and wetted to help limit dust generation.
  d) Charging of flue dust should cease once the agglomerator is operational. Until agglomerated material is available any dry flue dust should be wetted before handling and should be stored wet in an area protected from cross drafts.

- Improved local exhaust ventilation should be provided at the skip hoist loading station by attaching the existing exhaust ducts to the new exhaust system and tapering the entries to the exhaust pickups on each side.

- Mechanized materials handling equipment (Bobcat) should not travel into other smelter areas tracking mud, etc.

Tuyere Punching.

- A hood above all tuyeres to control emissions when tuyere covers are removed was recommended.

Blast Furnace Charging.

- Improved local exhaust ventilation of the blast furnace top should be provided by attaching the new ventilation system to the charging hoods at the top of the furnace. Capture characteristics at the charging hood should be evaluated to determine whether improved enclosure or otherwise altered hood designs are necessary.

Slag Tapping.

- The slag tapping hood should be repaired to correct deformations in ductwork.

- Better enclosure of the taphole and receiving vessel should be provided.

  a) The hood front and side pieces should be made to fit together better reducing gaps where sparks, etc. may escape.
  b) The hood sides and front should fit more tightly around the slag receiving vessel to better contain sparks, splashes, etc. which drop to the floor at the base of the receiving vessel.
  c) A means of clearing the slag taphole with the front portion of the slag tapping hood in its lowered position should be investigated. Possibly a small opening in the hood front would allow sufficient access.

- The slag tapping work area should be protected from strong wind currents. Doors leading to the yard area should be closed during windy conditions.

- The filled slag receiving vessel should remain under the slag tapping hood until it has cooled sufficiently to prevent fuming.

Crude Metal Tapping from Blast Furnace.

- Initially the new exhaust ventilation system should be hooked up to the existing local exhaust hoods provided for the lead well, launder and molding line.

- Improved local exhaust ventilation enclosures should be designed and installed for the crude metal tapping/pouring operation.

  a) A stationary hood could be constructed which would enclose the lead well, launder and molding line. This hood could be provided with inspection or access panels to service the launder, etc. The launder could remain stationary with the ingot molds passed in front of the pouring station by means of a rolling molding car at the pouring station. The hood over the pouring station should accommodate the mold being poured and the one cooling after pouring.
  b) A movable hood similar in some respects to the existing three hood system could be constructed. This movable hood should be of one or two piece construction. The hood system should be self supporting (supported on pivots, etc.). The hood(s) should enclose the lead well, launder, the mold being poured and the mold in the cooling position.

c) A combination of fixed and movable hoods could be
    applied. Any movable hood should be self support-
    ing. The hood system should enclose the mold being
    poured and the one in the cooling position. Transi-
    tions between hoods should overlap with no gaps in
    local exhaust coverage.

Kettle Refining.

•   Local exhaust ventilation hoods should be designed and
    installed for the refining kettles.

    a) Sufficient enclosure should be provided to capture
        heated air from kettle firing and fumes from the
        surface of the molten lead in the kettle.
    b) The local exhaust hood should be designed to permit
        charging of the kettle with large crude lead ingots.
    c) The local exhaust hood should be designed to permit
        insertion of a mechanical stirring apparatus used
        during drossing. Sufficient enclosure during dross-
        ing and stirring should be provided to control fumes
        and particulate emanating from the molten lead.
    d) The local exhaust hood should accommodate the pigging
        operation. Swing away doors may provide reasonable
        access for the molten lead pumping device.

Depending on the final design of refining kettle hoods,
varying configurations of additional local exhaust
ventilation for the skimming operation will be necessary.

    a) Fumes from kettle skimmings ("pies") should be
        controlled near their source.
    b) The present practice of building pies on a pallet
        should be reviewed to determine whether a more com-
        pact method of handling skimmings can be devised.
    c) Preferably, the auxiliary local exhaust ventilation
        for skimming should take advantage of the enclosure
        (swing away doors, etc.) of the kettle refining hoods.
        It may be possible to control skimming emissons by
        attaching local exhaust ductwork to the swing away
        portion of the hood thus eliminating the need for
        extra equipment.
    d) Any additional local exhaust ventilation for skim-
        ming operations should be easily operated by
        employees. Complicated ductwork, hoisting, and
        moving systems should be avoided.

Removing skimmings from the kettle could be accomplished
by a long handled implement supported at a pivot point at
the center of the handle. The increased handle length
would give more mechanical advantage to the skimming
operator than using a flat point shovel. Also the skimming operator would no longer have to reach out over the
molten lead thereby decreasing his potential exposure to
fumes.

Pigging.

- Exhaust enclosure for the pigging machine should be designed and installed on the machine.

    a) Of primary importance is the portion of the pigging machine where the reservoir of molten lead and pouring of lead into the ingot molds occurs. An enclosure of the "hot" end of the pigging machine could be made with hinged side access doors. The hood could be supported by metal extensions from the machine framework.

    b) The hood could be made a permanent part of the traveling pigging machine. Drops of flexible ductwork from the overhead local exhaust ductwork could be attached to the hood when the machine was positioned beside either refining kettle.

Building Enclosure.

- During windy weather, dross drafts which can blow through the smelter building should be minimized by closing the large sliding doors. This will help to minimize entrainment of contaminants out from under the local exhaust hoods.

Respiratory Protection Program.

- Plant management should review the respiratory protection requirements of 29 CFR 1910.134 and 29 CFR 1910.1025. Appropriate updating and improvement of the existing program should be instituted.

Specific attention should be given to the following items:

    a) A respiratory protection training session should be presented to employees to refresh their knowledge with regard to the fitting, use, care, and limitations of the respirators they are required to wear.

    b) A determination should be made that all smelter employees can obtain a proper face fit from the respirators in use. All respirators utilized should carry appropriate NIOSH approvals.

    c) A determination should be made that all employees who are required to wear respirators can perform their normal duties with the respirator properly fitted and adjusted. Employees who cannot effectively breathe through the respirator during normal work activity should be identified. A different form of respiratory protection (i.e., powered air-purifying respirator) may be supplied or the employee transferred to an area where respirators are not required.

    d) Special emphasis should be given to impressing the employees that respirators must be worn at all times while working in the smelting building and yard area.

    e) Employees should be instructed to prevent contamination of the respirator when it is removed at breaks or during lunch. Paper towels or soft clean cloths should be provided to wipe off the respirator facepiece before it is reworn after a break.

Hygiene Facilities and Practices.

- Smoking and consuming of food and beverage in the smelter building and yard area should be strictly prohibited. Smoking and chewing materials should not be carried by employees into the smelter building and yard area.

- Use of the existing smelter lunchroom facility should be discontinued. Smelter employees should be required to use the main plant lunchroom or outside food establishments.

- Smelter employees should not enter lunchroom facilities with protective work clothing or equipment unless surface lead dust has been removed by vacuuming, downdraft booth or other cleaning method. Alternatively, smelter employees could be provided with overalls which can be worn over their work clothing and removed before entering lunchrooms, breakrooms, or food establishments. Clean smocks could be provided to put on over work shirts and trousers.

- A shoe/boot cleaning station should be provided at the entrance to the locker room and to areas where food and beverages are consumed.

- Smelter employees should be required to thoroughly wash their hands, forearms, face and neck before consuming food or drink. As a minimum, hands and face should be washed before smoking or chewing materials are utilized.

- Smelter employees should be provided with clean work clothing each day.

- The smelter employee locker room should receive frequent janitorial service (mopping, etc.) on all shifts of the working day.

- Smelter employees should shower and change into street clothing at the conclusion of the workshift. Work clothing should not be taken home. It should be deposited in closed containers.

Employee Training and Information.

- Employees should be informed in organized sessions of the following:

  a) Content of 29 CFR 1910.1025.
  b) The hazards associated with exposure to lead and how they can be controlled.
  c) The purpose, proper selection, fitting, use and limitations of respirators.
  d) The purpose and description of the medical surveillance program.
  e) The engineering controls and work practices associated with controlling employee exposure to lead associated with a particular job.
  f) The contents of existing compliance plans.
  g) The dangers of chelating agents.

691

- Privately each employee should be told what his measured lead-in-air exposure is (without regard to use of respiratory protection) and what the results of his blood lead monitoring are. These results should be explained in terms of compliance with OSHA standards and related to potential adverse health effects.

Housekeeping.

- A smelter-wide cleanup should be undertaken. Lead dust and mud should be removed from yard surfaces, building surfaces, equipment, etc. Settled particulate which has accumulated on building structural members should be removed.

- Dry sweeping and shoveling should be discontinued. Shoveling of wet materials or vacuum methods are preferred.

Medical Monitoring.

- The existing blood lead monitoring program should be continued and the medical monitoring requirements of 29 CFR 1910.1025 should be implemented.

Recordkeeping.

- The recordkeeping requirements of 29 CFR 1910.1025 should be implemented.

PLANT HEALTH PRACTICES AT CANADIAN ELECTROLYTIC ZINC

P. Krick

Canadian Electrolytic Zinc Limited

Abstract

The worker in the primary zinc industry may be exposed to several heavy metals, gases, toxic agents, noise and heat.

Conscious of the possible effects on the employees' health, Canadian Electrolytic Zinc has implemented a systematic surveillance program during the past five years.

The purpose of this paper is to give a brief overview of the environmental monitoring and medical surveillance practices which are in use at this plant.

Introduction

Establishing individual medical surveillance and environmental monitoring, while being a social obligation, also produces major benefits. Good health contributes to lower absenteeism, better job performance and improved productivity.

Environmental monitoring quantifies the airborne exposure of the worker to a hazard. It measures the continuous effectiveness of the environmental control equipment and provides information of changes in process or of equipment functioning. Medical monitoring quantifies the individual susceptibility to that hazard. It provides an ongoing picture on the worker's state of health as it relates to the exposure. It also permits the evaluation of the individual risk and helps to determine if a temporary or permanent transfer from the job is necessary as a preventive measure.

Environmental and medical monitoring are complementary. In addition, education of employees on health hazards, good hygiene and work habits are an integral part of our surveillance program.

Overview of Occupational Risks in the Zinc Electrowinning Process

The zinc electrowinning process includes several steps. Each step uses or generates potentially dangerous chemical compounds. The worker must be protected. This is done through well designed equipment, personal protection and education.

For instance, in the zinc melting and casting area seven (7) occupational risks have been currently identified: dust, ZnO fumes, NH_4Cl fumes, CO and CO_2 gases, heat and noise.

The ventilating equipment installed in this area consists mainly of fans, hoods and stacks which exhaust the fumes, dust and heat. Acoustic barriers reduce the noise generated by the dross rod mill and the zinc atomizing gun. The personal protection consists of ear muffs while working on the casting machines and aluminized coats while drossing the furnaces. However, the worker must know that burning zinc with a torch will lead to zinc oxide fumes, and that loading a closed railcar with a forklift requires a fan to dissipate CO and CO_2.

It is now a common practice to refer to TLV's when talking about industrial inplant air standards. Scientific data have been collected and analyzed for many industrial chemicals and metals. Periodic revisions are made as new facts become available. A typical example is the case of arsenic and its compounds. Early in the last decade the TLV was set at 500 $\mu g/m^3$. Later it was proposed to bring it down to 50 $\mu g/m^3$. NIOSH is now proposing a 2 $\mu g/m^3$ TLV.

But still, the TLV's must in many cases be considered only as indicators of possible danger. An example related to our zinc plant, where the absolute reliance on a TLV is inadequate, is the dental erosion due to acid mist in the electrolysis section. The 1 mg/m^3 TLV leaves certain susceptible employees open to dental erosion after extended periods of exposure. TLV's should be considered, therefore, as guidelines only, and not as bench marks.

Table I shows the different steps of our zinc smelting process, their associated occupational hazards and the number of men who may be exposed to them.

TABLE I   OCCUPATIONAL RISKS IN PRODUCTION AND SERVICE DEPARTMENTS

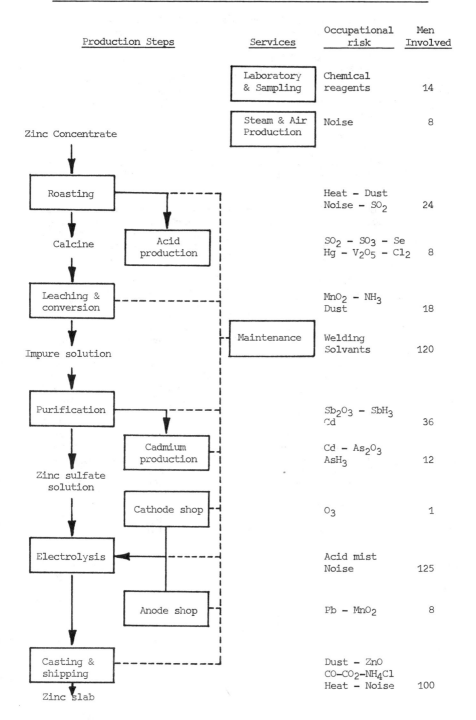

Production Steps	Services	Occupational risk	Men Involved
	Laboratory & Sampling	Chemical reagents	14
	Steam & Air Production	Noise	8
Zinc Concentrate			
Roasting		Heat − Dust Noise − SO_2	24
Calcine	Acid production	SO_2 − SO_3 − Se Hg − V_2O_5 − Cl_2	8
Leaching & conversion		MnO_2 − NH_3 Dust	18
Impure solution	Maintenance	Welding Solvents	120
Purification		Sb_2O_3 − SbH_3 Cd	36
	Cadmium production	Cd − As_2O_3 AsH_3	12
Zinc sulfate solution	Cathode shop	O_3	1
Electrolysis		Acid mist Noise	125
	Anode shop	Pb − MnO_2	8
Casting & shipping		Dust − ZnO $CO-CO_2-NH_4Cl$ Heat − Noise	100
Zinc slab			

Medical and Environmental Monitoring

By medical monitoring it is possible to determine accurately the concentration of toxic elements present in the blood and the urine.

Blood and urine sampling frequency depends on the toxic element itself and on the previous results. For instance blood Cd samples are taken monthly for the workers in the cadmium section, while blood Pb samples are taken every three (3) months for the people working in the anode shop. For cases in which the previous month's test reached 60 µg Pb/100 g blood, the sampling period is reduced to one month.

Urine analysis is useful as a screening method for determining the relative exposure. In most cases urine analysis is less dependable than blood analysis. The uniformity of the concentration throughout the circulatory system and the independance from the liquid intake of the body make the blood test more representative.

As shown in Table II, air monitoring frequency also varies with the type of element or gas. For instance, the sampling is done once every six (6) months for ozone while it is continuous for arsine. As stated earlier, air monitoring is a tool for measuring exposure levels and for indicating the need for engineering improvements. In certain cases the interim use of personal protection devices is required.

TABLE II OCCUPATIONAL RISKS AND RELATED HEALTH EFFECTS

Occupational risk	Health effect considered	1979 NIOSH Recommendation for exposure limit	Environmental Monitoring Frequency
Acid mist	Dental erosion, pulmonary irritation	1 mg/m^3	8 hours
As	Lung and lymphatic cancer	2 µg/m^3	Continuous
Cd	Lung and kidney	40 µg/m^3	1 month
Cl_2	Eye, airway irritation	1.5 mg/m^3	-
CO	Heart	40 mg/m^3	6 months
CO_2	Respiratory effects	18 g/m^3	6 months
Dust	Respiratory effects	10 mg/m^3	6 months
Heat	Heat illness	Action level: 26°C	8h when required
Hg	Central nervous system	50 µg/m^3	-
Mn	Nervous system	5 mg/m^3	6 months
NH_3	Airway irritation	35 mg/m^3	-
Noise	Hearing damage	85 dB	1 year
O_3	Lung	.2 mg/m^3	6 months
Pb	Kidney, nervous system	100 µg/m^3	1 month
Sb	Heart and lung	.5 mg/m^3	Continuous
Se	Skin, respiratory track	.2 mg/m^3	-
SO_2	Respiratory effects	1.3 mg/m^3	6 months
SO_3	Respiratory effects	N/A	6 months
Solvents	Skin	N/A	-
V_2O_5	Eye, skin and lung	.05 mg/m^3	As required
Welding fumes	Metal fume fever, lung	N/A	-
ZnO	Metal fume fever	5 mg/m^3	6 months

Description of Some Specific Health Practices

Over the past five years we have obtained toxicology expertise from the Université de Montréal and established our medical protocols using the latest worldwide information. Protocols were established for all the risks found at C.E.Z. However, many protocols are not in active use because of the occasional nature of the exposure. In order to understand the nature of our protocols, the case of the three (3) most important is presented.

Case of Acid Mist

Acid mist is generated in the cellhouses due to O_2 evolution on the anode and some H_2 evolution on the cathode. Gases and mist are exhausted by a battery of fans located at the far end of the rows of cells. Foaming agents such as licorice are added to produce a foam blanket which inhibits the escape of acid mist from the surface of the electrolyte. A cell operator is usually exposed to 0.4-0.7 mg/m^3 acid mist. However, before the new ventilation system was installed in 1974, the mist was evacuated through roof monitors. During this period air mist levels were usually in the high 0.8-1.2 mg/m^3 range due to the mixing with the ambient air.

Acid mist irritates mainly the nose and eyes but eventually causes dental erosion. This dental erosion is directly related to the amount of time an operator has his mouth opened allowing direct impingement of the acid mist on the teeth.

With regard to acid erosion of dental hard tissues, a standard method of classification was used:

Etching: Dull, ground-glass appearance with a line of demarcation from the normal shiny enamel. No loss of contour has occurred.

Grade 1: Loss of enamel only. Tooth may have a polished, rounded or bulbous appearance.

Grade 2: Loss of enamel with involvement of dentin.

Grade 3: Loss of enamel and dentin with exposure of secondary dentin.

Grade 4: Loss of enamel and dentin resulting in pulpal exposure.

Often the erosion is combined with attrition due to missing anterior teeth, but attrition is not taken into account when the dentist evaluates the erosion. Most of the operators' teeth are eroded. Etching is generally visible after 3-4 years, grade 1 after 5-8 years and grade 2 after 8-12 years. But again, depending on individual lip configuration and the breathing habit, ie the quantity of acid mist inhaled through the mouth, etc., the severity varies from one worker to the other.

It is our considered opinion that the acid mist level in closed cell-house buildings cannot be lowered much below .5 mg/m^3. Hence the effects of cummulative exposure remain. Cleaning the teeth with a kind of neutralizing paste is one possible approach. However, wearing of a light paper mask would give an efficient protection.

It should be underlined that the use of foaming agents to provide a foam blanket on the surface of the cells has the disadvantage of creating explosions from time to time with instantaneous noise levels reaching 140 dB. This happens when a spark reaches the gas bubbles while removing the cathodes for stripping. To protect against hearing loss, the practice is the wearing of ear muffs by all cell operators.

Case of Arsenic and its Compounds

Arsenic trioxide was used before 1975 in the conventional two stage purification. With the introduction of the continuous Sb_2O_3 purification process arsenic trioxide is only used in the cadmium section during the cobalt removal from the repulped purification cake. The main dangers for the operator are contact or ingestion of As_2O_3 or inhalation of arsine gas. As_2O_3 is shovelled into a ventilated reaction tank. During the cobalt removal process some arsine gas is formed, at times, due to the strong reducing medium in the tank. Arsine, however, can reach the working atmosphere only if the ventilation fails or if the reaction tank overflows.

Environmental monitoring is maintained continuously by an automatic sampler which gives an alarm when the concentration reaches 150 $\mu g/m^3$. In case of a power failure, everyone evacuates the building and returns only if the arsine tests, taken by a foreman wearing a "Scott Air Pak", prove negative.

The medical surveillance includes quarterly As-urine, As-blood, hemoglobin and hematocrit tests, plus a nose evaluation to detect irritation. This last test can be done by a nurse. An electrocardiogram and hemogram are also taken quarterly. Spirometric examination and blood multi-12 are performed on a yearly basis. To date no arsenic poisoning has been indicated. A false alarm was caused by the high As level in the seafood consumed by an employee before a urine sample was taken.

In the case of acute poisoning by arsine or stibine, a procedure has been established to provide the attending physician with expert consultation from the Toxicology department of Santa Cabrini Hospital in Montreal. In the case of a doubtful arsine leak, the workers are given a prepared form describing the symptoms of mild arsine poisoning and whom to contact.

Case of Cadmium

Cadmium is an important by-product of the zinc refining industry. At C.E.Z. the purification cake is repulped and the cobalt is removed. A liquid-solid separation results in a copper cake containing also Zn, Cd, As and Co and a filtrate containing Cd. Cd is precipitated by cementation with zinc dust. The majority of the sponge which was precipitated is washed and then compressed in a briquetting machine. There is very little cadmium exposure in this process because leaching, filtering and briquetting are wet operations. The danger for the operator arises when dry sponge is required by a customer. Dust is generated when shovelling the dried cadmium sponge into drums. This shovelling is done usually at the end of the shift, after which the operator removes his clothing and takes his shower. Masks are obligatory.

Environmental sampling is done at different locations of the cadmium section on a monthly basis. It demonstrated that the critical area is the cadmium sponge shovelling location. The medical protocol is summarized in Table III. The difficulty with Cd monitoring lies in the fact that there are very few indications of cadmium exposure before there is irreparable renal damage. Some of our older workers, who have been in the cadmium section when the controls were not as strict as now, show Cd-blood and Cd-urine analyses consistently higher than the average of the workers. These levels are respectively in the 1.0-1.5 $\mu g/100$ g blood and 10-20 $\mu g/g$ creatinin ranges. The practice is that these workers are removed from the cadmium section. They are transferred to other jobs as a preventive measure in spite of the fact that proteins and beta-2 microglobulin levels are normal. This is an intuitive executive decision. However, with the scientific progress in the neutron activation field, it is expected that the absolute levels of cadmium in kidney can be measured. This could become, in the near future, a reliable technique to determine cadmium exposure before the critical level is reached.

TABLE III   CADMIUM MEDICAL PROTOCOL

Cycle	Biological and parachemical parameters
Monthly	1. Cd-B
	2. Cd-U /g creatinin
	3. Qualitative test of urine proteins
Quarterly	1. Hemoglobin
	2. % hematocrit
	3. % blood sedimentation
	4. Beta-2 microglobulin
	5. Blood protein electrophoresis
	6. Qualitative test of blood proteins
Yearly	1. Lung radiography
	2. Spirometric examination: - vital capacity - peak flowrate
	3. Electrocardiogram
	4. Blood multi-12

Conclusion

In order to reduce the risk of industrial disease to workers from occupational hazards, a safe working environment must be provided and verified by environmental monitoring. A parallel effort must be made by the workers to maintain high standards of personal hygiene and work practices. Medical protocols must be followed to monitor these standards and practices and insure that the employees' health is not in danger.

MODERN WATER TREATMENT PRACTICE AT THE ZINC
SMELTER OF THE ST. JOE ZINC COMPANY

Gary M. Gulick
Charles A. Brockmiller
St. Joe Zinc Company
Smelting Division
Monaca, Pennsylvania

The St. Joe Zinc Company completed in early 1979 a "lime and settle" water treatment plant for controlling zinc and cadmium metal discharges in the process water stream from the zinc smelting operation located near Pittsburgh, Pennsylvania. The successful start-up of the treatment facility was the last step in a complex 8-year program to bring zinc smelter discharges into compliance with EPA NPDES effluent limits. The compliance effort also required upgrading existing process equipment, installing settleable solids removal and spill control facilities, segregating "clean" water discharges from contaminated discharges, and closing the loop on an existing effluent quality control facility.

The problem of achieving compliance was approached in an orderly manner and solved in five phases by applying time-proven technology to specific aspects of a complex and challenging problem. The St. Joe effluent compliance effort demonstrates the significant environmental improvements which can be achieved by "older" metallurgical facilities when given challenging but achievable compliance schedules and when the governing regulatory agency has some flexibility in setting effluent limits so that the goals set can be attained. The overall zinc smelter compliance program resulted in a 30% reduction in the quantity of effluent discharged, a 90% reduction in the total suspended solids and cadmium discharged, and a 95% reduction in zinc discharged to the Ohio River.

Introduction

St. Joe Zinc Company operates a primary zinc smelter employing the electro-thermic process to recover zinc from zinc sulfide ore concentrate (1). The smelter, located near Pittsburgh, Pennsylvania, has a rated capacity of 226,800 metric tons per year (MTPY) of zinc equivalent production. The primary products are zinc metal in various sizes of slabs, ingots and anodes, and multiple grades of zinc oxide produced by both the American and French processes. In 1978 a modestly sized zinc dust plant, employing the Imperial Smelting Corporation Ltd. electrothermal process, was started up at the zinc smelter to provide a domestic source for a zinc product with a rapidly expanding market (2). Major by-products include 317,500 MTPY of 66° Baumé sulfuric acid and about 450 MTPY of cadmium metal. The wide variety of pyro and hydrometallurgical and chemical processes employed in the zinc smelter provide many challenging environmental problems.

Recently the zinc smelter started up a "lime and settle" effluent treatment plant as a final step in bringing smelter effluent discharges into compliance with effluent limits set by the United States Environmental Protection Agency's (EPA) National Pollutant Discharge Elimination System (NPDES) program (3). The program to achieve compliance spanned about eight years and was carried out in five main phases. The basic steps in the phased compliance program were:

1. Installation of a facility to minimize total suspended solids discharge and to control spills which would adversely impact the quality of the effluent being discharged.

2. Segregation of the smelter sewer system so that slightly contami-nated process water (noncontact cooling water) could be handled separately from severely contaminated process water.

3. Installation of new shell-and-tube heat exchangers in the acid plant scrubber liquor circuit to replace cascade type lead tube coolers thus decreasing the potential for noncontact cooling water contamination resulting from process liquor leakage from the old style coolers.

4. Minimization of process water discharges by installing an electro-thermic furnace off-gas washer system recycle water facility.

5. Installation of a "lime and settle" facility to treat the process water effluent which could not be recycled or eliminated.

A detailed description of the design and operation of the "lime and settle" facility will be presented. However, since the compliance program evolved over a period of time, a review of the steps in the development of the compliance program will first be made to provide perspective for other industrial dischargers with similar problems.

Background

Under mandate from the Clean Water Act, the EPA was charged with developing a national program for control of effluents from municipal and industrial sources. Formulation of the program occurred in the early 1970's. To provide a clear reference point for evaluating the compliance program executed by the zinc smelter, 1970 will be used as a base year.

In 1970 the zinc smelter used a total of about 113.6 million liters per day (lpd) (30 MGD) of process water. Through the use of closed-loop recirculating systems and multiple uses of single pass water, the 1970 zinc smelter effluent discharge was limited to about 45.4 million lpd (12 MGD). All process water to be discharged flowed through a single sewer system which also collected and discharged storm water runoff. The typical quality of the zinc smelter effluent discharged in 1970 is summarized in Table I. While the 1970 zinc smelter effluent quality was far less than our present effluent limits, the 1970 effluent quality was substantially in compliance with ORSANCO* limits as administered by the Commonwealth of Pennsylvania Department of Environmental Resources (DER).

Table I. Typical Zinc Smelter Effluent Quality - 1970
Concentrations in mg/l

Parameter	Average	Range
Total Suspended Solids (TSS)	6.50	3.6 - 38.4
Zinc - Total	8.33	4.2 - 27.1
Zinc - Soluble	5.19	1.8 - 14.8
Cadmium - Total	0.32	0.17 - 0.75
Cadmium - Soluble	0.26	0.10 - 0.70
pH	7.2	3.0 - 8.1

In January 1971 the zinc smelter was informed of revisions in ORSANCO limits. Two of the revised limits significantly impacted zinc smelter effluents. The effluent limit for cadmium was decreased to ≤ 0.01 mg/l from a level of ≤ 0.50 mg/l. And all dischargers were to "substantially complete removal of settleable solids." This latter restriction was defined only as "removal to the lowest level attainable with current technology." Since cadmium discharges at the time averaged about 0.30 mg/l, the zinc smelter was presented with a formidable challenge in the area of effluent quality control.

To meet the challenge, St. Joe extensively reviewed zinc smelter water uses and discharges and studied effluent quality control alternatives available. The investigation of effluent quality control alternatives was focused in three areas:

1. Development of an end-of-line treatment process to control cadmium discharges and total suspended solids.

2. Development of design criteria for additional facilities to minimize total suspended solids discharges.

3. Assessment of the feasibility of converting the existing zinc smelter water supply-use-discharge system to a closed loop total recycle system.

---

*ORSANCO - Ohio River Valley Water Sanitation Commission - An organization whose members were states bordering the Ohio River and whose purpose was to develop uniform standards for sanitary and industrial discharges so that all users of the Ohio River would have equal opportunity for the use of good quality water and equal responsibility for the maintenance of that quality.

None of the effluent quality control investigations provided the zinc smelter with a clear path to compliance with the cadmium and total suspended solids limits imposed by ORSANCO.

Because of the large volume of effluent, end-of-line treatment focused on chemical addition systems. Laboratory treatability studies indicated that a form of the "lime and settle" process used for years in the mining industry was the most likely candidate for controlling cadmium discharges (4, 5). Operation of a 3.15 liter per second (lps) (50 GPM) pilot plant in 1971 to treat a slip-stream from the zinc smelter discharge resulted in significant decreases in both zinc and cadmium discharges. However, cadmium concentrations of $\leq$0.01 mg/1 were not routinely achieved (see Table II) and the overall effluent quality varied greatly, most likely as a result of extreme variations in the quality of the zinc smelter effluent at the time. Two additional problems associated with the operation of a full-sized "lime and settle" treatment facility were foreseen. First, the high pH (>9.0) required for cadmium removal required neutralization with sulfuric acid prior to discharge--such sulfuric acid neutralization would increase the effluent sulfate content above specified limits. Second, if equalization facilities prior to the treatment plant were not installed, the plant would have to be able to respond instantly to a wide variation in flow since the process sewer system also served as the storm sewer system for the zinc smelter.

Table II. Typical "Lime and Settle" Pilot Plant Effluent Quality Concentrations in mg/1

Parameter	Average	Range
Total Suspended Solids	No Data	
Zinc - Total	0.41	0.04 - 0.55
Zinc - Soluble	0.13	<0.01 - 0.44
Cadmium - Total	0.04	<0.01 - 0.28
Cadmium - Soluble	0.02	<0.01 - 0.08
pH	9.6	7.7 - 11.4

Development of design criteria for additional settleable solids control involved constructing a pilot settling pond with the same proportions as the anticipated full-sized facility and evaluating available filtration equipment. The results of tests conducted utilizing a slip-stream from the zinc smelter effluent discharge indicated that a pond with six hours' retention time provided optimum solids removal effecting about a 60% reduction in total suspended solids. Increasing the retention time or filtering the pond effluent did not measurably increase the removal of solids from the zinc smelter effluent.

Ideally the best form of effluent quality control is to eliminate the discharge. On this basis a detailed study of the zinc smelter water uses and effluent discharges was conducted to assess the feasibility of converting the existing distribution-discharge system to a total recycle system. The conclusion of this study was that the wide variety of metallurgical and chemical processes employed in the zinc smelter could not be accommodated in a closed loop system without significant revisions to many smelter processes. The concept, although perhaps technically possible, was not economically feasible. Data collected relative to the quantity and quality of discharges from various elements in the zinc smelter process, however, provided the key to achieving effluent discharge quality compliance.

Analysis of effluent discharges from various elements in the zinc
smelter process (Table III) revealed that about 297 lps (4700 GPM) of the
existing 536 lps (8500 GPM) smelter discharge was essentially noncontact
cooling water and, thus, should require no treatment prior to discharge.
Of the remaining 240 lps (3800 GPM), 202 lps (3200 GPM) was discharged
from process gas washers used to scrub the last traces of zinc from electrothermic furnace off-gases. The washer stream already passed through a
series of settling ponds for recovery of the metal values prior to discharge
to the sewer system. The effluent from the furnace washer settling ponds
accounted for about 57% of the total suspended solids, 90% of the zinc, and
47% of the cadmium discharges in the smelter effluent.

Table III. Typical Stream Analyses of 1970 Zinc Smelter Effluents

Parameter	Noncontact Cooling Water	Process Water	
		Furnace Washers	All Others
Flow, lps	296.5	202	37.8
Zinc, mg/l	<0.5	20	10
Cadmium, mg/l	<0.05	0.40	2
Total Suspended Solids, mg/l	<1	10	31

Investigation revealed that with the installation of a clarifier ahead of
the existing settling ponds--with polymer addition to accelerate settling--
and then using the existing ponds to polish the clarifier effluent, the
furnace washer system could be close looped with only about a 9.5 lps
(150 GPM) blowdown stream to the existing sewer system. Therefore, if the
furnace off-gas washer system was close looped and the 297 lps of clean
water was discharged through a separate sewer system, a 47.3 lps (37.8 lps
all other + 9.5 lps blowdown) stream of process water would remain which
would require treatment prior to discharge.

Utilizing the information developed in the studies, St. Joe proposed
a four-phase compliance plan to the DER as follows:

Phase 1

Immediately construct a settling pond to reduce the total suspended
solids discharged from the smelter. The facility would consist of a
16 million liter (4,225,000 gallon) capacity primary pond to settle
suspended solids and a 7.6 million liter (2,000,000 gallon) auxiliary
pond into which the smelter discharge could be diverted in the event
of a spill which would adversely affect effluent quality (9<pH<6).

Phase 2

Segregate the sewer system into noncontact cooling water discharges and
process water discharges. The noncontact cooling water discharges
would bypass the settling facility installation. (The Phase 1 and 2
proposals are shown schematically in Figure I.)

Phase 3

Install necessary clarification and pumping facilities to substantially
remove the furnace off-gas washer system effluent from the zinc smelter
effluent discharge.

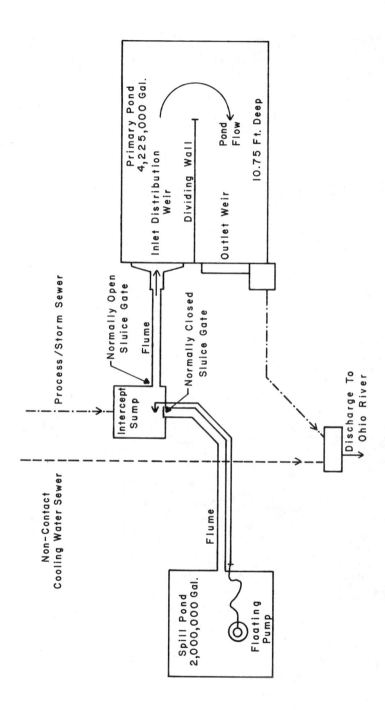

FIGURE I. ZINC SMELTER EFFLUENT SOLIDS REMOVAL AND SPILL CONTROL FACILITIES

Phase 4

At the conclusion of Phases 1, 2, and 3, perform treatability studies on the process water discharge and install a treatment process using the best available technology to reduce cadmium levels to the lowest achievable concentration.

It was anticipated that the proposed program would take three to four years to complete. The DER accepted the compliance plan and St. Joe began construction. The impact of the completion of Phases 1, 2, and 3 on zinc smelter effluent quality is shown in Table IV. The actual facilities are shown in Figures II, III, and IV. The revised pond inlet and discharge pumping facilities to accommodate the "lime and settle" treatment plant are visible in Figure III.

Table IV. Improvement of Zinc Smelter Effluent Quality With Completion of Phased Compliance Program

Parameter	1970 Quality mg/l	kg/d	At Completion of Phases 1 & 2 mg/l	kg/d	% Red.	At Completion of Phase 3 mg/l	kg/d	% Red.
Total Suspended Solids	6.50	301.2	2.52	116.6	61	2.28	67.6	78
Zinc - Total	8.33	385.6	5.65	261.7	32	1.90	56.2	88
Cadmium - Total	0.32	15.0	0.30	14.1	6	0.10	3.2	79
Zinc Smelter Discharge, lps	536		536		0	344		36

Figure II. Intercept Sump Facility - Discharge to Primary Pond at Upper Left; Discharge to Spill Pond at Upper Right

Figure III. Primary Settling Pond for Control of Settleable Solids

Figure IV. Auxiliary Pond for Spill Control

Through this same period in time, the EPA under mandate from the Clean Water Act was formulating effluent limits and guidelines for various types of municipal and industrial dischargers. The zinc smelter participated in a number of surveys and studies relating to nonferrous smelter water usage and effluent quality. A lengthy period of negotiations followed during which NPDES effluent limits for the zinc smelter were evolved. The final form of the zinc smelter effluent limits with two stages of compliance is shown in Table V. During the negotiations with EPA, the zinc smelter continued with the compliance program developed for the DER. At the end of Phase 3, the anticipated reductions in contaminants had been achieved. One additional problem which became evident was inconsistent quality of non-contact cooling water. The cause of this was traced back to the periodic leaking of coolers in the acid plant operation. As a result of this discovery, the zinc smelter proceeded with a program to replace the old coolers with new shell-and-tube heat exchangers, which had a leak-free operating history, as part of the overall compliance effort. With Phases 1, 2, and 3 of the original compliance plan completed and an NPDES permit with fixed effluent limits and phased compliance dates in hand, the zinc smelter now began the Phase 4 program to determine the most effective treatment for the process water stream and bring the zinc smelter effluent quality into compliance with NPDES limits.

Table V. St. Joe Zinc Company Zinc Smelter
NPDES Effluent Limits

Parameter	Time Interval			
	January 2, 1975-June 30, 1977		July 1, 1977-January 2, 1980	
	30-Day Average	Daily Max	30-Day Average	Daily Max
Cadmium, kg/d	5.73	13.1	3.25	6.50
Zinc, kg/d	219	874	28.4	85
Total Suspended Solids, kg/d	132	528	81*	161*
Arsenic, mg/l	-	0.05	-	0.05
Mercury, mg/l	-	0.005	-	0.005
pH	6-9	-	6-9	-
Temperature, °C	-	40.6	-	40.6

*Limits for total suspended solids only applies to process water discharge stream.

Process Water Treatment Plant Design

An intensive study was again conducted in 1976 to characterize zinc smelter effluents so that the best available technology could be defined for control of the process water stream. To complement the efforts of St. Joe personnel, an engineering consulting firm with a strong background in industrial effluent treatment was retained to conduct treatability studies; recommend a treatment process; and participate in the preparation of equipment specifications. After the treatability studies were completed, the consensus of those involved in the investigation was that a "lime and settle" process was the best available at the time. Cadmium removal was still proving to be a problem so cotreatment with sodium sulfide was recommended to improve cadmium removal.

Theory of "Lime and Settle" Process

Lime treatment has been the standard in the mining industry for many years and is, thus, extensively documented. In general lime is added to the effluent stream being treated until a pH of 9.0 or higher is achieved. At the elevated pH, the solubility of zinc and cadmium is reduced and metal hydroxides are formed. By the addition of flocculating agents to promote agglomeration, large particles can be formed which are amenable to settling.

In the case of cadmium, the solubility is not sufficiently reduced at a pH of 9.0 to achieve mandated effluent limits. The addition of sodium sulfide was recommended so that additional cadmium could be precipitated as a metal sulfide. Extensive literature exists indicating that, at or near neutral conditions, metal sulfides have much lower solubilities than the metal hydroxides (6, 7). Subsequent operating experience indicated that the precipitation of cadmium as cadmium sulfide was not effective in decreasing zinc smelter cadmium discharges.

External Factors Affecting Treatment Plant Design

Since the "lime and settle" process involved pH adjustment and clarifier operation, the need for relatively constant treatment plant flows to achieve consistent pH values and stable clarifier operation was recognized. At the time the zinc smelter process water discharge had increased to an average of 56.8 lps (900 GPM). Instantaneous flows throughout a day might vary from 25.2 lps (400 GPM) to 75.7 lps (1200 GPM), depending on the cycles of various operations within the smelter. Additionally, the zinc smelter process water sewer system also served as the storm sewer system. This necessitated the additional design criteria that the treatment plant be able to handle the rain water runoff from a 100-year storm.

To provide the constant flow necessary for good treatment plant operation, the primary settling pond, installed in Phase 1 of the compliance program, was to act as a surge facility between variations in the process/storm sewer water flow and the treatment plant. As originally designed, influent to the pond was distributed uniformly into one side of the pond, flowed around a dividing wall, and overflowed a 12.2 meter (40 ft) weir at the discharge of the pond (see Figure I). The head over the weir was about 1.5 centimeters. The nominal depth of the filled pond is 3.3 meters (10.75 ft). Therefore, at the start-up of the water treatment plant, flow to the plant would be at the maximum design rate until the pond level was lowered to a depth of 1.2 meters (4 ft). At that time the plant flow would be reduced to the average process water flow rate.

With the pond level at 1.2 meters of depth, approximately 10.7 million liters (2,830,000 gallons) of storage capacity is available. When heavy rains occur, the pond level will rise while the treatment plant flow is gradually increased in step-wise increments until the rain stops or the pond level starts to fall. As the pond level again approaches the 1.2 meter depth, treatment plant flow is decreased gradually to match process water flow. In the event of problems in the water treatment plant requiring a shutdown for maintenance, the pond storage capacity provides about two days of effluent retention at normal flows.

709

To protect the treatment plant from low pH excursions, the spill pond, also part of Phase 1 in the compliance schedule, was retained. If a pH of less than 6.0 is detected in the intercept sump, process water sewer flow is automatically diverted to the spill pond until the cause of the excursion is found and corrected. Caustic soda is then added to the spill pond until the pH rises to 7.0 or higher. The contents of the spill pond are then pumped back to the intercept sump for discharge into the primary pond.

Figure V shows the primary pond at the start-up of the water treatment plant. Note that the water level is already about 1.5 meters below the original discharge weir on the left so that all process water is passing through the treatment plant prior to discharge. Also note the floating inlet flume extending into the pond being used in place of the original inlet distribution weir. As the pond level is lowered, hypalon pond liner material cushioned by a sand bed is exposed. If the original inlet distribution weir were continued in service, the impingement of pond influent onto the exposed liner might result in displacement of the underlying sand bed which could lead to rips or tears in the liner material. The floating flume prevents direct impingement and thus protects the watertight integrity of the pond. The waste transfer pumps were also placed on a floating platform as a means of keeping the pond in operation during water treatment plant construction.

Figure V.   Water Treatment Plant Waste Transfer Pumps

General Treatment Plant Operation

As shown in Figure VI, process water is pumped from the primary pond at a rate of 37.8-75.7 lps (600-1200 GPM), as conditions warrant, to the first of two 22,700 liter (6000 gallon) capacity continuously agitated neutralization tanks. The pH is raised to within about 0.2 pH units of the desired pH level by the addition of a 5% lime slurry. The first tank overflows by gravity to the second neutralization tank where a 5% lime slurry is added until the final pH level required is achieved. To provide nuclei for particle growth, a 1.6 lps (25 GPM) stream of underflow solids slurry is pumped from the clarifier to the second continuously agitated neutralization tank. The overflow of the second neutralization tank goes to the clarifier.

Prior to entering the clarifier, a polymer solution (coagulant aid) is injected into the process water stream to promote particle growth and settling. The clarifier is a solids contact design with a process stream retention time of 5 to 10 hours, depending on the treatment plant flow rate. The effluent from the clarifier flows by gravity into a 22,700 liter (6000 gallon) capacity filter feed tank ahead of the multimedia filters. The filter feed tank is used to smooth out variations in flow to the filters and also provides an overflow bypass in the event of a filter failure. From the filter feed tank, water is pumped through the multimedia filters into a 60,600 liter (16,000 gallon) capacity effluent storage tank. Two of the three multimedia filters are in use at the same time with the third on standby. Filter switching is done automatically on a timed cycle. The effluent storage tank provides a source of clean water for backwashing the multimedia filters, general treatment plant washing, lime slurrying, and polymer batching. Overflow from the effluent storage tank is discharged to the Ohio River. Filter backwash water as well as general wash-down water is discharged to the primary pond for retreatment prior to final discharge.

The plate and frame filter press is operated as necessary to maintain a uniform depth sludge bed in the clarifier. Current operation requires 3 to 4 filter press cycles per week. The filtrate is returned to the clarifier while the filter cake is discharged into a mobile container for disposal in a licensed industrial landfill. Figures VII, VIII, IX, and X show key pieces of water treatment plant operating equipment.

Unit Operation

Lime Storage System. Hydrated lime is delivered in 16-18 metric ton (18-20 short tons) loads by trucks with pneumatic unloading capability. Prior to initiating the truck unloading cycle, sufficient water is added to the lime storage tank (71,900 liter capacity) to result in a 40% lime slurry. During the unloading operation, water sprays are operated in the tank to prevent dusting. The lime storage tank (center foreground Figure VII) is continuously agitated. Care must be exercised in mixing the lime slurry since a slurry of >45% solids results in a paste which does not mix or flow predictably.

Lime Slurry System. For neutralizing the incoming process water stream, a 5% lime slurry is used. The lime slurry batches are made automatically in 1140 liter (300 gallon) units. A low level probe in the continuously agitated lime slurry dilution tank activates the cycle. A valve in the pipe from the lime storage tank opens and allows 40% lime slurry to enter the pump tank. When the equivalent of approximately 45 kilograms (100 pounds) of lime has been timed into the dilution tank, the valve is closed. Water is then added to the dilution tank until a high level probe is satisfied. Each time the batch cycle is activated, the event is recorded on a counter.

711

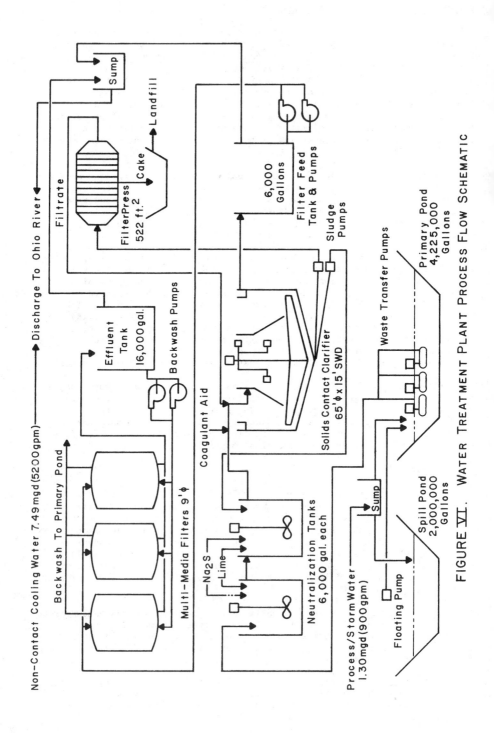

FIGURE VI. WATER TREATMENT PLANT PROCESS FLOW SCHEMATIC

712

Figure VII. Process Water Treatment Plant Facility

Figure VIII. Equipment Arrangement - Neutralization Tank in Upper
Right - Control Room in Upper Left - Multimedia
Filters in Lower Right

Figure IX.   Multimedia Filters

Figure X.   Plate and Frame Filter Press

By recording each lime slurry batch, a lime-use record is maintained. Current levels of operation require the use of about 320 kilograms (700 pounds) per day of lime.

Polymer Addition System. A batch of polymer solution (coagulant aid) is prepared daily by the treatment plant operator. The strength of the solution is determined by the water treatment plant flow rate. Extensive jar tests have indicated that Calgon WT-3100 polymer is the most effective coagulant aid for treating zinc smelter effluents. In preparing a polymer solution, no more than 9 kilograms (20 pounds) of polymer is diluted in 1520 liters (400 gallons) of water as a higher concentration of polymer results in a highly viscous solution which cannot readily be handled by the injection system. The polymer solution is injected into the neutralization tank effluent at a constant rate of one liter per minute. The desired dosage for optimum particle growth is in the range of 1.0-1.5 mg of polymer per liter of process water.

Clarifier Operation. Acceptable clarifier operation has been achieved by maintaining the sludge bed at the lowest practicable level. If an underflow sludge recycle stream were not required to provide nucleation material for the process water stream being treated, the clarifier could be operated with no sludge bed. The underflow sludge recycle system requires the clarifier to be operated with a sludge bed of sufficient depth to keep the clarifier underflow cone full of sludge.

Multimedia Filter Operation. Three multimedia filters, each about 3 meters (9 ft) in diameter, are utilized to polish the clarifier effluent. Two filters are on line at a time with the third being backwashed or on standby. Provision has been made in the existing multimedia pipe header system to accommodate additional filter equipment in the event additional process water treatment plant capacity should be required. The filter media consists of a 30.5 centimeter (12 inches) deep bed of sand covered by a 45.7 centimeter (18 inches) deep bed of 24 mesh anthracite coal. The filters may be backwashed on a regular cycle or whenever a specified filter pressure drop is reached. The filters are currently being operated on a 16-hour time cycle with a pressure drop override in the event a clarifier upset causes solids carryover to prematurely plug the filters.

Clarifier Sludge Filter Press Operation. The operating cycle of the filter press is determined by sludge accumulation in the clarifier, which can be correlated with lime usage. Presently one filter press cycle is required for every 11 lime slurry batches prepared. The filtrate is returned to the clarifier. The filter cake, about 0.85-1.0 cubic meters weighing 1590 kilograms (30-35 cubic feet - 3500 pounds) of 50% by weight solids material, is dropped into a mobile container for disposal in a local licensed industrial landfill. The filter cake, which contains about 35% zinc on a dry basis, is not recycled to the zinc smelter because the water treatment chemicals used are not compatible with the electrothermic process.

Staffing. The operation of the process water treatment plant is the responsibility of the zinc smelter acid plant personnel. No additional personnel have been required to support the operation of the water treatment plant. The acid plant operator makes a regular walk-through inspection of the water treatment plant about four times per shift as part of his normal routine. The acid plant control room has a remote annunciator panel to alert the operator to any abnormal conditions. Since stable operation has been achieved, the process water treatment plant has required only minimal attention.

Water Treatment Facility Start-up Problems

When the process water treatment plant was started up at the beginning of 1979, effluent quality was not good enough to comply with NPDES limits. The problem seemed to be the result of solids carryover from the clarifier since the floc did not attain sufficient size for settling. This was related to insufficient particulate matter in the influent stream for proper nucleation. Initially there was no underflow sludge pumping back to the clarifier influent stream since no provision had been made for sludge recycling in the original treatment facility design. The carryover from the clarifier resulted in short multimedia filter cycles due to filter bed plugging.

In an attempt to provide material for nucleation, a bentonite slurry was made using the lime slurry solution as a carrier. The addition of the bentonite material resulted in an incremental improvement in effluent quality, but the improvement was not sufficient to achieve NPDES compliance. Finally by using a spare clarifier sludge filter press feed pump, clarifier underflow sludge recycle was initiated and the improvement in water treatment plant effluent quality was immediate and dramatic.

After initiating clarifier underflow sludge recycle, cadmium control remained a modest problem. Increasing the sodium sulfide dosage did not achieve any recognizable incremental reduction in effluent cadmium concentrations over lower dosages. Finally acting on speculative comments in the report summarizing the operation of the pilot "lime and settle" treatment plant many years before, the addition of sodium sulfide was terminated. Although cadmium in the water treatment plant effluent was then reduced, several other changes were made at the same time, so it is difficult to assess how much impact terminating the sodium sulfide had on reducing cadmium levels. Even if the sodium sulfide was not interferring with the reaction taking place, it offered no benefit in achieving additional reduction in cadmium levels in the zinc smelter effluent and, therefore, was not required. The precipitation of cadmium carbonate appears to be the primary reaction for the removal of cadmium from zinc smelter process water discharges. The carbonate ion is available for reaction since the zinc smelter water supply is relatively hard. In periods of heavy rain, the addition of a soda ash to the incoming process water may prove necessary to maintain the concentration of carbonate ions at a level sufficient to control cadmium discharges.

During the first few months of process water treatment plant operation, problems also occurred in determining the proper sludge depth to be maintained in the clarifier to achieve optimum unit operation. On the basis of jar tests and experience with similar installations, the consultant had recommended a 2.1 meter (relative to the clarifier side wall) bed depth. The purpose of this was to provide solids contact with the incoming effluent to promote particle growth and eliminate the need for underflow recycle. The 2.1 meter sludge bed depth was thought possible because the treatability studies found that the resulting settled material remained light and fluffy with minimal bed compression. In actual operation, however, the bed not only compressed, but formed a cake which the clarifier rake could not break. After the first four months of operation, it was necessary to drain the clarifier and manually remove the cake. Since then the clarifier has been operated maintaining only a bed of sufficient depth to provide material for underflow recycle. The differences in the predicted sludge characteristics and the actual sludge are most likely the result of the actual influent containing lower concentrations of contaminants than the water used in the treatability studies. The inference is that the clarifier is conservatively sized and is probably capable of handling an equal volume process stream with substantially higher concentrations of contaminants.

Treatment Plant Performance

Since stable operation has been achieved, the process water treatment plant has proved very effective in reducing zinc and cadmium levels as indicated by typical analyses shown in Table VI.

Table VI. Typical Process Water Treatment
Plant Performance Concentrations
in mg/l

Parameter	Influent	Effluent
Zinc - Total	22.3	<0.66
Cadmium - Total	0.61	<0.06

The treatment plant effectively removes >97% of the incoming zinc and >90% of the cadmium. The "lime and settle" process is capable of achieving consistent zinc concentrations in the treatment plant effluent regardless of zinc loading in the influent stream. Extremely high levels of zinc, however, tax the capacity of the neutralization equipment and require excessive sludge filter press operation. Control of cadmium has been sufficient to consistently meet NPDES permit requirements, but the quality of the effluent relative to cadmium has been more variable than that of zinc. This is because cadmium solubility is more sensitive to variations in pH than zinc in the pH range in which the water treatment plant operates. The "lime and settle" process has proven to be an extremely versatile operation for controlling zinc smelter process effluent quality. With a properly controlled feed stream, the "lime and settle" process appears capable of consistently meeting current zinc smelter effluent limit requirements.

Smelter Compliance

With the changes in cooling equipment in the acid plant operation and the completion of the water treatment plant, the smelter effluent discharge was brought into compliance with current EPA NPDES effluent limits as shown in Table VII. Total suspended solids discharges have been reduced by nearly 90% from 1970 levels while zinc and cadmium have both been reduced by greater than 90%. As a result the smelter effluent discharges are well within current NPDES limits. To effect these reductions, considerable effort was necessary requiring a complex five-phase program to achieve compliance:

1. Installation of a facility to minimize total suspended solids discharges and control spills (Figure 1).

2. Segregation of the zinc smelter effluents into noncontact cooling water and process/storm water sewer systems.

3. Installation of new heat exchangers in the acid plant operation to minimize the potential for noncontact cooling water contamination.

4. Minimization of process water discharges by installing a furnace off-gas washer recycle water facility.

5. Installation of a "lime and settle" process to treat remaining process water.

Table VII. Typical Zinc Smelter Effluent Quality

	1970 Smelter Effluent		1979 Smelter Effluent			Current NPDES Limits
	mg/l	kg/d	mg/l	kg/d	% Red.	Avg
Cadmium	0.32	15.0	0.04	1.31	91	3.25 kg/d
Zinc	8.33	385.6	0.48	15.70	96	28.40 kg/d
TSS	6.50	301.2	1.12	36.63*	88	81.0 kg/d*
Arsenic	<0.05	-	<0.005	-	-	0.05 mg/l Max
Mercury	<0.005	-	<0.0002	-	-	0.005 mg/l Max
pH	7.2	-	7.8	-	-	6-9
Temp, °C	<40.6	-	<40.6	-	-	40.6 Max
Smelter Discharge, lps	536		378		29	

*Current NPDES limit applies only to process water discharge; smelter effluent value reflects combined zinc smelter discharge.

The zinc smelter NPDES compliance effort required a capital expenditure of about $4,780,000 with the "lime and settle" process/storm water treatment facility accounting for $2,160,000 of the total. The estimated annual operating and maintenance cost for the "lime and settle" facility is $150,000. The total zinc smelter annual direct operating and maintenance costs for all the facilities installed as part of the NPDES compliance program are about $450,000. To insure continuous compliance with NPDES limits, an extensive effluent discharge monitoring system is operated which adds an additional indirect cost of $150,000 per year to the cost of compliance.

The installation of facilities to control zinc smelter effluent discharge quality may be complete, but the maintenance of compliance requires a continuing effort on the part of many groups. Operating departments within the smelter must wisely use water and insure discharges go into the proper sewer system. Engineering personnel must carefully review water uses in new projects and incorporate water use conservation measures into new facilities and minimize discharges through reuse of process water. Water treatment plant operating personnel must maintain the physical plant in good operating condition; optimize the process to achieve maximum contaminant reduction; and, along with environment control personnel, continuously monitor the actual performance of the treatment facility.

References

1. Lund, R. E., et al., "Josephtown Electrothermic Zinc Smelter of St. Joe Minerals Corporation," pp. 549-580, AIME World Symposium on Mining & Metallurgy of Lead & Zinc, Vol. II, Cotterill and Cigan, ed.; AIME, New York, NY, 1970.

2. Newton, D. S., "The Electrothermal Process for Zinc Dust Production at Imperial Smelting Corporation (Alloys) Ltd.," pp. 995-1006, AIME World Symposium of Mining & Metallurgy of Lead & Zinc, Vol. II, Cotterill and Cigan, ed.; AIME, New York, NY, 1970.

3. United States Environmental Protection Agency, "Development Document for Interim Final Effluent Limitation Guidelines and Proposed New Source Performance Standards for the Zinc Segment of the Nonferrous Metals Manufacturing Point Source Category," EPA 440/1-75/032 Group 1, Phase 2, February 1975.

4. Shome, Shanjoy, "Environmental Considerations and the Modern Electrolytic Zinc Refinery," pp. 280-282, Lead-Zinc Update, Rausch, Stevens, and Mariachei, ed.; AIME, New York, NY, 1977.

5. Patterson, J. W., Wastewater Treatment Technology, first ed., pp. 247-258, Ann Arbor Science Publishers, Inc., Ann Arbor, MI, 1975.

6. Larsen, H. P., et al., "Chemical Treatment of Metal-Bearing Mine Drainage," J. Water Poll. Cont. Fed., 45 (8) (1973), pp. 1682-1695.

7. Shimoiizaka, J., "Recovery of Xanthate from Cadmium Xanthate," Nippon Kogyo Kaishi (Japan), 88 (1972), pp. 539-543.

MODERN WATER TREATMENT PRACTICES

AT THE HERCULANEUM SMELTER OF THE ST. JOE LEAD COMPANY

Donald H. Beilstein

&

Arthur E. Stanze

In August 1978, the St. Joe Lead Company started up a waste water treatment plant at its Herculaneum Smelter. The treatment plant removes the heavy metals from up to 850 gallons per minute of waste water prior to discharge to the Mississippi River. The water treatment plant also cleans 1000 gallons per minute of dirty scrubber water from the sinter plant's ventilation scrubbers. The clean scrubber water is recirculated back to the scrubbers.

Figure 1. - View of Water Treatment Plant from the north. From left to
right are the SVG filter building, the reactor clarifier,
the lime silo and the Agidisc filter building. Behind the
SVG filter building is the superstructure of the thickener.

The Herculaneum Smelter of St. Joe Lead Company started operation
the last decade of the 19th Century. The Smelter is situated on the
west bank of the Mississippi River in a rural area, approximately 30
miles south of the city of St. Louis. The Smelter is a classical lead
smelter operating a sintering machine, blast furnaces, drossing plant,
and pyrometallurgical refinery. Production levels are 230,000 short
tons per year of refined lead.

In the first half of the 20th Century, water usage was unrestricted,
and discharge was made to the Mississippi River following the settling
of solids from the effluent stream. In 1971, a study was made by
Singmaster and Breyer, who made recommendations for elimination of many
minor discharges, consolidation of waste water streams, and isolation
of particular streams. These recommendations eventually led to two
primary discharge streams from the Smelter. Sanitary waste was treated
as a separate item and is routed to a municipal sewage treatment installation. In the mid-1970's it became apparent that the quality of effluent
from the two remaining discharge points was unsatisfactory or would be
shortly unsatisfactory, pursuant to EPA regulations. At that time, a
study was undertaken to determine the best means of purification of the
remaining streams to a sufficient degree, to allow them to be discharged
under an existing NPDES Permit.

At present one of the effluent streams is completely treated prior
to discharge, and the remaining stream is in the process of being
eliminated by diversion to the existing water treatment plant. When
the final diversion is complete, there will be a single discharge stream
to the Mississippi, which will meet all requirements per the NPDES
Discharge Permit and should be very close to compliance with the 1983
regulations.

What follows is a description of the present Water Treatment Plant, which should be sufficient for treatment of all the effluent waste water from the Herculaneum Smelter. The plant can best be described as two separate circuits. The first circuit treats waste water prior to discharge to the Mississippi River. The second circuit is a closed scrubber water cleanup circuit for the sinter plant's ventilation scrubbers.

Waste Water Circuit

In the waste water circuit, the concentrations of the lead, cadmium, and zinc are reduced below 0.2 ppm Pb, 0.4 ppm Cd, and 1.0 ppm Zn. This is accomplished by adjusting the pH of the waste water to 9 with hydrated lime. Lead, cadmium, and zinc precipitate as hydroxide and are removed from the waters by settling and deep bed filtration.

Waste Water Flow

Pump stations, installed in the drain system, intercept water that at one time flowed out to the river. Clean water, from such applications as cooling, is pumped back to the plant water systems reservoir. Dirty, corrosive water, from such sources as the acid plant and copper dross furnace granulation system, are pumped to the blast furnace slag granulation system. The acid plant effluent is partially neutralized so that when it is combined with the caustic dross furnace granulating water, the resulting pH will be about 7.

The water pumped to the Waste Water Treatment Plant comes either from a sump which catches overflow from the blast furnace slag granulation system, or from the plant water reservoir. Automatic level controls at the granulation system overflow sump switch back and forth between the two sources. Discharge valves at both sets of pumps are set to deliver the same flow rate to the Waste Water Treatment Plant.

Waste water enters the treatment plant by flowing into a 1000 gal. forebay tank. (See Fig. 2 and 3) In the forebay, the waste water is mixed with reactor clarifier underflow. The pH of the resulting mixture is measured continuously and fed to the pH control circuit in the control room.

Figure 2 - Forebay, Parshall flume, and reactor tank.

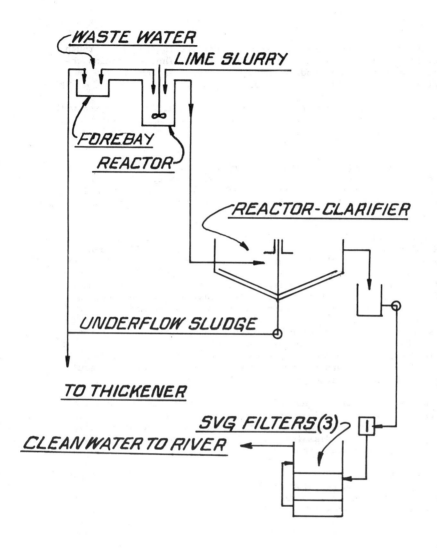

Figure 3 - Waste Water Circuit Flow Sheet.

The waste water/underflow mixture overflows the forebay through
a Parshall flume into a 4800 gal. reactor tank. The head behind the
flume is measured continuously and a flow signal is sent to the pH
control circuit in the control room.

In the reactor tank, the waste water and underflow are mixed with the
required amount of lime slurry. The tank is stirred by a twin blade
propellor agitator. The resulting mixture in the reactor tank overflows
to the reactor clarifier.

Further mixing, precipitation, contact with recirculated solids,
and settling of precipitated solids is provided by an EIMCO Type HRB
Solids Contact Reactor Clarifier. (See Fig. 4) The unit, which is
sized for a flow of 850 gpm, is 40 ft in diameter with a side water depth
of 11.75 ft, retention time is about 150 min.

A two-armed rake brings solids to the center of the tank for either
interval recirculation, or withdrawal through the underflow cone. A
conical skirt around the rake drive shaft forms about a third of the
tank's volume into a reaction zone in which settled solids are recirculated.
Inside the reaction zone a recirculation drum surrounds the rake drive
shaft. The inlet pipe comes through the side wall of the tank, through
the reaction well skirt, to introduce the waste water mixture into the
recirculation drum. A turbine at the top of the recirculation drum
draws up settled solids which have been raked to the center of the tank.
The solids mix with the incoming mixture of waste water, lime slurry,
and recirculated underflow as they pass through the turbine. Both the
rake and turbine drives are variable speed. The rake drive is equipped
with high torque overload protection.

Waste water leaving the reaction zone flows up through a bed of
solids. The solids settle, forming a zone of clear water at the top of
the tank. The clarified waste water overflows the reactor clarifier
through a system of radial launders. The pH of the overflow is measured
continuously and sent to the pH control system.

Three EIMCO type SVG deep bed granular media filters remove almost
all the solids which overflow the reactor clarifier. The 12 ft
diameter units operate in parallel. Each filter is capable of filtering
450 GPM of water, allowing two filters to handle the maximum flow through
the plant while the third is backwashing. Dual filter media is used,
two ft of graded authracite on top of 1 ft of graded sand. The upper
half of the filter is a storage compartment for back wash water.

Waste water from the reactor clarifier overflow is pumped to a
splitter box. The box divides the water between the filters. The elevation of the box limits the pressure head on the filter bed. The waste
water underflows the splitter box and enters the filter above the media.
As the water flows down through the anthracite and sand, reactor clarifier
solids are trapped in the voids of the bed. Filtered water leaves the
media through special nozzles installed in the plate which holds the
media. A transfer pipe leads the filtered water to the upper storage
compartment. The overflow from the storage compartment meets the requirements of the NPDES and is sent to the Mississippi River. Table I shows
a typical set of assays for the reactor clarifier and SVG filter effluent.

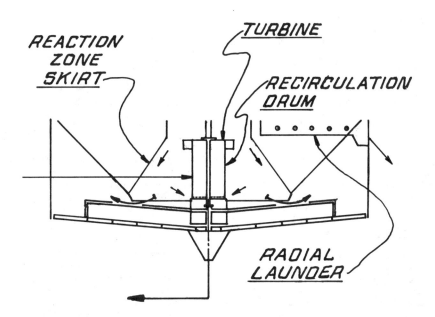

REACTOR-CLARIFIER

1979 P.J.J.

Figure 4 - Reactor Clarifier

Table I. Typical Analysis of Waste Water Circuit Streams

	pH	(1) ppm Pb	(1) ppm Cd	(1) ppm Zn	(2) ppm Solids
Waste Water	7.4	1.65	1.0	27.0	8
R/C Overflow	9.0	1.5	0.11	2.6	21
R/C Underflow	10.3	577	146	3514	10,090
SVG Overflow	8.9	0.17	0.01	0.11	5

(1) Pb, Cd, Zn figures are totals of both liquid and solid phase.

(2) Filterable solids.

The filters are programmed to back wash after 24 hours or when the headloss across the media reaches 15 ft W.G. The backwash cycle consists of filter compartment isolation, partial drain down, air scour and back wash at the rate of about 1700 gpm. The sand and anthracite were very clean when inspected after six months of operation.

pH Control

The pH control circuit has 3 inputs; the forebay pH, underflow plus waste water flow, and reactor clarifier overflow pH. The forebay pH and flow signals are combined and sent to an adder - subtractor. The reactor clarifier pH signal goes to a nonlinear pH controller. The output of the pH controller is sent to the adder - subtractor. The output from the adder - subtractor goes to the SCR drive which controls the speed of the DC motor which drives the lime slurry metering pump.

The forebay and flow signals provide a feed forward signal which allows the system to compensate for drastic changes in the amount or pH of the waste water. The reactor clarifier overflow pH signal provides the feed back necessary for the system to maintain the overflow at a constant pH.

Preliminary bench testing indicated that a pH of 9 in the reactor clarifier overflow was necessary to achieve the required lead removal. Plant experience has confirmed this.

Lime Slurry Preparation

Bulk hydrated lime is stored in a silo. Below the hopper of the silo is an agitated lime slurry tank. A level controller on the tank initiates an automatic makeup cycle which refills the tank with water and lime at preset rates to give a 5% by weight slurry. The lime slurry is pumped up to the metering pump and back through a return line to the lime slurry tank. The speed of the metering pump motor is controlled by the pH control circuit.

Waste Water Circuit Operation

The waste water treatment plant requires one operator per shift.
The operator's most important duties in the waste water circuit are to
keep close watch on the lime slurry makeup system and control the solids
level in the reactor clarifier. The lime slurry system is the most
troublesome part of the plant. Constant attention and frequent manual
corrections are required to keep a constant flow of 5% slurry going to the
metering pump.

Control of the level of solids in the reactor clarifier is a manual
operation. Two sample pipes go through the side of the tank and end 3
inches above and below the desired level. The operator diverts the underflow recirculation to the scrubber circuit's thickener to remove enough
solids to keep the top pipe clear and the lower pipe dirty. The amount
of solids produced is highly variable, but follows the forebay pH very
closely. We have found that a relatively high level of solids was
required to produce a clear overflow. This leaves the plant very
susceptable to thermal upsets. When the temperature of the incoming
waste water changes too rapidly, (20 deg F in an hour) the bed will
erupt in several spots and solids will carry over to the SVG filters.
High temperature also causes a shift in the equilibrium of the lead
precipitation reaction resulting in higher lead content of the SVG
overflow.

Scrubber Water Circuit

Scrubber Water Flow

Dirty water flows by gravity from the sinter plant's scrubbers to
a collection sump. These scrubbers are used for hygiene ventilation
of transfer points in the sinter circuit. They are DUCON UW-4 scrubbers
(See Fig. 5) Wash down and cooling water are also collected in this
sump. Dirty water is pumped from this sump to the feed well of the
thickener. The current flow rate is 1100 gpm. The solids content
varies from 1000 to 5000 ppm depending on sinter plant operation. The
system was designed to recirculate 2000 gpm of scrubber water and the
capacity will be increased as new scrubbers are brought on stream in the
sinter plant. Clean scrubber water overflows the thickener with a solids
content ranging from 5 - 25 ppm. Excess water overflows the surge tank
back to the plant water reservoir.

The thickener is an 80 ft EIMCO Type B Thickener with 15.5 ft side
water depth. (See Fig. 6 and 7) The two arm rake requires 12.5 min
for one revolution and can be raised 3 ft from the bottom position.
The rake drive is equipped with automatic torque control. Thickened
solids are withdrawn through a central discharge cone and pumped to the
filter. Our experience is that a velocity of 6 ft/sec is necessary to
transport the slurry.

727

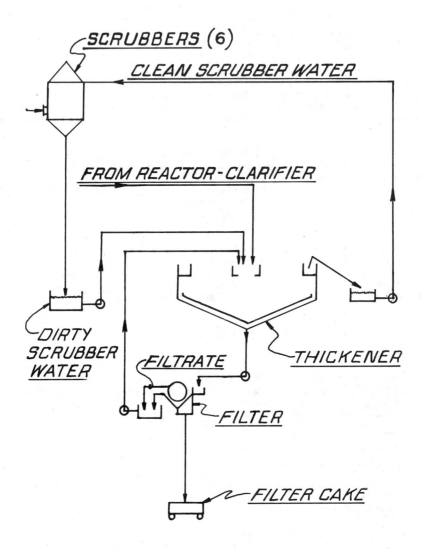

SCRUBBERS (6)

CLEAN SCRUBBER WATER

FROM REACTOR-CLARIFIER

DIRTY
SCRUBBER
WATER

FILTRATE

THICKENER

FILTER

FILTER CAKE

SCRUBBER WATER CIRCUIT

1979 P.J.J.

Figure 5 – Scrubber water circuit.

Figure 6 - 80' EIMCO Type "B" Thickener.

Figure 7 - Thickener device. Torque on the rake is logged once an hour.

Figure 8 - EIMCO 6' x 3 Disc Agidisc Filter.

729

Solids Removal

The solids are removed with an EIMCO 6 ft diameter x 3 disc Hy-Flow Metallurgical Agidisc Filter. (See Fig. 8) The unit has 150 square ft of filtration area. A variable speed drive allows disc speeds of from 0.1 to 1.0 revolution per minute. The vacuum pump is rated at 650 cfm against 20 in Hg. A paddle shaft agitator at the bottom of the vat keeps the heavy solids in suspension. When fed a slurry of +2.0 specific gravity (+61 wt % solids) the filter will produce 4.5 dry tons per hour of filter cake with a moisture content between 10 and 15 wt % H_2O.

The filter cake discharges through a rubber chute to a railroad car. The filter cake is blended back into the sinter machine feed. The filtrate, overflow from the vat, gland, and cooling water, and floor washing collect in a sump and are pumped back to the thickener.

Operation

The filter can handle about twice the solids as are collected in our scrubbers at this time. In order to keep the underflow specific gravity up, solids are allowed to accumulate in the thickener for about 3 days. Then a filter run is started and the filter is run till the underflow becomes so thin that filter cake formation is hurt. There have been no problems with using the thickener as a storage tank since we started the practice of raising the thickener rakes the full three feet between filter runs. Once, pump problems caused a five day delay with no problems. Cake formation is highly variable and seems to tie back to how well the sinter plant is running.

THE PROBLEMS OF TIN

John H. Harris
Technical Adviser
Centre for Natural Resources, Energy and Transport
United Nations, New York*

Tin is an indispensible metal. Resources of tin are limited in areal distribution. Exploration for new sources presents special difficulties as do the mining and treatment of tin ores and the smelting of tin concentrates. The available mineral processing technology is far from satisfactory and the metallurgy of tin smelting is imperfectly understood. Recent innovative proposals for alternative pyrometallurgical and hydrometallurgical processes compete with traditional methods. Further problems exist in the marketing of tin, and the metal is the only one subject to international control by agreement.

The Interplay of these factors is broadly reviewed.

---

* The views expressed in this paper are those of the author and do not necessarily reflect those of the United Nations.

Introduction

Tin is a rare commodity. Its ores have a limited geographic distribution and the total amount traded annually is of the order of only two hundred thousand tons. Nevertheless, it is one of the metals essential to modern social and technological welfare and, for the size of its market, plays an unusually large role in the international economic and political scene. It is one of the few mineral substances mined offshore as well as from land-based deposits. It is the only metal subject to global control through the United Nations sponsored International Tin Agreement.

Around forty percent of the world's production of tin goes on to tin-plate, ninety percent of which goes into food and drink packaging (6, Vol. IV, p.11-44). The next most important uses are in solder, bearings, and other alloys vital to today's electrical, electronic, and engineering achievements--in printing metals, in chemical intermediates, catalysts, and biocidal compounds and, amongst tin's contributions to the arts, pewter. Remove the tin in today's applications and all transport by land, sea, and air would be halted, all production of electricity eliminated, all electronic equipment stilled, and a vast amount of preserved food irretrievably lost.

Beyond the traditional pan, there are few technical aids to prospecting for tin. There are many economic and political factors which militate against investment in tin. The mining of tin presents distinctive and unusual problems; there is no real specific means of separating tin minerals from ores or contaminants. Much of the chemistry and metallurgy of tin smelting remains to be defined and the marketing and trading of the product has its own peculiar characteristics and difficulties.

Against this background, this paper attempts to chart the linkages between each stage and the others and to suggest that the participants in each stage must benefit by better understanding of the whole range of techniques and problems.

Resources and Exploration

Tin is a rare commodity, a relatively scarce element with a crustal abundance of 1.7 parts per million (ppm) compared with 94 ppm for zinc, 12 ppm for lead, and 63 ppm for copper (1). A remarkable aspect of the known resources is the apparent dominant position for the circum-Pacific deposits, particularily those in Southeast Asia, China, Siberia, and Bolivia. These deposits can, however, be classified into a number of different types, which differ again from those in Australia, Europe, and Africa. Relatively recently the location of tin deposits in Brazil and eastern Canada have made it clear that not only the Pacific but also the Atlantic has tin on both its borders (2). More recently still newer discoveries of tin in Czechoslovakia (3) (4) and India (5) have indicated that the possibilities of locating new resources of tin have not yet been exhausted.

In a paper delivered to the Fourth World Conference on Tin, Hosking (6) referred to the need for exploration for new tin resources and to the many problems which confront such exploration. He suggested that it is necessary to establish as far as possible the patterns of tin occurrences, within metallogenetic provinces, mining fields, individual deposits and, on both a macroscopic and microscopic scale, in the ore itself. He pointed out the problems which could arise in allowing classification of mineral deposits to dominate exploration thinking and referred to one tentative basic classification of tin deposits from which he had already had to depart when classifying, for example, the primary tin deposits of Malaysia.

The Conference concluded that the problems of coordinating the knowledge of the genesis of tin ores which has so far been gained and of creating a classification of tin provinces and tin deposits within those provinces have so far proved insoluble. Clearly, further investigations in these respects would enhance the possibilities of better delineation of world tin resources. Further, detailed knowledge of the characteristics of tin ores, the tin-bearing minerals contained therein, the associated minerals, and the nature and characteristics of the gangue are essential for the efficient design and operation of the mining, beneficiation, smelting and refining processes which are involved when the ore deposits are developed.

The author has referred elsewhere to the importance of this information in dealing with the problems of process design for tin placers (6).

Considering the low crustal abundance of the metals it is not surprising that economic deposits have been characterized as geologic freaks. Tin deposits are no exception. Catavi, in Bolivia, the world's largest tin mine, would never have been discovered if it had been covered by the sea of lava, which, in the event, lapped only around its foot and now conceals who knows what deposits which remain to be discovered - an intriguing problem for the geophysicist. The vast placers of Southeast Asia exist only because the tin, overwhelmingly as cassiterite, was disseminated into readily weathered limestones and sandstones which under the influence of humid tropical climate and favorable flat topography gently and conveniently shed their content of cassiterite (which, of all minerals, is one of the most resistant to weathering) into limited flat areas, eminently convenient for exploration and exploitation. This is not to say that exploration for - and of - placers is a simple matter. Placers are complex geologic formations, and indeed, the mechanisms of their evolution are still imperfectly understood (7). The placers of the Andean deposits in Bolivia pose special problems in prospecting and working due to steep topography and bouldery, morphology, the latter largely due glaciation.

The major mineral of tin is the oxide, cassiterite but it is often a minor mineral in its ores, the content of zinc as sphalerite in lode ("hard rock") ores being often far higher as at Colquiri in Bolivia, Wheal Jane in Cornwall, and in Nigeria (8). It is also accompanied by lead, bismuth, arsenic and antimony, by tungsten, tantalum, niobium and rare earths, and by titanium and zinconium minerals. Although the presence of these other elements causes problems in beneficiation and smelting, some of them particularly copper, zinc and arsenic, may be useful as "pathfinder" elements if found in anomalous amounts during geochemical exploration in known tin provinces. In tin placer areas, it is not uncommon to find zones where ilmenite predominates over cassiterite to the extent of four or five to one. Outstanding problem ores, difficult to locate and to treat, are typified by the Renison and Cleveland deposits in Tasmania, where the gangue is largely iron sulphides and liable to spontaneous combustion in the ore-bins.

A major problem in evaluating deposits of tin ores, apart from estimating recoverability, is the definition of cut-off grade. The question of recoverability of tin will be referred to later. It is generally conceded to be poor and to lead to large losses; but large losses, in a sense, are also sustained because some portions of known deposits are not worked at all, although their tin content could be profitably won, because the tin content is so low that, with the equipment installed, the target of production set by design or by policy cannot be met. Today some placers could be profitably worked with a tin content of 50 ppm but are not because the production target of the mine and the capacity of its equipment demand a feed grade not less than 200 ppm. The economics of installing additional equipment and the design of operations to re-mine for recovery of the previously rejected ore then become additional problems, often not readily solved. Thus, there is often a tendency to turn to "high-grading" (8).

Nevertheless, attention has been drawn to the potential value of specific search for large low-grade bodies of tin ore analogous to copper porphyries. Taylor (9) has deduced that deposits for "hard rock" ore containing quantities of the order of 40 million tons and operated at rates of the order of 10,000 tons per day could be worked profitably at a grade of 0.2 percent Sn.

As referred to above, there cannot be said to be a unique philosophy of exploration for tin deposits, especially "hard rock" deposits. Nevertheless, since tin deposits are generally associated with acid igneous rocks, there is general interest in the problem of the tin content of such rocks and in the comparison of, for example, tin-bearing and tin-barren granites (2b) (10) (11). There is likewise interest in the ages of such rocks and in plate-tectonic theories of their emplacement (12) (13) (14) (15).

Reference has also been made to the utilization of geochemical exploration techniques. Some useful results have been achieved in recent years in the study of problems arising in such work (6) (16) (17) (18) (19).

Turning now to problems of exploration of tin placers one of the greatest to be listed might be that of access. Swampy and overgrown as they often are, heavy mechanical drilling equipment is often out of the question. Manual labour with light, sectionalized drills such as the Banka drill has been found to be the only possible solution in the history of such undertakings. The mechanization of such a drill while striving for portability has had some limited success in recent years. The drilling is performed by turning and forcing a casing into the ground and bailing out its contents, length by length. The problems are: how to achieve 100 percent recovery of the core, how to avoid bailing beyond the bottom of the casing, what to do when a layer of running sand begins to rise in the hole or when encountering a boulder, how to recover stuck casing and many others. Estimating the tin contents of the sample brings more problems. There is no chemical field assay for tin except a colorimetric test which can be used for geochemical prospecting but has not been applied successfully for evaluation of drilled samples. Conventional analysis at the ppm levels encountered is imprecise and in any event the samples are wet and bulky, militating against transport. The solution is panning and evaluation of the pan concentrate. Losses of fines inevitably ensue, and it has been shown by the author that losses of coarser fractions also surprisingly occur (20). Complex combinations of panning, separation of constituents with or without the aid of heavy liquids, grain-counting, weighing and chemical analysis eventually yield a figure for tin present which can be related to recoverability from its related volume of ground. Underestimate is a general rule, leading to figures of recovery during mining in excess of 100 percent. The author has experimented with the use of portable isotope X-ray fluorescence for the analysis of field concentrates and has initiated research into the use of magnetogravimetric methods for the separation of cassiterite from its associated heavy minerals, so far with only limited success (21). The Southeast Asia Tin Research and Development Centre, formed with the aid of the United Nations, is preparing to test the possibility of down-the-hole analysis for tin by neutron activation analysis (22).

Finally, turning to exploration for tin offshore, the problems become compounded by the obvious factors of lack of visibility and access. The first approach was inferential, following a placer on land down to the coast and then on out. This was followed by tentative drilling with Banka drills on pontoons to follow a contact zone out to sea or to seek submerged river channels. More recently, sophisticated geophysics elucidated the sub-bottom profiles by seismic reflection and refraction aided by side-scan sonar. The work, confined by drilling, has shown that marine placers are not only submerged alluvials but also eluvials formed by weathering in situ of intruded mineralizations prior to submergence by a rising sea level subsequent to the last glaciation period. Geophysics has indicated its capability of identifying the unconsolidated sea-bottom sediments, the depth and nature of the bedrock, and the locality and characteristics of the intrusives; and new offshore tin fields are being discovered by these means combined with intuitive and investigative geological innovations (23-29), once the problems of operation, navigation, positioning, and interpretation of results have been overcome. (Note: Following on the International Symposium, Geology of Tin Deposits, held in Kuala Lumpur, Malaysia in March 1978 by the Geological Society of Malaysia a series of papers dealing in general with the subject and also with the particular deposits of Malaysia, Indonesia, Burma, Cornwall, Brazil, South Africa, India, Thailand, Nigeria, Australia and Bolivia, will be published in 1979 as Bulletin 11, Geological Society of Malaysia.)

Mining and Processing

The methods used for underground mining of "hard rock" tin ores do not differ in any major respect from those used for other ores. It is not proposed to catalogue them here or to comment on the problems involved except for two. One of these affects the succeeding process of recovering the cassiterite. This mineral is brittle and consequently easily shattered. It is therefore important, in designing methods of extracting the ore, to introduce drilling and blasting procedures which avoid over-breaking and the production of excessive fines. All the tin minerals are difficult to recover in sizes below 10 micrometers. This applies particularly in the case of cassiterite, a problem mineral, for which, it must be conceded, there is no specific method of recovery in any size range.

The other problem encountered is the emission from the granitoid intrusives of radon gas, the decay products of which, radon daughters, are a radioactive health hazard which has to be combated by efficient ventilation practice (30).

While the underground mining methods used do not vary much in principle from country to country the details are often strikingly different. In mountainous Bolivia, for example, the ore-bodies were entered by adit from the mountain-side and mined <u>upwards</u> into the mountain. The visitor would be surprised to note the reversal of the expected process of going down the mine.

Open-pit mining for hard rock tin, as at Ardlethan, New South Wales, is rare. The Ardlethan tin operation has been well described in detail by Shanahan (6). The main problem of mines of this type having low-grade ore is the accurate sampling and analysis required for grade control. The accurate determination of tin by chemical analysis has always been a problem (31-35) which has been alleviated to some extent when assaying for lower tin contents by the more recent availability of X-ray fluorescence analysis and atomic absorption spectrometry. The former has to cope with interference from matrix effects which with any given ore can be dealt with by some known methods but, on a trial and error basis, varied with changes in the ore. The latter is also subject to interference by other elements. Thus, it is extremely difficult to maintain a metallurgical balance, both feed and tailings assay being subject to quite wide ranges of discrepancy.

Other open-pit mines such as Sungei Besi in Malaysia and Pemali in Indonesia are "soft rock" mines where deep-weathering allows for excavation on benches without the use of explosives but, thereafter, the decomposed ore is treated almost by placer methods. The problems there have been in effecting the excavation of wet and stocky ore by means of power shovels and bucket-wheel excavators and in successfully disintegrating the ore and de-sliming it prior to the processing sequence. The hydrocyclone, for de-sliming and de-watering of placer ore-pulps began to be introduced in the middle 50's but is still not always used when indicated, or effectively used when finally installed.

Tin placers are mined by dredging or by hydraulicking. In the first case, the dredges (mostly bucket-line but a few cutter-section dredges are also at work) excavate the spoil and deliver it to a treatment plant on board, which makes a rough concentrate to be sent ashore for final dressing in the "tin-shed." In the second case, the ore is worked in an open pit by sluicing down the sides with high-pressure monitors. Ground is sometimes broken ahead of the monitors by drag-line or bulldozer. The slurry so formed is pumped out of the pit by gravel pumps to a treatment plant. Such a mine is called a gravel-pump mine. The term "hydraulic mine" used to be used to classify a mine where the ore was sluiced down by monitors and then pumped out of the mine by hydraulic elevators, powered by high-pressure water piped down from the hills. These elevators have now been superceded by diesel-or electrically-operated gravel pumps which are easier to install and operate and to move from location to location as the mine progresses.

The pulp elevated by the pump flows over a screen to remove over-size (3/4 - 1 inch) and boulders and then down a sluice up to 200 feet long divided into lanes six feet wide (in Malaysia, Thailand and Burma) or into jigs (in Indonesia). The losses in sluices used to be very high largely because the high rates of flow carried away the fine (say minus 150-mesh), but also because the lanes were run too long before cleaning and were then cleaned out by manual labour. Curiously enough, the photograph usually used to illustrate this type of mine is still one which shows a large number of labourers cleaning out one of the old-type sluices or palongs. This is a thing of the past and should be replaced by the photograph showing the more efficient but less glamorous modern equipment described by Jacket-Simpson (6) which evolved from methods pioneered by the author and co-workers, in particular the late Leow Yan Sip (20) (6). Improvement in the operation of the Indonesian jigs still awaits the adoption of the newer methods. Whereas the gravel pump mines used to recover scarcely more than half of the tin in placers the figures quoted by Jacket-Simpson show recoveries of 80 percent where the ground contains substantial amounts of cassiterite finer than 150 mesh and up to nearly 95 percent in other cases. Leow Yan Sip has reported even lower losses.

In the dredging industry the recovery picture is less clear. In the first place a dredge is unable to extract payable ground at the lowest part of the placer if the bed-rock formation is extremely uneven or pinnacled or below the digging depth of the dredge. Then come the difficulties of on-board treatment. The excavated material is first passed through a revolving screen where the oversize material is washed clean by high-pressure water-jets the remainder of the tin-bearing ground passing through. Should the ground contain heavy clay, a problem arises because the clay lumps may not disintegrate but, on the contrary, pick up grains of already liberated cassiterite and roll with them out of the system. The underflow from the screen amounts to 7000-10 000 ppm and this flow has to be divided into several streams and passed to the recovery plant, which usually consists of many banks of primary jigs. The problem here is to adjust matters so that the high speed flow does not carry values into the tailings. The difficulty of sampling these flows (strangely, mechanical samplers are never seen) makes any estimate of their efficiency suspect. Large circular jigs have recently been introduced to replace the smaller rectangular or trapezoid jigs formerly used and one report by McDonald claims high recovery, 95-97 percent (6). The same paper refers to other problems in the development of large-capacity dredges which include not only mechanical and operational design problems but also the emergence of an optimization factor indicating an economic limit to the size to which a dredge can be built. Cheng and others (6) have also referred to the problems of preparing and developing a placer property efficiently and adequately for initiation of large-scale dredging. In the discussion on these papers, the merit of de-sliming and de-watering the feed to the jigs, in order to improve recovery with fewer jigs, was mentioned but ensuing engineering complications have up to the present resulted in failure to find a means to adopt this proposal in general. Only one dredge so far has been built on this principle. It has had a chequered history and no reports on its performance have been published.

Dredge design is, however, not static. Problems arising in the mining or various deposits induce changes in design. The thin layer of tin-bearing sediment lying offshore Takuapa, West Thailand is one example. There were unique problems in exploring this deposit and developing the side-trailing suction dredge to mine it, as described by McDonald and Wong (36). Deiperink and Donkers also describe an innovative design of sea-going bucket-line dredge with a flexibly suspended ladder to cope with sea-state and wave action (26). A bucket-wheel cutter suction design has also been considered. Hewitt has further thoughts on future possibilities including a continuous bucket line and multi-grab configurations (37).

The rough concentrates from the dredges are worked up ashore by jigging and tabling to give clean concentrates up to 74 or 75 percent Sn in content, while the middlings are dried and separated by magnetic and high-tension (electrostatic) means with a final removal of sulphides by flotation. These processes yield a further concentrate of cassiterite and marketable by-products or ilmenite, monazite, xenotime, zircon, columbite and stouverite. Gravel pump mines, usually very small companies, separate only the high-grade concentrate and sell the middlings to a contractor. The problems in middlings treatment are, mainly, first that some of the cassiterite is magnetic and gets lost into the ilmenite product and, secondly, that it is not possible to separate all the niobium and tantalum from the cassiterite. Some is included mechanically and some is enclosed within the cassiterite grains and goes to the smelter, where it appears in the slags. Where, however, the concentrations of niobium and tantalum, particularly the latter, are high enough the slag becomes marketable as a raw material for the hydrometallurgical extraction of the two elements. Process design for tin placers has been discussed by the author elsewhere (6). While executing a United Nations project in Indonesia, the author and his team built several new treatment plants which solved problems of recovery.

It is in the processing of lode ores that the greatest problems are encountered. As stated previously, there is no real specific means of recovering cassiterite. In the case of the other major metals, there are available selective flotation and metallurgical processes which readily produce high recoveries. Even the major industrial minerals can be recovered cleanly by specific processes. All these processes are geared to specific properties of the minerals. Cassiterite, however, has only one specific property of any use and that is its specific gravity. That, however, does not suffice for clean separations since most of the associated minerals have specific gravities in high ranges and are often liberated at sizes coarser than that of the accompanying cassiterite, thus reducing the differential. The cassiterite crystal is not hydrophobic, at ordinary temperatures and pressures it is insoluble and inert to most chemical attack. Hence, it does not readily adsorb a flotation collector so that its recovery by flotation is far from outstanding and often appears fortuitous rather than specific.

Nevertheless, a great deal of work has gone into flotation research. Many hundreds of references could be cited including numerous patents. Only four are listed here (38-41), mainly dealing with recent successful use of cassiterite flotation in operating mines. The term "successful" has to be applied with qualifications since the concentrates produced are always of low grade and the stage recoveries are not high. The reason is, partly, that cassiterite slimes will not float and they hinder flotation of other sizes. Coarse cassiterite also will not float. So flotation feeds have to be scalped and de-slimed. Losses in the slimes can be 10-15 percent, or even more, of the tin in flotation feed. The principal successful collectors are succinamates, alkyl arsenates, and alkyl phosphonates, but their varying success with different ores gives the impression that the flotation characteristics are due not to the cassiterite itself but to point defects in the cassiterite lattice to which the collector latches on. In many cases, the use of flotation could be challenged by reference to an alternative gravity approach. Nevertheless, flotation does not provide a route to recovery of ultrafines and revival of oil flotation in a new guise is now approaching the problem of selective recovery of cassiterite from previously rejected slime fractions (42) (43).

There is also a new theory by Schulze on the upper particle size of flotability in general which probably has relevance to the cassiterite problem (44), while a paper by Jamieson (45) on physical factors affecting recovery rates in flotation gives rise to thoughts on the effect of reducing bubble size.

It is therefore understandable that much attention is being paid to the gravity concentration of cassiterite fines. First it must be stressed that overgrinding of ores should be avoided in order to reduce the production of fines. Finch and Matwijenko (46) point out that a grinding circuit closed by a cyclone can cause, by reason of the concentrating action of the cyclone, liberated heavy mineral to return to the mill. Attention is being paid to screening the cyclone underflow to recover liberated mineral which should have reported to the overflow, or even to closing the circuit with a screen (55).

New equipment has been invented and marketed which gives promise of higher recoveries of fine cassiterite down to 5 micrometers in size (47) (48) (49) (51), while work on the behaviour of various materials as collecting surfaces shows that there is a variation and that the pH of the pulp also has an effect (50).

A modern gravity mill for a complex tin ore is described by Shanahan (6). Ottley (52) refers in some detail to the problems of gravity concentrator design and operation. Bignell (54) considers it possible to predict and assess gravity separator performance from heavy liquid data. A word of caution is necessary in considering heavy medium separation for early rejection of waste. Commonly, specific gravities of 2.7 - 2.9 are used for the medium, and at this density a floating (i.e. reject) particle may easily contain several percent of tin, unliberated. In such cases, finer crushing and jigging should be investigated as an alternative.

With all the problems of processing tin ores still before us we turn with some interest to Holland-Batt's proposals (53) for design of gravity concentration circuits by mathematics.

Smelting and Refining

Until the middle of this century, the metallurgical problems of tin smelting received little publicity. The main problem experienced by the smelting companies was that of overcapacity and competition for the supply of concentrates. Consequently, the smelting kept their processes to themselves and virtually nothing appeared in the technical literature. Research, if any, was not made public and there was no patent literature on the main carbo-thermic process. It was generally understood that cassiterite was readily reduced to tin metal, at moderate temperatures, by carbon, that a two-stage process was required, involving the circulation of iron and that the tin loss in the final slag was high. The charges levied by the smelting companies included a unitage deduction (deduction of one or more units from the actual assayed percentage tin content of the concentrates) to allow for this loss. This and further penalties for "impurities" irritated producers, from countries lacking their own smelters, who had to ship their concentrates to distant points for smelting. They eventually decided to own their own smelters and solve their own technological problems and thus stimulated scientific activity into research on the existing methods and invention and development of new ones.

The fact of the matter is that tin smelting technology from the earliest times was built up around virtually pure cassiterite concentrates which are very readily reduced to metal. By carbon reduction, about half of the tin content can be recovered almost pure; but, if the reducing power is raised and reduction continued, the remainder of the tin will be contaminated with iron if such be present. The standard free energies of iron and tin oxides are so similar that this is unavoidable. Armstrong and Wheelwright describe (6, Vol. III, p. 89-110) how early smelters overcame this problem by partial smelting to recover a first-run pure metal and then smelted and re-smelted the slags with added carbon to yield a metal carrying about 95 percent tin which was refined by liquation. Final slags, still quite rich had to be discarded. Sooner or later they must have found how to use the liquation dross by returning it to the first stage smelt; probably as a result of using scrap iron in their slag smelting, and thus the two-stage smelt in reverberatory furnaces as used today and described by Armstrong and Wheelwright (6) or in rotary furnaces as described by Batubara and Mackey (6, Vol. III, p. 149-177) or in electric furnaces (66) has developed. The first stage produced tin with a limited amount of iron which can be drossed out by blowing with air and, after liquation, returned to the circuit. The rich slag is then intensively reduced to yield a tin-iron alloy until reduction will proceed no further.

These are not solid-state reactions; the cassiterite is reduced not by carbon but by carbon monoxide. The carbon has first to be oxidized in order to achieve this gaseous reduction and one of the problems of tin-smelting is to achieve this by correct adjustment of CO/CO_2 ratio in the furnace atmosphere. The author and co-workers found that oxygen fed precisely to the zone of reduction could produce a marked increase in rate of reduction and, hence, in furnace throughput but little use seems to have been made of this discovery. The alloy is returned to the first stage. The iron eventually has to leave the circuit in the final slag and not be allowed to build up in intermediates. There is an equilibrium between the iron/tin ratio in the alloy and that in the slag which was decribed by the author and his co-worker, Hallett (56) in 1967. This revealed for the first time the factor controlling the tin loss in final slag. It was shown that over a range of iron-tin ratios corresponding with the miscibility gap in the tin-iron system, slags in contact with iron-tin alloys would have an iron-tin ratio of between 10:1 and 12:1, the higher the iron the lower the ratio. Beyond the miscibility gap, where the iron content of the iron-tin alloy exceeds 60 percent, the iron-tin ratio in the slag rises dramatically. John Boxall, at Williams Harvey tried to take advantage of this by running a third stage of reduction but the alloy formed had a melting point too high for the furnace capability and set solid, so the process had to be abondoned. Hence, the tin loss in final slag is controlled by the amount of iron coming in to the circuit. Many attempts were made to devise means of removing iron from the circuit especially at the Williams Harvey and Texas City smelters which had to cope with high-iron concentrates from Bolivia. Texas City had cheap acid and could leach out much of their iron. Williams Harvey had not and many wondrous pyrometallurgical, hydrometallurgical and electrometallurgical routes were tried before it had to be conceded that, taking into account the energy content of the alloy when acting as a reductant in the first stage, it just did not pay to treat it by any other means. So, until the arrival of fuming, the tin lost in final slags stayed lost. If the iron intake dictated 25 percent Fe in the final slag, the Sn was an irrevocable 2.5. This, of course, was the reason for the unitage deduction - the smelters could not pay for tin they could not recover.

In a reducing atmosphere, cassiterite will react with sulphur or sulphides to form stannous sulphide which is volatile and which in contact with air, will burn to stannic oxide in the form of fine "fume" (actually a dust) which can be collected. This process has been used and is being updated (65) for treating low-grade sulphidic tin concentrates. The fume is then smelted in the same way as cassiterite concentrate. Around 1940 Phelps-Dodge utilized the reaction for fuming tin from slags by injecting pyrite and fuel into the slag in a converter. The process was abandoned there for reasons of cost and difficulty of maintaining regular injection, but taken up again at Williams Harvey where the author simplified the process by adding the pyrite batchwise directly into a re-designed vessel solving the problem of injection. Thus, the problem of the iron/tin ratio in the final slag was solved. It no longer applied.

Wright (6) has explained the chemistry of the process and referred to the problem which can arise when too high a CO_2/CO ratio caused the formation of magnetite and the consequent explosive reaction with pyrite. Wright also compares the economics of two or three stage smelting with that of a single-stage smelt followed directly by fuming of the rich slag (thus eliminating the re-cycle of alloy) and concludes that, for a new smelter the latter is probably preferable.

In conventional smelting, tin escapes from molten slag as volatile stannous oxide, SnO, at normal pressure and has to be recirculated. The author and his team in the Consolidated Tin Smelters Ltd. research department investigated the greatly increased rate at which the oxide was evolved at reduced pressure and pioneered a vacuum fuming process for which a patent was obtained. This method avoids atmospheric pollution since there are no emissions. A problem here, keeping the slag hot enough since the rapid evaporation of SnO induces rapid cooling, may have been overcome by Barrett, Howie, and Sayce (60) who employ electrical heating on a thin layer of slag continuously renewed in a rotating furnace.

Wright also describes the potential for chloride volatization and Esdaile and Walters, in the same volume (6, Vol. III) made new proposals along the same lines. There has been reluctance to bring these proposals to commercial scale until recently when the Texas City smelter commissioned a chloride volatilization process using a top-blown rotary converter. Industrial results are awaited with interest. Use of this process with total recycle of chlorine removes the environmental objection of sulphide fuming processes which emit sulphur oxides.

It must be recognized that these new processes and others which will doubtless be described or referred to at this meeting arise because of the inability of mineral beneficiation processes to extract cassiterite cleanly and with high recovery from its gangue and associated minerals. The cleanest possible concentrates on the one hand leave the problem minerals in the middlings on the other and the striving for high grade (to save transport and smelting charges) leaves values in the tailings.

To smelt middlings or low grade concentrates containing all the arsenic, antimony, bismuth, copper, lead, zinc and whatever requires a complexity of processing including electrolytic refining, a high cost process, from which the impurities now emerge as values in anode muds, or vacuum refining, a less expensive process, developed by the author and C. B. Gee and the team at Williams Harvey. See also Ref. (67). The tailoring or processes to treat the little piles of intermediate which grow along the way is an endless task. Problem flowsheets either lead them circulating endlessly or dangling loosely at the end of a line going nowhere - obviously impossibilities. See also Ref. (70).

Such complexities have induced efforts to recover tin from its ores and concentrates, especially low-grade concentrates, by a variety of hydrometallurgical and other methods of which the more recent are listed in the references (63) (64) (68) (69), but none is able to cope with the situation which arises because all the impurities are co-extracted and introduce the same problems of dealing with intermediates. Of course, ways can always be found to deal with the problem materials, sometimes on a break-even basis or, when all else fails, selling them at scrap prices to another smelter whose circuit is capable of absorbing them.

There is clearly a need in a country like Bolivia which has complex ores and is now planning a complex smelter to integrate the mining, processing, smelting, and marketing of its production in order to obtain the most cost-effective mix and the maximum value added. When the United Nations was assisting Bolivia in setting up the Instituto de Investigaciones Minero-Metalurgicas in Oruro 1966-1970 and Peter Wright, the author, and Luis Pommier were, in turn Director, the team began just such an investigation and, with inputs from Davey (57) (58) and others, put forward the outline of such a scheme which, according to Lema (59) may now be gathering momentum.

Environmental Problems

Public objections to mining usually focus on unsightly surface works, waste piles, tailings areas, and muddied rivers. These are common enough objections in populated areas like Cornwall but totally absent in the remote mountains of Bolivia. Operations in the placers of Southeast Asia, however, are much more an object of public dismay since they compete for land with agriculture, forestry, husbandry and habitat. Worked-out areas are not reclaimed except in small isolated instances and then, perhaps, too soon.

The problem is that alluvial ground is never completely worked out. At whatever cut-off grade exists at a given time economics will determine that lower grades will be left in the ground. Unworked layers exists at depths not reached by early dredges, or in pockets between pinnacles. When rising metal prices and ever more efficient methods lower the cut-off grades such ground will be re-worked, sometimes again and again. Mining law allows small leases - too small to be worth reclaiming - and even prohibits the use of the land for any use except mining, so the miner has no incentive to reclaim and farm it.

The United Nations has commenced a study of this problem and has come up with a promising solution, basically as follows: At any given period of mining the ground will be worked to the limit within the bounds of current economics and available technology. Without great difficulty the tailings can be laid behind the advancing operation in a condition suitable for immediate fertilization and planting of legumes and other animal feed crops. Within a short time ranching can commence. The benefits clearly outweigh the costs so we can have environmental reclamation of the mine site at no cost to the miner. If the miner would like to carry on and profit from the farming, so much the better (but Government must amend the law accordingly). The farm can follow the mine as the mine progresses along the placer until the placer is exhaused and becomes all farm. Should, in the future, the price of the metal or the technology for its recovery make the residual tin, which is still in the ground, a payable proposition, then mining could start again virtually without disrupting the ranch because it could progress forward into the farmland and simultaneously reclaim it behind exactly as it did in the first place. This "total" land use could even conceivably go through more than two cycles.

The other environmental problem which is frequently debated is actually a matter of conservation related to avoidance of waste of tin, namely the waste of the tin can. Tin is, of course, re-cycled by reclamation from off-cuts of tin-plate arising from can production, but reclamation from used cans is plagued by problems of the cost of collection and cleaning prior to stripping and recovery of the tin. Work on this continues and there are likely to be increases in such reclamation in favorable circumstances (71).

In the meantime, an artistic lady in Connecticut has discovered unsuspected beauty in the used can and by cutting and shaping produces spectacular transfigurations into utensils, ornaments, and jewelry and so contributes to recycling and avoidance of waste (72).

Marketing and Consumption

Tin, as with other commodities, responds to the law of supply and demand. Its price will rise when it is in short supply and fall when plentiful. Its price will also fluctuate due to speculation and its value as a hedge against inflation, currency fluctuations, and business risks. Nevertheless, a plot by Legoux and Diethrich (6, Vol. IV, p. 85-97) of the monthly average price in terms of constant currency units over a long period shows that swings in price are relatively shallow. Tin on this basis has behaved more quietly than copper, lead, and zinc until, in 1973, external causes resulted in a sharp rise. This steadiness is attributable to early tin agreements and to the UN-sponsored International Tin Agreements which commenced in 1956. It is not intended to go into the history of these Agreements here. They are well documented by Fox (74) and others (77). Suffice it to say that tin is still the only metal produced under international agreement, that the agreement is durable and that it achieves stability. It is a pity, however, that it has been classed as a cartel (75). It is not. Its members include both producers and consumers and, as such, by definition cannot be a cartel. It is a genuine agreement, in principle and practice, in the long-term interest of producers and consumers. Apart from this another influence on the stability of the market is the U.S. stockpile. Herkotroeter (6, Vol. IV, p. 57-81) makes a case for the stockpile having had a stabilizing effect but points out that uncertainty about sales, changes in GSA policy and related matters "have at times been sources of uncertainty in the market and, as such, causes for considerable price fluctuations in the shorter term". See also (73) (78).

There are two sets of problems in the marketing of tin. One is posed by the marketing of concentrates from the mine to the smelter and the other by the marketing of tin from the smelter to the consumer.

Since the smelters have always been so secretive and since practically nothing was known about the physical chemistry or costs of their processes their unitage deductions, treatment charges, and penalites for impurities were grounds for suspicion and mistrust. Also, since there is no producer price and all valuation of concentrates is done on the basis of Penang or IME prices, the producers tend to resent their dependence on the market.

The Penang market grew out of the needs of multitude of small-scale gravel pump mines in the Peninsula for a regular cash flow by means of spot cash on their production of concentrates. The price is fixed daily (79) by the Penang smelters (private corporations, independent of the miners and with no captive mines) who equate bids by traders for metal with offers by mines of concentrate for sale and then sell physical metal at the highest single price at which all offers and bids can be paired. Bids above the cut-off price pay only that price. There is no futures market. However, the seller has the option of pricing his delivery during any of the next 28 days. This appeals to the gambling instinct of the tin-miner. There is, however, no gamble in the metal which has to be delivered within 60 days. It is usually available in one to three weeks and may not be re-sold on the market.

The London Metal Exchange (LME) grew from the requirements for forward pricing of metals to cover the period of shipping of concentrates to smelters and smelting to metal for delivery. It, therefore, deals in cash and three months transactions and can thus be used for hedging and speculation. It is highly volatile as it needs to respond to sudden changes in the spot position and threats to the forward position posed by floods, breakdowns, strikes, or the like.

There is a further market in New York, but this is essentially only a traders' market where the traders fix among themselves the prices at which they believe they can make a profit on estimated spot, nearby and forward deliveries. This is a more sensitive market than the LME but on both there can be hectic scrambling for spot metal when as happens from time to time there is a delay in shipping metal already paid for and due for delivery.

The producers have felt themselves at a disadvantage in dealing with these smelter and market conditions and are aiming now at their own smelters and, in addition, at trading in their own metal (76). All the producing countries now have their own smelters. Indonesia is now vertically integrated and has its own selling organization. In Malaysia the largest producer, the Malaysian Mining Corporation (71.4 percent Government, 28.6 percent Charter Consolidated) which produces two-thirds of Malaysia's output of concentrates has by-passed Penang by having its production toll-smelted and taking over the marketing operation. This reduced the amount of tin available on the Penang market and pushed tin prices to record levels.

Malaysia's final aim is to have its own futures market in Kuala Lumpur on the lines of the LME. This awaits solution of a number of problems including the free convertibility of the Malaysian ringgit.

Bolivia's problem is to adjust the amount paid by the national smelter to the miner for concentrates by means of complex calculations based on values formerly obtained by export adjusted for freight charges and other dues.

So, as far as it goes, this recital has touched on the main problems of the tin industry today. We can be sure that, by tomorrow if they have gone away, they will have been replaced by others.

References

(1) K. L. Harris, Tin, Mineral Commodity Profiles, MCP-16, July 1978. United States Department of the Interior, Washington, D.C. See also: Slater, D., comp.-Tin, Mineral Resources Consultative Committee Mineral Dossier, No. 9. London (H.M.S.O.), 1974.

(2) (a) F. Hermann, Tin ore occurrences on both sides of the Atlantic Ocean, Berg-u. Huttenm. Mh., Vol. 122 No. 2a, February, 1977, pp. 59-63. (b) Also: T.D. Smith and A. Turek; Tin-bearing potential of some Devonian granitic rocks in S.W. Nova Scotia. Miner. Deposita, Vol. 11, No. 2, 1976, pp. 245-284. (c) Also: W. Pohl, Structural content of tin and tungsten mineralization in Rwanda. Berg-u. Huttenm. Mh., Vol. 122, No. 2a, February 1977, pp. 59-63. (d) Also: John McMahon Moore, Exploration prospects for stockwork tin-tungsten ores in SW England - Min. Mag., Lond,, Vol. 136, No. 2, February 1977, pp. 97-103. (e) Also: M. S. Pessoa de Souza, and P.V. Zalan, Cassiterite and wolframite occurrences in the south of the Serra Dourada, Goias State. - Min. Metal., Vol. 40, No. 385, April 1977, pp. 4-9. (f) Also: S. Banerjee, Tin in the base metal sulphide deposits at Geco, Manitouwadge, Ontario, Canada. Bull. Geol. Surv. India. A., No. 40, 1974. (g) Also: R.R. Potter, Tungsten, molybdenum join exploration targets in New Brunswick. - Nth. Miner, Vol. 63, No. 51, 2 March, 1978, p. C.10, C.13. (h) Also: W. M. Stear, The strata-bound tin deposits and structure of the Rooiberg fragment. - Trans. Geol. Soc. S. Afr., Vol. 80, No. 2, May/August, 1977, pp. 67-78. (i) Also: O. K. Bwerinofa, The Southern Province Tin Belt, Zambia. In: Utilization of mineral resources in developing countries. Conference, Lusaka, Zambia, 2-5 August, 1977, Vol. 2 Min. Mag. Lond., Vol. 137, No. 5, November, 1977, pp. 533-537.

(3) M. Treger, and I. Matula, New indications of tin mineralization in the Spissko-gemerske Rudohorie Mountains. Geol. Pruzkum, Vol. 19, No. 9, 1977.

(4) B. Kasak, and I. Matual, New information obtained from the research of the (tin-bearing) Gemeride granites in the Delava-Penisko-Majzlová area. (Spissko-gemerské Rudhorie Mountains). Miner. Slovaca, Vol. 9, No. 2, 1977, pp. 81-91.

(5) Tin discoveries in India. Min. Mag., Lond., May, 1976, p. 461; Min. Journ., Lond., June 23, 1978, p. 478; unpublished United Nations reports.

(6) Fourth World Conference on Tin. Kuala Lumpur, 1974. The International Tin Council, London.

(7) J. Adams, et al. Basin dynamics, channel processes and placer formation: A model study - Econ. Geol., Vol. 73, 1978, pp. 416-426.

References (Cont'd)

(8) P. Bowden and J. A. Kinnaird, Younger granites of Nigeria - a zinc-rich tin province. - Trans. Instn. Min. Metall., B., Vol. 87, May, 1978, pp. B66-B69. (b) See also: Y. A. Kettaneh and J. P. N. Badham, Mineralization and paragenesis at the Mount Wellington Mine, Cornwall. Econ. Geol., Vol. 73, No. 4, June/July, 1978, pp. 486-495. (c) Also: K. Atkinson, Wheal Jane - Br. Geol., Vol. 4, No. 3, July, 1978, pp. 47-48. (d) Also: J. A. Wells, In: The Development and operation of cassiterite flotation at Wheal Jane since commissioning. (Reference to mineralogical analysis, cassiterite 1.6 percent, sphalerite 7.7 percent, combined sulphides 34.1 percent). ·Simposio internacional del Estano. La Paz, Bolivia, November, 1977.

(9) R. G. Taylor, Observations on large low-grade tin ores, with special reference to Australia. - Trans. Instn. Min. Metall., A., Vol. 86, January, 1977, pp. A18-A27.

(10) D. I. Groves, and T. S. McCarthy, Fractional crystallization and the origin of tin deposits in granitoids. - Miner. Deposita, Vol. 13, No. 1, 1978, pp. 11-26.

(11) Haapala, - The controls of tin and related mineralization in the Rapakivi granites of Southeast Finland. - Geol. Foren. Stockh., Forh. 99, pp. 103-142.

(12) J. F. Evernden, et al. Potassium-argon ages of some Bolivian rocks. - Econ. Geol., Vol. 72, No. 6, September/October, 1977, pp. 1042-1061.

(13) M. T. Jones, et al. Age of tin mineralization and plumbotectonics, Belitung, - Econ. Geol., Vol. 72, No. 5, August, 1977, pp. 745-752.

(14) J. D. Bignell and N. J. Snelling, Geochronology of Malayan granites. - Overseas Geol. Miner. Resour., No. 47, 1977.

(15) A.H.G. Mitchell, Geosynclinal and plate-tectonic hypotheses: significance of late-orogenic Himalayan tin granites and continental collision. Instn. Min. Metall. Lond., 1978 (Paper 37, 11th CMMC, Hong Kong, 1978).

(16) H-J. Schneider, et al. Correlation of trace element distribution in cassiterites and geotectonic positioe of their deposits in Bolivia. - Miner. Deposita, Vol. 13, No. 1, 1978.

(17) W.W-S Yim, Geochemical exploration for offshore tin deposits in Cornwall. Instn. Min. Metall. Lond., 1978, (Paper 31, 11th CMMC, Hong Kong, 1978).

(18) G. H. Teh, K. W. Fung, and F. G. Hosking, The analysis of fluoride in tropical soils by selective ion electrode methods and its possible application to the search for sub-outcropping tin deposits in Peninsular Malaysia. - Bull. Geol. Soc. Malaysia, No. 8, December, 1977, pp. 151-158.

References (Cont'd)

(19)   Lenthall and Hunter, The geochemicstry of the Bushveld granites of
       the Potgietersrus tin-field. Precambrian Research, Vol. 5, pp.
       359-400.

(20)   J. H. Harris,   Serial Gravity Concentration: A new tool in
       mineral processing. - Trans. Instn. Min. Metall., Vol. 69, 1959-
       60, pp. 85-94, 295-318, 607-609.

(21)   T.B.M Rabelink and J. H. Harris,   Unpublished reports.

(22)   The Southeast Asia Tin Research and Development Centre, Ipoh,
       Malaysia.   Communication by the Chief Technical Adviser appointed
       by the United Nations.

(23)   G.J.J Aleva   - Aspects of the historical and physical geology in
       the Sunda Shelf essential to the exploration of submarine tin
       placers. - Geologie Mijnb., The Hague, Vol. 51, No. 2, March/April
       1973, pp. 79-91.

(24)   Paul F. Scholla and Associates.   Tin Deposits of Thailand.
       Bangkok, 1977.

(25)   G.J.J Aleva,   Exploration for placer tin deposits offshore
       Thailand.   Instn. Min. Metall., London. (Paper No. 1, 11th CMMC,
       Hong Kong, 1978).

(26)   F.J.H. Dieperink and J. M. Donkers,   Cassiterite deposits near
       Pulan Tujuh, Indonesia, and equipment developed for their mining.
       In: Oceanology International 78:   4th international conference
       for the offshore industries.   Technical session C. Instn. Min.
       Metall., London and BPS Exhibitions Ltd. 1978, pp. 16-21.

(27)   A.J.A. van Overeem,   Sonic underwater surveys to locate bedrock
       off the coasts of Billiton and Singkep, Indonesia. - Geologie
       Mijnb., Vol. 39, 1960, pp. 464-471.

(28)   G.J.J. Aleva, et al.   A contribution to the geology of part of the
       Indonesian Tinbelt: The sea areas between Singkep and Bangka
       islands and around the Karimata Islands. - Bull. Geol. Soc.
       Malaysia, No. 6, 1973, pp. 257-271.

(29)   E.H. Bon, Exploration techniques employed in the Pulau Tujuh tin
       discovery. - Trans. Instn. Min. Metall., A., Vol. 88, January
       1979, pp. A13-A22.

(30)   C. J. Dungey, et al.   An investigation into control of radon and
       its daughter products in some Cornish mine atmospheres. - Trans.
       Instn. Min. Metall., 88, April 1979.

(31)   J. A. Corbett and K. Riley,   An investigation of the discrepancy
       in results obtained using different methods for determining tin in
       some tin concentrates. - Invest. Rep. CSIRO, Aust., No. 116,
       December, 1976, p. 10.

(32)   K. Volker,   Methods of tin determination in tin ores of the
       Erzgebirga - Neue Hutte, Vol. 22, No. 8, August, 1977, pp. 451-
       453.

References (Cont'd)

(33) P. A. Wright, In: Discussions and Contributions. Tin: mining, metallurgy and economics. Trans. Instn. Min. Metall., A, April, 1977, p. 78.

(34) B. Lister, Second inter-laboratory survey of the accuracy of ore analysis. - Trans. Instn. Min. Metall., B, Bol. 86, August, 1977, pp. B133-B148.

(35) G. Kraft, The problem of sampling tin ores and concentrates. - Erzmetall, Vol. 31, No. 2, February, 1978, pp. 53-56.

(36) G.C.R. McDonald and Wong Kok Tong. Exploration and development of shallow coastal tin deposit by suction dredging at Takuapa, West Thailand. - In: Oceanology International 78, C. Offshore Mineral Exploitation. Instn. Min. Metall., Lond., 1978, pp. 7-15.

(37) J. A. Hewitt, Present and future trends in capabilities and design of sea-going tin dredges. - Ibid, pp. 22-27.

(38) Tin ore processing. Third Symposium on problems of tin processing, 1972. Freib. Forschft., No. A551, 1975.

(39) A. G. Moncrieff and P. J. Lewis, Treatment of tin ores. - Trans. Instn. Min. Metall., A, Vol. 86, April 1977, pp. A56-A60.

(40) S.R.J. Perkins, Ore treatment at Renison Ltd. In: Papers presented at the Tasmania Conference 1977. Australas, Inst. Min. Metall., Parkville, May 1977, pp. 129-139.

(41) R. H. Goodman, et al. A study of competing process requirements in a complex tin concentrator. Ibid pp. 141-154.

(42) A. G. Zambrana, et al. Recovery of minus ten micron cassiterite by liquid-liquid extraction. - Int. J. Miner. Process, Vol. 1, 1974.

(43) A. G. Zambrana and F. Arguedas, The recovery of ultrafine particles from Colavi deposit. - Simposio Internacional del Estano, 1977, La Paz, Bolivia. Ministerio de Mineria y Metallurgia.

(44) H. J. Schulza, New theoretical and experimental investigations on stability of bubble/particle aggregates in flotation: A theory on the upper particle size of flotability. - Int. J. Miner. Process., Vol. 4, No. 3, September 1977, pp. 241-259.

(45) G. J. Jamieson, Physical factors affecting recovery rates in flotation. - Miner. Sce, Eng., Vol. 9, No. 3, July 1977, pp. 103-118.

(46) J. A. Finch and O. Matwijenko - Individual mineral behaviour in a closed grinding circuit. - CIM Bull., Vol, 70, No. 787, November 1977, pp. 164-172.

(47) R. O. Burt and D. J.Ottley, Fine gravity concentration using the Bartles-Mozley concentrator. - Int. J. Miner. Process., Vol. 1, 1974, pp. 347-366.

References (Cont'd)

(48) R. O. Burt, Development of the Bartles Cross Belt concentrator for the gravity concentration of fines. - Int. J. Miner. Process., Vol. 2, 1975, pp. 219-234.

(49) R. O. Burt, On stream test-work of the Bartles Cross Belt concentrator. - Min. Mag., Lond., 137, 6 December, 1977.

(50) R. O. Burt, A study of the effect of deck surface and pulp pH on the performance of a fine gravity concentrator. - Int. J. Miner. Process., Vol. 5, 1978, pp. 39-44.

(51) A new slime concentrator - the vibrating-shaking vanner. - Acta Metal. Sin., Vol. 13, No. 1/2, 1977, pp. 35-45.

(52) D. J. Ottley, Gravity concentrator design and operation - developments and problems. - Min. Mag., Lond., Vol. 138, No. 1, January 1978, pp. 33-43.

(53) A. B. Holland-Batt, Design of gravity concentration circuits by use of empirical mathematical models. - Paper 21, 11th Commonwealth Mining and Metallurgical Congress, Hong Kong, 1978. Instn. Min. Metall. Lond.

(54) J. D. Bignell, Prediction and assessment of gravity separator performance from heavy liquid data. - Paper 51, Ibid.

(55) Mining Annual Review, London, 1978, p. 237.

(56) J. H. Harris and Hallett, G.D. Discussion of Davey, T.R.A, and Floyd, J.M. Slag-metal equilibria in tin smelting. Proc. Australas Inst. Min. Metall., No. 223, 1967, pp. 75-80.

(57) T.R.A Davey and F.J.F. Flossbach, Tin smelting in rotary furnaces. - J. of Met., N.Y., May 24, 1972, pp. 26-30.

(58) T.R.A. Davey and J. E. Joffre, Vapour pressures and activities of SnS in tin-iron mattes. - Trans. Instn. Min. Metall., C. Lond., Vol. 82, No. 3, September, 1973, C145-C150.

(59) J. Lema, G. Martinez, and J. Morales, Optimizacion tecnico economica para el tratamiento de minerales de estano. - Simposio Internacional del Estano, Ministerio de Mineria y Metalurgia, La Paz, Bolivia, November, 1977.

(60) M. F. Barrett, F. H. Howie, and I. G. Sayce, Oxide fuming of tin slags by use of electrical heating. - Trans. Instn. Min. Metall., C, Vol. 84, December 1975, C231-C238.

(61) R. Rawlings and U. J. Ibok. Gaseous reduction of cassiterite. - Trans. Inst. Min. Metal., C. Lond., Vol. 83, September 1974, C186-C190.

(62) Ibid. - Cementation of tin by aluminium. - Trans. Instn. Min. Metall., C. Vol. 85, March 1976, pp. C45-C48.

References (Cont'd)

— (63) I. J. Bear and R.J.T. Caney, Selective reduction of a low-grade
        cassiterite concentrate. - Trans. Instn. Min. Metall.C, Vol. 85,
        September 1976, pp. C139-C146.

— (64) Ibid. Extraction of tin from selectively reduced tin calcines as
        chloride. - Trans. Instn. Min. Metall., C, Vol. 86, March 1977,
        pp. C37-C40.

  (65) Erich A. Muller, Volatilization of tin from sulphur-bearing low-
        grade concentrates in the cyclone furnace. - Erzmetall, Vol. 30,
        No. 2, February 1977, pp. 54-60.

  (66) H. A. Uys, The metallurgy of tin smelting in a submerged-arc
        furnace. - J.S. Afr. Inst. Min. Metall., Vol. 77, No. 6, January
        1977, pp. 121-125.

  (67) R. Kammel and J. Mirafzali, Investigations on a process of
        selective vacuum refining of crude tin. - Erzmetall, Vol. 30, No.
        10, October 1977, pp. 437-444.

— (68) G. Holt and D. Pearson, Hydrometallurgical process for recovery
        of tin from low grade concentrates. - Trans Instn. Min. Metall.,
        C., Vol. 86, June 1977, pp. 67-C81. Also note by Wright, P.A., C,
        Vol. 86, September, 1977, p. C165.

— (69) D. Pearson, et al. Development of a hydrometallurgical process
        for tin recovery from low-grade concentrates. Trans. Instn. Min.
        Metall., C, Vol, 86, September, 1977, pp. C140-C146; and December
        1977, pp. C175-C185.

  (70) E. Muller and P. Paschen, Trends in smelting and refining of
        impure and complex tin concentrates. - In: Complex metallurgy
        '78. M. J. Jones, ed., London, Instn. Min. Metall., 1978, pp.
        82-90. See also Trans. Instn. Min. Metall., C., Vol. 88, March
        1979, p. C61, comment by N.J.B. Pocock and P. C66, comment by
        P.A. Wright.

  (71) W. R. Laws and E. Morgan, In: Waste in the process industries:
        The papers of a conference, London: Mech. Eng. Pub. for the
        Institution of Mechanical Engineers, 1977, pp. 7-13.

  (72) Lucy Sargent, (1) "Tincraft for Christmas", William Morrow and
        Co., Inc., New York, 1969. (2) "Tincraft", Simon and Schuster,
        New York, 1972. (3) "A Beginners Book of Tincraft", Dodd Mead
        and Co., New York, 1976. Reviewed in Tin and its used, Int. Tin
        Res. Inst., No. 118, 1978, p. 10-11.

  (73) W. Fox, Some thoughts on U.S. stockpile disposals. - Tin, Lond.,
        Vol. 46, August 1973, p. 272, p. 275.

  (74) Ibid. Tin: The working of a commodity agreement. London, Mining
        Journal Books Ltd. 1974.

  (75) K. W. Clarfield, et al. eds. Eight mineral cartels: The new
        challenge to industrialized nations. New York, McGraw Hill
        Publications Company, 1975.

References (Cont'd)

(76) N.J.B. Pocock, also P.A.A. de Koning, In: Tin: mining, metallurgy and economics. Trans. Instn. Min. Metall. A, Vol. 86, April 1977, pp. A76-A83.

(77) C. L. Gilbert, The post-war tin agreements: an assessment. - Resour. Policy, Vol. 3, No. 2, June 1977, pp. 108-117.

(78) G. B. Fitch, U.S. stockpile holds key to tin's price prospects. - Tin Int., Vol. 50, No. 10, October 1977, pp. 365-366.

(79) Anon. Workings of Penang. Metal Bulletin Monthly, London, September 1976, pp. 41-45.

DEVELOPMENTS IN THE SMELTING AND REFINING OF TIN

S.C. Pearce

Paul Bergsøe & Søn A/S
Glostrup
Denmark

Developments that have taken place in the field of tin smelting and refining are reviewed in this paper. In particular the development of the fuming process is traced and different flow sheets are discussed. The vacuum distillation process is reviewed as a means for refining tin.

Introduction

Historically tin has always been a comparatively expensive metal in relation to its fellow base metals, copper, lead and zinc. In the main this is due to the fact that it occurs relatively sparsely in nature and has not always been the easiest of metals to recover in mineral form.

Although predominantly occurring in nature as its oxide, SnO_2, cassiterite, this mineral is found in two distinct sets of circumstances. Firstly, cassiterite is found as distinct particles embedded in the rock as lodes associated with pyrite, arsenopyrite and in some places chalcopyrite. These ores are usually mined from considerable depths and as brought to the surface often average only 0.5-1.0% Sn. Due to the inert chemical nature of cassiterite, concentration techniques have mainly centred around gravity methods, and although it is not the intention in this paper to discuss these processes, it should be mentioned that recoveries in general are poor. Hence, as will be seen later, some thought has been given in recent years as to the point at which traditional concentration techniques should end and smelting techniques take over.

Tin is also found in alluvial deposits where distinct particles of cassiterite are present mechanically rather than chemically combined with the surrounding terrain. By washing very large tonnages of material, the small quantity of cassiterite particles contained in it can be recovered in a comparatively pure form assaying from 65% to 75% Sn.

Whilst probably three-quarters of the world's supply of tin is produced by smelting and refining concentrates from the alluvial deposits of four main areas, Thailand, Malaysia, Indonesia and Central Africa, the bulk of the remainder comes from the mining of a large number of lode deposits in Bolivia, the Soviet Union, the People's Republic of China and Australia. The concentrates from lode deposits, being of complex composition, are not readily treated by conventional methods and with the price of tin attractively high at the present time, much interest has been shown in new methods developed over the past twenty-five years to obviate the difficulties.

Tin Smelting

The simple, classical method of tin smelting has been known for hundreds of years and was described by Agricola (1). The flowsheet for this method is shown in Figure 1, from which it can be seen that the concentrates are smelted with a limited amount of reducing agent to give a crude tin, containing as little iron as possible, and a rich slag containing the bulk of the iron. This rich slag is then smelted at a higher temperature and under stronger reducing conditions to give a reject slag and a tin-iron alloy or hardhead as it is known. This hardhead plus refinery drosses are then returned to the first stage of smelting.

As pointed out by Wright (2) this circuit is fundamentally sound, but needs close control to work under optimum conditions. This optimisation was also studied by Davey (3) and a mathematical model established. Both should perhaps have emphasised that rigid discipline is necessary on the part of smelter managements not to be tempted into smelting disproportionately higher quantities of concentrates than rich slag when the tin price rises, or to attempt a greater degree of reduction whilst slag smelting in an effort to reduce the amount of tin discarded. Both result in an excessive quantity of iron and therefore tin being retained in the circuit. No doubt this has been one of the factors underlying the interest shown in the literature concerning methods of treating hardhead (4, 5, 6, 7, 8).

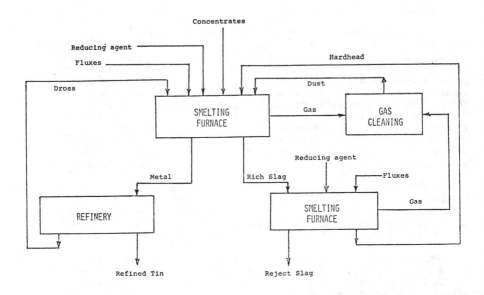

Figure 1.   The conventional tin smelting circuit.

The Chemistry of the Smelting Process

The chemistry of tin smelting appears extremely simple at first sight when summarised by the classical reaction,

$$SnO_2 + C = Sn + CO_2 \quad \dotsfill \quad (1)$$

This is, however, an over-simplification as the true reaction must take place in a number of stages. At the beginning of the process when the charge is heating up, carbon from the reducing agent will react with carbon dioxide of the furnace atmosphere to give carbon monoxide,

$$C + CO_2 \rightleftharpoons 2\,CO \quad \dotsfill \quad (2)$$

This carbon monoxide reacts at the surface of the solid cassiterite particles to produce tin and carbon dioxide,

$$2\,CO + SnO_2 \rightleftharpoons Sn + CO_2 \quad \dotsfill \quad (3)$$

Once sufficient liquid tin has been produced for the droplets to coalesce, they drain away by gravity, driving this reaction to the right. In most types of reducing furnace it is common to work with an open tap-hole during this early heating-up period to allow the tin to drain from the furnace.

As the temperature rises further, silica, which is a constituent of nearly all tin concentrates, reacts with cassiterite under reducing conditions to give stannous silicate,

$$SnO_2 + CO + SiO_2 \longrightarrow SnSiO_3 + CO_2 \quad \ldots\ldots\ldots\ldots \quad (4)$$

A further constituent of all tin concentrates is iron which is reduced from the ferric to ferrous condition when it also reacts with silica to form ferrous silicate,

$$Fe_2O_3 + CO + SiO_2 \longrightarrow FeSiO_3 + CO_2 \quad \ldots\ldots\ldots\ldots \quad (5)$$

Once the stannous and ferrous silicates have fused with any fluxes and formed a liquid slag, carbon monoxide ceases to be a very effective reducing agent. Carbon, still in the solid condition then becomes the predominant reductant and is capable of reducing both stannous silicate to tin and ferrous silicate to iron. This metallic iron, plus any introduced in the charge, can reduce tin from stannous silicate according to the reversible reaction,

$$SnSiO_3 + Fe \rightleftharpoons FeSiO_3 + Sn \quad \ldots\ldots\ldots\ldots \quad (6)$$

This is the equilibrium established at the end of each smelting cycle. It should be noted that as a considerable proportion of the tin produced in the heating-up stage has been drained from the furnace, it is only that metal remaining in the furnace at the final tapping time that comes into equilibrium with the slag.

At equilibrium the following relationship exists,

$$\frac{(Fe)}{(Sn)} = Constant \times \frac{(FeSiO_3)}{(SnSiO_3)} \quad \ldots\ldots\ldots\ldots \quad (7)$$

which can also be expressed as,

$$\frac{\% \ Fe \ in \ metal}{\% \ Sn \ in \ metal} = Constant \times \frac{\% \ Fe \ in \ slag}{\% \ Sn \ in \ slag} \quad \ldots\ldots \quad (8)$$

We can rewrite this in two ways, either,

$$\% \ Fe \ in \ metal = K \times \frac{\% \ Sn \ in \ metal \ \times \ \% \ Fe \ in \ slag}{\% \ Sn \ in \ slag} \quad . \quad (9)$$

or,

$$\% \ Sn \ in \ slag = K \times \frac{\% \ Sn \ in \ metal \ \times \ \% \ Fe \ in \ slag}{\% \ Fe \ in \ metal} \quad . \quad (10)$$

Although the absolute value of the equilibrium constant varies both with the metal composition and with temperature, for any specific set of conditions a definite value can be obtained. The actual numerical value is difficult to ascertain from the published literature since it is not always clear whether the iron content of the crude tin that is reported is related to the total tin produced or just to that portion remaining in the furnace at at tapping time. Nevertheless, an equilibrium does exist and cannot be avoided in a liquid tin - liquid slag system.

It will be seen from equation 9 that to produce a crude tin as low as possible in iron content, it is theoretically necessary to produce a metal low in tin, a slag low in iron and high in tin content. However, if the iron content of the metal is to be low, the tin content, which is the remainder, must be high and again all the iron which does not report in the metal, must report in the slag. Hence the only course open to the smelter management is to choose a limited amount of reductant in the charge which, of necessity, gives a high tin content in the slag. It is thus quite common to hear this slag termed "rich slag", and a second stage of smelting is required to recover its tin content.

A study of equation 10 shows that to obtain the lowest possible tin content in the reject slag from the second stage of smelting, it is desirable to produce a metal low in tin content and high in iron, together with a slag low in iron content. The iron content of the reject slag is however fixed in that it must contain all the iron arriving with the original concentrates, and it is therefore desirable to produce a metal, in this case hardhead, with a high iron content. By doing this and with good management it is possible to produce a reject slag from the classical two-stage smelting circuit with an iron:tin ratio of 10:1.

Three-Stage Smelting

A three-stage smelting process was devised by Boxall to obtain better iron:tin ratios in the final reject slag and was used to great effect for some years by Williams, Harvey & Co. Ltd.

In the three-stage smelting process, the tin concentrates were smelted to give a crude tin and a No. 1 slag of 15-20% Sn. The No. 1 slag was smelted to give a hardhead of approximately 30% Fe and a No. 2 slag of 4-5% Sn. The No. 2 slag was in turn smelted to give a hardhead of 60-70% Fe content and a reject slag of 0.5-1.0% depending on the iron to be eliminated from the circuit. Iron:tin ratios of 20:1 were obtainable. A certain amount of difficulty was experienced with the final stage of the process as a layer of high iron alloy would readily freeze on the cooler portions of the hearth. This required very strict rotation of charges to ensure that any such build-up was kept to a minimum.

A variation of this three-stage process is practised today by certain smelters in the Far East in which gypsum is added in the final stage. During this reduction, the calcium sulphate becomes reduced to calcium sulphide by the reaction,

$$CaSO_4 + 2 C \longrightarrow CaS + 2 CO_2 \quad \ldots\ldots\ldots\ldots (11)$$

The calcium sulphide formed reacts with stannous silicate to give calcium silicate and stannous sulphide,

$$CaS + SnSiO_3 \longrightarrow SnS + CaSiO_3 \quad \ldots\ldots\ldots\ldots (12)$$

As stannous sulphide is volatile at the working temperature of the furnace, it distills and burns in the furnace atmosphere to tin oxide,

$$SnS + 2 O_2 \longrightarrow SnO_2 + SO_2 \quad \ldots\ldots\ldots\ldots (13)$$

The tin oxide, together with that normally produced, is collected in a baghouse and returned to the circuit with new concentrates. This process has the ability to produce reject slags with 0.4-0.5% Sn.

Submerged Combustion Smelting Process

An interesting variation of the classical two-stage smelting process is that proposed by Floyd (9) and which is now undergoing large scale trials at the works of Associated Tin Smelters Pty. Limited, in Sydney, Australia. The first stage of smelting in which tin concentrates with by-products are reduced to tin and a rich slag is carried out in the conventional manner. The rich slag is then transferred by a crane ladle to a tall, cylindrical vessel, lined with refractories, for further processing. A simple lance, consisting of a thick-walled steel pipe equipped with an oil atomising nozzle near its tip, is lowered into the bath of molten slag and the oil is burned with a small deficiency of air beneath the surface. Fine coal is fed onto the surface of the violently agitated slag to give reducing conditions, and a rapid reduction occurs with the production of hardhead. Some tin is also eliminated as fume during processing. The process has been criticised as wasteful in fuel (10) due to the fact that to achieve the necessary CO/CO_2 and H_2/H_2O ratios for tin reduction from slags, a lot of unburned gases with a useful energy content escape from the vessel. However the equipment required is relatively simple and no heat is wasted in melting down solid slags as in conventional slag smelting. It is therefore to be hoped that the results of the present trials will be published before long.

Tin Fuming Process

There is no doubt that the greatest contribution to tin smelting in recent years has been the development of a satisfactory process for the fuming tin from slags. Because of the steadily rising price of tin there has been a corresponding rising pressure on smelters to achieve better recoveries. This has applied particularly to smelters of tin from lode concentrates where the slag fall is much higher, due to a greater gangue content, than at smelters dealing with alluvial concentrates. As methods of gas cleaning have improved over the years, losses of tin in slags have replaced the chimney losses as the largest single factor, due to the equilibrium between iron and tin in the reject slags. The higher the iron content of the incoming concentrates, a correspondingly higher loss of tin in reject slags had become to be regarded as almost inevitable.

Any process which could obtain a clean separation of tin from iron would therefore be of great value in improving recoveries and the volatilisation of tin from slags has been found to be the most practical way of achieving this.

The most volatile compounds of tin for practical purposes are the halides, stannous oxide, SnO and stannous sulphide, SnS. The boiling points of the halides at atmospheric pressure are reported by Sidgwick (11) as follows,

$SnCl_2$	$SnBr_2$	SnI_2	$SnCl_4$	$SnBr_4$	SnI_4
$652^{\circ}C$	$619^{\circ}C$	$720^{\circ}C$	$114^{\circ}C$	$203^{\circ}C$	$346^{\circ}C$

From the economic point of view, only the chlorides are of importance at the present time, when considering the halides.

The boiling point of stannous oxide is reported by Veselovskii (12) to be approximately 1530°C and the boiling point of stannous sulphide is reported by Richards (13) to be 1230°C.

The volatilisation of tin as stannous oxide was discussed as long ago as 1929 by Kohlmeyer (14) and Feiser (15), but they arrived at the general conclusion that as SnO was always in combination with silica as stannous silicate $SnSiO_3$, this method of volatilisation held little promise. However Wright (16) reported the results of some experiments at Capper Pass & Son Ltd., using a blast furnace in a similar manner to that reported by Jensen (17) but in which lump limestone was used instead of the sulphidising agent, gypsum. As a similar performance was obtained without the sulphidising agent, it was concluded that stannous oxide, SnO, and not stannous sulphide, SnS, was the volatile species of tin produced.

Much of the early work on the volatilisation of tin from slag as stannous sulphide was centred around the use of long kilns but was abandoned due to poor results and constant problems with accretions. The first successful process for treating tin slags on a commercial scale was that developed by Phelps Dodge (18) (19). This process utilised a converter of the Pierce-Smith type, having a capacity of 5 tons of liquid slag. It was fitted with special tuyeres each consisting of three concentric tubes. The centre tube fed a suspension of pyrites in air, the middle fed atomised light fuel oil and air, whilst the outermost tube carried air only. Liquid slag was fed in from a crane ladle and the blowing time was that required to add the necessary quantity of pyrites. The reject slag from this process contained 0.5% Sn.

The air/oil ratio has to be maintained at approximately 80% of that required for complete combustion in order to maintain reducing conditions within the bath. One of the first reactions when the pyrites comes in contact with the slag is dissociation into iron sulphide and sulphur,

$$FeS_2 \longrightarrow FeS + S \quad \dots\dots\dots\dots\dots\dots (14)$$

The elemental sulphur volatilises rapidly and probably does not contribute greatly to the process. With the strong agitation provided by the combustion gases, the iron sulphide is distributed in the slag where it reacts with tin silicate,

$$SnSiO_3 + FeS \longrightarrow FeSiO_3 + SnS \quad \dots\dots\dots\dots\dots (15)$$

Tin sulphide vapour is carried to the surface of the bath by the bubbles of combustion gases and oxidises in the furnace atmosphere,

$$SnS + 2 O_2 \longrightarrow SnO_2 + SO_2 \quad \dots\dots\dots\dots\dots\dots (16)$$

The chemistry of the process is complicated by the fact that some of the reactions due to burning oil within the bath are slow and promote the formation of magnetite, which then requires an excess of pyrites to reduce it again,

$$FeS + 3 Fe_3O_4 \longrightarrow 10 FeO + SO_2 \quad \dots\dots\dots\dots\dots (17)$$

The chemistry of the process is discussed by Wright (20) in greater detail.

The chief difficulty with a vessel such as a converter is the extremely heavy wear that takes place in the vicinity of the tuyeres, and this led the tin smelting industry to look at furnaces of the type used for vola-

tilising zinc from lead blast furnace slags. One of the first recorded
installations was at a Russian smelter in 1957 and was described by
Murach et al (21) in some detail. This was probably the forerunner of the
furnace described by Kolodin (22) and illustrated in Figure 2,

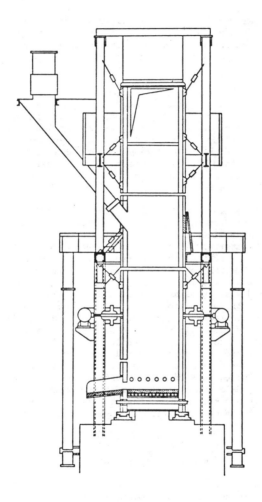

Figure 2.   The stationary tin fuming furnace as
             designed by Kolodin.

The Kolodin type of furnace, with some modifications, has become the most widely used type of furnace for the volatilisation of tin from slags today. It consists basically of a water-jacketted shaft with a water-cooled cast iron hearth. Kolodin's furnace was approximately 1.2 metres x 2.1 metres in cross section and was equipped with six tuyeres on each of the shorter sides.

The tuyeres were used to blow a coal/pyrite mixture into the bath of liquid slag a little under 1 metre in depth, with a given quantity of air. Kolodin mentions that due to incomplete combustion in the furnace newer furnaces would be fitted with combustion chambers and possibly waste heat boilers.

Today most fuming furnaces have been modified to burn oil instead of coal, as this is capable of greater control, and the pyrites is added to the surface of the bath to avoid the air necessary to inject it through the tuyeres. As will be appreciated many of the important reactions taking place in the fuming process occur at the gas bubble-liquid slag interface. To optimise these reactions it is therefore advisable to use a larger number of small tuyeres than a few large ones. For a given volume of combustion air this results in a large number of small bubbles with a considerably greater surface area.

It should not be imagined from this brief description that the sulphide volatilisation of tin from slags is a cheap process or without its difficulties. Pyrites is a mineral commonly contaminated with lead, arsenic and zinc. If such material is used these elements are volatilised with tin and will report in the dust recovered after the exhaust gases have been cleaned. This then entails either a separate circuit for smelting dust or the risk of contaminating a much larger bulk of metal if smelted with fresh concentrates. The emission of sulphur dioxide can also be a problem depending on the location of the smelter. The bulk of the sulphur present in the pyrites added is converted to sulphur dioxide and becomes diluted down with the products of combustion. It therefore has a nuisance value, but is insufficient in quantity to justify conversion into acid. An interesting possibility is that proposed by Wright (24) in which sulphur dioxide is extracted from roaster and/or fuming furnace gases and introduced into special tuyeres in a fuming furnace with a certain amount of fuel and air to provide a thermally balanced mixture. If this proposal is found to work on a large scale it not only reduces the pollution risk from sulphur dioxide, but also eliminates the contamination frequently found due to impure pyrites.

Chloride Volatilisation

Volatilisation of tin as the chloride, whether in the stannous or stannic form, has probably attracted more research work from institutions and academic sources than the combined work on all other systems put together. The attraction of such low boiling point compounds, has produced a large number of processes, despite the practical problems involved. Not only are the corrosion problems very severe, but control of the furnace atmosphere has also to be precise. The only known application of chloride volatilisation is at the Texas City tin smelter described by King and Pommier (24). This involves the use of a Kaldo type of furnace, capable of close atmosphere control, with the addition of calcium chloride to promote the production of stannous chloride. The long history of the Texas City smelter with hydrochloric acid leaching methods has, no doubt, given them considerable knowledge of the corrosion problems involved.

Smelting Flowsheets

Before leaving the subject of smelting, it is interesting to note the various combinations of smelting and volatilisation furnaces that are either in use or proposed to be used at plants under construction today. Smelting in accordance with the classical flowsheet can be carried out in a reverberatory, rotary or electric smelting furnace and the choice tends to be dictated far more by economic circumstances than by technical factors. Thus at the smelters in Central Africa, such as the Zaire smelter and that under construction in Rwanda, isolated by thousands of miles from a source of fuel oil, the electric furnace is a natural choice due to the availability of electric power. In Malaysia with oil available but expensive electricity, the reverberatory furnace is favoured. Rotary furnaces have not found favour mainly due to excessive lining wear and limited ability to meet the temperature requirements.

As soon as the grade of the concentrates falls into the 30-50% Sn range due usually to association with pyrites and arsenopyrite, roasting becomes a necessary pretreatment. Again as the volume of slag increases, fuming becomes necessary to effect a tin-iron separation and thus avoid the losses inherent in the classical two-stage smelting process. This is illustrated in Figure 3, which shows a simplified flowsheet of the ENAF smelter in Bolivia described by Lema Patiño (25). Fuming replaces the second stage of smelting, and tin oxide filtered from the gases, which is substantially free from iron, is recirculated to the first stage of smelting in a reverberatory furnace.

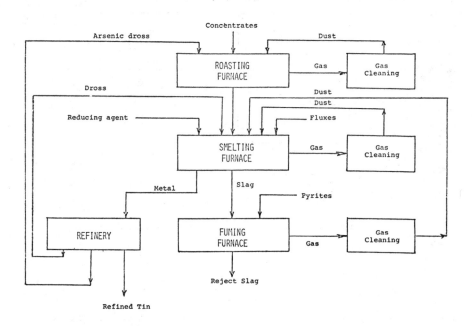

Figure 3.   Smelting circuit for medium-grade concentrates.

When the tin content falls still further into the 5-25% Sn range, there comes a point at which a tin-iron separation becomes necessary at an earlier stage in the flowsheet. This can be achieved by fuming the concentrates without roasting to produce a tin oxide dust, free from iron, which is then readily converted to metallic tin by conventional means. This is illustrated in Figure 4. This also perhaps indicates how dependent such techniques are on the efficiency of the gas cleaning equipment used. Any process which seeks to convert an expensive metal, such as tin, first into a gaseous phase and then into a dust, must depend heavily on a high efficiency in the gas cleaning system chosen. This must also be designed to withstand corrosion from gases which at times contain considerable amounts of sulphur dioxide.

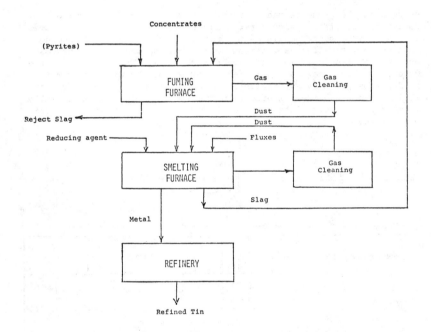

Figure 4.   Smelting circuit for low grade concentrates.

Refining of Tin

Impurities in tin concentrates tend to fall into two categories, the first being those that report predominantly in the metal and therefore need removal during the refining of the crude metal, and the second those that report predominantly in the slag during smelting. This latter group consists mainly of the compounds of tantalum, niobium, wolfram, titanium and zinc. Wolfram does become partially reduced in the later stages of smelting and can be found in the iron-rich layers which build up as accretions on the hearth of stationary furnaces. Those impurities which report mainly in the metal are copper, antimony, arsenic, lead, bismuth, silver and gold. The two principal exceptions are sulphur and iron. Because of the volatility of stannous sulphide, sulphur is usually removed prior to smelting by roasting, unless direct volatilisation of tin from the concentrates is envisaged as in Figure 4. In all smelting processes, as distinct from fuming, iron divides between the metal and the slag, but because of its importance in refining it will be treated here as an impurity reporting in the metal.

Iron Removal

Iron is probably the only impurity found in the crude metal irrespective of the type of concentrate being smelted. It has been the practice in some high grade smelters to blow air through the molten tin which has the effect of gathering iron into a scum on the surface. This is not a true oxidation but more akin to a flotation process. Nevertheless substantial quantities of tin are oxidised with the result that the iron scum seldom contains more than a few percent of iron. This dross is often liquated in an attempt to concentrate the iron further, but the average liquation dross seldom exceeds 8-10% Fe. At a few smelters the crude tin is cast into large ingots and these are charged into a liquation furnace. As the pure tin drains away the iron-tin dross is left behind and with correct working can be enriched to 12-14% Fe with progressive heating. At some other smelters the hot liquid metal centrifuge is used to concentrate the iron into a crystalline by-product. The crude tin is allowed to cool down to a little above its melting point in refractory lined containers. The clean metal is then decanted carefully into the first of the refinery kettles, care being taken to prevent any solid material being carried with it. The remaining slush of liquid tin and solid material is then poured into an adjacent small kettle. This is then heated to 550°C to convert as much FeSn$_2$ compound into FeSn, and the solids separated from the liquid using the centrifuge. When little solid material remains behind in the small kettle it is cooled to a little above the melting point of tin and after allowing time for any tin-iron crystals to settle, the liquid tin is pumped to the large kettle. Tin is collected in the large kettle until full when it is cooled down to a few degrees above its melting point. It is then pumped to the next large kettle in the refinery leaving behind a small quantity of scum and crystals adhering to the sides. This small amount of remaining material is allowed to solidify around a steel hook and is then transferred to the small kettle. In this way only tin containing less than 0.005% Fe is produced and iron-tin crystals, which at the Bolivian smelter average 23-25% Fe. Only the material in the small kettle, amounting to 5-10% of the total, is required to pass through the centrifuge.

Copper Removal

Copper is reduced to very small proportions by the additon of sulphur to the molten tin at a little above its melting point. The scum produced is then dried by stirring in sawdust or palm oil to produce a dry, black powder which can then be skimmed off.

Arsenic and Antimony Removal

Arsenic, antimony, and any nickel present, are most efficiently removed with the classic aluminium process. However, whilst metallurgically efficient, this process produces drosses containing the impurities in the form of AlAs, AlSb and AlNi. The compound AlAs, if allowed to become wet, generates a poisonous gas arsine AsH_3 and very strict handling procedures are necessary. If nickel is not a difficulty, many smelters now prefer to use sodium for the removal of arsenic and antimony as this is safer even if slightly more costly.

Lead Removal

Lead is a common impurity in tin and if present only in small quantities can be most economically removed utilising the fact that at low temperatures lead chloride is formed preferentially to stannous chloride in the presence of chlorine. Unfortunately the reaction between stannous chloride and lead is a reversible reaction and it can be shown that the amount of lead in the salt phase is proportional to the amount of lead in the tin with which it is reacting (26). It is therefore advisable to adopt a counter-current procedure by dividing the de-leading operation into a series of stages and moving the salts in a counter-current direction to the flow of tin.

Bismuth Removal

Bismuth, if present in small quantities, can be removed from tin by the conventional Kroll-Betterton type of process using sodium or calcium in conjunction with magnesium. However the concentration of bismuth in the by-product is low, seldom exceeding 2-3%, and for this reason many smelters prefer to eliminate bismuth prior to the smelting operation by using a chloridising roast.

Electrolysis

A different picture is presented if large quantities of lead and bismuth, either singly or together, have to be removed from the crude tin. Then a choice has to be made between treating the tin in an electrolysis plant or the more newly developed vacuum refining process, and the decision is influenced largely by the amount of silver and gold present. In practice, these precious metals can only be removed by electrolysis, although it was once claimed that gold could be selectively removed from tin by the use of aluminium (27).

The electrolysis of tin is carried out in several different ways and the processes involved can be classified broadly into two types, those using acid electrolytes and those using alkaline. The acid electrolytes are usually based on stannous sulphate to which materials such as phenol or cresol-sulphonic acids have been added. Tin is a metal which in the stannous condition is very prone to irregular deposition and addition agents, such as glue, are necessary to produce a reasonably smooth cathode. Acid electrolysis is not very tolerant of impurities without frequent removal of the anode slimes to prevent passivation. Detailed descriptions of plants operating this type of electrolysis have been published by Mackey (28) and by Weigel and Zetzsche (29).

Alkaline electrolytes, such as sodium stannate and sodium thiostannate, are much more tolerant of impurities and require little or no addition agent to produce smooth cathodes. However, they have the disadvantage that they must be used hot (>80°C) and as the tin is deposited from the tetravalent condition twice as much current is consumed per ton of tin as with bivalent electrolyte.

The great disadvantage of both types of electrolysis process is the lock-up of tin metal estimated by Wright (30) to be of the order of 25 tons of tin for every ton of tin produced daily with bivalent electrolytes, and 50 tons when using tetravalent tin electrolytes. It is mainly for this reason that perhaps the most significant development in recent years has been the practical application of vacuum distillation to tin refining.

Vacuum Distillation

The principles involved have been known for many years, but it is only in recent times that suitable materials have been available from which to construct practical equipment. Thanks to the spin-off from atomic energy research, very dense graphite is now available which will contain molten tin at temperatures up to 1400°C without penetration. A vacuum of 10 microns and below is now readily attainable and by choosing a specific time/temperature combination a reasonable degree of selectivity can be obtained. Table I shows the effect of distillation time on

Table I

Element	Input Metal	Distillation time (minutes)						
		2	3	5	7	10	15	30
Sn	87.3	Bal.	Bal.	Bal.	Bal.	Bal.	Bal.	Bal.
Sb	1.5	1.4	1.5	1.3	1.2	0.9	0.7	0.7
Cu	3.2	3.3	3.5	3.6	3.2	3.6	3.6	3.4
Pb	3.9	1.2	1.0	0.3	0.07	<0.01	<0.002	<0.002
Bi	1.1	0.44	0.35	0.10	0.04	<0.002	<0.002	<0.002
As	2.7	1.47	1.3	0.86	0.60	0.15	0.05	0.04

a crude metal held at 1200°C under a vacuum of 10 microns. As will be seen lead and bismuth distill off very quickly, whilst arsenic is a little slower and antimony very much slower. Table II shows the effect of heating a batch of similar metal from cold under a vacuum of 10 microns and holding at various temperatures for 15 minutes.

Table II

Element	Input Metal	Temperature °C				
		940°	1075°	1220°	1255°	1315°
Sn	87.3	Bal.	Bal.	Bal.	Bal.	Bal.
Sb	1.5	1.5	1.4	0.8	0.45	0.2
Cu	3.2	3.4	3.6	3.6	3.6	3.6
Pb	3.9	3.4	0.5	<0.01	<0.002	<0.002
Bi	1.1	1.0	0.34	<0.01	<0.002	<0.002
As	2.7	2.58	0.99	0.13	0.03	0.03

Thus it will be seen that vacuum distillation can produce metal complying with ASTM B339 Grade A standard with respect to lead, bismuth and arsenic at reasonable temperatures and within a short time, but that this does not hold true for antimony. Other work has shown that a temperature of 1400°C is necessary for removal of antimony in a reasonable time, but that at this temperature tin itself has a significant vapour pressure and the distillate is spoiled by its high tin content. In fact higher concentrations of antimony and better recoveries are possible with standard thermometallurgical processes.

Commercially Available Systems

Two commercial systems based on different approaches to vacuum refining are currently available, one developed in the U.S.S.R. and the other by the joint Bergsøe-Redlac Engineering organisation. Both were recently reviewed by Müller and Paschen at the international symposium "Complex Metallurgy '78" (31). The Russian system requires the crude tin to flow by gravity down a spiral launder and the volatile impurities are condensed as a liquid on a central vertical cooled surface. Both the purified tin and the distillate flow out of the vacuum chamber through separate barometric legs. In the Bergsøe-Redlac system the tin flows almost horizontally through the vacuum chamber and the volatile impurities migrate vertically to a water-cooled condenser positioned immediately above the molten tin surface, where they condense as a solid phase. The condenser is rotated and the condensate is removed mechanically. It is then conveyed out of the vacuum chamber through a vacuum lock, still in the solid condition. The significance of this difference in approach is considerable in respect of the behaviour of arsenic. In the Russian system the surface temperature of the condenser is of the order of 350°C and provided that the condensate consists essentially of lead, bismuth and tin, these elements form a low melting point liquid which flows easily under gravity. If however arsenic is present intermetallic compounds of tin and arsenic would form with melting points considerably in excess of 350°C. To prevent high melting point compounds interfering with the liquid flow, the surface temperature of the condenser would have to be raised considerably. However this would then produce the situation in which the temperature of the condenser would be above the sublimation point of arsenic in a vacuum better than 100 microns. In other words arsenic cannot be condensed in these circumstances and any arsenic contained in the crude tin has to be removed before feeding it to the vacuum chamber. No such restriction applies if the condensate is a solid and is removed from the vacuum chamber in the solid condition. Condensates high in arsenic, can be reprocessed to produce metallic arsenic with a purity in excess of 99% As.

The great advantage of the vacuum refining process is the small lock-up of tin involved and except in cases in which significant quantities of silver and gold are present, it is very likely that some of the existing electrolytic plants will eventually be replaced by vacuum distillation.

768

References

(1)   Georgius Agricola, De re metallica. (Translation by Hoover H.C. and Hoover L.H.) Mining.

(2)   P.A. Wright, Proceedings of Symposium on Advances in Extractive Metallurgy and Refining. The Institution of Mining and Metallurgy, London, October 1971, Paper 13.

(3)   T.R.A. Davey, Proceedings of Symposium on Advances in Extractive Metallurgy and Refining. The Institution of Mining and Metallurgy, London, October 1971, Paper 14.

(4)   W. Kroll, Erzmetall, 19 (1922), p. 317.

(5)   G.C. Pearce, British Patent 733, 956 (1955)

(6)   J.A. Ruppert and P.M. Sullivan, U.S. Bur. Mines Rept. Invest. 6529, (1964).

(7)   D.A. Wilson and P.M. Sullivan, U.S. Bur. Mines Rept. Invest. 5756, (1961).

(8)   C.W. Jensen, Mining Magazine, London, 82 (1950), p. 73.

(9)   J.M. Floyd, Fourth World Conference on Tin, Kuala Lumpur 1974, Vol. 3, p. 179.

(10) D. Young, Lecture, Symposium on Metallurgy, Bandung, February 1979.

(11) N.V. Sidgwick, Chemical Elements and Their Compounds, Oxford Univ. Press, Oxford, 1950.

(12) V.K. Veselovskii, Zh. Priklad. Khim, 16 (1945), p. 397.

(13) A.W. Richards, Trans. Faraday Soc., 51 (1955), p. 1193.

(14) E.J. Kohlmeyer, Metall und Erz, 26 (1929), 62.

(15) H. Feiser, Metall und Erz, 26 (1929), 269.

(16) P.A. Wright, Extractive Metallurgy of Tin, Elsevier Publishing Co., 1966, p. 120.

(17) C.W. Jensen, Mining Magazine, London, 81, (1949, p. 337.

(18) W.H. Osborne, U.S. Patent 2,261,559, 1941.

(19) W.H. Osborne, U.S. Patent 2,304,197, 1942.

(20) P.A. Wright, Fourth World Conference on Tin, Kuala Lumpur 1974, Volume 3, p. 33.

(21) N.N. Murach et al, Metallurgy of Tin, Moscow, 1964 (Translation into English by N.W. Litwinov), Chapter 13.

(22) S.M. Kolodin, Vetoritznoe Olova (Secondary Tin), Moscow, 1964, p.207.

(23)   P.A. Wright, International Symposium on Tin, La Paz, 1977

(24)   E.B. King and L.W. Pommier, International Symposium on Tin, La Paz, 1977.

(25)   J. Lema Patiño, Fourth World Conference on Tin, Kuala Lumpur, 1974, Vol.3, p. 111.

(26)   P.A. Wright, Extractive Metallurgy of Tin, Elsevier Publishing Co., 1966, p. 142.

(27)   G.E. Behr and L.H. Schroeder, U.S. Patent 2,296,196, 1942.

(28)   T.S. Mackey, Journal of Metals, June 1969, p. 32.

(29)   H. Weigel and D.Z. Zetzsche, World Mining, July 1974, p. 32.

(30)   P.A. Wright, Extractive Metallurgy of Tin, Elsevier Publishing Co., 1966, p. 151.

(31)   E. Müller and P. Paschen, Complex Metallurgy '78, The Institution of Mining and Metallurgy, 1978, p. 82.

TIN SMELTING AT BERZELIUS, DUISBURG

Friedrich Karl Oberbeckmann

Manfred Porten

"Berzelius"Metallhütten-GmbH
Duisburg W.-Germany

Tin smelting by electric furnace

Since 1917 Berzelius is operating a tin smelter and a tin-lead-smelter at Duisburg. Both lines are presented in a short summary, special attention being drawn to the smelting by electric furnace. Plant experience on this process as well as the results are reported. Careful control in the preparation of charges of materials, which contained varying levels of impurities, has resulted in production of a raw metal suitable for pyrometallurgical refining, the 2000 kVA-furnace used has interchangeable crucibles and meets environmental requirements. As a result of these investigations the resistance furnace at Berzelius is used as a two stage process in the first of which a slag containing 10 % tin is obtained. Further treatment of this slag in campaigns in the same furnace yields a final slag containing only 1 % tin. The process is very flexible and can treat big variety of raw material.

In 1917, the Tin Smelter Berzelius in Duisburg started production on the same site as Metallgesellschaft had previously established a zinc smelter in 1905.

With a production capacity of about 5000 tpy, the tin smelter in Duisburg is the only one of its kind in the Federal Republic of Germany. Berzelius tin is marketed in three brands, registered at the LME: M Standard, M Special, and Rose Tin.

Apart from the production of refined tin, the Duisburg tin smelter has manufactured solders from the very beginning of its commercial operation. The annual production of 6,000 - 8,000 t of soldering tin in the form of blocks, bars, wire, and sheet covers more than 40 % of the demand of the Federal Republic.

In the following, the metallurgical and process concept of the Berzelius tin smelter will be described for the two production routes employed.

A. Solder Smelter

The starting material for soldering tin production is a tin-lead alloy obtained according to the Berzelius rotary kiln process. With this process, Sn-Pb-Zn-containing oxides, the so-called mixed oxides, are conditioned in a charge preparation unit by adding coal as reductant and soda as flux, followed by pelletizing. The pellets are charged to a 24 m long rotary kiln which has a diameter of 2.20 m (inside diameter 1.85 m). The charge travels through the kiln, which is inclined by 2.5 % in its longitudinal axis and rotates slowly at 1 rpm.

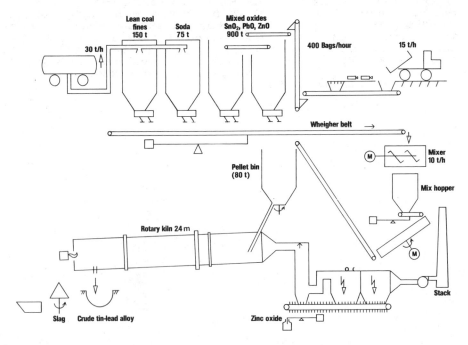

Figure 1: Flow sheet of the Berzelius soft solder plant.

An oil burner arranged at the discharge end of the kiln is provided to heat the charge by counter-current gas flow. In the reaction zone, the oxides of tin, lead and zinc are reduced. The zinc vapour produced is burnt to form zinc oxide, which is collected in an electrostatic precipitator. The liquid tin-lead alloy collected at the discharge end of the kiln, being tapped from time to time. The slag of low tin content (- 0.5 % Sn) flows through the retention ring and is granulated.

Table I: Rotary kiln data

Technical Data:	Length	24 m
	Diameter (outside)	2.20 m
	Diameter (inside)	1.85 m
	Firing (light oil)	100 l/h
	Lining	12 m magnesite, 12 m fire bricks
Charge:	Throughput	70 tpd
	Mixed oxides	35 tpd
	Lean coal fines	14 tpd
	Soda	7 tpd
Discharge:	Crude metal	15 tpd
	Soda slag	15 tpd
	Zinc oxide	12 tpd
Waste gas:	Rate	10,000 m^3 NTP/h
	Gas cleaning system	3-chamber electrostatic precipitator (Lurgi)
	Raw-gas dust content	30 – 50 g/m^3 NTP
	Clean-gas dust content	< 5 mg/m^3 NTP

The crude Sn-Pb metal recovered is conveyed to the conventional pot refinery with centrifuges or, if it contains precious metals, it is submitted to H_2SiF_6 electrolysis. It is then utilized for the manufacture of commercial solders.

The intermediate products by refining are oxidized and then either leached to separate Cu or roasted for As oxide recovery. The residues of the two processes are reduced in short rotary furnaces.

Depending on the demand and market situation, the short rotary funaces may also be used to smelt ashes and residues to recover Sn-Pb alloys. The short rotary furnace slag is fumed and the mixed oxide in the process is fed to the rotary kiln plant.

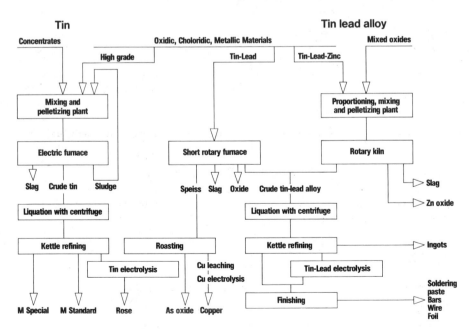

Figure 2: Flow sheet of the Berzelius tin smelter.

773

B. Refined Tin Smelter

The traditional feed materials for the refined tin smelter had consisted, for more than 50 years, of complex low-grade concentrates, mainly from Bolivia, which were submitted to a three-stage smelting process including concentration, reduction and refining. Concentration and reduction were carried out in shaft furnaces. In close cooperation with our traditional suppliers, we have restructured our works in parallel with the development of the smelting technology in Bolivia.

Since 1976, we have been applying a double-stage smelting process for primary raw materials, with reduction and refining.

After comparing the reduction units which could be used for that purpose, such as shaft furnace, revolving reverberatory furnace, reverberatory furnace and electric furnace, with regard to throughput, waste gas rates, dust formation, attainable temperatures, consumption of energy and carbon for reduction as well as the bricklining life, we decided in favour of the electric furnace. This plant section will be described and a report on the first few years of operation will be given.

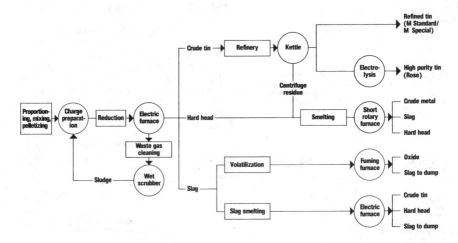

Figure 3: Process flow diagram of electrical furnace tin smelting.

Charge preparation and charging

The electric reduction furnace charge is composed as follows: up to 75 % concentrates, up to 50 % flue dust and up to 20 % each consists of ash, sludge, Ca stannates and intermediate products.

The physical characteristics of the preproducts cover a wide range, from an extremely fine to a coarse size (below 1 to 50 mm), from dry to moist (1 % - 40 % H_2O) and they may contain metals.

There are only a few preproducts, such as ash, which can be charged and smelted individually. Extremely fine concentrates and flue dust will lead to increased dust formation in the furnace and give rise to deflagration as a result of poor permeability to gases. In view of their high moisture content, sludge and filter press residues cannot be charged to

774

Table II: Tin bearing materials

	Sn %	Sb %	Pb %	Cu %	Zn %	Fe %	Bi %	Ag g/t	Range of physical condition
Concentrate	46,2-73,9	<0,01-2,9	<0,01-7,1	<0,01-0,2	<0,1- 3,0	0,2 -10,8	<0,01-0,2	<0,5-84	fine to coarse
Fumed oxides	49,0-72,6	<0,1 -0,4	0,6 -6,4	<0,1 -0,3	<0,1-13,4	0,13- 8,7	<0,01-0,04	10-3000	fine to lumpy
Stannates	36,3-40,3	0,5 -2,1	0,2 -1,2	0,01-0,1	<0,1- 0,3	0,05- 0,1	0,01	<10-30	
Residues	9,0-81,8	<0,1 -0,2	0,4 -9,8	0,08-1,2	0,3-10,2	0,2 - 8,0	<0,01-0,03	<10-1090	metal-containing
Slimes	42,0-75,5	<0,1 -0,3	<0,1 -4,0	0,01-0,4	<0,1- 2,9	0,29-11,0	<0,01-0,1	<10-450	fluid to pasty

the furnace directly because they would cake together and therefore require pretreatment.

When processing complex preproducts with strongly varying contents, as is the case in our plant, it is indispensable to provide for adequate control of composition of the mix because it is imperative for smooth electric furnace operation that the charge behaviour in the furnace should be as constant as possible.

In order to provide for optimum burden preparation, we installed plant for weighing, mixing and pelletizing, because the analysis, conductivity and moisture content of the mix are as important as the aim of producing green pellets of sufficient strength from the preproducts available without having to add binders such as bentonite or cement.

The pelletizing plant consists of 7 storage bins for concentrates, sludge, dust, additives and reductants. The bin discharge is controlled with a programme providing for exact proportioning to a belt with incorporated weigher. An elevator feeds the material to the mixer (concrete mixer design). From this, the mixed charge passes via a weigh feeder to the pelletizing disc of 3.20 diameter. The disc has a capacity of about 10 t/hour.

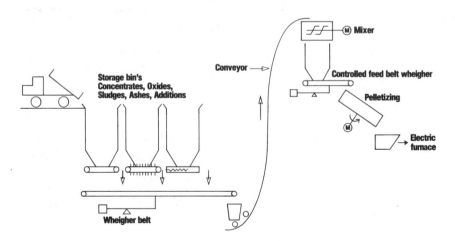

Figure 4: Flow sheet of the proportioning, mixing and pelletizing plant.

Figure 5: View of the proportioning, mixing and pelletizing plant.

After storage for 48 hours or drying at 110°C, the pellets attain sufficient strength for charging. In the top of the electric reduction furnace they are dried further, increase in porosity which, coupled with the intimate mixing of the reagents, results in a quick and complete reduction. Depending on the furnace operation, small additions may be introduced such as sand or limestone chips, by adding reductants such as coke breeze or pea coke, and by adding stubstances designed to increase the conductivity such as hard heads.

The programme-controlled transport of the materials to the furnace is effected by means of buckets with a holding capacity of two tons, suspended from a trolley and emptied into three charge hoppers of 1.3 m³ each, arrangend above the electric furnace. By means of hydraulically operated needle valve, the materials are charged from these into the furnace at intervals of about 30 minutes.

1.)
50% Concentrate
20% Fumed oxides
6% Filter Cake   } Pellets
10% Residues
11% Coke Breeze
3% Limestone Chips } Correction

2.)
60% Fumed oxides
8% Leancoal fines   } Pellets
20% Sludge
4% Pea Coke   } Correction
8% Tin-Iron Dross

Figure 6: Charge composition

Table III: Characteristics of the electric furnace

Furnace vessel:	2 Interchangeable crucibles		**Waste gas system:**	Wet scrubber, desintegrator, Theisen system	
	Outside diameter 4 m			Gas rate	$100 - 1000 \text{ m}^3$ NTP/h
	Inside diameter	2.90 m		Raw-gas temperature	573 – 1073 K
	Height	3 m		Furnace pressure	0,1 mbar
				Dust content of raw-gas	$50 - 150 \text{ g/m}^3$ NTP
Lining:				Dust content of clean-gas	0.02 g/m^3 NTP
	Bottom, magnesite	0.7 m		Clean-gas temperature	308 – 343 K
	Side walls, magnesite	0.5 m		Scrubbing water temperature	298 – 338 K
	From 1.5 m height, fire day	0.3 m		Scrubbing water rate	3 l/m^3
	Roof, fire days			Cooling and scrubbing water	
	electrode pitch circle	1260 mm		rate in circuit	40 m^3/h
				Total scrubbing water rate	80 m^3
				Rotor speed	1475 rpm
				Pressure generation	35 mbar
Cooling:				Power requirements	32 KW
	Contact jaws				
	Roof passages				
	Waste gas duct		**Waste gas:**	Rate	500 m^3 NTP/h
	High-power ducts	20 m^3/h		Temperature	311 K
	Transformer			H_2	5% by vol.
				CO	28% by vol.
				CO_2	32% by vol.
Electrics:				N_2	15% by vol.
	Name plate rating	2 MVA		SO_2	100 mg/m^3 NTP
	Primary voltage	10 kV, 50 cps		Solids	3 mg/m^3 NTP
	Secondary voltage	50 – 135 V, 18 steps			
	Electrodes	3 graphite			
	Electrode diameter	405 mm			
	Current intensity/electrode	12 KA			
	Electrode lift	2 m, hydraulically			
	Electrode control	current intensity or impedance control			

The electric furnace used is a standard furnace as applied in ferrous metallurgy, modified according to our "know-how".

It has two exchangeable air-cooled crucibles, accessible from all sides. The cross sectioned area is about 6 m² with an inside diameter of 2.90 m and an inside height of 3 m.

In the lower part, which is in contact with the melt, the bricklining consists of magnesite. The roof has eleven water-cooled openings and is fabricated from shaped fire bricks and ramming mass.

Furnace off gases pass into an inclined water cooled tube in which part of the entrained dust drops out of the gas stream and returns into the furnace. The gas passes to a cleaning system consisting of a mechanical wet scrubber (Theisen).

The cleaned gas which contains CO and H_2 is flared off. Additionally the system contains an emergency stack and a safety valve.

Figure 7: Roof openings

Figure 8: Electric furnace control panel

Operation

The electric furnace is operated continuously. The furnace at all times contains a charge column consisting of unsmelted material (approx. 1.0 m), a pasty zone in which the tips of the electrodes are immersed (approx. 80 cm) and one layer each of slag (40 cm) and metal (5 cm) above the bottom. The electrodes must be at a height of about 1 m from the floor level. If they are arranged deeper, it may occur that cavities will form underneath the charge column during tapping, and if they are arranged at a higher level, the top material will become too hot and fuse together. The electrode position can be influenced by the particle size of the reductant (coarser material will produce a higher conductivity and the electrodes will rise at constant power consumption of the furnace), by the reductant rate, which controls the degree of reduction and electrode burning, by the conductivity and temperature of the charge column. Metallic components of the charge and high temperatures of the charge column will have the effect of increasing the conductivity, in the same way as FeO contents in the slag. The least favourable power consumption of the furnace will occur when the electrodes are immersed in a slag low in SnO and FeO (1 - 5 %) with silica in excess (SiO_2 35 %, CaO 25 %).

The metal is tapped every 1.5 - 3 hours and the slag, every 0.5 - 2 hours.

One tapping launder is arranged at metal bath level and the other one at slag level.

Figure 9: Furnace crucible with tappings

In order to ensure economic refining of the smelted crude tin, the pre-products are separated, mainly according to the companion elements Pb, Bi, Ag, and then reduced in the two crucibles.

Figure 10: Furnace top

Changing of crucibles takes, as a general rule, 12 hours. The crucible being put into operation retains the slagged layers from the preceding smelting period and is furnished with an ignition triangle.

Processing of the Electric Furnace Products

The solidified metal in   ton blocks is melted   in refining pots of 10 tons holding capacity, liquated at decreasing temperature and the segregated phase is separated from the adhering tin by means of centrifuge. Separation of Cu, As, Sb and Pb is effected conventionally in pots of 25 t capacity or by electrolytic refining with a sodium sulphide electrolyte to remove Sb, Pb, Bi and Ag.

The centrifuge residues and the hard heads from the electric furnace, above all the top layer forming on the solidifying metal melt, are smelted in the electric furnace or revolving reverberatory furnace to concentrate the iron. The remaining hard head with about 50 % Fe and 30 % Sn can be smelted with lead-containing preproducts or treated with FeSi to separate the tin.

The slag from the first stage of the electric furnace and from the revolving reverberatory furnace is blown or smelted   in the electric furnace to reduce its metal content. The flue dust from the wet scrubber is recycled to the process.

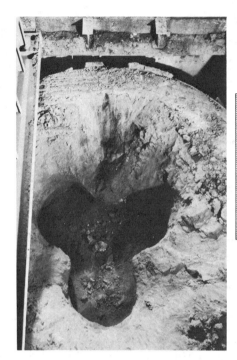

Figure 11: View inside the furnace

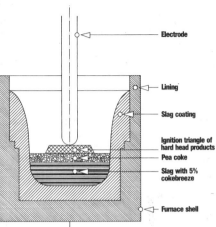

Figure 12: Ignition triangle

Table IV: Product analyses

	Sn %	Pb %	Zn %	FeO %	CaO %	Al$_2$O$_3$ %	SiO$_2$ %
Slag, type 1	1,9- 7,9	<0,02	<0,2-8,0	2,1- 6,2	27,8-37,4	6,3-11,2	26,8-37,8
Slag, type 2	4,0-19,1	0,04-0,6	2,3-5,2	16,0-23,5	15,9-26,6	2,2-10,9	25,6-32,9
Slag, type 3	0,3- 1,0	<0,02	0,3-0,5	0,3- 2,8	37,9-50,4	6,4-13,5	37,1-39,3

	Sn %	Sb %	Pb %	Cu %	As %	Zn %	Fe %
Filter Cake	32,1-47,1	0,1 -2,0	0,08-12,5	<0,1 - 3,4	1,4-4,2	5,0-33,2	1,5- 2,0
Hard head	40,0-58,9	0,1 -1,8	1,2 - 4,1	0,08- 1,5	0,8-7,1	<0,2- 3,9	5,1-26,2
Crude tin	78,4-99,3	0,03-2,8	0,05- 9,9	0,03- 3,4	0,1-2,0		0,1- 2,2
Dross	13,4-91,5	0,1 -4,5	0,1 -14,2	0,10-17,9	1,1-6,3	0,1- 0,6	0,2-18,6

	Sn %	Sb %	Pb %	Cu %	As %	Fe %	Bi %
M Standard	99,79	0,04	0,05	0,008	0,008	0,003	0,04
M Special	99,88	0,02	0,05	0,006	0,006	0,003	0,03
Rose	99,94	0,03	0,003	0,004	0,004	0,003	0,002

Operational Experience

We have run several series of tests in order to gather data for metallurgical and economic optimization of furnace operation.

Table V: Average operating results

Charge	1 Concentrate		2 Concentrate pellets		3 Oxide pellets		4 Slag	
Feed rate	%	t/d	%	t/d	%	t/d	%	t/d
Throughput Tin bearing material		39,8		31,4		28,0		43,0
Tin	59,5	23,7	57,0	17,8	44,6	12,5	11,6	5,0
Iron	5,0	2,0	5,9	1,8	2,2	0,6	13,7	5,9
Carbon	12,6	5,0	10,0	3,1	9,3	2,6	1,5	0,6
Additions		0,9		2,0		3,7		0,4
Total throughput		45,7		36,5		34,3		44,9
Discharge	%	t/d	%	t/d	%	t/d	%	t/d
Crude tin		17,2		14,4		14,1		2,5
Slag		17,0		10,1		8,9		31,1
Tin in slag	19,1		8,3		6,6		1,2	
Iron in slag	23,5		14,7		5,4		4,7	
Hard heads		1,1		2,5		1,6		3,6
Sludge (dry)		1,8		1,6		1,8		0,4
Consumptions								
Energy kWh/t	750		950		950		550	
Electrode consumption kg/t	2,0		2,3		2,5		1,0	

1. Pure, unpelletized concentrates:
The charge consisted of fine concentrates with Sn contents around 60 % and iron contents of about 5 %. We added sawdust to loosen up the charge. Smelting did not give rise to any problems, but the Sn contents in the slags were up to 25 % and also the iron contents. The slag had a high conductivity.

2. Pelletized concentrates with small portions of sludge or flue dust, at Sn contents of 57 % and Fe contents of 2 - 10 %. At a reduced throughput compared with the former case, the tin contents in the slag were below 10 %. The energy consumption was higher because the pellets had to be dried in

the furnace.

3. Pelletized flue dust with sludge and ash:
The tin contents were about 40 - 50 % at lead contents of around 5 %. High
volatile contents in the sludge and ash led to irregular furnace operation
with deflagrations. The throughput of material to be smelted was below 30 tpd
and the total throughput including additives reached 35 tpd.

4. Slag smelting:
The slags contained 10 - 15 % Sn. It was possible to de-tin the slag down to
0.8 % at throughputs above 40 tpd in continuous operation.

Discussion

Based on systematic charge preparation it is possible to balance the
multiplicity of analytical and physical differences in the various primary
and secondary materials and to optimize their use from both metallurgical
and economic viewpoints.

In the course of operation it has been demonstrated that all secondary
preproducts available at present with medium to high metal contents, such
as ash, flue dust, filter slurry, etc., can be smelted in an electric
furnace despite their substantial chemical and physical differences,
provided that the charge is adequately prepared. It is a prerequisite,
however, that the necessary conditions for smooth operation of the pelletizing unit and electric furnace are fulfilled by appropriate material
mixtures.

The electric furnace offers optimum flexibility.

Outstanding features are the high temperatures that can be achieved,
the utilization in terms of time and durability of the lining.

The extremely low waste gas rates compared with other reduction
processes underline the environmental harmlessness of the process.

A comparison of the units used for tin reduction is shown in table VI.

Table VI: Comparison of tin reduction furnaces

	Rotary Furnace	Reverberatory Furnace	Electric Furnace
Smelting time h/t charge	0.7	0.48	0.48
Temperatures °C	1100	1350	1500
Throughput t/m^{2}d	1.36	1.0	6.6
Furnace operation days/year	300	300	330
Refractory life, months	8	8 2 years: walls and roof	>36
Fuel, oil, l/h	200	200	–
Reductants kg/t metal	330	370	200
Power consumption kWh/t metal	208	126.9	1580
Cooling water m^3/t metal	5.88	0.2	18
Tin Recovery %	98.5 – 99.0	99.0	99.5
Furnace Size m^2	8 x 3.60	4 x 12	6
Waste gas m^3/t charge	6500	16000	250
Dust, max.,%	10	10	5

References

M. Porten, A Ueberschaer: "Untersuchungen zum Recycling zinnhaltiger Produkte und oxidischer Stäube mittels Versuchs-Elektro-Widerstandsofens", Forschungsbericht NTS 50, BMFT, Juli 1978

H.W. von Reis: "Das Berzelius Verfahren, ein Sonderverfahren im Drehrohr", Erzbergbau Metallhüttenwesen 11 (1958), p 196

Th. S. Mackey, K. Batubara: "Rotary Furnace Smelting of Indonesian Tin Concentrates", Tin International, 6 (1974) pp 165-176

N.N. Murach, N.N. Sevryukov, S.I. Pol'kin, and Yu.A. Bykov: "Metallurgy of Tin", Translation of Russian book, National Lending Library for Science and Technology, England (1967)

O.M. Katkov: "Tests on a Technique for the Continuous Smelting of Tin Concentrates in an Electric Furnace", Tsvetnye Metally, The Soviet Journal of Non-Ferrous Metals, N.Y., 12 (1971) pp 28-31

C. Ferrante: "Das elektrische Schmelzen von Zinnerzen", pp 96-154, Handbuch der technischen Elektrochemie, Akademische Verlagsgesellschaft Leipzig (1956)

J.B. Huttl: "Electric Furnace Smelts Sullivan Tin Concentrate", reprinted from Engineering an Mining Journal, Jan. 1948

M. Gamroth, S. Wirosoedirdjo, P. Paschen: "Zum Stand der Anlagentechnik in der Zinnmetallurgie", Metall 13, (1977) 9, pp 999-1004

P.A. Wright: "Extractive Metallurgy of Tin", Elsevier Publishing Company, New York, 1966

P. Paschen: Aktuelle Probleme der Zinnmetallurgie". Erzmetall 29 1976, pp 14 - 18

T. Davey, F. Flossbach: "Tin Smelting in Rotary Furnaces", Journal of Metals, 5 (1972) pp 26-30

H. Roth, G. Rath: "Betrieb der neuen Elektro-Roheisenöfen bei Mysore/ Indien", DEMAG-Publikation aus SEA (5) Quarterly, Apr. 1973

J.L. Patiño: "Fuming of Tin Slags", Fourth World Conference of Tin, Kuala Lumpur, Malaysia 1974

J. Puttick: "Note on the Electric Smelting of Tin Concentrate at Iscor Works, Vanderbylpark", Journal of the South African Institute of Mining and Metallurgy, 69, (1968-69) pp 139-45

H.A. Uys: "The Metallurgy of Tin Smelting in a Submerged Arc Furnace", Journal of the South African Institute of Mining and Metallurgy, 1 (1977), pp 121-125

E. Mueller, P. Paschen: "Entwicklungstendenzen beim Schmelzen und Raffinieren komplexer und verunreinigter Zinnkonzentrate", Erzmetall 32 (1978) 6, pp 266-272

DEVELOPMENT OF THE MATTE FUMING PROCESS FOR TIN

RECOVERY FROM SULPHIDE MATERIALS

K.A. Foo
Aberfoyle Ltd.
Melbourne, Australia
&
J.M. Floyd
C.S.I.R.O. Division of Mineral Engineering

The matte fuming process previously described by Foo and Denholm was given trials on a small pilot scale using the submerged lance injection SIROSMELT furnace system. Two sulphide feed materials from Tasmanian sources were used in the trials. For the first series of trials Queen Hill pyritic tin ore, which could not be economically upgraded by mineral dressing, was charged as lump material to the furnace. In the second series of trials Cleveland copper concentrate containing a significant level of tin which is only partly recovered in conventional copper smelting, was charged as pelletized material incorporating a siliceous mill tailing as flux. For both materials indicated tin recoveries in fume were greater than 90%. Problems with small scale operation resulted in very significant dilution of the fume with carry-over feed but it is not expected that this fume dilution would occur to a significant extent on a large scale.

Design criteria for a large pilot plant are indicated.

INTRODUCTION

The winning of tin from its ores has traditionally been achieved by a combination of simple gravity concentration processes for separating the heavy cassiterite from lighter gangue minerals. Over 70% of the world's tin is recovered from the secondary alluvial deposits of Malaysia, Indonesia and Thailand by the use of simple devices such as palongs (sluices) and jigs. However, a significant and increasing amount of tin is recovered from hard rock sources found mainly in Bolivia, Australia, South Africa and U.K.

These deposits require a substantial mineral processing effort to recover the cassiterite successfully, because of the need to liberate the brittle cassiterite from its host gangue and other minerals such as sulphides without creating unrecoverable, minus 5 micron "slimes". Tin recovery from these sources varies from 20-80% and averages about 50% in Bolivia and 60-70% for the other countries. Recovery can be related directly to the natural grain size of cassiterite in the ore, the amount of gangue and accessory minerals of high specific gravity and the methods employed for concentration. Consequently, fine grained, mineralogically complex ores cannot be treated successfully with gravity separation and therefore cassiterite flotation has emerged as a supplement to gravity processing. However flotation itself has limitations in that cassiterite finer than 5 microns and coarser than 30 microns cannot be recovered efficiently or economically. Furthermore, the reagents used require removal of sulphides before cassiterite flotation and the process rarely produces concentrate above 25% Sn grade.

Lode ores are becoming increasingly complex, tin grades are slowly decreasing and recovery methods, both traditional and contempory are proving inadequate. Furthermore, the demand and price of tin has increased markedly to record levels because current supply is inadequate and future production shortfalls are anticipated.

The use of fuming processes to overcome the limitations of conventional concentration and smelting techniques and to improve tin recovery, is receiving increasing attention.

This paper is concerned with one such development which has evolved from testwork on a very fine grained, complex pyritic tin ore from Queen Hill, near Zeehan in Tasmania. This material could not be economically concentrated using conventional methods such as gravity separation and cassiterite flotation but laboratory fuming trials using the matte fuming concept gave high recoveries[1].

Following the successful treatment of Queen Hill ore, the work was extended to products from the tin concentrator at Cleveland Tin Ltd., Tasmania. The tin content of the table tails and copper concentrate used in the trials is not at present recovered in the concentrator or copper smelters but initial laboratory trials showed high tin recoveries using matte fuming.

This paper discusses the development of the process in a small pilot plant scale of 50 kg/hr using SIROSMELT technology[2,3].

PROCESS DESCRIPTION

Matte Fuming

Matte fuming is a simple pyrometallurgical process. Crushed ore or pelletised tin and copper bearing products are continuously fed to the surface of a molten iron sulphide or copper iron sulphide matte overlain by a thin layer of iron silicate slag at 1200°C. When the system is blown with air, which in the SIROSMELT furnace is delivered through a steel lance submerged in the matte, intense reaction occurs with efficient liquid-gas mixing so that reaction rates involving tin sulphidisation and volatilisation, matte oxidation and slag formation are very rapid. Autogenous heating can be achieved with air as oxidant if sufficient iron sulphides are present in the feed. The presence of FeS at high activity ensures that solid magnetite formation does not occur so that discard slags are fluid.

Residual matte is maintained in the bath to collect copper, gold and silver and is tapped as required. Alternatively, for copper production the matte can be blown to blister copper if desired.

Tin oxide fume is collected in a baghouse and waste gases are disposed of by methods that meet environmental standards.

Sirosmelt System

The SIROSMELT furnace system has been developed as a versatile and cheap unit for carrying out a wide range of operations. The furnace is refractory lined and uses a simple lance for injecting air or fuel-air mixtures (which may be oxygen-enriched) into baths of matte, slag or metal. A single furnace can therefore carry out either smelting, converting, refining or slag cleaning operations and can be operated either in batchwise or continuous mode to suit the type of operation required. Fuels used in pilot plants have included oil, natural gas, and various types of coals.

The very high degree of turbulence generated in the SIROSMELT furnace has allowed very rapid smelting rates of up to 2 tonnes $m^{-3} h^{-1}$ to be achieved.

The Pilot Plant

The 50 kg pilot plant furnace used in the present investigation is illustrated in Figure 1.

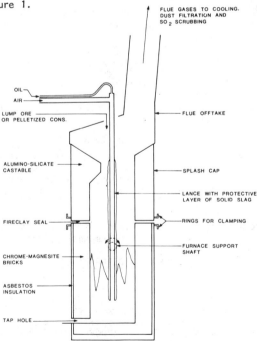

Fig. 1 - Details of the 50 kg pilot furnace.

The furnace had an internal diameter and volume of approximately 0.31m and 0.072 m^3 respectively. It was lined with chrome-magnesite refractory bricks backed by insulation. The lance was lowered and raised on a chain-driven holding arm which also operated a lance position transducer permitting remote indication of lance position and activation of oil, air and solid feed flows at given lance positions.

Oil and air flows were controlled using variable area flow meters and solids were fed into the mouth of the furnace from a belt feeder.

Flue gases passed through a water-cooled heat exchanger, a baffle box and a baghouse and were finally scrubbed free of SO_2 with aqueous sodium carbonate in an absorption tower before venting to atmosphere.

Liquid matte and slag were tapped into cast iron moulds during the Queen Hill ore smelting trials. When the lighter slag phase rather than matte was to be tapped the furnace was tilted away from the tap hole, which was opened in a high position. A furnace suspended on trunnions was used to pour off slag or matte into the moulds during fuming trials on the copper-tin bearing feed from Cleveland.

Feed Materials
Typical analyses of Queen Hill ore, Cleveland copper concentrate and table tailings are shown in Table 1.

Table 1. Typical Chemical Composition of Feed Materials

Component	Queen Hill Ore	Cleveland Concentrate	Table Tail
Sn	1.6 wt. %	4.0 wt %	0.7 wt. %
Fe	30.0	32.8	22.8
S	24.0	37.9	1.6
SiO_2	23.3	0.8	39.0
TiO_2	1.6	-	-
CO_2	6.0	1.6	7.8
Cu	0.05	21.3	0.06
Other	13.3	1.0	27.9

The Queen Hill ore is rich in pyrite and contains cassiterite as the main tin mineral, which has a grain size of less than 20 microns. Other components of the ore include 0.25% Pb, 0.05% As, 0.12% Mn, 0.05% W, 0.02% Sb and 0.5% F. The lump ore fed to the plant varied in size from 2 cm down to fines.

The Cleveland Copper Concentrate contains copper as chalcopyrite and tin as both stannite and cassiterite. The concentrate requires silica for fluxing during smelting and a 1:1 mixture of concentrate with the siliceous Table Tailing shown in Table 1 was pelletized with 5% cement binder and air dried for feed to the furnace. The concentrate also contains 0.1% Pb, 0.7% Zn and 0.02% As while the other constituents of Table Tailings include 9.3% Al_2O_3, 7.8% CaO, 4.6% F and <0.005% As.

Pilot Plant Operation

A starting bath was provided by melting sulphide materials and slag in a rotary furnace and tapping into the pre-heated SIROSMELT furnace. The lance was lowered into the furnace with the oil flow required to maintain the temperature at 1250°C and an excess of air over that required for complete combustion of the oil to provide for the desired level of oxidation of the sulphides. For the Queen Hill Ore the excess air was sufficient for complete oxidation of ore fed while for Cleveland Copper Concentrate pellets it was sufficient to produce a matte grade of 40% Cu. When the slag level reached furnace capacity, slag was removed, leaving the matte bath, and smelting was continued.

A number of control techniques was employed to ensure that target conditions were being achieved. These included simple dip-rod checks of matte and slag depth, disposable-tip emf cell checks on the oxygen potential of slag, infra-red sulphur dioxide determinations on the gas samples taken from within the furnace and rapid x-ray fluorescence analysis of matte samples for copper content. All of these proved useful, but they will not be described in detail here.

Results of Trials on Queen Hill Ore

The important constituents of the batches of Queen Hill ore used in the three trials are shown in Table 2. They are lower in grade than the typical Queen Hill ore sample shown in Table 1 because the ore for these trials was nonselectively mined from an old adit. Nevertheless it is considered that the trials are representative of the results that would be expected from run-of-mine ore.

Table 2. Analyses of Feeds to Trials

Lump ore to Queen Hill material trials

Trial No.	% Sn	% Fe	% S	% SiO$_2$
A/50/1	0.93	30.0	24.0	23.3
A/50/2	0.43	18.9	18.1	45.4
A/50/3	0.44	31.7	35.7	27.6

Pellets to Cleveland Copper Concentrate Trials

Trial No.	% Sn	% Fe	% S	% SiO$_2$	% Cu
CT/50/1	2.5	26.5	26.5	18.5	10.7
CT/50/2	"	"	"	"	"
CT/50/3	1.54	23.5	16.6	20.2	9.08

The operating details and product weight and analysis from the three trials are shown in Table 3. The rotary furnace feed materials for producing the starting bath were tin-free. The duration of trials shown in the Table is the time that the lance was operating and does not include time involved in tapping operations, correcting faults, etc. The total elapsed time for trials was 25 to 50% more than the injection time.

The volume of smelting air was calculated by subtracting the theoretical air requirement for complete combustion of the oil (10.8 m^3/kg) from the total volume of air injected. Because of variability in the properties of the oil used and difficulty in measuring oil and air flows accurately, the smelting air figure is only approximate.

Table 3. Operating Details and Results of Trials on Queen Hill Ore

Trial No.	Feed to Rotary Furnace		Feed to SIROSMELT		To lance		Smelting Air	Slag produced		Matte produced		Fume		Smelting Rate *2
	Pyrr. kg.	Slag kg.	Ore kg.	Duration *6 h	Oil *1 kg.	Air m³	m³	kg.	% Sn	kg.	% Sn	kg.	% Sn	kg/m³h.
A/50/1	30.0	10.0	90.0	2.87	28.7	393.2	83.2	105.9 *3	0.06	- *4	-	4.7	14.4	436
A/50/2	30.0	10.0 *5	69.0	2.05	23.2	280.9 *7	33.9	94.8 *8	0.07	4.7	0.12	3.9	7.7	467
A/50/3	25.0 *10	10.0	450 *12	11.32	116.0	1546 *11	295.9	344.8 *9	0.02	24.7	0.03	12.9	10.1	552

*1 Theoretical air for complete combustion is 10.8 m³/kg.
*2 Internal volume of furnace = 0.072 m³
*3 Two taps of slag.
*4 A matte bath was maintained throughout the run, of depth 2 to 11 cm but this was removed in a final oxidation with no ore being fed.
*5 Rotary furnace tapped cold and most slag remained in the furnace.
*6 Total time of injection. The time of ore feeding was less.
*7 Oxygen added at a rate of 4.5 m³/h for the last 10 minutes ($\frac{O_2}{O_2 + N_2}$ = 0.25)
*8 A single tap at conclusion of run. Weight includes splashed slag, lance cover, etc.
*9 Slag tapped 7 times.
*10 5 kg of commercial black iron sulphide also added.
*11 0.57 m³ of oxygen used over a 12 minute period at an enrichment of $\frac{O_2}{O_2 + N_2}$ = 0.23 to increase temperature before one tap.
*12 18 kg of silica added with ore.

The ratios of smelting air to weight of ore smelted are compared with calculated ratios allowing for initial and final mattes in Table 4. The calculated figures are for the major reactions (ignoring non-stoichiometry):

$$Fe\ S_2 = FeS + S_2 \tag{1}$$

$$FeS + \frac{3}{2}\ O_2 = FeO + SO_2 \tag{2}$$

$$S + O_2 = SO_2 \tag{3}$$

and do not include the small contribution of ferric oxide in slag and matte and sulphidization of tin. Two figures are shown which are for reactions (2) and (3) or for only reaction (2). This represents two possible situations in the smelting of Queen Hill ore depending on whether labile sulphur is oxidised within the furnace or in the flue system. Comparison of the figures in Table 4 suggests that labile sulphur produced by pyrite decomposition was not oxidized within the furnace.

Table 4. Experimental and Calculated Ratios of Smelting Air Volume to Ore Weight for Queen Hill Trials

Trial No.	Experimental Ratio m^3/kg	Calculated Ratio m^3/kg	
		Labile S used	Labile S not used
A/50/1	0.92	–	–
A/50/2	0.49	0.75	0.45
A/50/3	0.66	1.48	0.89

Table 5 is the tin material balance for the trials. The inferred tin recoveries were obtained by assuming that the only tin loss from the process was in the slag. The inferred recoveries for trials A/50/1 and A/50/3 were very high at 93% and 96% respectively. The lower recovery of 77% for A/50/2 may be caused by the low grade of the ore and its low smelting air requirement (Table 4). The high inferred recovery for trial A/50/3 is particularly significant because a large quantity of material was smelted in an extended trial involving seven cycles of smelting and tapping operations.

Table 5. Tin Material Balances for the Queen Hill Trials

Trial No.	Tin In (kg)	Tin Out (kg)			Tin Loss (kg)	Inferred Tin Rec. (%)
		slag	matte	fume		
A/50/1	0.84	0.06	–	0.68	0.10	93
A/50/2	0.30	0.07	0.07	0.30	(-0.08)	77
A/50/3	1.98	0.07	0.01	1.30	0.61	96

The tin collected in the first two trials was acceptable in view of the small quantity of fume involved and the impossibility of shaking bags completely clean. During the last trial the baghouse was losing fume due to the need for maintenance and this is the cause of the large tin loss.

The low grade of fume was caused by fine ore particles being carried away in the flue gases during the feeding operation. This is due to a limitation of small scale operation in that feed enters the furnace through the small flue off-take hole which also contains the lance. Slag accretions on the lance and the sides of the off-take occurred during the trials and required breaking away. At times ore particles up to quite a large size were fluidised in the flue off-take. On a larger scale, suitable location of separate ports for charging, lancing and flue off-take would overcome the problem of elutriation of feed in the flue gases.

Typically the slag contained 40% SiO and 35% Fe and possessed a low Fe^{3+}/Fe^{2+} ratio. The matte typically contained 64% Fe and 1.5% SiO_2.

Results of Trials on Cleveland Copper Concentrates

The important constituents of the two batches of pellets used in the three trials are shown in Table 2. The concentrate whose analysis is shown in Table 1 was used for Trials CT/50/1 and CT/50/2 while the concentrate for Trial CT/50/3 contained 21.6% Cu, 37.1% S, 32.9% Fe and 2.76% Sn. The operating details and product weights and analyses from the three trials are shown in Table 6. As for the Queen Hill ore trials the time shown is the time of operation of the lance and elapsed times for the trials were longer by 10 to 25% due to the time involved in tapping and clearing accretions. As explained earlier the calculated smelting air volumes are only approximate.

The ratio of smelting air volumes to pellet weights are compared in Table 7 with the calculated ratios, which includes the air to upgrade the matte from the rotary furnace to the final matte grade. The experimental ratios are closer to the values calculated assuming that labile sulphur is not oxidised in the furnace.

Table 8 shows materials balances for tin and copper in the three trials. The copper level in slag shown in Table 6 is the lowest analysis for slag samples selected to be matte-free. Since, on this small scale, settling and separation of matte from slag was not satisfactory, the copper level in the bulk of slag was generally higher due to suspended matte prills. Thus the copper balance contains a column for copper in total slag that includes copper in suspended matte prills. This figure was obtained from bulk samples of all of the slag moulds, together with material swept up after the trial and that forming the lance cover.

The copper balance shows only a small loss of copper, which would have occurred due to small scale problems such as incomplete tapping of furnaces. The copper recovery inferred from the copper in matte-free slag is 97 to 98%.

The large loss of tin from these trials shown by the material balance occurred because, for runs CT/50/1 and CT/50/2, problems were experienced in commissioning a new gas handling system and baghouse so that fume was not collected adequately. The new bags retained a coating of fume despite vigorous after-shaking. For CT/50/3 the fan was not functioning properly and much fume and flue gas was lost at the flue off-take point because of inadequate suction. Assuming that the tin in matte and slag represented the only loss the inferred recoveries varied from 99 for CT/50/1 to 93 for CT/50/3.

Table 6. Operating Details and Results of Trials on Cleveland Copper Concentrate

Trial No.	Feed to R.F. Conc. kg.	Feed to R.F. Cement kg.	Feed to R.F. Conv. slag kg.	Feed to SIRO. (Pellets) kg.	Duration h.	To lance Oil kg.	To lance Air m³	Smelting Air m³	Slag Produced kg.	Slag Produced % Sn	Slag Produced %Cu	Matte Produced kg.	Matte Produced % Sn	Matte Produced %Cu	Fume kg.	Fume % Sn	Fume %Cu	Smelting Rate kg/m³h
CT/50/1	38	2.0	10	100*4	2.83*1	33.1	381	23.5	24.0*2	0.02*2	0.40	0.9*2	0.04	39.5	0.2*3	19.2	4.7	491
CT/50/2	38	2.0	10	79	2.27	27.0	326	34.4	80.5	0.08	0.65	30.0	0.24	41.4	1.8	5.0	9.1	483
CT/50/3	50	2.5	10	142	3.30	29.4	400.3	82.8	136.5	0.20	0.47	36.5	0.36	52.2	5.3	22.3	2.6	598

*1 Smelting period followed by a white metal blow and a blister copper blow.
*2 Matte and some slag left in the furnace for converting blows.
*3 New bags used for this trial – little fume collected.
*4 5 kg of Table Tails added during smelting because the rotary furnace was tapped cold and little slag flowed.

Table 7. Experimental and Calculated Ratios of Smelting Air Volumes to Pellet Weight For Copper Concentrate Trials

Trial No.	Experimental Ratio m³/kg pellets	Calculated Ratio m³/kg pellets Labile S used	Calculated Ratio m³/kg pellets Labile S not used
CT/50/1	0.24	0.83	0.56
CT/50/2	0.44	0.88	0.61
CT/50/3	0.58	0.68	0.53

Table 8. Tin and Copper Balances for Copper Concentrate Fuming Trials

Tin Balance

Trial No.	Tin In (kg)			Tin Out (kg)				Tin loss (kg)	Inferred Tin Rec. %
	R.F	SIRO.	Total	Slag	Matte	Fume	Total		
CT/50/1	1.52	2.50	4.02	0.04	0.02	0.04	0.09	3.93	99
CT/50/2	1.52	1.98	3.50	0.06	0.07	0.09	0.23	3.27	98
CT/50/3	2.00	2.06	4.06	0.27	0.13	1.22	1.62	2.44	93

Copper Balance

Trial No.	Copper In (kg)			Copper Out (kg)					Copper loss (kg)	Inferred Cu Rec.
	R.F.	SIRO.	Total	Selected slag	Total slag*2	Matte	Fume	Total		
CT/50/1	8.09	10.70	18.79	0.35*1	-*1	-*1	0.01	-	-	98
CT/50/2	8.09	8.45	16.54	0.52	2.82	12.42	0.16	15.40	1.14	97
CT/50/3	10.80	12.78	23.58	0.64	4.14	19.05	0.14	23.33	0.25	97

*1 Assumed weight of slag and matte before converting blows are 87 kg and 47 kg respectively.

*2 Includes clean-up material from around furnace, lance cover etc.

Contamination with fine feed material was the main cause of the low tin content of the fume and its high copper content. Other components of the fume from CT/50/3 included 2% SiO_2, 9% Fe, 6% S, 14% Zn and 0.5% As. The fume was dark grey in colour but contained only 1-2% of carbon.

Slags typically contained 25% SiO_2, 8% Al_2O_3, 8% CaO, 40% Fe and 0.5% Zn.

GENERAL DISCUSSION

The trials indicated high recoveries of both tin and copper from the materials treated but did not yield sufficiently accurate data to indicate oxygen utilisation. High specific smelting rates of about 0.5 tonnes m^{-3} h^{-1} were found. Results indicate that the labile sulphur in the materials was not oxidised in the furnace but this sulphur would be burnt to utilize its heat of combustion in a larger plant by blowing air into the furnace above the bath.

Calculations of heat requirements for a process are largely dependent on the particular data used and assumption made. For the fuming of Queen Hill ore calculations indicated that the process should be autogenous with air if the sulphur content in feed is about 25%, and evidence of autogeneity was in fact found in the first trial. With a constant oil injection rate of 10 kg/hr the feed rate and smelting air rate were doubled with a resultant temperature increase from a constant 1250°C to 1350°C over a period of about an hour while the matte depth remained constant.

An indication of refractory life cannot be obtained from the present work, which was carried out concurrently with other smelting trials. The

rate of refractory attack is very dependent on scale factors such as
surface area to volume ratios and the degree of turbulence at the furnace
walls. Generally the rate of attack on larger plants has been much less
than for the 50 kg plants and it is considered that refractory life similar
to that for the walls of copper converters distant from the tuyere line
could be expected.

The grade of fume produced in the present trials was very low, but it is
considered that this was due to small scale operating problems. Fume grades
of 60% Sn have been achieved in previous small scale crucible experiments(1).
Impurities which volatilise to a significant extent during operations
include As, Zn, Bi and Pb, and at high levels these will mostly report to
the fume, requiring treatment of the fume for their removal before smelting
to produce tin. The fume could also possess significant sulphur levels,
which could result in a relatively high fume recycle in a smelting operation.

It is apparent that the reaction path which is used in this work is
more complex than would occur with direct combustion of feed material. The
material is dropped into the furnace in lump or agglomerated form. Before
it enters the matte phase where its main oxidation occurs the material heats
up and melts in its own micro-environment so that sulphidation and volatilisation processes have time to occur away from the presence of highly oxidising
species. Thus constituents with more volatile sulphides, such as Sn and Pb
should be more effectively volatilised than in systems where the material is
conveyed into the furnace as fine material suspended in air. On the other
hand suspension smelting of sulphides generates very high particle
temperatures during oxidation so that this may favour volatilization of
oxidized species.

Large Pilot Plant
Success on the 50 kg scale has encouraged Aberfoyle Ltd. to investigate
the feasibility of a pilot plant that could generate data to enable an
evaluation of the potential of a large commercial plant. A study was made
of the smelting technology that could be used for commercial operations and
all fuming and smelting furnaces available for tin and copper production
were evaluated. The choice of furnace types is limited to refractory-lined
converters or suspension-settling furnaces(4) by the need to maintain a
matte phase for the recovery of copper and precious metals. Not withstanding
this, the possible use of water-jacketted fuming furnaces(5) was considered
for feed materials free of copper and precious metals. The converter option
was chosen because of its greater simplicity, its ability to accept lump
feed prepared by a primary crushing operation and its ability to operate
without supplementary fuel. Of the possible converter types, the SIROSMELT
furnace was chosen because of its simplicity and robustness and proven
capabilities on the 50 kg scale.

Continuous operation was preferred to batch operation on the grounds of
reduced labour and operating costs, production of flue gases with a constant
SO_2 tenor, and better control of low-level emissions. This entails the
installation of a settling furnace for separation of matte from slag.

A scale of 4 tonnes/hour was chosen because it represents a significant,
but not adventurous increase in scale above those SIROSMELT furnaces in
existence. Furthermore, a 4 tph plant would be one fifth to one tenth the
size of a commercial plant. If multiple units were to be used in a commercial
plant, the pilot plant could be considered as a commercial module.

A series of engineering and design studies have been completed and a
flowsheet and plant design has been developed and costed. Fig. 2 is an
indicative material balance for the copper concentrate - middlings table

tailing blend and Fig. 3 is a simplified flowsheet of a fuming plant.
Fig. 4 is a sketch of the proposed furnace and forehearth area of the pilot
plant.

Evaluation of possible sites for the pilot plant is underway, the choice
lying between establishment at the Cleveland mine, or at existing smelters
in Australia or overseas.

The Queen Hill prospect is undeveloped and could not be considered
capable of supplying a large amount of feed for some time. Nevertheless,
matte fuming provides the long term incentive to develop this prospect
commercially.

Environmental and Economic Considerations
Adequate disposal of sulphur dioxide is of course the major environmental problem and studies have been completed on all known ways of solving
it.

Water absorption to form sulphurous acid which was then neutralised
with lime and mixed with mill tailings prior to ponding was given detailed
laboratory study but the high operating costs associated with this route
have disqualified it. Sulphuric acid manufacture is uneconomic and
undesirable on the proposed scale.

After evaluating a number of alternatives, dispersion of SO_2 to the
atmosphere via a tall stack has emerged as the only viable answer for the
Cleveland site. It follows that the major advantage of piloting the
process at an existing smelter is use of installed SO_2 handling equipment.

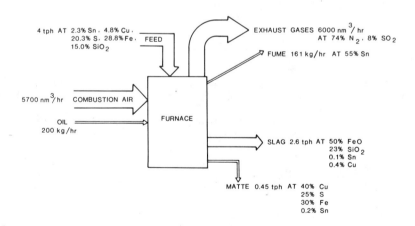

Fig. 2 - Mass balance for 4 tph matte fuming plant treating
tin-bearing copper concentrate.

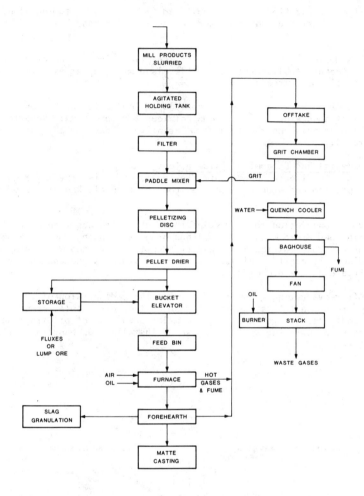

Fig. 3 - Simplified flowsheet of the matte fuming plant
treating tin-bearing copper concentrate.

An important part of any economic analysis is an examination of the
underlying risks, and possible rewards. The major risks associated with the
proposal are:

. the high scale-up ratio from 50 kg/hr to 4 tph and beyond
. the risk of failure of lances
. the potential refractory problem.

The scale-up ratio is high, but confidence has been gained by the recent
successful reduction of copper slags in a 1 tonne batch furnace at Mt. Isa
Mines and of tin slags at Associated Tin Smelters in Sydney in a 4 tonne
commercial SIROSMELT plant. Essentially, the proposed Aberfoyle furnace will
be about twice the volume of the ATS furnace, so it is considered that the
actual scale-up in terms of knowledge about the furnace is not excessive.

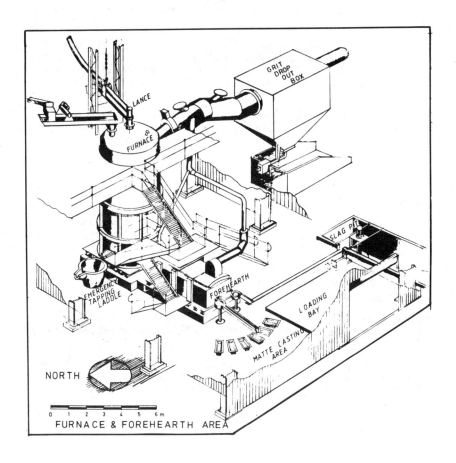

Fig. 4 - Details of the furnace and forehearth area
of the 4 tph plant.

Lance scale-up is considered a fair risk, and if necessary multiple
lances can be used. Refractories can best be chosen on the basis of prior
copper smelting experience, but with the intense reactions and mixing in the
confined space of the vessel, wear rates will be significant and refractory
consumption is expected to be high, but not excessive.

Benefits of the project are related to the new dimension it adds to tin
production methods and the incentive it gives to develop marginal tin
deposits at recoveries of greater than 90% compared with much lower
recoveries from mineral dressing processes.

It is envisaged that a centrally placed commercial plant could treat a
variety of products such as Queen Hill material, upgraded Cleveland tailings
and mill products, and Cleveland copper concentrates.

One important factor to consider is the interplay between the sulphur
content of the feed and its inherent savings in fossil fuels in relation to
the cost of disposal of SO_2 gas. Capital and operating costs are markedly

influenced by the amount of sulphur in feed and the method and cost of disposal.

A recent capital cost estimate of a fully engineered 4 tph matte fuming plant, including all housing and services, came to A\$4.5 million. A major component of the capital cost was the gas cleaning and SO_2 disposal system. Indicated operating costs were approximately \$40/tonne of feed.

ACKNOWLEDGEMENT

The authors wish to thank Aberfoyle Limited for permission to publish this paper. We also thank individuals in the C.S.I.R.O. who assisted in the trials, in particular W.T. Denholm, N.C. Grave and B.W. Lightfoot.

REFERENCES

1. K.A. Foo and W.T. Denholm, "Matte fuming of tin from sulphidic ores", paper to the International Tin Symposium, Bolivia, 1977.

2. J.M. Floyd, "The submerged smelting of tin slags - A new approach to lower-grade concentrate smelting", in Fourth World Conference on Tin, Kuala Lumpur, 1974, N.L. Phelps ed., Vol. 3, pp.179-190.

3. J.R. Siemon and J.M. Floyd, "The treatment of liquid tin slags by submerged combustion-reduction", paper to the International Tin Symposium, Bolivia, 1977.

4. Jorge Lema Patino, "Fundacion de Minerales de Estaño de Baja Ley en Bolivia", paper presented at International Tin Symposium, La Paz, Bolivia, Nov. 1977, 32p.

5. V.G. Brovkin, B.F. Verner, V.V. Kostelov, B.P. Derevensky, V.N. Kostin and S.N. Suturin, "Method of processing tin- and bismuth-bearing materials", British Patent, 1391572 (1975), 4p.

INNOVATIONS IN TIN USE

Joseph B. Long

Tin Research Institute, Inc.

Columbus, Ohio

Tin has a very long history and this fact, coupled with its high intrinsic value, means that applications of the metal have been refined by the study and experience of centuries and by the pressure of economic conditions. In general, expansion of tin usage is stimulated by the improvement of existing tin products, by the invention of new applications and by increases in the world demand for consumer goods and services.

The vast number of alloys and chemical compounds which include tin find a variety of commercial applications. Work at the International Tin Research Institute and other research facilities has been aimed at modifying the properties of these alloys and chemicals already in application thus improving their competitive position by extending their range of applicability; finding new systems with novel or improved properties which lend themselves to existing uses; or devising new processing techniques which may provide future markets. These new developments will be discussed as well as the role of tin in light of today's changing technology.

Introduction

Tin as used in modern technology is more often chosen on its technical attributes rather than on aesthetic or other grounds. World concerns as to the state of environment, pollution control, energy requirements, consumer safety, and product reliability have moved many technologists to restudy their material requirements. Because of its unique properties and versatility of use many new horizons for tin are appearing as the result of definitive research and development programs.

Tin in Storage Batteries

One of the advancing technologies has been in the development of more efficient storage batteries. Lead use in batteries was nearly 700,000 tons in 1978 (1). It appears that the most significant factor influencing the surge in demand for batteries in 1976 and 1977 was the severe cold weather conditions experienced during the winter months. Millions of batteries failed or deteriorated during these periods. This was coupled with an aggressive marketing campaign by some national retailers resulting in an increase in replacement battery shipments of 22 percent in 1976 and an additional 7 percent increase in 1977. In addition, sales in the battery sector were influenced by an increase in automotive production as well as greater acceptance and demand by consumers for "maintenance-free" batteries.

"Maintenance-free" batteries are designed so that gassing, and hence water-loss, is substantially reduced. Composition of the battery grid alloy has an influence on the gassing rate of these batteries. It has been demonstrated that gassing rates of automotive batteries based on the lead-calcium-tin alloy system are lower than the conventional antimonial leads used as grid materials. Using this factor, investigators have been able to estimate water-loss rates during overcharge and in service. Temperature is an important consideration during overcharge because with increasing temperature, there is a corresponding increase in water-loss. The ternary lead-base alloy grids containing calcium and tin have superior water-loss characteristics compared to the 2.5 percent antimonial alloys and only about one-third of the water-loss, when compared to grids of 6 to 12 percent antimony, balance lead.

Other properties of the lead-calcium-tin grids have demonstrated decided advantages over the binary lead-antimony alloys. The costs and amounts of alloying materials, mechanical properties and corrosion properties of the ternary alloy are far superior to the lead-antimony grid alloys. The lead is hardened by the small additions of calcium providing enhanced handling during processing, and increasing creep and strength properties. Tin additions increase castability, and not only improve electrochemical properties, but also provide fully aged mechanical properties which involve more complex precipitation reactions.

Battery grids are manufactured by gravity-fed casting techniques or by fabrication of wrought alloy strip. The wrought grids are made from punched, roll forged or expanded metal which is produced from a continuous casting machine (Fig. 1).

Fig. 1 – Continuous casting machine for battery grid alloys.

The lead-calcium-tin alloys for battery grids appear to be a very significant development in the battery manufacturing industries. Potential uses are not only for automobiles and trucks but in such applications as portable television, cordless tools, home security and emergency lighting systems. Notwithstanding, it also represents a potentially important new use for tin.

Carbon Reinforced Babbitts

In 1839, Isaac Babbitt patented his axle box which was lined with an alloy of tin, copper and antimony. The same classes of bearing alloys, now called babbitts, are still used in machinery of all types. Modern axle bearings are used to support journals for railroad rolling stock and are often designed with a tin-bronze shell backing with a tin-base white metal as the antifriction lining.

Increased loads and elevated temperature conditions arising during severe service of babbitted bearings causes the lining to creep and spread over the edge of the bearing shell. With a view of overcoming problems associated with the wiping and degradation of soft bearing alloys, consideration has been given to the strengthening of babbitts by reinforcement with fibers. Carbon was the obvious choice for study because of the low friction properties of the material at elevated temperatures and stress. Unfortunately, carbon is not wetted by tin and even electroforming techniques involving infiltration did not yield a coherent material. A coating unit was devised in which titanium iodide is vaporized and the vapor is allowed to react with carbon filaments to form titanium carbide. The iodine released during the reaction is recycled and used to form more titanium iodide. The titanium carbide coated strands of carbon are then thinly coated in a molten bath of tin containing 0.5 percent titanium. The tin bath temperature is about 800 to 1250°C.

After tinning, white metal is cast around the carbon filaments, and a tape of the material up to 2 inches wide is provided. Up to 40 percent

(by volume) of fibers can be incorporated in the white metal, but 10 percent
has been found to provide adequate strength (Fig. 2). The finished carbon
reinforced white metal inserts are finally positioned laterally across a
bearing shell, and pressed into place to provide the bearing surface.

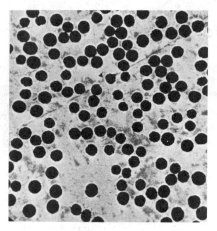

Fig. 2 – Carbon reinforced white metal.

To date, carbon fiber-reinforced babbitts have not shown wear or
wiping of bearing surfaces under adverse operating conditions. Tests will
be continued and eventually this development may lead to an extended use
for tin-babbitted bearings.

Tin Stabilizers for PVC

U. S. shipments of plastic pipe totaled about 702 million pounds in 1970.
Eight years later, production of plastic pipe had grown to 2.3 billion pounds
(2). This growth has been lead by gains in the production of polyvinyl
chloride (PVC) pipe. About 1.8 billion pounds of PVC pipe were produced
in 1978 which represents a growth of 350 percent above the 1970 level.

Organotin chemicals, used as PVC stabilizers, have shown growth
rates which follow the advancing production of PVC pipe. In order to
understand the role of organotins in the manufacture of extruded plastic
pipe, a description of the breakdown of PVC during extrusion is necessary.
Stimpfl (3) has stated that during the extrusion of PVC resins in pipe manu-
factures, extremely high rates of shear are provided resulting in temperatures up to $215^{\circ}C$. This heat causes a hydrochlorination reaction at allylic
or tertiary chloride sites with the formation of a double bond. As degradation of the plastic continues, a chromophoric conjugated structure develops
which leads to blackening of the polymer. As little as 0.1 decomposition of
the polymer will produce this discoloration. The dehydrochlorination reaction is autocatalytic and is accelerated further by the presence of oxygen
or contamination such as iron, residual polymer catalysts, or suspension
agents.

To prevent discoloration of the plastic, tin-base stabilizers such as dibutyltin or dioctyltin sulfur containing compounds are added to inhibit the formation of the conjugated double bonds. These compounds are usually added in liquid form to provide for stabilization and heat resistance of the PVC resins. Levels of 0.5 parts per hundred of the organotins are used in the resin mix.

Recently sulfur containing methyltin stabilizers have been developed as stabilizers for PVC piping and fittings for drain, waste and vent systems and piping for irrigation and gas distribution. These powerful stabilizers are used in concentrations as low as 0.85 parts per hundred of resin when single screw extruders are employed and up to 0.25 parts per hundred of resin when twin screw extrusion equipment is used (Fig. 3).

Fig. 3 – Single screw extrusion of PVC pipe.

Sintering P/M Parts

Excellent growth has been predicted for powder metallurgy products (4). The compounded annual growth of the P/M industry over the past ten years has increased faster than basic steel, primary aluminum, and iron and steel castings. In 1978, iron powder shipments to P/M fabricators showed almost 200 thousand tons and an annual growth rate of 8 to 10 percent through the next five years is predicted. The research arm of the tin industry was one of the first groups to recognize the potential of the P/M market using sintered iron components. Programs were initiated in the 1960's to determine the feasibility of using tin as an aid to the sintering of iron powders.

Experiments at the International Tin Research Institute demonstrated the beneficial effects of making additions of tin powders to iron compacts (5). These studies of the sintering behavior of iron powders showed that 2 percent of tin powder accelerates the sintering process. Later it was demonstrated that by use of 2 percent tin plus 3 percent copper powders, sintering temperatures could be reduced. Sintered iron-tin-copper alloys exhibit uniform strength characteristics as high as iron with 7 to 10 percent copper mixtures which are ordinarily sintered at temperatures near 1150°C. Mixtures containing tin were fully sintered at 950°C. Furthermore, the addition of tin to iron-copper compacts reduced the dimensional change (viz. growth) of sintered parts. This had the effect of maintaining closer control over the finished dimensions of the sintered part and thus, the quality and cost effectiveness of the products were improved considerably.

The economics of producing sintered iron parts with tin and copper additions was recently published (6). A piston and connecting rod for a refrigerator compressor were formerly made of cast iron. The cast iron parts for the piston weighed about 58 g and considerable machining was required to provide a finished piston weighing about 34 g. A German firm produced a sintered shape (Fig. 4) which was close to the geometry of the finished part. Many of the machining steps were eliminated and material costs were reduced by providing a sintered preform which weighed only 40 g.

Fig. 4 – Sintered iron refrigerator piston made from Fe-Cu-Sn powders.

When the part was first made, iron and copper powder mixes were used, but the sintered part was difficult to machine and antifriction properties were poor. When the piston was fabricated with iron powders plus

4 percent copper and 1 percent tin, excellent machinability and sliding properties were realized. In addition, sintering time and furnace temperatures were reduced to 20 minutes at 950°C. This amounts to considerable economies in operations since a sintering time of 60 minutes at 1150°C was necessary when mixes of iron and copper were employed. Overall, the production costs were 15 percent lower when tin-copper-iron alloys were used as compared to the production of the same part cast in gray iron. Energy costs for production of 1 kg of sintered iron parts are higher by about 65 percent than those for 1 kg of cast iron. However, the cost differences are more than compensated by the lower weight of the sintered parts as compared to gray iron castings. In addition, lower machining costs for the sintered pistons allowed savings of about 10 percent.

In the sintering of iron parts, zinc stearates are commonly added to provide lubrication during the pressing of parts. The time to burn off these lubricants is often inordinately long. Failure to remove pressing lubricants during sintering results in the inclusion of these compounds within the compact matrix causing porosity and subsequent weakening of finished parts. Research has shown that tin additions to iron powder mixes increases the green (compacted) strength of pressed parts. The metallic lubricity of the powdered tin allows reduction in the amount of zinc stearate pressing lubricant to 0.5 percent, which is less than half the amount normally used. Therefore, burn-off of lubricants is significantly reduced and contamination of sintering furnace atmospheres and degradation of the interior of furnaces is decreased.

Recently Madeley and MacKay reported on their work concerned with the properties of iron/solder compacts produced by warm pressing techniques (7). Warm pressing is defined as compaction of powders at temperatures between ambient and the temperature at which the iron powder matrix materials are capable of retaining work-hardening. Mixtures of -100 mesh Swedish sponge iron and 7 percent of -300 mesh atomized and prealloyed 60 tin-40 lead powder were compacted at pressures of 770 MN/m^2 (50 tons/in^2) into dies preheated at 450°C. Pressure was maintained while the solder powder was molten. After cooling, the iron particles were bonded together by the solder and a continuous network was established around the ferritic iron boundaries. Compacts showed very little porosity. Successful cohesion of the iron/solder parts during warm compaction is believed to be due to the attrition and rupture of oxide films on the iron and simultaneous wetting by the molten solder. Significant increases in tensile strength and transverse rupture strength of iron-solder compacts are attributed to the formation of $FeSn_2$ crystallites within the sintered mass. High matrix strength values are obtained since the initial work hardening of the iron powders obtained during the low-temperature sintering cycle, is below the annealing point for the iron particles.

Tin in Nuclear Applications

There is an increasing anxiety in regard to world energy requirements. Most persons readily acknowledge that nuclear power would provide vast sources of energy, but development of nuclear facilities is often blocked by individuals or groups who are concerned about problems of environmental safety, the cost of establishing proper controls, and recycling or storage of nuclear wastes (Fig. 5).

Fig. 5 – The core of a typical nuclear reactor.

From present knowledge, it appears that tin may, in part, provide a viable solution to the world's energy problems and reactor-related concerns. Researchers at Stanford University have developed the Actinide-Nitride Fueled (ANF) reactor system which is simple in concept. According to the developers, the system is self-correcting with a minimum of waste products and low radiation hazards. The ANF concept involves a graphite lined core which contains molten tin under a low positive pressure of nitrogen. Some of the nitrogen is dissolved in the molten tin. When uranium fuel is introduced into the tin, it combines with the nitrogen forming uranium nitride. This compound is more dense than the tin and therefore, it settles to the bottom of the reactor vessel. When the uranium nitride reaches a critical mass, fission begins and the reactor is in operation. If the temperature from fission of the radioactive material increases above planned operational levels, the uranium nitride dissociates and uranium is returned to the system, reducing the critical mass. Thus, the equilibrium reaction provides the ANF reactor system with stability against temperature excursions or runaway reactions. If temperature falls below the operating level, the reaction between uranium and the dissolved nitrogen will increase and the fission reaction will, in turn, be increased.

In the ANF system, the control of the critical mass of uranium nitride can be accomplished in any of three ways, namely, varying the pressure of nitrogen in the reactor vessel, increasing the amount of uranium or dilution of the uranium nitride by increasing the supply of molten tin. Additional control of the system can be made by use of neutron absorbing rods which can be inserted into the high flux region of the reactor.

The use of tin in the ANF system is advantageous for a number of reasons. At high temperatures, tin reduces the chemical activity of the uranium by the tendency of the system to form stable intermetallic compounds which compete for the uranium in solution. The reduction in the chemical activity of uranium promotes the reversibility of the nitriding-denitriding reaction.

The high boiling point of tin is employed in the ANF reactor concept. Studies have shown that a temperature of 1475 to $1800^{\circ}C$ is necessary to ensure precipitation of UN in preference to U_2N_3 which is precipitated more slowly. This temperature range is below the vaporization point for tin. In addition, certain lanthanide and transition metal impurities dissolved in molten tin as fission products are not seriously coprecipitated with the uranium nitride since they form low density nitrides which float on the molten tin surface and can be easily removed. Alternately, their concentrations in the solvent tin are sufficiently low and they remain in solution with other non-nitride forming elements.

Tin also provides the means for recycling of irradiated nuclear fuels by allowing the precipitation of any dense actinides such as those of plutonium. These are reusable fuel components and amount to more than 95 percent of the reactor charge. Once these components are precipitated from the molten tin, they can be drawn off and finally returned to the main reactor.

The small amount of wastes resulting from operation of the ANF reactor can be encased in slabs of cast tin and stored safely without the difficulties normally associated with the storage of nuclear wastes. Presently, much of the atomic waste products are stored in steel drums which are susceptible to corrosion and leakage, with subsequent contamination of the environment.

Although the ANF reactor has not reached the commerical stage of development, it would appear that the principles of its operation are sound. Perhaps tin will provide the means in the future for a fail-safe method of nuclear energy generation.

Can Lacquers Containing Tin

Tinplate has enjoyed the position as the preeminent packaging material for food and beverage for over 165 years due to the fact that it has undergone continual technological improvements. Tin coated steel is adaptable to the changing requirements of the packaging industry. Tinplate manufacture is the world's leading use for tin and over the past 25 years, the use of this material has increased steadily in spite of severe competition from other packaging materials. However, the average percentage by weight of tin coatings has fallen. The main reason for the drop in the average tin coating thickness per unit of tinplate used is because of a shift toward the greater use of lighter coated grades and the increasing use of can lacquers.

Recently it has been estimated that 60 percent of tinplates used for packaging are lacquered (8).

Tin coatings on containers serve the useful functions of protecting the product from anerobic spoilage reactions or biologic contamination and improving solderability of can side-seams. During the early stages of storage of foodstuffs in an unlacquered can, the oxygen remaining in the can during processing acts as a depolorizer for tin dissolution. Thus, the large area of tin surface exposed inside unlacquered cans acts as an effective sacrificial anode against dissolution of any iron exposed through pores in the tin coating and also against lead dissolution from contact of the pack with the soldered side-seam. Lacquering of the interior surfaces of tin cans may lead to undesirable color or taste changes in the pack and increased dissolution of iron or exposed solder. When there is no requirement for tin migration from can surfaces to preserve the organoleptic properties of the food, lacquered tinplated can be used. In addition, when a lead-base solder seam would provide an inadmissable source of contamination, pure tin solders have been used.

A can lining has been developed which combines the advantages of both the lacquer coating on thin tin coatings and plain tinplate cans (9). Spherical tin powders have been incorporated into can lacquers for application to thinly coated sheets of tinplate before the can is made. After oven stoving, the tin-loaded lacquered tinplate sheets are sheared and made into cans by conventional can making equipment. The finished cans have lacquered coatings which have tin particles exposed through the can enamels. It has been demonstrated in practical trials that if the proper size of the tin powder particles in relation to lacquer film thickness is made, sacrifical protection by the exposed tin is assured.

Work is in progress to determine whether the loading of tin powders in can lacquers will be sufficient to allow the migration of tin to preserve the organoleptic properties of some foods. A variety of foods and fruit juices have been packed in containers which have tin pigmented lacquers. Preliminary results indicate that the program will be successful.

Controlling Orchard Pests

Mankind wages a constant battle against insects and other organisms which destroy his agricultural products. Chemists are searching for new classes of compounds which agricultural pests have not built-up an immunity and still do not adversely affect the environment.

Certain organotin compounds are finding ready acceptability in agricultural applications as fungicides and acaricides (10). These compounds have relatively low mammilian and plant toxicity as well as a favorable environmental breakdown pattern. A new product based on the organotin bis (trineophytin) has proved to be an effective control over a wide range of plant feeding mites which cause bronzing of leaves or defoliation of citrus and other fruit trees. The commercial formulation is a wettable powder which can be dispersed readily in water and applied by conventional sprayers (Fig. 6). The compound kills adult and nymph forms of plant-feeding acarine organisms on contact, but at the same time it shows low

Fig. 6 – Spraying an organotin acaricide to
control orchard pests.

toxicity to beneficial mites and no toxicity to honeybees. The tin compound
has proven effective in controlling pest infestations which have acquired resistance to conventional phosphate and chlorinated hydrocarbon pesticides
which have now been banned by Federal agencies.

This organotin pesticide may be helpful to the home gardiner. Although
the main use thus far has been in controlling orchard pests, interesting
results are being obtained with greenhouse crops, flower beds and other
ornamental plants (Fig. 7).

Organotin Compounds as Antifoulants

One of the most promising uses of organotins appears to be in the field
of antifouling paints. Throughout history there has been a need to protect
ship hulls from marine fouling organisms. Arsenic and sulfur compounds
were formulated in wax and tar-base coatings to protect ship bottoms as
early as 300 BC (11). Soon after the first immigrants established Plymouth
Colony in 1620, copper sheathing was employed to protect hulls from the
attachment of sea organisms. With the advent of steel hulls, copper
sheathing could not be used alone because the copper-iron electromotive
cell provided an area for corrosion. Wood sheathing, used in a sandwich
construction with copper sheathing, prevented galvanic attack on iron hulls.

Fig. 7 – Ornamental plants treated with organotins.

The first copper oxide paints were developed about this same time and many of the systems employing copper oxide are in use today. Mercuric oxide paints were also employed from about 1908 to the mid 1940's but the effectiveness of these paints, as well as copper-based antifoulants depend upon controlled disintegration or exfoliation of the paint film. This leads to localized breakdown of films and eventually to discontinuous coatings and roughness of surface finishes.

It has been demonstrated that marine sliming bacteria and algae in estuarine locations provide a favorable environment for growth of barnacle cyprids (Fig. 8). There is ample opportunity for fouling of submerged hull surfaces where vessels are docked or moored in tropical waters. Fouling is significant in reducing speed and increasing fuel consumption of vessels underway. It has been estimated that a vessel of 80,000 tons displacement and having a speed reduction of only 0.15 knot due to fouling, would cost owners an extra $12,000 a year in fuel costs.

Suitable antifouling systems which incorporate tri-n-butyltin fluoride or bis (tributyltin) oxide have been developed which provide control over a wide range of fouling organisms (Fig. 9). These compounds are formulated into plasticized vinyl coatings. Controlled release of T.B.T.O. or T.B.T.F. succeeds by diffusion from the plastic film which does not exfoliate and therefore, maintains a smooth surface. In addition, diffusion is controlled by low solubility of these compounds in sea water. An additional benefit is that organotins provide a plasticizing action on the film.

Fig. 8 – Panels submerged in sea water. Panel on the
left is treated with an organotin antifoulant.
Panel on the right is untreated.

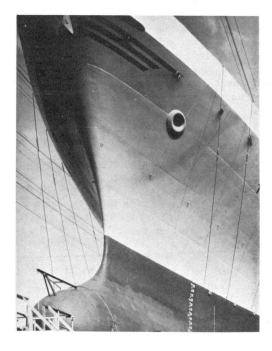

Fig. 9 – A ship hull painted with an organotin containing paint.

In the long term, tri-organotins used as antifoulants are important from the environmental point-of-view since residues in sea water are eventually degraded in nature to non-toxic inorganic tin compounds (i. e., stannic oxide) by successive cleavage of the three tin-carbon bonds. This fact together with the slower and predictable leaching rate of tin chemicals, point the way toward still another important use for tin.

Tributyltin Compounds As Wood Preservatives

The American Wood-Preservers' Association has issued its standard which is concerned with oil-borne preservatives for wood (12). Bis (tri-n-butyltin) oxide, commonly called T. B. T. O., is regarded as an effective preservative for wood. Pentachlorophenol is advocated in the same standard but environmental concerns may eliminate this compound as an effective wood preservative.

In any paint or preservative system for wood, solvents are important because these compounds control viscosity of the coating and provide the carrier media for the preservative. Petroleum derivatives and water are the major solvents and while many organotins are soluble in petroleum-derived solvents, the trend has been to limit the amount of organic solvents and to formulate various systems with water.

Water solubility of tributyltin compounds is very low (i. e., 0. 001 percent) and therefore work was programmed to find a suitable tributyltin compound which is sufficiently water soluble to allow the use of aqueous solvents. The patent literature reveals that tributyltin methane sulphonate or ethane-sulphonate have limited water solubility (13). Therefore, efforts have been directed to include a series of these tributyltin compounds in water-base formulations. A number of tributyltin alkylsulphonates have been employed (14). These compounds provide solubilities in water of 0. 8 to 1. 5 percent which is a useful range for the prevention of attack on timber by white and brown-rot fungi.

Tin Oxide Oxidative Catalysts

Mixed oxide systems based on tin have shown to be excellent oxidative catalysts. A tin oxide/copper oxide catalyst developed at the International Tin Research Institute's Laboratories in Perivale, England has demonstrated the usefulness of this system in the conversion of carbon monoxide to carbon dioxide at low temperatures. The reaction is continuous and is enhanced by the presence of water vapor. A direct and somewhat novel application of this development is in its use as a constituent in filters for cigarettes. A smoker encounters considerable quantities of carbon monoxide. The use of a tin oxide catalyst in these filters to convert the poisonous monoxide to harmless carbon dioxide may offer an exciting new application for tin. Commerical tests are currently underway to evaluate this idea.

Flame Retardants

Wool is generally regarded as the least flammable of the commonly used textile fibers because it has a high moisture content, high ignition temperature, low heat of combustion and a low flame temperature. New consumer legislation (15) prohibits the sale of flammable carpets, and although untreated conventional all-wool carpets will usually pass tests for flammable products, some long-pile (shag) wool carpets may fail flame tests because of their low pile density which allows a flame to spread because of the ready access of oxygen to the material. Many chemical compounds have been tested for their effectiveness as flame retardants for wool. Recent work at the Wool Research Organization of New Zealand has shown that tin salts offer some distinct advantages as flame retardants.

In the manufacture of certain wool shag carpets, the fibers are combed and brushed to provide a fur-like appearance. During this operation, the fibers are sprayed with a solution of stannic chloride, a fluoride complexing agent and polishing chemicals dissolved in an isopropyl alcohol/water mixture. This spray treatment relaxes the fibers and enhances their luster. They are then combed under a hot (200C) revolving metal drum which is fitted with combing teeth. The heat from the polishing drum is sufficient to evaporate the alcohol solvent and "fix" the tin containing fire retardant.

The tin oxide is concentrated in the top 10 mm of the wool pile. Flammability tests have shown that only the very tips of the wool pile are burned since they are in direct contact with air (Fig. 10). Below this level there is little access to air and combustion is prevented. Thus, tin solutions are only applied at the surface of the nap of the wool leading to significant economies when compared to other methods of applying flame retardant chemicals.

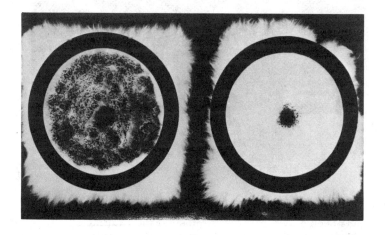

Fig. 10 — Flame test results on untreated wool (left)
and wool treated with tin compounds (right).

Research has shown that the active flame retardant tin species present in unwashed wool is likely to be hydroxypentafluorostannate di-anion, $SnF_5(OH)^{2-}$. However, upon washing, SnO_2 is formed by decomposition of the fluoro-tin complex (Figs. 11 and 12).

Fig. 11 – Scanning electron microscope photograph of untreated wool fibers.

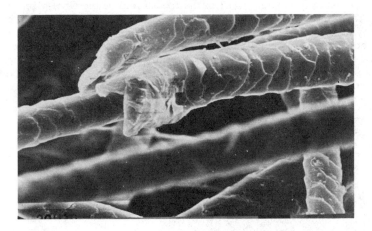

Fig. 12 – Scanning electron microscope photograph of tin-treated wool fibers.

Summary

This has been only a cursory glimpse at the exciting and dynamic innovations in tin use. It is increasingly apparent as we move towards the 21st century, that tin will continue to play an important role in technology to the benefit of all mankind.

References

(1) Gail W. Brown, "Battery Manufacturers Surprised by Size of Lead Price Increses," American Metal Market, 87 (55) (1979) pp. 12, 25.

(2) William J. Storck, "Plastic Pipe Use Growing Despite Problems," Chemical and Engineering News, 57 (12) (1979) pp. 15-18.

(3) R. J. Stimpfl, "Old and New of PVC Stabilization," Popular Plastics, May, 1973, 33.

(4) Robert E. Harvey, "Nutured by Automakers P/M Products Come of Age," Iron Age, 222 (2) (1979) pp. 32-34.

(5) R. Duckett and D. A. Robins, "Tin Additions to Aid the Sintering of Iron Powder," Metallurgia, 74. October, 1966, pp. 163-167.

(6) F. J. Esper and G. Leuze, "Economical Sintered Parts Through Suitable Alloy Selection," Powder Metallurgy International, 10 (3) 1978, pp. 148-149.

(7) D. J. Mandeley and C. A. MacKay, "A Preliminary Study of the Properties of Iron/Solder Compacts Produced by Warm Pressing," Powder Metallurgy International, 7 (4) (1975) pp. 170-171.

(8) A. Kleniewski, "Polarization Resistance Measurements as a Guide to the Performance of Lacquered Tinplate," British Corrosion Journal, 10 (92) (1975) pp. 91-98.

(9) Malcolm E. Warwick, "Role of Tin Powder Incorporated into Lacquers Applied to Tinplate Containers," British Corrosion Journal, 12 (4) (1977) pp. 247-252.

(10) Colin J Evans, "A New Organotin Miticide," Tin and Its Uses, 110, (1976) pp. 6-7.

(11) J. Engelhart C. Beiter and A. Freiman, Paper presented at the Controlled Release Pesticide Symposium, University of Akron, Akron, Ohio, September 1974.

(12) American Wood Preservers' Association Standard P8-77.

(13) R. Suzuki, Y Kuriyama and H. Shioyama, Japanese Patent 18,489 (1976).

(14) S. J Blunden, A. J. Crowe and P. J. Smith, "The Preparation of Some Water Soluble Tributyltin Biocides," International Pest Control, July/August (1978) pp. 5-7.

(15) Federal Register, 35, 74, 6211 (1970).

R&D PROGRAMS TO MEET NEW CHALLENGES FOR LEAD AND ZINC

Dr. S. F. Radtke and Dr. M. N. Parthasarathi
International Lead Zinc Research Organization, Inc.
New York, New York

The paper begins with a brief discussion of trends in consumption of lead and zinc, the growth rates in different periods and changes in patterns of end usage. The two metals had overcome many crises in the past and had acquitted themselves well in spite of the challenges and competition from other metals and materials. Research and development programs in major areas of application such as die casting, galvanizing, battery, cables, chemicals, etc., are discussed. The accent on these programs is very much market oriented and through the various R&D programs, it is planned not only to defend some of the old and established uses which are being challenged but also penetrate into new markets. The market pressures have proved to be a blessing in disguise as new potentialities for the two ancient but versatile metals are being recognized and new vistas of market opportunities are being opened through research and development programs.

Introduction

Lead and zinc are ancient metals. The former has been in usage for over 2,000 years and the latter for several centuries. They have survived the vicissitudes of fortune over hundreds of years and they are now poised for a new era of growth and challenges. This paper will attempt to trace and discuss the growth trend of the two metals and how they survived past crises. This paper will also discuss how in the recent past, research and development programs have brought about the survival of these metals. It also proposes how lead and zinc should get organized for the future in the light of an unprecedented competition from more modern materials.

Three Periods of Growth

The history of lead and zinc can be broadly grouped into three periods. The first period, up to the Second World War, was one of steady but slow growth of the two metals. This was a period when lead, zinc, copper and, of course, along with steel, were the only available common metals. Their usage was developed over several centuries. The metals were, in fact, put to, what one may call, traditional usage. They had no real competition from other materials. In spite of the lack of competition, the growth rates of lead and zinc were relatively low. This was because the pace of industrialization also was slow, so the demand for the two metals was only moderate. Furthermore, there was virtually no research or development or any effort to innovate new uses. Whenever new uses were discovered, such as, for instance, galvanizing, they were evolved rather accidentally and their usage was developed slowly over the years by trial and error methods.

The next period was the post second war era ending approximately with the onset of the first oil crisis. During this period, there was relatively high industrial activity which resulted in a much greater growth for the two metals. Although the growth rates were much greater than in the earlier period, they were not as impressive as the rate of industrialization because

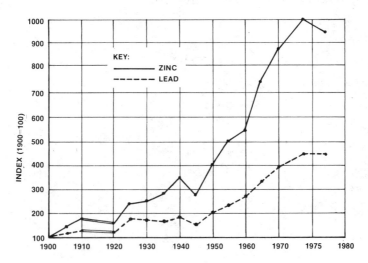

Fig. 1 - Indices of Consumption of Lead and Zinc
(1900 = 100)

of the keen competition from modern metals, like aluminum and other materials, like plastics. During this period, the pattern of consumption changed following the trends in industrialization, but both metals were for the first time meeting tremendous challenges and competition from other metals and materials.

Table 1.

Consumption (000 Metric Tons)

Year	1900	1938	1950	1960	1965	1970	1974	1977
Lead	851	1423	1683	2212	2725	3216	3683	3680
Zinc	455	1276	1828	2454	3333	3901	4562	4224

% Growth Rates/Year

Year	1900-38	1950-60	1960-65	1965-70	1970-74	1974-77
Lead	1.4	2.8	4.2	3.5	3.5	Negligible
Zinc	2.8	3.0	6.3	3.2	3.4	Negligible

Changes in Pattern of Consumption

Before the Second World War, the principal uses of lead were in cable sheathing followed by sheet and pipe for building, pigments and lastly batteries. After the Second World War, with the growth of the global automobile industry, lead batteries became the most important application and lead cable sheathing lost considerable ground. Sheet and pipes were relegated to a relatively insignificant position. For the first time, the price of lead became an unfavorable factor, and plastics made some permanent inroads in the cable industry in a relatively short time. Although plastics would have eventually supplanted lead cable sheathing in many applications, it happened somewhat sooner because of the acute shortage of lead during the Korean War. The process of substitution was certainly hastened by the nonavailability of the metal rather than the high prices prevalent at the time. While lead was losing a very lucrative market in cables, it was also gaining one in the automobile industry through lead acid batteries and lead in gasoline.

On the zinc side also, there were changes in the pattern of consumption. Before the Second World War, the important application of zinc was brass followed by sheet in building and galvanizing. The post war era saw the rapid growth of two principal uses of zinc; namely, galvanizing and die casting. The rapid growth of die casting can be mainly attributed to the growth in the automobile industry.

Crises for Zinc

While the growth of zinc was rapid during this period, it was also a period of great challenge. Zinc faced its first crisis as a result of premature failure of zinc die castings. This crisis was very serious and would have wiped out the usage of zinc in the automobile industry if the problem relating to tramp elements was not solved. At a later date, zinc met another crisis when the automobile industry demanded guarantees for plated finishes. Once this crisis was successfully solved, it looked as if zinc was about to settle down for a long period of uninterrupted growth. However, with the advent of plastics, challenges from this new material were slowly being recognized as threats. Plastics soon became, in fact, a very serious challenge.

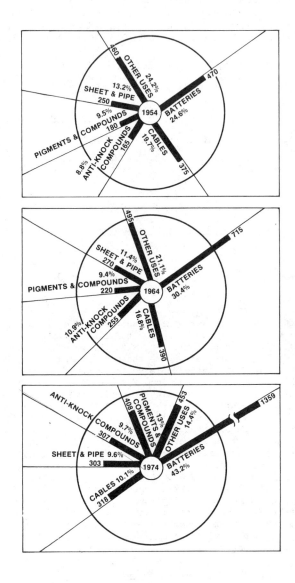

Fig. 2 - Changes in the Pattern of Consumption
for Lead

824

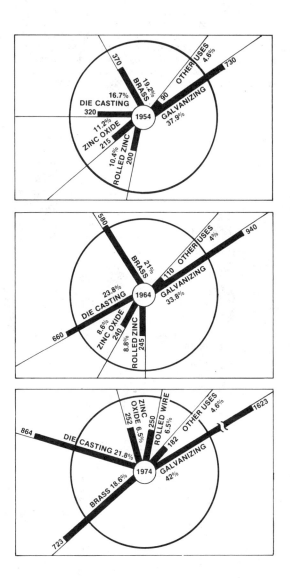

Fig. 3 - Changes in the Pattern of Consumption
         For Zinc.

825

Zinc in coating enjoyed a period of rapid growth but after a couple of decades of unchallenged growth, galvanized coatings began facing competition from several sources, most notably aluminum and plastics. While the markets for galvanizing were not seriously affected, the threat continues.

The next period is the post oil crisis period, which began in 1974, to the present. The crisis was aggravated by one of the worst recessions on a global scale, one from which both the metals have not completely recovered. Zinc suffered most from this recession and while it has yet to recover completely, there is a threat of a new recession.

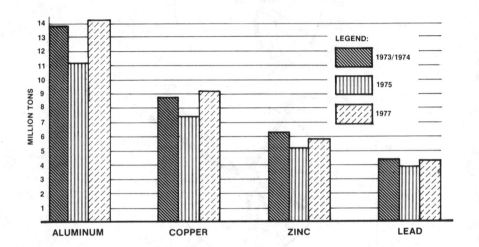

Fig. 4 - Bar Charts relate to peak consumption figure 1973-74, figures during 1975 recession and 1977 figures. They show fall in metal consumption and subsequent recovery.

Post Oil Crisis Challenges

During this period, both the metals went through a most depressing situation of low prices and threatened markets although lead recovered recently. The oil crisis forced many changes in the automobile industry and these changes have affected zinc more than any other metal.

The zinc and lead industries are not unaware of the future difficult situation but are prepared to meet the new challenges. In fact, even as early as 20 years ago, the lead and zinc industries anticipated the rough road ahead for the two metals and realized that only through research and development could the industry hope to develop new markets or defend the old and established ones. The only way to achieve this aim was to identify the shortcomings or limitations of the two metals and either remove them or mitigate their adverse effects through R&D.

It became evident, for example, that die casting had a serious contender in aluminum and plastics. It also became clear that these two materials would replace zinc markets to a certain extent. The question then was - Is it possible to retain the markets by any improvements in the process, and secondly, can zinc increase its market share through new developments, or alloys, or improvements?

Shortcomings Overcome by R&D

At this stage, it would be useful to discuss briefly the shortcomings of zinc in terms of the relevant markets.

1. Zinc has a relatively higher density than aluminum or plastics. Zinc die castings could not be used in certain applications because they did not have the creep resistance.
2. Zinc could not compete in the general foundry area because there were no suitable gravity casting alloys. Die casting techniques were costly and they were prone to mistakes and consequently losses.

Let us now deal with these shortcomings and discuss how zinc was able to sharpen its competitive edge and how it is now poised to extend its markets. Zinc did recognize it had a serious handicap in its relatively high density. This certainly was a handicap, but then zinc had better mechanical properties of strength and ductility and it had favorable casting characteristics like good fluidity, low melting point and good castability. The question was whether one could utilize these favorable characteristics to compete with aluminum and plastics. This was the basic foundation of the thin wall technology research program, and while it appeared that zinc die castings could be redesigned so as to compete effectively with aluminum and plastics, it became evident that a new concept of technology had to be developed. The success of this program hinged around the development of a well defined, simple and easily adaptable "foolproof" technology. Die design was an important component of this technology and it was believed that unless the required technology was perfected, and the know-how easily transferred, it would be impossible to achieve the objective of competing with plastics in a reasonable time. This was no easy task since die design in die casting had always been more of an art than a science. The then existing available guidelines and principles were derived on an empirical basis. Therefore, even for the best die designer, the die design could lead to unexpected problems in getting a good casting, with attendant delays and unnecessary expense. Therefore, it was considered impossible for a newcomer to the field of die casting to aspire to design a die.

Thin Wall Technology for Die Casting

The research program on thin wall die casting was based on an unusually simple approach. In pressure die casting, there are two important aspects; namely, (1) the flow of liquid metal into the die, and (2) the extraction of heat from the die - in other words, metal flow and heat flow. It was believed if both these parameters could be well understood and controlled, there should not be any difficulty in developing some well defined simple principles so as to arrive at a scientific die design. These principles should take into consideration the various common defects in die casting and should set them right or attempt to eliminate them altogether so as to arrive at a relatively simple and perfect design. With that approach, the objective was to attempt to produce a casting with the thinnest possible dimensions that would do the job.

The thin wall technology as developed by ILZRO/Battelle program has the following achievements: (1) extremely thin walls, as low as 0.4 mm cross section, can be prepared by adopting this technology. Normally when we speak of thin walls, we do not have any specific thickness in mind, but we are aiming at the thinnest sections possible for a certain application, which is certainly attainable through thin wall technology; (2) The rules governing die design are simple and are based on scientifically developed principles of fluid flow and thermal flow; (3) It would enable even a newcomer or, as

Fig. 5 - Untrimmed and trimmed-plated die castings using thin wall technology concepts.

Fig. 6 - Photograph shows the computer TRS80 Microprocessor video display and cassette tape deck. Once the simple concepts of thin wall technology are understood, the computer will perform the otherwise long and laborious calculations in a short time.

a matter of fact, a total stranger to the field of die design to produce a viable die that will work exceedingly well without any need of extended trials; (4) The occurrence of common defects can be minimized or totally eliminated; (5) Considerable metal savings can be achieved if the principles of thin wall technology are followed; (6) Productivity increases because of lesser metal used; (7) A great improvement can be obtained in the finished appearance of the die castings.

As a step to further simplification and make the die design calculation quick and easy and free of hassle, a simple computer program has been developed and with the easy to follow directions, it would be possible to obtain all the die design data in a short time. Previously, it could take several days or hours before such data can be computed.

New Creep Resistant Alloy

The conventional alloys have excellent mechanical properties at room temperature. However, their creep resistance is not adequate for certain structural applications where even moderately high temperatures as are prevalent as, for example, under the hood in an automobile where temperatures range between 100-200°C. Therefore, research was aimed at developing a die casting alloy with superior creep resistance, particularly for engineering applications in automotive and other fields. To meet this requirement, a new alloy, ILZRO 16, was developed. This is a cold chamber die casting alloy of zinc which is eminently suited for relatively high temperature applications.

Pore-Free Die Casting

Coming back to conventional die castings, it was always considered impossible to produce a die casting without any gas porosity. Porosities long have been accepted as inevitable in die castings, whether zinc or aluminum, because the air contained in the cavity gets entrapped as a result of turbulence within the liquid metal and remains embedded in the castings. Such porosities never posed any serious problems in zinc since the castings were always designed with adequate thicknesses or often overdesigned. Furthermore, if the parts are not required to be used at relatively high temperatures or not welded or not heat treated, porosities can be tolerated. However, the fact that conventional die casting has this serious and basic handicap made the use of die castings restrictive to only certain applications. Often, die cast parts have to be impregnated so as to seal the porosities.

Gas porosity can be almost totally eliminated by the Pore-Free Die Casting process patented by ILZRO. This process, which was developed in partnership with Nippon Light Metal Company, consists of the purging of the die with a reactive gas, like oxygen, prior to filling so that when the metal is injected into the cavity, the oxygen would react with the metal and be converted into its oxide, leaving the casting free of gas porosity. Pore-Free Die Casting extends the die casting process into applications which have long been unavailable to die castings. Furthermore, because of the absence of porosity, the castings are superior in properties and they do not need impregnation, a process routinely applied in many critical castings. The castings can also be welded and heat treated and, above all, can be used in relatively high temperature applications.

Research relating to die casting is not confined to the process and die design. The die casting machine is an expensive investment and it is important to make sure that it is efficiently utilized. The establishment of a PQ^2 diagram has gone a long way in achieving this objective (p is the metal pressure and Q is the flow rate; usually, p is plotted on a linear scale and Q on a square scale.)

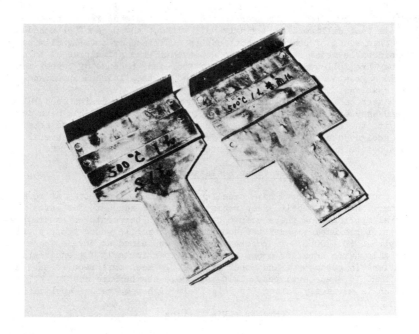

Fig. 7 - Pore-Free Die Casting on the left shows absence of porosities. Conventional die casting on the right shows
presence of considerable porosities as revealed by blister
tests (heat castings at 500°C for one hour) to which both
castings were subjected.

Computer Program for Cost Calculation

Another area of research in die casting worth mentioning is the use of
computers to calculate the cost of a die casting. In the highly competitive
market of the present era, accurate and speedy estimates of cost are essential.
Previously, and to a large extent even now, estimation of the cost of die
casting was based on direct material cost or part weight. This often led to
unpleasant surprises. There is such a large number of factors in the estimation of the cost of die castings that the use of a computer would be beneficial in arriving at a quick and accurate estimate and in avoiding costly
mistakes. This system could also be used as a cost control tool.

To meet this need, ILZRO developed what is called a DIECAP program by
which a computer is utilized to provide a quick estimate of the cost of die
castings and of the various cost elements involved.

Thus, all the components of the die casting research program are aimed
at making the process of die casting more scientific, less empirical, more
efficient and foolproof. Above all, the process is made so much simpler that
it no longer requires master craftsmen with long experience and it is now
easy to train people to a high level of competence with simple training.

Three New Gravity Casting Alloys

Until recently, one of the limitations of zinc was that its alloys were
exclusively confined to pressure die casting. There were no gravity casting
alloys. As a result, zinc could not enter the profitable general foundry

Table II.

PROPERTIES OF THE THREE GRAVITY CASTING ALLOYS

PROPERTIES	Zn-27% Al 2% Cu-0.01% Mg — Sand Cast	Zn-27% Al 2% Cu-0.01% Mg — Heat Treated	Zn-11% Al 0.75% Cu-0.02% Mg — Sand Cast	Zn-11% Al 0.75% Cu-0.02% Mg — Perm. Mold	Zn-8% Al - 3% Cu 0.03% Mg — Sand Cast	Zn-8% Al - 3% Cu 0.03% Mg — Perm. Mold
Physical						
Density (lb/cu.in.)	.181		.218		.22	
Electrical Conductivity (% IACS)	28		25		25	
Thermal Expansion (68-212°F) (in./in./F°)	14.4×10^{-6}		15.5×10^{-6}			
Pattern Maker's Shrinkage (in./ft)	5/32		5/32	1/8		1/8
Melting Range (°C)	380 - 493		380 - 432		387 - 393	
Mechanical[4]						
Tensile Strength (lb/in² x 10³)	58-68	45-47	40-45	45-47	38-42	38-44
Yield Strength 0.1% (lb/in² x 10³)	49	40	30	31	34	
Yield Strength 0.2% (lb/in² x 10³)	53					
Elongation (% in 2 in.)	3-6	8-11		1.3	1	
Shear Strength (lb/in² x 10³)	41-43	32-33	36-38		42-44	
Hardness BHN (500, 10, 30 sec.)	110-120	90-100	100-125		100-115	
Hardness V.P.N. (5 kg)	135	105				
Impact Strength(1) (ft-lb at 20°C)	11	19				
Impact Strength(2) (ft-lb at 20°C)	2					
Creep Strength(3) (lb-in² x 10³)	≥ 10	≈ 8	2.8		2	
Creep Rate at 20,000 lb/in² stress, (% in 1000 hrs. at 20°C)	0.1		4.95		3.16	

(1) ¼ in. square ASTM unnotched samples.
(2) Charpy V-notch.
(3) Stress to produce a steady creep rate of 0.01% per 1000 hrs. (1% per 11.4 years) at 20°C.
(4) Range of properties reflects change upon ageing for 10 days at 95°C.

Data From a Paper Presented at the 16th Annual Conference of Metallurgists of the Met. Society of CIM, August 1977 by E. Gervais et al.

market which had been dominated by aluminum to a large extent and brass, bronze, cast iron, etc. Zinc alloys have excellent mechanical properties and are relatively low in cost. It was believed, therefore, that it would be very easy for zinc alloys to capture a significant portion of the foundry market provided it was possible to develop suitable gravity casting alloys. The potential markets or applications where zinc alloys would be successful would have the following characteristics: (1) strength and ductility in the cast and heat treated condition; (2) Costs could be kept to a minimum by reduced machining or eliminating it altogether; (3) A good, easily applied plated finish would be an advantage.

There are now three gravity casting alloys and it has been demonstrated that these three alloys can compete with almost all of the general foundry alloys. The combination of excellent mechanical properties and relatively low cost of the material do make the gravity alloys very attractive. The foundry alloys and their properties are presented in the Table.

New Finishes and Processes

The success of zinc gravity and die castings to a significant extent is often interlinked with the finishing process; most importantly, with electro-plating. The cost of electroplating is a substantial part of the cost of the finished castings. Furthermore, in recent times, the electroplating industry has come under very strict pollution control regulations. In order to reduce costs and considering the likelihood of more stringent regulations regarding the disposal of plating effluents, ILZRO embarked on a comprehensive research program using trivalent chromium solutions. Trivalent chromium plating has many advantages. The research on trivalent chromium plating, in addition to developing a viable plating process, has developed a process for directly plating chromium on zinc without the intermediate stages of copper and nickel under-coats. A process of putting a pleasing black plated finish has also been developed.

Galvanized Coating

The biggest market for zinc on a global basis is galvanized coatings. Galvanized coatings have reigned supreme in the corrosion protection market and they are expected to hold this preeminent position for a long time. However, in recent years, several other materials, notably aluminum and plastics, have made inroads in certain specialized applications. With the demand for sophisticated and tailor-made coatings, galvanized coatings will lose a small share of the market. If the zinc industry intends to increase the market for galvanized coatings, vigorous research and development will be needed. Search is in progress for new zinc rich alloys which would enhance the life of the coatings.

The Industrialized Housing System

Considerable research and development has been carried out to extend the use of galvanized coating in large and profitable markets, such as the housing and building industry. All over the world, the building and housing industry is using zinc coated material; but the object of the ILZRO R&D program is to innovate a new approach to building and housing which would result in a greatly enlarged use of zinc. This has been achieved in the ILZRO industrialized housing system which uses sandwich panels as the major component. The sandwich consists of two galvanized sheets with polyurethane foam in the core. This system has shown the way to great cost reduction without compromising quality. The system is strong, durable, maintenance-free and has much less labor content so that considerable cost savings can be achieved.

Fig. 8 - The ILZRO Industrialized Housing System during a stage of erection. This system uses galvanized steel sandwich panels containing foam insulation.

Fig. 9 - Finished house of the ILZRO Industrialized Housing System.

Improvements in Process and Product

While the housing system is a move to penetrate into new markets, considerable effort is being utilized to strengthen the present market. This relates to two aspects; the process of galvanizing, and product research. The former would permit galvanizing of high silicon steels, the proportion of which will be considerably increasing in the coming years. Several routes are being investigated for handling high-silicon steels and it turns out that

high-silicon steels will not pose any difficult problems. A number of research
projects are in hand to investigate problems relating to paint adhesion, such
as coping with the spangled structure of the galvanized coating, corrosion
problems in automobile joints between galvanized steel and bare steel. These
and similar problems will continue and very stringent demands are made by
the users in as much as they demand very highly specialized and tailor-made
coatings suitable for particular applications.

Zinc Oxide as UV Stabilizer in Plastics

Zinc oxide has been losing markets in the paint industry but it is growing stronger in the rubber market. However, zinc oxide, on the whole, has
not shown any great growth trend. This material has been identified in a
recent ILZRO research program as a very good stabilizer in plastics against
ultraviolet radiation. In outdoor applications, plastics tend to become dis-
colored or breakdown by embrittlement unless adequately stabilized against
ultraviolet light degradation. Research and tests have demonstrated that zinc
oxide is a good stabilizer.

Engineering Development Program for the Auto Industry

There has been a great change in the outlook of the automobile industry
all over the world and more so in the U.S. Previously, the industry used to
specify materials, particularly in the days in the U.S. when most of the
accent was placed on styling. But with the advent of plastics, the auto industry began to vacillate between plastics and zinc particularly where only
styling was the main consideration and plastics did make some inroads in the
zinc market. Following the EPA mileage mandate and the need to downsize the
car and lighten every part, the auto industry moved away from the convention
of specifying any material. They, in general, now appeared to favor that
material which, in addition to meeting predetermined performance specifications and tests, would have the most attractive cost-weight ratio. There is
intense competition among all materials because of the large potential size
of the market. As a result of these developments, all materials have intensi-
fied the R&D effort and, furthermore, they extended their efforts to production
of prototypes to demonstrate the function of the part. The zinc industry has
not lagged behind in this challenge and is now committed to a comprehensive
engineering development program to produce prototype parts of not only light-
weight zinc parts but redesigning components which hitherto have been produced
in other materials. An example is presented in Figure 10.

Lead Battery Research

The battery market has emerged as the most important market for lead in
spite of the numerable power storage systems that are being investigated. Lead
acid battery systems will flourish for a long time to come. The lead acid
markets will grow steadily; however, there is intense competition among the
various systems and several promising ones are on the horizon. The lead industry, conscious of the tough competition ahead, has embarked on a comprehensive
fundamental research program aimed at producing new forms of lead oxide which
is the starting material in the production of lead acid battery grids. In this
research program, the role of dopants to control the microstructure and gross
structure of the battery plate is being investigated.

Lead Cable Research

Lead cable sheathing had lost considerable ground from aluminum and more
notably from plastics. The reason for the loss of the market was not from
technical considerations but rather on economic grounds. An effort was made
to minimize the loss of market for lead by thinning the cable sheath.

834

Fig. 10 - Camshaft Sprocket on the left is made of sintered iron.
It is redesigned as a zinc die casting shown on the
right. Weight savings effected - approximately 50%.
The creep resisting alloy, ILZRO 16, was used. The alloy
does not have adequate wear resistance in this application.
Several successful wear resisting surfaces have been identified.

In spite of considerable cost savings as a result of reduction in thickness of lead sheathing, plastic and aluminum sheaths were still cost attractive
in certain ranges. However, these materials and notably plastics have been in
service only for a limited period and the long term life and performance
characteristics have not been established as clearly as in the case of lead.
In the use of materials, these are important considerations since the cost of
replacement of cables as a result of failures or repairs and maintenance is
very expensive. Some utilities have had some unpleasant and unexpected experiences with some of the newer materials so that lead cable sheathing appears
to be favored where reliability is of paramount importance. But, the lead
industry is conscious that lead cable sheathing should be made more favorable
in cost. Therefore, research is aimed at reducing costs of not only manufacture but also of maintenance. One such project relates to improving the
efficiency and decreasing the cost of the process of splicing and sealing
joints in medium voltage power cables where lead is still preferred.

Lead Chemicals

The lead industry is aware that it can no longer depend on its traditional
markets, and for its survival, the metal has to seek new markets through research.
One such example is in the use of lead chemicals as additives in asphalt compositions for stabilizing them against thermal and oxidative degradation during

preparation, application and outdoor service on highways. Lead diamyl dithio-carbamate has been identified as a promising additive and field trials are being planned to assess the economics on an industrial scale.

Conclusion

In conclusion, it must be said that in a short paper such as the above, it was not possible to cover all aspects of the various research and development activities that are taking place to search for new marekts for lead and zinc and defend the old and established ones. Lead and zinc industries do recognize the challenges of the present and future, and that only by concert-ed R&D efforts alone, it can maintain even a modest growth rate.

ZINC-BASED COATINGS

PROCESS, PRODUCTS AND MARKETS

T.F.Kirk, F.W.Ling, W.R.McClure, T.E.Weyand

St. Joe Minerals Corporation
Pittsburgh, Pennsylvania

Summary

The strength, formability, low cost and availability of steel have been responsible for its utilization as a building and fabricating material. However, left unprotected, it will corrode in most environments. Zinc-based coatings have offered both an effective barrier and galvanic defense in the protection of steel from corrosion.

The use of zinc in coatings accounted for 46% of the U.S. 1978 total zinc consumption of 1.03 million metric tons. The foremost types of zinc coating are: continuous sheet hot dip galvanizing and electrogalvanizing (58%), after fabrication hot dip galvanizing (31%), and zinc paints (11%).

During the 1970's, performance requirements for products that utilized zinc coatings changed due to energy, government and processing considerations. Zinc coating technology responded to these new demands with modified coating processes, new coating alloys and advanced products. Because many of these developments commenced commercial operation in the late 70's, their full impact on zinc consumption has yet to be realized.

The performance and economic requirements of zinc coating markets have never been more stringent than at present. Coincident with this challenge, the opportunities for zinc coatings over the next decade will be at their greatest. Zinc's proven record for well over 150 years of successfully protecting countless steel products, combined with the new technology advances in the 70's indicates that the zinc coatings industry will respond to the changing market and product demands with improved quality coatings for the 80's.

Introduction

It has long been recognized that steel is heavily utilized as a building or fabricating material because of its great strength, good formability, low cost and availability. If left unprotected, however, it corrodes rapidly in most environments. Of the various methods available for preventing or minimizing the atmospheric corrosion of steel, the most commonly used is coating of the steel surface.

Zinc coating systems offer two forms of corrosion protection: barrier protection plus electrochemical or galvanic protection. A general guideline is that the rate of uniform corrosion of iron is more than 25 times that of zinc. In addition, when the coating is penetrated via scratches, fabrication processes, or during normal wear, zinc's secondary defense bridges these gaps electrochemically to continue to protect the steel by galvanic action.

Zinc's ability to economically provide both barrier and galvanic protection to steel is why zinc coatings continue to represent the largest segment of total zinc consumption. In 1978, the U.S. consumption of zinc was 1.03 million metric tons which represented 23% of the free-world consumption of zinc (1). In the U.S., 470,000 metric tons representing 46% of the U.S. consumption were used in the coating of steel. Shown in Table I are the major zinc coating markets.

Table I. U.S Zinc Consumption by Application Process 1978 (2)

	Continuous Galv.	After Fabrication Galv.	Zinc Paints	Total
Zinc Consumption (metric tons)	270,000	147,000	53,000	470,000
Percent	58	31	11	100
Steel Coated (metric tons)	5,830,000	1,840,000	N/A	7,670,000+

Throughout the 1970's, a mature zinc coating industry has been faced with new challenges based on more stringent economic and performance demands of the marketplace. This has resulted in significant market shifts and new products such as one sided galvanized sheet, Zincrometal and Galvalume. This paper will review the new markets, products and processes of the 1970's, several of which have yet to realize their full commercial impact. Also, possible problems and opportunities for the industry in the 1980's will be identified.

A review of the traditional processes and products will be presented followed by a discussion of the market needs that stimulated new zinc coating technology and products.

Process and Product Review

Continuous Sheet Galvanizing

Essentially, all continuous sheet galvanizing operations include high temperature or chemical cleaning, followed by coating either in a bath of liquid galvanizing alloy (hot dip) or by immersion in a zinc salt electroplating solution (electrogalvanizing) (3).

Hot Dip. The high temperature continuous galvanizing processes, such as the Sendzimir process (4), subject the sheet to a high temperature oxidation to burn off various organic rolling oils, followed by an elevated temperature reducing operation carried out in an atmosphere furnace generally containing a high hydrogen concentration. Typically, sheet temperatures within the reducing furnace are within the range of 700 to 800°C, thereby reducing the oxide layer on the surface as well as fully annealing the steel substrate. The sheet is then partially cooled before entry into the bath via a furnace extension snout containing the protective atmosphere of the reducing furnace.

The chemical continuous galvanizing processes are characterized by off-line annealing and in-line alkaline degreasing and acid pickling. After cleaning, the sheet is coated with an aqueous ammonium-chloride bearing flux, dried and preheated slightly before entering the galvanizing bath. The sheet temperature increases to that of the molten zinc bath, approximately 455°C, so that the sheet reacts with the molten zinc to form the metallurgically bonded galvanized coating.

After exiting from the molten bath, the coating in both processes is normally wiped to the desired thickness with steam/air knives or coating rolls (5), and then cooled continuously at a rate dependent on the final coating spangle requirements.

Electrogalvanized. This process consists of high speed alkaline and acid bath cleaning, either or both of which may be electrolytic (6). In the electroplating unit, zinc anodes are used in 65°C acidic solutions. Either zinc sulfate or zinc chloride plus other salts are used to improve conductivity and plating performance.

During and after the above processes, additional in-line chemical or mechanical processing is often applied. This may include a special steam quench immediately after exiting the molten zinc bath to physically obtain a 'killed' minimum spangle surface or a small in-line annealing furnace immediately above and after the galvanizing pot to provide heat for accelerated alloying of the zinc to produce a zinc-iron alloy (galvannealed) coating. Further treatments may include a hot dilute chromic acid rinse applied as a conversion coating, phosphate coating to improve paintability, tension leveling to improve sheet flatness or an in-line temper pass to achieve an extra-smooth nonspangled surface.

The continuous hot dip galvanized sheet products consist of cold rolled (as rolled or as annealed) steel substrates with coatings that fall into two broad classifications: regular and alloyed. The bulk of the coatings can be described as the regular type which have a minimal iron-zinc intermetallic layer due to short bath immersion times and small aluminum additions to the galvanizing bath (Figure 1).

Coating

Substrate

Figure 1 - Cross Section of Regular Continuous Hot Dip
Galvanized Steel

The alloyed or galvannealed coatings, although not as ductile and formable as the regular coatings, offer enhanced paint adherence in the final product with no significant loss in either corrosion resistance or galvanic protection. This coating is shown in Figure 2.

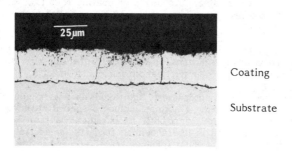

Coating

Substrate

Figure 2 - Cross Section of Alloyed Continuous Hot Dip
Galvanized Steel

The electrogalvanized product exhibits a uniform zinc coating without any intermetallic layer and achieves excellent coating adherence even to the point of substrate fracture. Another advantage of this low temperature process is that the substrate retains the batch annealed qualities required in many forming operations. This structure is shown in Figure 3.

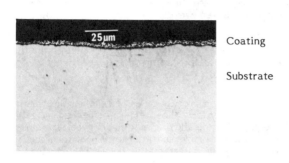 Coating

Substrate

Figure 3 - Cross Section of Continuous Electrogalvanized Steel

Table II summarizes the metallic bath composition and coating thickness, and indicates the effect of minor bath additives.

Table II. Continuous Galvanized Bath Composition and Thickness (7)

| Element | Bath Composition | | Remarks |
	Electro-Galvanize	Hot Dip	
Zinc	(Anode) 99.99%	>97.65%	---
Aluminum	---	.1-.25%	Suppresses iron-zinc alloy layer development
Lead	---	< 1%	Promotes large spangles and enhances bath fluidity
Antimony	---	<.1%	Promotes large spangles with lower coating relief
Cadmium, Tin	---	< 1%	Promotes frosty spangles
Coating Thickness	.0025-.025 mm	.005-.060 mm	
Specification ASTM	A-591	A-525	

After Fabrication Hot Dip Batch Galvanizing

This hot dipping process involves the immersion of a steel or iron fabrication or casting into molten zinc to provide a corrosion resistant zinc coating. The operation is divided into three steps: surface cleaning, fluxing, and coating. The purpose of the surface cleaning is to remove the contaminants and oxides from the steel surface, and can be accomplished either chemically or mechanically. Fluxing is performed to ensure an oxide-free surface prior to the galvanizing operation and can be conducted in three ways: a preflux, a top flux, and a modified top flux. The operations using these techniques are usually referred to respectively as dry, wet, and modified wet hot dip galvanizing (8).

The final stage of the galvanizing process takes place when the clean steel is immersed into the molten zinc. After the steel has reached a galvanizing temperature of approximately 455°C, the alloying reaction between the zinc and the steel initiates. This involves interdiffusion and interaction between the zinc and steel which produces various zinc-iron intermetallic layers on the steel surface. After a period of time, usually two minutes or more, the galvanized steel is withdrawn from the bath at a rate which allows sufficient drainage of the excess molten zinc from the surface. A typical after-fabrication hot dipped galvanized coating is shown in Figure 4.

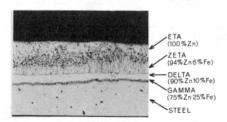

Figure 4 - Cross Section of After Fabrication Hot Dipped
Galvanized Steel

The following, Table III, summarizes a typical coating bath composition and coating thickness.

Table III. After Fabrication Galvanized Bath Composition and Thickness (9)

Element	Bath Composition	Remarks
Zinc	>98.0%	---
Lead	$\leq$1.6%	Increases molten zinc drainability
Cadmium	< .5%	---
Iron	< .05%	---
Aluminum	.002%	Promotes shiny surface
Coating Thickness	.07-.2 mm	
Specification ASTM	A-123, 153, 386	

Zinc Paint Systems

Zinc paints when properly formulated perform two functions (10); initially they are electrically conductive, protecting cathodically. Then the exposed surface of the zinc reacts with water, oxygen and carbon dioxide in the air to form insoluble zinc corrosion products, which fill the pores of the film. The resulting coating then acts as a barrier which, in turn, protects the underlying free zinc in the film, allowing it to remain dormant until the film is damaged. When damaged, the cycle is repeated.

There are two types of zinc pigmented paints which can be used for corrosion control: zinc dust-zinc oxide and zinc-rich coatings (11). Both types offer excellent rust inhibitive properties, adhesion, and abrasion resistance. The former are ideal for both prime and finished coat applications, while the latter serve as excellent primers.

Zinc dust-zinc oxide paints normally possess high covering power and can hide backgrounds of almost any other color. Zinc-rich coatings are effective when steel is subjected to high humidity and water immersion. Zinc paint systems can be broken down into industrial, automotive spray and coil coatings; each class offering particular corrosion protection characteristics, and each requiring its own unique method of steel surface preparation before painting.

Zinc paint systems are applied either before or after fabrication (12). Continuous roll coating of coiled steel sheet occurs at high speeds and is automated. Zinc paints can be applied after fabrication by spraying, dipping, or brushing. In all cases, the steel substrate surface must be clean to ensure paint adherence. Unlike the galvanizing process, no metallurgical reaction or bond occurs so the paint adherence is dependent upon mechanical attachment. Typical one-coat thicknesses can vary from .02 to .20 millimeters depending upon specific application conditions.

Market and Product Shifts

Continuous Galvanizing

The amount of zinc consumed on steel sheet products was 270,000 metric tons in 1978, or 58% of the total zinc utilized in coatings. This represented a 32% increase over the 204,000 metric tons of zinc consumed in 1970. This growth reflects a strong end-user demand for galvanized sheet steel. The customers for zinc in continuous hot dip and electrogalvanized sheet steel are mainly the steel mills and, for this reason, the AISI Continuous Hot Dip and Electrogalvanized Sheet Steel Shipments have been used to depict shifts in the marketplace for zinc. The years 1970 and 1978 are representative and shown in Table IV.

Table IV. Continuous Hot Dip and Electrogalvanized Sheet Steel
Shipments by Market Classification -- U.S. (13)

(000 Metric Tons Steel)

	1970		1978	
Market Class	Tons	% Total Steel	Tons	% Total Steel
Contractor's Products	1,780*	40.0	1,545	26.0
Automotive	870	20.0	1,765	30.0
Construction	345	8.0	990*	18.0
Agriculture	245	6.0	640	11.0
Appliances	375	9.0	315	5.0
Electrical Machinery	120	2.0	150	3.0
Other Domestic & Commercial Equipment	140	3.0	160	3.0
Machinery	60	2.0	80	1.0
Containers, Packaging & Shipping	55	1.0	50	1.0
All Other	370	9.0	135	2.0
Total Steel	4,360	100.0	5,830	100.0
Tons of Zinc Consumed (000 Metric)	204		270	

*1970 Culvert sales are included within contractor's products (330,000 metric tons)
 1978 Culvert sales are included within construction (317,000 metric tons)

The increase in galvanized steel shipments represents an increase in annual zinc consumption of about 66,000 metric tons in 1978 compared to 1970. The categories contractor's products (after culvert adjustment), appliances, electrical machinery, etc. have remained relatively constant for the 8-year period. The major growth has occurred in automotive (100%), construction (95%), and agriculture (160%). It is in these areas that significant changes in market and product demand have occurred during the 1970's, and new, more stringent demands are anticipated in the 1980's.

Automotive. The automotive segment's consumption of galvanized steel doubled during the 8-year period representing an additional 895,000 metric tons of galvanized steel per year in 1978.

In the early 1970's, the automotive industry suffered a dramatic loss in consumer confidence because of the frequency and magnitude of steel corrosion which penetrated the exterior bodies in relatively new cars. To minimize the severity of this condition, the automotive industry demanded that the steel sheet be zinc coated to protect against corrosion and that this new system be compatible with existing painting, fabricating, joining, and finishing practices.

The regular G90 two-side galvanized steel sheet was provided for use in nonexposed areas. The increase since 1970 in regular two-side galvanized product has consumed an additional 20,000 metric tons of zinc per year. However, the commercial galvanized surface was not capable of meeting the standards imposed by the automobile industry.

Detroit had requested a 'one-side' galvanized product since the 1960's, but not until 1970 through 1972 was the need and demand severe enough to justify the development of new technologies and the capital investment to convert existing galvanizing lines to produce this new product. The ideal one-side galvanized sheet product was expected to offer protection of zinc on the inside of the steel panel, while leaving an exposed drawing quality exterior which could be fabricated, joined, and painted in the normal fashion.

Several hot dip one-side galvanizing techniques were proposed and investigated, including a standing wave process (14) developed by one zinc supplier; St. Joe Minerals Corporation. Since 1974, three continuous hot dip one-side coating processes achieved commercial operation in the U.S. Each is unique, thus yielding slight variations in final coating and substrate characteristics. In 1978, these three processes accounted for 195,000 metric tons of hot dip one-side product which consumed approximately 4,900 metric tons of zinc:

Stop-Off Method (15). This product is produced by Republic Steel in a continuous process utilizing a stop-off coating to protect one side of the steel sheet from reacting with the molten zinc in the galvanizing bath (see Figure 5). This first attempt at a one-side product is not used in exposed panel applications, but has been successfully employed in less critical areas of the automobile.

Coating

Substrate

Figure 5 - Cross Section of One Side Galvanized Steel - Stop-Off Method

Electrostrip Method. In this 'Unikote' (16,17) process developed by National Steel, the sheet is galvanized on both sides in the conventional continuous hot dip manner. After exiting from the bath, one side of the product is wiped via air knives to yield a minimum coating thickness. The zinc on this thin coating surface is then electrolytically removed and replated on the opposite side on top of the thicker hot dip galvanized coating (see Figure 6).

Coating

Substrate

Figure 6 - Cross Section of One Side Galvanized Steel - Electrostrip Method

Meniscus Method. 'Zincgrip OS' (18), developed by Armco Steel, is produced in a continuous hot dip line by exposing only one side of the sheet to the zinc molten bath (see Figure 7). The high surface quality of this product is due to nitrogen shrouding and has permitted its utilization in high visibility areas.

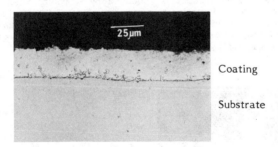

Coating

Substrate

Figure 7 - Cross Section of One Side Galvanized Steel - Meniscus Method

Several other successful one-side and pseudo one-side products have been developed to meet the automotive challenge:

U.S. Steel developed the 'Carosel' electrolytic process (19,20) and Sharon Steel modified an existing electrogalvanizing line to produce a continuous electrogalvanized 'one-side' coated steel sheet product. These products exhibit not only a uniform and spangle-free zinc coating, approximately .013 millimeters (Figure 8), but also better formability compared to a hot dipped process. In 1978, 295,000 metric tons of one-side electrogalvanized product were shipped, representing consumption of about 4,800 metric tons of zinc.

Coating

Substrate

Figure 8 - Cross Section of One Side Electrogalvanized Steel

Inland Steel developed 'Paint-Tite B' (21) which is produced by differentially wiping the galvanized sheet and subsequently heating in a controlled fashion such that only the thinly coated surface is completely transformed into an iron-zinc alloy layer while the heavily coated surface remains essentially unaffected (see Figure 9). This product offers improved paint adherence while retaining significant corrosion protection on both sides of the sheet. In 1978, 11,000 metric tons of Paint-Tite B were shipped which consumed approximately 300 metric tons of zinc.

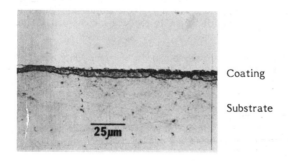

Coating

Substrate

Figure 9 - Cross Section of 'Paint-Tite B' Coated Steel

All of the above processes required significant lead times due to technology development and equipment conversion. As a result, the automotive demand was, and continues to be, in excess of the steel industry's ability to produce these new zinc coated products.

The continuing gap between supply and demand for one-side galvanized has been filled by the zinc paint industry, which developed a two-stage, continuous process which yields a one-side zinc-rich painted steel sheet. First, a 'Dacromet' coating, approximately .0025 millimeters thick, is applied to the base steel sheet and heated to $175^{\circ}C$. Then, the second stage 'Zincromet' is applied and cured at $290^{\circ}C$ to achieve a total coating thickness of .014 millimeters. This 'Zincrometal' (22) coil coated product has achieved wide acceptance for many automotive applications as an alternative to one-side galvanized. In 1978, 1.1 million metric tons of this product were consumed, utilizing approximately 5,500 metric tons of zinc.

The automotive needs for corrosion protection started to change dramatically in late 1978, so that a new dilemma is now facing the automotive industry. Government regulations continue to demand increased fuel economy which is presently being met through additional downsizing and weight reduction measures. The use of thinner gauge steel sheet will require new and improved corrosion resistant coatings. In addition, product liability considerations and the need for increased car life continue to stimulate the automotive industry. Furthermore, the government is requesting, and threatens imposition of a corrosion warranty that would give not only 10-year's perforation protection, but also a 5-year cosmetic exterior corrosion guarantee (23).

These factors have precipitated a change in the expected zinc coated steel sheet requirements for this market. It is generally agreed that a high quality, non-zinc rich paint system can only provide 2 to 3 years of cosmetic exterior rust protection. Thus, zinc will be needed on both sides of the sheet to ensure longer lasting protection.

If these projections are true, a new opportunity for zinc could exist in the form of a switch in demand to a high quality two-side galvanized sheet. However, this transition can only occur if the fabrication, joining, finishing, and painting problems associated with zinc on the exterior sheet panel can be resolved. At the present time, considerable development work is being performed to produce a high quality, formable, two-side, minimum spangle galvanized sheet with low coating relief. Testing is ongoing to determine the feasibility of producing a two-side nitrogen-shrouded and wiped product. In addition, one-side electrogalvanizing lines are being evaluated to determine required modifications to produce a competitive automotive two-side product.

A caveat must be considered in this demand transition. Only about one-third of the domestic galvanizing capacity has the necessary sheet width ranges required, i.e., those above 122 centimeters. Thus, should these projections become fact, the steel industry would still be unable to supply the projected short-term requirements of the automotive industry. To this end, industry sources indicate that at the beginning of 1979, three North American steel companies were evaluating the feasibility of installing new, wide, flexible continuous hot dip galvanizing lines capable of meeting the long-range demand and quality requirements.

In order to supplement the limited supply of acceptable two-side galvanized product, the Zincrometal coil coating lines are being studied and evaluated to determine the optimal two-side coating system. To the coaters' advantage, over 700,000 metric tons of new Zincrometal capacity is expected to be on-stream in the U.S. by 1980.

848

Once clearly defined, the steel and coil coating industries appear willing to respond to the automotive industry demands with two-side and one-side galvanized products as well as two-side and one-side Zincrometal. Because the zinc loadings for these products vary from 5 to 50 kilograms per metric ton of steel, the opportunity could represent anywhere from 12,000 to 35,000 metric tons per year of additional zinc consumption by 1985.

Construction and Contractor's Products. Considered jointly, the steel tonnage shipped to these markets has increased by 20% to 2.54 million metric tons per year from 1970 to 1978. This represents an overall increase in zinc consumption of about 20,000 metric tons.

One key technical development affecting the zinc consumption in this market occurred in steel substrate metallurgy in the early 1970's. In the metal roofing industry, the production of a more formable 'full hard' recovery annealed sheet product permitted an overall reduction in sheet gauge, a redesign in the building panels, and an increase in on-site fabricability. A result of the decrease in sheet thickness was a proportional increase in zinc loadings per ton of steel; since 1972, the average amount of zinc per ton of galvanized sheet roofing has increased by almost 17 kilograms per metric ton.

The consumer's continued desire for a longer lasting product has led to the utilization of painted galvanized roofs which carry a 20-year warranty. The color availability was particularly appealing with high-sloped roof designs found on most buildings.

A new coating, 55% aluminum, 44% zinc, 1% silicon, called 'Galvalume' (24,25) was developed by Bethlehem Steel and commercialized in the mid-1970's. Product claims for Galvalume include a best of both worlds approach in that it provided not only the barrier protection of aluminum coatings, but also the galvanic protection of zinc coatings at a cost less than that for aluminized sheet. Figure 10 illustrates a typical cross-section of the Galvalume product which exhibits an aluminum-rich dendritic and a zinc-rich interdendritic phase. In recent years, building design has also featured lower sloped roofs which do not require an aesthetically pleasing colored surface, i.e., where paint is not needed. The improved corrosion resistance of Galvalume as compared to galvanized sheet has accounted for its acceptance in this market. Approximately 91,000 metric tons of this product were shipped in 1978, utilizing approximately 1,500 metric tons of zinc, but simultaneously displacing anywhere from 2,000 to 2,500 metric tons of zinc in galvanized steel (26).

The market demand for improved corrosion resistance stimulated by competitive products such as Galvalume and Aluminized coatings has provided the impetus for the development and evaluation of other types of zinc-aluminum alloys rich in zinc.

Continuous galvanized sheet still exclusively enjoys the wall panel and deck end-use applications. Accordingly, modifications are being made to ensure a better, more competitive zinc product. The finished painted surface is of prime importance in the wall panel market. The coating has been improved by the addition of antimony, which lowers the spangle relief and ultimately provides a smoother finished surface.

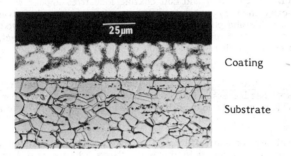

Coating

Substrate

Figure 10 - Cross Section of 'Galvalume' Coated Sheet

At the present time, it is not expected that Zincrometal will be a competitive factor in these construction market segments. The large amount of handling and potential damage due to scratching prevents Zincrometal's participation.

The production of galvanized culvert stock represents a significant portion of the total shipment in these market classifications. In 1970, the Federal Government sponsored an intensive highway construction program which included a heavy reliance on new culvert stock. Although this new construction has recently slowed, it has been replaced by primary and secondary road maintenance projects which also utilize significant culvert stock. Throughout the period 1970 to 1978, the overall shipments of culvert stock remained fairly constant.

Continued growth of 2 to 4% is expected in the construction and contractor's products markets for zinc coated steels.

Agriculture. The agricultural market class has enjoyed an average annual growth rate in excess of 12%. A consistently increasing demand for galvanized steel sheet in this market classification has been in the preengineered metal farm building and shelter applications. It is anticipated that this market will remain a strong consumer of galvanized sheet throughout the 1980's.

Other Classifications. In total, the remaining classifications have been relatively stable throughout the 1970's. The desire to conserve energy will force downsizing and weight reduction of the equipment and packaging items contained in these market classes. This downsizing will be accomplished by a strong consumer and governmental interest in corrosion protection. The National Bureau of Standards, in conjunction with Battelle Institute, conducted a study on the economic effects of corrosion in the U.S. which concluded that metallic corrosion costs approximate $70 billion per year. Therefore, on the basis of energy conservation and corrosion costs, it is anticipated that the demand for continuous galvanized sheet will increase through the 1980's.

The overall zinc consumption is not only dependent upon the growth rates of the different market classifications. Another factor affecting overall consumption is the amount of zinc utilized per ton of steel. This can be broken into three significant components: substrate thickness, coating thickness, and coating weight control. For example, the installation of coating weight control systems between 1972 and 1977 decreased consumption by 20,000 metric tons of zinc per year.

After Fabrication Hot Dip Galvanizing

The customers for zinc in the after fabrication market are the almost 400 batch hot dip galvanizers. The amount of zinc consumed in this segment was 147,000 metric tons in 1978, or 31% of the total zinc utilized in coatings for corrosion protection. This consumption was slightly lower than the 150,000 metric tons consumed in 1970, indicating a stable market for zinc. A breakdown of this market by market classification for the years 1970 and 1978 is shown in Table V.

Table V. After Fabrication Hot Dipped Galvanized Steel by Market Classification -- U.S. (27,28)

(000 Metric Tons Steel)

Market Classification	1970		1978	
	Tons	% Total Steel	Tons	% Total Steel
Electric Utilities	390	21.0	400	22.0
Fabricated Wire	400	22.0	380	21.0
Light Construction	220	12.0	230	12.0
Heavy Construction	200	11.0	215	12.0
Highways	240	12.0	210	11.0
Agriculture	110	6.0	125	7.0
Recreation	125	7.0	115	6.0
Marine	15	1.0	15	1.0
Metal Containers	20	1.0	15	1.0
Other	140	7.0	125	7.0
Total Steel	1,860	100.0	1,830	100.0
Tons of Zinc Consumed (000 Metric)	150		147	

The galvanizing sector depends primarily on the products for electric utilities, fencing and packaging, light and heavy construction, and highway safety. The demand for these products tends to rise and fall with the status of the economy. The electric utility industry utilizes galvanized steel for transmission and distribution towers, and pole line hardware. Galvanized tubing has been used on potable water systems for a number of years with good success. Waste water treatment plants, irrigation systems, and playgrounds are also large users of galvanized tube and pipe. The fence and packaging industries maintain a relatively constant demand for galvanized wire. Galvanized guardrails, overhead sign supports, light poles and chain link fences provide long-lasting safety and protection for the highway systems. Over the past 13 years, more than 250 galvanized bridges have been installed on federal, state, and county road systems.

The major consumers of structural members are the utility, construction, and highway market classes. These account for 60% of the total after fabrication galvanized steel shipments. A significant trend to reduce weight in structural members has led to the specification of thinner gauge, higher strength silicon steel sections.

Silicon causes increased reactivity in the galvanizing bath resulting in a thicker, more brittle, and less aesthetically pleasing coating (29). From a processing standpoint, these effects increase costs and ultimately reduce the competitiveness of galvanizing. Traditional methods for controlling this reaction have relied upon reducing immersion time or decreasing bath temperature. Both of these techniques can result in production delays and excessive costs. Zinc suppliers are continuing to evaluate types of additions that can be made universally to galvanizing baths which can control this reaction. Thus far, only one system, 'Poly-galva' (30), has been commercially introduced in the U.S. Others are expected to follow.

Although only 5% of the products hot dip galvanized are fabricated from silicon bearing steels, this percentage is expected to increase. Economic solutions to the silicon steel galvanizing problem could increase zinc consumption by 5,000 to 7,000 metric tons per year.

Galvanized steel framing is growing in popularity for use in single and multi-family housing because of increasing labor and wood building component costs. The galvanized system features high strength and proven economy with lasting protection against rust, rot, and termites. With little or no technological changes, galvanized steel studs and joists can be directly substituted for traditional wood products. This was recently demonstrated by the International Lead Zinc Research Organization (ILZRO) with an actual installation. The Zinc Institute and supporting companies have initiated an effective educational program targeted for manufactured housing producers. Preliminary indications are optimistic and a potential 23,000 metric tons of increased zinc consumption during the period 1983 to 1987 is considered reasonable.

In industrial applications, galvanizing has been proven successful in pulp and paper mills, refineries, power plants and other large industrial complexes for protecting equipment and structures. Of the 14 million metric tons of steel used annually in the construction industry, only about 10% is hot dip galvanized at present. When protective coatings for structural steel are considered, the primary selection has traditionally been paint. Hot dip galvanized has received some consideration but, basically, it is labeled a specialty coating for use in very corrosive environments. The case histories on galvanized applications are fairly abundant, all showing that any initial premium for galvanizing is more than offset through maintenance savings over the life of the structure. The problem is that designers and specifiers of construction materials are not sufficiently aware of galvanizing's technical and economic performance. St. Joe and the American Hot Dip Galvanizers Association have initiated aggressive marketing programs utilizing both mass media and direct sales techniques to place this information in the hands of the specifiers. Along with complementary and reinforcing programs sponsored by galvanizers and other zinc producers, the galvanizing industry appears ready to play a dominant role in the coatings industry of the 80's. It is estimated that these educational programs will increase zinc consumption by 23,000 metric tons per year over the 5-year period 1983 to 1987.

The results of these technical and marketing programs will not impact the after fabrication hot dip galvanizing industry for approximately 3-5 years. Thus, zinc consumption is projected to be stable through the early 1980's and increasing in the middle and latter parts of the decade.

Paints

As previously mentioned, zinc oxide and zinc dust are the primary ingredients in zinc paint used to protect steel from oxidation. In 1978, this market segment accounted for approximately 11% of the total zinc consumed in zinc coatings. A breakdown by market classification is shown in Table VI.

Table VI. Zinc Pigment Consumption in Corrosion Resistant Paints by
Market Classification -- U.S. (31)

(000 Metric Tons ZnE)

Market Class	1978	
	Tons	(%)
Zinc-Rich Coil Coatings	12.3	23.3
Automotive Spray Coatings	13.4	25.3
Industrial Zinc Coatings	27.3	51.4
Total	53.0	100.0

Industrial Coatings. Inorganic zinc-rich coatings are unsurpassed by any other paint system in the corrosion protection of steel. They result in a harder film, which has excellent abrasion resistance, outstanding solvent resistance, and good temperature stability over conventional primers. This system requires careful preparation of the steel surface to ensure a good bond.

Organic zinc rich coatings require less critical surface preparation, permit greater variation in application techniques, are less sensitive to varying climatic conditions and are more flexible and resistant to chemical environments. The most often mentioned drawbacks with organic zinc systems are flammability, blistering, and low heat resistance (32). Furthermore, with lead being virtually eliminated as a pigment for paint, designers will be forced to specify zinc rich systems rather than the older, more familiar lead primers. This general trend has already begun in Florida and California, where recently changed bridge and highway specifiers opted for zinc rich coatings. It appears that this general trend will extend into the federal and state bridge renovation programs as well. A 7% annual growth rate in this application is anticipated.

Coil Coatings. Zincrometal accounts for the majority of the products in this category. All available coil coating capacity is currently being used. Additional lines are being built to allow for growth in the future. The present capacity of 1.1 million metric tons per year will be increased to 1.8 million metric tons per year in 1980. A portion of the Zincrometal production may be converted to a two-side coating in response to the automobile industry demands. The additional demand combined with a switch to two-side coatings could result in 8,000 metric tons per year of increased zinc consumption.

Spray Coatings. Spray zinc paints for automobiles have also benefited from Detroit's increasing use of zinc coated steel. When the application is performed properly, an excellent corrosion resistant coating results.

The major benefits of these spray zinc coatings are the economic advantages compared to Zincrometal and galvanized steel. However, this lower initial cost is offset by operator variability, which is inherent in the application. The quality of the coating is ultimately related to the quality of the operator. Also, spray coatings don't penetrate to some areas of formed steel parts, such as door seams and trunk lids.

The spray zinc coatings are being attacked from an environmental standpoint. The solvents used in some zinc coatings are considered harmful, which results in additional labor problems for assembly plants.

The major automotive companies agree that the use of spray zinc coatings will decline in the future. The problems associated with incomplete coverage of tightly formed parts, labor variability, and solvent emissions will combine to reduce the use of these coatings.

In total, the future for zinc paints is positive with an annual growth rate of 5% per year expected through the 1980's.

In conclusion, the use of zinc coatings for corrosion protection of steel is rapidly changing zinc from a commodity to a product. The future of zinc in coatings will depend on the zinc coating industry's ability to respond to specific market needs with new products. Zinc's successful record combined with recent technical and marketing advances indicates that zinc coatings will enjoy increased utilization in the 1980's.

References

1. Non-Ferrous Metal Data-1978, American Bureau of Metal Statistics, Inc.; New York, N.Y.

2. Annual Review-1978 U.S. Zinc Industry, Zinc Institute Inc. New York, N.Y.

3. Harold E. McGannon Editor, The Making, Shaping and Treating of Steel, 8th ed., p.979; United States Steel Corporation, Pittsburgh, PA, 1964.

4. J. J. Butler, D. J. Beam, and J. C. Hawkins, "The Development of Air Coating Control for Continuous Strip Galvanizing: I," Industrial Heating (1970) pp. 1985-1986.

5. Harold E. McGannon, The Making, Shaping and Treating of Steel, 9th ed., p.992; United States Steel Corporation, Pittsburgh, PA, 1970.

6. R. M. Burns, W. W. Bradley, Protective Coatings for Metals, 2nd ed., p. 100: Reinhold Publishing Company, New York, N.Y. 1955.

7. Harold E. McGannon, The Making, Shaping and Treating of Steel, 9th ed., p.1033; United States Steel Corporation, Pittsburgh, PA 1970.

8. The Galvanizing Manual, St. Joe Minerals Corp. Monaca, PA, 1974.

9. Ibid., p. 20.

10. Masciale, M. J., "Use Zinc Rich Coatings - But Use Them Wisely", paper presented at the International Corrosion Forum, Toronto, Ontario, April, 1975.

11. D. C. Nevison, "Zinc Paints for Maintenance," American Paint and Coating Journal, (December, 1977) p. 42.

12. A. W. Kennedy, A. Osawa, and T. Sato, "A Systems Approach," Modern Paint and Coatings, (August, 1978).

13. "Shipments of Steel Products by Market Classification," American Iron and Steel Institute, AISI 6C, 1970, 1978.

14. Weyand, T.E.; "A Standing Wave Method for One Side Galvanizing." A paper presented at Z.I. Galvanizing Committee Meeting, Birmingham, AL, April, 1978.

15. A. F. Prust, "Manufacture of Hot-Dipped One-Side Galvanize," paper presented at SAL Automotive Engineering Congress, Detroit, MI, January, 1971.

16. Unikote, A National Steel Corporation product brochure, 10-77-10M.

17. L. W. Austin, and G. W. Bush, "Unikote," paper presented at SAE International Automotive Engineering Congress, Detroit, MI, February, 1977.

18. A. F. Gibson, H. F. Graff, and M. B. Pierson, "Zincgrip OS Armco's One-Side Galvanized Steel," paper preesented at ZI Galvanizers Committee Meeting, Birmingham, AL, April, 1978.

19. D. T. Carter, "A Practical Process for Production of Galvanized Sheets with Coating on One Side," Iron and Steel Engineer, (October, 1971), p.54.

20. Notes from a speech given by R. F. Higgs, "Production and Use of One Side Electrogalvanized Steel for Improved Corrosion Resistance of Automobile Bodies," presented at the spring meeting of the Pittsburgh Electrochemical Society, May, 1979.

21. J. A. Kargol, "Paint-Tite B An Alternative to One-Side Galvanized Steel," a paper presented at the ZI Galvanizers Committee Meeting, Birmingham, AL, April, 1978.

22. ZINCROMETAL FACT SHEET, a product brochure by Diamond Shamrock, MC-Z-IC, June 30, 1977.

23. Lake, P.B. "Galvanized Consumption in the Automotive Industry," ZI/LIA, 1978 Annual Meeting, April 11, 1979, Drake Hotel, Chicago, IL.

24. GALVALUME, A product brochure by Bethlehem Steel Corporation, folder 3236, July, 1977.

25. G. J. Harvey, "Fifty Five Percent Al-Zn Alloy A New Hot-Dipped Coating for the Protection of Steel Sheet," Metals Australia, (September, 1976).

26. Lake, P.B. "Galvalume" ZI/LIA 1978 Annual Meeting, April 11, 1979, Drake Hotel, Chicago, IL.

27. Annual Review - 1970 U.S. Zinc Industry, Zinc Institute, Inc. New York, NY and conversation with Mr. L. E. Gage.

28. American Hot Dip Galvanizers Association, "Distribution of Galvanized Products Survey's 1978," Washington, D.C.

29. Proceedings of a seminar on, "Galvanizing of Silicon-Containing Steels," sponsored by International Lead Zinc Research Organization, Inc. (ILZRO) and Centre DeRecherches Metallurgiques (CRM), Liege, Belgium, May, 1975.

30. Polygalva-A Process for Galvanizing Rimmed and Silicon Contained Steel", AM&S Europe Ltd., Bristol England, November 1978.

31. Internal communication, "Market Analysis of Domestic Zinc Pigment Usage," St. Joe Zinc Company, Pittsburgh, PA 1978

32. op. cit. Nevison, D.C.

PRODUCT DEVELOPMENT - AN EFFECTIVE TOOL IN DEVELOPING

NEW MARKETS FOR ZINC DIE CASTINGS

Gene O. Cowie, P. E.

Pioneer Engineering and Manufacturing Company, Inc.
Warren, Michigan

The recent proliferation of plastics and plastic fabricating technology has caused plastics to replace relatively high density materials, such as zinc, in a large number of applications. This paper looks into the changes which have recently taken place in the automotive industry where losses to zinc have been particularly heavy. The forces that caused these changes are analyzed, and product development is identified as a means of halting and reversing the trend away from zinc. The product development process is described, and the industry wide product development efforts in zinc die casting are discussed using selected case histories. The state-of-the-art of product development is discussed, emphasizing the contribution of the zinc industry. The paper concludes with a scenario of the product development process in the near future and the opportunities for zinc die castings.

Introduction

The automotive industry has been a major market for zinc die castings for several decades. It has been an excellent market for zinc, providing an endless stream of product opportunities outside the vehicle, inside, and under the hood. While constant change has been the rule, the changes were predictable. Completely new body styles came typically at two or three year intervals with annual "facelifts" between; as a result, each year a large number of decorative parts were discontinued, and an equivalent number of new parts were introduced. These annual changes impacted upon the individual firms that supplied the decorative parts, since they had to compete for new business each year. But the changes were not felt by the zinc producers, because material selection was governed by economics and durability, and zinc held a clear advantage over all other materials. Therefore it was a foregone conclusion that zinc would be specified and used. The advantage enjoyed by zinc was so great, in fact, that there was no incentive to reduce metal content thru improved manufacturing technology. That technology would actually have been counter productive to zinc producers and alloyers because it would have reduced their markets. But six years ago the automotive industry experienced a sudden, massive rearrangement of its product objectives. Weight suddenly became a major, sometimes a dominating factor, while cost and durability became less important. The changes were so radical that they seemed, to the zinc industry, to defy common sense. Zinc components were being discarded while the automotive industry was paying higher prices for inferior products.

The Automotive Material Revolution

It is important to understand the materials revolution and its impact on the automotive industry because it is affecting a major market for zinc. Furthermore, the same thing can happen in any other industry, just as suddenly, and with the same consequences. Under mandate from the federal government, the automotive industry was charged with the responsibility of approximately doubling the efficiency of the automobile, in terms of miles per gallon, while sharply reducing certain emission components, in ten years. Since major product changes in that industry require a minimum of three years lead time, the industry was left with seven years or less to redevelop a mass produced product that had been developing for over seventy. To complicate matters, the mandated efficiency and emissions goals are largely antagonistic; most of the measures that improve one tend to aggrevate the other. And other standards, such as crashworthiness and reparability are adding to the conflict.

But there is one rather obvious way to help emissions and efficiency together; make cars smaller and lighter. As a result, the automotive industry plunged into a massive program of downsizing with two major thrusts; redesign and material substitution. The industry was well equipped for the redesign task, since designing and building new automobiles is the very thing they are geared to do. However, they did not have an adequate task force of materials engineers to handle the massive materials substitution job that was required. Therefore engineers from other areas, with marginal or inadequate materials experience, were suddenly pressed into service to do the best they could while learning on the job. They learned their lessons well and in relatively short order, but the learning period was extremely turbulent for the automotive industry and its suppliers. The first instincts of the newly commissioned materials task force was to replace high density materials with low density materials. Iron and steel gave way to aluminum and plastics, and zinc suffered wholesale replacement by plastics. It was a compound loss for the so-called high density metals: less material per vehicle, and a lower percentage of what there was.

The metals industries had difficulty in realizing the extent of the threat, and that it was no longer business as usual. It seemed impossible that their alloys, which had dominated certain applications, were suddenly being abandoned in favor of more costly and less reliable materials. It seemed improbable that the trends could go very far. But by the time economics finally began to reassert itself, major markets had been lost. A few applications, lost in an overswing to low density materials, were regained. However, most major applications have not been regained and will not be in the forseeable future.

The affected metals industries reacted by reevaluating the potential of their available alloys and processes, developing new alloys with improved properties, and developing new processes. The steel industry, for example, developed high strength low alloy (HSLA) steels so that the auto designers could reduce vehicle weight while continuing to use steel. And they helped to develop the companion fabricating technology. Then they carried their story to the automotive industry. The automotive industry's response to this thrust was almost as shocking to the steel industry as the materials revolution had been. It was graphically summarized by a representative of one of the major steel producers, reflecting on the response by the automotive industry during the year that followed his initial presentation on HSLA steel. He said to an assembly of body engineers, "Gentlemen, we've been underwhelmed". Nobody had beaten a path to their door. The reason was that they had failed to grasp the plight of the automotive designer who was, and still is, caught up in a crash program. He rarely has time for the classic product development process--calling in the materials consultants, weighing the merits of several material and design options, and systematically choosing the optimum. He is usually scurrying about trying to get the job done yesterday in a schedule that is telescoped so tightly that often production tooling is being fabricated by the time he sees his first prototype parts. He needs something that will work on the first go around.

The plastics industry had learned what to do about this dilemma and were doing it with great success. They weren't promoting materials; they were producing parts. They weren't talking about what their materials could do, they were demonstrating with ready to test prototypes delivered to the engineer's desk top. The plastics industry had literally relieved the automotive engineer of the bulk of his product development task. The relief was welcomed, and the automotive engineers bought plastics.

The Response of the Metals Industries

Various metals industries sensed the situation and began to get into the product development business. A major nickel producer developed a bumper system, featuring chrome plated steel, to protect their market in chrome plating. A major steel producer developed a number of weight reduction items, including their own version of a bumper system, and a brake master cylinder, which was intended to recoup an application that had been converted from iron to aluminum and plastic.

The zinc industry, which had suffered extensive losses of its automotive market to plastics, had a number of materials and processes at its disposal that were suitable for use by the automotive industry. The best known was thinwall die casting, which had been developed by Battelle Colombus Laboratories under contract to the International Lead Zinc Research Organization, ILZRO. The technology was made available to the die casting industry in 1974, enabling zinc parts to be made lighter in weight, often at less cost, with no sacrifice in essential component characteristics. The technology was quickly adopted by some die casters to produce decorative components. Substantial weight reductions were achieved, and the automotive industry accepted and appreciated the effort. But thinwall technology was not being used to any extent for functional, load carrying components. The zinc industry had also developed a number of other processes and alloys with excellent potential for the automotive market. For example ILZRO 16, a high temperature, creep resistant, cold chamber die casting alloy had demonstrated its properties in marine outboard engine cylinder blocks. Foundry alloys 12 and 27, which possess some of the properties of die cast Zamak 3, were available for fabricating prototypes of die castings at low cost and in relatively short time. And zinc anodizing was an established process for improving the wear and corrosion resistance of zinc components. But none of these innovations were being used by the automotive industry; they simply did not have the time to adapt them.

The zinc industry recognized the needs of the automotive industry and began to get involved in product development development programs for the purpose of designing, fabricating and delivering functional prototypes into the hands of the automotive engineers. Some of this activity is being carried on by zinc producers, working in conjunction with aggressive die casters. A large part of the work is being done on an industry wide basis via an ILZRO sponsored program.

The Product Development Process

The complete product development process is schematically portrayed in Figure 1, in an array of three columns and seven rows. The first of the three vertical columns defines the activities that are performed, the second indicates the people who normally perform those activities, and the third indicates the resources--both intellectual and physical--that are utilized. In brief, the columns indicate what is being done, who is doing it, and the resources involved. The seven horizontal rows indicate the discreet levels of activity or tasks that are performed. These consist of first, identifying opportunities with potential; second, appraising the market and profit potential; third, synthesizing several possible solutions; fourth, determining the optimum solution and reducing it to design drawings; fifth, fabricating a working prototype; sixth, pretesting the prototype to assure that it has a reasonable chance of success; and seventh, delivering the prototype to the customer who conducts his own tests, and ultimately accepts the product. The feedback loops arise from those junctures at which a need for revision is identified; in those cases activity is redirected to the appropriate level, and a part of the cycle is repeated. The user, such as the automotive engineer, is quite prominent in this process. The product objectives come from him, and he is periodically given the opportunity to review the development work, often initiating feedback loops. Although these reviews and the attendant feedback loops increase the development time, they will in the long run save time and money by assuring that the development efforts have been directed toward the proper objectives.

Although all applications in some sense go through all seven development levels, the relative importance of any level, and the point at which the zinc industry becomes involved vary considerably from one application to another. Decorative applications, for instance are frequently picked up by die casters as late as the fifth, or prototyping level since the automotive companies have traditionally performed the first four.

THE PRODUCT DEVELOPMENT PROCESS

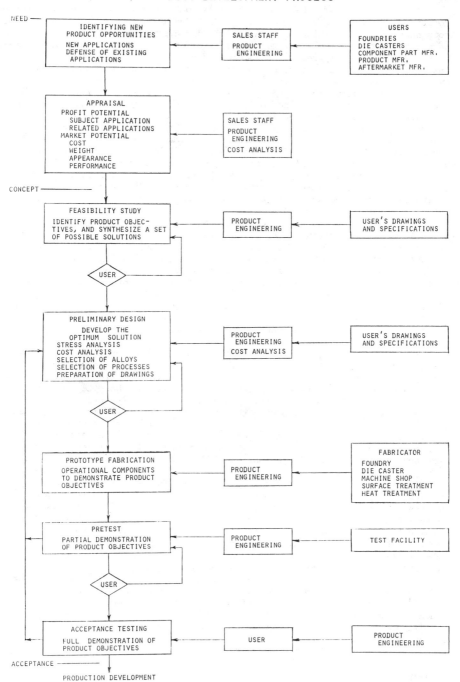

Decorative Vs. Functional Components

It would be helpful here to distinguish between two classes of components: decorative and functional. From the product development viewpoint, decorative components are those which provide a desired configuration or appearance, but have little if any structural requirements beyond their own system of attachment. They invariably receive some decorative surface treatment such as paint or plating. Decorative components are usually conceived and designed by the automotive companies ready for production tooling. Decorative die castings are usually prototyped in a single cavity production die, and acceptance testing is concerned only with the appearance and dimensional accuracy of the casting. Bezels, medallions and moldings are examples of decorative parts. Functional parts are those which must carry prescribed loads or withstand a specified number of performance cycles. They may or may not receive a decorative surface treatment. Door handles and window cranks, which must withstand the normal loads and predictable abuse imposed in operating the vehicle's doors and windows, are classified as functional by the product designer. The distinction drawn here is different from that typically drawn by die casters who equate decorative with bright finish.

As mentioned prior, the automotive companies normally bring decorative zinc die castings through the preliminary design stage ready for prototyping. The challenge to the die caster is to produce the part as economically and reliably as possible. This often means casting thinner walls to improve his competitive cost-weight position without sacrificing surface quality or increasing his rejection rate. The process primarily requires the application of good die casting technology.

Selected Case Histories

The rear fender extension, shown in Figure 2, is an example of a thinwall decorative casting. This unit was jointly developed by a zinc producer and a die caster with thinwall capability, to challenge the plastic unit, which in turn had displaced a relatively heavy zinc die casting. Despite the fact that this is a relatively large casting with a surface area of approximately 774 sq. cm. (120 sq. in.) average wall thickness was held to .76 mm (.030"); yet the exterior surface is of high enough quality to accept paint or bright chrome plating. The configuration of the plastic unit and the attachment points were duplicated. The light weight and outstanding surface quality were achieved by judicious application of thinwall die casting technology.

Functional components present an entirely different development challenge. The zinc industry must usually enter the product development process at the first stage by meeting with the product designer to learn of his needs and review his designs in process, then visualizing a better way of meeting the product objectives with zinc die castings. This usually means ignoring the way the component is being proposed or made, and concentrating on what it is supposed to do. You must then find the best way to do the job with the materials and processes at your disposal. This requires activities such as analysis of the applied loads, determination of the working stresses, and the utilization of new alloys or processes, culminated by the design of a totally new component that will perform at the optimum combination of weight and cost. But design is not enough; since most customers have a healthy skepticism of new design concepts, the burden of proof rightfully rests upon the one who proposes innovation. Therefore it is usually necessary to fabricate prototypes to demonstrate that the part can perform according to the user's specifications. Furthermore, since the user's test facilities are usually occupied with his own designs, it often becomes necessary to conduct

862

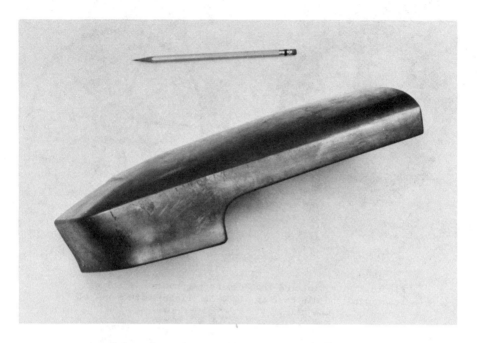

Figure 2 The rear fender extension is an example
of a relatively large zinc die casting with thin
walls and excellent surface quality. Courtesy of
Gulf + Western.

extensive pre tests on the component so that the customer has evidence that
it is worth his time and effort to test it himself. To repeat a prior
statement, you must literally do the bulk of the customer's product
development task for him.

An example of a functional component is the camshaft sprocket, shown
on the left in Figure 3, which is currently made of sintered iron and weighs
.66 Kg (1.45 pounds). The sprocket receives a rubber belt and is part of a
system which transmits approximately 3 Kilowatts (4 horsepower) from the
crankshaft to the overhead camshaft of a four cylinder internal combustion
engine, while maintaining a precise angular relationship between the two.
Since a variety of metals and some plastics are used in similar sprockets,
this application appeared to have potential in zinc. A comparative cost
analysis revealed that no advantage could be gained in zinc by merely repli-
cating the existing configuration and attempting to reduce wall thicknesses.
A significant reduction in material content was necessary, and that meant a
total redesign of the sprocket. Further study indicated that the sprocket
configuration had been determined by fabricating constraints, and that it was
not optimum for the function it was to perform. But the versatility available in zinc die casting made optimum design in zinc possible; therefore the
sprocket was redesigned. The heavy rim of the iron unit, which incorporates
the teeth, was redesigned into a thinwall shell and the 6.9 mm (.270 in.)
thick flat disc was changed to a 1.5 mm (.060 inch) thick disc which
developed the necessary strength and rigidity via five convolutions.

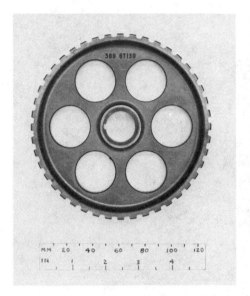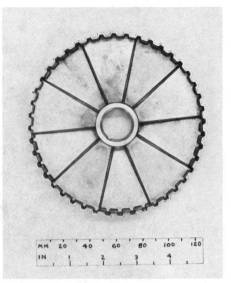

Figure 3   The sintered iron camshaft sprocket (left) was
redesigned and die cast in ILZRO 16 (right) with a weight
reduction of one half.

Calculations based on the design study indicated that the zinc die casting,
shown on the right in Figure 3, would have approximately one half the volume
of material of the sintered iron unit, would weigh one half as much, and be
economically competitive. The sprocket was therefore fabricated in ILZRO 16
which was specified due to the anticipated heat and stress.

   The automotive manufacturer who uses the sprocket was presented with the
die cast prototypes. And although he was impressed with the weight reduction,
he declined testing since his available facilities were committed to more
urgent work, relating to compliance with government mandated standards.
Furthermore he had not specified a test procedure, because existing sprockets
were apparently exhibiting unlimited durability. The belt manufacturer, who
was deeply concerned with durability, was consulted, and his test cycle of
1000 hours at 5000 engine RPM (equivalent to approximately 135,000 Km
(85,000 miles)) at a temperature of 82°C (180°F) was adopted with the consent
of the user. In order to duplicate the loads and belt configuration
encountered on the vehicle, a scrapped engine was procured so that the actual
camshaft, sprockets and belt could be used. The irrelevant engine parts were
removed and relevant parts were reconditioned. The engine test fixture was
driven thru pulleys, attached to the rear of the crankshaft, by a 3.75 Kw
(5 hp) electric motor. The required sprocket temperature was maintained with
a 375 watt infrared lamp.

   The test cycle revealed that the sprocket strength was adequate, but
abrasion resistance was inadequate. The rubber belt bars were causing heavy
wear on the flanks of the sprocket teeth due to the sliding action that
occurred at disengagement. Therefore a number of surface treatment processes
were evaluated, and four were found that would provide adequate wear
resistance. As a result of this work, a light weight zinc sprocket with
adequate performance is now available to challenge cast iron.

It is important to recognize that throughout the development work no attempt was made to evaluate or duplicate the strength or durability of the iron sprocket, since the iron sprocket is overdesigned in this application. Numerous worn out engines with good-as-new iron sprockets testify to this fact. Instead, the manufacturer's performance requirements were identified and met. This point is important. Many parts now in service are over-designed and may often be replaced with parts which, though inferior by comparison, are still more than adequate for the application.

Another example of a functional component is the trunk latch shown in Figure 4 which must not only perform over the expected life of the vehicle,

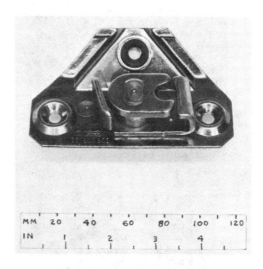

Figure 4 The trunk latch is an all steel component
with stringent wear and strength requirements.

but also secure the trunk lid in the body opening in the event of a rear end collision. To meet these objectives, the vehicle manufacturer specifies that the latch must withstand 25,000 cycles of latching and unlatching, and one time loading in various modes up to 2,500 pounds. The base, which is the major structural member, was being made of three steel stampings and two steel screw machine parts as shown in Figure 5. A cost study indicated that a substantial cost reduction could be achieved by combining these five parts into a single piece zinc casting. The base was therefore redesigned and prototyped as shown in Figure 6. The zinc base plate weighs approximately the same as the steel, and the anticipated cost reduction gives it a clear advantage over steel.

Once again no attempt was made to replicate the shape or strength of the steel assembly. The zinc unit was designed to interface with existing vehicle components and perform the required functions in a way that is best suited to zinc die casting. Because of the high cost and lengthy time schedule required to fabricate a die, the base plate was prototyped by gravity casting in Alloy 27, then heat treated to attain an ultimate tensile strength and ductility equivalent to die cast Zamak 3. Prototypes suitable for structural testing and functional operation were thus produced in a few weeks at moderate cost.

Figure 5 The base plate for the steel trunk latch is a costly assembly of five pieces.

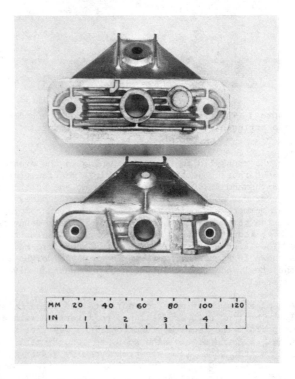

Figure 6 The zinc casting integrates the essential features of the steel base plate into a single casting. The prototype was gravity cast in alloy 27.

Identifying products made from other materials and converting into zinc die castings is one important thrust of the product development program. Another, and equally important, is to identify and defend those zinc die castings which may be lost to other materials. The window regulator handle shown on the right in Figure 7 is an example. The unit was already a zinc die casting which was performing satisfactorily. However, the mood of the automotive industry clearly indicated that this item was likely to give way to plastic because of its weight. On an absolute scale, a component weighing .123 Kg (.27 lb.), and accounting for .5 Kg (1.1 lb.) vehicle weight does not seem like much to worry about. However, on a comparative scale, plastic cranks had been prototyped at approximately one fourth that weight, and a 75 per cent weight reduction was attractive to weight conscious automotive designers. Because of the threat from plastic, a die cast zinc crank was developed that weighed half as much as the existing crank, and cost less to produce. The light weight prototype protected zinc by presenting a weight-cost profile with a more competitive position against plastic. It may seem counter productive to a zinc producer or alloyer to develop a component which reduces zinc consumption by fifty percent, but experience in the automotive market has taught that if you stand pat, you run a high risk of a one hundred percent reduction.

This very significant weight reduction required much more than a face-lift of the existing handle which had been in use for several years and was grossly overdesigned in some respects. In this case the customer had well defined criteria. The new unit was to retain the appearance of the existing unit as much as possible, interface with existing parts, transmit prescribed loads, and function under specific conditions for a given number of cycles. Simultaneously reducing weight and meeting these criteria required an entirely new design perspective. Therefore the handle was perceived as a structural component with decorative features to be made by die casting, rather than as merely a decorative die casting. The knob was structurally connected to the hub, consistent with the loads imposed in service and on test. A totally new section configuration was developed to provide maximum strength with minimum metal. The hub wall thickness was substantially reduced in anticipation of improved metal quality derived from improved die gating. Purely decorative features were identified and their wall thick-nesses reduced to a minimum, based on metal flow and surface finish criteria. The prototype crank is shown on the left in Figure 7.

While prototypes were being made, a finite element analysis was performed to determine whether this powerful computer aided design tool could be correlated with the actual die cast part. Finite element analysis predicts performance under load by computing displacements and stresses in a mathe-matical model of the component. Construction of the model is currently performed by an experienced analyst, and it is the only labor intensive aspect. The process also requires a mathematical model of the stress-strain relationship and Poisson's ratio for zinc under both elastic and plastic deformation. While some degree of correlation was attained in this exercise, the correlation was not totally satisfactory due to gas porosity in the casting, which cannot yet be modeled, and a lack of stress-strain data on thinwall die cast Zamak 3 sections in the as-cast condition.

Trends In Product Development

The state-of-the art of product development is moving toward increased computerization, which is being developed at a rapid pace by the electronics industry. Computer aided design, CAD, is now being used by big business, and the price is being brought down within the range of smaller firms. With this tool, a designer sits down at a computer console, instead of a drawing board,

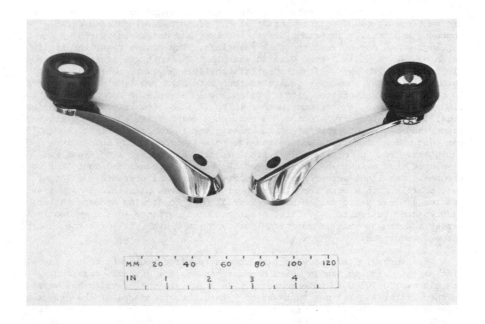

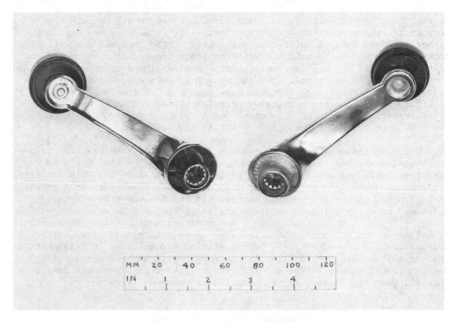

Figure 7 The prototype zinc regulator handle shown on the left offers the same appearance as its production counterpart on the right, but the weight is one half. Total redesign was required to achieve the weight reduction while meeting the manufacturer's structural criteria.

and punches in data instead of drawing lines. A total design including complete views, intricate projections, and difficult sections is developed by the computer in a small fraction of the time required on a drawing board; revisions are accommodated with equal facility. And the computer obligingly prints out a hard drawing of the part when the design is finalized.

Finite element analysis, mentioned earlier, gives the product engineer the ability to explore the effect of in-service loading in perhaps one-tenth the time and expense required to cut a die, cast parts, and run tests. This technology has been in use longer than CAD. And the software to meld CAD with finite element analysis is now under development. When this software is available, the computer will take the data from the CAD program and construct a model of the component, eliminating the major labor intensive phase of finite element analysis.

Computer aided manufacturing, CAM is now being used to control metal cutting machine tools in the manufacture of a wide variety of components. This technology has already been linked with computer aided design into a complete CAD-CAM system which allows for completely computerized design and manufacturing control of a component in substantially less time than by conventional means, without drawing a line on paper. Engineering drawings are produced by the CAD-CAM system, almost as afterthoughts, for visual record. And the CAD-CAM system can readily be modified to produce a die cavity, with runners, gates and overflows.

The zinc industry is contributing to this technology in ways that will directly benefit product development in zinc. The thinwall system has been computerized, using low cost, readily available hardware, to eliminate the drudgery and sharply reduce the time required to design the gates, runners and overflows. The location of water lines has also been expedited with a program that can be run on low cost, hand held calculator. Gas porosity, which has always been a major concern in die casting, is reduced by the low turbulence metal flow in dies made according to the thinwall system, and it can be totally eliminated by using the pore free die casting system. Control or outright elimination of gas porosity will assure that the die casting under test and in service will perform as predicted by finite element analysis. And the DIECAP program, which is also computerized, allows the die caster to predict his costs in advance with a high degree of accuracy.

The Future For Product Design - A Scenario

Within only a few years, these technologies will be developed and inter-connected to the point where a zinc die casting can be designed, analyzed for cost and performance under load, a die fabricated, and prototypes cast in the space of a few weeks. This fast turnaround from concept to prototype will be appreciated and accepted by the product engineer who is harassed by tight schedules. And the parts made by this system will stand an excellent chance of meeting his performance tests at the optimum of weight and cost on the first go around. The automotive industry, and a lot of other industries, will have some excellent reasons to buy zinc.

But there is one missing link in this scenario of a fast moving, highly efficient computerized development program for zinc die castings. Finite element analysis requires a complete data matrix defining the mechanical properties of zinc die castings under a wide variety of conditions, because the output of any process is only as good as its inputs. It is therefore no longer adequate to tell the product designer that die cast zinc has an ultimate strength up to 282 K Pa (41,000 psi), in some cases. He needs to know the exact strength for his particular case, which means data such as the stress-strain relationship to failure in tension, compression and shear; the fatigue strength and endurance limit; and the creep properties. When the product designer is told that an alloy can perform satisfactorily at a temperature of 82°C (180°F), he wants the same data all over again at that temperature. When the zinc industry publicizes the fact that thin wall sections of .76 mm to 1.14 mm (.030" to .045") have properties superior to sections of 3.05 mm (.120"), the product designer, to take advantage of those properties, needs data relevant to those thicknesses; data derived from 6.35 mm (.25 in.) diameter bars is no longer adequate. The new found capability for casting more intricate configurations and thinner walls in zinc has attracted real interest; at the same time it has attracted a lot of questions which have identified a need for a lot more basic information about the metal. Without this vital link, the scenario will not materialize, and a lot of potential markets will be lost.

Conclusion

The zinc industry, throughout its various periods of growth, has established an outstanding track record for rising to meet the needs of its various markets. The industry has also displayed an outstanding capability for responding to crises which have threatened its markets. It now faces a new challenge, to develop a data matrix that will meet the needs of the forthcoming era of computerized design. Without this data, zinc die castings cannot remain competitive with other materials in the arena of functional components. With this data, the product opportunities and the market areas for zinc die castings will be unlimited.

NICKEL/ZINC BATTERY: A PROMISING CANDIDATE

FOR ELECTRIC VEHICLE PROPULSION

N. P. Yao and J. F. T. Miller
Argonne National Laboratory
9700 South Cass Avenue
Argonne, Illinois 60439

A rechargeable nickel/zinc battery is one of the three near-term candidates, along with improved lead-acid and nickel/iron batteries, which are
under development for electric and hybrid vehicles application. Commercial
viability of these batteries will be greatly enhanced if concomitant improvements in performance, cycle life, and cost reduction are achieved, and their
manufacturing technology demonstrated. The nickel/zinc battery has the highest specific energy and specific power of the three candidates, but its
demonstrated cycle life falls short of the acceptable level for the application. This paper addresses the developmental status of nickel/zinc battery
technology and the expected progress. Battery materials requirements, especially nickel and zinc, will be projected in the light of unit manufacturable
battery and a set of electric vehicle market scenarios.

Introduction

A rechargeable nickel/zinc battery is one of three near-term candidates, along with improved lead-acid and nickel/iron batteries, that are under development for electric and hybrid vehicle applications. Nickel/zinc batteries have the potential of doubling the energy density and power density of present day lead-acid batteries. Yet the development time for nickel/zinc batteries will be comparatively shorter than that required for the more exotic advanced batteries such as zinc/chlorine, sodium/sulfur, and lithium/iron sulfides systems.

As a result of the Electric and Hybrid Vehicle (EHV) Research, Development, and Demonstration Act passed by Congress in 1976, the U.S. Department of Energy (DOE) has initiated an EHV Program designed specifically to advance EHV technology and to demonstrate and promote EHV market potential. As part of the program, DOE has awarded cost-sharing contracts totaling over $25 million to eight battery manufacturing firms for improving the performance and life, while lowering the cost of Ni/Zn, Ni/Fe and lead-acid batteries. Argonne National Laboratory - as a field project office for DOE, manages, directs, and coordinates the industrial battery contracts. As a part of any full-scale commercialization scheme, material resource requirements and their future availability must be recognized as an important concern. Therefore, the objectives of this paper are twofold. First, it will provide an introduction to the nickel/zinc battery system, giving an overview of the present technical status and describing the technical problems which need be overcome. Secondly, the materials requirements for Ni/Zn batteries will be assessed based upon various projections of the electric vehicle market up to the year 2000. A comparison of the critical battery material requirements with the projected demand and availability of these materials will be presented.

Nickel/Zinc Battery System

The historical development of the Ni/Zn battery system has arisen from and is based upon the technologies of the Ag/Zn and Ni/Cd batteries. Ag/Zn cells, characterized by high energy and power densities, were developed for specialized military applications and have found considerable use when the requirements for reduced weight and reliability justify their high cost. However, the high cost of Ag/Zn cells precludes their use in electric vehicle applications. Ni/Cd cells, presently in use for aerospace applications as well as for limited consumers' products, are equally high cost and furthermore, the use of Cd in the large quantities required for EV commercialization would be intolerable. Consequently, the positive nickel electrode of Ni/Cd has been mated with the zinc negative electrode of Ag/Zn to form a near-term candidate which is potentially cost-effective with acceptable performance for EV propulsion. As such, the nickel/zinc system is a relatively new and emerging technology.

The basic electrochemistry of the Ni/Zn battery system can be described by the following electrode reactions:

$$2 \text{ NiOOH} + 2H_2O + 2e^- \underset{C}{\overset{D}{\rightleftarrows}} 2 \text{ Ni(OH)}_2 + 2OH^- \qquad E_o = 0.48 \text{ volts} \quad (1)$$

$$\text{Zn} + 2OH^- \underset{C}{\overset{D}{\rightleftarrows}} \text{ZnO} + H_2O + 2e^- \qquad E_o = 1.25 \text{ volts} \quad (2)$$

Adding these two electrode reactions yields the overall cell reaction:

$$2 \text{ NiOOH} + \text{Zn} + \text{H}_2\text{O} \underset{C}{\overset{D}{\rightleftharpoons}} 2 \text{ Ni(OH)}_2 + \text{ZnO} \qquad E_0 = 1.73 \text{ volts} \qquad (3)$$

Based solely on the electrical capacity and weight of the reactants only, a theoretical energy density of 345 Wh/kg is predicted. However, the practical energy density obtainable in a completed battery for any system is typically only 1/6 to 1/4 of the theoretical energy density, due to the fact that utilization of reactants may be somewhat less than complete and owing to the need for current collectors, separators, terminals, and cell cases. Using the aforementioned rule-of-thumb, reasonable energy densities expected for Ni/Zn batteries lie in the range of 60-90 Wh/kg.

Parasitic reactions other than those given above occur in actual Ni/Zn batteries. These include the gassing reactions which occur near the end of charge. At the nickel electrode, the relevant parasitic reaction is:

$$2 \text{ OH}^- \overset{C}{\rightleftharpoons} 1/2 \text{ O}_2 + \text{H}_2\text{O} + 2\text{e}^- \qquad (4)$$

while at the zinc electrode, the parasitic reaction is:

$$2 \text{ H}_2\text{O} + 2\text{e}^- \overset{C}{\rightleftharpoons} \text{H}_2 + 2 \text{ OH}^- \qquad (5)$$

Other important reactions are those by which ZnO is dissolved in the hydroxide electrolyte to form zincate ions by the possible two-step process:

$$\text{ZnO} + \text{H}_2\text{O} \rightleftharpoons \text{Zn(OH)}_2 \qquad (6)$$

$$\text{Zn(OH)}_2 + 2 \text{ OH}^- \rightleftharpoons \text{Zn(OH)}_4^{-2} \qquad (7)$$

The feature of ZnO solubility in KOH solutions is the source of many of the inherent problems associated with the Ni/Zn system. Non-uniform dissolution and deposition of zinc during cycling leads to a redistribution of zinc (shape change) and to the formation of zinc dendrites which can cause cell failure by internal shorting. Additional reactions are those responsible for corrosion of the electrodes, particularly the zinc electrode which can be described by:

$$\text{Zn} + \text{H}_2\text{O} \rightarrow \text{ZnO} + \text{H}_2 \qquad (8)$$

or

$$\text{Zn} + 2\text{H}_2\text{O} + 2\text{OH}^- \rightarrow \text{Zn(OH)}_4^{-2} + \text{H}_2 \qquad (9)$$

The gassing reactions - overcharge and corrosion - prevent the operation of cells in the sealed condition unless additional measures - such as the use of catalysts or third and fourth auxiliary electrodes - are incorporated to promote gas recombination.

The position of the Ni/Zn system in relation to other battery types is shown in Figure 1, where theoretical energy densities have been plotted as a function of the equivalent weight of the reactants and the thermodynamic voltages. One advantage of the nickel/zinc combination is a high thermodynamic voltage (1.73 volts). This feature, coupled with a modest equivalent weight, produces a theoretical energy density that is one of the highest among aqueous systems amenable to near-term commercialization. While high-temperature battery systems are appealing from a theoretical viewpoint, it has been found that a rough proportionality exists between the desirability of a particular battery system and its difficulty of development (1).

Electric vehicles typically require battery systems that are capable of storing 20-30 kWh of energy. Considering the battery alone, the largest cell capacity would generally result in the greatest energy density. On the other hand, a small cell capacity and a greater number of cells would create a higher operating voltage for better motor efficiency. As a compromise, EV battery capacities range from 200-500 Ah with operating voltages of 50-120 volts.

The number of electrodes per cell, *i.e.*, available electrode surface area per cell, is determined by the peak power requirements of the application. From solely a cost and energy density point of view, the fewest number of thickest possible electrodes would be desirable. However, in order to meet the power requirements of EV applications, it is found that 10-16 electrodes of each type (nickel and zinc) are necessary. Under peak demand, they are designed to carry current densities of 30-60 ma/cm^2 or 20-40 amps per electrode.

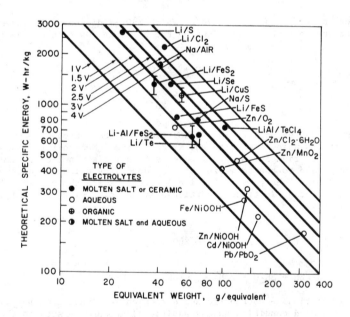

Figure 1. Theoretical Specific Energy as a Function of Equivalent Weight and Cell Voltage for Various Battery Systems

Typical manufacturing processes for Ni/Zn batteries are illustrated by the simplified flow chart shown in Figure 2. Fabrication of the nickel electrode involves two steps - one with the manufacture of the porous nickel substrate or plaque and the other with the impregnation of the plaque with the nickel hydroxide active material. The plaque therefore serves both as a current collector and as a structural containment for the active material. The plaque is formed by sintering nickel powder in a reducing atmosphere to form a 75-85% porous substrate. Embedded within the plaque is a suitable grid material. A perforated sheet of nickel, nickel-plated steel, or nickel plated copper is often used for the grid. The nickel powder is of high purity and obtained commercially by the thermal decomposition of nickel carbonyl gas. Conventional plaque fabrication has relied upon a continuous wet slurry method that is normally limited to less than 0.1 cm in thickness. In order to achieve greater energy density and reduce material costs, a semi-continuous dry loose-powder method is being developed which is capable of producing plaques with thicknesses of up to several millimeters.

The impregnation of sintered plaques is carried out by vacuum immersion into a nickel nitrate or sulfate solution, filling the pores of the plaque. The nickel salt is subsequently converted to nickel hydroxide by any of several chemical, thermal, or electrochemical techniques. Multiple impregnation steps may be required to reach the desired nickel hydroxide loading level. Final porosities are typically 30-60%. Historically, cobalt hydroxide has been incorporated into the active material in amounts of 1-10% to increase capacity and life (2). Barium hydroxide reportedly has a similar effect and has sometimes been used instead (2). Other hydroxides are presently being investigated as substitutes for cobalt.

An alternative process of nickel electrode manufacture is being developed for the non-sintered, plastic-bonded electrode (3). In this process, a mixture of nickel hydroxide ($\sim$60%), cobalt hydroxide (1-10%), graphite ($\sim$30%), and a plastic binder (1-8%) are molded and pressed to form strips. A suitable grid is subsequently sandwiched between two such strips and the three pieces are then pressed into an electrode. Replacement of the sintered nickel plaque by graphite as a conductive diluent eliminates over half of the nickel used in the electrode and total electrode weight is also reduced.

Zinc electrode manufacture begins by mixing together powders of Zn, ZnO, additives and plastic binder. Zinc, predominantly in the form of oxide, accounts for most of the mass, binder is 1-3% while the percentage of additives varies considerably among manufacturers. A pressed electrode is then constructed in a preform-and-marry process similar to that used for the non-sintered nickel electrode. Because the electrical conductivity of zinc is much greater than that of nickel hydroxide, no conductive diluent is used in the zinc electrode.

Utilization efficiency of materials during manufacturing is typically 95-99% for Ni and Zn. This occurs because scrap nickel from the plaque fabrication can be re-used in the preparation of the nickel nitrate solution for impregnation. Likewise, scrap from the preformed zinc electrode strip can be rolled into the next electrode. Scrap rates for separator materials are higher however (typically 10%).

A breakdown of material requirements and material costs by components in a 500 Ah Ni/Zn cell, based on sintered type of nickel plaque, is presented in Table I.

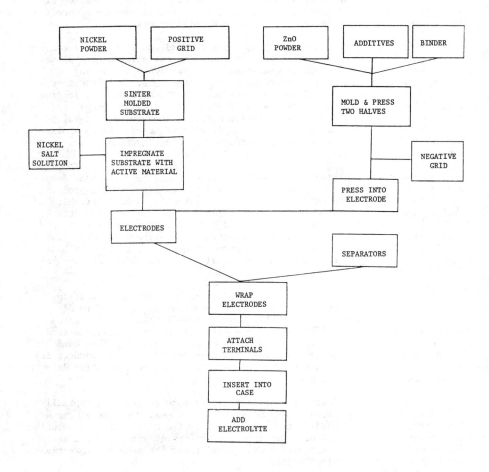

Figure 2. Typical Manufacturing Steps for Nickel/Zinc Batteries

Table I. Detailed Materials and Cost Breakdown* Per Ni/Zn Cell**

Components	kg/Cell**	Scrap kg/Cell	Total kg/Cell	Price $/kg	Total Materials $/Cell
Nickel Electrodes					
Substrate Ni Powder	1.3636	0.0207	1.3843	5.61	7.77
Grids	0.0738	0.0015	0.0753	19.80	1.49
Tabs	0.0509	0.0084	0.0593	10.78	0.64
Impregnation Solution	4.39	–	4.39	0.88	3.86
Zinc Electrodes					
ZnO	1.2945	0.0196	1.3141	0.97	1.26
Binder	0.3236	0.0049	0.3285	0.55	0.18
Grids	0.0855	0.0018	0.0873	19.80	1.73
Tabs	0.0573	0.0095	0.0668	10.78	0.72
Separators					
Substrate	0.483	0.054	0.537	1.10	0.59
Inorganic/Organic Mix	0.432	0.023	0.455	0.84	0.38
Cell Assembly					
Electrolyte (KOH)	–	–	–	–	0.22
Case and Cover	–	–	–	–	1.30
Terminals	–	–	–	–	0.59
Vent	–	–	–	–	0.15
Total					20.88

*At a production level of 13,000 25 kWh batteries per year

**Based on a 500 Ah cell

Source: Adapted from Reference 4, pp. 101–104.

Recycling of battery materials is expected to occur whenever battery production levels are sufficient to warrant the processing necessary for such recovery. Several promising processes for the recovery of nickel and zinc have been identified (3). The technical feasibility of each process is estimated to have a 80-100% chance of technical success. A typical process calls for the separate dissolution of positive and negative plates in sulphuric acid, dissolving the metallics and leaving plastic, binder, graphite and other inert material to be filtered out. Treatment of the respective sulfate solutions with DEPHA (di-2 ethylhexyl phosphoric acid) or a liquidion-exchange material can selectively extract Zn or Cu leaving (Ni,Co)SO$_4$ or ZnSO$_4$. The (Ni,Co)SO$_4$ could then be recycled into battery manufacture or used in the production of stainless steels. Likewise, ZnSO$_4$ could be used in electrolytic zinc production or used in galvanizing/plating applications. It is therefore not necessary that recovered materials be returned to battery manufacturing. Estimates of the probable percentage of nickel, cobalt, and zinc recoverable from spent batteries range from 73-95% (3, 4, 5, 6). It is even conceivable that whole sintered nickel electrodes could possibly be recycled after a simple cleaning operation. Likewise, cell cases, terminals, and packaging hardware might also be recycled.

Major Research & Development Problems for Nickel/Zinc Batteries

A number of major battery manufacturers are presently actively pursuing the development programs for Ni/Zn batteries, using both corporate as well as government-sponsored funds. Major DOE contracts have been awarded to Gould, Inc., Yardney Electric Company, Energy Research Corporation, and ESB Incorporated for the development of Ni/Zn technology. Worldwide interest has accelerated in recent years and numerous new programs have been initiated in Europe and Japan (7). General Motors/Delco Remy recently announced their decision to commercialize an EV in the 1980's using a nickel/zinc battery. Eagle-Picher Industries Inc. is also developing Ni/Zn batteries (6).

The present status of nickel/zinc EV battery technology is summarized in Table II. State-of-the-art Ni/Zn cells have specific energies of 60-70 Wh/kg. Demonstrated results, obtained by the National Battery Test Laboratory (NBTL) at Argonne National Laboratory (ANL), have surpassed 63 Wh/kg for modules supplied by Gould and Yardney. The peak power capability appears to be a function primarily of current collection design and limited to a lesser extent by the electrochemistry of the cell reactions. For selected grid designs, 15 second peak power has been demonstrated to be as high as 150 W/kg at 80% depth-of-discharge, although a value of 130 W/kg may be more typical. The ability to maintain peak power capability throughout a deep discharge is impressive. Battery cycle life remains at present the most deficient area of Ni/Zn performance. Present lifetime capability has demonstrated about 120-200 deep cycles for full-sized cells. Due to scale-up problems and other unforeseen features which have caused premature cell failure, results obtained to date at NBTL/ANL have been limited to 120 deep cycles. Many vehicle applications, however, would require a mix of predominantly shallow discharges and relatively few deep discharges. The lifetime of nickel/zinc batteries under this type of cycling conditions is not yet established, but presumably would be much better than for repeated deep discharges. Charge/discharge turn-around energy efficiencies are typically 75%. The cost of Ni/Zn batteries is very dependent upon the production level. Present costs, based on a production rate of forty 25 kWh units/year, are estimated to be $550/kWh. At a production level of 3,000 units/year per manufacturer in 1985, costs are projected to be $100/kWh; and at 100,000 units/year, projected costs would be $70/kWh. Key battery development goals for 1984 are listed in Table II and compared to the present status. Of these, the cycle life goal appears to be the most difficult to attain. The probability of achieving the energy and performance goals is high.

Table II. Present Status and Development Goals for
Ni/Zn Electric Vehicle Batteries[a]

	1979	1984
Energy Density[b] (Wh/kg)	65	85
Peak Power Density[e] (W/kg)	130	160
Sustained Power Density[d] (W/kg)	90	100
Energy Efficiency (%)	75	75
Cycle Life[e]	120	800
Battery Price ($/kWh)	550[f]	250[g](70)[h]

a) Values for 1979 are based upon cell data; 1984 estimates
   are for assembled batteries of 20-30 kWh capacity
b) At a three-hour constant current discharge rate
c) 15 sec. peak power at 50% state of battery charge
d) Sustained power at 1/2-hour rate at 50% state-of-charge
e) Life cycle testing to 80% depth-of-discharge
f) Selling price at a production level of 40 batteries per year
g) Selling price at a production level of 500 batteries per year
h) Selling price at a production level of 100,000 batteries per year

R&D problems areas currently regarded as having the most significant
impact on barriers to commercial development of the nickel/zinc battery are
1) short cycle life, 2) cost reduction, 3) sealed-cell development, and 4)
thermal management. The lifetime problem is jointly related to separator
degradation and the zinc electrode characteristics of dendrite formation and
shape change. The development of microporous separators has reduced the
dendrite problem somewhat. Early work on microporous separators was focussed
on cellulosic plastics such as cellophane. However, these were found to be
gradually degraded by the cell environment. More advanced inorganic/organic
composite and cross-linked organic separators have proven to be more hydroxide resistant. Current efforts are concentrated on developing stable, zinc-
free separators which have low ionic resistivity and are available at low
cost (less than $0.10/ft^2). A number of additives - both to the zinc
electrode and to the electrolyte - are also being investigated in an attempt
to extend cycle life. The role of the additives is to alter the characteristics of zinc dissolution and deposition in such a way to suppress dendrite
formation and shape change. By providing dispersed nucleation sites, some
additives reduce the extent of zinc redistribution. Others reduce zincate
solubility in the electrolyte. Still other additives affect zinc deposition
characteristics by raising or lowering the overpotential for the competing
hydrogen-evolution reaction. A recent novel approach to reducing shape
change has been an attempt to immobilize or otherwise confine zinc species
within an inert electrode material. This latter work remains in an early
exploratory research stage. Taken collectively, these new concepts offer
substantial reason to believe that improvements in cycle life will be forthcoming.

An alternative solution to the cycle life problem is based upon the use
of a vibrating zinc electrode (8) being developed at ESB Incorporated. By
vibrating the Zn electrode during charging, lifetimes of 1400 cycles in full-
sized cells have been obtained without the need for a microporous separator.

However, such a system is inherently more complex and bulky, resulting in lower volumetric energy densities. The future of the vibrating electrode battery will depend upon the degree to which progress is made in extending the lifetime of conventional Ni/Zn systems.

Efforts in cost reduction are focussed primarily on the nickel electrode as it represents the cost dominant element in the battery (over half of all materials costs). Energy Research Corporation and General Motors believe that the only route to commercial success is through the development of a non-sintered plastic-bonded electrode which they are pursuing. Indeed, the non-sintered nickel electrode has less than half the materials costs of conventional aerospace nickel electrodes. However, the plastic-bonded electrode has not yet achieved the performance and life capabilities of the sintered version. Gould, Yardney, and Eagle-Picher are developing advanced fabrication techniques for sintered electrodes which they believe, coupled with the superior performance of such electrodes, will reduce the life-cycle costs sufficiently to make the sintered electrode become as economically attractive as the non-sintered nickel electrode.

Sealed cell development is being pursued by Yardney, General Motors, and Energy Research Corporation as a means to reduce or eliminate maintenance (watering) requirments and possibly to extend life. To keep hydrogen and oxygen pressures low in sealed-cells, auxiliary electrodes or pocket catalysts are being employed to promote gas recombination. Early results are favorable, but additional work will be required before cell capacities, which are limited during charge by gas pressure build-up, are equal to those of comparable vented cells.

In the nickel/zinc system, as in many alkaline batteries, heat generation occurs during discharge, including a significant entropy contribution. For EV applications, this presents a definite problem. Temperature increases of 20-30°C at the battery interior are observed after three hour discharges. The effects on cycle life of such temperature increments and associated thermal gradients may present a serious and perhaps under-appreciated at present limitation on cell life. Adequate temperature control may be difficult to achieve in compact, high specific energy batteries unless appropriate steps are taken - either by cell design, by proper packaging of cells, or by active cooling methods - to provide protection from overheating the battery. Additional studies have shown that charge acceptance of the nickel electrode is quite sensitive to temperature and that effective charging cannot proceed without excessive gassing unless the cell temperature is below 50°C.

Table III. Estimated Requirements of Materials Per Ni/Zn EV Battery

Material	kg/kWh
Nickel	3.50
Zinc	1.90
Cobalt	0.069
Copper	0.086
KOH	3.96
Plastic	0.85

Material Resource Requirements for Application to Electric Vehicles

Estimates of material requirements for future Ni/Zn batteries, expressed in kg of material per kWh of capacity, are given in Table III. Because the Ni/Zn technology is in the development stage, considerable latitude exists in the design of batteries, creating a broad range of estimates for the type and quantity of materials required. Additives, such as cobalt, may be reduced or even eliminated by substitutes. Values given in Table III are based upon a battery having sintered nickel electrodes, nickel grids and tabs, a negative to positive theoretical capacity ratio of 2.5, and a cobalt hydroxide/nickel hydroxide ratio of 5%.

Based upon 100% utilization of positive active material and a cell voltage of 1.6 volts, a minimum of 1.37 kg of nickel is required per kWh. However, nickel is also used in the substrate (for sintered electrodes), in grids and tabs, and for plating cell terminals and connections. Thus, the total nickel required is about 3.5 kg/kWh. If nickel-plated tabs and grids are used, the total nickel requirements can be reduced by 10%. If plastic-bonded nickel electrodes are used instead of sintered ones, the substrate can be eliminated, and although utilization may be reduced somewhat, nonetheless, total nickel requirements would be cut in half.

The theoretical minimum amount of zinc required is 0.76 kg/kWh. But in order to avoid capacity decay and passivation effects, most zinc electrodes will contain from two to four times that theoretically required. The vibrating zinc electrode of ESB, Inc. is an exception and is designed to contain only 20% more than theoretical. A cost breakdown of components is given in Table IV. Raw materials costs, together with the manufacturing cost of converting the raw materials into battery components, are presented.

Table IV. Manufacturing Cost Estimate*

	Materials $/Cell	Conversion** $/Cell	Total $/Cell
Nickel Electrode	16.39	10.54	26.93
Zinc Electrode	6.37	3.00	9.37
Hardware	2.20	-	2.20
Separator	1.30	0.98	2.28
Electrolyte	0.46	-	0.46
Assembly	-	6.62	6.62
$/Cell:	26.72	21.14	47.86
$/kWh:	32.78	25.94	58.72

---
*Based upon a production level of 1,300 25 kWh batteries per year and 500 Ah cells.

**Manufacturing cost of converting the raw materials into battery components.

Source: Reference 4, Page 102

881

Estimates of electric vehicle market forecasts differ quite markedly because of uncertainties about future EV performance and cost, future energy availability, and other exogenous parameters. Predictions of EV passenger car usage in the year 2000 range from 1.8 to 30 million vehicles (9). Table V presents the range of estimates for total EV's, EV sales per year, and Ni/Zn battery sales per year predicted under various low, medium, and high growth-rate market scenarios (10, 11). From these scenarios, it is apparent that Ni/Zn is portrayed as an intermediate-term battery system, capturing a dominant segment of the EV market beginning in the late 1980's and continuing past the end of the century.

Table V. Electric Vehicle Market Scenarios

Year	SRI* Study	ANL** Low	ANL** Medium	ANL** High
Total Electric Vehicles (thousands)				
1985	–	54	85	168
1990	–	211	410	955
2000	–	3000	8000	24000
Total Electric Vehicle Sales Per Year (thousands)				
1985	10	16	32	70
1990	60	69	154	422
2000	600	961	2900	8340
Total Ni/Zn Battery Sales Per Year (thousands)				
1985	1.2	0.4	21	21
1990	36	7.3	128	260
2000	300	123	2260	5310

*Reference 10

**Reference 11

Material resource requirements for the production of one million 25 kWh EV batteries per year are presented in Table VI and compared with the projected annual U.S. demand in the year 2000 for such materials. For zinc and copper, the impact of the additional demand would be minimal. In the case of nickel and cobalt, the introduction of one million Ni/Zn batteries per year would have a significant impact on demand, leading to a greater dependence on foreign imports. However, the availability of nickel would probably not limit the production of Ni/Zn batteries below their natural market demand in the low and medium scenarios provided sufficient notice is given to nickel suppliers to allow for adequate planning.

Table VI. Battery Material Requirements and Projected U.S. Demand In the Year 2000

Materials	Required for One Million 25 kWh Ni/Zn Batteries (10^6 kg)	U.S. Demand (without EV's) Projected for 2000 (10^6 kg)	Increase of Demand For Production of 10^6 Ni/Zn Batteries/Year (%)
Nickel	87.5	349*	25.0
Zinc	47.5	2767*	1.7
Cobalt	1.72	20*	8.6
Copper	2.15	4600**	0.05

*Reference 12

**Reference 13

Conclusion

The nickel/zinc battery is a promising candidate for commerical electric vehicle application. Because of its intrinsically high specific energy and specific power, its popularily among EV developers is steadily growing. But major improvements in cycle life must still be obtained. The technology of the Ni/Zn systems is a new and emerging one, and though varied approaches, achievement of development goals is envisioned. Successful technical accomplishments and the availability of needed materials will not in themselves guarantee the full-scale commericalization of Ni/Zn EV batteries. The degree of market penetration will also hinge upon many exogenous parameters such as consumer acceptance, federal energy and environmental policy, or the development of an alternate competing transportation system and energy technology.

Acknowledgement

This work was performed under the auspices of the U.S. Department of Energy, Division of Energy Storage Systems.

References

1. E. Behrin, et al., Energy Storage Systems for Automobile Propulsion, Lawrence Livermore Laboratory, Report UCRL-52553, Volume 2, 1978, p. 31.

2. S. Uno Falk and A. J. Salkind, Alkaline Storage Batteries, p. 54 Wiley & Sons, Inc., New York, NY, 1969.

3. Energy Research Corporation, Design and Cost Study of Nickel-Zinc Batteries for Electric Vehicle, ERDA Report No. ANL-K-76-3541-1, Energy Research Corporation, October, 1976.

4. Gould Incorporated, Development of a Nickel-Zinc Battery Suitable for Electric Vehicle Propulsion, ERDA Report ANL-K-3558-1, Gould Report No. 762-003-1, February 1977.

5. Yardney Electric Division, <u>Final Report-Design and Cost Study - Zinc/ Nickel Oxide Battery For Electric Vehicle Propulsion</u>, ERDA Report No. ANL-K-76-3453-1, Yardney Electric Reference No. 21686, October, 1976.

6. Eagle-Picher Industries, Inc., <u>Design and Cost Study for Nickel/Zinc Battery Manufacture, Electric Vehicle Propulsion Batteries</u>, ERDA Report No. ANL-K-77-3542-1, Eagle-Picher Industries, Inc., June, 1977.

7. T. Takamura and T. Shirogami, "New Development in Sealed Ni-Cd and Ni-Zn Cells in Japan in 1977," pp. 152-154 in <u>Progress in Batteries and Solar Cells</u>, A. Kozawa, et al., eds.; JEC Press Inc., Cleveland, 1978.

8. Otto von Krusenstierna, "High Energy Long Life Zinc Battery for Electric Vehicles," pp. 303-319 in <u>Power Sources 6</u>, D. H. Collins, ed.; Academic Press, New York, NY, 1977.

9. U. S. General Accounting Office, "The Congress Needs to Redirect the Federal Electric Vehicle Program," GAO Report No. EMD-79-6, April, 1979, p. 5.

10. E. M. Dickson and B. L. Walton, "A Look Down the Road - Beyond 2000, A Scenario of Battery Vehicle Market Evolution," <u>Electric Vehicle News</u>, May, 1978, pp. 27-29.

11. Private communication, Martin J. Bernard, Argonne National Laboratory, December, 1978.

12. U. S. Bureau of Mines, <u>Mineral Facts and Problems</u>, 1975 Edition, U. S. Government Printing Office, Washington, 1976.

13. H. J. Schroeder, "Copper - Mineral Commodity Profiles - 1977," MCP-12, Bureau of Mines, U. S. Department of the Interior, June, 1977.

DEVELOPMENTS IN LEAD-ACID BATTERY TECHNOLOGY

Dr. Schrade F. Radtke
International Lead Zinc Research Organization, Inc.
New York, New York

Although the lead-acid battery continues to be the dominant force in the storage battery field, aggressive research is required to maintain that position. The paper reviews major ongoing research programs to improve energy density, and deep-discharge capabilities and to provide longer cycle life. Major new research to develop lead oxides which are consistent in quality, high in performance, economical to produce, and superior to battery oxides produced by present techniques is discussed and results to date reviewed. The needs of such potential markets as load leveling, peak-power shaving and electric vehicles are examined in relation to the role to be played by the lead-acid battery.

Introduction

As the world moves into a new decade, there is increasing reference to "superbatteries" that are expected to unseat the lead-acid battery as the undisputed king of storage batteries on a cost-effective basis. The most optimistic forecast for the challengers has this changeover taking place by around 1985. To paraphrase an old saying: The reports of the lead acid battery's imminent demise have been frequent and generally exaggerated.

Those not familiar with the lead-acid battery apparently are not aware of the great efforts being expended by the lead industry and by the battery industry worldwide to improve the performance levels of the lead-acid battery while maintaining or reducing costs. Lead producers and battery manufacturers are not resting on their laurels, like contented cats; rather, they are running lean and hungry to maintain their dominance. Nor are they taking their competition for granted or underestimating the possibilities of other battery material combinations.

No matter how remote the competitive threats may be at this time, the lead industry, through its cooperative research arm, International Lead Zinc Research Organization, Inc. (ILZRO), and the individual member companies, aided by the further efforts of battery manufacturers, is conducting research and development programs that surpass all previous efforts in intensity and scope.

Three Big Markets for Batteries

The principal research in the late 70's and on into the 80's is aimed at developing such major new end-use markets as peak-power shaving and load leveling for electric utilities and electric vehicle propulsion. A survey by the Mitre Corp. (1) on behalf of the U.S. Department of Energy (DOE) in the late 1970's indicated that a base-case national market penetration for lead-acid batteries in peak-power shaving will account for 572 megawatts of peak power per year by 1995. Given a discharge capability of five hours, this would mean a lead consumption of over 114,000 tons for that year, on the basis of nearly 40 tons of lead per megawatt hour of capacity. A typical peak-power shaving plant might have a capacity of 40 megawatt hours (MWh) that is, 10 megawatts discharged over a four-hour period. The estimate of nearly 40 tons of lead per megawatt hour, actually 67,500 lb/MWh, is derived from an average of figures obtained from five battery companies (2).

A prior, related ILZRO project to encourage the building of a plant to demonstrate the use of the lead-acid battery for peak-power shaving by electric utilities led to cooperative work with the Energy Research and Development Administration (ERDA) and the Electric Power Research Institute (EPRI). These efforts were culminated in a late 1978 announcement by the DOE that it planned to go forward with a demonstration plant employing a lead-acid battery having approximately 50 MWh capacity. This five-to-seven-year project will require an investment of from $800,000 to $1-million each year, all by the U.S. government. EPRI previously had recommended that 10% of U.S. generating capacity should be in the form of energy storage.

The first progress report on the Battery Energy Storage Test (BEST) Facility, issued in early 1979, noted that a "shakedown" battery with a capacity of 1.8 MWh over a 10-hour period could be operational by early 1980 for use in checking and testing all baseline facility equipment (3). The report recommended the use of four strings of cells that could be placed in either series, parallel, or series-parallel circuit configurations, with each string having a nominal voltage rating of 250 V. In competitive bidding, C & D Batteries Division of Eltra Corporation was awarded a subcontract to provide a deep-cycling, lead-calcium grid, lead-acid battery with hydrogen recombiners. Six cells will be packaged together to form a module with pre-wired monitor and control leads. The particular design selected will reduce or eliminate toxic gas and hydrogen evolutions and provide an easily installable package.

After the lead-acid modules are checked out over a period of a few months, advanced batteries, such as sodium-sulfur, lithium-molten salt, and zinc-chlorine, will be installed and each tested for up to 1-1/2 years. The hope is that the BEST facility will produce three to six satisfactory battery possibilities by 1981 or 1982 which would then be operated at various levels over the next 10 years. The facility is being built and will be operated by the Public Service Electric and Gas Company of New Jersey. It will serve as a U.S. center for testing advanced batteries and ac-dc power-conditioning equipment in modes of operation that are anticipated for commercial units in utility power systems.

A complementary project which is scheduled for start-up by 1983 is the Storage Battery Electric Energy Demonstration program (4). This will involve use of improved lead-acid batteries operating in a commercial-size plant that will be at least 10 times larger than the BEST facility and employ conventional utility operating and mainenance personnel to respond to actual utility load demands. This demonstration unit will examine the impact of the battery on power systems, including how well its output can be controlled, its dynamic response, its effect on the system's stability, and other questions.

Battery Requirements

Among the key objectives that advanced batteries must achieve to be effective for load-leveling use is a minimum life of 10 years, a charge-discharge cycle of 2000, and a cost of around $40 in U.S. dollars per kWh (1978 base). This compares with an industrial lead-acid battery's readings of eight years service life, 1750 life cycles, and a cost of about $80/kWh (1978 base) (4).

A major U.S. automobile producer has advised ILZRO that it estimated the U.S. market at from two to five million electric passenger vehicles by 1988. The DOE estimates that 3 million to 24 million electric vehicles could be in use in the year 2000, frequently using eight million as a realistic estimate (4). The Council also did a cost study of operating costs in 1979 and found that it costs about 2 cents per mile to drive an electric car, using a national average figure for electricity costs, which vary from place to place. An automobile which averages 15 miles to a gallon and using gasoline at $1 a gallon, costs three times as much. With gasoline costs rising steadily, the cost advantage for EV's becomes more impressive (5).

These estimates are based largely on the concept that EV's would be used as a second or third family car, for short trips to work or for shopping. Studies show that over 98% of all auto trips in the U.S. cover less than 160 km (100 miles), so there is a potential market for 30 million EV's, the number now used as second or third cars by a family, provided that cost and performance are satisfactory (4).

Again, considerable research is being conducted to develop batteries or battery-hybrid systems for EV's. But where some new designs surpass the lead-acid battery in one respect or another, they fail in some other equally important parameter. For example, higher power and quick recharge require many thin electrodes in each cell (maximum surface area for electrochemical reactions), while the lightest weight and smallest volume are achieved through fewer, thicker electrodes.

It should be noted that many of the new "superbatteries" being tested have major unresolved questions on safety. Some of them include battery-active materials which are molten liquids that react violently when mixed. If the potential hazards, especially during accidents, were not eliminated by design, it is doubtful that regulatory approval would ever be granted to permit use of such batteries.

R & D Programs are Broad Based

It is for these and other reasons that ILZRO and the individual lead-producing companies look to the lead-acid battery as the choice of battery to be used in automobiles, both conventional and battery-powered, well through the decade of the 1980's, and probably beyond. They are supporting their beliefs by a broadly based research and development program to improve the lead-acid battery so that it will remain in the forefront, regardless of the competition.

Table I compares the major characteristics of the lead-acid battery with those of the major competitive candidates.

Table I. Load-Leveling Batteries: Candidates,
Characteristics, and Demonstration Dates

	Lead-Acid	Sodium-Sulfur	Lithium-Metal Sulfide	Zinc-Chlorine
Operating temperature, $^{\circ}C$	20-30	300-350	400-500	25-50
Electrolyte	Sulfuric acid	Ceramic	Molten Salt	Aqueous zinc chloride
Design modular energy density, Wh/kg*	20	44	53	84
Utilization of active material during charge/dischare, %	25	85	80	100
Current density (5-h discharge), $m/A/cm^2$	10-15	100	30	40
Active materials, cost, $/kWh	8.5	0.49	4.27	0.74
Cell life goal, cycles	2000	1300 (30 Wh)	1000 (150 Wh)	800 (1.7 kWh)

* These figures represent different cooling arrangements: water cooling for lead-acid, gas gooling for sodium-sulfur and lithium-metal sulfide, and a refrigeration system for zinc chloride. Liquid cooled designs for sodium-sulfur have about 40-50% higher energy densities per unit weight and unit volume than corresponding values for gas cooling.

One of ILZRO's projects is being carried out at the U.S. Naval Research Laboratory, where ILZRO has two associates, Dr. S. M. Caulder and Mr. A. C. Simon (6). They have been concerned in the past few years with improving the lead-acid batteries required for electric vehicle propulsion and for load leveling. The EV program is a joint venture with the U.S. Army electro-chemical laboratory (MERADCOM) at Ft. Belvoir. Batteries that have been cycled at MERADCOM according to an electric vehicle regime are analyzed at NRL, using optical and scanning electron microscopy and x-ray diffraction techniques to determine whether any changes in battery components could be related to the high current pulse discharge type of duty cycle.

In their latest reporting period, Caulder and Simon studied 6V EV batteries, having a nominal capacity of 135 Ah that were subjected to 290A, 167-Hz pulse discharges. A 50% duty cycle was used, giving an average discharge current of 145A. A second battery system was cycled and discharged at 145A d.c.

The results of the examination on battery plates after cycling, with and without pulsed discharge, indicated that pulsing might be responsible for a definite accumulative and detrimental effect seen in pulsed positive plates, and that it might also slightly change the negative in a manner that could be beneficial. The fact that about the same amount of corrosion seemed to have occurred in the pulsed battery as in the unpulsed battery, after only half as many cycles, was probably also an effect of the pulsing. Although these results agree with what might be expected, further tests were scheduled to corroborate the results, which should be considered preliminary until then.

In another phase of their research, Caulder and Simon conducted a fundamental study concerning the atomic structure of the lead dioxide active material and how changes in this PbO_2 structure affect its electrochemical activity. They succeeded in establishing the atomic structure, firmly paving the way for further study of the changes that occur in the atomic structure of PbO_2 taken from the active material of cycled positive electrodes. They noted some errors and differences in earlier studies. An investigation of alpha-PbO_2 is underway with the ultimate goal of forming a solid structural foundation to study actual mixed oxide battery electrodes at various stages of activity.

Research to Improve the Active Material

In the late 1970's, ILZRO launched a major three-project program on the assumption that of the six principal components needed in the production of a lead-acid storage battery, the active materials, usually a mixture of high-purity lead oxides or a blend of litharge with about 20% red lead, represented the most promising potential for a major breakthrough in better performance.

It was decided to have two projects in which the objective would be the synthesis of new lead oxides, one by direct oxidation and the other by chemical and electrochemical methods. After characterization of the oxides from such standpoints as particle size, shape, purity, morphology, and reproducibility, the new lead oxides would be evaluated for suitability of the product for general application and particularly for load leveling, vehicle propulsion and other deep-discharge applications.

This research course was selected because available materials do not provide superior deep discharge capacity while at the same providing good charge-discharge cycle life. Also, current manufacturing techniques result in lead oxides which are not always reproducible with identical electrochemical characteristics. This variability, however, can be compensated by adjustments in the battery manufacturing process. For deep-discharge batteries, on the other hand, the ILZRO objective is to produce lead oxides that have uniform, consistently reproducible characteristics, combined with high-charge acceptance, high energy density, and extended cycle life.

To achieve these objectives, the ILZRO research effort includes a detailed examination of selected dopant materials. Dopants are chemical additives to the lead oxide which, hopefully, will change the crystal morphology, increase the electrochemical capacity, improve the paste utilization, and extend the cycle life -- in short, meet the battery requirements of the three major new markets.

The dopants selected initially had favorable properties for compatibility with the lead oxide structure. Other significant factors in developing an optimum lead oxide include chemical composition, particle size and particle size distribution, particle shape, and the crystal system.

Over 100 Samples Synthesized

Over 100 experimental sample lead oxide production trials were completed by the direct oxidation method through mid-1979, with 14 of these samples being evaluated in the third, companion project. All of the lead oxides produced in these processes are spherical in shape and have a surface area which is higher than commercially available Barton pot or ball mill oxides. Two sample oxides produced by this method are shown in Figure 1. Work through 1980 will involve production of 10 new samples of doped oxides, including some containing more than one dopant, and characterization of the failure mode of 10 previously cycled battery electrodes. Research also is continuing to determine the effect of dopants on oxide characteristics.

Testing of the early undoped samples showed them to be similar to fumed lead oxides made by other methods. Fume oxides typically have high initial capacities which drop off to normal capacity levels within a few cycles. By varying the processing conditions, the oxides produced ranged from 100% orthorhombic lead oxide to 25% orthorhombic-75% tetragonal lead oxide. The cause of the change in crystal structure had not been determined by early 1979, but the results did show that significant variations in lead oxide properties could be obtained (7).

In the second project to develop new lead oxides, the work involves synthesis by three methods: (1) chemical precipitation; (2) polymorphic

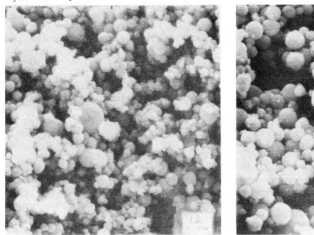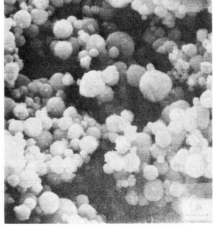

8000 X   8000 X

Fig. 1 - Scanning electron micrographs of doped lead oxides, one at 0.02% (left) and the other at 1.15%. Oxides are all spheroidal, but particle sizes vary.

transformation; and (3) electrolysis. Both doped and undoped sample oxides have been obtained by all three methods and selected samples tested.

Making Lead Oxides Chemically

Fine, morphologically characteristic alpha-type oxides were obtained by use of the chemical precipitation and polymorphic transformation methods. Controlled agitation accelerated the polymorphic transformation process and changed the powder morphology, as shown in Figures 2 and 3. With the electrolytic method, lead oxides also were obtained efficiently, but not directly in the electrolysis cell, as had been expected originally.

Fine, non-doped tetragonal lead oxides exhibited a good deep-discharge cycle performance comparable to that of conventional battery oxides. However, some doped samples also gave poor results. Both the type of dopant and level used have been varied in this project. Through mid-1979, a total of 16 dopant candidates had been evaluated, of which five were battery tested (8).

Oxide Test Center

Since the two concurrent investigations began, a total of 36 oxides had been tested through mid-1979 in the evaluation test center, with reports on 31 of these issued by that time. A portion of the test center is shown in Figure 4. A standardized set of testing procedures had been established to produce three kinds of information on the new oxides: (1) a limited amount of physical and chemical data; (2) performance data in moderately high-rate, deep-discharge cycling in the paste of the positive plate on a commercial lead-antimony, book-mold grid; and (3) a limited amount of diagnostic data on the active material. These test procedures, however, were modified in

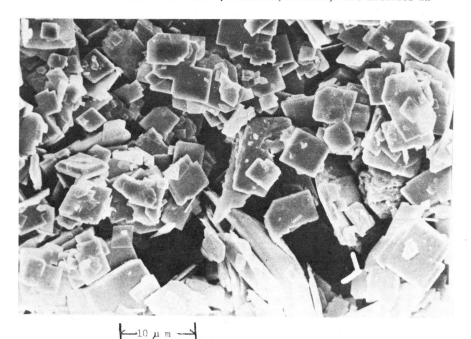

$\vdash$—10 μ m —$\dashv$

Fig. 2 - Undoped PbO oxide prepared by chemical method, showing platelet structure.

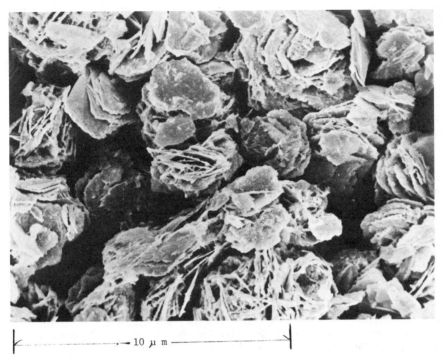

$\longmapsto$ ———10 μ m ———$\longmapsto$

Fig. 3 - Doped PbO sample prepared by chemical method shows change in structure.

Fig. 4 - Automated battery cycling test facility used in evaluating new forms of lead oxide for deep-discharge applications.

some areas as part of the continuing effort to optimize the test procedure. These modifications, for example, included the lowering of the curing temperature for non-leady oxides from 85°C to 40°C.

The latest available report on this project verified that very fine oxides give unusually high initial capacity which cannot, as yet, be maintained over a large number of deep-discharge cycles. The initial structure of the active material and its subsequent morphological alteration during cycling were under study.

Four reference oxides have been characterized chemically and physically and subjected to a deep-discharge cycling test. The reference oxides selected are a fume litharge, calcined (yellow) litharge, and two forms of leady litharge. Test results of the new oxides are compared with results compiled from the reference oxides. Plans were to conduct analyses of tested plates to produce a failure analysis of sample oxides. It also was proposed that battery companies, beginning in 1980, would test selected oxides for their suitability in deep-discharge application, using their preferred test method.

In the basically preliminary work through this period, no experimental oxide had been found superior in overall performance to the commercially available oxides. One of the experimental oxides of the very small particle size gave an excellent initial capacity but had an inferior cycle life. A later oxide, a very finely divided leady oxide, demonstrated a high capacity and good cycle life, but further considerations, especially economic, had yet to be determined whether this oxide is a valid prospective oxide for deep-discharge cycling batteries (9).

Research on Grids and Grid Alloys

ILZRO research also has included much work on improvement of grids and grid alloys. One major project in this area is directed to the problem of early battery failure due to corrosion of the positive grid and the role of antimony in this failure. In one of these projects, the researchers found that the major corrosion product of pure lead in sulfuric acid is alpha-PbO_2 and its microstructure is distinctly different from that on lead-antimony alloys (10).

These and other findings led to a new objective in ILZRO research: to find a way to achieve good cycle life without antimony in the grid. The presence of antimony in the lead-acid battery increases the cost of the grid alloy, gives rise to self-discharge, may allow evolution of toxic stibine gas, and reduces the conductivity of the grid. Eliminating antimony for deep-discharge batteries requires assurance that the active material will adhere to the grid in its absence. Therefore, establishing the mechanism whereby antimony achieves adhesion of the active material may permit replacement of antimony by some less objectionable material.

Corrosion Mechanism Clarified

In the work to date on this project, the corrosion mechanism of pure lead and the antimonial alloy have been clarified. In sulfuric acid, the anodic corrosion of the pure lead proceeds with the solid-state oxidation of the lead. Through the formation of intermediate lead oxides (tetragonal PbO, PbOx) alpha-PbO_2 is preferentially formed. In the corrosion of an antimonial alloy, however, antimony continues to dissolve out of the antimony-rich phases and the corrosion film is rich in beta-PbO_2.

In the latest reporting period, to mid-1979, the relationship between the premature capacity-loss and the distribution of $PbSO_4$ around the grid, as well as the relationship between the properties of the grid alloy and the buckling of a positive electrode made from such alloy, were studied. Results of the electron probe microanalyzer (EPMA) showed that the distributions of $PbSO_4$ around the grid of pure Pb and a Pb-0.09% Ca alloy are richer than those of a Pb-Sb alloy discharged for the same time at the same cycle, which is at the end of life of such antimony-free cells. On the other hand, results of the scanning electron microscope (SEM) showed the morphology of such an antimony-free electrode is quite similar to that of an antimonial electrode.

Macroscopic observation results of a positive electrode showed that although the plate with an antimonial grid did not show any buckling at all during cycling and that the distribution of $PbSO_4$ around the grid made of such an electrode was poor during cycling, the other plate with non-antimonial alloy grid showed a severe buckling on cycling and the distribution of $PbSO_4$ around the grid made of such an electrode was rich. In particular, the plate with a Pb-4% Cd alloy grid, which had the shortest cycle life, also showed the largest degree of buckling and the richest distribution of $PbSO_4$ around the grid.

These findings lead to the conclusion that such difference in the distribution of $PbSO_4$ was related to the difference in the properties of the grid alloy, such as alloy structure, a stress-strain curve, a corrosion morphology and corrosion production. The study, therefore, has been aimed at clarifying the relationship between the formation of $PbSO_4$ and the positive electrode behaviors of antimony-free grid alloys which have properties similar to those of lead-antimony grid alloys (11).

Conclusion

In spite of the forecasts of the imminent demise of the lead-acid battery, the lead industry's research and development efforts, including those conducted by ILZRO, will support the continued viability of the lead-acid battery through the 1980's. This total effort will be bolstered by the further research work being conducted by the battery manufacturers. The latter have found, and will continue to find in the 1980's, that no other presently known combination of battery materials matches the cost/performance characteristics of the "old reliable," the lead-acid battery, particularly for starting-lighting-ignition applications. And despite its longevity, expected improvements will place the lead-acid battery on the threshold of vast new markets -- peak-power shaving, load leveling, and electric vehicle propulsion -- in the decade of the 1980's. It should prove to be an exciting period for the lead industry.

References

1. "Lead Acid Battery: Market Assessment," by Metrek Div. of the Mitre Corp. MTR 7593, Revision 1, for the U.S. Dept. of Energy, Contract No. EC 77-C-01-5025, April, 1978.

2. "Lead-Acid Batteries for Utility Application: Workshop II," Electric Power Research Institute, Special Report, EPRI EM-399-SR, March, 1977.

3. "Battery Energy Storage Test (BEST) Facility, First Progress Report," EM-1005, Research Project 25502, U.S. Department of Energy, Contract No. EY-76-C-02-2857, February, 1979.

4. J. R. Birk, K. Klunder, and J. C. Smith, "Superbatteries: A Progress Report," pp. 49-55, IEEE Spectrum, March, 1979.

5. "Electric Cars: They're Slow, But So Are Gasoline Lines," New York Times, July 1, 1979.

6. S. M. Caulder, P. D'Antonio, and A. C. Simon, "Improving Vehicle Propulsion and Load Levelling Batteries," ILZRO Project No. LE-255, Progress Report No. 6, Jan.-Dec., 1978.

7. R. F. Dvorak, G. A. Parker, and J. P. Ryan, "Synthesis of Lead Oxides - Direct Oxidation," ILZRO Project No. LC-269, Progress Report Nos. 5 and 6, January, 1978 - June, 1979. Industrial Confidential.

8. Y. Kuroda and Y. Okada, "Synthesis of Lead Oxides - Chemical and Electrochemical Methods," ILZRO Project No. LC-271, January, 1978 - June, 1979. Industrial Confidential.

9. T. G. Chang and J. A. Brown, "Evaluation of Battery Oxides," ILZRO Project No. LE-272, "Progress Report Nos. 4 and 5, January, 1978 - June, 1979. Industrial Confidential.

10. S. Hattori, et al., "The Role of Antimony in the Positive Plate," ILZRO Project No. LE-253, Final Report, December, 1977 - June, 1978.

11. S. Hattori, et al., "Antimony-Free Grids for Deep Discharge," ILZRO Project No. LE-276, Progress Report No. 3, December, 1978 - June, 1979.

PHYSICAL METALLURGY OF LEAD ALLOYS

FOR EV BATTERIES

W. O. Gentry
Globe-Union, Inc.
5757 N. Green Bay Avenue
Milwaukee, Wisconsin 53201

Y. A. Chang
Materials Department
University of Wisconsin-Milwaukee
Milwaukee, Wisconsin 53201

ABSTRACT

Propulsion batteries for electric vehicles promise to be a significant market during the next decade. In the near term, the battery
system most likely to be utilized is the lead-acid storage battery.
How long this market will be dominated by the lead-acid battery system,
in preference to other theoretically more attractive systems, depends
on the battery manufacturers' innovation with respect to the design
of the battery system.

The electrochemical characteristics of batteries for EV applications
are very briefly reviewed to draw attention to the characteristic differences in the performance of antimonial and nonantimonial grid alloys,
particularly the known problems associated with the use of each.
Possible solutions to these problems are sketched and the (arguable)
conclusion reached that EV battery systems in the early '80's are
likely to be designed around antimonial lead alloys.

The literature on the physical metallurgy of antimonial lead alloys
published since Hofmann's treatise is reviewed, attention being paid
to those aspects important in their manufacturing and use in electric
vehicles.

The prospects for the electric vehicle look increasingly promising. There exists today an automotive industry which has developed over the past eighty years, sometimes leading but largely responding to the demands of society, a society whose interest is in the personal transportation the automobile provides and not in the details of its propulsion system. Clearly, an industry which has been satisfying a societal need as evidenced by its many years of high profitability will strive to continue doing so no matter what perturbations are placed on its ability to do so. Thus if the automotive industry is to continue into an era of accelerating hydrocarbon fuel costs, it must bring to the consumer market suitably designed vehicles regardless of whether the engine be internal combustion, external combustion, or electric.

Three important aspects of an energy source for a vehicle are its availability, price, and performance. There can be little question but that gasoline because of its high energy content per unit weight and its portability has been a most efficient fuel for the automobile. But as its price increases and its availability decreases other fuels will become, first, acceptable and, finally, necessary. The recent difficulties in gasoline availability and increasing price here in the United States make the near term commercialization of electric vehicles look more and more reasonable even given the questions as to their performance. One can visualize the time in the future when that average American household will use an EV as a family car for accomplishing the multitude of chores done daily and will rent a liquid fueled car for those occasional longer trips.

If the introduction and wide scale commercialization of electric vehicles is likely, an important question is just which secondary battery systems will be used. It can be argued that this battery system will be lead/acid both in the near term and in the longer term. The argument, again, would be based on its availability, price and performance.

One of the more favorable details of the availability of lead over that of the constituents being used in the developing battery systems is that there is in place a functioning secondary smelting industry to re-cycle the spent batteries. We can assure the people who worry about such things that adoption of lead/acid powered EV's will not result in cluttering the world with used battery packs nor grossly deplete the world's non-renewable resources.

The price of lead is also an item in favor of the use of lead/acid EV batteries. And until the recent run-up in its price, the same was true for its stability. In considering the economics of EV battery systems, two further points should be considered. The first is that the price the consumer will see will be the life-cycle cost of the vehicle. This is a pronounced negative from the point of view of the lead producers. This life-cycle cost will depend on the cost of the electricity used to re-charge the batteries, clearly beyond the control of the lead producers, and the life-time of the battery pack, a performance characteristic which cannot be directly controlled by the lead producers. The second economic point to be considered is that the price of the battery system must include its engineering. This is a pronounced positive to the lead producers. The lead/acid battery manufacturers have many, many years experience in manufacturing somewhat similar cells; and it is not difficult to see how that improves the relative position of this battery system compared to its many newer, but un-engineered, competitors.

But a final word on price. Today's lead/acid battery manufacturers are not as inflexible in their self image as the lead producers might wish. These manufacturers may actually consider themselves to be in the battery business rather than lead/acid battery business, and be busy investigating other battery systems. So in thinking about price, two figures you might concern yourselves with as marketeers of lead are its cost per unit electrical energy and the volatility, that is the time-derivative, of this number. And if your marketing people have the perspicacity to now normalize these figures to those for nickel, you will have a good feel for the relative competitive position of lead in the EV battery market.

Even in the area of performance, the outlook is promising for the lead/acid system. There, of course, is that very well-known chart of EV propulsion battery performance that shows the lead/acid system well below the performance of a large number of other systems. What should be stressed in evaluating that oft reproduced data is that the curve for the lead/acid EV battery is drawn in the <u>present</u> tense but those of the others, in the <u>future</u> tense; the lead/acid EV battery system exists. Another interesting characteristic that should be looked at is how the performance of the various systems has been changing over the past few years. Table I is reproduced from a recent report reviewing EV battery systems (1).

Table I. Change in Projected Energy Storage Characteristics Since 1977

Battery System	Change in Peak Power Capability (%)	Change in Specific Energy (%)
Pb/acid	+69	+5
Ni/Fe	+27	0
Zn/Cl_2	+57	-14
Ni/Zn	+ 7	-11
Na/S	- 4	-6
Li-Al / FeS_2	+ 7	0

As can be seen, the improvements in the lead/acid system are generally a great deal better than those which have occurred in the other systems; and this is true even though much of the proprietary activity being pursued around the world on the lead/acid system has not yet been reported.

The prospects for the electric vehicle are promising as are the prospects that it will be powered by a lead/acid battery system. Note that this battery system will be a carefully engineered and manufactured product meant to help satisfy a societal need in an economic manner. For that need to be met in the future by the lead/acid system will require the cooperation of the lead producers in maintaining the system's current competitive position in availability, price, and performance.

The lead/acid battery industry is an old and well-established industry, one which can be characterized as mature in that its continued technological developments are more evolutionary than revolutionary. The markets historically addressed have been more sensitive to the electrochemical than to the metallurgical performance of the product so it is not surprising that, in spite of the tremendous amount of metallic lead the industry consumes, the majority of its technical personnel are more chemically than metallurgically oriented. While the interesting and important developments of the past several years in maintenance-free grid alloys and novel grid manufacturing processes have seen changes in the technical work force of the battery manufacturers, it is still generally true that they rely heavily on the expertise of their suppliers for guidance on the metallurgical details of their production. Because these details will play a more significant role in battery systems for electric vehicles, if for no other reason than because the systems will be far larger and more costly than individual batteries, the technical support of the lead producers will become more important. This paper is meant to provide some insight into the physical metallurgy of lead alloys for EV batteries to help in providing that support.

The most important metallic structure in the battery is the battery grid. These are almost always prepared from alloyed lead. The reason pure lead is not used is its poor mechanical properties in a manufacturing environment. Given a particular battery grid design and particular manufacturing procedure, it is possible to actually calculate the minimum yield strength (not ultimate tensile strength) and minimum elongation to plastic instability (not elongation to fracture) necessary for the grid alloy to have. It is even possible to calculate the maximum creep rate acceptable in the alloy for a given type of battery service. The numbers one calculates generally preclude pure lead or only solution strengthened lead.

Lead alloys considered for use in preparing battery grids can be divided amongst just two families, those containing antimony as one of their constituents and the non-antimonial lead alloys. This is because these two families of alloys are characteristically different in their electrochemistry. In particular, the antimony if present retards the degradation of the positive plate through loss of its active material. This active material is an electro-chemically active lead oxide which discharges as the battery is used to lead sulphate, and it is well known that in cyclic service the battery's life is limited by the softening of the active mass or by loss of electrical continuity between it and the grid. The degree to which the material is retained is dependent on the amount of antimony present in the grid; it is known that the deep cycle life of a lead/acid battery is improved as the antimony content of the positive grid is increased towards six weight percent. Much research, a good deal of it proprietary, is being done to identify the mechanism by which the antimony improves both the adhesion of the active material to the grid wires and its own cohesion and hence the battery's cycle life.

This interest in the role of the antimony is because its use in the positive grid leads to a number of undesirable side effects which become increasingly severe as the concentration of antimony is increased. These include water loss by electrolysis, corrosion of the positive grid during overcharge, an increase in the cells internal resistance because of the poorer electrical conductivity of the antimony, and the cost of the alloy. Experience has shown that the corrosion problem is somewhat ameliorated by additions of tin and arsenic to an antimonial lead alloy.

In the negative plate, the active material is a spongy lead which discharges on use, again, to lead sulphate. Here, retention of the active material is not a problem and antimony need not be used. In fact, there is a good electrochemical reason - namely, the influence of the composition of the negative upon the corrosion rate of the positive plate -- for excluding antimony entirely from the negative. However, since some alloy must be used because of the mechanical strength needed during manufacturing, the negative grid would likely be a low antimony alloy again with some tin and arsenic.

For ease of processing these antimonial alloys, particularly during the casting operations, small concentrations of copper and sulfur are maintained in these antimonial alloys.

Thus, a review of lead alloys for near-term EV batteries can be limited
to the lead-antimony-tin system with some concern being taken for the affects of arsenic, copper, and sulfur additions. The review has been further
limited to the applicable literature published since Hofmann's treatise (2).

Equilibrium Phase Diagrams

Antimonial lead alloys contain the following alloy elements: Sb, As, Sn,
Cu, and S. These elements are added either for improving the mechanical
and physical properties of the alloys or their casting quality. In order
to understand and eventually to improve the performance and the processing
of these alloys, it is desirable to have a thorough knowledge of the phase
relationships of the multi-component alloy system. Complete information is
not available in the literature concerning the phase diagrams of the six-
component systems. Limited amount of information is available for the quaternary Pb-Sb-Sn-As system. In the following, the phase diagrams of the
eight binaries: Pb-As, Pb-Cu, Pb-S, Pb-Sb, Pb-Sn, As-Sb, As-Sn, and Sb-Sn are
presented first, then the ternaries: Pb-As-Sb, Pb-As-Sn, Pb-Sb-Sn, and
As-Sb-Sn; and lastly the quaternary Pb-As-Sb-Sn.

The intermediate phases given in Table II appear in the Pb-S, As-Sn, and
Sb-Sn binaries. The structure data of PbS are from Chang, Neumann and
Choudary (3); the structure of the As-Sn binary are from Hansen and Ander-
ko (4); and those of the Sb-Sn binary are from the Metals Handbook (5).

Table II. Structure and Designation of Intermediate Phases

Designation	Composition	Symmetry	Symbol	Prototype
η	PbS (galena)	fcc	B1	NaCl
θ	Sn_4As_3	rhombohedral	--	--
κ	SnAs	fcc	B1	NaCl
β, β'	SbSn	rhombohedral	--	--

Binary Systems

Pb-As. Lead-Arsenic. The diagram shown in Fig. 1 is based on the recent
investigation of Hutchison and Peretti (6) from pure Pb to 8 wt% As and
the information given in Hansen and others (4,8,9). The hypothetical melting
point of 808°C for As at 1 atm is from Hultgren, Desai, Hawkins, Gleiser,
Kelley and Wagmen (7). The extrapolated liquidus to pure As is shown as a
dashed curve.

Pb-Cu. Lead-Copper. The diagram shown in Fig. 2 is from the Metals
Handbook (5).

Pb-S. Lead Sulfur. The diagram shown in Fig. 3 is from the Metals
Handbook (5). However, the data concerning the liquid phase may be incorrect as discussed by Choudary, Lee and Chang (10).

Pb-Sb. Lead-Antimony. The diagram shown in Fig. 4 is from the Metals
Handbook (5). The more recent investigation of Clark and Pistorius (11) on
the effect of pressure on the melting behavior confirm the existing phase
diagram at 1 atm as shown in Fig. 4.

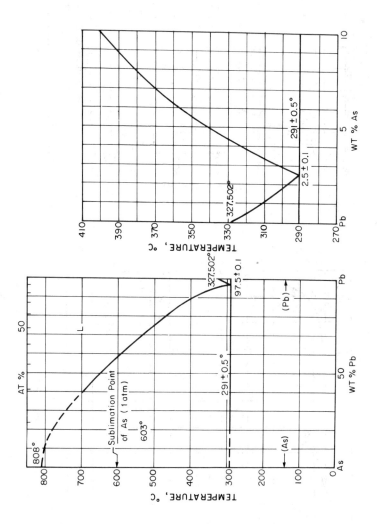

Fig. 1. Phase diagram of Pb-As

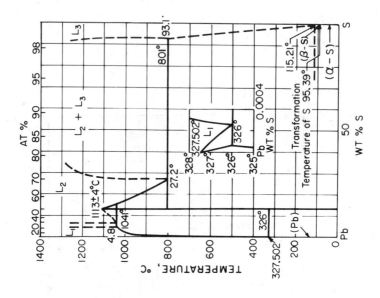

Fig. 3.   Phase diagram of Pb-S

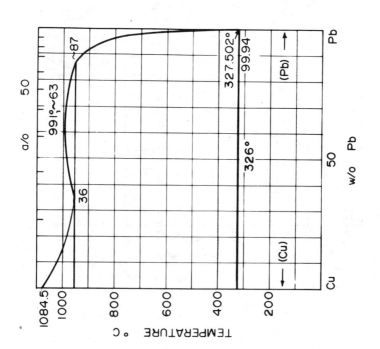

Fig. 2.   Phase diagram of Pb-Cu

904

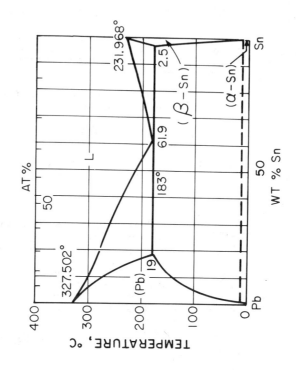

Fig. 5. Phase diagram of Pb-Sn

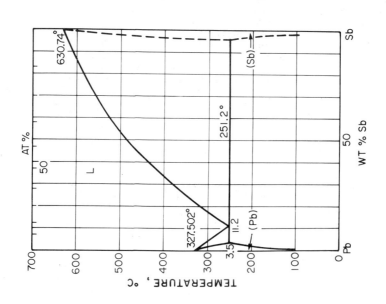

Fig. 4. Phase diagram of Pb-Sb

Pb-Sn. Lead-Tin. The diagram shown in Fig. 5 is from the Metals Handbook (5).

As-Sb. Arsenic-Antimony. The diagram shown in Fig. 6 is from Hansen and Anderko (4) except that the liquidus and solidus are extrapolated to the melting point of As given by Hultgren et al. (7).

As-Sn. Arsenic-Tin. The phase diagram from pure Sn to SnAs (κ) as shown in Fig. 7 is based on the data of Peretti and Paulsen (12); that from κ to pure As are from Hansen and Anderko (4) except that the liquidus and solidus curves are extrapolated to the melting point of As at 1 atm according to Hultgren et al. (7).

Sb-Sn. Antimony-Tin. The phase diagram as shown in Fig. 8 is from the Metals Handbook (5).

Ternary Systems

Pb-As-Sb. Lead-Arsenic-Antimony. The only information available in the literature is concerned with the liquidus in the Pb-rich corner as given by Hoffman (2). Fig. 9 shows the liquidus projection and the three binaries.

Pb-As-Sn. Lead-Arsenic-Tin. This system has been investigated by Hutchison and Peretti (13, 14, 15). The data given in Tables III and IV as well as the liquidus and the isothermal sections presented as Figs. 10-14 are deduced by us on the basis of their data. Some of the diagrams given in (15) are not consistent with each other.

Pb-Sb-Sn. Lead-Antimony-Tin. The phase diagram data of this system were evaluated by Brewer and Chang as reported in the Metals Handbook (5). The liquidus and the isotherms 240°C and 189°C shown in Figs. 15-18 are from (5) with some minor modifications. Fig. 15 shows the liquidus in the ternary and the phase diagrams of the three binaries while Fig. 16 shows only the ternary liquidus but in greater detail. Tables V and VI summarize the four-phase equilibria and the reaction diagram for this system. However, the compositions of the four-phase equilibrium I, as shown in the liquidus and given in Table V are those of Kogan and Semionov (16) which fall in between the values of Iwase and Aoki (17) and Weaver (18). The compositions of the four-phase equilibria at 240° and 189°C given in the two isotherms are those summarized in Table V. The compositions of the other equilibria are estimated based on the liquidus and isopleths given in the literature. These equilibria are shown as dashed lines. (5) gave four isopleths; however, the isopleths 14% Sn and 15% Pb as given are not consistent with the given liquidus and isotherms. The 4% Sn would be consistent if 4% Sn is taken to be ternary eutectic composition. The five isopleths 4% Sn, ~3.7% Sn, 14% Sn, 15% Pb and 90% Pb given here as Figs. 19, 20, 21, 22 and 23 are consistent with Figs. 15-18.

As-Sb-Sn. Arsenic-Antimony-Tin. Fig. 24 shows an estimated liquidus of the As-Sb-Sn system. The isotherm 385°C shown in Fig. 25 is based on the data of Kosovinc (19) and Kosovinc, Frank and Schubert (20).

Pb-As-Sb-Sn. Lead-Arsenic-Antimony-Tin.

Relatively little information is available in the literature on the phase diagram of the Pb-As-Sb-Sn quaternary system. Kerr (21) and Sudocha and Kerr (22) studied the solidification microstructure of some Pb-As-Sb-Sn alloys, but insufficient information is available to construct a meaningful phase diagram.

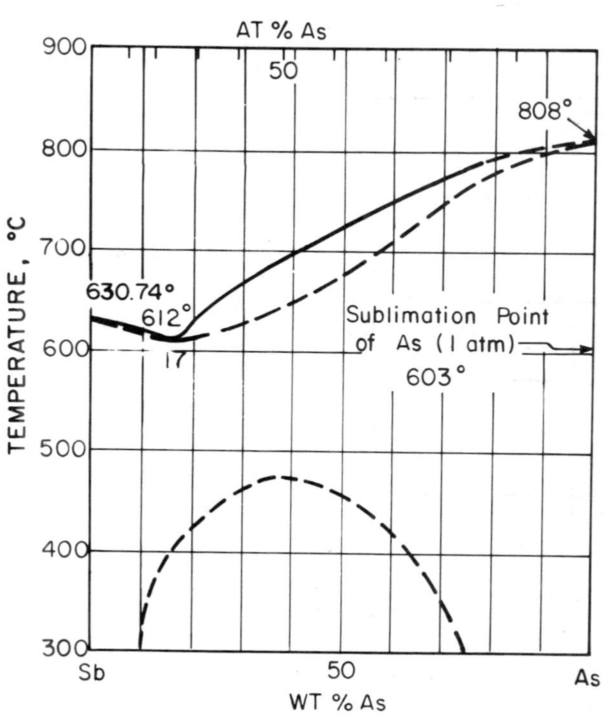

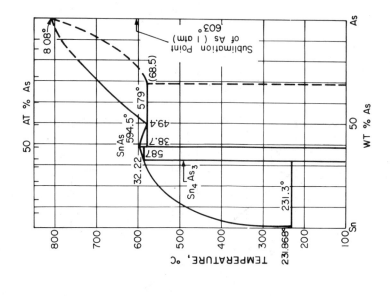

Fig. 7. Phase diagram of As–Sn

Fig. 6. Phase diagram of As–Sb

907

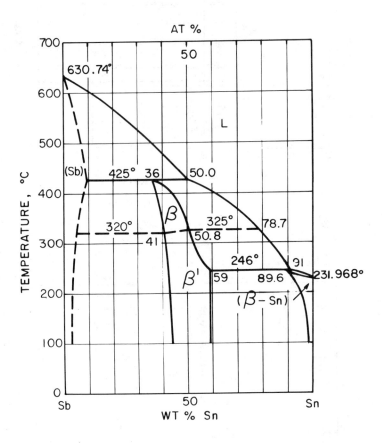

Fig. 8.   Phase diagram of Sb-Sn

908

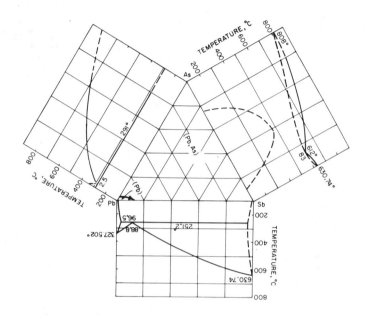

Fig. 9.　Liquidus projection of Pb-As-Sb and
phase diagrams of the three binaries

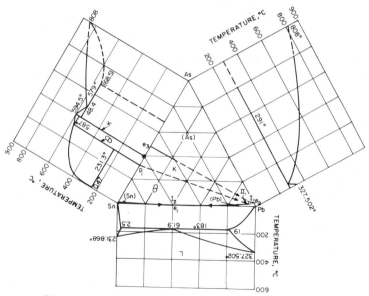

Fig. 10.　Liquidus projection of Pb-As-Sn and
phase diagrams of the three binaries

Table III. Four-Phase Equilibria in the Pb-As-Sn System

Reaction	Temp, °C	Co-existing Phases	Composition of Phases		
			Wt% Pb	Wt% As	Wt% Sn
II_1: $L + \kappa = \theta + (Pb)$	311	L	98.1	0.6	1.3
		κ	~0	~38.5	~61.5
		θ	~0	~32.4	~67.6
		(Pb)	~100	~0	~0
I_1: $L = (As) + \kappa + (Pb)$	290	L	97.2	2.62	0.18
		As	~0	~100	~0
		κ	~0	~38.5	~61.5
		Pb	~100	~0	~0
I_2: $L = (Sn) + \theta + (Pb)$	182.2	L	~38.1	~0	~61.9
		(Sn)	~2.5	~0	~97.5
		θ	~0	~32.4	~67.6
		(Pb)	~81	~0	~19

Table IV. Reaction Diagram for the Pb-As-Sn System

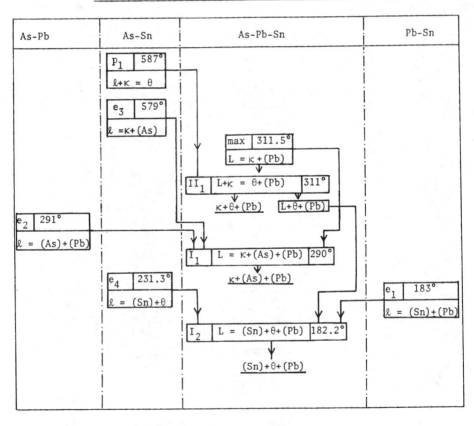

910

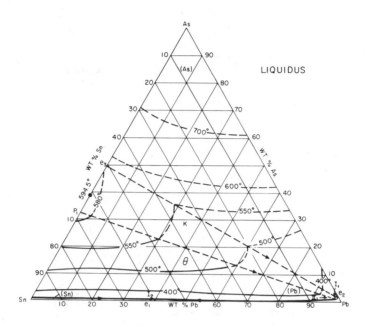

Fig. 11. Liquidus projection of Pb-As-Sn

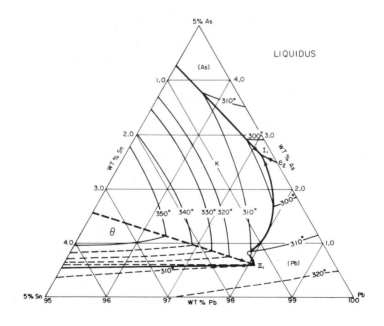

Fig. 12. Liquidus projection of Pb-As-Sn
in the Pb-rich corner

911

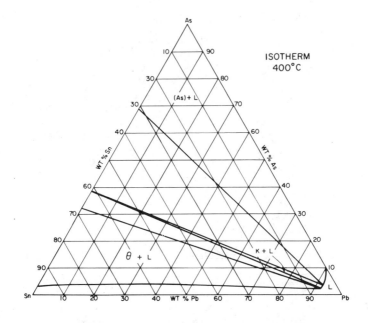

Fig. 13.   400°C isotherm of Pb-As-Sn

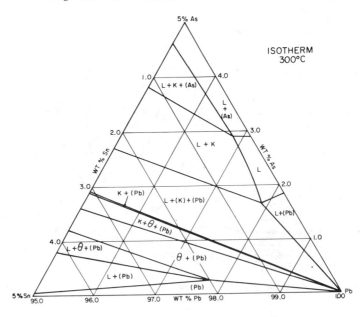

Fig. 14.   300°C isotherm of Pb-As-Sn
in the Pb-rich corner

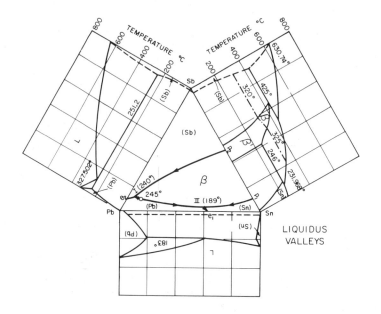

Fig. 15.   Liquidus projection of Pb-Sb-Sn and
phase diagrams of the three binaries

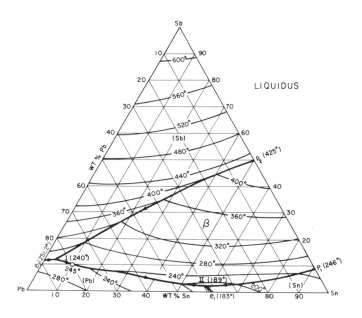

Fig. 16.   Liquidus projection of Pb-Sb-Sn

913

Table V.   Four-Phase Equilibria in the Pb-Sb-Sn System

Reaction	Temp,°C	Co-existing Phases	Composition of Phases		
			Wt% Pb	Wt% Sb	Wt% Sn
I: L = (Pb)+(Sn)+β	240	L	84.3 ± .5	12.0±0.5	3.7±0.3
		(Pb)	95.5	2.0	2.5
		(Sb)	3.0	92	5.0
		β	13.0	58	29.0
II: L+β = (Pb)+(Sn)	189	L	40.0	2.5	57.5
		(Pb)	85	0.5	14.5
		(Sn)	2	7	91
		β	5	45	50

Table VI. Reaction Diagram for the Pb-Sb-Sn System

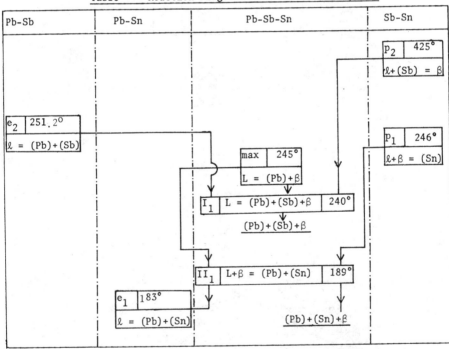

914

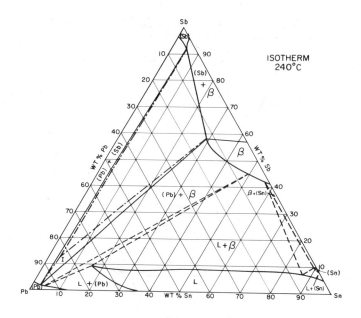

Fig. 17. 240°C isotherm of Pb-Sb-Sn

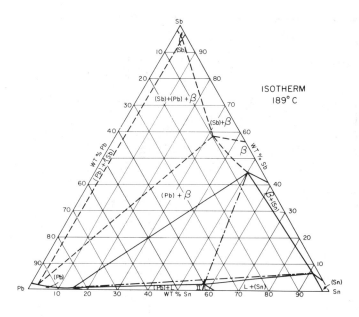

Fig. 18. 189°C isotherm of Pb-Sb-Sn

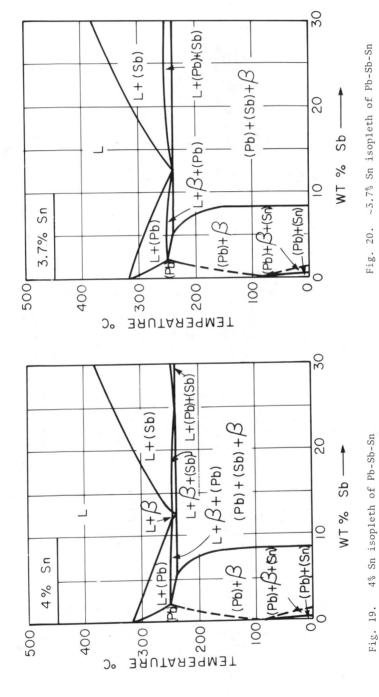

Fig. 20. ~3.7% Sn isopleth of Pb-Sb-Sn

Fig. 19. 4% Sn isopleth of Pb-Sb-Sn

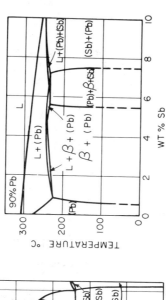

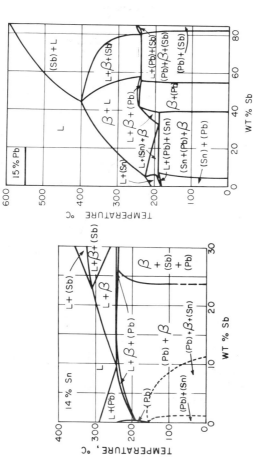

Fig. 21.   14% Sn isopleth of Pb-Sb-Sn   Fig. 22.   15%Pb isopleth of Pb-Sb-Sn   Fig. 23.   90% Pb isopleth of
Pb-Sb-Sn

917

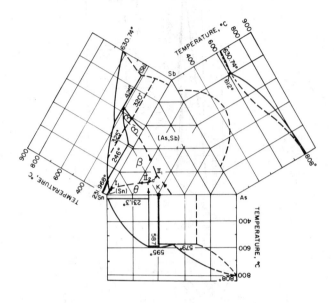

Fig. 24.   Liquidus projection of As-Sb-Sn and
phase diagrams of the three binaries

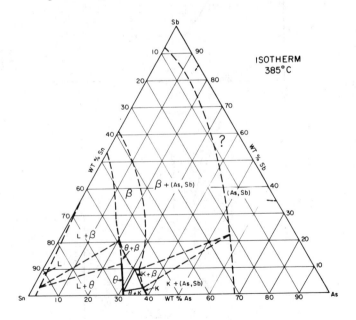

Fig. 25.   385°C isotherm of As-Sb-Sn

Non-Equilibrium Phase Diagrams

Although equilibrium diagrams are important in the understanding of the metallurgy involved in the processing of the lead alloy grids, the actual casting operations take place at such a rate that equilibrium conditions are never seen. Published information on the non-equilibrium phase distribution of antimonial leads is very sparse. And, because the details of the non-equilibrium solidification of a given alloy are so dependent on the rate of heat extraction, what information is available is likely applicable only to the particular casting operation studied. This makes generalizations difficult.

Fortunately, there are some non-equilibrium phenomena which do not occur in antimonial leads. For example, no evidence was seen (23) of the formation of meta-stable intermediate phases in splat-cooled lead-antimony alloys. And Russian workers, studying the occurrence of non-faceted antimony in a number of antimony-binary alloys (24), found the antimony in Pb-Sb alloys to always form in a faceted morphology independent of composition or cooling rate (25).

Powell and Colligan (26) have reported that the eutectic temperature in Pb-Sb alloys is depressed and the eutectic composition shifted towards antimony. This happens because the unfaceted primary lead is a poor nucleant for the antimony. They also report (27) work on Sn-Sb alloys where the undercooling is influenced by the nucleating ability of SbSn for tin.

The degree of undercooling in casting both Pb-Sb and Pb-Sn alloys was seen by Youdelis and Iyer (28) to be strongly dependent on the melt composition, being greatest at the eutectic compositions.

Davis (29) has pointed out that the actual local composition in a non-equilibrium eutectic alloy will not in general be the eutectic composition, and that this actual composition is calculable. For example in a Pb-Sn alloy, the tin-rich node would be expected to have a composition of 73 w/o Sn rather than the 61.9 w/o Sn under equilibrium conditions. The local composition of a solidifying casting is also influenced by the difference in density between the primary lead and the tin-rich liquid. It has been seen (30) that significant macrosegregation occurs on solidification because of fragmentation of dendrites and their subsequent sedimentation. This density difference has also been used (31) to explain the formation of pipes in lead-alloy castings and, by extension, may play a role in the formation of the "crystals" often seen on poor castings of antimonial alloys.

Another common type of non-equilibrium phase distribution seen in alloys in a eutectic system is the "halo" of second phase surrounding the primary particle. This halo has been ascribed (32) to the variation in the growth rates of the two phases with undercooling and solute content. The small halos observed in the Cu_2Sb-Sb system were interpreted to mean that a large zone of coupled growth existed.

The well known, though not well understood, improvement in the casting characteristics of antimonial leads on adding tin has been studied by Heubner and his associates (33). They conclude that the effect is due to easier movement of the solidified and fragmented lead dendrites in the solidifying alloy and to the mechanical properties of the oxide skin at the surface of the melt. This conclusion is supported by recent analytical work (34) which found that the surface of cast battery grids are tin-rich, possibly due to the presence of a superficial tin oxide layer. The improvement is not due to any large change in viscosity on adding tin; Thresh and

919

Crawley (35) have found that the isothermal viscosity of Pb-Sn melts varies
linearly by about a factor of two between pure lead and pure tin, but that
the viscosity at the liquidus temperatures of lead-rich Pb-Sn alloys is approximately constant. It is also unlikely that the improvement is due to
some strong effect of the tin on the alloy's thermal expansion behavior;
measurements (36) have shown that the coefficient of expansion in Pb-Sn alloys to be a simple linear function of composition, decreasing about 25%
on going from pure lead to pure tin. It would be interesting to extend
the work of Kumar and Sivaramakrishnan (37), who found that lead-antimony
melts can be considered as dispersions of lead- and antimony-rich clusters,
to see if tin has any influence on the cluster size.

Because of the ease with which they are handled experimentally, Pb-Sn
alloys are a favorite for the investigation of solidification phenomena.
Some of this work is of significance to grid casters. Verhoeven and his
co-workers at Iowa State (38, 39, 40) have observed four different crystallographic realtionships between the solidifying lead and tin solid solutions
in eutectic alloys; the actual orientation relationship depended on the
eutectic's growth rate. In alloys on the tin-rich side of the eutectic, they
found that both graphite and MoS_2 reduced the nucleation rate of the dendrites which formed at discontinuities on the mold walls. In similarly
tin-rich binary alloys, Jackson(41) studied the dendrite to dendrite-eutectic
transition; he proposed a model in which the solidified structure depends
on the temperature ahead of the growing solid, the composition of the liquid,
and the temeprature gradient in the liquid. In more slowly cooled Pb-Sn
castings, a cellular profile rather than the dendritic is expected. Coulthard and Elliott (42) have measured the cell sizes in lead alloys with small
additions of tin and found them to be inversely related to the product of
their growth rate and the temperature gradient in the liquid.

Chell and Kerr (43, 44) have looked at the solidification structures of
eutectic and monovariant Pb-Sn-Cd alloys and concluded that the single
phase dendrites they observed were due to the failure of both phases to be
nucleated simultaneously and because the liquid composition was not within
the zone for coupled growth at the given growth rate. However, they point
out, the mechanical properties of the resulting dendritic samples may be as
good as those expected from an equiaxed structure because the many changes
of crystallographic orientation from dendrite branching are similar to the
orientation variations seen across the grain boundaries in an equiaxed
sample.

An attempt to modify the structure of binary Pb-Sb alloys by the use of
ultrasonic vibrations (45) showed relatively little change at the low antimony concentrations of interest to grid casters.

Much more information on solidification processing is available. An
excellent reference is Flemings' (46) text on the subject. An interesting
and less mathematical treatment by Ohno (47) is available.

The utility of mentally, if not yet actually, extrapolating the equilibrium phase diagram information into a non-equilibrium phase diagram is that
it permits the metallurgist to account for the great process sensitivity seen
in casting lead alloys. Quite large differences are possible in the microstructures of the cast alloys, and these differences become apparent in
the grids' mechanical and corrosion properties during the fabrication and use
of the battery.

Effects of Copper and Sulfur

It is well known by solder users that copper readily reacts with Pb-Sn alloys, and the reactions and their rates have been studied (48). It is also well known by lead smelters that sulfur is an appropriate addition to de-copper lead melts and that small levels of tin (or silver) are beneficial although not necessary for substantial copper removal. Australian workers (49) have shown that the reaction between the sulfur and lead is inhibited by the tin so that its reaction with the copper to form CuS can occur; this compound then provides further decoppering by reacting to form Cu_2S unless the melt is poorly stirred in which case the copper returns to solution and PbS is formed.

It is interesting, then, that the solution to the hot-cracking problem seen upon the industry wide introduction of low-antimony grid alloys was to use small additions of copper plus sulfur to the alloys as grain refining agents. Heubner and his coworkers (50) found that either element provided some refinement, sulfur performing somewhat better than copper, but that the greatest grain refinement came when both were used. They argue that this refinement was due to either an inoculating effect by PbS, Cu_2Sb, and $CuSbS_2$ or due to these small additions reducing the dendritic growth rate allowing more time for additional dendrites to nucleate.

Age Hardening in Pb-Sb Alloys

An interesting metallurgical experiment would be to add small but increasing amounts of various solid solution elements to high purity lead to see which of them might be the most effective solution strengthening agent. When this is done (51) what is seen is that the yield-strengths of the resulting alloys increase linearly with increasing atom-fractions of the solutes' solubility limits regardless of which element was added. Each solid solution element is equally effective; but, unfortunately, the maximum strength obtainable from solid solution strengthening is insufficient for processing battery grids. The age-hardening of supersaturated alloys is important in battery manufacturing because it is commonly used to provide that necessary strength.

In a long series of papers (52-59), researchers at Spain's National Center for Metallurgical Research (CENIM) have reported their studies of the age-hardening in dilute Pb-Sb alloys. The utility of their first studying dilute alloys rather than grid alloy compositions was simply that those alloys could be solution treated and brought into equilibrium whereas the more heavily alloyed commercial alloys could not. After the samples were quenched from the solution treatment, the subsequent precipitation was monitored by hardness and electrical resistivity measurements and by metallography. They find an activation energy for the precipitation process of about 0.47 ev which they argue is what would be expected if the rate determining step were antimony diffusion in the lead matrix. The precipitates are plate shaped. Their degree of dispersion is dependent on the temperature at which nucleation occurs after the solution treatment. If the single phase alloy is quenched to a low temperature and held there long enough for the plates to nucleate, the precipitates will be more numerous and the rate of precipitation increased although, of course, the size of the plates will be reduced.

More recently they reported (60) a somewhat similar study in which the two binary alloys investigated had antimony contents above the solubility limit and, hence, could not be solution treated. In these more grid alloy like compositions the age-hardening was modified by the manner in which the castings were cooled in the mold. With one set of specimens the mold was water cooled; with the other set, air cooled. They found that the age-hardening was most rapid in the slowly cooled lower antimony, Pb 4 w/o Sb, alloy. The age hardening was slowest for the slowly cooled higher antimony,

Pb 6 w/o Sb, alloy. The age-hardening for the two rapidly cooled alloys was approximately the same. They attribute this behavior to the way in which the antimony concentration in the primary lead dendrites is affected by the cooling rate. They conclude by noting that the antimony content of the alloy used for manufacturing cast battery grids is by no means the most important variable influencing the mechanical properties of cast grids.

While the tensile yield strength is the most important strength parameter of a grid alloy during the fabrication of a battery, after fabrication the alloy's creep strength is more important. The creep strength determines how the positive battery plate will respond to the relatively small stresses applied to the grid by the volumetric change of the corrosion film which forms on its surface during charging, a stress sufficient to cause "grid growth" by creep deformation over a period of time leading to failure with a poorly designed alloy.

To improve the creep properties of antimonial leads, Tilman, Crosby, and Neumeier (61) used the plastic deformation from multiple hot-rolling passes to develop a substructure and precipitate distribution which conferred particularly good creep strength. This was seen even in two-phase alloys in which the occurance of localized phase boundary sliding (62) would have been expected to result in poor creep properties.

The Spanish investigators reported (59) on the effect of plastic deformation immediately after the quench from the solution treating temperatures of their low antimony samples. They found that the initial age-hardening rate was reduced. Borchers and Eyl (63, 64) have looked at a similar alloy which was stressed merely by quenching from the solution treating temperature. They noticed that in the regions which had been plastically deformed in yielding to the thermal stresses on quenching, the precipitates were non-coherent. In regions of the specimens which had been only elastically strained, the precipitates were coherent.

Effects of Arsenic

There is a marked effect of small additions of arsenic on the age-hardening of antimonial leads; both the rate at which the hardening occurs and the maxima in the alloys are increased. Typical data were reported by Williams (65) in 1966, Borchers, Nijhawan, and Scharfenberger (66) in 1974, and Tsumuraya and Nishikawa (67) in 1975. These last used electrical resistivity to follow the precipitation in solution treated low-antimony alloys and then differential scanning calorimetry to study the reversion, or re-solution, of the precipitates as the temperature was raised. They conclude that the enhancement is due to the 0.01 w/o As addition increasing the number of precipitate particles.

Overaging

A transformation responsible for much of the overaging or re-softening seen in age-hardenable antimonial and non-antimonial alloys is the cellular reaction (68). In this form of discontinuous precipitation a large angle grain boundary migrates; and, as it does, the uniformly distributed precipitated plates formed during the age-hardening are dissolved at the grain boundary and reprecipitated behind it into a coarser lamellar structure having relatively low hardness. This transformation was seen again in low-antimony leads recently (69, 70). Borchers and Nijhawan (69) reported the significant observation that the transformation is retarded by silver additions.

922

Conclusion

As part of the discussion of one of the classic papers (71) on electric vehicles, published back in 1903, a commentator with admirable politeness says that he "considered that the paper was an interesting one, on an interesting subject; and in view of the previous failures of accumulator-traction, he thought it had required a great deal of courage on the part of the author to attack the problem as he had done. The author was an enthusiast: and were not enthusiasts apt to be a little rosy in their view of what could be done? Manufacturers also were enthusiasts, and looked at the matter from an optimistic point of view."

Three-quarters of a century later, the enthusiasm remains.

References

1. E. Behrin, et.al., "Energy Storage Systems for Automobile Propulsion: 1978 Study", UCRL-52553, University of California (Livermore, Cal., 1978).

2. W. Hofmann, Lead and Lead Alloys, Springer-Verlag (New York, 1970).

3. Y. A. Chang, J. P. Neumann and U. V. Choudary, INCRA Monograph VII, Phase Diagrams and Thermodynamic Properties of Ternary Copper-Sulfur Metal Systems, The International Copper Research Association, Inc., New York, 1979.

4. M. Hansen and K. Anderko, Constitution of Binary Alloys, 2nd Edition, McGraw-Hill, New York, 1958.

5. Metals Handbook, 8th Ed., Vol. 8, Metallography, Structures and Phase Diagrams, American Society for Metals, Metals Park, Ohio, 1973, p. 333 and p. 374.

6. S. E. Hutchison and E. A. Peretti, J. Less-Common Metals, 1973, 30, 306.

7. R. Hultgren, P. D. Desai, D. T. Hawkins, M. Gleiser, K. K. Kelley and D. D. Wagman, Selected Values of the Thermodynamic Properties of the Elements, American Society for Metals, Metals Park, Ohio 1973.

8. R. P. Elliott, Constitution of Binary Alloys, First Supplement, McGraw-Hill, New York, 1965.

9. F. A. Shunk, Constitution of Binary Alloys, Second Supplement, McGraw-Hill, New York, 1969.

10. U. V. Choudary, Y. E. Lee and Y. A. Chang, Met. Trans., 1977, 8B, 541.

11. J. B. Clark and C. W. F. T. Pistorius, J. Less-Common Metals, 1975, 42, 59.

12. E.A. Peretti and J.K. Paulsen, J. Less-Common Metals, 1969, 17, 283.

13. S. E. Hutchinson and E. A. Peretti, J. Less-Common Metals, 1973, 31, 101.

14. S. E. Hutchinson and E. A. Peretti, J. Less-Common Metals, 1974, 34, 107.

15. S. E. Hutchinson and E. A. Peretti, J. Less-Common Metals, 1973, 31, 321.

16. V. A. Kogan and A. A. Semionov, Zh. Fiz. Khim., 1963, 37, 802.

17. K. Iwase and N. Aoki, Kinzoku-no-Kenkyu, 1931, 8, 253.

18. F. D. Weaver, J. Inst. Metals, 1935, 56, 209.

19. I. Kosovinc, Mining and Met. Quarterly, 1969, 29, 343.

20. I. Kosovinc, K. Frank and K. Schubert, 1968, 22, 690.

21. H. W. Kerr, J. Inst. Metals, 1971, 99, 238.

22. J. P. Sandocha and H. W. Kerr, J. Inst. Metals, 1972, 100, 315.

23. P. Ramachandrarao, P. K. Garg, and T. R. Anantharaman, Indian J. Tech., 1970, 8, 263.

24. I. V. Salli, L. V. Samoilenko, and V. Z. Kolinskaya, Sov. Phys. Chrstallogr., 1974, 18, 823.

25. I. V. Salli, V. Z. Dolinskaya, and L. V. Samoylenko, Fiz. Metal. Metalloved., 1974, 38, 132.

26. G. L. F. Powell and G. A. Colligan, Met. Trans., 1970, 1, 133.

27. G. L. F. Powell, G. A. Colligan, and A. W. Urquhart, Met. Trans., 1971, 2, 918.

28. W. V. Youdelis and S. P. Iyer, J. Inst. Metals, 1973, 101, 176.

29. K. G. Davis, Can. Met. Quart., 1978, 7, 93.

30. T. E. Strangman and T. Z. Kattamis, Met. Trans., 1973, 4, 2219.

31. N. Streat and F. Weinberg, Met. Trans., 1972, 3, 3181; but see comments by Hebditch and Hunt, Met. Trans., 1973, 4, 2474.

32. M. F. X. Gigliotti, G. A. Colligan, and G. L. F. Powell, Met. Trans., 1970, 1, 891; ibid. 1970, 1, 2046.

33. U. Heubner, G. Rudolph and A. Ueberschaer, Z. Metallkde., 1976, 67, 277.

34. C. Sanchez and P. Ruiz, Personal communication, 1979.

35. H. R. Thresh and A. F. Crawley, Met. Trans., 1970, 1, 1531.

36. L. J. Balasundaram and A. N. Sinha, J. Appl. Phys., 1971, 42, 5207.

37. R. Kuman and C. S. Sivaramakrishnan, J. Matl. Sci., 1969, 4, 383.

38. J. D. Verhoeven and E. D. Gibson, Met. Trans., 1972, 3, 1893.

39. J. D. Verhoeven, D. P. Mourer, and E. D. Gibson, 1977, Met. Trans., 1239.

40. D. P. Mourer and J. D. Verhoeven, J. Crystal Growth, 1977, 37, 197.

41. K. A. Jackson, Trans. Met. Soc. AIME, 1968, 242, 1275.

42. J. O. Coulthard and R. Elliott, J. Inst. Metals, 1967, 95, 21.

43. M. F. Chell and H. W. Kerr, Met. Trans., 1972, 3, 2002.

44. M. F. Chell and H. W. Kerr, Met. Trans., $8A$, 520.

45. A. V. Sapozhnikov, Met. Sci. Heat Treat., 1972, 10, 74.

46. M. C. Flemings, Solidification Processing, McGraw-Hill (New York, 1974).

47. A. Ohno, The Solidification of Metals, Chijin Shokam (Tokyo, 1976).

48. M. A. H. Howes and Z. P. Saperstein, Weld. J., 1969, 48, 805.

49. I. S. R. Clark, L. A. Baker, and A. E. Jenkins, Trans. Inst. Min. Met., 1973, $82C$ (796), C1.

50. U. Heubner, I. Muller, and A. Veberschaer, Z. Metallkd., 1975, 66, 74.

51. W. O. Gentry and D. Marshall, (Globe-Union, Inc.), Unpublished Research.

52. M. Torralba, J. J. Regidor, and J. M. Sistiaga, Z. Metallkde., 1968, 59, 184.

53. M. Aballe and M. Torralba, Rev. Metal. CENIM, 1969, 5, 385.

54. M. Torralba, J. J. Regidor, and J. M. Sistiaga, Rev. Metal. CENIM, 1969, 5, 706.

55. J. J. Regidor, M. Aballe, M. Torralba, and J. M. Sistiaga, Rev. Metal. CENIM, 1970, 7, 447.

56. M. Torralba and J. J. Regidor, Rev. Metal. CENIM, 1971, 7, 304.

57. M. Aballe and M. Torralba, Rev. Metal. CENIM, 1972, 8, 360.

58. M. Aballe, J. J. Regidor, M. Torralba, and J. M. Sistiaga, Rev. Metal. CENIM, 1973, 9, 95.

59. M. Aballe, J. J. Regidor, J. M. Sistiaga, and M. Torralba, Z. Metallkde., 1972, 63, 564.

60. M. Torralba, M. Aballe, and J. J. Regidor, J. Power Sources, 1979, 4, 53.

61. M. M. Tilman, R. L. Crosby, and L. A. Neumeier, Improved Properties of Lead-Antimony Alloys by Thermomechanical Treatments, U. S. Bureau of Mines (Washington, 1975).

62. A. Eberhardt and B. Baudelet, J. Matl. Sci., 1974, 9, 865.

63. H. Borchers and J. Eyl, Metall., 1977, 31, 959.

64. H. Borchers and J. Eyl, Z. Metallkde., 1977, 68, 612.

65. J. D. Williams, Metallurgie, 1966, 74, 105.

66. H. Borchers, S. C. Nijhawan, and W. Scharfenberger, Metall, 1974, 28, 863.

67. K. Tsumuraya and S. Nishikawa, J. Jpn. Inst. Met., 1975, 39, 916.

68. K. N. Tu, Met. Trans., 1972, 3, 2769.

69. H. Borchers and S. C. Nijhawan, Metall, 1975, 29, 465.

70. M. Torralba and M. Aballe, Rev. Metal. CENIM, 1978, 14, 25.

71. H. F. Joel, Proc. Inst. Civil Eng., 1903, 152, 2.

MATERIAL REQUIREMENTS FOR LEAD ALLOYS

AND OXIDES IN THE 80'S

A. Eckert, J. Pierson, and W. Tiedemann

Globe-Union Inc.
Milwaukee, Wisconsin

The most significant detrimental effects of impurities in lead alloys and oxides for use in automotive type lead-acid batteries are those which increase the rate of undesirable reactions: excessive gassing during charging, and self-discharge during rest periods in the duty cycle. Requirements for analytical methods for identification and quantization of trace elements in these materials are becoming more demanding, requiring capabilities for: 1) increasing number of samples to be analyzed, 2) analysis for lower concentrations, 3) more elements included in the analyses, 4) shorter turn around times for the analyses to be performed.

Electroanalytical methods have many capabilities that indicate their superior applicability for these purposes; they are rapid, highly sensitive to small concentrations of ions of interest, and they are based on phenomena that are fundamentally similar to reactions in a lead-acid battery. These methods are promising particularly for certain special elements (e.g., S, Se, Te, Sb, As) where other methods have limited capability. The most popular approach now is to use direct reading emission spectroscopy; by this means, acceptable turn around times are possible, but it is not always possible to include all elements of interest and to meet all of the desired lower detection levels. Mass spectroscopy can meet most turn around time requirements, and can probably analyze for any element required, but it is only moderately successful in meeting the lowest detection levels.

Projected analytical requirements tax capabilities for greater detection sensitivities rather than higher precision and accuracy, but extremely rapid, high volume capabilities significantly beyond the present state of the art are not indicated. For the most demanding analyses, costly, skill limited procedures are the only useful methods in sight.

Introduction

Since the invention of the lead-acid battery 120 years ago by Gaston Plante', contaminants have been blamed for inadequacies in the electrochemical system's performance. Organic and metallic impurities in minute quantities have been shown to have deleterious effects ranging from increased self-discharge and lowered capacity to premature corrosion of the grids, shedding of the active material and attack of the separators. Extensive experimental work in research laboratories within the battery and lead industries have identified many of the problem-causing contaminants and established acceptable concentration levels. As a result of these studies, Material Specifications have been prepared by battery manufacturers for all of the components used in making batteries.

In recent years the relative importance of impurities has been increasing due to the reduction in total amount of materials in the battery (more sensitive to changes) and the increase in performance required of the remaining materials. Automotive batteries for the 1980's will be significantly smaller and lighter than their predecessors. Most will also be of the sealed-but-vented maintenance-free type with no provision for replenishment of water in the electrolyte. Because this new constraint on the lead-acid battery increases dramatically the effects and tolerance levels of impurities on the gassing behavior of the Pb and PbO_2 electrodes, lead-acid battery materials must be reviewed and respecified.

The typical performance specifications for maintenance-free batteries incorporates not only initial electrical and life tests but also some form of definition of the term "maintenance-free." A representative specification for the maintenance-free aspect of such a battery is "no water addition required during the normal life of the battery and no failure due to water loss over the service period--typically four years or 50,000 miles (80,450 Kilometers)."

System Overview

An overview of the lead-acid battery from electrochemists' point of view reveals two basic facts: first, the system is thermodynamically unstable (decomposition of H_2O into H_2 and O_2, see figure 1). Both H_2 and O_2 are generated several hundred millivolts before the conversion of $PbSO_4$ to Pb and PbO_2 occurs. However, as we know, thermodynamics tells us about the possibility of reactions occurring, nothing is stated concerning the rate (kinetics) at which these reactions proceed. Decomposition of H_2O into H_2 and O_2 is acceptable due to the relatively low rate of reaction (slow kinetics). However, it is widely known that kinetics of various reactions can be increased several orders of magnitude by addition of trace amounts of impurities as reviewed in 1974 by Pierson et al. (1).

Second, any constituent in the lead-acid system other than Pb, H_2O and H_2SO_4 is considered to be an impurity; however, other constituents are needed to produce viable products for the market place. The appearance of some of these materials in the battery can be viewed as a potential danger to the apparent thermodynamic stability of the system. In addition, the location of the impurities (solution, grids, Pb and PbO_2 materials) can be very critical to stability of the system.

As mentioned above, the kinetics of H_2 and O_2 gas evolution on Pb and PbO_2, respectively, are very slow, however, impurities such as Pt, Te, Se, etc., and traditional grid-lead alloying ingredients such as Sb and As are known to deposit on the lead electrode and serve as sites for increased rates of H_2 evolution. One can view this effect as a corrosion cell (self dis-

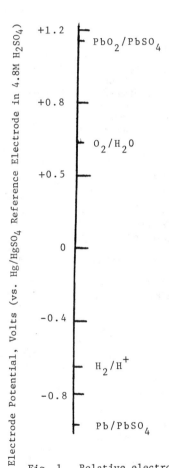

Fig. 1. Relative electrode potentials for the
decomposition of H_2O into O_2 and H_2 and
the conversion of $PbSO_4$ to Pb and PbO_2.

charge) in which $Pb \rightarrow Pb^{++} + 2e^-$ and $2H + 2e^- \rightarrow H_2$ occurs on the surface. In addition, these sites will significantly increase H_2 generation during battery overcharge.

The dissociated water must be replaced if the battery is to perform properly and have an acceptable service life. The first task encountered by the battery engineer in designing a "sealed" maintenance-free product therefore is to reduce or eliminate the antimony and arsenic from the grid alloy. This has been accomplished by replacing these strengthening ingredients in the alloy with electrochemically innocuous materials such as calcium and more recently strontium in a patented lead alloy, reported in 1979, by Weinlein et al. (2).

Increasing the rate of H_2 evolution also alters the relative distribution of the cell voltage during charge (decreasing the relative voltage on the lead electrode and increasing the voltage on the positive). The increased voltage on the positive electrode increases the rate of corrosion on the lead current collector. Increased corrosion by itself shortens battery life, but, depending on the specific type of alloy constituents, corrosion may release impurities which may also increase the rate of H_2 evolution when deposited on the lead electrode. This behavior can dramatically accelerate battery failure. Therefore, we must be aware of not only the initial level and type of impurities but also where they are used in the system.

It appears that there is a need to know what are the desirable, acceptable and maximum concentration of the various impurities which adversely affect battery life. It is therefore imperative that analytical procedures for rapidly and accurately detecting impurity type and concentration be developed in conjunction with electrochemical techniques in order to cross correlate chemical analysis with electrochemical H_2 gassing tests. In depth studies are presently underway to develop both chemical and electrochemical techniques to detect impurity type, concentration and acceptable level.

Chemical analysis of lead, lead alloys, and lead oxide for the manufacture of automotive type lead-acid batteries is performed to support engineering and electrochemical research activities and to support operations activities, which include raw material quality control, product quality assurance, and manufacturing plant trouble shooting.

Analytical requirements for support of engineering and research activities, extremely varied and constantly changing, do not fit neatly into a discussion of requirements for support of operations. Some of the research service requirements later form the basis for methods used to support operations.

Analyses for material and product quality control are generally performed with a minimum of pressure in scheduling. There is a trend toward somewhat more exacting requirements for precision and accuracy, and for keeping certain trace contaminants below prescribed levels of concentration, but analyses for support of operations activities can almost always be performed by state of the art methods.

Plant problem related analyses are the most demanding and costly, but also are the most urgently needed. They will be discussed in more detail than the others.

Discussion

Sources of Impurities

Let us discuss briefly how impurities enter the lead-acid system and which areas might be more sensitive to impurities. The lead received is either of secondary or primary source. The secondary lead derived primarily from salvaged batteries and lead battery parts can be expected to contain a wide and potentially harmful source of impurities due to the variety of lead alloys being employed by various battery manufacturers. While a measure of impurities is commonly taken as ppm of the total sample, more important is the surface area of the sample. For example, 10 ppm of a specific impurity may be quite acceptable for the alloy used to make grids for the lead electrode but may be totally unacceptable for lead-oxide used in the Pb and PbO_2 electrodes. It is the impurity level which is exposed to the solution-solid interface that affects the rate of H_2 evolution.

It is therefore unwise to specify a critical level of a specific impurity without knowing in which part of the system it will be encountered. We can also expect many impurities which enter the system via the electrolytic solution to resist identification because they will be deposited on the lead electrode leaving the solution concentration below routine detection limits. Furthermore the analysis of the solid lead active material may also resist identification since the impurities may only be present on the surface and therefore represent an undetectable quantity when results of the bulk sample analysis are expressed in the form of ppm.

Having reduced or eliminated the major sources of electrochemical contamination the next task is to determine the acceptable (yet realistic) impurity limits that should be specified for materials used in the battery.

A test procedure reported in 1974, by R. L. Bennett (3) for determining gas generation and therefore water loss was utilized to define the effect of various impurities at different concentrations on the maintenance-free aspect of batteries. The test calls for maintaining a fully charged cell at 2.35 $\pm$ .001 volts and 125°F (51.7°C) for a period of four hours while collecting and measuring the volume of all gasses generated during this period. The results of a study of the impact of twenty-four contaminant ions on gas generation of lead-acid cells are reported in 1975, by Pierson et al. in "Power Sources 5" - "Proceedings of the 9th International Power Sources Symposium" (1). Six of the elements investigated (antimony, arsenic, cobalt, manganese, nickel and tellurium) increased the gas generation rate when present at concentrations of ten parts per million or lower (see Table I).

Battery Material Specifications

Material specifications for the principal ingredients in a battery must be based on a combination of gas generation (or current acceptance) data and historical evidence concerning effects other than gassing.

Primary and secondary lead is purchased in various forms by battery manufacturers.

Oxide. Uncalcined or leady litharge has been utilized in the manufacture of battery plates for many years. This material, commonly called battery oxide, grey oxide or lead dust is generally made by one or two processes, the ball mill process or the Barton pot process. The "oxide" typically contains about 75 percent lead monoxide and 25 percent unoxidized metallic lead. Material specifications for oxide call out metallic lead content, apparent density, reactivity (acid absorption), crystal morphology and acceptable impurity levels.

All of these physical and chemical properties apply to standard and maintenance-free products but impurity levels become more stringent in maintenance-free batteries. Oxide becomes the porous positive and negative active material which participates in the charge-discharge reactions in the battery and therefore it must be assumed that any contaminant contained in

Table I. Maximum Allowable Contaminant Concentrations

Contaminant	Maximum Allowable Concentration in Electrolyte (P.P.M.)					
	5000 or saturation	500	160	3	1	.1
Aluminum		X				
Antimony					X	
Arsenic					X	
Barium	X					
Bismuth		X				
Cadmium	X					
Calcium	X					
Cerium		X				
Chlorine	X					
Chromium		X				
Cobalt					X	
Copper		X				
Iron			X			
Lithium	X					
Manganese				X		
Mercury	X					
Molybdenum		X				
Nickel					X	
Phosphorous	X					
Silver		X				
Tellurium						X
Tin	X					
Vanadium		X				
Zinc	X					

932

this material is totally leachable and capable of migrating to those sites where they exhibit their detrimental effects. Therefore those materials which have been identified as increasing gas generation must be controlled within prescribed limits for maintenance-free oxide and the lead from which the oxide is made. This kind of restriction can cause processing difficulties since antimony in low concentrations has been used as a spiking agent to increase the rate of oxidation in the Barton oxide manufacturing process.

Grid Alloy. The lead alloys which are to become the raw materials for grid production are usually procured in either pig or strip form. Pig lead is used to produce cast grids while strip material is fabricated into expanded or wrought grids. In either case specification of impurity levels is critical to the performance of maintenance-free products.

Composition of the positive grid is of particular importance since this grid is subjected to corrosive attack during its entire life thereby releasing any impurities contained in the alloy. The negative grid, on the other hand, does not undergo oxidative corrosion during normal service and so the release of contaminants from this source is less likely.

Expander. Additives blended into the negative active material to maintain porosity during service are typically a blend of carbon black, lignin and barium sulfate with leady oxide as a carrier. Some commercial expanders have incorporated current acceptance enhancers such as nickel into their formulation. Other mix formulations have called for addition of small quantities of tellurium for the same purpose. These two materials (nickel and tellurium) were found to have the most profound influence on gas generation of any of the twenty-four contaminants tested by Pierson et al. (1975) in the previously referenced experiments (1). Expanders formulated specifically for maintenance-free batteries will become prominent in the 1980's.

Analytical Service Requirements

Analyses of lead and its alloys and its oxides to serve the needs in the lead-acid battery industry require versatility in the application of analytical chemical principles and in the type of equipment used. Areas of application requiring chemical analyses may be grouped into three categories:

- Engineering and Research
- Material and Product Control
- Manufacturing Operations

The areas of application have differing needs with respect to:

- types of sample analyzed
- number of analyses made on a given sample
- elements for which analyses are made
- levels of concentration at which analyses are made
- precision and accuracy
- turn around time

Equipment and facilities required range from conventional analytical laboratory facilities through X-ray and emission spectroscopic instrumentation to clean room type facilities and sophisticated surface analysis and neutron activation and mass spectrometric apparatus.

Engineering and Research Applications. These applications are interesting because of their variety and because they often anticipate future analytical requirements for control and operations areas. They range from elemental analyses of alloys and trace element composition of pure lead and lead oxide to highly demanding and sophisticated ultratrace and surface analyses.

Elemental analyses include those similar to conventionally required material control analyses; they are performed by classical and X-ray flourescence methods, as reported in 1969, by Fillmore et al. (4). The surface analyses (to depths of 1 μM or less) require instrumentation, as described in 1979 by Evans (5), ranging from a probe attachment on a scanning electron microscope to an electron microprobe apparatus to electron and ion probe instruments such as ESCA, Auger, and SIMS.

An example of a current problem is the need to identify and quantify trace element impurities on the surface of some battery electrodes on which excessive gassing is observed during charging of the battery. Wang (1979) has found that certain elements can cause this reaction when present in quantities as low as 10^{-12} Geq/mM^2 (6). A typical offending element, Sb, with an equivalent mass of about 40, would need to be identified and measured at about 40 picograms per mM^2. If the surface could be stripped, a 100 mM^2 (10 x 10 mM) would yield 4 nanograms. For practical analysis, a mass of electrode would be taken; probably it would be 1 mM thick. At a density of about 10 G/cM^3, the mass of the sample would be about 10 G; the concentration of Sb would then be 4 nG/G, or 4 parts per billion. This is a challenging problem in chemical analysis. Analyses at this level require clean room type facilities, specialized equipment, and special skills in operation, as described in 1976, by Zief and Mitchell (7). Forsberg et al. (1975) and Hansen et al. (1978) show how electroanalytical methods offer advantages to this type of analysis (8, 9): 1) They can be used to strip certain elements from the electrode surface into a solution simulating the battery electrolyte, and they are sensitive to small concentrations of many of the resulting ions, 2) inversely, they can be used to deposit certain ions from very low concentrations onto an analytical electrode on which they can then be detected, identified, and measured.

The analytical service capabilities needed now to support research suggest the capabilities that will be required in the future to support manufacturing quality control and operations; in fact, there already is demand for some of them. However, to project what capabilities will be needed in the future to support research activities is at best speculative.

Material and Product Control Applications. Analytical service for material and product control is required to maintain lead alloy compositions within specified ranges and to maintain certain trace element contaminations below prescribed concentration levels in corroding lead and lead oxide. The service requirements are fulfilled satisfactorily by state of the art methods. Anticipated upgrading can probably be accomplished by using analytical quality evaluation procedures to use the present methods to their full capabilities. Additional upgrading may be realized as a fringe benefit of new capabilities needed for process troubleshooting and maintenance.

There are conventional ASTM standardized methods for control of lead alloys (10). Precision can be improved by use of systematically repeated analyses of reference materials plotted on control charts and compared between laboratories of vendors and users of the materials as discussed by Eckert and Mongan (1975), The ASTM (1963 and 1974), and Lewis (1974) (11, 12, 13, 14). The present, usually uncontrolled, analyses have precision that is consistent with accuracy obtained from standardizations made with laboratory reagent grade materials. Additional precision gained by systematic analytical quality evaluation procedures may justify the development of standard reference materials by the National Bureau of Standards.

The cost of analyses for routine control of trace level contaminating elements is not justified for routine use. However, the cost is moderate for analyses by direct reading spectroscopy if the cost of the equipment can be ignored. The need for such equipment for manufacturing operation trouble-shooting is discussed in the next section; if such equipment is available for that purpose, it will probably be used also for more frequent material control analyses.

Manufacturing Operations Applications. Analytical service support to manufacturing operations needs to have the capability--anticipated by the nature of needs for service to research--to find trace level impurity elements in battery electrolyte (4 M/L H_2SO_4) and electrode materials. Included are twenty to forty elements that are known or suspected to cause excessive gassing when a battery is being charged, or to accelerate the self discharge of a battery when it is not being charged and is not giving power. Concentrations of interest range from fractions of a per cent to fractions of a part per million. Problems could develop in manufacturing plants that would require immediate answers.

Of several possible approaches to this requirement, two are of particular interest: 1) Electroanalytical methods, and 2) Direct Reading Emission Spectroscopy. Mass spectroscopy is attractive because of its capability to measure small quantities of all elements. However, its high cost, limited sensitivity and moderate turn around time make it a poor third competitor. Other possible approaches have even more significant limitations. Neutron activation analysis and surface analysis methods (such as ESCA, Auger, SIMS) may be useful in some special instances, but for general application they are much less likely competitors.

Electroanalytical methods, (e.g., anodic stripping voltammetry), are especially interesting for this application because the samples are being analyzed to detect materials that promote reactions similar to those on which the analytical methods are based--electrochemical reactions that promote gassing and self discharge at the electrodes. Anodic stripping voltammetry can be associated with a process that concentrates the material for which the analysis is being made. Electroanalytical methods have the greatest detection sensitivities (lower detection limits) for ions in solution. These methods are rapid, and require only modest investment in instrumentation. Their principal limitations include specificity, the number and nature of ions for which an analysis can be made on a single sample, and the rigid limitations on the composition of the material being analyzed. They require more knowledge about the sample than can always be available for manufacturing plant samples, and more skill in manipulation than is often available in manufacturing plants. Electroanalytical methods seem to be most promising for special applications where the capabilities of other, more generally applicable methods (such as spectroscopy) are limited.

The most popular approach now to trace level contaminant analyses is by direct reading emission spectroscopy. One spectroscopic instrument manufacturer's user list includes eight instruments designed to analyze lead for trace element composition located in plants owned by two lead-acid battery manufacturers. The instrument manufacturer's sales literature lists lead analysis capabilities for sixteen trace elements from minimum concentrations of 1 to 10 ppm. Instruments of this type cost from K\$ 50 to 100, depending on the number of spectral lines included, the type of excitation equipment, and other options. The design of these instruments is derived from thirty years of use in the steel industry; they have been updated, and they are readily customized for other applications such as analyses of lead and lead alloys. Readout is simultaneous for all elements included; turn around time can easily be within $\frac{1}{4}$ to 1 hour when required; instrumental time can be less than 5 minutes per sample. The lower detection limits are adequate for many present requirements.

Principal limitations of direct reading emission spectroscopy, in addition to cost and flexibility once the instrument is built, are the 40 to 60 elements which can not be readily included (some not at all), and the lower detection limits (in some cases as high as 0.1%;--in some cases parts per billion or less may be required). Some of the more interesting additional elements can be included with newly available vacuum spectrometers. Parts per billion analyses will require completely new laboratory and instrumental facilities and skills. Relative merits of other approaches have been listed.

Direct reading emission spectroscopy is a late comer to the battery industry because the analytical needs for battery and battery material production have been modest, and the number of analyses required has been small compared with steel industry needs. The more stringent requirements are brought about by maintenance-free battery technology which can not tolerate more than minimal gassing during the charging of a battery in service. Even these requirements might not justify the cost of this approach just for routine quality control, but the urgent need for immediate results for many trace element compositions in a plant troubleshooting situation more readily justifies the investment in facilities; the incremental cost for using the instrument for routine work is then modest and more easily justified.

Projection. Required capabilities for analytical service in the battery industry are developing toward greater detection sensitivities for identification of increasingly lower concentrations of impurities in battery electrode materials and electrolyte. If the research based estimate that a few parts of one of the offending elements per billion parts of sample is confirmed, the only question will be cost justification. There is little or no basis of expecting that significantly higher precision and accuracy, faster turn around time, nor the capacity to perform analyses on greater numbers of samples will be required beyond what a straightforward sophistication of presently available methods and equipment can attain. The prospects are that the anticipated developments in analytical capability requirements toward identification of increasingly lower concentrations will probably be attained only by means of costly, laborious, slow, and skill limited procedures which have to be tailored to handle one or no more than a few elements at a time, and that have to be performed in expensive, well maintained clean room type facilities with research technician level skill.

Conclusions

1. Increase in the use of electroanalytical methods is likely, especially for support of research activities.

2. Increased use of surface analysis methods is highly probable, especially for support of research activities.

3. Investment in direct reading emission spectroscopic equipment seems to be inevitable for service to manufacturing operations.

4. Material and product quality control will probably need the benefit of upgraded precision and accuracy that can be attained by systematic programs of analytical quality evaluation.

5. Material and product quality control will probably benefit from the use of direct reading emission spectroscopy that will be justified by the need for its capability to support manufacturing operations.

6. The acceptable limits of various metallic impurities will depend on their location in the system (e.g., positive grid, negative grid, positive oxide, negative oxide, solution, separator).

7. A great deal of effort must be expended to develop engineering specifications from the test procedures which are presently under investigation.

References

(1) J. R. Pierson, C. E. Weinlein, and C. E. Wright, "Determination of Acceptable Contaminant Ion Concentration Levels in a Truly Maintenance-Free Lead-Acid Battery," pp. 97-108 in Power Sources 5, D. H. Collins, ed.; Academic Press, London, New York, 1975.

(2) Conrad E. Weinlein, John R. Pierson, and Dennis Marshall, "A New Grid Alloy for High Performance Maintenance-Free Lead-Acid Batteries," pp. 67-77 in Power Sources 7, J. Thompson, ed.; Academic Press, London, New York, 1979.

(3) R. L. Bennett, "Maintenance-Free Batteries: Definitions, Special Tests, and Advantages," paper presented at Maintenance-Free Battery Symposium: BCI Jubilee Convention, London, England, May 13-16, 1974.

(4) C. L. Fillmore, A. C. Eckert, Jr., and J. V. Scholle, "Determination of Antimony, Tin, and Arsenic in Antimonial Lead Alloys by X-Ray Flourescence," Applied Spectroscopy, 23 (1969), pp. 502-7.

(5) Charles A. Evans, Jr., "Surface and Thin Film Compositional Analysis: Description and Comparison of Techniques," Anal. Chem., 47 (1975) pp. 818A-20A, 22A, 24A, 26A-29A; "...: Instrumentation," ibid. pp 855A-856A, 858A-62A, 864A-66A.

(6) Private Communication, C. L. Wang, Globe-Union Inc., Milwaukee, WI, July, 1979.

(7) Morris Zief and James W. Mitchell, Contamination Control in Trace Element Analysis, Wiley-Interscience, New York, 1976.

(8) Gustaf Forsberg, Jerome W. O'Laughlin, and Robert G. Megargle, and S. R. Koirtyohann, "Determination of Arsenic by Anodic Stripping Voltammetry and Differential Phase Anodic Stripping Voltammetry," Anal. Chem. 47 (1975) pp. 1586-92.

(9) Wilford N. Hansen, C. L. Wang, and Thomas W. Humphreys, "A Study of Electrode Immersion and Emersion," J. Electroanyl. Chem. 93 (1978) pp. 87-98.

(10) See for example:
    a) ASTM Committee E-3 on Chemical Analysis of Metals, "Standard Methods for Chemical Analysis of Pig Lead," ASTM Designation E 37-76 (1976), American Society for Testing and Materials, Philadelphia, PA, 1976.
    b) ASTM Committee E-2 on Emission Spectroscopy, "Standard Method for Spectrochemical Analysis of Pig Lead by the Point to Plane Spark Technique," ASTM Designation E 117-64 (1976), ASTM, Philadelphia, PA, 1976;
    c) ASTM Committee E-3, "Standard Photometric Methods for Chemical Analysis of Lead, Tin, Antimony, and Their Alloys," ASTM Designation E 87-58 (1978), ASTM, Philadelphia, PA, 1978.

d) ASTM Committee E-3, "Standard Methods for Chemical Analysis of White Metal Bearing Alloys," ASTM Designation E 57-60 (1978), ASTM, Philadelphia, PA, 1978.

(11) Alfred C. Eckert, Jr., and Dennis M. Mongan, "Working Reference Materials for Lead Contamination Analyses of Air and Water," pp. 545-51 in Methods and Standards for Environmental Measurement, William H. Kirchoff, ed.; National Bureau of Standards Special Publication 464, Washington, D.C., 1977.

(12) ASTM Committee E-11 on Quality Control of Materials, ASTM Manual for Conducting an Interlaboratory Study of a Test Method, ASTM Special Technical Publication No. 335, American Society for Testing and Materials, Philadelphia, PA, 1963.

(13) ASTM Committee E-3 on Chemical Analysis of Metals, "Standard Recommended Practices for Conducting Interlaboratory Studies of Methods for Chemical Analysis of Metals," ASTM Designation E 173-68 (1974), American Society for Testing and Materials, Philadelphia, PA, 1974.

(14) Lynn L. Lewis, "Analytical Chemistry and Consumerism in the Automobile Industry," Anal. Chem., 46 (1974) pp. 866A-868A, 870A, 872A, 874A, 876A-877A, 879A.

SOLDERING, THE ESSENTIAL LINK IN ELECTRONICS

W. G. Bader
Bell Laboratories
Murray Hill, New Jersey   07974

Soldering has traditionally been the most widely practiced and the most economical method of providing permanent, reliable electrical connections. Soldering technology has progressed considerably in response to the needs of the rapidly advancing electronics industry. The role of soldering in modern electronics is discussed in this paper. The factors that must be considered to achieve good quality solder joints for electronic applications are examined. Developments in solders and soldering processes for the fabrication of electronic assemblies are reviewed.

Introduction

Soldering has been the most widely practiced method of providing permanent electrical connections in the electronic industry. Over the years considerable investigations have been conducted to develop alternate methods of joining metals to provide electrical connections. A few methods have been developed but their use has been limited to specific metals or particular joint member geometries, i.e., thermocompression bonding of gold plated surfaces and wire wrapped connections of copper wires to square terminal posts. Soldering, however, has continued to provide quality electrical connections between most metals in a multitude of designs, both economically and reproducibly.

Solders have been used for over two thousand years and perhaps because of this longevity the importance of soldering in modern electronics is occasionally taken for granted. The fact remains that soldering is essential in the production of the sophisticated assemblies that constitute modern electronics. Since the majority of connections between electronic devices and electrical wiring are accomplished by soldering, it can be considered to be the essential link in the electronic chain. It is this role of soldering in electronics that is reviewed in this paper.

Historical Perspective

Prior to the second World War, the majority of soldered electrical connections were made manually with hand held tools, primarily soldering irons with copper bits. The copper bits were either heated electrically or by a gas flame. Solder was supplied as flux cored wire or bar stock. Craftsman acquired considerable speed and skill in making soldered connections and as a result soldering was viewed as more of an art than as an engineering process. This view changed with the rapid advances in soldering technology that occurred in response to the needs of the electronic industry.

In the last four decades the electronic industry has experienced a continuous evolution in technology and a rapid expansion in the range and scope of applications. It is likely that this age of modern electronics started in the late 1940's. The invention of the transistor marked the beginning of solid state electronics and a multitude of new components, devices, and assemblies. The development of printed circuitry introduced a structure suitable for automatic assembly, mass soldering techniques, and miniaturization in electronic packaging. Finally, the introduction of commercial television provided an expanded market for electronic assemblies that encouraged the sophistication and specialization of electronic instrumentation and equipment. Obviously, enormous advances have been made in electronics over the last thirty years. The space program of the 1960's spurred many developments in electronic technology. Electronic devices have progressed from discrete transistors and diodes to hybrid integrated circuits and on to LSI technology where tiny chips contain the equivalent of 15,000 transistors. Computer circuitry and memory that once filled an entire room now performs more functions and is located on a chip smaller than a postage stamp. Computers are utilized in most all industrial operations. Indeed, electronic markets encompass nearly every segment of the economy including communications, manufacturing, space, medicine, entertainment, the home, and the office. During this period of rapid expansion in electronics, soldering technology has advanced in both innovative processes and scientific principles. The advancements in soldering can be observed by examining the component parts of the process used to obtain sound soldered joints.

Soldering Fundamentals

Soldering can be defined simply as the joining of two metal members with a relatively low melting alloy. A more exact definition is provided by the American Welding Society where soldering is "A group of welding processes which produce coalescence of materials by heating them to a suitable temperature and by using a filler metal having a liquidus not exceeding 450°C and below the solidus of the base materials. The filler metal is distributed between the closely fitted surfaces by capillary attraction" (1). The filler metals are usually called solders and the formation of a sound soldered joint requires that the molten solder:

1) wet the base metals
2) flow and spread to the joint entrance
3) fill the joint by capillary action

The general procedures followed in producing a soldered joint are:

1) designing the joint to provide clearances that permit capillary action by molten solder
2) preparing the joint surfaces by cleaning or precoating with a solderable metal or nonmetallic coating
3) applying a soldering flux to remove residual contaminate films and facilitate wetting
4) applying heat and solder to the joint surface. The application may be simultaneous or separate depending on the soldering process
5) cooling of the completed joint
6) removing the flux residues, if required, and cleaning the joint assembly
7) inspection of the completed joint

In addition to these procedures, a number of other factors must be considered if sound, reliable soldered joints are to be produced.

Metallurgical Factors

Wetting

Wetting of the solid metals by molten solder is essential in the soldering process to provide spreading and flow of solder to the joint area. Wetting of solid metal surfaces by liquid metals is a complex subject, however, in soldering, alloying of the solder with the base metals has a significant effect in the wetting process. In soldering there is some degree of mutual solubility between the solder and the base metals. Generally, an intermetallic compound is formed at the joint interface or the base metal takes solder into solution (2). In common tin lead solders, tin is the active constituent that readily alloys with most base metals and forms intermetallic compounds, i.e., Cu_6Sn_5 with copper and Ni_3Sn_4 with nickel Figure 1.
It is this alloying that provides rapid wetting in the soldering operation and promotes spreading of the molten solder. Indeed, with automated rapid soldering operations, the rate of wetting is quite important and test instruments are used to provide these data for various solder-base metal combinations.

The bond that is provided by alloying is limited only by the strength of the solder alloy. It should be recognized, however, that wetting and alloying are not necessary to provide a bond. Indium solders on glass or pure lead solidified on clean iron will provide bonds where wetting and alloying have not occurred. In many soldering operations, oxides and contaminent films interfere with the wetting and alloying reactions and

941

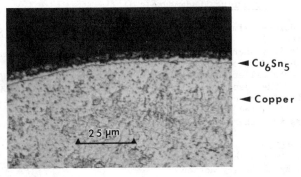

Fig. 1 Cu_6Sm_5 layer at solder to copper interface.

hence a flux is employed to remove these films and permit intimate contact between the molten solder and the base metals.

Dissolution

Dissolution of basis metals by molten solder can seriously affect the formation and integrity of soldered joints (3). Dissolution behavior in the soldering process has grown significantly more important in present day electronic assemblies. Miniature components and circuitry utilize thin film layers and coatings on non-metallic substrates. Dissolution effects can range from complete solution of metallic coatings with loss of adherence to the substrate, see Figure 2, to the reduction in thickness of metal coatings

Fig. 2 Thin film substrate after immersion in molten solder.

and impairment of soldered joint properties. Unfortunately, the metals that are easily wetted and readily alloy with molten solder are more likely to exhibit the highest dissolution rates. These effects are most pronounced with precious metals such as gold and silver that are frequently used as coatings in semiconductor device assembly. The kinetics of dissolution of solids in liquid metals are well defined (4,5). Investigations of the dissolution rate behavior of various metals in molten solders have provided valuable data useful in the selection of metal coatings, solder alloys, and the temperature and time of the soldering process (6).

Intermetallic Compound Formation

Intermetallic compounds are formed at the joint interface of most solder-base metal combinations and in a few cases, solid solutions as with tin lead solders on cadmium, lead, or zinc. In normal soldering operations, the compound layer is extremely thin and soldering time and temperature are kept to a minimum to prevent excessive growth. Although the compounds are by nature hard and brittle, in thin layers they have little if any effect on the properties of the soldered joint. A noteable exception is the rapid reaction of gold with tin in solders and the formation of $AuSn_4$. In instances where thick gold coatings have been soldered, fractures have occurred between the $AuSn_4$ compound and the residual gold coating. This problem is overcome by limiting the gold coating to ~ 1 µm to permit complete dissolution of the gold and wetting of the under layer by solder. The intermetallic dispersed in the solder does not impair the solder properties if the gold content is less than 4 w/o (7). Prolonged heating at soldering temperatures can cause excessive compound formation with other base metals. Additionally, exposure of soldered joints to elevated temperatures in service conditions will cause compound growth by solid state diffusion. Investigations of the growth rates of intermetallic compounds have provided data that permit the calculation of compound thickness for service conditions at elevated temperatures and the time for a coating to be consumed (8). These data are particularly useful with thin and thick film devices and components where degradation of the joint strength is encountered after elevated temperature exposure (9).

Solder Alloys

Tin-lead alloys are the most widely used solders for applications in electronic assemblies. Innovations in designs that utilize a variety of metals and metal coatings have necessitated the development of special solders for unique applications and a variety of service environments. However, the majority of soldered electrical connections, particularly with printed circuit assemblies, are made with eutectic or near eutectic tin-lead solders. The most useful properties of solder alloys are their low melting characteristics. The binary tin-lead solders furnish a variety of useful compositions with a solidus as low as 183°C and a liquidus as high as 327°C. This range of compositions permits the selection of solders with the most desirable properties for specific applications. Some solders used in electronic applications are shown in Table 1.

In general, solders do not provide high mechanical strength but do possess good ductility. The tensile strengths of bulk tin lead solders at ambient temperature range between 4,000 and 8,000 psi depending on tin content and the shear strengths are closely related to the tensile values. In many applications, the creep behavior of solders is the most important design criteria but most solders exhibit low creep strength. Values of stress to produce creep rates of 0.0001 mm/mm/day at ambient temperature for solders with 30 to 50 w/o tin are ~ 120 psi (1). Fatigue properties of solders and soldered joints are equally important in electronic designs. Present day designs incorporate a wide variety of materials with differing coefficients of thermal expansion. During assembly and service life, thermally induced stresses in soldered joints may cause failures due to low cycle fatigue Figure 3. Investigations of soldered joints after thermal cycling have provided an understanding of the failure mechanisms and supplied corrective measures (10,11).

Soldered joint strengths can be appreciably different from those of bulk solder because of the joint design and the soldering conditions. While

TABLE I - SOLDERS FOR ELECTRONICS

Composition					Solidus °C	Liquidus °C	General Comments
Sn	Pb	Sb	Ag	In			
62	38	--	--	--	183	183	Wave and dip soldering of printed
60	40	--	--	--	183	190	circuit assemblies
50	50	--	--	--	183	216	Hand soldering of wiring
30	70	--	--	--	183	255	Low tin content solder
5	95	--	--	--	270	312	Higher temperature solders
10	90	--	--	--	268	299	with improved ductility
1	97.5	--	1.5	--	309	309	
95	--	5	--	--	233	240	Intermediate solders with high
96.5	--	--	3.5	--	221	221	creep strength
62	36	--	2	--	179	189	For soldering to silver surfaces
48	--	--	--	52	118	118	Low melting solder
--	50	--	--	50	180	208	Reduced dissolution of gold

944

Fig. 3 Cracked soldered joints on printed circuit assembly.

most soldered joint members are designed to be mechanically secure prior to soldering, many current designs employ solder as the strength member. Recent investigations of the strength of solders and soldered joints have shown that mechanical properties are greatly dependent on the joint design and the rate of straining (12). These data are particularly relevant to the practical design of electronic assemblies.

The effects of impurities in solders on the properties of solders is not completely understood and this uncertainty is reflected by the variations in compositional requirements of solder specifications. Impurities are present in solders as supplied by the manufacturer and are acquired as the result of dissolution of the members that are soldered with wave and dip processes. The primary concern with impurities has been the effects on the wetting properties of the solders. These effects have been investigated and show that aluminum, phosphorus, and zinc have the greatest detrimental effect in trace quantities (13).

Fluxes

Soldering fluxes are solid or liquid materials used in the soldering process to facilitate wetting of the base metals by molten solder. They perform this function by removing oxides and other tarnish films from the joint members and by providing a cover to prevent oxidation of the joint area and solder during the heating and cooling period.

Traditionally, soldering fluxes have been functionally classified as either corrosive or noncorrosive depending on their chemical activity before soldering and the nature of their post-soldering residues. For some time rosin was the only noncorrosive soldering flux accepted for electrical and electronic applications where freedom from corrosion and high insulation resistance were of prime importance. Rosin is composed primarily of abietic acid and while active when molten is inert after it has cooled and solidified. The residues after soldering are noncorrosive, nonhygroscopic and nonconducting. However, rosin is a relatively mild flux and because of its weak fluxing action, a number of rosin base fluxes were developed that incorporated chemical agents to improve their fluxing action. This class of fluxes are designated as activated rosins and typically contain .5 to 1% halide activators. The majority of compositions are proprietary and the corrosivity of their residues has been the subject of considerable debate.

945

In the past fifteen years, water soluble fluxes have received a great deal of attention for electronic applications (14). The corrosivity and potential hazards of the flux residues are recognized, however, the advocates of water soluble fluxes contend that with proper design of assemblies and rigorous water cleaning of assemblies after soldering, product reliability can be maintained. The advantages that are claimed are superior fluxing action and the elimination of solvents used to remove rosin residues. It is likely that the acceptance and use of water soluble fluxes will be a controversial matter for some time.

Solderability

Solderability is not a property of a material but a term used to describe the ease with which a surface can be wetted by solder under conditions of a manufacturing process. The solderability of joint members is of prime importance if sound soldered joints are to be produced both rapidly and consistently, Figure 4. The development of solderability test methods

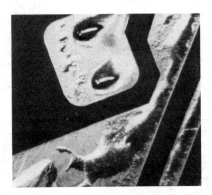

Fig. 4 Poor solderability of printed circuit board.

and the evaluation of coatings to provide solderable surfaces have been the subject of numerous investigations. Solderability test methods have been developed that measure the time to achieve wetting, the extent of solder spread, the degree of capillary penetration and the completeness of coverage of dipped specimens (15). The test methods are used to evaluate solders, fluxes, and base metal coatings and also as an inspection procedure of materials and components.

While many base metals can be satisfactorily soldered after proper cleaning, components and parts of electronic assemblies are usually protected with metallic coatings that are readily wetted by molten solder. The coatings are applied by electrodeposition, a hot dip process, evaporation or sputtering. The coatings most frequently used for good solderability are tin-lead solder, tin, gold, and silver. Other metals such as copper and nickel are used as undercoats when soldering to refactory metals or nonmetallics is required. In the soldering process the final coating may melt, dissolve rapidly or remain intact. Minimum coating thickness have been established to provide and maintain solderability after storage (16,17,18). Organic coatings, usually rosin lacquers, are employed as solderable coatings on printed wiring boards. The shelf life of these coatings is not as satisfactory as with metallic coatings.

Soldering Process

Soldering processes are often designated by the heating method used in the soldering operation. For some processes solder alloys are preplaced at the joint areas prior to the application of heat while in other methods the molten solder supplies the heat required to produce the soldered joint. The soldering process may be divided into two categories for electronic applications:

1) Those that produce a single soldered joint. The process may be manual or automated.
2) Those that produce multiple joints simultaneously. These processes are frequently automated and are described as mass soldering.

While there are a number of soldering processes, only those of major importance in soldering electronics will be described.

Hand Soldering

Soldering irons are probably the most familiar tools used to produce soldered electrical connections. Most soldering irons are electrically heated and a great number utilize copper bits protected with an iron coating to prevent erosion of the copper by molten solder. They are used in soldering operations of small assemblies and for touch-up and repair of printed circuits after automated mass soldering operations.

Reflow Soldering

Reflow soldering is a term used to describe soldering processes used to solder lead wires and tapes to printed circuit land areas. The land areas are usually coated with a sufficiently thick layer of solder to form the joint. Heat is produced by the electrical resistance of the lead by employing two electrodes, one on each side of the joint area. The process is similar to parallel gap welding operations. In some cases, heat is supplied by conduction by using a bar that is heated electrically or by infrared. Flat pack integrated cirucits and flat ribbon cables are frequently soldered with this process.

Dip Soldering

Dip soldering is a process whereby assemblies are fluxed and then dipped into a molten bath of solder. The immersion depth is sufficient to provide solder and heat to the joint areas. The process can be automated and modern systems use a conveyor system that moves printed wiring boards over fluxing stations and then over a solder bath so that the parts contact the molten solder surface. The units are call drag soldering systems and provide a mass soldering process for electronic assembly.

Wave Soldering

Wave soldering is accomplished by pumping molten solder from a reservoir through a shaped slot to produce a continuous wave. Printed circuit boards or other work pieces are moved by a conveyor system over a fluxing station and across the crest of the solder wave. It is the principal process used to produce soldered joints on printed circuit assemblies. In many soldering systems of this type, fluxing, soldering, and cleaning stations are incorporated into automated in-line machines.

Condensation Soldering

Condensation soldering is a new technique where the latent heat of vaporization of a saturated condensing vapor is used to provide the heat required for soldering. It is most useful for large assemblies that require a large number of soldered joints to be made simultaneously with uniform heating and cooling of the assembly (19). Printed circuit assemblies with more than 30,000 joints are soldered by this method. Solder alloy and flux are applied to the joint areas and the cold assembly is fixtured and loaded into the condensation chamber where the soldered joints are accomplished. This method has also been useful in soldering operations of thin film hybrid integrated circuit assemblies.

Other Considerations

A number of other factors must be considered if soldered electronic assemblies are to function reliably under the intended service conditions. These include post solder cleaning of flux residues and adequate inspection of completed assemblies. With the continuing trend toward miniaturization, both of these factors have become more difficult. However, with careful design for soldering, proper selection of materials and good process control, the attainment of sound, reliable soldered connections can be assured with minimal inspection.

Summary

Soldering is an essential segment of the electronics industry. Solder provides the connections that link electronic devices and circuitry into a multitude of products and equipment. The technology of soldering has advanced at a rapid pace with innovative processes that produce more than 30,000 joints simultaneously on a single printed circuit board. It is likely that soldering will continue to be the primary method of providing electrical connections.

REFERENCES

1. Soldering Manual, 2nd ed., p. 1; American Welding Society, Inc., Miami. FL., 1977.

2. J. F. Lancaster, The Metallurgy of Welding, Brazing and Soldering, p. 128; American Elgevier Publishing Company Inc., New York, NY., 1965.

3. W. G. Bader, "The Dissolution and Formation of Intermetallics in the Soldering Process," paper presented at TMS-AIME Fall Meeting, St. Louis, MO, October, 1978.

4. E. A. Moelwyn-Hughes, The Kinetics of Reactions in Solution, 2nd ed., p. 366, Clarendon Press, Oxford, 1947.

5. J. M. Lommel and B. Chalmers, "The Isothermal Transfer from Solid to Liquid in Metal Systems," Trans. of the Metallurgical Society AIME, 215, 1959, p. 499-508.

6. W. G. Bader, "The Dissolution of Au, Ag, Pd, Pt, Cu, and Ni in a Molten Tin-Lead Solder," Welding Journal 48 (12), 1969, p. 551s-557s.

7. C. J. Thwaites, "Some Aspects of Soldering Gold Surfaces," Electroplating and Metal Finishing, 1973, Aug. p. 10-14, September, p. 21-26.

8. P. J. Kay and C. A. MacKay, "The Growth of Intermetallic Compounds on Common Basis Metals with Tin and Tin-Lead Alloys," Trans. of the Institute of Metal Finishing, Vol. 54, 1976, p. 68-74.

9. W. A. Crossland and L. Hailes, "Thick Film Conductor Adhesion Reliability," Solid State Technology, Feb. 1971.

10. Roger N. Wild, "Fatigue Properties of Soldered Joints," Welding Journal, 51, 11, 1972, p. 521s-526s.

11. E. R. Bangs and R. E. Beal, "Effect of Low Frequency Thermal Cycling on the Crack Susceptibility of Soldered Joints," Welding Journal, 54, 10, 1975, p. 377s-383s.

12. C. J. Thwaites and W. B. Hampshire, "Mechanical Strength of Selected Soldered Joints and Bulk Solder Alloys," Welding Journal, 55, 10, 1976, p. 323s-327s.

13. M. L. Ackroyd, C. A. MacKay, C. J. Thwaites, "Effect of Certain Impurity Elements on the Wetting Properties of 60 tin-40 lead Solders," Metals Technology, February, 1975, p. 73-85.

14. Alvin F. Schneider, "Water-Soluble Organic Fluxes for the Electronics Industry," Insulation/Circuits, February, 1979, p. 31-34.

15. J. B. Long, "A Critical Review of Solderability Testing," Journal of the Electrochemical Society, 122, No. 2, 1975, p. 25c-32c.

16. C. J. Thwaites, "The Solderability of Some Tin, Tin Alloy, and other Metallic Coatings," Trans. of Institute of Metal Finishing, 36, 1959, p. 203.

17. W. G. Bader and R. G. Baker, "Solderability of Electrodeposited Solder and Tin Coatings After Extended Storage," Plating, 60, No. 3, March, 1973, p. 242-246.

18. M. L. Ackroyd, "A Survey of Accelerated Aging Techniques for Solderable Substrates," Tin Research Publication 531, 1977.

19. W. K. Comella, "Quality Assurance for ESS: Advanced Measures for Advanced Technology," Bell Laboratories Record, Vol. 56, 11, December, 1978, p. 289-295.

ROTARY KILN SMELTING OF SECONDARY LEAD

Richard C. Egan, M. Vikram Rao & Karl D. Libsch

NL Industries, Inc.
Hightstown, N. J.

This paper describes the layout and operation of the Pedricktown Plant of NL Industries, Inc., located at Pedricktown, New Jersey, approximately 20 miles from Philadelphia. A description of the complete operation will be given commencing with the decasing facilities, describing the kiln smelting operation and finishing with the refining and casting. The emphasis, however, will be placed upon the kiln smelting which represents a significant departure from conventional smelting, which is predominantly composed of blast furnaces, reverberatory furnaces and short rotary furnaces. The long rotary furnace has some similarities with the short rotary in its chemistry, but in all other respects it is a substantially different technology from any other secondary lead process.

The kiln smelting described in this paper represents a modification of the process employed by Preussag AG, Goslar, Germany, from whom the basic technology was licensed by NL. Some of the major comparisons of this technology with conventional smelting are discussed at the end of the paper.

Battery Decasing

The Pedricktown decasing plant separates the received batteries into acid, grid, large case and separator, middlings fractions and battery paste. Most of the battery acid drains when the battery is received. The battery is then shredded and screened at 5/16". The +5/16 fraction contains grid and large case and separators. These are separated by a magnetite sink float process. The -5/16 contains some non-paste materials which are removed and report to the middlings fraction. The remaining paste slurry is thickened, filtered and dried to make the battery paste product. Figure 1 shows the flow of battery material.

Incoming battery trucks report to the truck dump. There the cab and the trailer are lifted to 60° from the horizontal and the batteries spill out to a receiving area. While the truck is being lowered it is washed with spray water to remove any battery acid. Figure 2 shows the truck dump. Batteries on skids and loosely packed on flat bed trucks cannot be unloaded by the truck dump. These are removed by pushing with a payloader or by fork lift at a separate receiving dock. The sides of the loosely packed flat beds cannot withstand the rush of batteries on dumping and break out creating damage to the vehicle and danger from falling batteries.

The drop during dumping fractures most of the battery cases and a large amount of the acid drains immediately. This flows through a swale into a receiving acid sump and is pumped to storage tanks. The acid is disposed of by waste disposal firms who pump from these tanks to their trucks for transport to their treatment facilities.

An International 515 payloader is used to remove the batteries from the receiving area into one of five receiving bins. These are constructed primarily of concrete with acid brick in areas where most of the drainage occurs. We have observed that acid attack on concrete is not severe until puddling begins.

Batteries to be decased are dumped into the shredder receiving hopper (Figure 3) by a second IH 515 payloader. The shredder is a Saturn Model No. 52-32, 100-HP with MR 375 low speed high torque motor and specially constructed acid resistant alloy cutter teeth. The two opposing rows of teeth rotate at approximately 30 RPM. The blades are driven by a hydraulic system which automatically reverses direction when torque increases above a certain level. This feature allows the hopper to be emptied even when pieces of tramp iron fall into it. The shredder shreds between 1200-1600 batteries per hour depending on the condition of the teeth. We have found it necessary to rebuild them about every six weeks. If this is not done the battery rate drops sharply. Shredder product is nominally minus 2-inch with some plates quite a bit larger in two dimensions.

A 30 inch wide conveyor belt feeds the shredded product through a chute to a 6 feet diameter trommel (Figure 4). This unit is driven by a 25 HP electric motor and has a solid receiving section 4 feet long and a 12-1/2 feet long screening section. The receiving section of the trommel acts as a gentle grinding area in which the material grinding against itself tends to free the metallic grid of any paste adhering to it. The holes in the screen are 5/16 inch in diameter. Water is fed into the receiving section through the chute at a controlled rate and also is sprayed internally on the screen and externally on top of the screen. The trommel rotates at 10.7 RPM and is tilted downward at 7° to the horizontal. In our operating experience we have never had the trommel blind with separators.

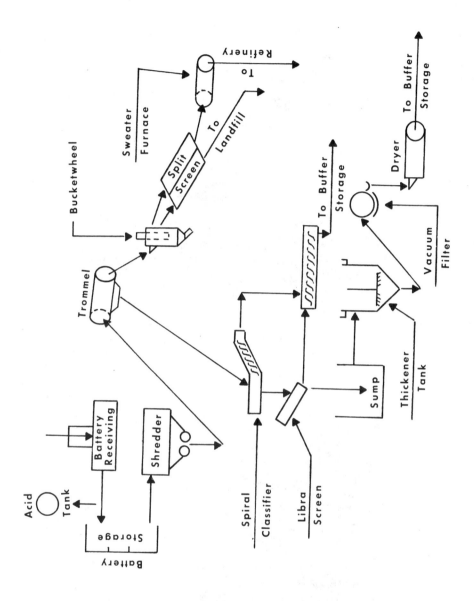

FIG. 1

PEDRICKTOWN
DECASING
FLOW SHEET

Figure 2 - Truck Dump

Figure 3 - Shredder Receiving Hopper

956

Figure 4 - Trommel

The 5/16" overs fall through a chute into a magnetite heavy media separator. The media is kept between 1.6 and 1.8 specific gravity. The unit is a Humboldt bucketwheel separator and has approximately 1.25 square meters of separating surface. The case and separators float on top of the bath over a discharge chute onto a split screen. The lead grid sinks to the bottom of the bath, is picked up by a revolving bucketwheel, brought to the top and discharged down a chute to the opposite side of the split screen. The separation is complicated by the fact that often times small 1"x1/16" pieces of grid sink past the bucketwheel. For this reason, the unit has been modified so that media and these fines discharge continuously below the wheel itself.

The magnetite used in the sink float is reclaimed and reused. This reclaiming process starts on the split screen. Its first portion is not sprayed. Here much of the heavy media drips off the separated products and drains immediately by gravity into a receiving cone. The heavy media pump then pumps it back into the separator. The heavy media pump is a Wemco Model 6X4WD with 40 HP and is driven at 714 RPM. The farther end of the split screen is washed with sprays which remove the rest of the magnetite from the separated products. They flow by gravity through the screen into the light media receiving cone and into the light media pump. It transfers the light media to a magnetic separator which reclaims the magnetite from the water. The discharge from the magnetic separator drops by gravity into the heavy media cone.

The heavy media sinks fall from the split screen through a chute onto a conveyor belt which delivers them to the sweater furnace. This is a rotating tube furnace 4 feet in diameter and 20 feet long, which is fired from the discharge end by one burner which also has the ability to cool by blowing air through the nozzle without flame. Grid metal melts in the furnace. Off-gases are treated in a baghouse. About 85% of the feed reports as lead and 15% as dross. The lead buttons are removed for use directly in the refining area. The dross is transported to the kiln area for smelting. We have found that the amount of plastic in the grid portion influences the operation of the sweater greatly. Excessive plastic requires the unit to be cooled.

The washed case and separator material is delivered by conveyor to a storage pile. We are currently selling this material to a firm who is reclaiming its plastic content and extruding this to make plastic product for sale and manufacture of specialty items.

All the -5/16" which goes through the trommel screen is eventually smelted. The equipment used subsequently for processing removes the water used in screening in the trommel screen.

The first step in water removal is to remove the unpumpable material to form a middlings fraction. This is done using the spiral classifier followed by a flat screen. The spiral classifier removes that material which sinks so quickly that it cannot be pumped. The water content is approximately 15% on discharge. The overflow contains the battery paste and -5/16" overs case and separators. The latter are removed on a flat screen with approximately 44 square feet of surface area. Both the spiral classifier and flat screen discharges go to a long screw which delivers the middlings to the buffer storage room for kiln smelting.

The screen unders are battery paste slurry with between 10-15% solids. These fall by gravity to a 50 HP pump which transfers them to a thickener. The thickener is 30 feet in diameter by 9-1/2 feet deep and has an automatic torque lifting sensing device on the rake.

The thickened battery paste flows out of the bottom of the thickener through one of two 5 HP Moyno pumps which are made using a chrome-plated tool steel type rotor, natural rubber type stator, and pump the battery paste through a 1-1/2" line to a rotary drum filter. The slurry is usually between 60-70% solids once it leaves the thickener. The line to the filter contains a nuclear density gauge which continually measures the density of the slurry. The filter is an EIMCO filter 10 x 12 feet in size, rests in a tub, and is fitted with an agitator to keep the slurry in suspension. It is operated without an overflow and material is fed to it intermittently to keep the level in the filter within about a foot of the top of the tub. The filter is run between 19-22" mercury vacuum. The vacuum system consists of a filtrate receiver connected through a barometric seal to an output sump and a 60 HP Siemens vacuum pump. The filter drum typically has a cycle time of 3 minutes and the cake is released by using a scraper blade and a small amount of snap blow air. The cake is usually released at about 10-15% moisture. It is usually about 1/4" to 1/2" thick.

The filter cake falls into a receiving screw, and is conveyed to a screw which feeds it to a 4 feet diameter by 24 feet long rotary drum drier. The rotary drum drier is fitted with a scraper to prevent accumulation of material inside the drum. The dried material contains 6-8% moisture and consists of nodularized material (pellets) usually between 1/2" and 1/4" in diameter and containing 15-20% fines. The discharge is conveyed by a ribbon screw conveyor to the buffer storage building. The drier off-gasses are treated in a Ducon scrubber to remove particulates.

The thickener serves as the plant's water reservoir. The water for the sprays for the trommel screen, the spiral classifier, the libra screen and the split screen is supplied by the thickener overflow. It is then pumped to the sprays or if there is excess water to a holding inventory tank. The plant has numerous valves for floor wash down and cleanup all of which are done with water from the thickener overflow. All washup water eventually ends in the bottom floor of the building and drains from there into a sump pump which pumps it over the magnetic separator and then into the thickener.

The plant is operated 3 shifts a week, 5 days a week on a 24-hour basis. The department employs a superintendent, three foremen and six hourly personnel per shift. One of these is used to load the batteries into the shredder; a second to monitor the equipment before the heavy media section; the third to monitor the operation of the bucketwheel separator; a fourth to monitor and operate the thickener, filter and drier; a fifth as a utility man to provide for relief breaks and lunches; and the sixth to provide for absentee replacement and cleanup.

The most significant costs in operating the plant are depreciation, maintenance, and the cost of disposing of waste acid and decasing effluent. The need for maintenance is continual. This facility uses many pieces of equipment in one line and without much surge. We have been able to obtain 70% operating factor during the last year and believe it will be possible to improve this to 80-85%. That improvement depends on careful preventative maintenance. We pay very careful attention to the amount of gland water used on our pumps as we find that attention pays off in greatly reduced effluent haulage cost.

Kiln

The Pedricktown rotary kiln is a 177 foot long, 10 foot diameter used cement kiln converted to lead smelting specifications (Figure 5). The furnace is inclined slightly to the horizontal and contains 392 tons of

Figure 5 - Kiln

refractory made up of alumina brick in the feed and preheat zones, burned magnesite-chrome brick in the reaction zone, and the entire length is backed up by 3 inches of insulating firebrick. The discharge end of the kiln is narrowed for 6 feet 3 inches with the basic brick to form a dam (corbel) to allow for the collection of metal. Bricked working diameter of the furnace is 8 feet at the feed end and 4 feet at the discharge end.

The kiln is fired by a 35 million BTU oil burner positioned at the discharge end and angled forward enough so that the flame strikes the edge of the bath just behind the corbel (Figure 6).

The rotary kiln sits on three large concrete piers, and each pier has two trunnion rollers upon which the kiln tires rest. The roller bearings are lubricated through a recirculating oil system, which has provisions for maintaining the oil temperature at 100-120°F. The kiln is rotated by means of a large ring (bull) gear attached to and encircling the outside of the kiln. This gear is driven by a pinion drive gear, which is in turn driven through a water-cooled magnetic coupling clutch by a fixed speed A.C. electric motor. Rotational speed variation through a range of 0.3 to 1.1 revolutions per minute is obtained by varying the current to the magnetic clutch. An auxiliary gasoline engine is provided as a backup source in case of electrical failure.

Lead-bearing feed material is stored in an enclosed bin that is kept under negative pressure. A bridge crane with a 1-1/2 yard capacity bucket operating within the building delivers the feed to a 15-ton capacity hopper. The hopper is emptied by the action of an apron feeder pulling material away from the opening at the hopper bottom. The apron feeder conveys the material to the kiln feed conveyor, a 150 foot long completely enclosed belt conveyor set at a 17° angle. At the top of the conveyor, the material is discharged into the kiln via an angled feed chute or a feed screw, depending on the physical nature of the material being used.

Petroleum coke and soda ash are stored in enclosed bins having a capacity of 260 and 409 tons respectively. Each of the bins is equipped with individual piping and baghouse systems for blowing coke and soda ash from pressurized delivery vehicles into the storage bins. Both bins have vibrating bottoms that assist the movement of the bin contents into screw conveyors which are operated by variable speed D.C. motors. The variable speed screws convey to a common flux feed screw, which in turn discharges onto the aforementioned kiln feed conveyor. Cast iron chips are fed from a 10 ton V-bottomed hopper and, as they settle, are pulled from the bottom of the hopper by an apron feeder and deposited into the flux feed screw. The cast iron chips are mixed with the petroleum coke and soda ash in the screw and are discharged onto the kiln feed conveyor.

The rotary kiln metallurgical gas stream is cleaned by an 18 cell Norblo baghouse powered by a 150 HP fan capable of discharging 38,000 CFM. Each cell contains 78 acrylic fiber bags having a permeability of 40 ft^3/ft^2/min. The kiln is drafted from the charging end through a large gravity settling chamber known as the "fuchs". The dust that drops out in the fuchs funnels into a screw conveyor, which carries it back to the kiln feed conveyor by way of the recovery loop. The recovery loop is comprised of a horizontal screw conveyor, a rotary valve, an inclined screw conveyor, and a bucket elevator. The gas stream from the fuchs is first pulled through a vee duct cooling tube and then through the balloon flue (Figure 7), expanded duct with a screw conveyor mounted on the bottom, prior to entering the baghouse.

ROTARY KILN-FLOW DIAGRAM

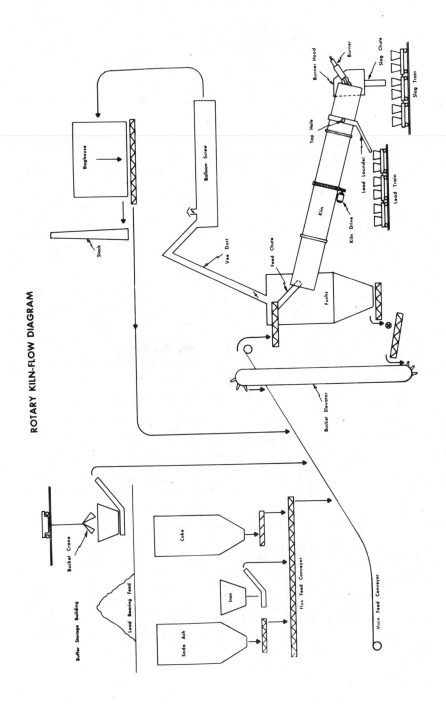

Figure 6 - Kiln Burner

Figure 7 - Vee Duct
Balloon Flue

Mounted on the balloon flue is a bleed air damper, its function to allow outside cool air to flow in to maintain the proper baghouse inlet temperature. The damper is operated by a Leeds and Northrup controller and is capable of operating in either the manual or automatic mode. Negative pressure in the kiln is controlled by a damper mounted on the outlet side of the fuchs at the base of the vee duct.

A 24-cell Norblo baghouse handles the sanitary gas cleaning requirements for the kiln and the refining department. Sanitary hoods are provided over the lead launders at the discharge end of the kiln (firing hood), over the slag and lead trains, and over the lead tapping section. The dust that is collected from the kiln operation, both from the metallurgical and sanitary systems, is returned to the kiln feed conveyor through a screw conveyor system.

The slag, which forms on the surface of the metal, overflows the corbel continuously through a chute and is collected in slag pots. Two slag pots sit atop each car of a five car cable-drawn train. Lead is tapped periodically by stopping the kiln, breaking open the tap hole, then rotating the kiln until the tap hole points downward. Lead is collected by means of refractory lined launder chute which directs the flow into botton molds placed on a cable-drawn lead train.

All controls and instrumentation that support the rotary kiln operation are housed in one of two air-conditioned, dust-free control rooms. The feed control room (Figure 8) is located within the enclosed negative pressure bin and contains the crane controls, the start/stop controls and indicator lights for all feed and recovery system screws and belt conveyors, the rheostats that control the variable feed screws and conveyors and the controls for the pneumatic coke and soda ash handling system. The kiln control room, located in the tapping area, houses all kiln and baghouse temperature indicators and recorders, start/stop controls and indicator lights for kiln systems and draft control instrumentation (Figure 9).

Start Up

The time requirement for kiln start up varies depending on the condition of the refractory. In the case of a newly rebricked kiln, the fuchs and firing hood are dried out for a two-day period by placing salamanders inside them. At the end of this period, the main burner is turned on and the furnace temperature is raised 100°F per hour to 1200°F. Once the temperature has reached 1200°F, it is held for an additional 12 hours to permit the brickwork to get uniformly hot out to the steel shell. At the end of this "soak" period, the kiln temperature is again raised 100°F per hour until the temperature reaches 2400°F. During this last 12-hour period, the kiln is rotated one-quarter turn every 15 minutes. At the end of this heatup period, feeding begins at approximately 35% of capacity for 12 hours, after which it is increased to 65% for 12 hours and thereafter to designed capacity. Start up with used refractory involves heating the kiln without rotation to process conditions over an 8-10 hour period. When the shell temperature reaches 250°F at the tap hold, the kiln is rotated 180°F every half hour for approximately 4 hours. At this point, feeding begins and proceeds as described above.

Technical Training

The training of personnel for both the kiln and decasing operation was accomplished by the central technical training department via a Criterion

Figure 8 - Feed Control Room

Figure 9 - Kiln and Baghouse
Temperature Indicators & Recorders

Referenced Instruction Program. CRI training is a course of instruction in which the subject matter is taught a small portion at a time with the requirement that the trainee demonstrate his understanding of that one small portion before proceeding to the next higher learning step. The program proved to be very effective; operators were well versed with the equipment, contingency procedures and operating perimeters, and above all no accidents were suffered during the three-month start up period.

Manpower Requirements

The kiln is operated on a three shift per day, seven day per week schedule. A tapper, charger, slag handler, lead train operator, baghouseman and relief operator per shift comprise the hourly work.

A smelting manager is responsible for the overall operation and has four shift foremen to supervise the furnace.

Maintenance

Ninety percent of all kiln maintenance and repair is performed by the plant maintenance department. One maintenance mechanic per shift is assigned to the kiln, each performing a total of 91 equipment checks and lubrications. Rebricking and the large fabrication and/or repair jobs that are conducted during scheduled shutdowns are contracted out.

Since the kiln start up in April of 1978, there have been four shutdowns to make repairs to or replace refractory. Refractory experiences are outlined in Table 1.

Operation

As is the case with all furnace operations, there are certain important aspects which must be monitored and controlled via instrumentation and operator observation. With the kiln operation, this can be broken down into four major headings: the feed system, burner system, off gas system, and the kiln. Each major heading is then sub-divided as follows:

The feed system, as previously mentioned, is made up of two primary feed systems to include lead bearing feed and flux feed. Both systems, as will be explained further later, are set by a potentiometer and re-calibrated at each feed rate change. Once calibrated, the controls are locked at the panel board and a reading is taken once per hour to insure accuracy.

The burner system consists mainly of oil pressure valves and air rate control dampers, which are set for maximum burner efficiency, and monitored via gauges at each critical point. In addition, an oil rate meter is utilized to monitor fuel consumption and aid us in maintaining an energy efficient system.

The kiln off gas system is equipped with draft control, temperature monitoring and environmental controls. To enable us to monitor temperatures, we have installed a 12 point recorder which records fuchs temperature, top and bottom baghouse temperatures and stack temperature. Individual temperature meters have also been installed to monitor the baghouse inlet and burning zone temperature. To cover draft requirements and environmental needs, we have incorporated a recording meter and photohelixes for draft monitoring, a combustion meter for gases, and an opacity meter for monitoring

Table 1

Kiln Refractory History

	Rebricking Period	Short Tons Produced During Campaign	Cause of Failure	Scope of Repairs
Initial Lining	Installed 8/77	--	--	--
First Rebricking	8/22-9/5/78 (15 days)	7,898	Normal deterioration of refractory. Shell temp. in reaction zone rose to critical levels.	Replaced entire corbel plus 14' of reaction zone. Due to severe spalling, portion of alumina brick beyond reaction zone was replaced with magnesite chrome. 13' of alumina brick in feed zone was replaced due to crushing & radial shifting.
Second Rebricking	10/27-11/2/78 (7 days)	5,141	Corbel deterioration with top dam bricks broken away and lead penetrating through lower courses.	Replace corbel.
Third Rebricking	4/7-4/22/79 (15½ days)	19,241	Normal deterioration of refractory.	Replaced entire corbel & reaction zone. Improvements made include: (1) Added insulating brick under steel retaining rings for greater insulation in critical areas. (2) Tighter refractory placement with reduced longitudinal expansion spacings. (3) Tighter refractory joints with mortar fill in critical and vulnerable locations. (4) Straightened & aligned nose castings for greater support.
Fourth Rebricking	5/6-5/9/79 (4 days)	937	Excessive temperatures created refractory deterioration in top courses of corbel bricks. Problem could not be remedied by spot repair.	Replaced corbel with even tighter expansion joints.
	Through 7/22/79	12,117		

967

stack emissions.

The kiln itself is equipped with rotational speed control and a monitoring meter, a drift monitor alarm, and cooling water and lube oil temperature meters for monitoring the bearings.

Several other critical functions not covered by instrumentation as such are safeguarded by alarm systems. Typical examples of these alarms are: vee duct high temperature, baghouse inlet high temperature, flame failure, clutch cooling water low pressure, kiln drive motor stopped, etc.

As with any new furnace, unexpected operational problems develop that require perseverance on the part of the furnace staff and hourly personnel to analyze and overcome procedurally. A discussion of some of these difficulties follows.

The original feed system at start up, for the rotary kiln was a gravimetric system using weighbelts combined with various types of transition conveyors. This system was, and still is, considered the optimum system for charging the kiln. However, early in the start up, it became apparent that the system, as designed, was too sophisticated for reliable day-to-day operation in a smelter environment. It wasn't long before the weighbelts were taken out and the feed system was converted to a simple volumetric basis.

Operating a volumetric feed system required that an accurate means of calibrating feed rate be developed. Our present system utilizes a variable potentiometer which controls the speed of each individual feed conveyor. This potentiometer is calibrated by the simple expedient of running the system at a particular setting for a set period of time with the feed diverted to a drum. The feed thus collected is weighed and the rate is calculated. This is done a number of times at different settings and the results are plotted. Using the reference calibration curves thus obtained, it is a simple matter to change charge composition when required.

During the 16 months of kiln operation, the composition of the charge to the kiln has varied considerably. Some of this was expected, and planned campaigns to run particular lead-bearing feeds have been scheduled. However, some of it is due to the natural variation that occurs in scrap lead feed. The kiln is a dynamic smelting system and, while not as sensitive as a blast furnace, changes in charge composition must sometimes be made.

Care must be taken however to distinguish between those conditions which are transient in nature and do not require charge adjustments and those which do. Adjustments to the ratio of fluxes are most common, particularly soda ash and coke. These adjustments are based on several factors: slag chemical analysis, slag pool appearance and general kiln parameters. Adjustments in the rate of feed is usually dependent on the type of scrap being fed to the kiln.

The role of the burner in kiln smelting is only now beginning to be fully appreciated. The burner's main role is to initiate the smelting reactions at start up through the ignition of the coke in the charge. Once these reactions are underway, the oxidation of the coke and the exothermic heat from the lead oxide reduction supplies the bulk of the necessary heat to the kiln. The burner thereafter is mainly used for slag control. Minor adjustments are made in the burner to keep the slag within a specific viscosity range. These adjustments can involve changes in flame characteristics and/or angular position.

Slag viscosity control is important for the clean separation of lead and slag. A viscous slag can carry a significant amount of lead prill out of the kiln. It also causes a transport problem - building up on the corbel face and clogging the slag chute.

Too thin a slag can signal the beginning of problems within the kiln. The flux ratio may be incorrect or the smelting zone may be moving. The smelting zone can and does shift within the kiln. Factors such as feed rate, coke level, draft control and feed type can cause these shifts. Movement of the smelting zone can be followed by monitoring the temperature readouts at each end of the kiln.

The kiln draft is controlled by the reciprocal adjustment of the fuchs damper located in the vee duct and the bleed in air damper on the balloon screw. Automatic controllers are provided for these two dampers. However, their operation must be periodically monitored to see that they stay within the effective controller range. Since there are almost an infinite combination of settings for these two dampers which can yield the same draft and temperature conditions within the kiln system, they have a tendency to drift until they are out of control. The control of these two functions is critical for maximum kiln production.

As previously mentioned, the rotary kiln at Pedricktown is a used concrete kiln. It is mounted, out in the open, on three concrete support piers. Because of the combined weight of the refractory and the lead charge (about five times denser than concrete) and the marginal support, extra care must be taken to keep the kiln from warping and/or drifting. The extreme load also makes lubrication of the trunions and bearings extremely critical.

As a result, once the kiln is fired up and up to operational heat, it is never fully stopped for any length of time (about 15 minutes maximum). An idling procedure has been instituted wherein the kiln is slowly rotated or given a periodic quarter turn during any temporary "hold" condition. Should the hold be prolonged, the kiln is emptied of slag and lead and gradually cooled down while being rotated.

Start up of the kiln proceeds differently depending on whether or not the start up follows a rebricking. Following a rebricking, the kiln is heated up very slowly to drive off the moisture on the refractory. As the bricks expand and lock in place, the kiln is rotated slowly and the heating is continued until it is up to near operation temperature. When this point is reached, feeding is commenced.

The original lead tapping procedure called for a receiving kettle mainstay with a manually rotated launder chute filling button molds as a backup. This concept was abandoned prior to start up due to the anticipated difficulty of pumping "dirty" bullion out of the receiving kettle.

A lead carrousel was designed and used successfully for some 10 months. However, mechanical difficulties became an increasing problem. Currently a lead train, virtually identical to the one used for slag, has replaced the carrousel and is operating satisfactorily.

Charge

Daily kiln tonnage fluctuates significantly as a function of charge. Three basic charges are utilized, depending on antimony requirements in the generated metal: pellets/middlings, plant scrap, and occasionally 3:1 grid

metal/pellets. The pellets/middlings charge has produced a daily high of
225 tons but normally averages 200 tons per day with bullion at a >1.25%
antimony level. A high of 288 tons has been realized on plant scrap, the
average daily tonnage being 225 with an average antimony content of 6%.
The 3:1 grid metal/pellets charge averages 180 tons per day, the high
tonnage being 200 and the average antimony content runs 3%.

Laboratories

The chemical laboratories at Pedricktown are located in the main office
building, while the assay room is located in a wing off the refining building. The lab is equipped with a Baird-Atomic 1 meter, Spectromet 1000,
direct reading spectrometer, a Varian Techtron Model AA6 atomic absorption
spectrophotometer with a M-80 gas box, a Leco Model 532-000 sulfur analyzer,
a Leco Model 761-100 carbon analyzer and the conventional wet chemical
facilities.

The laboratory's major emphasis is on kettle process control samples
for which the spectrometer and atomic absorption spectrophotometer are most
highly utilized. Incoming drosses, batteries and metal scrap are assayed
as required. Daily samples of kiln slag and periodic samples of feed materials and fluxes provide control for the kiln operation.

Kiln Chemistry and Reactions

During the design and construction of the Pedricktown kiln, a computer
model of the process was constructed in order to better define the process
parameters. We will not here go into any details regarding the model, since
this will follow as the subject of a separate paper. However, in order to
get up the main feature of the model, namely a mass and heat balance
simulation of the kiln, we had to arrive at an understanding of the kiln
reactions as functions of distance from the feed end. This understanding
is described below.

The feed to the kiln comprises the lead-bearing portion of the charge,
composed of the output from the decasing operation, combined with other
lead scrap, finely divided petroleum coke, case iron chips, and soda ash,
to name the principal constituents. The amounts of coke, iron and soda are
adjusted to suit slag chemistry and also temperature. Equations 1 through
8 describe the basic smelting reaction.

$$3\,C + 2\,PbSO_4 + \frac{Na_2CO_3}{Flux} \longrightarrow 2\,Pb^0 + Na_2S_2O_3 + 4\,CO_2 \qquad (1)$$

$$5\,C + 2\,PbSO_4 + \frac{2\,Na_2CO_3}{Flux} \longrightarrow 2\,Pb^0 + 2\,Na_2S + 7\,CO_2 \qquad (2)$$

$$PbSO_4 + 2\,C \longrightarrow PbS + 2\,CO_2 \qquad (3)$$

$$C + 2\,PbS + \frac{2\,Na_2CO_3}{Flux} \longrightarrow 2\,Pb^0 + 2\,Na_2S + 3\,CO_2 \qquad (4)$$

$$PbS + \frac{Fe^0}{Flux} \longrightarrow Pb^0 + FeS \qquad (5)$$

$$PbO_2 + C \longrightarrow Pb^0 + CO_2 \qquad (6)$$

$$2\ PbO + C \longrightarrow 2\ Pb^0 + CO_2 \qquad (7)$$

$$\frac{Na_2CO_3}{Flux} \longrightarrow Na_2O + CO_2 \qquad (8)$$

It can be seen that the coke acts primarily to reduce the lead oxides to metallic lead and to reduce the sulfate to sulfide. Additionally, coke also undergoes the following reaction:

$$C + O_2 \longrightarrow CO_2 \qquad (9)$$

and we have found in practice that this reaction plays an important part even though heat is provided at the front end of the kiln with a burner. The burning of the coke is experienced over the entire length of the kiln even up to the point where the coke may be seen burning on the slag layer as slag pours over the corbel. The reactivity of the coke is an important factor, and we have found petroleum coke to be most suitable. The sodium carbonate reduces the lead sulfide to lead, and the other reaction product varies from Na_2S all the way to Na_2SO_4 with several different intermediate compounds such as sodium thiosulfate also being found. Since the predominant species are sodium thiosulfate and Na_2S, the proportions of these two constituents are used as indicators of the state of oxidation of the slag. The small amounts of sulfur in the coke and other incidental sources also react with sodium carbonate in the same way. An important function of the soda is also to reduce the levels of sulfur oxides in the off gases. This "gettering" action causes the sulfur dioxide in the off gases to be well below 200 parts per million. Some unreacted soda also escapes with the flue dust and is returned to the kiln when the dust is recycled. The airborne sodium carbonate also exerts "gettering" action upon some chlorides that happen to be present in the air stream. These chlorides are generally in the form of HCl evolved from the burning of PVC separators.

The action of the cast iron is to act synergistically with the soda to reduce the lead sulfides to lead. The proportions of iron and soda are adjusted so that the compositions of the slag are such as to be in the low melting portions of the eutectic valleys.

The heat for the kiln reaction comes from both the coke and the fuel oil being burned. In the early portions of the kiln where the materials are solid, heat transfer from the hot gases to the solids is affected by the rate of rotation, since increasing the rate increases the turning over of the surface solids, thus transferring the heat to the layers below. At the tapping end of the kiln, the temperature is controlled primarily by adjusting the burner. Although from the standpoint of entrained metallic lead it is preferable to have slag as fluid as possible, an intermediate viscosity is preferred in order to minimize the refractory wear.

The majority of the slag is composed of the sulfides and oxides of sodium, iron and lead, together with silica, alumina and other oxidic constituents present in the feed materials. Also present to levels up to 20% is unburned coke. Of the compounds mentioned above, the sulfides predominate; although in the case of sodium, there also exist thiosulfates. The

971

presence of the high concentration of sulfides allows this material to be
alternately called a matte rather than a slag. The oxides in sulfides will
not separate out into two different layers as they do in the blast furnace.
The nonmetallic lead present in the slag is extremely insoluble in water
because the predominant specie is PbS. The slag is hygroscopic and this has
to be taken into account when running slag analyses. Furthermore, due to
the presence of the entrained carbon, this slag must be handled and disposed of using special techniques, which are simple but absolutely necessary.
A discussion of the methods employed to properly handle slag is contained
in a patent that is expected to be issued shortly.

Refining and Casting

The Pedricktown refinery was constructed in 1972 and was not altered
during the kiln construction and decaser revamp in 1977-78. The refinery
consists of twelve 90-ton kettles arranged in two groups of six with an
isleway and 100 mold Wirtz casting machine separating the two. Plans
originally called for reserving one group of six kettles for soft lead
production and the other group for antimonial lead production. Due to
higher demand for soft lead and soft lead based grid alloys, nine kettles
are now designated for "soft lead use only". Soft lead is manufactured
from low antimony kiln metal by oxidation with air, intermetallic impurity
removal and caustic-sodium-nitrate treatment. Antimonial lead is produced
from 2-1/2 to 3% kiln metal or sweater metal.

The refining kettles are constructed from ASTM-A285C firebox steel,
1-1/2 inches thick. They have straight sides, 38 inches length and an
elliptical bottom. All kettles are stress-relieved at 1100°F by the
manufacturer. Kettle settings consist of a steel shell lined with 3 inches
of insulating castable refractory and 13-1/2 inches of firebrick.

The floor of each setting is made up of a layer of 3 inches castable
refractory, a course of 4-1/2 inch brick and a layer of sand. Fumes and/or
unburned combustion gases in the setting are carried off by a ventilating
flue system.

Each kettle is equipped with a North American Mfg. Co., Model No. 5795-
54-8A nozzle-mix type oil burner rated at 10 million BTU/hour and a full
flame safety system.

Drossing is conducted using an automatic drossing machine which
operates under ventilated kettle hoods.

All mixers are mounted on kettle hoods and are of right angled drive
design, driven by belts and powered by 30 HP motors. Three bladed impellers
are used exclusively.

Lead pumps are comprised of two types: transfer-pumps which are used
to move molten lead from kettle to kettle and casting-pumps. All pumps are
of centrifugal design and are driven by fixed AC electric motors. The
casting pump that feeds the Wirtz casting machine is made to operate as a
variable speed pump by varying the frequency at which the motor operates.
This is accomplished by plugging the pumps into a "black box" speed control
panel and dialing the speed required.

Discussion

Comparison of Rotary Kiln Smelting with Blast Furnace

We will discuss here some of the major differences between these two types of smelting. Although both methods employ continuous smelting, in the kiln process the slag is tapped continuously while the metal is tapped periodically, whereas in the blast furnace the situation is exactly reversed. Major differences, too, lie in the chemistry and the ease of operation. The rotary kiln is, without question, a simpler furnace to operate in that individuals can be trained expeditiously; and there is required a minimum of experiential knowledge of the type required in the blast furnace. This is partly because there is a higher degree of automation in this furnace and partly because the kiln is a more stable furnace in that departures from normalcy do not generally have catastrophic effects.

In a blast furnace, furnace shutdown, or lost time burn down, can be caused by the following situations: sow formation due to a high iron condition, excessive unsmelted material due to low iron or coke, improper charge distribution, hands and accretions in the shaft, extended elevation of the smelting zone and high back pressure. The last item can have a variety of causes, including presence of too many fines on the charge. In contrast, the kiln operates efficiently with fines; and an extended shutdown can only be caused by a few events.

In recent times, the extensive use of PVC separators has affected blast furnace operations adversely because the volatile lead and iron chlorides form a reflux loop in the furnace resulting in operating difficulties and also lead lost in slag (or lead chloride). In the kiln, the chlorides easily react with the soda and the resulting NaCl product is not detrimental as a slag constituent.

Overall manpower requirements are lower on the kiln when compared on a tonnage basis. Also the tasks are simpler and less intensive in that a majority of the operations are on a timed schedule, whereas in a blast furnace the slag tapping is often required to be on the basis of visual observations (requiring experience), and on the same furnace tapping intervals can vary from 10 minutes to 30 minutes depending upon furnace conditions.

Acknowledgements

The authors wish to thank NL Industries, Inc. and Preussag AG for permission to publish this paper. Gratefully acknowledged is also the assistance provided by John Cestaro, Joe Reichard, and Rosemary Bauer.

FIVE YEARS'UTILIZATION OF THE SHORT ROTARY FURNACE IN

THE SECONDARY SMELTING OF LEAD*

Jacques Godfroi

Penarroya, Trappes, FRANCE

INTRODUCTION

The aim of this paper is to show how, following changes in the socio-economic context, Penarroya radically modified its secondary smelting technology.

The methods employed and the results obtained after five years of operation during which many improvements were implemented, are discussed.

* This work was supported by Penarroya which is part of the IMETAL Group of companies.

HISTORY

In order to better appreciate the reasons behind the choice of the short rotary furnace for secondary lead smelting, it is first necessary to briefly review the history of our Secondary Smelting Division.

This dates back to 1924 when under the name of Métaux & Alliages Blancs, two plants began operation in France, one located at Saint-Denis near Paris, and the other at Lyon. These plants produced lead, copper and aluminium from scrap. The reduction step was performed in blast furnaces which permitted not only to treat battery plates, but also scraps with little contained metal or containing other metals. For rich material such as type metal dross, aluminium chips or copper and bronze scrap, rotary furnaces were used.

In addition, a reverberatory furnace was employed at Saint-Denis to produce soft lead from battery plates, while various furnaces were used for refining at Lyon.

Subsequently, in 1965, the purchase of a secondary smelter at Escaudoeuvres in the North of France, which treated only battery plates and lead dross, gave us the opportunity to gain further experience in operating the rotary furnaces of the time. These units were small capacity, long body furnaces, with the gas discharging at the opposite end from the burner.

By 1969, in an expanding market, it became more and more difficult to provide a sufficient quantity of battery plates which at that time came from whole batteries, broken manually by our suppliers or wrecked by a partially mechanized process (guillotine shear) in our plant.

For this reason, it was necessary to develop our operations along two separate lines. The result was a completely automatic battery wrecking system of industrial size (20t/h) and also a process for treating whole batteries in our blast furnaces.

With these furnaces, using oxygen enrichment, we achieved good productivity and specific fuel consumption ; for example, at Saint-Denis, an average daily lead production of 60 t for a furnace area at tuyere level of 1.60 square meters with a whole battery charge of 60 %.

Coke consumption was reduced from 170 to 100 kg and the lead loss dropped from 21 to 15 kg per ton of lead bullion produced.

However, we encountered gas treatment problems, as the blast furnaces were equipped with wet scrubbers which produced offgases containing a lead concentration of more than 200 mg/Nm3.

Although our Saint-Denis and Lyon plants were initially located in industrial zones, by 1970 these had become increasingly incorporated in an urban environment subjected to greater control by government regulations.

At this stage, our experience in secondary smelting included the use of small rotary furnaces, reverberatory furnaces and blast furnaces, with and without whole batteries. A mechanized battery wrecker had also become operational.

The next step was therefore to choose the metallurgical process for our future development.

REASONS FOR THE CHOICE OF THE SHORT ROTARY FURNACE

The basic factors governing our choice were as follows :

1. Due to the closedown of more and more coal mines in France, coke had become expensive. It was therefore risky to become too tied to this type of fuel outside the steelmaking regions.

2. The enormous rise in antimony prices in late 1969 had considerably modified the Sb, Sn, and As contents of the alloys used for batteries.

   It was thus important to have a very flexible process capable of satisfying a large range of requirements.

3. In the aftermath of 1968, continuous operation became less acceptable, especially near large cities where the sources of used batteries, the battery plants and our secondary smelters are located in France.

4. The only type of filter which could meet the government standards for stack emissions (< 30 mg/Nm3) was the bag filter.

Based on the above combination of factors, Penarroya opted for the rotary furnace, because :

   a) it is not dependent on any particular type of fuel, as the burner can be fired with oil, gas, coal or any pulverized fuel.

   b) its batch operation allows to change the charge composition at will, to meet the bullion analysis requirements.

   c) its thermal inertia, rapid fill and emptying capabilities, make it possible to shut the furnace down on weekends with minimum loss of productivity.

Another factor supporting this choice was the development of a mechanized battery wrecker which allows to eliminate the chlorine contained in the PVC separators before the reduction operation.

BATTERY WRECKAGE

The primary aim of the wrecking operation is to get rid of as much of the waste material contained in the batteries as possible, that is to say, the electrolyte, casings and separators, without any human intervention except for loading, monitoring and cleaning.

Because of the changing trends in battery alloys, we also decided to separate the antimony metal from the grids and the soft lead oxides and sulfates contained in the battery paste.

Our wrecker was designed to handle 20 t/h of batteries, whatever the type of casing, with consistent results in terms of product quality and metal recovery.

The batteries are first crushed and then sent to a screening installation. The material obtained is separated according to particle size as follows :

. The fine fraction coming mainly from the battery paste is transported by conveyor to the furnace charge preparation section.

. The coarse particles are either sent through a second crusher to reduce the material to < 60 mm, or through a screen producing the same result.

The product obtained, i.e. a mixture of battery casings, separators and metal grids, is then treated in hydraulic separators in which the lighter materials (waste) float, are recovered and then disposed of, while the heavy materials sink to the bottom of the tank.

The distribution of the products obtained in the battery wreckage operation is as follows :

	% of the weight obtained	% of contained metals
a) Fines...........................	64 %	70 %
b) Lead-bearing mud.................	5 %	3 %
c) Metallics.......................	17 %	26,5 %
d) Waste material (casings)........	14 %	0,5 %

It can be seen that the metal recovery of the operation is 99.5 %.

a) The fines have the following composition :

	H_2O...................	7 to 11 %
on dry product	Pb...................	74 to 80
	Sb...................	1 to 2
	S...................	4 to 5
	C...................	3 to 4
	SiO_2................	0.5 to 1
	Cl_2................	0.01 to 0.10

The last three elements come from a fraction of the waste material crushed in the primary crusher and their analyses depend on the type of crusher used. It can be seen that the plastic separators are fairly well eliminated which will avoid a concentration of chlorides in the flue dust.

These fines are combined with by-products of the metallurgical process (flue dust and lead bullion) to produce a lead bullion containing 1.7 to 1.9 % Sb, which corresponds to the composition requirements of the new battery alloys. This is also a very suitable base metal from which soft lead can be obtained economically.

b) The lead-bearing mud is composed of paste entrained with the coarse particles after the separation of the fines. It is recovered from the water of the hydraulic separators or from the final washing of the waste material.

The mud is allowed to settle to the bottom of the clarifier and the clear water is recycled back to the hydraulic separation process. The lead-bearing mud is then removed and mixed with the fines to be used as furnace feed.

c) The metallics contain less than 5 % of casing material and are treated directly in the furnaces or remelted continuously in kettles.

d) The waste material consists primarily of separators and cases in which the lead contained (approx. 2 %) is mainly due to impregnation. This should disappear, however, with the generalized use of polypropylene.

The polypropylene is presently disposed of as there is not enough employed in France today to justify its recovery. However, we are studying this possibility along with the use of this material as fuel for firing the rotary furnace burners.

The material is separated in upward circulating water and because of the wide range of specific gravities (0.8 to 1.5 for waste material, 4 to 6 for metallics) the process is not influenced by the type of casing or the amount of water.

The fact that heavy media are not used for separation is an advantage.

Wrecker capacity   :   20 t/h of batteries ($\pm$ 2)

Installed power   :   270 kW

Power consumption   :   11 to 15 kWh/t of batteries

Screen heating   :   900 thermies/h

Manpower   :   3 to 4 persons/shift.

ROTARY FURNACE REDUCTION

Although the decision to use rotary furnaces for reduction was taken in 1970, this did not become a reality until 1972 and 1974.

We chose the large capacity furnace since it permits to mechanize the feed and tapping operations. In addition, the looping of the flame which exits just above the burner, provides a greater heat efficiency.

The first furnace of this type was installed in 1972 in our Escaudoeuvres smelter to replace the two small existing furnaces. Its external dimensions are : $\emptyset$ 3.5 m, length 4 m.

In 1974, a second unit, 3.6 m in diameter and 4 meters in length, was added.

The same year, the blast furnace at Saint-Denis was shut down. The Lyon plant was phased out and replaced by the new Villefranche plant which is equipped with two short rotary furnaces, 3.6 meters in diameter and 5 meters in length or approximately 8 m3 of working capacity each.

As the Saint-Denis plant no longer has a reduction furnace, it uses the metallics from the new battery wrecker and other antimonial scrap to produce alloys containing < 4 % Sb. The fines from its wrecker operation are sent to the two other secondary smelters for use in the production of low antimony alloys.

FURNACE CHARGE PREPARATION

The chemical reactions occurring in the rotary furnace are well known. The aim is to reduce the oxide, remove the sulfur and dispose of the waste material in the form of a slag which melts at a relatively low temperature. This is a compromise between the specific consumption of the fuel used, the duration of the batch operation, the life of the refractories and the recoverable metal contained in the slag.

This temperature is approximately 950°C and we use the $SiO_2-FeO-Na_2O$ ternary system to determine the composition of the slag and the reagent requirements.

The reagents consist of iron chips or millscale and sodium carbonate for sulfur removal, anthrasite fines as reductants and eventually sand if silica addition is necessary.

The reagents and lead-bearing materials are stored in bins which all discharge onto a central weigh belt.

From the control room, the various charge constituents can be added alternately, but the rotating speed of the furnace during charge is too slow to provide an adequate mixing.

Furthermore, the presence of dry pulverulent material in the charge causes an intense emission of dust in the furnace. This delays the use of the burner at full capacity to prevent the production of dust outside the furnace. During this period, the heat exchange between the flame and the charge is also considerably reduced.

All of this lengthens the cycle time, decreases furnace productivity and increases the specific consumption of fuel.

For this reason, we have introduced charge pre-mixers. The first ones were installed at Escaudoeuvres in 1977, in the form of a vertical rotary drum with 10 t working capacity.

The use of this device has improved productivity by about 11 % and decreased fuel oil consumption by 15 %.

With the main aim of protecting the in-plant environment, at Villefranche we have recently installed a pelletizing system for fine material, i.e. battery fines, skimmings and drosses, flue dust, lead-bearing mud, fluxes.

The granulated form of the mixture obtained has permitted to automate the storage and furnace charging operations and realize further productivity gains.

First used in connection with the recovered baghouse dust, pelletizing has cut flue dust production in half :

Without pelletizing	Flue dust : 7 to 8 % of the Pb bullion produced
With pelletizing	Flue dust : 3 to 4 % of the Pb bullion produced.

FURNACE PRODUCTIVITY

Our major concern during the past five years of operation of these new furnaces has been to improve productivity continuously. This has been made necessary by the rising costs of oil, reagents and labor.

Various measures have been taken to achieve this objective, including :

- premixing of the charge
- use of more efficient burners than those originally installed
- mechanization and automation of the drilling of metal and slag tapping holes
- treatment of flue dust by pelletizing.

We have also devoted considerable effort to improving metallurgical reactions and kinetics, which has resulted in an optimization of furnace operation, charge volume, flux quantities and the length of the cycles.

	Production growth
1975 :	35 t/day/furnace
1976 :	43 t/day/furnace
1977 :	50 t/day/furnace
1978 :	55 t/day/furnace
1979 (first half)	59 t/day/furnace

METAL BALANCE

The large productivity increase realized has, however, since 1978 been accompanied by a greater loss of lead in slag, which rose from 13 to 23 kg per ton of bullion produced.

In contrast, the antimony balance has improved, since a loss of only 1.1 % is presently observed.

After the design and installation of the necessary mechanical and heating equipment, we are currently making metallurgical investigations to decrease the lead losses economically while maintaining the production levels achieved.

The sulfur balance shows that more than 80 % of this element is fixed in the slag, 7 % in the lead bullion drosses, 11 % in the dust and less than 2 % escapes in the furnace offgases. After dilution with ventilation gases, an SO_2 concentration of 130 mg/m3 is found in the stack discharge.

HEAT BALANCE

The heat balance is as follows :

Heat from fuel oil...... 52 %	Water vapor................ 7 %
Heat from carbon combustion.............. 44 %	Decomposition of carbonates 4 %
Oxidation of the iron... 4 %	Reduction and desulfurizing 41 %
	Sensible heat of gases..... 36 %
	Sensible heat of metal and slag..................... 5 %
	Heat losses............... 3 %
	Errors.................... 4 %
100 %	100 %

The proportion of heat lost in the furnaces offgases has evolved as follows :

1975	61 %
1976	51 %
1978	36 %

This is due to the improvement in the heat transfer as a result of premixing of the charge, in burner efficiency and in furnace operation.

A 25 % savings on fuel oil consumption per ton of bullion produced has resulted.

CONSUMPTION

Per ton of lead bullion produced :

Fuel oil........................	92 kg
CO_3Na_2........................	62 kg
Iron...........................	50 kg
Coal...........................	65 kg
Labor..........................	1 hour 12 min. including supervision and maintenance
Electricity....................	45 kWh for Reduction
	105 kWh for Filtration

GAS FILTRATION

The furnace offgases are mixed with the ventilation gases from the charging and tapping areas and sent to the baghouse.

Each furnace is individually equipped with a bag filter cleaned by counter-current air and by hitting the bags.

The sizing of the baghouse was based on a ratio of 100 square meters of filtering area/cubic meter of gas to be treated per second, to have a filtering speed of less than 1 centimeter/second through the cloth. This allows to maintain the full filtering efficiency even during maintenance when 1 cell out of 4 is down.

All measurements of dust emission whether made by ourselves, or by official laboratories, give less than 20 mg/Nm3 of dust, or less than 1.50 kg/h per furnace. The lead concentration in the dust is slightly more than half.

For Villefranche, the standards specify $<$ 30 mg/Nm3.

We have been using the same bags for five years. They are washed and repaired every two years.

The dust collected has the following analysis :

Pb..........................	60 %
Sb..........................	0.7
Sn..........................	0.2
S...........................	6
C...........................	4
Cl..........................	4
Na..........................	5

As a result of pelletizing, dust production has declined from 8 % to 3 % of the lead bullion produced.

AUTOMATION

With the development and use of automatic kettle skimmers, we were able to install a dust-tight collecting system for all plant fine materials including lead bullion dross, refining by-products, flue dust and battery fines. This system discharges to the charge preparation area where the fine material is granulated. Granulator feed is controlled automatically by computer.

This means that the following operations are now fully automated : transport of material starting with battery wreckage, furnace charge preparation, furnace charging, metal and slag tapping, transfer of the bullion in liquid form to the refining area, and skimming the kettles. As most refining operations are performed by air or gas injection, operators are only needed for triggering of the desired sequences, lead pumping and control.

Although the primary purpose of automating, which has continuously developed at Villefranche since its opening in late 1974, was to improve working conditions and remove workers from exposure to dust emissions, productivity gains have also resulted by decreasing idle time between cycles.

For the plant as a whole, productivity has evolved as follows :

Saleable ton produced per day	1978 = 1975 x 1.42
Hours per ton	1978 = 1975 x 0.69

IN-PLANT ATMOSPHERE

All measurements performed inside the plant give lead airborne concentrations of less than 150 μg/m3.

Furnace.......................... 34 to 135 μg/m3

Refining......................... 56 to 97 μg/m3

Tapping.......................... 32 to 104 μg/m3

ENVIRONMENT

Stack discharge

. Dust

The particulates discharging to the atmosphere are in the range of 3 to 15 mg/Nm3 depending on the cycle phase, with a weighted average of 10 mg/Nm3, or approximately 0.75 kg of Pb per hour for an annual production of 30,000 tons with the two Villefranche furnaces.

. SO_2

The SO_2 concentration in the offgases discharging to the atmosphere is 130 mg/m3.

. Environmental fallout

There are five permanent stations installed at distances of 100 to 800 m from the stack. Dust value is on the order of 1 mg/m2/24 h. The lead concentration in the ambiant air at these same points is $<$ 1 μg/m3.

Water

All industrial and rain water is collected and treated in an effluent treatment plant with a capacity of 50 m3/h. The clean water obtained is recycled.

CONCLUSION

The review we have just made of Penarroya's 55 years of experience in producing lead from secondary sources, including five with short rotary furnaces in our new Villefranche plant, has led us to conclude that flexibility must be the main characteristic of the secondary smelting industry.

This notion must be foremost in our choice of technology and equipment layout so as to successfully cope with economic and social change.

The combination of several factors including the separation of battery constituents in our battery wrecker, the flexibility of rotary furnace operation and our selective refining capabilities have allowed us to adapt to changes in the supply and demand situation which have been particularly marked in the present decade.

The mechanization of materials handling is an essential component for depollution of the in-plant atmosphere. It also contributes to productivity gains.

In order to guarantee the in-plant airborne lead concentration of $< 100 \mu g/m3$ prescribed by EEC regulations, we need to make some further improvements in our raw materials storage.

Finally, the considerable part of the capital investment we have devoted to pollution control should be noted :

        in 1974   : 34 % of the initial investment

        in 1979   : 44 %, with additional environmental investment.

The power consumed for environmental protection represents 62 % of the total plant power consumption.

It is therefore obvious how important it is to improve productivity constantly in order to limit these costs.

REVERBERATORY FURNACE-BLAST FURNACE SMELTING OF BATTERY SCRAP AT RSR

R. David Prengaman
General Manager Research and Development
RSR Corporation
1111 W. Mockingbird Lane
Dallas, Texas 75247

RSR operates 5 lead recycling plants in the U.S.A. Each smelter is designed to produce a maximum amount of fully refined pure lead from recycled
lead acid batteries. The reverberatory furnace-blast furnace combination
approach to smelting battery scrap allows for the production of maximum
amounts of pure lead.

In the RSR process, the batteries are crushed, drained of acid, the
lead values separated from the case materials and separators, and the plastic cases recycled. The lead values from the battery wrecker are smelted
in a reverberatory furnace where the alloying elements are oxidized to the
slag, sulphur is eliminated, and bullion containing very low levels of impurities is recovered for refining into pure lead.

Through controlled additions of reducing agents the lead is reduced
from the slag, producing a high antimony slag and very low antimony bullion.
The high antimony slag is reduced in a blast furnace under conditions to
promote maximum recovery of the antimony arsenic, and tin in a high antimony bullion. The high antimony bullion can be used to produce any
antimonial lead alloys desired.

Introduction

RSR operates 5 lead recycling plants throughout the United States. These plants are located in Middletown, New York; Indianapolis, Indiana; Seattle, Washington; Los Angeles, California; and Dallas, Texas.

The raw material feed to the RSR lead recycling plants is almost wholly automotive batteries. The plants handle only small amounts of battery plant scrap, drosses, and residues from customers' plants.

Historically lead recyclers operated blast furnaces and produced antimonial lead for resale to battery customers. Small amounts of pure lead were produced, mainly by kettle softening. Higher antimony alloys than that produced from blast furnace bullion were produced by adding antimony ore or high antimony softening skims to the blast furnace.

In the early 1970's with the advent of the maintenance-free battery, RSR changed from emphasis on production of antimonial lead to maximum pure lead production. While maximum pure lead production was aimed for, the recycling plants would have to remain flexible so that production could be switched between pure lead and antimonial lead alloys as the markets dictated.

At the same time, RSR developed special refinery techniques for pure lead which permitted RSR to offer high purity recycled refined pure lead comparable to that produced by primary lead smelters. Because of the extensive refining, RSR pure lead has been used for the most demanding requirements such as the TEL and oxide for maintenance-free batteries.

To maximize pure lead production, RSR optimized the reverberatory furnace-blast furnace combination for smelting of battery scrap. This paper deals with the complete recycling process at RSR from receipt of whole automotive batteries through the smelting and refining processes. RSR recovers and recycles as much of the battery as possible including lead and battery case. Figure 1 is a flow sheet describing the operation of the RSR lead battery recycling process.

Battery Receiving

Most batteries are received at the RSR recycling plants in trucks. A small amount are received by rail. In order to unload the trucks as rapidly as possible and with minimal labor, RSR devised a truck dumper for unloading batteries. AS seen in Figure 2, batteries are unloaded at the Dallas plant using the truck dumper. The truck dumper at Dallas has proved so successful that additional truck dumpers at the other plant locations are planned.

In the truck dumping process, the tractor, trailer, and entire load of scrap batteries are secured to the truck dumper. The dumper is elevated to a 30-45° angle, and the entire load of batteries is discharged from the truck into a receiving area. After the batteries are discharged, the dumper is lowered to the initial position and the truck is permitted to drive off the dumper. The entire unloading procedure requires about 5 minuts per truck. Trucks are thus not delayed for long periods of time waiting to be manually unloaded.

Figure 1. RSR Reverberatory Furnace-Blast Furnace Smelting Process

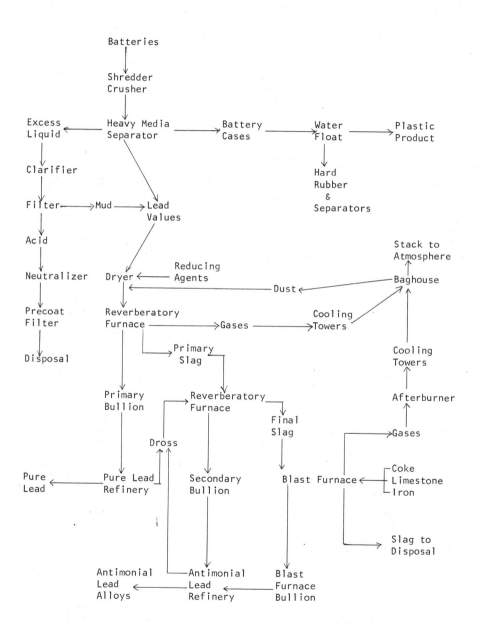

In the bulk unloading of the batteries, some small amount of acid may be spilled during the dumping process. To prevent damage to the trailers, the trucks then proceed to a wash area where the trailers are thoroughly washed to remove all traces of acid.

The battery receiving and storage area is located beneath the truck dumper. The receiving area is 30 meters X 20 meters and can store 1000 tonnes of batteries. The receiving area is paved with acid-proof concrete. The concrete is sloped to allow acid from batteries cracked or broken during the dumping to flow to a sump where it is pumped to the acid handling system. In past years when batteries contained hard rubber cases, many of the batteries were broken and much acid was liberated during the dumping operation. With the introduction of the polypropylene-cased battery, very few batteries are broken during dumping.

Battery Wrecking

Historically battery wrecking has been done manually with axes or by using saws or guillotines to cut open the batteries. The opened batteries were dumped by hand. This required a great deal of labor and exposed the workers to acid mist. During the 1960's, several mechanical methods were employed to wreck batteries and recover the lead-bearing materials. These include processes developed by Bunker Hill (1), Tonolli (2), and Stolberg (3). These battery wreckers have the following features:

1. Mechanically breaking or crushing the batteries,
2. Separating the acid from the solids,
3. Separating the lead-bearing portions from the cases and separators,
4. Greatly reducing manual labor, and
5. In some processes, separating metallic lead from the lead compounds.

RSR operates a modified Tonolli battery wrecking process at the Dallas plant and modified Bunker Hill battery wreckers at the other plants.

With the introduction of polypropylene-cased batteries in the late 1960's, battery wrecking has become much more difficult. The older hard rubber batteries would break upon impact. The new polypropylene battery cases do not break upon impact. Because the polypropylene-cased batteries are difficult to wreck, RSR has instituted a process for crushing or shredding the whole battery. The batteries from the receiving area are scooped up by a large front-end loader and dumped into an apron feeder where they are raised up by a belt feeder, and dropped into the crusher-shredder.

Various plants at RSR are experimenting with different types of crushers and shredders. These include double roll and single roll impact crusher, and knife shredders. Regardless of the type of shredder used, the batteries must be broken down into small enough pieces to liberate the lead-bearing portion of the battery from the case and separators. Particularly troublesome is liberation of the through-the-wall intercell connectors in polypropylene batteries.

The shredder at the Dallas plant consists of two opposing rotating shafts driven by 110 kilowatt motors containing 33 knives spaced 37.5mm apart. The knives are heat-treated stainless steel. All internal parts of the shredder are of stainless steel to prevent attack by the sulphuric battery acid. The knives must be kept very sharp to assure shearing of the polypropylene case, and achieve liberation of the lead-bearing portion from the case and separators. To prepare good feed for the reverberatory furnace effective crushing and shredding of the batteries is required.

The crushers or shredders are completely enclosed to prevent acid spray during the wrecking process. In all the plants, the case materials and separators are floated away from the metal-bearing parts of the shredded battery in a heavy media separator.

The heavy media consists of a slurry of the dilute H_2SO_4 of the battery and the finely divided lead oxides and sulphates from the shredded battery. The agitated slurry, at a specific gravity between 1.3 and 2.0 allows the metallic lead portions of the grids, posts, and connectors; and large pieces of lead oxide and sulphate active material to fall to the bottom of the separator where they are continuously removed. The separators, plastic, and hard rubber cases float. The agitation of the slurry prevents the rising float material to be trapped and pulled beneath the surface by the heavier lead particles.

The float portion is carried over onto a screen where the heavy media slurry is drained and washed from the separators and cases. The washed float product is conveyed to the plastic recovery system.

Plastic Recovery System

In the RSR plastic recovery systems, the shredded case material is further reduced in size by an impact hammermill granulator to less than 25mm. The granulated product is fed to a second sink float unit using water as the floating media. The polypropylene case material, along with small amounts of other plastics from the tops and vents of the batteries, floats on the agitated water bath. The separators and hard rubber cases, having densities greater than water, sink to the bottom of the unit where they are continuously removed.

The polypropylene fraction is dewatered, washed, dried, and sold to plastic recycling companies. The plastic product is shown in Figure 3, consisting of small multicolored pieces averaging 15mm. Currently polypropylene-cased batteries represent 50-60% of the batteries obtained by RSR. Within several years the percentage of polypropylene-cased batteries is expected to increase to greater than 90% of the recycled batteries. At these percentages recycling the polypropylene becomes economically attractive.

The hard rubber cases mixed with small pieces of lead and separators constitute the sink portion of the plastic recovery system. The hard rubber can be used as a source of carbon reducing agent in the furnace. RSR is currently investigating processes to utilize the hard rubber battery cases as fuel.

Battery Wrecker Product

The battery wrecker sink product consists of grids, posts, internal connectors, and the large pieces of battery active material ($PbSO_4$, PbO_2, PbO, and finely divided metallic sponge lead). The battery wrecker material is continuously removed from the battery wrecker and placed into piles from which the excess H_2SO_4 is allowed to drain. The battery wrecker product is shown in Figure 4.

Figure 2. Truck Dumper

Figure 3. Polypropylene
Plastic Product

Figure 4. Lead-Bearing Battery
Wrecker Product

The excess dilute H_2SO_4 overflows the heavy media sink float unit into a clarifier. Here the fine suspended lead oxide and sulphate particles sink and are compacted into a sludge called battery wrecker mud. The mud from the clarifier is combined with the battery wrecker sink material to consolidate all the lead-bearing materials for furnacing. This material consists of the feed to the reverberatory furnace.

Sulphur Removal System

The battery wrecker material contains some unreacted H_2SO_4, metallic lead, $PbSO_4$, PbO_2, and a small amount of organic material from entrained cases, separators, and fibers. The typical battery wrecker product is shown in Table I. The battery wrecker material is a very uniform product because of the nature of the battery wrecking process which combines large numbers of batteries and thoroughly mixes the output. The composition of the battery wrecker material varies only slightly from plant to plant and is more a result of slight changes in battery composition between geographical areas than battery wrecker differences.

Table I. Composition of Battery Wrecker Material (WT%)

Element	Battery Wrecker Material	Typical	Sulphur Converted
Pb	68-75	71	68-75
Sb,As,Sn	2.1-2.5	2.3	2.1-2.5
C	1.5-2.5	1.8	1.5-2.5
SiO_2	1-3	1.8	1-3
H_2O	6-10	8.5	6-10
S	4-6	4.6	<.4

The battery wrecker material consists of 4-6% sulphur mainly in the form of $PbSO_4$ and entrained H_2SO_4. During furnacing, the $PbSO_4$ liberates SO_2 as is shown in the reverberatory furnace reactions section. Using a patented process developed by Acoveno and Freudiger (4), RSR is currently operating a pilot plant to remove the sulphur from the battery wrecker material prior to furnacing. The process uses a solution of $(NH_4)_2CO_3$ to convert the $PbSO_4$ to $PbCO_3$, producing a solution of soluble $(NH_4)_2SO_4$. By removing greater than 90% of the sulphur from the battery wrecker feed material prior to furnacing, the SO_2 in the furnace off gases can be decreased to such low levels that the current SO_2 emissions levels may be reached without the need for scrubbers or high stacks.

Acid Recovery and Treatment

The acid from the receiving area acid sump, excess acid from the battery wrecker sink float clarifier, and water used to wash acid from the equipment, trucks, or grounds is collected and pumped to the acid treatment plant.

The dilute, clarified acid and wash waters are treated with NH_3, Na_2CO_3, or $Ca(OH)_2$, depending on location, to neutralize the acid. The neutral, clarified solution with controlled pH is then pumped to a precoat vacuum filter to remove any traces of suspended lead or heavy metal compounds. The final neutral, filtered, water can then be disposed of into the sanitary sewers.

RSR is currently investigating the possible recovery of the H_2SO_4 for use as dilute acid for pickling applications, concentrating the acid to produce a high grade product, or producing a saleable sulphate product.

Reverberatory Furnace Operation

The reverberatory furnaces at RSR average approximately 2.4 meters wide X 10m long at the slag line. The large melting furnace is fired by two burners, each producing 9-12 gigajoules/hr. using oil or natural gas at stoichiometric ratio with oxygen enriched air.

The standard furnace construction uses magnesite brick below the slag line, chrome magnesitic brick at the slag line, and high alumina brick on the upper walls. The suspended arch roof is constructed using basic brick in areas where slag may splash and high alumina brick elsewhere. The floor arch is heavy-duty fireclay brick.

The drained battery wrecker material is blended with recycled flue dust and carbon in a batchhouse. The blended charge is fed to a dryer which reduces the moisture to less than 4%.

Carbon is added to the charge as a reducing agent in the form of coke breeze, coal fines, or hard rubber battery case material, in an amount up to about 5% by weight of the total charge. The dryers are natural gas or oil fired rotary kilns.

The premixed, dried charge is fed continuously to the furnace by a closed belt conveyor and a vibrating feeder system. This feeder drops the charge onto a pile of unsmelted material in the furnace between the two burners. A mechanical stoker with a travel of about 450-600mm pushes the pile of material into the furnace at a rate of 2-8 cycles/minute.

As the material is pushed into the furnace, the pile is heated by radiation and convection from the hot furnace gases and walls. The lead melts, and the carbon reacts with the oxides and sulphates. Most of the reactions take place in the first half of the furnace. The remaining portion of the furnace is mainly a settling area for separation of the slag from the metal, although some smelting is done in this zone.

The slag consists of a thin, fluid layer atop the heavier lead layer. The slag is tapped continuously from the furnace onto an endless belt slag caster. The slag is cast into flat, shallow molds to maximize heat transfer, and promote cooling and solidification of the slag. The solidified slag is discharged into a slag bin.

While the slag is tapped continuously, the bullion is tapped from the furnace when the metal level builds up to the height that small amounts appear in the slag. At this point, about 4000-6000Kg of bullion are tapped from the surface into either water-cooled hog molds approximately 1000mm X 1000mm X 600mm, or into a kettle. Sufficient lead is tapped to draw the lead in the furnace down by 25-50mm below the slag discharge tap.

The furnace is operated on a campaign basis of 24 hours per day, 7 days per week. During periods of short raw material availability, the furnaces may be operated on a 5 day/week basis with the furnace banked during the weekend.

Rerunning Primary Slag

During the smelting of battery wrecker material, the furnace produces low (.2-.7%) antimony bullion and a primary slag containing 80% metallic oxides.

The primary slag is stored and reprocessed in the reverberatory furnace in a separate slag campaign. The slag is blended with up to 5% carbon in the form of coal fines, coke, or hard rubber battery case material, and is fed back to the furnace in the same manner as the battery wrecker material. In addition to the slag, softening skims from the soft lead refinery are added to the furnace charge as they are produced.

In the slag smelting run, about 60% of the lead contained in the slag is recovered as a low (.8-2.5%) antimony bullion. The remaining lead and alloying elements are concentrated in the slag. Because of the low lead content and high antimony content of the final slag, it must be processed in the blast furnace.

The composition of primary slag and bullion, as well as final slag and second run bullion, are shown in Tables II and III.

Table II. Composition of Reverberatory Furnace Products (WT%)

Element	Primary Bullion	Second Run Bullion
Sb	.2-.7	.8-2.5
As	.001-.01	.01-.05
Sn	<.001	.001-.01
Cu	.09-.12	.09-.12
Fe	.02-.05	.02-.07
Bi	<.02	<.010
Ag	002	002
S	0.1-.5	0.1-.3

Table III. Slag Composition (WT%)

Element	Primary	Final
Pb	65-75	50-65
Sb	7-10	10-18
As	.8-2.0	.8-1.6
Sn	1-2.5	.8-3
Fe	1.1	1-3
CaO	1.2	1.5-3
SiO$_2$	4-8	8-14
S	1.5-3	.4-1.5

Reverberatory Furnace Smelting Reactions

The major aim in reverberatory furnace smelting is the reduction of the lead compounds to metallic lead bullion, and at the same time oxidation of the alloying elements in the battery grids, posts, straps, and connectors to produce a slag containing virtually all the alloying elements.

The following reactions take place in the reverberatory furnace:

1. $PbSO_4 + C \rightarrow Pb + CO_2 + SO_2$

2. $PbO + 1/2C \rightarrow Pb + CO_2$

3. $4Sb(m) + 3PbSO_4 \rightarrow 3Pb + 3SO_2 + 2Sb_2O_3$

4. $2Sb(m) + 3PbO \rightarrow 3Pb + Sb_2O_3$

5. $Sn(m) + PbSO_4 \rightarrow Pb + SO_2 + SnO_2$

6. $Sn(m) + 2PbO \rightarrow 2Pb + SnO_2$

7. $3As(m) + 3PbSO_4 \rightarrow 3Pb + 3SO_2 + 2As_2O_3$

8. $2As(m) + 3PbO \rightarrow 3Pb + As_2O_3$

Some of the lead sulphate is reduced by carbon or carbon monoxide to metallic lead evolving CO_2 and SO_2 as shown in reaction #1.

The major part of the $PbSO_4$ is converted by oxidizing the melted grid metal according to equation 1A. In this reaction, termed the "lead eater reaction", one mole of $PbSO_4$ reacts with 4 moles of molten lead to produce 4 moles of PbO and one mole of PbS, which enter the slag. The antimony, arsenic and tin contained in the grid metal are also oxidized in a similar manner. Reaction 1A is responsible for the PbS which can be found in the primary reverberatory slag. Most of the PbS is oxidized in the highly oxidizing slag via reaction 1B to SO_2 and metallic lead. To complete reaction 1,

the remaining two moles of PbO are reduced to lead with Carbon in reaction
1C. The complete reaction is thus reaction 1.

$$1A. \quad PbSO_4 + 4Pb \rightarrow 4PbO + PbS$$

$$1B. \quad 2PbO + PbS \rightarrow 3Pb + SO_2$$

$$\underline{1C. \quad 2PbO + C \rightarrow 2Pb + CO_2}$$

$$1. \quad PbSO_4 + C \rightarrow Pb + CO_2 + SO_2$$

Thus, not only is the $PbSO_4$ a major oxidizing agent in the reverberatory
furnace, but the PbO generated from reaction 1A also aids in the oxidation
of the alloying elements antimony, arsenic, and tin from the melted metallics
as shown in reactions 4,6, and 8.

Because of the $PbSO_4$ and subsequently generated PbO, the furnace conditions are reducing to lead and oxidizing to the alloying elements in the
battery grids. The product of the furnace is a bullion containing very low
levels of antimony and almost no arsenic and tin, and a free-flowing litharge
(PbO) slag containing virtually all the alloying elements. The slag consistency depends on the self-fluxing action of the PbO rich slag.

Flue Gas Handling System

The off gases from the reverberatory furnace exit the furnace through
a refractory lined flue. The gases pass into a series of water-jacketed
cooling towers. In the cooling towers, some dust drops out and the gases
are cooled by contact with the water-cooled steel jacket walls. The exit
gases are further cooled in a long steel gas trail by dilution air, or by
the use of water sprays if necessary, so that the gas is cooled sufficiently
to pass through the dust collecting bags without damaging them.

The typical dust collection baghouse used at RSR are the mechanical
shaker type. The baghouse operates at 1.5:1 air to cloth ratio . The
typical baghouse system consists of 6-8 cells using teflon bags 127mm in
diameter and 3 meters long, and can handle about 1150 m3/min. of the furnace gases.

The flue dust is shaken from the bags and is continuously returned to
the batch house where it is blended back into the furnace feed. Because
the dust is handled continuously, there is no measurement of the amount
generated, but by continuously feeding the dusts back into the furnace
charge, there is no buildup of dust in the system.

Blast Furnace

The operation of the blast furnace at the Dallas, Texas plant was described by Murph & Pinkston (5). This blast furnace is of circular cross-
section 152 cm in diameter, and 3.7 meters between the tuyere level and the
top of the charge. The water-jacketed steel bosh slopes at an angle of 15°.
The upper portion of the furnace is a straight cylindrical water jacket.
The lead is discharged from the crucible continuously through a siphon into
a holding kettle. The slag is tapped intermittently. Blast air is provided
via a positive displacement blower through a 300mm d bustle pipe. The furnace has 10 tuyerers 50mm in diameter. The blast is enriched with oxygen.

The furnace is blown at a rate of 28.3 m^2/min. The blast air is not heated.

Blast Furnace Charge

The blast furnace utilizes the final reverberatory slag as feed material. The charge to the furnace differs from conventional blast furnace practice because of the very high antimony content, and low lead content of the input slag.

The typical charge for the blast furnace, slag composition, and bullion composition is shown in Table IV.

Table IV. Composition of Blast Furnace Feed and Products (WT%)

Furnace Charge

Final Slag	60-80
Coke	8-10
Limestone	8-10
Silica Rock	2-10
Iron	2-10

Blast Furnace Slag

Lead	1-3
CaO	22-30
SiO_2	25-35
FeO	19-29
S	1-5
As	.5-1.0
Sn	.5-1.0

Blast Furnace Bullion

Sb	15-25
As	1-3
Sn	1-3
Cu	.1-.2
Fe	.05-.2
S	.2-.5
Bi	<.01
Ag	<.002

The charge is prepared based on instructions from the central laboratory based on the final slag analysis. All materials fed to the blast furnace are automatically weighed.

The final reverberatory slag has a size range of 30-100mm. The coke has a preferred size range of 30-100 mm. The iron is added in the form of shredded automobile scrap 30mm X 120mm X 1mm. The thin iron reacts much faster than large thicker pieces normally used.

Operation

Under normal operating conditions, the furnace produces equal amounts of disposable slag and high antimony content bullion. The major problem in

running the blast furnace using the final high antimony slag as feed material is to prevent the loss of substantial amounts of antimony and arsenic as
a speiss in the furnace. The feed to the furnace is generally low in sulphur which allows less iron to be used in the charge. Even at higher sulphur
levels, the iron must be kept under 10% of the charge, because at higher iron
levels, it will react with the antimony and arsenic to form a speiss. Control
of the blast furnace to retain antimony values was presented at the 1977 AIME
meeting in Atlanta by Weiss, Prengaman & Freudiger (6).

Because of the reduced iron content and increased CaO content of the
slag, there is an increase in the lead content of the disposable blast furnace slag. This lead content of 1-3% is a relatively small part of the total
lead feed to the system, because most of the lead is recovered in the reverberatory furnaces prior to blast furnacing. The higher value of the antimony,
arsenic, and tin dictates maximum recovery of these elements in the blast
furnace bullion.

The bullion contains between 15-25% antimony with tin and arsenic contents of about 1-3%. The high antimony bullion is shipped to the refinery
to produce antimonial lead alloys. The high antimony content of the bullion
permits maximum flexibility to produce antimonial lead alloys from 0.5% to
20% by combining the reverberatory bullion with the blast furnace bullion.
The high antimony blast furnace bullion alloys provide an outlet for the
antimony in the form of high antimony alloys, and permits the plants to concentrate on the production of pure lead and pure lead alloys. The slag and
any matte present is tapped into large cast iron slag pots and allowed to
solidify. This slag is disposed of in an approved landfill.

Blast Furnace Gases

The off gases from the blast furnace contain about 20% CO. The CO is
burned in an afterburner which maintains sufficient excess air for complete
CO combustion. A burner maintains the temperature above 800°C at all times
to assure combustion of the CO. The off gases pass from the afterburner to
cooling tanks similar to those on the reverberatory furnace. Here the gases
are mixed with sanitary air and cooled in water-jacketed cooling towers.
The dust is filtered in a mechanical shaker baghouse containing 1440 acrylic
bags 127mm in diameter and 3 meters long for a total cloth area of about
2700 m^2. The baghouse provides 99.99% dust removal. The clean filtered air
is vented to atmosphere via a tall (91.5m) stack. The dust collected in the
blast furnace baghouse is conveyed to the batch house where it is blended
with the reverberatory furnace feed.

Lead Refining

Pure Lead Refinery

The lead recycling plants at RSR produce large quantities of high purity refined lead. The primary bullion from smelting battery wrecker material
contains only small amounts of impurities which must be removed to produce
pure lead as seen in Table II. The production of pure lead takes place in
100 Tonne kettles.

Blocks of primary reverberatory furnace bullion 1700-2500 Kg are transferred to the pure lead refinery and loaded into the primary melting kettle.
Here the bullion is melted and drossed to remove the nickel, iron, and suspended oxides and sulphides. The lead is oxidized by stirring in scrap oxide,

or using the oxide generated on the surface of the lead to remove the antimony, arsenic, and tin as a dry dross. This dross, along with the iron and
nickel breakdown dross, is returned to the reverberatory furnace as it is
produced. Thus there is no buildup of drosses to be handled and cause dusting, and the slag is somewhat enriched by the removed metallic oxides.

The major impurity in reverberatory furnace bullion other than antimony,
is copper. Because of the very low tin content of the reverberatory furnace
bullions, sulphur cannot be used to remove copper. In past years, the copper
was removed by the addition of metallic aluminum. Aluminum is no longer used
because of the danger of arsine or stibine generation from the drosses if
the antimony and arsenic has not been completely removed from the lead prior
to the aluminum treatment. Because of this danger, the lead is currently
decopperized using phosphorous.

In order to be used for battery oxide for maintenance-free batteries, or
for the production of non-antimonial battery grid alloys, the pure lead is
treated via the modified Harris Process to remove traces of tellurium. To
assure maximum cleanliness, the fully refined lead is pumped from the final
refining kettle to a casting kettle. Here the lead is given a final cleaning to remove all traces of suspended oxides, and is cast into either 30Kg
pigs, ingots, or 900Kg blocks. The typical chemical analysis of RSR pure lead
is shown in Table V.

Table V. Typical Chemical Analysis of RSR Pure Lead

Element	WT %
Sb	<.001
As	<.001
Sn	<.0003
Cu	<.001
Fe	<.001
Ni	<.001
Ag	<.002
Bi	<.015
Zn	<.001
Te	<.00
Pb	>99.98

The pure lead thus produced is suitable for the production of TEL, oxide
for maintenance-free batteries, and pure lead alloys such as Pb-Ca or Pb-Sr
alloys for maintenance-free battery grids.

Antimonial Lead Refinery

In addition to the pure lead refinery, each lead recycling plant has an antimonial lead refinery. Depending on the particular alloy to be made, the refinery may use the primary bullion or second run bullion. These may be used as is, or blended with some of the high antimony blast furnace master bullion to produce antimonial lead alloys from 0.3% to 20% antimony. Generally, the refining plants produce mostly pure lead or low antimony (<3% antimony) alloys.

Most of the antimonial lead alloys produced today range between 1.5 and 3.0%, and are used in the production of low gassing batteries. The major problem in manufacturing antimonial lead alloys is control of the micro-alloying elements. Prior to 1975, there was little concern about alloying elements other than antimony, arsenic, and tin. The antimony content of reverberatory bullion was merely adjusted by the blast furnace bullion. The tin and arsenic contents were adjusted using high tin and arsenic master alloys produced in special blast furnace campaigns. The metal was decopperized, drossed, and was cast into pigs.

With the advent of the low antimony alloys, the control of the refining of antimonial lead alloys became much more critical. The low antimony alloys must contain nucleating agents to innoculate the metal, and produce a uniform fine grain size which prevents brittleness. The major nucleating agents are copper, sulphur and selenium (7).

To make these nucleating agents effective, the amount of each must be controlled, as well as the alloying and pouring temperatures. Manufacturing the alloys is particularly difficult because copper can easily be removed from the alloys using sulphur. This removes not only the copper, but also the sulphur from the alloy.

Since the nucleating agents are also dross formers, the amount added must be accurately controlled to prevent excessive drossing at the battery companies. To accurately control the amount and types of nucleating agents in the antimonial lead alloys, RSR has developed a refining technique to produce consistent, dross-free alloys.

Reverberatory bullion is combined with blast furnace bullion, and the arsenic and tin contents are properly adjusted by blending in the melting kettle. The kettle is then drossed, an a controlled amount of nucleating agents are added.

The metal temperature is then brought to the casting temperature. To prevent reaction between the copper and sulphur, the pouring temperatures must be very closely controlled. The actual pouring temperature will depend on the concentration of nucleants desired in the metal. Figure 5 shows the structure of the metal with uncontrolled pouring temperature, while Figure 6 shows the structure of a low antimony alloy cast under controlled temperature to maintain the nucleants at the proper concentration. The metal in Figure 5 will produce brittle grids while the metal shown in Figure 6 will produce good grids.

Metal Casting

Pure lead, or lead alloys, are cast into 30Kg pigs using Wirtz endless belt casting machines. 900Kg blocks are cast into water-cooled block molds 600mm X 600mm X 290mm. The lead is removed from the block molds via a vacuum lifting device.

Figure 5. Structure of a low antimony battery grid alloy without proper control of nucleants. Note the large elongated crack prone grains. Magnification 240X after reduction.

Figure 6. Structure of RSR-R275 low antimony battery grid alloy with proper control of nucleants. Note the uniform, rounded crack resistant grains. Magnification 240X after reduction.

Refinery Quality Control

The central chemical laboratory at each of the refineries controls the refining process. Metal samples are analyzed on a direct reading spectrometer connected to a mini-computer. The mini-computer gives a direct readout of the chemical analysis of the metal at each step of the refining and alloying process. Based on the in-line analysis, very close control over the composition of both pure lead and alloys can be achieved. Using the rapid analysis, the process control technicians accurately compute reagent requirements, thus reducing costs, as well as amounts of dross to be recycled.

Conclusion

The reverberatory-blast furnace process for smelting battery scrap is an effective method to produce maximum amounts of pure lead necessary for the new non-antimonial battery grid alloys. At the same time the process is flexible enough to produce any antimonial lead alloys required and may be used to concentrate antimony in a readily available form.

Acknowledgement

The author wishes to thank RSR Corporation for permission to publish this paper. Also, the assistance provided by Michael G. Weiss and Rose Adams is acknowledged.

References

1. M.E. Elmore, U.S. Patent 3,393,876 (1968).

2. G. Tremolada, U.S. Patent 3,456,886 (1969) and 3,614,003 (1971).

3. Reinhard Fischer, "Treatment of Lead Battery Scrap at Stolberg Zink A.G.", AIME World Symposium on Mining & Metallurgy of Lead and Zinc, Volume II, 984 (1970).

4. F. Acoveno and T.W. Freudiger, U.S. Patent 3,883,348 (1975).

5. B.D. Murph and J.A. Pinkston, "Current Blast Furnace Practice at Murph Metals Southern Lead Company Smelter", Paper presented at 99th AIME Annual Meeting, Dallas, Texas, February 1970.

6. M.G. Weiss, R.D. Prengaman, and T.W. Freudiger, "Retention of Antimony Values in Secondary Lead Blast Furnace", Paper presented at 106th AIME Annual Meeting, Atlanta, Georgia, March 6-10, 1977.

7. R. David Prengaman, "Structural Control of Low Antimony Alloys for Battery Grids by the Use of Nucleants", Independent Battery Manufacturers Association 40th Convention, Chicago, Illinois, October 12-14, 1977.

SECONDARY LEAD SMELTING

AT BRITANNIA LEAD COMPANY LIMITED

John D. Taylor & Phillip J. Moor

Britannia Lead Co Ltd
Northfleet, Kent, England

The recycling of lead acid batteries has by tradition been carried out by a breaking operation in which unwanted components are discarded, followed by smelting in a rotary or reverberatory type furnace.

The Blast Furnace has been developed as an alternative approach in which a significant proportion of complete batteries, drained of acid, can be smelted without the need for mechanical processing. Another major advantage is that the ancillary gas cleaning and hygiene equipment can be scaled to a continuous steady operation rather than peak loads being associated with the other batch operations.

Smelting is carried out in a continuously tapped Blast Furnace fitted with an external forehearth. Waste gases are combusted in an afterburner and the dust recovered in a baghouse. Dust is agglomerated by melting to produce a slag which can be hygienically handled.

Introduction

Britannia Lead Co Ltd commenced refining lead at Northfleet, Kent in 1931. The raw material was crude lead containing silver from Mount Isa Mines Ltd, Queensland, Australia and throughput has gradually increased to the current level of 150,000 tonnes of lead per annum. In 1974 design work started on a new smelter and refinery to process reclaimable lead bearing materials in the form of used lead acid batteries and various residues. Construction began in 1975 and was completed in 1977. Design production rate was 70 tonnes of crude lead per day.

The plant equipment consists of a blast furnace, short rotary furnace, six refining kettles and a moulding machine. Ancillary equipment includes an afterburner for blast furnace gases, a filter baghouse and a 94 metre stack. The plant layout is shown in Fig. 1.

Materials

The materials used in the blast furnace are as follows:

a. **Cased Batteries and Battery Plates**

These are purchased from scrap dealers in the southern half of England. The physical quality of purchased battery plates has been poor and the purchase of whole batteries is now preferred. Some batteries are charged whole and others are broken on a toll basis near the refinery. Mechanical handling and stockpiling of the plates is minimised to prevent break up and degradation. Treatment of old stockpile material has resulted in poor furnace operation due to low permeability of the charge, high back pressures and carry over of dust into the afterburner.

b. **Softener Skim**

High antimony softener skim from the primary refinery (No. 1) is treated to recover lead and provide antimony for the higher antimonial alloys. Softened skim from No. 2 Refinery will also be re-processed and will assist in making high antimony alloys with stringent silver specifications.

c. **Lead Residues**

Materials arriving from processes ex customers' plants.

d. **Foul Slag ex Rotary Furnace**

Foul slag will be treated to recover lead and antimony.

e. **Dry Skims and Caustic Skims from Refining**

These materials are re-circulated as they are produced.

f. **Chalk**

Chalk is used as a flux and is abundant and cheap in Kent.

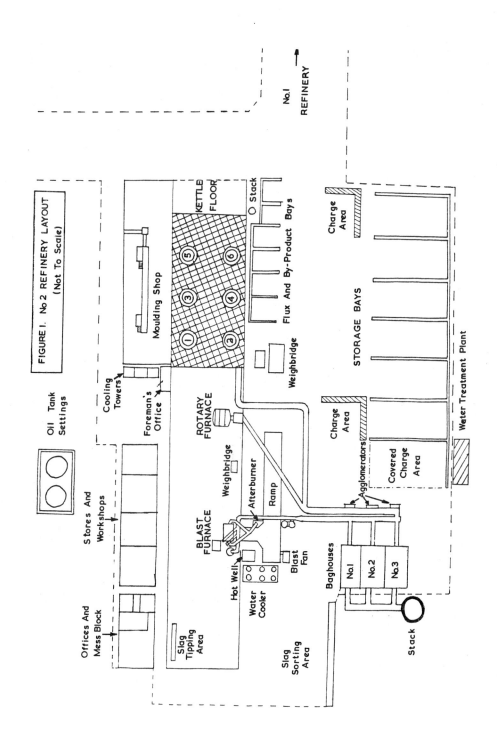

FIGURE 1. No 2 REFINERY LAYOUT (Not To Scale)

g.   Iron

The scrap iron is added to:-

i.   React with the sulphur in acid and lead sulphate to form an
iron sulphide matte.

ii. React with the alumino-silicates in the battery case
materials to form a slag.

Note - Cast iron borings and millscale have been used instead of "frag
iron" but lead to increased furnace pressures.

h.   Metallurgical Coke

The source is Bedwas Coke Ovens and size range preferred is
25-75 mm.

i.   Return Slag

The return slag content of the charge is varied according to the
smelting rate such that a reasonably constant amount of slag is
produced each day with the desired level being about 90-100 tonnes per
day.

Note - The materials flow sheet is shown in Fig. 2 and typical assays
of all materials are shown in Appendix I.

Blast Furnace and Ancillary Equipment

The plant was originally constructed according to the SB design of
Paul Bergsoe & Son under their consultancy. Operation with this system
from July 1977 to May 1978 resulted in 17 short campaigns and two fires in
the Bag Filter House. The design was then changed to incorporate a Roy
Continuous Tapping System and to re-route ducting so that all furnace top
gases must go through the afterburner (see Fig. 3).

Shaft and Forehearth

The furnace hearth is 4 metres long by 1.3 metres wide and slopes 50 mm
downwards to the taphole. It is curved across the width so that the centre
is 150 mm lower than the sides. The lower shaft wall consists of 22 water
cooled steel jackets with eight tuyeres on each side and one tuyere in the
centre jacket at the end opposite the taphole. The tuyeres were originally
100 mm in diameter but have been reduced to 50 mm by fitting inserts and
inclined 12° downwards. Each jacket is fitted with a thermometer showing
water temperature and water flow is individually controllable. The water
cooled stainless steel tapping breast has an opening 275 mm wide by 150 mm
high.

The upper shaft is lined with magnesite bricks and is 1.1 metre wide by
4 metres long. The shaft widens at the top to accommodate an inverted U-
shaped duct for fume and waste gas removal (the plant name is "dog kennel").

Two pneumatically operated doors close the front of the charging opening. Under normal operating conditions all the furnace top gases are drawn
in through the "dog kennel". In addition sufficient air to combust these
gases is drawn in through the charging opening and down into the dog kennel.
The off-takes situated at each end of the dog kennel join above the charging
ramp and a single duct then carries the gases to the afterburner.

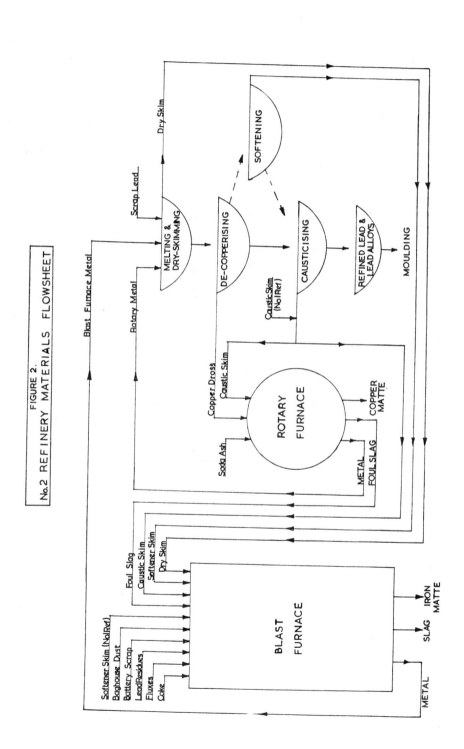

FIGURE 2.
No.2 REFINERY MATERIALS FLOWSHEET

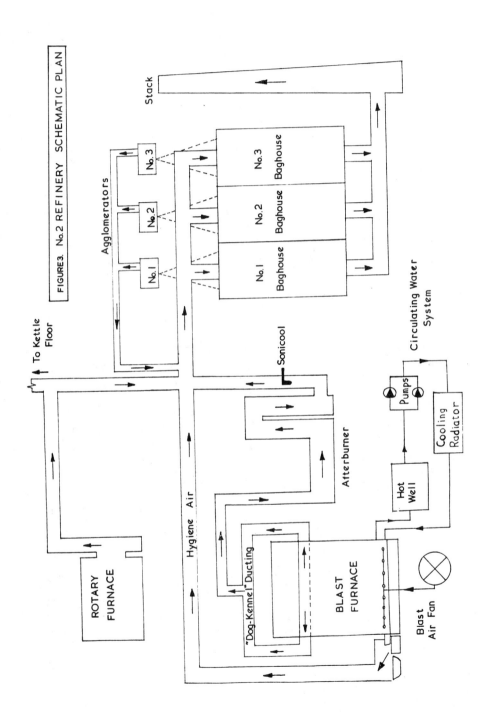

FIGURE 3. No. 2 REFINERY SCHEMATIC PLAN

The tapping area of the furnace consists of a chrome magnesite lined weir box and V notch weir whose height is adjusted by placing or removing brick splits. The weir height is adjusted to give sufficient depth to give a gas-tight seal on the furnace while permitting free flow through the tapping breast for metal and slag. The weir box is covered by a refractory lined plate through which a gas burner projects.

The slag and metal flow from the weir into a rectangular forehearth lined with Kaiser Permanente castable concrete. In one corner is an underflow weir through which metal flows into a lead well and then overflows down a covered Y-shaped launder. The metal is directed alternately into one of two 2-tonne refractory lined pots in a sand lined pit and from there transported by overhead crane to the kettle floor.

The forehearth cover is lined with high alumina castable refractory and a gas burner fires through it to the area behind the slag spout. A door is provided for cleaning the lead well.

Slag and matte are tapped into 3-tonne capacity cast steel pots.

The hygiene ventilation around the tapping area consists of five extraction hoods, i.e. over weir, metal pots and launder, slag pot, slag spout and over the lead well door. The five flues merge together into a common flue of 0.9 metres diameter which joins the afterburner exit gas flue prior to the baghouse inlet. The ventilation gases are currently used as the cooling medium for baghouse temperature control.

Waste Gas System

In the original system the hood above the furnace charging area was connected by ventilation duct to the baghouse inlet flue and hence gases issuing from the top of the furnace could by-pass the afterburner (see Fig. 4). The duct has been disconnected and all the furnace top gases now pass into the afterburner where the carbon monoxide and inflammable materials from the battery cases and separators are burnt. The afterburner consists of a refractory lined cylinder with an 8.42 gigajoule per hour natural gas burner at the end nearer the furnace. The afterburner gas temperature is automatically controlled at $800^{o}C$ so as to ensure combustion has been completed. The temperature generated by the furnace top gas combustion is normally above $800^{o}C$ and when this is the case the burner is on minimum fire and acts as a continuous ignition burner. The gases exit via a U-shaped refractory structure where the final stages of combustion are completed and are then mixed with the ventilation gases. Problems have been experienced with Baghouse temperature control using hygiene gases. In order to improve control a Sonicool spray which can inject up to 30 litres per minute of water has been installed prior to the gas mixing point. The combined and cooled gases then pass along the ducting to the three baghouse sections.

Some entrained dusts fall out in the afterburner and are removed by tapping on line or during furnace shutdowns. A smaller burner is situated two-thirds of the way along the afterburner but has not proved of use and will be removed.

Flow through the afterburner is normally 11,000 m^3 per hour and retention time approx. 3.3 seconds. Control is effected by monitoring the temperatures recorded by thermocouples through the system, e.g. high temperatures late in the system are indicative of late burning and lack of sufficient air induced at the top of the furnace.

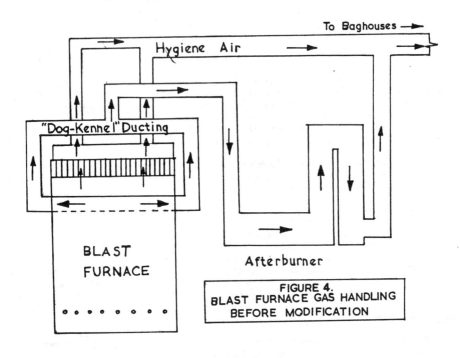

To Baghouses →

Hygiene Air

"Dog-Kennel" Ducting

BLAST FURNACE

Afterburner

FIGURE 4.
BLAST FURNACE GAS HANDLING
BEFORE MODIFICATION

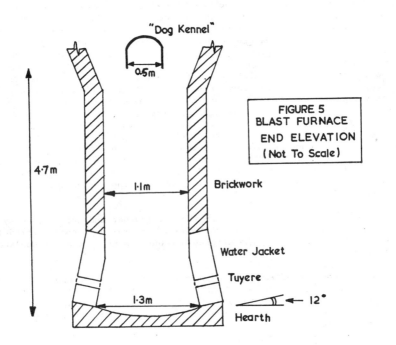

"Dog Kennel"

0.5m

4.7m

1.1m

1.3m

12°

Brickwork

Water Jacket

Tuyere

Hearth

FIGURE 5
BLAST FURNACE
END ELEVATION
(Not To Scale)

Baghouse and Agglomerator

The waste gases from the afterburner plus the hygiene air are filtered through a bag filter house before passing up a 94 metre stack to the atmosphere. The filter house consists of three Serck Visco Mark IV automatic suction type dust collectors. Each collector contains 18 compartments with 28 filter sleeves giving a total of 504. Cleaning is carried out by isolating each compartment in turn and then reverse blowing with scavenging air while the sleeves are mechanically shaken by an eccentrically mounted motor.

The dust is collected in a hopper at the bottom of the compartments and is continuously removed by screw conveyors to the appropriate one of three Bergsoe agglomerators. These consist of small sloping hearth furnaces on which a gas burner plays. The leady dusts are melted and flow to a conical slag pot and the slag produced is stockpiled for eventual re-treatment. Ducts from the agglomerator burner chamber and the outlet hood are connected to the Baghouse inlet manifold.

The Baghouse has a total filter capacity of 165,000 Nm^3/hr at 120^oC with filter bags of Orlon fibres on a Dralon skin. The total filter area is 15,909 m^2 and all three sections are used when running the blast furnace due to the method of gas cooling employed at present. The baghouse gas temperature is controlled at 120^oC by admitting dilution air and the use of the Sonicool spray. Emergency dampers are set to admit cold air when the temperature reaches 130^oC.

Water Cooling

Cooling water is pumped through a "Fin-Fan" radiator cooler and on through the furnace jackets at a total flow rate of 1,000 litres per minute and is collected in open troughs. From these it re-circulates to a hot well and flows by gravity to the pump inlet. The cooler can extract 7.5 gigajoules of heat per hour and temperature control is by means of switching on or off one or more of the six cooling fans. Commercial distilled water is used and is regularly dosed with a corrosion/scaling inhibitor. An emergency towns water cooling supply is also available and when this is in use the water is rejected to the drains.

Blast Air

The blast air was provided originally by two fans of 37 K.W. each at 2,950 r.p.m. The design capacity of each fan was 5,000 Nm^3/hr at 1,000 mm W.G. The pressure was found to be inadequate and the expedience of operating the fans in series was adopted and they then produced up to 6,000 Nm^3/hr at 2,000 mm W.G. A new fan has recently been installed which can produce 9,000 Nm^3/hr at 1,750 mm W.G.

Oxygen Plant

An oxygen plant was originally installed but is not currently in use. As originally supplied the oxygen delivery pressure was inadequate to inject the oxygen into the blast air system. The delivery has been changed to the inlet of the blast air blowers so the oxygen could be used. However, the total volume available is only 110 Nm^3/hr which would only produce a 2.0% enrichment and is unlikely to prove significant. In addition the enrichment would be even smaller as blast air volumes increase.

Furnace performance is steadily improving so no genuine base period information is available against which to hold an oxygen trial. When the maximum performance is achieved with air only then the effect of oxygen enrichment will be determined.

A natural gas air pre-heater was supplied to provide blast air at 500°C. The system is not now in use because the use of expensive natural gas to preheat blast air is doubtful economics.

Other Plant Equipment

Lead Kettle Floor

There are six hemispherical kettles of 3.66 metres diameter with a capacity of 130 tonnes each. Firing is by tangentially mounted natural gas burners rated at 4 gigajoules/hour maximum output and 1 gigajoule/hour on low fire setting. Kettle setting gases are vented by a 32 metre steel stack. Equipment includes 3 impeller type mixers of 30 K.W. capacity, 2 lead pumps for transfer and two for moulding, all of which are of 11 K.W. capacity. All the mixers are hooded and these and other hoods are linked in the kettle floor hygiene ventilation as required.

The building itself is 112.5 metres long by 18 metres wide and 14.6 metres high.

A 15 tonne overhead crane is used for all metal and primary slag handling. An additional crane will be installed in the near future.

Moulding Machine

The machine is a straight line belt which can mould 25 kg or 40 kg pigs at 30 tonnes or 45 tonnes per hour respectively. Moulding is carried out on a daywork basis.

Rotary Furnace

The furnace is of Schmitz & Apelt design and is 4 metres long by 3.6 metres diameter and is lined with chrome magnesite bricks. The natural gas burner has a maximum capacity of 12.6 gigajoules per hour and is inserted through the off-take flue at the rear of the drum. Charging is effected through the front by means of a rotary head fork lift truck using 1.5 tonne charge buckets. Throughput is approx. 60 tonnes per day of by-product materials and the furnace is campaigned to clear up any by-products when production/sales balance permits a blast furnace shutdown.

Plant Operations

A summary of operating statistics is shown in Appendix II.

Charge Preparation and Charging

Charge materials are made up, on daywork, into six or seven beds each 35 tonnes weight. The materials are weighed and then spread on a concrete pad using a Volvo front end loader. Each bed is 9 metres long, 1.5 metres wide and 1 to 1.5 metres high and consists of a "layer cake" in a specified order, i.e. return slag, coke, fluxes and leady materials. Correct and even distribution is critical to steady operation. A driver and weighman are employed each day on these duties. (See Appendix III).

Furnace charging is then effected by Volvo front end loader which takes slices off the end of the bed and this is a 24 hour continuous operation for one driver. The furnace is filled up as the charge descends and

1012

four Volvo bucket loads form a layer in the furnace.

Slag Handling

The slag and matte from the furnace stands for 12 hours and is then tipped by the overhead crane and transported outside by front end loader. The matte phase sets as a distinct button at the bottom of the pot and is removed from the overturned slag by the six claw petal grab of an Atlas excavator. The slag is broken to less than 300 mm and moved to a bay for re-cycling. When slag stocks rise some is shipped to No. 1 Refinery for use in its cupola. A driver and a sorter are employed for about 6 hours per day on these duties.

Furnace Tapping Area

Three men per shift are continuously employed in this area, i.e. Furnace-man, Assistant Furnaceman/Spoutman and Furnace Attendant/Weighman. The crane and its driver are also required for approx. 70% of the shift on an inter-mittent basis.

The general duties involved are attention to taphole, weir and fore-hearth as necessary, metal and slag pot changing, metal weighing, tuyere punching and general cleaning.

Kettle Floor Operations

Furnace metal is charged into one of four refining kettles until the level is 300 mm down from the top when the kettle burner is turned on. Charging then continues until the level is 100 mm down. A mixer is placed in the kettle, the dross is dried up at 450°C and then skimmed. The metal is sampled and analysed and the refining planned according to sales requirements.

The usual procedure commences with decopperising by cooling and sulphur treatment (1 kg per tonne lead).

Note - A minimum tin content of 0.03% is necessary for efficient decopperising. When the copper content is reduced to the desired level the temperature is raised to 480°C and caustic soda treatment given to remove tin and arsenic in a caustic skim. The metal is eventually transferred to one of the two moulding kettles where any final alloying adjustments and/or cleaning treatments are given before moulding.

A kettle softening procedure has recently been developed for antimony removal using mixed oxygen and air blowing. A permanent installation will be installed in the near future. The procedure has been successful to produce 99.97% lead from 3% antimony production from the furnace.

Problems and Modifications

Two main periods are covered.

A. Up to May 1978 when major modifications were made while the Bag-houses were repaired following a fire. The problems which existed in the original period were as follows:

i. The original furnace was fitted with an internal metal sump and goose neck siphon for lead removal and with intermittent tapping of slag and matte from four separate tapholes. Metal from the siphon ran via a swinging launder to two tonne metal moulds. Slag was tapped into slag pots of 800 kg capacity.

Problems soon became evident when iron matte third phase material separated out inside the furnace. The goose neck weir height was critical to matte removal and poor control led to the formation of a solid matte shelf across the furnace which led to the furnace being lost. In addition the repeated operation of the slag tapholes and the frequent fork lift movements to handle slag pots were hazardous and manpower intensive.

ii. The formation of the shelf inside the furnace was caused by a combination of the operating problems described previously and by the low blast volume which the furnace would accept.

iii. The original front end loaders used for charging had insufficient reach to allow the whole furnace to be charged. The bucket was also wider than half the length of the furnace so that overlap and some humping of the charge occurred in the centre. Uneven feeding caused poor blast distribution, channelling and eventual loss of tuyeres, with cold slag and lead conditions.

iv. The baghouse screw conveyors frequently overloaded and motors cut out and clearing these screws caused considerable hygiene problems.

v. The original oil burners of the agglomerators were not easily adjustable and the net result was baghouse dust going through unmelted or the ducts, etc. being overheated.

vi. Ventilation and Gas Cleaning System.
Fires occurred in the baghouses in February 1978 and May 1979. In the first fire a number of sleeves were burnt. The second fire was far more serious and four baghouse sections lost all the sleeves and suffered severe structural damage necessitating a re-build.

Both fires were caused by spontaneous combustion at the mixing point of the afterburner and furnace ventilation gases. They were brought about by furnace waste gases, containing combustible concentrations of carbon monoxide, by-passing the off-take flue and entering the ventilation hood situated over the furnace top. These gases of a combustible mixture were then ignited when they came in contact with the hot (800°C) afterburner exit gases.

The rotary furnace was operated using the remaining two baghouse sections until the baghouse repairs and furnace modifications were completed in November, 1978.

Modifications (May-November 1978)

i. The sump was bricked up to form a hearth sloping downwards gradually to the new taphole.

ii. The shaft was narrowed by adding a 23 cm layer of bricks to each side.

iii. The side jackets were tilted so that the shoulder was now vertical and the tuyeres angled down at 12° and 50 mm diameter inserts fitted to reduce tuyere area. The increased blast velocity plus the downward angle was intended to increase activity in the centre of the hearth and shaft. New end jackets were installed to match the configuration created by the side jacket movement.

iv.   A new tapping breast, weir and forehearth system was installed.
It was necessary to excavate a pit to house the metal pots and to extend the
metal launder so that the pit could be outside the piled raft on which the
furnace stands.

v.   Pneumatically operated doors were fitted to the charge bay openings.

vi.   The ducting from above the furnace charge bay was removed.   The
ventilation gases from the new hoods covering the weir, slag pot, metal pots
and launder were connected to the exit gases from the afterburner to supply
the cooling requirement for baghouse temperature control.   From this point
on the furnace top gases were exhausted through the afterburner.

vii.   The 254 mm Baghouse screws were replaced with 305 mm screws and
the driving motors increased from 2 kW to 5 kW.

B.   27 November 1978 to 18 June 1979

Following these modifications the furnace started up on November 27th
and operated until December 18th when it was shut down following industrial
action.   The campaign, while short, is regarded as the first reasonably
successful operational period.   A dispute over wage negotiations prevented
a re-start until 26th February 1979.   From that date the operation has been
more continuous with good production rates and an availability of 77%.   The
results are shown in Appendix II.   The operations have been encouraging but
operational problems and analysis of downtime causes have shown the need for
further modification.   Installation of these modifications incurred down-
time of 4% during the period under review.   They were as follows:

i.   Campaign 20 was terminated because dust "fall-out" in the afterburner peaked below a baffle wall creating a constriction.   The baffle has
been re-sited opposite a clean out door and a taphole installed.   Regular
tapping prevents excessive build up.

ii.   Erosion of the furnace gas off-take necessitated extensive
repairs during two shutdowns and terminated campaign 22.   The design has
been modified to make changing the off-take a simple operation and a spare
off-take is carried.   The present off-take is of mild steel and relatively
inexpensive.

iii.   The tuyeres at the back end of the hearth gave continuous
problems by slagging up due to insufficient hearth to tuyere clearance.   The
rear hearth level was lowered 50 mm to be level but the problem persisted.
The weir height was then dropped 50 mm leaving only 12 mm of clearance
between weir and forehearth.   The rear tuyeres now keep open provided there
are no other furnace problems.   An additional tuyere has been installed in
the centre rear jacket of the furnace and this has helped marginally in keeping the area operational.

iv.   The weir was found to be too wide with low liquid flow rates and
the slag, metal and matte tended to separate out causing matte build-up.
Brickwork wear also compounded the problem.   The weir was then narrowed and
lined with high quality chrome magnesite bricks.   A protective brickwork
facing has been installed in front of the tapping breast to prevent it being
accidentally oxygen lanced as occurred in campaigns 19 and 21.

v.   Ventilation hoods have been installed over lead well, slag spout
and forehearth metal outlet.   Fibre-glass cloths have been added to extend
the metal hood sides.

vi. The agglomerators have been converted to natural gas firing with easily variable flow rate. In conjunction with the installation of a larger duct system the previous problems have been mainly eliminated.

vii. The main limitation on operations and production is baghouse capacity because of its inability to cope with the large volumes of hot gases produced from the furnace, particularly when treating whole batteries. No other definite constraint has yet been determined although metal handling, charge preparation and kettle floor treatment have been stressed when metal production has exceeded 120 tonnes per day when treating high softener skim charges.

The baghouse capacity constraint will remain until the use of dilution air for cooling can be replaced by evaporative cooling but minor improvements have been made by:

a. Installation of a Sonicool spray in the duct after the afterburner. Flows of up to 30 litres per minute can be used but flows above that cause water carry over due to insufficient residence time.

b. A crossover flue with damper has been fitted between the kettle floor ventilation duct and the furnace baghouse duct. The kettle floor hygiene air is thus used as dilution air and all sections of the baghouse can be used for furnace gases. The main duct in front of the baghouse is too small for full advantage to be taken of this because the most distant sections throughput is limited by the high suction head.

Charge Variations

The furnace charge is based on battery plates and whole batteries and the relative proportion of these has a significant effect on the production rate.

The lead contained in plates is approximately 70% and in whole batteries averages 50%. High case battery charges obviously contain less metal. The major problem however is the large quantity of volatile inflammable gases produced from the cases which immediately brings up the baghouse capacity limitation factor. Trials have been held at varying proportions of whole batteries. At the high proportions steady furnace operation is difficult to achieve. The cases tend to partially melt high up in the shaft and cause bridging by agglomerating together. The precise proportion above which this tends to occur is not known but problems have occurred at 60% and above. The current tendency is to limit whole batteries in the charge to 40% maximum with normal levels at 30%. A lower limitation of the use of whole batteries and/or battery plates is the fact that charges should always produce sufficient inflammable materials to make afterburner combustion self sustaining and thus prevent natural gas waste.

The basic charge of battery plates and whole batteries currently produces a daily production varying between 70-100 tonnes of lead containing approximately 3% of antimony.

Alloys containing higher antimony contents are produced by incorporating softener skim in the charge. The antimony content of this skim ranges up to 25% and lead containing up to 14% antimony has been produced directly from the furnace with production rates well in excess of 100 tonnes per day. The limitation was the inability to make up sufficient charges per day as the low calorific value top gas produced left a reserve of baghouse capacity. The high antimony production has been moulded and stocked as rough moulds for blending with normal furnace production to make alloys as required.

Hygiene and the Environment

The lead in air levels in the refinery are monitored using a Millipore field type sampler utilising 37 mm 0.8 micron pore size cellulose acetate filters. The mean of all determinations so far is 70 ug/m with the majority of the readings in the 30-40 ug/m^3 region. The higher mean has been caused by occasional high figures when there has been equipment malfunction or furnace problems. They still compare favourably with the current British TLV of 150 ug/m^3. As continuity has improved and operations steadied the levels have decreased. The average level of determinations to date in the June 1979 campaign have been 45 ug/m^3. These figures will be still further improved by the modifications envisaged in the future.

In the period when the furnace was operating intermittently, large stockpiles of feed materials, by-products and furnace barrings were accumulated. Most of these have now been processed and the areas cleared. In future all the stocks will be contained in concrete bays and either covered or sprayed automatically with water to prevent wind dispersal. The bays and battery stockpiles drain to a collection sump with facilities for acid neutralisation and a pump and filter system for disposal.

A mechanical wheel washing system is regularly used for all vehicles and a road washer/sweeper vehicle cleans all roads and concrete surfaces daily.

Future

On campaigns 20 and 21 the rear tuyeres were lost and furnace operation continued at 5,000 m^3/hr on ten tuyeres utilising 60% of the furnace. Hearth and tuyere conditions were good and production unaffected. The experience indicates that 500 m^3/hr per tuyere gives excellent tuyere condition and activity in front of it. On a seventeen tuyere furnace this gives a total blast volume of 8,500 m^3/hr., i.e. 70% more than current levels. Extrapolating from campaign 23, the furnace shaft and hearth would appear to have a capacity to produce up to 170 tonnes/day of lead at this blast volume. In order to achieve this rate major modifications are required in the blast furnace ancillaries and the refining areas as described below.

An evaporative spray cooling tower will be installed to carry out 80% of the gas cooling by the injection of up to 120 litres of water per minute. Dilution air will still be provided but this will be the hygiene air required to ventilate the furnace tapping area and the kettle floor. The narrow section of the main baghouse flue will also be increased in cross sectional area to enable three baghouses to be fully utilised. The injection of large quantities of water into the gases going to the baghouses will lower the dew-point. To prevent condensation and corrosion it will be necessary to sheet in the lower baghouse area and insulate all areas of the screws, etc. where gas flows are low.

The existing ducts from the furnace to afterburner are needlessly long and have been damaged by high temperatures during furnace run downs. The current proposal for replacement is to use a gas off-take in which the gases are removed at one end only and follow the most direct route to the afterburner. The afterburner would then be extended to 4 metres closer to the furnace and additional doors fitted to clear the dust which falls out. In this way the retention time in the afterburner would remain the same with the higher blast rate and ensure complete burning.

Slag handling should not present a problem as the proportion of returned slag can probably be reduced to match the increase in newly formed slag.

Metal handling will require modification as the 2 tonne pots in use at present require an excessive number of movements. Metal pot size will probably be increased to 5 tonnes each and a larger metal pot pit will be installed.

A second overhead crane will be required as the existing crane is already heavily utilised.

Charge bed preparation will have to be carried out either on shiftwork or two vehicles be used on daywork. The daywork solution appears the most attractive as greater control over charge preparation can be exercised and one weighman could cope with the two vehicles. Two separate areas for bed preparation would be required to prevent congestion particularly as the shiftwork vehicle would still be charging the furnace.

Moulding would have to be carried out on two shifts when moulding 25 kilo pigs.

The use of flash agglomerators deals with the baghouse dust effectively in the short term. Unfortunately the processing of the agglomerator slag and/or baghouse dust is very difficult because of the volatile metal chlorides it contains. Indications are that recirculation tends to increase the chloride content creating a vicious circle. It is believed that chlorine must be bled out of the system. A wet treatment plant is under consideration in which the dusts will be slurried with a sodium carbonate solution to convert the chlorine present to soluble sodium chloride. The slurry would then be filtered and the residues recirculated and the liquid discarded after treatment to remove soluble metal salts other than sodium chloride.

Longer Term Future

In the longer term a system must be developed to smelt greater proportions and/or quantities of whole batteries. The plastic problem and agglomeration of the battery cases together will certainly be reduced when the proportion of polypropylene batteries increases. Once the new evaporative cooler is installed, trials of higher proportions of batteries will be held.

The other constriction of large quantities of inflammable gas produced may be solved in another manner. It may prove possible to use these gases to preheat blast air. A heat exchanger system seems attractive on first examination. However, tube heat exchangers are high maintenance items particularly at the $1,100^{\circ}C$ reached by combustion of the Blast Furnace gases and all the Imperial Smelting Furnace operators are converting to Cowper stoves. The modern design stoves would not be suitable as the small chequer openings would block up. The more reliable system would appear to be the use of dirty gas Cowper stoves as used on early iron blast furnaces with large openings designed for easy cleaning.

Other improvements envisaged are as follows:

- Improved instrumentation and siting including a continuous gas analysing system to optimise afterburner operation.

- Improved materials storage.

- Improved hooding and ventilation in the weir and forehearth area.

- Tilting launder to obviate need to remove hood to redirect metal flow in the Y launder.

- Redesign of hoods and handling system for slag metal pots to enable changing without loss of suction.

- Redesign of mixer hoods.

- Hooding and ventilation of tapholes on afterburner (occasional use only).

- Vacuum cleaner with piped inlets to all plant areas.

- Possible widening of the furnace shaft.

Assay Schedule

Material	Sb	Pb	FeO	CaO	SiO$_2$	Al$_2$O$_3$	ZnO	Na$_2$O	H$_2$O	S
Battery Cases	–	–	–	1.5	2.4	1.58	–	–	–	–
Chalk	–	–	0.12	42.2	0.57	0.1	–	–	16.0	–
Return Slag	–	1.5	29.3	17.8	32.1	8.6	1.4	2.7	–	–
Foul Slag	6.5	19.9	10.9	3.1	18.0	5.9	5.4	23.6	–	2.9
Iron Matte	–	7.7	60.5	–	0.5	0.4	–	–	–	22.4

	H$_2$O	Ash	Vola-tile	Fixed Carbon	FeO	CaO	SiO$_2$	Al$_2$O$_3$	S
Coke	15.1	9.9	0.9	89.2	0.9	0.4	4.0	1.9	0.7

	Pb	Sb	As	Cu	Na$_2$O	ZnO	Cl
Caustic Skim	17.5	9.3	2.0	–	Balance	–	–
Dry Skim	71.9	5.1	0.06	0.08	–	–	–
Softener Skim	53.7	25.3	2.2	0.01	–	2.9	–
Baghouse Dust	60.1	0.7	–	–	–	3.3	14.2
Lead Residues	72.4	2.0	0.009	0.002	–	–	–

	Sn	Sb	Cu	As	Bi	Ag
Blast Furnace Metal (Batteries)	0.02	2.9	0.08	0.007	0.01	0.005
Blast Furnace Metal (Softener Skim)	0.02	11.6	0.08	0.08	0.003	0.007
Rotary Furnace Metal	0.4	6.8	0.12	0.5	0.007	0.008

Furnace Operating Statistics

Campaign No.	Operating Days	Material Charged T.P.O.D.	Kg Coke per tonne Charged	Crude Metal T.P.O.D.	Percent Lead in Slag	Percent Antimony in Crude Metal	Kgs Coke per T.C.M.	Therms of Gas per T.C.M.
1977-78	121	98.0	89.0	36.6	Not available	Not available	237	Not available
18	20	161.5	62.9	62.0	1.8	2.5	164	"
19	7.7	145.7	90.0	50.6	1.6	2.8	260	36.9
20	30	178.7	76.5	68.9	1.7	4.5	198	16.6
21	18.7	186.5	82.0	86.9	1.8	5.5	176	17.0
22	22	175.0	76.6	75.0	1.3	5.4	179	19.0
23	7	201.7	70.8	100.8	1.4	3.0	142	12.0

Note – i) T.P.O.D. = Tonnes per operating day.
ii) T.C.M. = Tonnes of Crude Metal.
iii) Campaign 23 was still in progress when report was written.

Breakdown of Materials Charged
(all figures given in tonnes)

Campaign No.	Cased Batteries	Battery Plates	Softener Skim	Foul Slag	Lead Drosses	Scrap Lead	Dry Skim	Caustic Skim	Return Slag	Chalk	Iron	Fluor-spar	Coke
1977-78	1776	4542	279	-	63	675	-	-	3439	233	841	5	1050
18	526	1290	-	-	-	108	-	-	1039	51	215	-	203
19	199	460	-	112	-	10	-	-	367	18	72	.9	101
20	742	2070	526	-	16	-	-	-	1462	116	315	.8	410
21	496	1115	706	-	20	10	-	-	846	96	197	1.2	286
22	356	1744	382	-	25	-	47	10	1013	73	199	1.1	295
23	12	868	-	-	-	-	51	11	349	39	82	-	100

Note – The small quantity of cased batteries treated in campaign 23 was so that the furnace's capabilities at smelting battery plates could be demonstrated.

APPENDIX III

No. 2 Refinery Establishment

Process Personnel

Staff			Hourly Paid		
Manager	1		Shift	Furnaceman	4
Metallurgist	1			Spoutman	4
Day Supervisor	1			Furnace Attendant	4
Shift Foreman	4			Crane Driver	4
Relief Foreman	1			Baghouse	4
Day Foreman	1			F.E.L. Driver	4
Weighman	1			Kettle Floor	8
				Relief	8
Total	10			Total	40

Day

Moulding Crew	
Chargehand	1
Pourer	1
Minder	2
Fork Lift	1
Charge Bed Preparation	2
Slag Sorting	2
Relief	2
Total	11

The above manning requirements allow for Blast Furnace operation and
a smaller workforce is required when only the Rotary Furnace is operating.

Maintenance Personnel

Day			Shift		
Engineer	1		Fitters	4	
Foreman	1		Mates	4	
Electricians	2				
Fitters	3				
Mate	1				
Total	8		Total	8	

GRAND TOTAL 77

Vehicles For Blast Furnace operation only.

2 Volvo 1.5 m^3 bucket Front End Loaders.
1 Coventry Climax 4600 kg Fork Lift.
1   "   "   2300 kg   "   "
1   "   "   2300 kg Rotary Head Fork Lift.
1 Atlas Tracked Excavator with petal grab.

For Rotary Furnace operation only.

1 Coventry Climax 7000 kg Rotary Head Fork Lift.
1   "   "   2300 kg   "   "   "   "
1 Volvo 1.5 m^3 Bucket Front End Loader.

LEAD SMELTING, REFINING AND POLLUTION

S. Bergsøe and N. Gram

Paul Bergsøe & Søn A/S
(Denmark)

The various metallurgical furnaces that are used for secondary lead smelting are reviewed critically, taking into account their significance to subsequent refining operations and for recycling of byproducts, as well as the pollution potential of the individual flowsheets. In particular, the technical, economic and environmental advantages and disadvantages of different smelting systems are discussed.

Background

More and more lead is being used in lead-acid batteries and the world consumption of lead for this application now exceeds 1½ million metric tons per year. The dominating type of battery is the motorcar SLI battery, which in the US has an average lifetime between 2½ and 3 years. It can be shorter or longer depending on local conditions, but in any case, a comparable quantity of lead is contained in the batteries that are discarded each year, and it is obvious that this lead cannot be left in the environment but has to be recycled, mostly by secondary smelters. World statistics on secondary smelting are incomplete, but those figures that are available add up to a production of between 1.3 and 1.4 million metric tons of lead per year. The real figure is certainly higher and with primary lead smelting accounting for just over 3 million tons per year, the two trades are beginning to resemble each other in size.

In most countries they are also having troubles in common, arising from the demand for environmental protection and work hygiene improvement. The overall effect of the modern regulations is of course to increase the price of taking lead from the mines or from spent batteries back to the battery industry, but there is considerable disagreement as to the size of the impact from the new regulations. It is the purpose of this presentation to point out how the choice of technology for each step of the cycle must be evaluated for its effect on pollution and work hygiene, and how this will affect the price of lead recovery. We feel that the secondary lead industry is only beginning to realize how little of its existing plants can be used in the future, if we are going to meet today's regulations. And who knows how much they will be tightened tomorrow?

It has been claimed that the US lead industry is in a particularly bad situation because foreign competitors are not compelled to comply with such tight standards. It is perhaps a more correct statement that the country which has had the latest revision of its standard is in the worst position, so next time someone else will have a tighter standard to meet. It is going to be a continuous fight to keep less and less lead from spilling out of the process.

Most secondary lead smelting in the United States is carried out in two stages, as also described in another paper at this symposium (1). Starting from conventional battery scrap, the first stage is a non-reductive smelting or "sweating" of the charge in a reverberatory or rotary furnace. During this part of the process most of the antimony and other alloy constituents are oxidized and dissolved in molten lead oxide while lead metal can be tapped off with 0.2-0.3% antimony. The second stage can be carried out in the same furnace by giving the oxide slag a reductive smelting with addition of coal and fluxes to reduce the oxides to a lead-antimony alloy and to produce a final slag. If no special precautions are taken, the slag nevertheless contains too much lead to be dumped and most smelters prefer to carry out the second stage in a small blast furnace (often called a cupola furnace) which is fed with a mixture of battery scrap and the antimonial slag from the first furnace. The blast furnace slag can more easily be brought down to a level of lead that allows dumping.

The combination of a reverberatory or rotary furnace and a shaft furnace has up to now fitted very well with the needs of the battery industry. Lead from the first smelting stage can conveniently be refined to soft lead of sufficient purity to manufacture battery oxide, and that from the second stage needs little refining and alloy adjustment to become useful again as a grid casting alloy. The economy of the process is good with a comparatively small investment per ton of annual capacity.

Smelting in the conventional furnaces requires conventional battery scrap from which the casing material and the acid have been removed. With modern equipment the scrapping allows recovery of plastic casing material (polypropylene) in addition to the lead recovery, but whatever the degree of mechanisation and containment, the disintegration of spent batteries and the sorting of metallics from non-metallics etc. remains a potential source of pollution. It is our experience that it cannot be made occupationally safe at a reasonable cost. The charging and operation of conventional rotary, reverberatory, and blast furnaces, as still carried out in many places, also causes serious pollution outside and inside the plant. It is possible to retrofit hoods and exhaust systems, but the cost tends to be prohibitive. The same is true of the conventional extraction, handling and recycling of flue dust collected in baghouses from the furnace gases and sanitary exhaust.

New Approach

An entirely new approach is required today. The future smelter must be prepared to receive wet batteries and to process them all the way through to produce lead of a purity that suffices for making grid alloys with tin and calcium or with antimony as required. All processes on the way must be so contained that internal and external pollution can be controlled within narrow limits, and lead and other heavy metals cannot be allowed to escape in quantity via stack gases, in waste acid or effluent water, or on plastic material from separators and casings if such material is sorted out from the metalliferous fractions. Since there is no practical way of distinguishing between scrap batteries containing antimonial lead grids and those containing lead-calcium grids (and the latter type also usually contains connectors made of antimonial lead) the process must also be so designed that there is no risk of arsine and stibine generation.

Basic considerations are whether battery breaking should be included or avoided and whether the subsequent processes should be wet or dry (or a combination of the two). Since batteries are to be received with acid, there will be a wet stage to begin with and it is of course possible to continue processing in the wet to end up with electrowinning of high purity lead. Theoretically, there is then no handling of dry, powdery or dusty material and therefore no such source of pollution. However, it remains to be seen whether large scale wet processing of lead can be made safe in practice. In addition to the problems of cleaning large quantities of waste water, it is our experience that wet processes are not as inherently safe as they look. It is very difficult to avoid completely spilling and splashes of lead-containing liquid, and when they dry up on equipment, tools and work clothes they constitute a very dangerous source of lead dust in the breathing zone.

After some bad experience of this sort we decided to use as little wet processing as possible and to go for direct smelting of unbroken batteries. The wet batteries still have to be pierced or punctured to drain most of the acid, but apart from that everything is left in the casing and without further processing, it is taken to the smelting furnace.

Modern Furnaces

Two types of furnace appear to fit into the concept: The wide type of shaft furnace that we have developed into the so-called SB-furnace, and the modern deep rotary furnace described in a paper at this symposium (2). Although they are very different, both of these furnaces lend themselves to a good and efficient "containment" of the process, so that installations can be made to meet work hygiene regulations. As always, environmental protection calls for efficient filtration of furnace gas and exhaust air, but it is also necessary to design the reception and the handling of raw material and byproducts carefully. A particular advantage in this respect comes from continuous agglomeration of flue dust (3).

The deep rotary furnace (also called the short bodied rotary furnace) is fired from the rear end where the off-gases are also extracted around the air- or water-cooled burner. A burner using fuel oil, gas or coal dust can be fitted and in any case the flame is forced to cycle inside the furnace and therefore to deliver much more of its heat energy than is possible with conventional furnaces with the burner and exhaust at opposite ends. The furnace is charged through a door in the front end and during charging the burner can operate and the furnace can rotate slowly if desirable. The compactness of the furnace further enhances its good heat economy, and it can reach high temperatures without excessive fuel consumption. The furnace thus offers a wide selection of smelting procedures, including the two-stage smelting carried out within one and the same furnace. At the cost of extending the charge cycle it is possible to produce clean slag ready for dumping, and also to run a matte process in which most of the sulphur contained in the charge is converted to iron sulphide instead of escaping as sulphur dioxide.

In its present form the rotary furnace does not readily accept a large fraction of cased batteries in the charge, but we believe that there is scope for further development in this respect. One disadvantage of the rotary furnace compared with shaft furnaces, is its requirement for a rather costly lining at intervals. The length of the interval depends very much on the charge feed and the mode of operation, but would typically be between 3,000 and 8,000 tons throughput. When operated round-the-clock and when smelting battery scrap, the most common size of deep rotary furnace has an annual capacity of 8,000-10,000 tons of lead, and re-lining thus becomes necessary between once and twice per year.

The SB-furnace has been described earlier (4,5), but we would like to repeat that it is very wide in relation to its production and it has a very low and compact smelting zone. The combustion gases are well cooled on the way upwards through the shaft, and they are extracted below the cool surface of the charge. A large afterburner is fitted to deal with all combustible matter and to promote the reaction between lead oxide and sulphur dioxide to form lead sulphate that precipitates and is collected in the baghouse. Mixing the hot gases from the afterburner with the cold sanitary exhaust air allows a close control of gas speeds and temperatures, and produces a type of dust that is less sticky and is better to collect in a filter. From the baghouse hoppers the dust is fed continuously into a flash agglomeration furnace that converts it to a solid slaglike material that can be recycled without any pollution.

With its latest modifications the SB-furnace can accept an almost unlimited amount of cased batteries in the charge, but it is still recommended to smelt recycled dross, agglomerated flue dust and other scrap and residues concurrently. Such material typically constitutes 10-15% of the feed and the remaining 85% is then varying mixtures of old batteries and battery scrap. The old batteries must be cracked or punctured to drain most of the acid, but the furnace is tolerant regarding moisture and acid left in the charge. By means of the matte process in the furnace itself and the afterburner reaction mentioned above, the process can withhold more than 90% of the sulphur contained in the charge and thus maintain a low sulphur dioxide emission. The emission of particulate matter including lead flue dust, has to be controlled by means of baghouses, and of course both process gases and

and sanitary exhausts from point sources such as charging and tapping have
to be filtered either individually or mixed as in our case. We find that it
is necessary to treat at least 50 tons of gas for each ton of lead smelted.

Filtration of gas and air, however, is only half of the environmental
protection. The other half is to prevent fugitive emission. A major source
of fugitive emission used to be the handling of flue dust from baghouses,
a source that can now be controlled at a reasonable cost by means of continuous flash agglomeration. With such an agglomerator in use, no flue dust
is ever handled as such, except inside the closed hopper system with its
screw conveyers that feed the agglomerator. With proper engineering of
the flue and filter system, the need for cleaning and for exchange of bags
can be minimized.

Fugitive emissions also originate from the reception, unpacking,
transport, and handling of raw material in the plant and from the movement
of internal and external vehicles. In the future smelter it will be necessary to do almost everything under roof and to minimize handling. It may even
be necessary to refuse raw materials which are difficult to handle or which
require sampling. Controlling these fugitive emissions is going to be one
of the costly items.

What has been said about emission control applies to short rotary
furnaces as well as to the SB-furnace, but the fact that the latter is
operated continuously allows the most safe control. On the other hand, the
SB-furnace cannot perform two-stage smelting. It yields only one lead
product: a bullion or crude lead containing a mixture of the alloy constituents from the scrap batteries, with antimony, tin and arsenic usually
dominating. Among the common constituents calcium burns away and is lost
in the slag. In one way this is advantageous, because with no calcium in
the metal there is no risk of producing drosses with calcium antimonide and
calcium arsenide, which may later generate the poisonous gases stibine and
arsine, if the drosses become wet or moist.

Refining

The average antimony content in the crude lead is now between 2 and 3%
and in order to produce soft lead, this quantity has to be removed by softening, while the first step in two-stage smelting yields lead that only needs
softening from a much lower antimony level. However, antimony removal is
rapid to begin with and a batch of antimonial lead is quickly brought down
to less than 0.5% Sb. In addition, the average antimony content of scrap
batteries continues to decline and can be expected to approach 1% in a few
years time. Furthermore, some control of the antimony content in crude lead
is possible by running the SB-furnace in periods without any addition of
antimony-containing recycle material, and at other times increasing the
antimony content by recycling a large proportion of such material, in
particular antimony-rich slag from softening. Crude lead from such campaigns
can be refined to grid casting alloy without any need to remove antimony
(and tin).

To make soft lead, tin and arsenic must also be removed and by a special
process the valuable tin can be recovered for use in other alloys or in
solder. Crude lead from secondary smelting invariably also contains a small
amount of copper and some impurities, notably bismuth and silver that are
present in the World's circulating stock of old lead. They must all be
removed to produce "four-nines" lead, but bismuth can be left if the product
is going to be "Corroding Lead". Although the discussion has not been
brought to a conclusion, most producers of calcium-lead batteries prefer to

1027

use 99.99% lead that can be obtained from primary producers. In the long run calcium-lead batteries will also have to be made from recycled lead, and we are prepared to meet the requirements, in other words to remove the bismuth if it is really necessary.

Pollution

Producing 20,000-25,000 tons of lead per year (each), the SB-furnace at Glostrup, Denmark, has been in operation for 4 years and its sister furnace in Sweden for about 3 years. Unlike primary smelters, secondary smelters rarely have a constant feed mixture, but have to rely on the scrap that eventually becomes available. There is considerable variation between campaigns and even between days or weeks within a campaign, which is usually of several months' duration. The furnaces have readily accepted very different mixtures of scrap and residues, including materials such as wet battery mud, loose flue dust from other smelters, and complex metallurgical byproducts, in addition to all conceivable grades of battery scrap and old batteries, sometimes partly filled with acid.

The two installations have demonstrated that plants of their size can be made acceptable regarding environmental protection and work hygiene at the same time. The optimum size of a secondary lead smelter is not known, but since the collection and smelting of lead-containing scrap will always cause pollution, it is not desirable to make very large secondary smelters. Even with the best technology, increased production means higher emission and longer distances for scrap collection.

Sponsored by the US agencies EPA and NIOSH, a team of engineers from Radian Corporation made a study to measure stack emissions, lead-in-air figures and many other parameters in our SB-smelter at Glostrup, Denmark (6). In normal operation our main stack serves both the SB-furnace and another smelting department in which short bodied rotary furnaces are used to treat other materials, but during measuring periods the gases were separated so that the SB-furnace emission could be studied. The composition of the lead-bearing charge was held constant during the test period and comprised,

Cased batteries	39%
Battery scrap	39%
Battery mud	7%
Recycled flue dust	2.5%
Recycled dross	2.5%
Lead scrap and residues	10%

In spite of the constant mix the chlorine content of the stack gases varied quite widely, while emissions of lead, antimony and sulphur were reasonably constant as shown in Table I. The table also illustrates how the four

Table I. SB-Furnace Emission

	Total emission	Elemental balance		
		Metal	Matte	Smoke
Lead	0.04-0.12 kg/hr	98.9%	1.0%	0.003%
Antimony	0.5 -0.6 kg/hr	89.1%	10.1%	0.7%
Sulphur	3.3 -4.5 kg/hr[+)]	0.2%	93.0%	6.8%
Chlorine	1.6 -7.1 kg/hr	10 %	29 %	61%

+) Measured as sulphur dioxide

elements became distributed between the metal, the matte and the stack gases. Average compositions have been used to calculate the percentages, and they confirm that more than 90% of the sulphur is bound in the matte with only about 7% escaping as sulphur dioxide in the stack gases. In the case of chlorine, the percentages are more uncertain.

The SB-smelter was designed with a solid separation between "clean" and "dirty" rooms, the latter being used for storage and handling of raw materials. In such rooms, work is only carried out in the protection of a loading machine cabin provided with filtered air supply. On the "clean" side the furnace can be operated without use of respirators or other personal protection. There are efficient exhaust hoods available at all point sources, and when they are used properly, the lead exposure can almost be held below the magic figure of 100 microgram/m^3. Table II summarizes the lead-in-air figures measured by Radian Corporation and shows

Table II. Lead Exposure at SB-Furnace

Lead-in-air	Time weighted exposure			Percent exceeding	
	High	Low	Average	100 μg/m^3	50 μg/m^3
Indoor sampling	110	10	51.9 μg/m^3	8%	50%
Stationary samplers ("work area")	85	10	38.4 μg/m^3	None	36%
Portable samplers ("breathing zone")	110	14	67.6 μg/m^3	17%	67%
Outdoor sampling [+)]	18	8	13 μg/m^3	–	–

[+)] Few samples only.

that 92% of all measurements were within that limit, but only 50% within the lower standard of 50 microgram/m^3 that has been imposed by US Department of Labor. We believe it will be very difficult, if not impossible to meet such a standard in practical operation of a lead smelting furnace. In Denmark, we work under a standard of 100 microgram/m^3 which is within practical reach with a well designed and well maintained smelter. Good cooperation with employees and frequent instruction is also necessary to maintain the work hygiene standard.

About half of the samples were taken with stationary samplers and the other half with samplers carried by the employees, and Table II clearly shows the difference between the two groups. The sampler carried by a working person invariably catches more lead than a stationary sampler. The portable sampler is moved closer to the point sources and there is an extra contribution from work clothes.

It has to be added that no emergencies occurred during the test period and it was not necessary to carry out any maintenance operations other than normal cleaning of tuyeres, lead well and slag tapping holes. In case of emergencies and during furnace shut-down periods, the lead-in-air figures do reach higher levels so that it becomes necessary to use respirators.

References

(1) R.D. Prengaman, Contribution to TMS-AIME World Symposium, Las Vegas, Nevada, 1980.

(2) J. Godfroi, Contribution to TMS-AIME World Symposium, Las Vegas, Nevada, 1980.

(3) T.S. Mackey and S. Bergsøe, "Flash Agglomeration of Flue Dust", Journal of Metals, 29 (11) (1977) pp. 12-15. .

(4) S. Bergsøe and S. Pearce, "The Story About the SB Furnace", Antifriction Journal No. 4, 1976, pp. 30-35. (Published by Paul Bergsøe & Søn A/S).

(5) N. Gram, "Der Bergsøe-SB-Ofen, Aufbau und Betriebserfahrungen", Metall, 32 (9) (1978) pp. 942-945.

(6) R.T. Coleman and R. Vandervort, "Source Characterisation of the SB Battery Smelting Furnace", Report issued by Radian Corporation, Austin, Texas, 1979.

Subject Index

Zinc

Zinc-Lead ISF

Tin

Author Index